Springer Monographs in Mathematics

Springer
*Berlin
Heidelberg
New York
Barcelona
Budapest
Hong Kong
London
Milan
Paris
Singapore
Tokyo*

Ludwig Arnold

Random Dynamical Systems

With 40 Figures

Springer

Ludwig Arnold
Universität Bremen
Institut für Dynamische Systeme
Postfach 33 04 40
D-28334 Bremen, Germany
e-mail: arnold@math.uni-bremen.de

Library of Congress Cataloging-in-Publication-Data
Arnold, L. (Ludwig), 1937-
Random dynamical systems / Ludwig Arnold.
p. cm. -- (Springer monographs in mathematics) Includes bibliographical references and index.
ISBN 3-540-63758-3 (hardcover: alk. paper)
1. Stochastic differential equations. 2. Differentiable dynamical systems. 3. Ergodic theory. I. Title.
II. Series
QA274.23.A75 1998 519.2--dc21 98-27207 CIP

Mathematics Subject Classification (1991): 34F05, 60H10, 93E03

ISBN 3-540-63758-3 Springer-Verlag Berlin Heidelberg New York

This work is subject to copyright. All rights are reserved, whether the whole or part of the material is concerned, specifically the rights of translation, reprinting, reuse of illustrations, recitation, broadcasting, reproduction on microfilm or in any other way, and storage in data banks. Duplication of this publication or parts thereof is permitted only under the provisions of the German Copyright Law of September 9, 1965, in its current version, and permission for use must always be obtained from Springer-Verlag. Violations are liable for prosecution under the German Copyright Law.

© Springer-Verlag Berlin Heidelberg 1998
Printed in Germany

The use of general descriptive names, registered names, trademarks etc. in this publication does not imply, even in the absence of a specific statement, that such names are exempt from the relevant protective laws and regulations and therefore free for general use.

Typesetting: The Author's LaTeX input files have been edited and reformatted by Satztechnik K. Steingraeber, Heidelberg using a Springer LaTeX macro package
SPIN 10574441   41/3143-5 4 3 2 1 0 – Printed on acid-free paper

Preface

Background and Scope of the Book This book continues, extends, and unites various developments in the intersection of probability theory and dynamical systems. I will briefly outline the background of the book, thus placing it in a systematic and historical context and tradition.

Roughly speaking, a random dynamical system is a combination of a measure-preserving dynamical system in the sense of ergodic theory, $(\Omega, \mathcal{F}, \mathbb{P}, (\theta(t))_{t \in \mathbb{T}})$, $\mathbb{T} = \mathbb{R}^+, \mathbb{R}, \mathbb{Z}^+, \mathbb{Z}$, with a smooth (or topological) dynamical system, typically generated by a differential or difference equation $\dot{x} = f(x)$ or $x_{n+1} = \varphi(x_n)$, to a random differential equation $\dot{x} = f(\theta(t)\omega, x)$ or random difference equation $x_{n+1} = \varphi(\theta(n)\omega, x_n)$.

Both components have been very well investigated separately. However, a symbiosis of them leads to a new research program which has only partly been carried out. As we will see, it also leads to new problems which do not emerge if one only looks at ergodic theory and smooth or topological dynamics separately.

From a *dynamical systems point of view* this book just deals with those dynamical systems that have a measure-preserving dynamical system as a factor (or, the other way around, are extensions of such a factor). As there is an invariant measure on the factor, ergodic theory is always involved.

Our book is a "continuation" of that by Guckenheimer and Holmes [162] on *Nonlinear Oscillations, Dynamical Systems, and Bifurcations of Vector Fields*. In their own words (Preface, page xi), their book *"should be seen as an attempt to extend the work of Andronov et al. (i.e. the analysis of a single degree of freedom nonlinear oscillator, L. A.) by one dimension (i.e. by adding a small periodic forcing term, L. A.)"*. Specifically, they look at certain equations of the form $\dot{x} = f(\theta(t)\omega, x)$ in $\mathbb{R}^2$ where $\theta(t)\omega$ is periodic. We will go further and beyond the periodic "noise" to a general measure-preserving dynamical system to which the ordinary differential equation is coupled. In yet other words, we take the step from autonomous systems $\dot{x} = f(x)$ to nonautonomous systems, but of the special kind $\dot{x} = f(\theta(t)\omega, x)$, i.e. to those which are coupled to a dynamical "bath".

If the flow $\omega \mapsto \theta(t)\omega$ in the equation $\dot{x} = f(\theta(t)\omega, x)$ is a flow of homeomorphisms of a compact space we are in the realm of skew-product flows in the sense of Sacker, Sell and Johnson (see e.g. [187], [298], [299], [316]). We go beyond this again by stripping off all the topology from $\omega \mapsto \theta(t)\omega$, and instead adding an invariant measure – shortly, by going from "almost periodic" to "random".

We also extend and generalize Mañé's book [249] on *Ergodic Theory and Differentiable Dynamics*. He has a measure ρ invariant with respect to the flow $\varphi(t)$ of a deterministic vector field $\dot{x} = f(x)$ on a manifold M. Here, we have a measure μ on $\Omega \times M$ with marginal $\mathbb{P}$ on Ω invariant with respect to the skew-product flow $(\omega, x) \mapsto (\theta(t)\omega, \varphi(t, \omega)x)$, where $\varphi(t, \omega)$ is the solution flow generated by the random vector field $\dot{x} = f(\theta(t)\omega, x)$.

From a *probabilistic point of view* this book offers another look at the quite classical subject of random difference equations and of random and stochastic differential equations, i.e. ordinary differential equations driven by real or white noise.

During the last 20 to 30 years an impressive structure called "stochastic analysis" has been erected, part of which is a theory of differential equations with semimartingale (rather than only Gaussian white noise, or Wiener) driving processes, providing us with a unified theory of random and stochastic differential equations.

Around 1980 it was discovered by Elworthy, Baxendale, Bismut, Ikeda, Watanabe, Kunita and others (see e.g. [137], [53], [72], [178], [223]) that a stochastic differential equation generates "for free" a much richer structure than just a family of stochastic processes, each solving the stochastic differential equation for a given initial value. It gives us in fact a flow of random diffeomorphisms. We can now bridge the gap between stochastic analysis and dynamical systems by proving that a random or stochastic differential equation generates a random dynamical system.

This makes it possible to re-evaluate and improve all the classical results (which are based on one-point motions and Markov transition probabilities) on stochastic stability, existence of invariant measures, etc. by Kushner [225], Khasminskii [206], Bunke [84] and many others. In [8] I have described the extension of the horizon when going from Markov processes to stochastic flows and cocycles.

The present book also adds a new chapter to the volume by Horsthemke and Lefever [175] entitled *Noise-Induced Transitions* and re-interprets their findings: Their noise-induced transitions are nothing but bifurcations on the static level of the Fokker-Planck equation. We will also study bifurcation scenarios on the dynamic level.

The book closest to ours in spirit and content is the one by Kifer [207] on *Ergodic Theory of Random Transformations*. He, however, deals exclusively with the i.i.d. case, i.e. with the case of iterations of random mappings chosen independently with identical distribution. In this case the orbits in state space form a Markov chain. We go beyond that by allowing a stationary stochastic sequence of mappings to be iterated, keeping the i.i.d. case as an important particular case.

It is a characteristic feature of the theory of random dynamical systems that every problem involves some ergodic theory and ergodic theorems. The most crucial and most important ergodic theorem applies to the linearization of smooth random dynamical systems. It is traditionally called the *Multiplicative Ergodic Theorem* and was proved by Oseledets [268] in 1968. This theorem provides a random substitute of linear algebra and hence makes a local theory of smooth random dynamical systems possible. Without it the whole field (in particular this book) would not exist.

Structure of the Book As this is the first monograph on random dynamical systems, my main intention is foundational. This forces me to adopt a systematic, maybe sometimes even pedantic, style, and put my emphasis on theory rather than applications. I hope nevertheless to present a useful, reliable, and rather complete source of reference which lays the foundations for future work and applications.

Part I (Random Dynamical Systems and Their Generators) introduces the subject matter, settles the subtle perfection question, develops the theory of invariant measures (Chap. 1) and gives a (hopefully ultimate) treatment of the problem of which random dynamical systems have infinitesimal generators (Chap. 2).

Part II (Multiplicative Ergodic Theory) is the heart of the book. I first present and prove the classical Multiplicative Ergodic Theorem for products of random matrices in $\mathbb{R}^d$ (Chap. 3), then present its various modifications and the concept of random norms which turns out to be basic (Chap. 4). In Chap. 5 the multiplicative ergodic theory of related linear random dynamical systems obtained by taking the inverse, the adjoint, and exterior and tensor products is studied. The same is done with the systems induced by a linear random dynamical system on the unit sphere, the projective space, and Grassmannian manifolds, culminating in the Furstenberg-Khasminskii formulas for Lyapunov exponents. Finally, a multiplicative ergodic theorem for rotation numbers is proved (Chap. 6).

Part III (Smooth Random Dynamical Systems) addresses the three most fundamental problems regarding nonlinear systems. The first one is the construction of invariant manifolds (Chap. 7). We adopt the new method of Wanner [340] which provides a unified approach towards invariant manifolds and the Hartman-Grobman theorem. The second basic problem is the simplification of a random dynamical system by means of a smooth coordinate transformation (normal form problem) (Chap. 8). I finally present the state of the art of random bifurcation theory (Chap. 9) which is still in its infancy and is not much more than a collection of (numerical) examples.

Part IV (Appendices) collects some facts from measurable dynamics (Appendix A) and smooth dynamics (Appendix B).

VIII   Preface

This book is a research monograph which belongs to the richly structured interface of probability theory and dynamical systems. Therefore, substantial mathematical knowledge is required from the reader.

The book as a whole is probably not suitable for a course. There are, however, various possibilities to use subsets of the text as the basis of a graduate course or seminar or for private study. For example:

(i) The multiplicative ergodic theorem: Extract the basic definitions from Chap. 1, then read Chaps. 3 and 4.

(ii) Smooth random dynamical systems: Read Chaps. 1, 3 and 4, then choose one or several of the Chaps. 7, 8, or 9.

(iii) Stochastic bifurcation theory: Read Chaps. 1, 3 and 4, then go to Chap. 9.

Omissions The exclusion of the following topics from this book is partly compensated by some recent publications which augment this book and complete the overall picture of the subject.

I completely omitted topological dynamics of random dynamical systems and refer instead to the book by Nguyen Dinh Cong [261].

Fortunately, I do not need to include *Pesin's theory*, probably the deepest of recent developments in random dynamical systems, as it is beautifully and completely presented in the book by Liu and Qian [244].

With many scruples, I decided to omit the beautiful "geometry of stochastic flows" (see the work of Baxendale [56, 55, 59, 61, 62], Baxendale and Stroock [64], Carverhill [91], Carverhill and Elworthy [95], Elworthy [139, 140, 141], Elworthy and Rosenberg [144], Elworthy and Yor [145], Elworthy, Le Jan and Li [142], Elworthy and Li [143], Kunita [224: section 4.9] Li [234], Liao [236, 237, 238, 239], and the references therein). The subject is worth a book of its own.

I also did not include the theory proper of products of random matrices. On the one hand, this area with its elaborate methods and numerous applications is so vast that it would easily fill a volume in itself. On the other hand, the subject is already quite well-documented: Besides the books by Bougerol and Lacroix [77] and Högnäs and Mukherjea [172], there are numerous contributions, survey articles, and further references in the three proceedings volumes [104], [39], and [14].

Finally, I omitted infinite-dimensional random systems and instead refer the reader to the work of Crauel and Flandoli [117, 119], Flandoli and Schmalfuß [151], Mohammed [255, 256], Mohammed and Scheutzow [257], Schaumlöffel [301], in addition to many others.

Acknowledgements Beginning with my collaboration with Peter Sagirow in the late 1960's in Stuttgart, I have always maintained strong and extremely fruitful contacts with engineers, to whom I am indebted for numerous suggestions and problems. In addition to my nestor Peter Sagirow, I would like to explicitly mention Walter Wedig, S. T. Ariaratnam, Mike Lin, and N. Sri Namachchivaya. My contacts with them gave me the badly needed reassurence that I was doing something that would be useful.

Some of my former students and collaborators also deserve an extra special acknowledgement, as they took very active part in the unfolding of the subject and taught me much of what I know today: Wolfgang Kliemann, Volker Wihstutz, and Hans Crauel.

I would also like to thank all those who proof-read parts of preliminary versions, in particular Peter Baxendale, Thomas Bogenschütz, Hans Crauel, Matthias Gundlach, Stefan Hilger, Peter Imkeller, Peter Kloeden (who very efficiently served as my default "native English speaker"), Liu Pei-Dong, Hans-Friedrich Münzner, Nguyen Dinh Cong, Gunter Ochs, Klaus Reiner Schenk-Hoppé, Michael Scheutzow, N. Sri Namachchivaya, Björn Schmalfuß, Stefan Siegmund and Thomas Wanner.

I am grateful to Gabriele Bleckert and Hannes Keller who produced the figures. Gabriele Bleckert also did the extensive numerical studies of the noisy Duffing-van der Pol oscillator in Chap. 9.

Last but not least I would like to thank my wife Birgit for her understanding and support during the six very hard years of writing.

Bremen, May 1998 *Ludwig Arnold*

Contents

Part I. Random Dynamical Systems and Their Generators

Chapter 1. Basic Definitions. Invariant Measures 3
- 1.1 Definition of a Random Dynamical System 3
- 1.2 Local RDS 11
- 1.3 Perfection of a Crude Cocycle 15
- 1.4 Invariant Measures for Measurable RDS 21
- 1.5 Invariant Measures for Continuous RDS 26
  - 1.5.1 Polish State Space 26
  - 1.5.2 Compact Metric State Space 30
- 1.6 Invariant Measures on Random Sets 32
- 1.7 Markov Measures 37
- 1.8 Invariant Measures for Local RDS 40
- 1.9 RDS on Bundles. Isomorphisms 43
  - 1.9.1 Bundle RDS 43
  - 1.9.2 Isomorphisms of RDS 45

Chapter 2. Generation 49
- 2.1 Discrete Time: Products of Random Mappings 50
  - 2.1.1 One-Sided Discrete Time 50
  - 2.1.2 Two-Sided Discrete Time 52
  - 2.1.3 RDS with Independent Increments 53
- 2.2 Continuous Time 1: Random Differential Eqs. 57
  - 2.2.1 RDS from Random Differential Equations 57
  - 2.2.2 The Memoryless Case 64
  - 2.2.3 Random Differential Equations from RDS 66
  - 2.2.4 The Manifold Case 67
- 2.3 Continuous Time 2: Stochastic Differential Eqs. 68
  - 2.3.1 Introduction. Two Cultures 68
  - 2.3.2 Semimartingales and Dynamical Systems: Stochastic Calculus for Two-Sided Time 71
  - 2.3.3 Semimartingale Helices with Spatial Parameter 78

| | 2.3.4 RDS from Stochastic Differential Equations 82
| | 2.3.5 Stochastic Differential Equations from RDS 87
| | 2.3.6 White Noise 91
| | 2.3.7 An Example 98
| | 2.3.8 The Manifold Case 101
| | 2.3.9 RDS with Independent Increments 103

Part II. Multiplicative Ergodic Theory

Chapter 3. The Multiplicative Ergodic Theorem in Euclidean Space 111

3.1 Introduction 111
3.2 Lyapunov Exponents.................................... 113
    3.2.1 Deterministic Theory of Lyapunov Exponents 113
    3.2.2 Singular Values 117
    3.2.3 Exterior Powers 118
3.3 Furstenberg-Kesten Theorem 121
    3.3.1 The Subadditive Ergodic Theorem.................. 122
    3.3.2 The Furstenberg-Kesten Theorem for One-Sided Time 122
    3.3.3 The Furstenberg-Kesten Theorem for Two-Sided Time 130
3.4 Multiplicative Ergodic Theorem 134
    3.4.1 The MET for One-Sided Time 134
    3.4.2 The MET for Two-Sided Time 153
    3.4.3 Examples 159

Chapter 4. The Multiplicative Ergodic Theorem on Bundles and Manifolds 163

4.1 Temperedness. Lyapunov Cohomology 163
    4.1.1 Tempered Random Variables....................... 163
    4.1.2 Lyapunov Cohomology 166
4.2 The MET on Manifolds 172
    4.2.1 Linearization of a C^1 RDS...................... 172
    4.2.2 The MET for RDS on Manifolds 174
    4.2.3 Random Differential Equations..................... 179
    4.2.4 Stochastic Differential Equations 181
4.3 Random Lyapunov Metrics and Norms 186
    4.3.1 The Control of Non-Uniformity in the MET 187
    4.3.2 Random Scalar Products 191
    4.3.3 Random Riemannian Metrics on Manifolds 197

Chapter 5. The MET for Related Linear and Affine RDS 201

- 5.1 Inverse and Adjoint 201
- 5.2 The MET on Linear Subbundles 206
- 5.3 Exterior Powers, Volume, Angle 211
  - 5.3.1 Exterior Powers 211
  - 5.3.2 Volume and Determinant 213
  - 5.3.3 Angles 215
- 5.4 Tensor Product 218
- 5.5 Manifold Versions 221
- 5.6 Affine RDS 221
  - 5.6.1 Representation 221
  - 5.6.2 Invariant Measure in the Hyperbolic Case 223
  - 5.6.3 Time Reversibility and Iterated Function Systems 231

Chapter 6. RDS on Homogeneous Spaces of the General Linear Group 235

- 6.1 Cocycles on Lie Groups 236
  - 6.1.1 Group-Valued Cocycles and Their Generators 237
  - 6.1.2 Cocycles Induced by Actions 238
- 6.2 RDS Induced on S^{d-1} and P^{d-1} 241
  - 6.2.1 Invariant Measures 242
  - 6.2.2 Furstenberg-Khasminskii Formulas 251
  - 6.2.3 Spectrum and Splitting 260
- 6.3 RDS on Grassmannians 263
  - 6.3.1 Invariant Measures 263
  - 6.3.2 Furstenberg-Khasminskii Formulas 266
- 6.4 Manifold Versions 269
  - 6.4.1 Sphere Bundle and Projective Bundle 269
  - 6.4.2 Grassmannian Bundles 273
- 6.5 Rotation Numbers 277
  - 6.5.1 The Concept of Rotation Number of a Plane 277
  - 6.5.2 Rotation Numbers for RDE 285
  - 6.5.3 Rotation Numbers for SDE 293

Part III. Smooth Random Dynamical Systems

Chapter 7. Invariant Manifolds 305

- 7.1 The Problem of Invariant Manifolds 306
- 7.2 Reductions and Preparations 308
  - 7.2.1 Reductions 309
  - 7.2.2 Preparations 311
- 7.3 Global Invariant Manifolds 317

	7.3.1	Construction of Unstable Manifolds	318
	7.3.2	Construction of Stable Manifolds	337
	7.3.3	Construction of Center Manifolds	340
	7.3.4	The Continuous Time Case	343
	7.3.5	Higher Regularity	346
	7.3.6	Final Global Invariant Manifold Theorem	360
7.4	Hartman-Grobman Theorem		361
	7.4.1	Invariant Foliations	361
	7.4.2	Topological Decoupling	368
	7.4.3	Hartman-Grobman Theorem	373
7.5	Local Invariant Manifolds		379
	7.5.1	Local Manifolds for Discrete Time	380
	7.5.2	Dynamical Characterization and Globalization	386
	7.5.3	Local Manifolds for Continuous Time	393
7.6	Examples		396

Chapter 8. Normal Forms ... 405

8.1	Deterministic Prerequisites		405
8.2	Normal Forms for Random Diffeomorphisms		412
	8.2.1	The Random Cohomological Equation	412
	8.2.2	Nonresonant Case	417
	8.2.3	Resonant Case	419
8.3	Normal Forms for RDE		420
	8.3.1	The Random Cohomological Equation	421
	8.3.2	Nonresonant Case	425
	8.3.3	Resonant Case	426
	8.3.4	Examples	430
8.4	Normal Form and Center Manifold		433
	8.4.1	The Reduction Procedure	433
	8.4.2	Parametrized RDE	439
	8.4.3	Small Noise: A Case Study	442
8.5	Normal Forms for SDE		446
	8.5.1	The Random Cohomological Equation	447
	8.5.2	Nonresonant Case	450
	8.5.3	Small Noise Case	461

Chapter 9. Bifurcation Theory ... 465

9.1	Introduction		465
9.2	What is Stochastic Bifurcation?		468
	9.2.1	Definition of a Stochastic Bifurcation Point	468
	9.2.2	The Phenomenological Approach	471
9.3	Dimension One		477
	9.3.1	Transcritical Bifurcation	477
	9.3.2	Pitchfork Bifurcation	480

	9.3.3	Saddle-Node Case 482
	9.3.4	A General Criterion for Pitchfork Bifurcation 482
	9.3.5	Real Noise Case 489
	9.3.6	Discrete Time 490
9.4	The Noisy Duffing-van der Pol Oscillator 491
	9.4.1	Introduction. Completeness. Linearization 491
	9.4.2	Hopf Bifurcation 497
	9.4.3	Pitchfork Bifurcation 510
9.5	General Dimension. Further Studies....................... 518
	9.5.1	Baxendale's Sufficient Conditions for D-Bifurcation and Associated P-Bifurcation 518
	9.5.2	Further Studies 529

Part IV. Appendices

Appendix A. Measurable Dynamical Systems 535

- A.1 Ergodic Theory 535
- A.2 Stochastic Processes and Dynamical Systems............... 542
- A.3 Stationary Processes 545
- A.4 Markov Processes 548

Appendix B. Smooth Dynamical Systems 551

- B.1 Two-Parameter Flows on a Manifold 551
- B.2 Spaces of Functions in $\mathbb{R}^d$ 552
- B.3 Differential Equations in $\mathbb{R}^d$ 555
- B.4 Autonomous Case: Dynamical Systems 557
- B.5 Vector Fields and Flows on Manifolds 560

References 563

Index 581

Part I

Random Dynamical Systems and Their Generators

Chapter 1. Basic Definitions. Invariant Measures

Summary

In this first chapter we introduce the concept of a random dynamical system and study its invariant measures – objects which characterize the possible long term behavior of the system.

The definition of a random dynamical system or cocycle is given in Sect. 1.1 and basic properties are derived (Theorem 1.1.6). The local concept (allowing for explosion) is introduced in Sect. 1.2.

Sect. 1.3 deals with the key but tricky technical problem of "perfecting" a "crude" cocycle (as generated by a stochastic differential equation). We give a satisfactory answer in Theorem 1.3.2 and its two corollaries. This section can be omitted at first reading.

The basic Sects. 1.4 to 1.7 are devoted to the study of invariant measures of random dynamical systems. We describe invariance in terms of the factorization of the measure (Theorem 1.4.5). For a Polish state space we introduce a topology of weak convergence of measures which permits us to carry over the Krylov-Bogolyubov procedure (Theorem 1.6.4) and to prove that each continuous random dynamical system on a compact state space has at least one invariant measure (Theorem 1.5.10; for a useful generalization to a random compact set see Theorem 1.6.13).

In Sect. 1.7 we relate our general definition of an invariant measure to the classical one for Markov processes (Corollary 1.7.6).

Theorem 1.8.4 in Sect. 1.8 stating that all invariant measures for continuous random dynamical systems with state space $\mathbb{R}$ are random Dirac measures will be frequently quoted later as we will explicitly work out several one-dimensional examples.

Finally, in Sect. 1.9 we introduce the bundle version of a random dynamical system and provide several notions of isomorphism which are used later to identify different systems of similar structure.

1.1 Definition of a Random Dynamical System

Imagine a mechanism which at each discrete time n tosses a (possibly complicated, many-sided) coin to randomly select a mapping φ_n by which a given point x_n is moved to $x_{n+1} = \varphi_n(x_n)$. The selection mechanism is permitted at time n to remember the choices made prior to n, and even to foresee the future. The only assumption made is that the same mechanism is used at each step. This scenario, called a *product of random mappings*, is one of the prototypes of a random dynamical system (see Fig. 1.1).

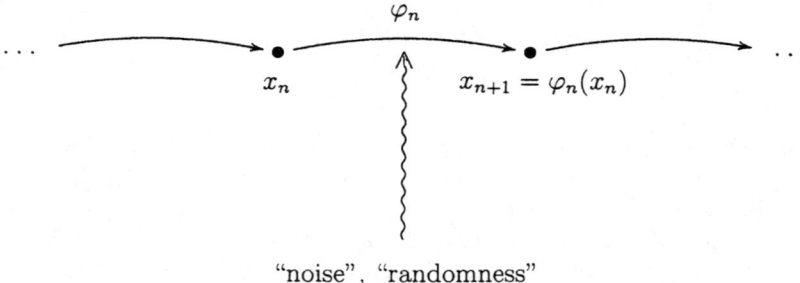

Fig. 1.1. Product of random mappings

In continuous time the random selection at time t would be from a set of differential equations (or vector fields) $\dot{x} = f(t, x)$, again with the stipulation that the statistics of $f(t, \cdot)$ be independent of t.

We will now give a formal definition of a random dynamical system which is tailor-made to cover the most important families of dynamical systems with randomness which are currently of interest, in particular random and stochastic ordinary and partial differential equations.

We will often describe the above situation by saying that a certain type of noise influences or perturbs a dynamical system.

A random dynamical system thus consists of two basic ingredients:

− a model of the noise,
− a model of the system which is perturbed by the noise.

Throughout the book, noise will always be modeled by a metric (i.e. measure-preserving) dynamical system in the sense of ergodic theory (see Appendix A for all basic notions and some examples), and the system will in most cases be modeled by a difference or a differential equation or its solution flow, respectively.

Dynamics studies those properties of a collection of self-mappings of some space which become apparent asymptotically through iteration. This collection of maps is (algebraically) always a semigroup, often a group $\mathbb{T}$ which we call *time*. Throughout the book, time $\mathbb{T}$ always stands for the following (additive) semigroups or groups:

- $\mathbb{T} = \mathbb{R}$: *Two-sided continuous* time,
- $\mathbb{T} = \mathbb{R}^+ := \{t \in \mathbb{R} : t \geq 0\}$ (sometimes $\mathbb{T} = \mathbb{R}^- := -\mathbb{R}^+$): *One-sided continuous* time,
- $\mathbb{T} = \mathbb{Z} := \{0, \pm 1, \pm 2, \ldots\}$: *Two-sided discrete* time,
- $\mathbb{T} = \mathbb{Z}^+ := \{0, 1, 2, \ldots\}$ (sometimes $\mathbb{T} = \mathbb{Z}^- := -\mathbb{Z}^+$ or $\mathbb{T} = \mathbb{N} := \{1, 2, 3, \ldots\}$): *One-sided discrete* time.

We will now give a hierarchy of definitions.

1.1.1 Definition (Random Dynamical System). A *measurable random dynamical system*[1] on the measurable space $(X, \mathcal{B})$ over (or covering, or extending) a metric dynamical system $(\Omega, \mathcal{F}, \mathbb{P}, (\theta(t))_{t \in \mathbb{T}})$ with time $\mathbb{T}$ is a mapping
$$\varphi : \mathbb{T} \times \Omega \times X \to X, \quad (t, \omega, x) \mapsto \varphi(t, \omega, x),$$
with the following properties:
 (i) *Measurability*: φ is $\mathcal{B}(\mathbb{T}) \otimes \mathcal{F} \otimes \mathcal{B}, \mathcal{B}$-measurable.
 (ii) *Cocycle property*: The mappings $\varphi(t, \omega) := \varphi(t, \omega, \cdot) : X \to X$ form a *cocycle* over $\theta(\cdot)$, i. e. they satisfy

$$\varphi(0, \omega) = \mathrm{id}_X \quad \text{for all} \quad \omega \in \Omega \quad (\text{if} \quad 0 \in \mathbb{T}), \tag{1.1.1}$$

$$\varphi(t + s, \omega) = \varphi(t, \theta(s)\omega) \circ \varphi(s, \omega) \quad \text{for all} \quad s, t \in \mathbb{T}, \quad \omega \in \Omega. \tag{1.1.2}$$

Here "$\circ$" means composition, which canonically defines an action on the left of the semigroup of self-mappings of X on the space X, i. e. $(f \circ g)(x) = f(g(x))$. This fact creates, as we will see, a certain "break of symmetry" in the theory.

If we want to emphasize that equation (1.1.2) holds identically, we call φ a *perfect* cocycle. We call φ a *crude* cocycle if (1.1.2) holds for fixed s and all $t \in \mathbb{T}$, $\mathbb{P}$-a. s. (where the exceptional set N_s can depend on s). We call φ a *very crude* cocycle if (1.1.2) holds for fixed $s, t \in \mathbb{T}$, $\mathbb{P}$-a. s. (where the exceptional set $N_{s,t}$ can depend on both s and t). ∎

Note that axiom (1.1.1) of Definition 1.1.1 is not redundant. However, if the mappings $\varphi(t, \omega) : X \to X$ are known to be invertible, (1.1.2) implies (1.1.1).

It is very useful to imagine an RDS move on the (trivial) bundle $\Omega \times X$, as Fig. 1.2 depicts: While ω is shifted by the dynamical system θ in time s to the point $\theta(s)\omega$ on the base space Ω, the cocycle $\varphi(s, \omega)$ moves the point x in the fiber $\{\omega\} \times X$ over ω to the point $\varphi(s, \omega)x$ in the fiber $\{\theta(s)\omega\} \times X$ over $\theta(s)\omega$. The cocycle property is also clearly "visible" on this bundle.

1.1.2 Definition (Continuous RDS). A *continuous* or *topological RDS* on the topological space X over the metric DS $(\Omega, \mathcal{F}, \mathbb{P}, (\theta(t))_{t \in \mathbb{T}})$ is a measurable RDS which satisfies in addition the following property: For each $\omega \in \Omega$ the mapping

---
[1] "Dynamical system(s)" and "Random dynamical system(s)" are henceforth often abbreviated as "DS" and "RDS", respectively.

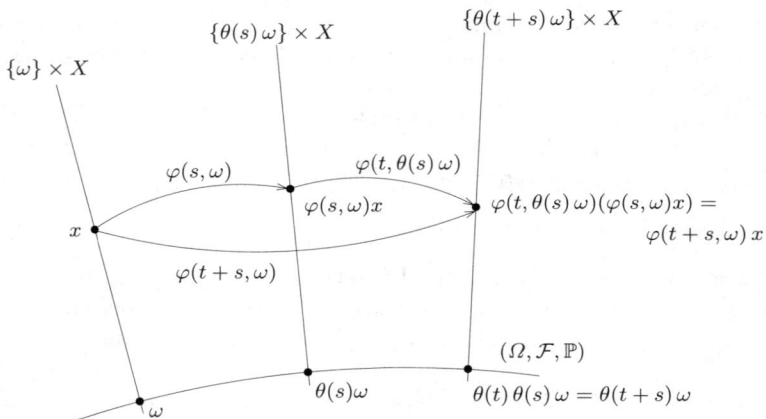

Fig. 1.2. A random dynamical system as an action on a bundle

$$\varphi(\cdot, \omega, \cdot) : \mathbb{T} \times X \to X, \quad (t, x) \mapsto \varphi(t, \omega, x),$$

is continuous. ∎

1.1.3 Definition (Smooth RDS). A *smooth RDS of class C^k*, or a C^k *RDS*, where $1 \leq k \leq \infty$, on a d-dimensional (C^∞) manifold X is a topological RDS which in addition satisfies the following property: For each $(t, \omega) \in \mathbb{T} \times \Omega$ the mapping

$$\varphi(t, \omega) = \varphi(t, \omega, \cdot) : X \to X, \quad x \mapsto \varphi(t, \omega, x),$$

is C^k (i.e. k times differentiable with respect to x, and the derivatives are continuous with respect to (t, x)). ∎

1.1.4 Definition (Linear RDS). A continuous RDS on a (for simplicity) finite-dimensional vector space is called a *linear RDS*, if $\varphi(t, \omega) \in \mathcal{L}(X)$ for each $t \in \mathbb{T}$, $\omega \in \Omega$, where $\mathcal{L}(X)$ is the space of linear operators of X. ∎

If we endow the vector space X with its natural manifold structure, then $\mathcal{L}(X) \subset C^\infty(X, X)$. Hence a linear RDS is automatically C^∞.

Notations: (i) We often omit specifically mentioning the underlying metric DS $(\Omega, \mathcal{F}, \mathbb{P}, (\theta(t))_{t \in T})$ (or abbreviate it as θ) and speak of an "RDS φ" (over θ), thus identifying an RDS with its cocycle part. Whenever we speak of a C^k RDS we assume $1 \leq k \leq \infty$.

(ii) We denote by $\mathcal{C}(X, X)$ or $\text{Homeo}(X)$ the semigroup or group of continuous mappings or homeomorphisms of a topological space X endowed with its compact-open topology. If X is a locally compact Hausdorff space, this is a Hausdorff topological semigroup or group, and the evaluation mapping $(f, x) \mapsto f(x)$ is continuous.

(iii) Finally, we denote by $\mathcal{C}^k(X,X)$ or $\mathrm{Diff}^k(X)$ the semigroup or group of C^k mappings or C^k diffeomorphisms of a manifold X, respectively, endowed with its compact-open topology (for the definition see Appendix B.2 for $X = \mathbb{R}^d$ and B.5 for the manifold case). This is a Polish topological semigroup or group, and the evaluation mapping $(f,x) \mapsto f(x)$ is C^k with respect to x. In the manifold case also $\mathrm{Homeo}(X)$ is a Polish group.

1.1.5 Remark. (i) If $\mathbb{T}$ is discrete, measurability of $(t,\omega,x) \mapsto \varphi(t,\omega)x$ is equivalent to measurability of $(\omega,x) \mapsto \varphi(t,\omega)x$ for each fixed $t \in \mathbb{T}$, continuity of $(t,x) \mapsto \varphi(t,\omega)x$ for each $\omega \in \Omega$ is equivalent to continuity of $x \mapsto \varphi(t,\omega)x$ for each fixed $(t,\omega) \in \mathbb{T} \times \Omega$, and the C^k smoothness of $x \mapsto \varphi(t,\omega)x$ is just with respect to x for each fixed $(t,\omega) \in \mathbb{T} \times \Omega$.

(ii) A measurable/continuous/C^k RDS with continuous time $\mathbb{T}$ is also a measurable/continuous/C^k RDS if restricted to discrete time $\mathbb{T} \cap \mathbb{Z}$.

(iii) A measurable/continuous/C^k RDS with two-sided time $\mathbb{T}$ is also a measurable/continuous/C^k RDS if restricted to one-sided time $\mathbb{T}^+ = \mathbb{T} \cap \mathbb{R}^-$.

(iv) We stress that we never allow our exceptional sets in the definition of a cocycle to depend on $x \in X$. In fact, it is one of the basic problems of a theory of RDS in an infinite-dimensional space X that

$$\varphi(t+s,\omega,x) = \varphi(t,\theta(s)\omega,\varphi(s,\omega,x))$$

often holds only outside a set of measure zero which depends on $s,t \in \mathbb{T}$ and on $x \in X$. See e.g. Skorokhod ([319: Chap. I] and [320: Chap. IV]) for examples of linear stochastic differential equations in Hilbert space.

(v) Example: Deterministic DS and DS in the sense of ergodic theory are particular cases of RDS. Indeed, if φ is independent of ω then the RDS decouples into a metric DS $(\Omega, \mathcal{F}, \mathbb{P}, (\theta(t))_{t \in \mathbb{T}})$ and a deterministic measurable/continuous/C^k DS φ on X. ∎

In case time $\mathbb{T}$ is a group, the underlying metric DS θ is invertible with $\theta(t)^{-1} = \theta(-t)$. Equations (1.1.1) and (1.1.2) then force the cocycle to be invertible too. More precisely, we have the following far-reaching consequences of the cocycle property.

1.1.6 Theorem (Basic Properties of RDS with Two-Sided Time).
Suppose $\mathbb{T}$ is a group (i.e. $\mathbb{T} = \mathbb{R}$ or $\mathbb{Z}$).

(i) Let φ be a measurable RDS on a measurable space $(X, \mathcal{B})$ over θ. Then for all $(t,\omega) \in \mathbb{T} \times \Omega$, $\varphi(t,\omega)$ is a bimeasurable bijection of $(X, \mathcal{B})$ and

$$\varphi(t,\omega)^{-1} = \varphi(-t,\theta(t)\omega) \quad \text{for all} \quad (t,\omega) \in \mathbb{T} \times \Omega, \tag{1.1.3}$$

or, equivalently,

$$\varphi(-t,\omega) = \varphi(t,\theta(t)^{-1}\omega)^{-1} \quad \text{for all} \quad (t,\omega) \in \mathbb{T} \times \Omega. \tag{1.1.4}$$

Moreover, the mapping

$$(t,\omega,x) \mapsto \varphi(t,\omega)^{-1}x$$

is measurable.

(ii) Let φ be a continuous RDS on a topological space X. Then for all $(t,\omega) \in T \times \Omega$ we have $\varphi(t,\omega) \in \operatorname{Homeo}(X)$. If

1. $T = \mathbb{Z}$, or
2. $T = \mathbb{R}$ and X is a topological manifold, or
3. $T = \mathbb{R}$ and X is a compact Hausdorff space

then $(t,x) \mapsto \varphi(t,\omega)^{-1}x$ is continuous for all $\omega \in \Omega$.

(iii) Let φ be a C^k RDS on a manifold X. Then for all $(t,\omega) \in \mathbb{T} \times \Omega$, $\varphi(t,\omega) \in \operatorname{Diff}^k(X)$. Moreover, $(t,x) \mapsto \varphi(t,\omega)^{-1}x$ is C^k with respect to x for all $\omega \in \Omega$.

Proof. (i) We use the fact that $g = f^{-1}$ if and only if $f \circ g = \operatorname{id}$ and $g \circ f = \operatorname{id}$. Put $t = -s$ in (1.1.2) and use (1.1.1) to obtain

$$\operatorname{id}_X = \varphi(-s, \theta(s)\omega) \circ \varphi(s,\omega) \quad \text{for all} \quad \omega, s.$$

Now use (1.1.2) for $s = -t$ and $\tilde{\omega} = \theta(t)\omega$ and relation (1.1.1) to obtain

$$\operatorname{id}_X = \varphi(t,\omega) \circ \varphi(-t, \theta(t)\omega) \quad \text{for all} \quad \omega, t,$$

yielding for $s = t$ equation (1.1.3). The mapping $(t,\omega,x) \mapsto \varphi(t,\omega)^{-1}x = \varphi(-t,\theta(t)\omega)x$ is measurable since it is the composition of the measurable maps $(t,\omega,x) \mapsto (-t,\theta(t)\omega,x)$ and $(t,\omega,x) \mapsto \varphi(t,\omega)x$. The measurability of the inverse mapping with respect to x follows from that of $(t,\omega,x) \mapsto \varphi(t,\omega)^{-1}x$ by freezing (t,ω).

(ii) Equation (1.1.3) says that the left-hand side is continuous with respect to x since the right-hand side is, by Definition 1.1.2. Thus $\varphi(t,\omega) \in \operatorname{Homeo}(X)$. In case 1 the continuity of $(t,x) \mapsto \varphi(t,\omega)^{-1}x$ is trivially satisfied. In case 2 observe that $(t,x) \mapsto (t,\varphi(t,\omega)x)$ is a continuous (by Definition 1.1.2) and bijective (by part (i) of this proof) mapping of $\mathbb{R} \times X$ onto itself. This mapping is thus a homeomorphism by Brouwer's theorem (see Dieudonné [127: p. 52]), so the inverse $(t,x) \mapsto (t,\varphi(t,\omega)^{-1}x)$ is continuous, in particular $(t,x) \mapsto \varphi(t,\omega)^{-1}x$ is continuous. Case 3: A continuous bijection of a compact space into a Hausdorff space is a homeomorphism (see Dunford and Schwartz [132: p. 18]). We apply this to $\psi(\omega) : K \times X \to K \times X$, where $K \subset \mathbb{R}$ is a compact interval and $\psi(\omega)(t,x) = (t,\varphi(t,\omega)x)$.

(iii) The C^k diffeomorphism property of $\varphi(t,\omega)^{-1}$ follows again from equation (1.1.3) and Definition 1.1.3. For the last statement we use the facts that the derivative of a diffeomorphism is nonsingular and that the derivative of the inverse is given by the formula

$$D\varphi(t,\omega)^{-1}x = (D\varphi(t,\omega)y\,|_{y=\varphi(t,\omega)^{-1}x})^{-1}.$$

Hence $(t,x) \mapsto D\varphi(t,\omega)^{-1}x$ is continuous since

– $(t,x) \mapsto D\varphi(t,\omega)x$ is continuous by assumption,

- $(t, x) \mapsto \varphi(t, \omega)^{-1} x$ is continuous by part (ii) of this proof (we are in the case where X is a C^∞ manifold), and
- $g \mapsto g^{-1}$ is continuous in $Gl(d, \mathbb{R})$.

Similarly for higher order derivatives. □

1.1.7 Remark. (i) It is somewhat surprising that under the assumptions of part (ii) of the above theorem the function

$$t \mapsto \varphi(t, \omega)^{-1} x = \varphi(-t, \theta(t)\omega) x$$

is continuous in t although $\varphi(t, \omega) x$ was assumed to be only measurable in ω.

(ii) Let φ be a continuous RDS with time $\mathbb{T} = \mathbb{R}$. If X is not locally Euclidean or compact Hausdorff we can in general not conclude that $(t, x) \mapsto \varphi(t, \omega)^{-1} x = \varphi(-t, \theta(t)\omega) x$ is continuous. This is due to the appearance of the shift operator $\theta(t)$ in formula (1.1.3) for the inverse. This suggests replacing the continuity requirement in Definition 1.1.2 by the following weaker (for discrete time: equivalent) one: For each $(t, \omega) \in \mathbb{T} \times \Omega$ the mapping

$$\varphi(t, \omega) : X \to X, \qquad x \mapsto \varphi(t, \omega) x,$$

is continuous. We still could conclude that $\varphi(t, \omega)$ is a homeomorphism. In fact, this weaker assumption suffices for most things we do with continuous RDS. The reason we stay with the stronger version of a topological RDS as given in Definition 1.1.2 is that we automatically obtain such continuous RDS when solving random or stochastic differential equations (see Sects. 2.2 and 2.3). ∎

1.1.8 Remark (RDS as a Skew Product). Given an RDS φ. Then the mapping

$$(\omega, x) \mapsto (\theta(t)\omega, \varphi(t, \omega) x) =: \Theta(t)(\omega, x), \quad t \in \mathbb{T}, \tag{1.1.5}$$

is a measurable DS on $(\Omega \times X, \mathcal{F} \otimes \mathcal{B})$ (exercise) which is called the *skew product* of the metric DS $(\Omega, \mathcal{F}, \mathbb{P}, (\theta(t))_{t \in \mathbb{T}})$ and the cocycle $\varphi(t, \omega)$ on X. Conversely, every such measurable skew product DS Θ defines a cocycle φ on its x component, thus a measurable RDS. We can consequently use "RDS φ", "cocycle φ" and "skew product Θ" synonymously. ∎

1.1.9 Remark (RDS have Stationary Increments). Let φ be a measurable RDS. For $n \in \mathbb{N}$ and $t_i \in \mathbb{T}$ for $i = 1, \ldots, n$, where $t_1 \leq t_2 \leq \ldots \leq t_n$, the X^X-valued random variables

$$\varphi(t_2 - t_1, \theta(t_1)\cdot), \ldots, \varphi(t_n - t_{n-1}, \theta(t_{n-1})\cdot) \tag{1.1.6}$$

are called the *(forward) increments*. If time is two-sided the cocycle property yields

$$\varphi(t - s, \theta(s)\omega) = \varphi(t, \omega) \circ \varphi(s, \omega)^{-1} \quad \text{for } t \geq s,$$

so that the quantities in (1.1.6) are indeed (multiplicative) increments.

The θ-invariance of $\mathbb{P}$ implies that the joint distribution of the increments does not change if the t_i are replaced by $t_i + h \in \mathbb{T}$, i.e. each RDS has *stationary increments*. In particular

$$\mathcal{L}(\varphi(t-s,\theta(s)\cdot)) = \mathcal{L}(\varphi(t-s,\cdot)) \quad \text{for all } t \geq s,$$

where $\mathcal{L}(\xi)$ denotes the probability distribution of the random variable ξ. Similarly for backward increments if time is two-sided. ∎

1.1.10 Remark (Backward Cocycles). A family $\psi(t,\omega)$ of mappings of a space X which satisfies all of the requirements of an RDS except that instead of the (*forward*) cocycle property (1.1.2) it satisfies the *backward cocycle property*

$$\psi(t+s,\omega) = \psi(t,\omega) \circ \psi(s,\theta(t)\omega) \tag{1.1.7}$$

is called a *backward cocycle*. Since composition ∘ is canonically a left action on X, backward cocycles are somewhat unnatural (or "acausal"), since in the first step one applies a mapping which depends on the second step.

Backward cocycles typically occur as companions of cocycles. For example, if $\mathbb{T}$ is two-sided then $\varphi(t,\omega)$ is a cocycle over θ if and only if

$$\psi(t,\omega) := \varphi(t,\omega)^{-1}$$

is a backward cocycle over θ, and, since φ is a cocycle (backward cocycle) over θ if and only if $\varphi(-\cdot)$ is a cocycle (backward cocycle) over θ^{-1},

$$\chi(t,\omega) := \psi(-t,\omega) = \varphi(-t,\omega)^{-1} = \varphi(t,\theta(-t)\omega)$$

is a backward cocycle over θ^{-1}. (Remark: χ can also be defined for one-sided time and non-invertible φ, but invertible θ as $\chi(t,\omega) := \varphi(t,\theta(t)^{-1}\omega)$, which is important for the study of random attractors, see for example Arnold, Demetrius and Gundlach [15], Crauel and Flandoli [117], Schmalfuß [307]).

The differences between the cocycle φ and the two backward cocycles ψ and χ in asymptotic behavior for $t \to \infty$ are quite dramatic. The reason is, in the bundle picture, that $\varphi(t,\omega)$ maps $x \in \{\omega\} \times X$ to $\varphi(t,\omega)x \in \{\theta(t)\omega\} \times X$ at time t, while $\psi(t,\omega)$ maps $x \in \{\theta(t)\omega\} \times X$ (which is moving with t) to $\psi(t,\omega)x \in \{\omega\} \times X$, the fixed fiber at time 0. Similarly, $\chi(t,\omega)$ maps $x \in \{\theta(-t)\omega\} \times X$ at time $-t$ to $\chi(t,\omega)x \in \{\omega\} \times X$ at time 0, see Fig. 1.3.

In general, in contrast to autonomous systems, it makes a big difference in the non-autonomous case (with which we are concerned here) between moving points from 0 to t, and moving them from $-t$ to 0. Only in the second case will the result be in the same fiber for all t, hence can be studied for $t \to \infty$. This is the reason why the backward cocycles ψ and χ will prove to be of fundamental importance for the construction of the invariant objects (measures, attractors) of the cocycle φ.

In Sect. 6.1 forward and backward cocycles are considered from the perspective of actions of groups on spaces.

In the deterministic (autonomous) case, forward cocycles and backward cocycles coincide, namely with a flow. ∎

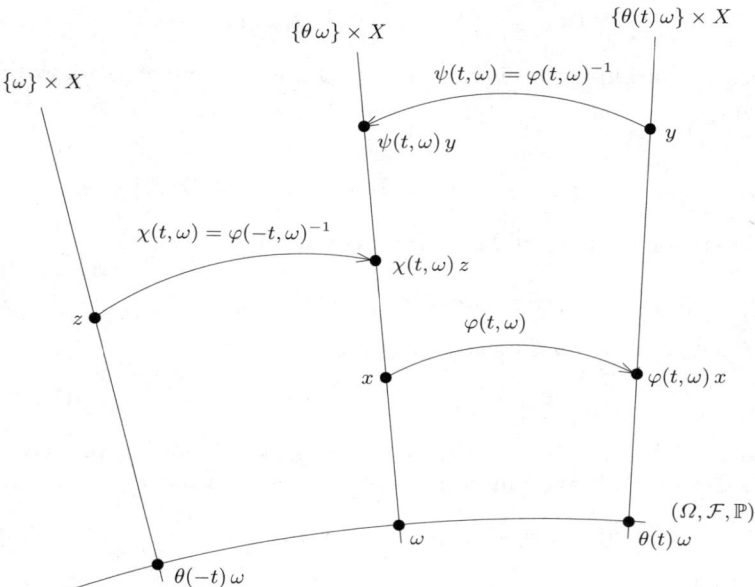

Fig. 1.3. A cocycle φ and corresponding backward cocycles ψ and χ

1.2 Local Random Dynamical Systems

It is well-known that a deterministic vector field in general only defines a local flow, due to the possibility of explosion (exit from the state space in finite time), see Appendix B.3 and B.4. Explosion is a continuous time phenomenon which originates from the fact that we can define a DS via an "infinitesimal generator" with the potential for explosion built in. In the random case we will find a similar situation: It will turn out that the solution of a random or stochastic differential equation will generally explode in finite time, so we need the concept of a local RDS which allows for this possibility. We restrict our attention to time $\mathbb{T} = \mathbb{R}$ and the continuous/C^k case.

1.2.1 Definition (Local RDS). Suppose $\mathbb{T} = \mathbb{R}$ and $(\Omega, \mathcal{F}, \mathbb{P}, (\theta(t))_{t \in \mathbb{R}})$ is a two-sided metric DS. A *local continuous/C^k RDS* over θ on a topological space/manifold X is a measurable mapping

$$\varphi : D \to X, \quad (t, \omega, x) \mapsto \varphi(t, \omega, x),$$

where $D \subset \mathbb{R} \times \Omega \times X$ is a measurable set, with the following properties: For all $\omega \in \Omega$

(i) The random *domain*

$$D(\omega) := \{(t, x) \in \mathbb{R} \times X : (t, \omega, x) \in D\} \subset \mathbb{R} \times X$$

is non-void and open, and

$$\varphi(\omega) : D(\omega) \to X, \quad (t, x) \mapsto \varphi(t, \omega, x),$$

is continuous/C^k (meaning that it is k times continuously differentiable with respect to x).

(ii) For each $x \in X$

$$D(\omega, x) := \{t \in \mathbb{R} : (t, \omega, x) \in D\} = \{t \in \mathbb{R} : (t, x) \in D(\omega)\} \subset \mathbb{R}$$

is an open interval containing 0, hence can be written as

$$D(\omega, x) =: (\tau^-(\omega, x), \tau^+(\omega, x)) \subset \mathbb{R}.$$

(iii) $\varphi(\omega)$ satisfies the *local cocycle property*:

$$\varphi(0, \omega) = \mathrm{id}_X \qquad (1.2.1)$$

and for all $x \in X$ and all $s \in D(\omega, x)$ we have the following property: $t \in D(\theta(s)\omega, \varphi(s, \omega, x))$ if and only if $t + s \in D(\omega, x)$, equivalently

$$D(\omega, x) = s + D(\theta(s)\omega, \varphi(s, \omega)x) . \qquad (1.2.2)$$

In this case we have

$$\varphi(t + s, \omega)x = \varphi(t, \theta(s)\omega)(\varphi(s, \omega)x). \qquad (1.2.3)$$

The local continuous/C^k RDS is said to be *global* if $D = \mathbb{R} \times \Omega \times X$. ∎

1.2.2 Remark. (i) $\tau^+(\omega, x)$ and $\tau^-(\omega, x)$ are, respectively, the forward and backward explosion times of the orbit $\varphi(\cdot, \omega)x$ starting at time $t = 0$ in the position x. The set $D(\omega, x) \subset \mathbb{R}$ is automatically open by part (i) of Definition 1.2.1 as a section of the open set $D(\omega)$.

(ii) A local RDS which is global in the sense of Definition 1.2.1 is an RDS in the sense of Definitions 1.1.2 and 1.1.3, respectively. The following conditions are obviously equivalent for a local RDS to be global:

– $D(\omega, x) = \mathbb{R}$ for all $(\omega, x) \in \Omega \times X$,
– $\tau^+(\omega, x) = \infty$ and $\tau^-(\omega, x) = -\infty$ for all $(\omega, x) \in \Omega \times X$,
– $D(t, \omega) = X$ for all $(t, \omega) \in \mathbb{R} \times \Omega$, where

$$D(t, \omega) := \{x \in X : (t, \omega, x) \in D\} \subset X$$

is the (in general possibly empty) set of initial values $x \in X$ for which the trajectories still exist at time t. As a section of $D(\omega)$, $D(t, \omega)$ is open. If $0 \leq t \leq s$, then by definition $D(t, \omega) \supset D(s, \omega)$ and $D(-t, \omega) \supset D(-s, \omega)$.

(iii) We have

$$t \in D(\omega, x) \iff x \in D(t, \omega),$$

since $D(t, \omega) = \{x : t \in D(\omega, x)\}$ and $D(\omega, x) = \{t : x \in D(t, \omega)\}$. See Fig. 1.4. ∎

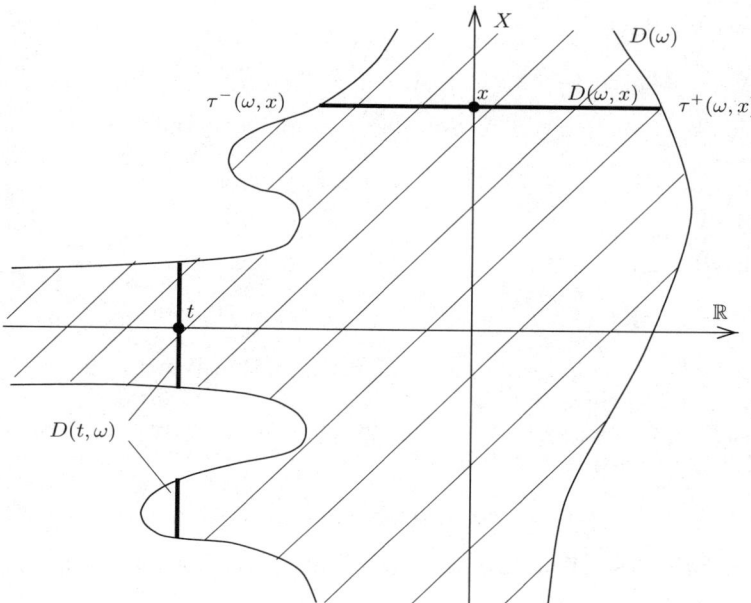

Fig. 1.4. The domain of a local random dynamical system

1.2.3 Theorem (Basic Properties of Local RDS). *Let φ be a local continuous/C^k RDS over the metric DS $(\Omega, \mathcal{F}, \mathbb{P}, (\theta(t))_{t \in \mathbb{R}})$ with time $T = \mathbb{R}$ on a topological space/manifold X. Then*

(i) τ^+ and τ^- are measurable.
Furthermore, for all $\omega \in \Omega$ the following assertions hold:
(ii) $x \mapsto \tau^+(\omega, x) \in (0, +\infty]$ is lower semicontinuous and $x \mapsto \tau^-(\omega, x) \in [-\infty, 0)$ is upper semicontinuous.
(iii) For all $(t, x) \in D(\omega)$

$$D(\omega, x) = t + D(\theta(t)\omega, \varphi(t, \omega)x).$$

(iv) For all $t \in \mathbb{R}$

$$\varphi(t, \omega) : D(t, \omega) \to R(t, \omega) = D(-t, \theta(t)\omega)$$

is a homeomorphism/C^k diffeomorphism and

$$\varphi(t, \omega)^{-1} = \varphi(-t, \theta(t)\omega) : D(-t, \theta(t)\omega) \to D(t, \omega). \qquad (1.2.4)$$

(v) If X is a manifold, then the mapping

$$(t, x) \mapsto \varphi(t, \omega)^{-1} x = \varphi(-t, \theta(t)\omega)x$$

is continuous/C^k.

Proof. (i) For $t > 0$ fixed,
$$\{(\omega, x) : \tau^+(\omega, x) > t\} = \{(\omega, x) : (t, \omega, x) \in D\}$$
is measurable as the section of a measurable set. Similarly for $t < 0$
$$\{(\omega, x) : \tau^-(\omega, x) < t\} = \{(\omega, x) : (t, \omega, x) \in D\},$$
hence the $\tau^\pm$ are measurable.

(ii) For fixed ω and $t > 0$,
$$\{x : \tau^+(\omega, x) > t\} = \{x : (t, \omega, x) \in D\} = D(t, \omega)$$
is open, hence $\tau^+(\omega, \cdot)$ is lower semicontinuous. Analogously for $\tau^-(\omega, \cdot)$.

(iii) This is (1.2.2) since $(t, x) \in D(\omega)$ if and only if $t \in D(\omega, x)$.

(iv) It suffices to prove the inversion formula (1.2.4). By part (iii) of Definition 1.2.1, we have for $t \in D(\omega, x)$
$$-t \in D(\theta(t)\omega, \varphi(t, \omega)x) \iff 0 \in D(\omega, x),$$
which is always the case. Thus the local cocycle property holds for t and $-t$, yielding
$$\varphi(0, \omega)x = x = \varphi(-t, \theta(t)\omega)(\varphi(t, \omega)x).$$
Similarly, with t and $-t$ interchanged,
$$\varphi(0, \omega)x = x = \varphi(t, \omega)(\varphi(-t, \theta(t)\omega)x).$$
We conclude from either of these two equations for $\varphi(t, \omega) : D(t, \omega) \to R(t, \omega)$ that $R(t, \omega) \subset D(-t, \theta(t)\omega)$ and $D(t, \omega) \subset R(-t, \theta(t)\omega)$. Taken together
$$R(t, \omega) = D(-t, \theta(t)\omega) \quad \text{for all } t, \omega,$$
from which the inversion formula follows.

(v) Fix ω and consider the mapping $(t, x) \mapsto (t, \varphi(t, \omega)x)$ from $D(\omega)$ into $\mathbb{R} \times X$. This mapping is continuous with respect to (t, x) by definition and bijective by part (iv) of the theorem. The set $D(\omega) \subset \mathbb{R} \times X$ is open by definition and connected since $D(\omega, x)$ is an interval containing 0. As we have restricted ourselves to the case where X is a manifold, we can (as in the global case of Theorem 1.1.6) appeal to the principle of the invariance of domain (cf. Dieudonné [127: p. 52]) which says in particular that the mapping $(t, x) \mapsto (t, \varphi(t, \omega)x)$ is a homeomorphism, so
$$(t, x) \mapsto \varphi(t, \omega)^{-1}x = \varphi(-t, \theta(t)\omega)x$$
is continuous. That it is C^k follows from the formulae for the derivatives of the inverse of a C^k diffeomorphism, as in the proof of Theorem 1.1.6. □

1.2.4 Remark (Globality Test for Positive Time). Whether a local system is global or not can be checked using positive time only as follows: Assume we have already verified that $\tau^+(\omega, x) = \infty$ for all $(\omega, x) \in \Omega \times X$, i.e. that $D(t, \omega) = X$ for all $t \geq 0$ and $\omega \in \Omega$. Then

$$\tau^-(\omega, x) = -\infty \text{ for all } (\omega, x) \in \Omega \times X \iff$$
$$\varphi(t, \omega) : X \to X \text{ is surjective for all } (\omega, t) \in \Omega \times \mathbb{R}^+,$$

i.e. $R(t, \omega) = X$ for all $(\omega, t) \in \Omega \times \mathbb{R}^+$. By Theorem 1.2.3(iv), this is equivalent to $D(t, \omega) = X$ for times $t \leq 0$.

We stress, however, that $\tau^+ = \infty$ and $\tau^- = -\infty$ are two independent criteria. Many nonlinear RDS relevant in applications are only forward global, but explode backwards in time, e.g. the noisy Duffing-van der Pol oscillator (see Sect. 9.4 and Schenk-Hoppé [303, 304]). ∎

1.3 Perfection of a Crude Cocycle

In Sect. 2.3 we will see that it follows from the uniqueness of the solution of a stochastic differential equation that the solution is a crude cocycle. The question is: how close is a (very) crude cocycle to a perfect one? More specifically, can a (very) crude cocycle φ be perfected, i.e. is there a ψ which is indistinguishable from φ (see Appendix A.2) for which the cocycle property holds identically? This is a technical point of great importance. For example, only for perfect cocycles does the formula

$$\varphi(t, \omega)^{-1} = \varphi(-t, \theta(t)\omega)$$

hold, and the skew product $\Theta = (\theta, \varphi)$ is a flow if and only if φ is perfect.

The strongest argument, however, in favor of perfection is the fact that the multiplicative ergodic theorem for a perfect linearized cocycle holds on an *invariant* set $\tilde{\Omega}$ of full measure (see Chap. 3). This makes constructions based on it (like invariant manifolds, normal forms, stochastic bifurcations etc.) much easier to handle.

Perfection techniques were first developed for crude multiplicative functionals by Walsh [334] and then applied to the perfection of crude helices by de Sam Lazaro and Meyer [123]. However, these methods are strictly limited to cocycles where $\varphi(t, \omega) \in (\mathbb{R}, +)$ (i.e. where composition is addition) and cannot be applied to cocycles consisting of mappings which are elements of more general groups. This was clearly seen by Bismut [72: p. 69] who doubted the possibility of perfection in general.

Perfecting a crude cocycle φ needs quite sophisticated changes on sets of probability zero in order not to arrive at contradictions, since the cocycle property couples those ω at which we have to change φ with those where it is already correctly defined. For example, assume that φ is an almost perfect cocycle, i.e. satisfies the cocycle property identically outside a null set $N \in \mathcal{F}$.

The first idea for constructing a perfect ψ would be to put $\psi(t,\omega) = \varphi(t,\omega)$ for all $t \in \mathbb{T}$ if $\omega \notin N$, and $\psi(t,\omega) = \mathrm{id}_X$ (say) for all $t \in \mathbb{T}$ if $\omega \in N$. Suppose for some $\omega \in N$ there exists an $s \in \mathbb{T}$ for which $\tilde{\omega} := \theta(s)\omega \notin N$. If ψ were a perfect cocycle we would have $\mathrm{id}_X = \varphi(t, \theta(s)\omega) \circ \mathrm{id}_X$, thus $\varphi(t, \tilde{\omega}) = \mathrm{id}_X$ for all $t \in \mathbb{T}$ which is typically a contradiction.

Fortunately, the perfection problem has a satisfactory solution in all relevant situations. For discrete time, a simple universal answer is possible.

1.3.1 Theorem (Perfection for Discrete Time). *Let φ be a very crude measurable/continuous/C^k cocycle over θ with discrete time $\mathbb{T}$. Then there exists a measurable/continuous/C^k cocycle ψ over θ which is perfect and indistinguishable from φ, i.e. there exists an $N \in \mathcal{F}$ with $\mathbb{P}(N) = 0$ and*

$$\{\omega : \psi(t,\omega) \neq \varphi(t,\omega) \text{ for some } t \in \mathbb{T}\} \subset N.$$

Proof. Let $N_{s,t}$ be the ω-set for which (1.1.2) does not hold for some fixed $s, t \in \mathbb{T}$. Let $N := \cup_{s,t \in \mathbb{T}} N_{s,t}$ and $\Omega_0 := N^c$. Consider the set of full measure

$$\tilde{\Omega} := \cap_{t \in \mathbb{T}} \theta(t)^{-1} \Omega_0.$$

If $\mathbb{T} = \mathbb{Z}$ then $\theta(s)^{-1}\tilde{\Omega} = \tilde{\Omega}$, i.e. $\tilde{\Omega}$ is invariant and

$$\psi(t,\omega) := \begin{cases} \varphi(t,\omega), & \omega \in \tilde{\Omega},\ t \in \mathbb{T}, \\ \mathrm{id}_X, & \omega \notin \tilde{\Omega},\ t \in \mathbb{T}, \end{cases}$$

is obviously a perfect cocycle, which is indistinguishable from φ since $\psi = \varphi$ on $\tilde{\Omega}$. Hence we can choose $N = \tilde{\Omega}^c$.

If $\mathbb{T} = \mathbb{Z}^+$ (extension from $\mathbb{N}$ to $\mathbb{Z}^+$ is always possible by putting $\varphi(0,\omega) = \mathrm{id}_X$) then $\theta(s)\tilde{\Omega} \subset \tilde{\Omega}$, i.e. $\tilde{\Omega}$ is forward invariant. Define the random variable

$$\tau(\omega) := \min\{t \geq 0 : \theta(t)\omega \in \tilde{\Omega}\},$$

which is the first entrance time of θ into $\tilde{\Omega}$. Note that (a) $\tau(\theta(t)\omega) = (\tau(\omega) - t)^+$ and (b) $\theta(t)\omega \in \tilde{\Omega}$ for all $t \in [\tau(\omega), \infty)$ (this statement is vacuous for $\tau(\omega) = \infty$). Define

$$\psi(t,\omega) := \begin{cases} \varphi(t,\omega), & \omega \in \tilde{\Omega},\ t \in \mathbb{T}, \\ \mathrm{id}_X, & \omega \notin \tilde{\Omega},\ 0 \leq t < \tau(\omega), \\ \varphi(t - \tau(\omega), \theta(\tau(\omega))\omega), & \omega \notin \tilde{\Omega},\ t \geq \tau(\omega). \end{cases} \quad (1.3.1)$$

If $\tau(\omega) = \infty$ then the second line of (1.3.1) applies. Since $\psi = \varphi$ on $\tilde{\Omega}$, ψ and φ are indistinguishable. It can be easily checked that ψ is a perfect cocycle of the same regularity category as φ (exercise). □

For continuous time, the perfection of a crude cocycle is a much more subtle problem. The following theorem of Scheutzow (see Arnold and Scheutzow [35]) and its proof were inspired by a theorem of Zimmer [353: Theorem B.9]) on the perfection of measurable cocycles.

We will discuss the perfection problem for the following more general setting:

1. The group $(\mathbb{R}, +)$ is replaced with a locally compact Hausdorff topological group $(G, *)$. Our proof relies on the fact that such a group has a Haar measure.
2. $\theta : G \times \Omega \to \Omega$, $(t, \omega) \mapsto \theta(t)\omega$, is an action of G (on the left) on the set Ω, i.e. it satisfies

$$\theta(e_G)\omega = \omega \quad \text{for all } \omega \in \Omega, \quad e_G \text{ the identity of } G,$$

$$\theta(t * s)\omega = \theta(t)\theta(s)\omega \quad \text{for all } t, s \in G, \omega \in \Omega.$$

3. The probability measure $\mathbb{P}$ invariant under θ is replaced with a σ-finite measure μ which is quasi-invariant under θ, i.e. $A \in \mathcal{F}$ and $\mu(A) = 0$ implies $(\theta(t)\mu)(A) = 0$ for all $t \in G$.
4. The cocycle as a family of self-mappings $\varphi(t, \omega) : X \to X$ of some space X is replaced with a *group-valued cocycle* over θ, i.e. a function $\varphi : G \times \Omega \to H$ with values in some topological group $(H, \circ)$ satisfying the perfect, or crude, or very crude cocycle property: For all $t, s \in G$

$$\varphi(t * s, \omega) = \varphi(t, \theta(s)\omega) \circ \varphi(s, \omega) \tag{1.3.2}$$

for all $\omega \in \Omega$, or $\omega \notin N_s$, or $\omega \notin N_{s,t}$, respectively. Note that (1.3.2) and the fact that we are in a group implies that

$$\varphi(e_G, \omega) = e_H, \quad e_H \text{ the identity of } H,$$

for all $\omega \in \Omega$, or μ-almost all $\omega \in \Omega$, respectively.

1.3.2 Theorem (Perfection for Continuous "Time" $\mathbb{T} = G$). *Assume that $(H, \circ)$ and $(G, *)$ are Hausdorff topological groups with a countable base of their topologies, with respective Borel σ-algebras $\mathcal{H}$ and $\mathcal{G}$. In addition, assume that G is locally compact, and let $(\Omega, \mathcal{F}, \mu)$ be a σ-finite measure space, $\theta : G \times \Omega \to \Omega$ be $\mathcal{G} \otimes \mathcal{F}, \mathcal{F}$-measurable such that*

$$\theta(e_G)\omega = \omega \quad \text{for all } \omega \in \Omega, \quad e_G \text{ the identity of } G,$$

and

$$\theta(t * s)\omega = \theta(t)\theta(s)\omega \quad \text{for every } s, t \in G, \omega \in \Omega.$$

Further, assume that μ is non-trivial and quasi-invariant under θ, i.e. $A \in \mathcal{F}$, $\mu(A) = 0$ implies $(\theta(t)\mu)(A) = 0$ for all $t \in G$ and let $\varphi : G \times \Omega \to H$ satisfy

(i) *φ is $\mathcal{G} \otimes \mathcal{F}, \mathcal{H}$-measurable;*
(ii) *$\varphi(\cdot, \omega) : G \to H$ is continuous for μ-almost every $\omega \in \Omega$;*
(iii) *Very crude cocycle property: For every $s, t \in G$ there exists $N_{s,t} \in \mathcal{F}$ such that $\mu(N_{s,t}) = 0$ and*

$$\varphi(t * s, \omega) = \varphi(t, \theta(s)\omega) \circ \varphi(s, \omega) \quad \text{for } \omega \notin N_{s,t}.$$

Then there exists $\psi : G \times \Omega \to H$ which satisfies (i),

(ii)' $\psi(\cdot, \omega) : G \to H$ is continuous for all $\omega \in \Omega$.
(iii)' Perfect cocycle property:

$$\psi(t * s, \omega) = \psi(t, \theta(s)\omega) \circ \psi(s, \omega) \quad \text{for every } s, t \in G, \ \omega \in \Omega.$$

In particular, $\psi(e_G, \omega) = e_H$ for all $\omega \in \Omega$.

(iv) φ and ψ are indistinguishable. Moreover, if $\varphi(\cdot, \omega)$ is continuous for all $\omega \in \Omega$, then

$$\tilde{N} := \{\omega : \varphi(t, \omega) \neq \psi(t, \omega) \text{ for some } t \in G\} \in \mathcal{F},$$

and $\mu(\tilde{N}) = 0$.

Proof. *Step 1:* Preparations: By (ii), there exists $N \in \mathcal{F}$, $\mu(N) = 0$, such that $\{\omega : \varphi(\cdot, \omega) \text{ is not continuous}\} \subset N$. Redefining φ on N to be identically equal to e_H, we see that we can assume that φ satisfies (i), (ii)' and (iii).

Let $G_0 \subset G$ be countable dense, and define $N_s := \cup_{t \in G_0} N_{s,t}$ for $s \in G$. Then $N_s \in \mathcal{F}$, $\mu(N_s) = 0$. Since every $t \in G$ is the limit of a generalized sequence from G_0, since $\varphi(\cdot, \omega)$ is continuous, and since the limit of a generalized sequence is unique in the Hausdorff space H, we have:

(iii)'' Crude cocycle property:

$$\varphi(t * s, \omega) = \varphi(t, \theta(s)\omega) \circ \varphi(s, \omega) \quad \text{for every } t, s \in G, \ \omega \notin N_s.$$

Hence we can, and will, assume that φ satisfies (i), (ii)' and (iii)''.

Let ν be a probability measure on $(G, \mathcal{G})$ which is equivalent to a Haar measure and assume without loss of generality that μ is a probability measure (for the construction of such an equivalent probability measure find a positive function which is integrable with respect to the original measure) (exercise).

Define

$$M := \{(s, \omega) \in G \times \Omega : \varphi(t * s, \omega) = \varphi(t, \theta(s)\omega) \circ \varphi(s, \omega) \text{ for all } t \in G\},$$

$$\Omega_0 := \{\omega \in \Omega : (s, \omega) \in M \text{ for } \nu\text{-a.a. } s \in G\},$$

$$\Omega_1 := \{\omega \in \Omega : \theta(s)\omega \in \Omega_0 \text{ for } \nu\text{-a.a. } s \in G\}.$$

Step 2: We show that $\Omega_0, \Omega_1 \in \mathcal{F}$, $\mu(\Omega_0) = \mu(\Omega_1) = 1$ and that Ω_1 is invariant under θ, i. e. $\omega \in \Omega_1$ implies $\theta(t)\omega \in \Omega_1$ for all $t \in G$.

Note that since φ is continuous, H is Hausdorff and, by the definition of a Haar measure, ν charges all non-empty open subsets of G, the set M remains unchanged if we replace "for all $t \in G$" by "for ν-a.a. $t \in G$".

The map

$$(s, t, \omega) \mapsto \varphi(t * s, \omega) \circ \varphi(s, \omega)^{-1} \circ \varphi(t, \theta(s)\omega)^{-1} =: A(s, t, \omega)$$

is $\mathcal{G} \otimes \mathcal{G} \otimes \mathcal{F}, \mathcal{H}$-measurable since for topological spaces $(T, \mathcal{T})$, $(U, \mathcal{U})$ with a countable base the Borel σ-algebra of $T \times U$ coincides with $\mathcal{T} \otimes \mathcal{U}$ (see Dudley

[131: p. 90]), hence continuity entails measurability. Since H is Hausdorff, $\{e_H\}$ is closed, hence Borel, and
$$A^{-1}(e_H) \in \mathcal{G} \otimes \mathcal{G} \otimes \mathcal{F}.$$
Using (iii)" and Fubini's theorem we get $(\nu \otimes \nu \otimes \mu)(A^{-1}(e_H)) = 1$. Again by Fubini's theorem, using
$$(s,\omega) \in M \iff \int_G 1_{A^{-1}(e_H)}(s,t,\omega)\,d\nu(t) = 1,$$
we obtain $M \in \mathcal{G} \otimes \mathcal{F}$, $(\nu \otimes \mu)(M) = 1$. Similarly, $\Omega_0 \in \mathcal{F}$ and $\mu(\Omega_0) = 1$. Further, $\Omega_1 \in \mathcal{F}$ follows from Fubini's theorem and the relation
$$\omega \in \Omega_1 \iff \int_G \int_G 1_M(s,\theta(u)\omega)\,d\nu(s)\,d\nu(u) = 1,$$
and $\mu(\Omega_1) = 1$ again by Fubini's theorem and also since ν is equivalent to a Haar measure.

Finally, Ω_1 is invariant under θ since ν is equivalent to a Haar measure.

Step 3: Define
$$\psi(t,\omega) := \begin{cases} \varphi(t*s^{-1},\theta(s)\omega) \circ \varphi(s^{-1},\theta(s)\omega)^{-1}, & \omega \in \Omega_1,\ \theta(s)\omega \in \Omega_0, \\ e_H, & \omega \notin \Omega_1. \end{cases}$$

Remember that for $\omega \in \Omega_1$, $\theta(s)\omega \in \Omega_0$ for ν-a.a. $s \in G$. We show that ψ is well-defined.

Assume that $\omega \in \Omega_1$, $\theta(s)\omega \in \Omega_0$ and $\theta(u)\omega \in \Omega_0$. By the definition of Ω_0 and since ν is equivalent to a Haar measure, we have for all $t \in G$ and $r \notin N^{s,u}$, where $N^{s,u}$ is a ν-null set,
$$\varphi(t*s^{-1},\theta(s)\omega) = \varphi(t*r^{-1},\theta(r)\omega) \circ \varphi(r*s^{-1},\theta(s)\omega) \tag{1.3.3}$$
and
$$\varphi(t*u^{-1},\theta(u)\omega) = \varphi(t*r^{-1},\theta(r)\omega) \circ \varphi(r*u^{-1},\theta(u)\omega). \tag{1.3.4}$$

Now for $r \notin N^{s,u}$ and consecutively using (1.3.3) for $t = 0$ to transform $\varphi(s^{-1},\theta(s)\omega)^{-1}$, (1.3.4) for $t = 0$ to transform $\varphi(r^{-1},\theta(r)\omega)^{-1}$, (1.3.4) to transform $\varphi(r*u^{-1},\theta(u)\omega)$, and finally (1.3.3) to transform $\varphi(r*s^{-1},\theta(s)\omega)^{-1} \circ \varphi(t*r^{-1},\theta(r)\omega)^{-1}$ we deduce that
$$\varphi(t*s^{-1},\theta(s)\omega) \circ \varphi(s^{-1},\theta(s)\omega)^{-1} = \varphi(t*u^{-1},\theta(u)\omega) \circ \varphi(u^{-1},\theta(u)\omega)^{-1}.$$
Hence ψ is well-defined.

Step 4: Obviously $\psi(\cdot,\omega) : G \to H$ is continuous for all $\omega \in \Omega$. (iv) follows since φ and ψ agree on the set $\Omega_0 \cap \Omega_1 \cap N_{e_G}^c$ of full μ-measure, where N_{e_G} is the exceptional set constructed in step 1 for $s = e_G$.

Step 5: We show that ψ satisfies the perfect cocycle property (iii)': Fix $s, t \in G$ and $\omega \in \Omega$. The assertion is clear for $\omega \notin \Omega_1$, so we assume $\omega \in \Omega_1$, hence $\theta(s)\omega \in \Omega_1$. Pick $u \in G$ such that $\theta(u)\omega \in \Omega_0$ and $\theta(u * s^{-1})\omega \in \Omega_0$ (this holds even for ν-a.a. $u \in G$). Then by the definition of ψ

$$\psi(t * s, \omega) = \varphi(t * s * u^{-1}, \theta(u)\omega) \circ \varphi(u^{-1}, \theta(u)\omega)^{-1},$$

$$\psi(s, \omega) = \varphi(s * u^{-1}, \theta(u)\omega) \circ \varphi(u^{-1}, \theta(u)\omega)^{-1},$$

and

$$\psi(t, \theta(s)\omega) = \varphi(t * s * u^{-1}, \theta(u)\omega) \circ \varphi(s * u^{-1}, \theta(u)\omega)^{-1},$$

so (iii)' follows.

Step 6: It remains to show (i) for ψ. Define

$$B(s, t, \omega) := \begin{cases} \varphi(t * s^{-1}, \theta(s)\omega) \circ \varphi(s^{-1}, \theta(s)\omega)^{-1}, & \omega \in \Omega_1, \theta(s)\omega \in \Omega_0, \\ e_H, & \text{otherwise.} \end{cases}$$

Then B is $\mathcal{G} \otimes \mathcal{G} \otimes \mathcal{F}, \mathcal{H}$-measurable (recall that $\Omega_0 \in \mathcal{F}$ and θ is jointly measurable). According to step 3, $\varphi(t * s^{-1}, \theta(s)\omega) \circ \varphi(s^{-1}, \theta(s)\omega)^{-1}$ does not depend on s as long as $\omega \in \Omega_1$ and $\theta(s)\omega \in \Omega_0$, the latter happening for ν-a.a. $s \in G$ by the definition of Ω_1. Since the countable base of H generates $\mathcal{H}$ and separates points, $(H, \mathcal{H})$ is isomorphic as a measurable space to a subset of $[0, 1]$ (see Zimmer [353: p. 194]). Consequently,

$$\psi(t, \omega) = \int_G B(s, t, \omega) \, d\nu(s)$$

for all t, ω, so by Fubini's theorem ψ is $\mathcal{G} \otimes \mathcal{F}, \mathcal{H}$-measurable.

If $\varphi(\cdot, \omega)$ is continuous for all $\omega \in \Omega$ then

$$\{\omega : \varphi(t, \omega) = \psi(t, \omega) \quad \text{for all } t \in G\} =$$
$$\{\omega : \varphi(t, \omega) = \psi(t, \omega) \quad \text{for all } t \in G_0\} \in \mathcal{F},$$

thus proving the last statement of (iv). □

1.3.3 Remark. (i) If $G = (\mathbb{R}, +)$ and in (ii) and (ii)' "continuous" is replaced by "right continuous" or "left continuous" (or "cadlag", "caglad", see Appendix A.2) then Theorem 1.3.2 remains true.

(ii) If $G = (\mathbb{R}, +)$ and a complete filtration $(\mathcal{F}_t)_{t \in \mathbb{R}}$ or $(\mathcal{F}_s^t)_{s \leq t}$ is given on $(\Omega, \bar{\mathcal{F}}^\mu, \mu)$ (see Sect. 2.3) and if in addition φ is adapted (i.e. $\varphi(t, \cdot)$ is $\mathcal{F}_t$-measurable, or $\varphi(t, \omega) \circ \varphi(s, \omega)^{-1}$ is $\mathcal{F}_s^t$-measurable), then obviously ψ is also adapted. Moreover if H is also metrizable, then both φ and ψ are in fact progressively measurable. ∎

We now apply this general theorem to our situation and obtain the following corollaries.

1.3.4 Corollary (Perfection of a Very Crude Continuous Cocycle). Let $\mathbb{T} = \mathbb{R}$. Let φ be a very crude continuous cocycle over the metric DS θ on the Hausdorff topological space X, where X is either

- compact metric, or
- locally compact and locally connected[2] with a countable base (e. g. a manifold).

Then φ can be perfected to a continuous cocycle ψ which is indistinguishable from φ.

Proof. By the definition of a continuous cocycle and by Theorem 1.3.2, $\mathbb{R} \ni t \mapsto \varphi(t,\omega) \in \mathrm{Homeo}(X)$ is continuous for all $\omega \in \Omega$. It is known (see Arens [3]) that, under the above conditions, $(H, \circ) = (\mathrm{Homeo}(X), \circ)$ endowed with the compact-open topology is a Hausdorff topological group which has a countable base (in fact, is even a Polish group). □

1.3.5 Corollary (Perfection of a Very Crude C^k Cocycle). Let $\mathbb{T} = \mathbb{R}$ and let φ be a very crude C^k cocycle, where $1 \leq k \leq \infty$. Then φ can be perfected to a C^k cocycle ψ which is indistinguishable from φ.

Proof. It is known that $\mathrm{Diff}^k(X)$, $1 \leq k \leq \infty$, endowed with the C^k compact-open topology is a Hausdorff topological group which has a countable base (in fact, is even a Polish group). For more details see Appendix B.5 or Baxendale [54: p. 21]. □

1.3.6 Remark. (i) If φ takes values in a subgroup H of $\mathrm{Homeo}(X)$ or $\mathrm{Diff}^k(X)$ (endowed with its respective relative topology) then the perfected ψ also takes values in H.

(ii) One-sided time $\mathbb{T} = \mathbb{R}^+$: The fact that we need to assume that time is a group leaves the perfection problem unsolved here for the case $\mathbb{T} = \mathbb{R}^+$. However, in case φ is generated by a random or stochastic differential equation, the latter driven by a continuous semimartingale helix, we can always assume (basically without loss of generality) that $\mathbb{T} = \mathbb{R}$ (see Sects. 2.2 and 2.3).

If a stochastic differential equation in $\mathbb{R}^d$ is driven by a semimartingale helix which is not necessarily continuous we have a genuinely one-sided situation, since the solution (semi-)flow only takes values in the semigroup $\mathcal{C}(\mathbb{R}^d, \mathbb{R}^d)$ or $\mathcal{C}^k(\mathbb{R}^d, \mathbb{R}^d)$. The perfection problem in this case is considered by Kager [193] and Kager and Scheutzow [194]. ■

1.4 Invariant Measures for Measurable Random Dynamical Systems

Let φ be a measurable RDS over θ. Suppose the probability measure μ on $(\Omega \times X, \mathcal{F} \otimes \mathcal{B})$ is invariant for the skew-product Θ corresponding to φ, i. e.

---

[2] A topological space is called *locally connected*, if the connected open sets form a base of the topology.

$\Theta(t)\mu = \mu$ for all $t \in \mathbb{T}$. As $\pi_\Omega \circ \Theta(\cdot) = \theta(\cdot) \circ \pi_\Omega$, where $\pi_\Omega : \Omega \times X \to \Omega$, $\pi_\Omega(\omega, x) = \omega$, is the projection onto Ω, we have

$$\pi_\Omega \mu = \theta(t)(\pi_\Omega \mu) \quad \text{for all } t \in \mathbb{T}.$$

Hence the marginal $\pi_\Omega \mu$ of μ on $(\Omega, \mathcal{F})$ is θ-invariant. As our philosophy is that noise is something given to us and not at our disposal, we require that $\pi_\Omega \mu = \mathbb{P}$, which leads to the following definition.

1.4.1 Definition (Invariant Measure for RDS). Given a measurable RDS φ over θ, a probability measure μ on $(\Omega \times X, \mathcal{F} \otimes \mathcal{B})$ is said to be an *invariant measure for the RDS φ*, or *φ-invariant*, if it satisfies

1. $\Theta(t)\mu = \mu$ for all $t \in \mathbb{T}$,
2. $\pi_\Omega \mu = \mathbb{P}$. ∎

Define

$$\mathcal{P}_\mathbb{P}(\Omega \times X) := \{\mu \text{ probability on } (\Omega \times X, \mathcal{F} \otimes \mathcal{B}) \text{ with marginal } \mathbb{P} \text{ on } (\Omega, \mathcal{F})\}$$

and

$$\mathcal{I}_\mathbb{P}(\varphi) := \{\mu \in \mathcal{P}_\mathbb{P}(\Omega \times X) : \mu \ \varphi\text{-invariant}\},$$

which are both convex sets, if we define $(\alpha \mu_1 + \beta \mu_2)(\cdot) := \alpha \mu_1(\cdot) + \beta \mu_2(\cdot)$ and $\Theta(t)(\alpha \mu_1 + \beta \mu_2) := \alpha \Theta(t) \mu_1 + \beta \Theta(t) \mu_2$.

1.4.2 Remark. (i) For two-sided time, it suffices to require condition 1 only for $t > 0$, since $\Theta(-t)\mu = \Theta(-t)\Theta(t)\mu = \mu$.

(ii) For discrete time, condition 1 follows from $\Theta(1)\mu = \mu$.

(iii) Invariant measures which are σ-finite, but not finite, are interesting in stochastic bifurcation theory, see Sect. 9.5. ∎

Note, however, that an RDS in general does not come equipped with an invariant measure (it comes only with a θ-invariant $\mathbb{P}$ on $(\Omega, \mathcal{F})$). In fact, finding and studying the invariant measures of an RDS which are lifts of a given $\mathbb{P}$ on $(\Omega, \mathcal{F})$ to $(\Omega \times X, \mathcal{F} \otimes \mathcal{B})$ will be one of the major tasks of our theory of RDS.

Note also that an invariant measure is not a product measure in general, see the examples below.

Suppose $\mu \in \mathcal{P}_\mathbb{P}(\Omega \times X)$. We call a function $\mu_\cdot(\cdot) : \Omega \times \mathcal{B} \to [0,1]$ a *factorization* (or *disintegration*, or *sample measure*) of μ with respect to $\mathbb{P}$ if

1. for all $B \in \mathcal{B}$, $\omega \mapsto \mu_\omega(B)$ is $\mathcal{F}$-measurable,
2. for $\mathbb{P}$-a.a. $\omega \in \Omega$, $B \mapsto \mu_\omega(B)$ is a probability measure on $(X, \mathcal{B})$,
3. for all $A \in \mathcal{F} \otimes \mathcal{B}$

$$\mu(A) = \int_\Omega \int_X 1_A(\omega, x) \mu_\omega(dx) \mathbb{P}(d\omega). \tag{1.4.1}$$

We write symbolically
$$\mu(d\omega, dx) = \mu_\omega(dx)\,\mathbb{P}(d\omega).$$

Condition 3 is also equivalent to the following one: For all $f \in L^1(\mu)$
$$\int_{\Omega \times X} f\,d\mu = \int_\Omega \left(\int_X f(\omega, x)\,\mu_\omega(dx) \right) \mathbb{P}(d\omega).$$

Introducing the sections $A_\omega := \{x : (\omega, x) \in A\}$, (1.4.1) can be written as
$$\mu(A) = \int_\Omega \mu_\omega(A_\omega) \mathbb{P}(d\omega).$$

The following proposition gives sufficient conditions for the existence and uniqueness of a factorization.

1.4.3 Proposition (Existence and Uniqueness of Factorization).
Let $\mu \in \mathcal{P}_\mathbb{P}(\Omega \times X)$. Suppose
(i) $\mathcal{B}$ is countably generated,
(ii) the marginal $\pi_X \mu$ on $(X, \mathcal{B})$ can be compactly approximated.
Then a factorization of μ exists and is $\mathbb{P}$-a.s. unique.
Conditions (i) and (ii) are satisfied for any μ if $(X, \mathcal{B})$ is a standard measurable space[3], in particular if X is a Polish space with its Borel σ-algebra.

For a proof see Gänssler and Stute [156: p. 196].

Let $\mathcal{E} \subset \mathcal{F}$ be a sub-σ-algebra and $\mu \in \mathcal{P}_\mathbb{P}(\Omega \times X)$. If the conditions (i) and (ii) of Proposition 1.4.3 are satisfied, then the factorization of the restriction of μ to $\mathcal{E} \otimes \mathcal{B}$ with respect to the restriction of $\mathbb{P}$ to $\mathcal{E}$ is said to be the *conditional expectation of μ with respect to $\mathcal{E}$*, and is denoted by $\omega \mapsto \mathbb{E}(\mu.|\mathcal{E})_\omega$. Note that
$$\mathbb{E}(\mu.|\mathcal{E})_\omega(B) = \mathbb{E}(\mu.(B)|\mathcal{E})(\omega) \quad \mathbb{P}\text{-a.s.}$$
for any $B \in \mathcal{B}$.

1.4.4 Lemma (Factorization of Image Measure).
Let φ be a measurable RDS on a standard space and let $\mu \in \mathcal{P}_\mathbb{P}(\Omega \times X)$. If θ is measurably invertible (e.g. if $\mathbb{T}$ is two-sided) then the factorization of $\Theta(t)\mu$ is given by
$$(\Theta(t)\mu)_\omega = \varphi(t, \theta(t)^{-1}\omega) \mu_{\theta(t)^{-1}\omega} = \varphi(-t, \omega)^{-1} \mu_{\theta(-t)\omega}, \quad \mathbb{P}\text{-a.s.}, \quad (1.4.2)$$
where the second equality holds for two-sided time.

---

[3] See Appendix A.1 for the definition of a standard measurable space.

Proof. We check equation (1.4.2) on product sets $F \times B$ where $F \in \mathcal{F}$, $B \in \mathcal{B}$. We obtain

$$\begin{aligned}(\Theta(t)\mu)(F \times B) &= \int_{\theta(t)^{-1}F} (\varphi(t,\omega)\mu_\omega)(B)\,\mathbb{P}(d\omega) \\ &= \int_\Omega 1_F(\theta(t)\omega)(\varphi(t,\omega)\mu_\omega)(B)\,\mathbb{P}(d\omega) \\ &= \int_\Omega 1_F(\omega)(\varphi(t,\theta(t)^{-1}\omega)\mu_{\theta(t)^{-1}\omega})(B)\,\mathbb{P}(d\omega),\end{aligned}$$

using the θ-invariance of $\mathbb{P}$ for the last equation. $\square$

The next theorem rewrites invariance of μ in terms of its factorization.

1.4.5 Theorem (Invariance in Terms of Factorization). *Let φ be a measurable RDS on a standard space $(X, \mathcal{B})$ and let $\mu \in \mathcal{P}_\mathbb{P}(\Omega \times X)$. Then*

(i) $\mu \in \mathcal{I}_\mathbb{P}(\varphi)$ if and only if for all $t \in \mathbb{T}$

$$\mathbb{E}(\varphi(t,\cdot)\mu.|\theta(t)^{-1}\mathcal{F})_\omega = \mu_{\theta(t)\omega} \quad \mathbb{P}\text{-a.s.} \tag{1.4.3}$$

(ii) If θ is measurably invertible (e.g. if $\mathbb{T}$ is two-sided) then $\theta(t)^{-1}\mathcal{F} = \mathcal{F}$ for all $t \in \mathbb{T}$, and $\mu \in \mathcal{I}_\mathbb{P}(\varphi)$ if and only if for all $t \in \mathbb{T}$

$$\varphi(t,\omega)\mu_\omega = \mu_{\theta(t)\omega} \quad \mathbb{P}\text{-a.s.} \tag{1.4.4}$$

Proof. (i) It suffices to check equation (1.4.3) for product sets $F \times B$, $F \in \mathcal{F}$, $B \in \mathcal{B}$. We have for those sets

$$\begin{aligned}(\Theta(t)\mu)(F \times B) &= \mu(\Theta(t)^{-1}(F \times B)) = \int_{\theta(t)^{-1}F} \mu_\omega(\varphi(t,\omega)^{-1}B)\,\mathbb{P}(d\omega) \\ &= \int_{\theta(t)^{-1}F} (\varphi(t,\omega)\mu_\omega)(B)\,\mathbb{P}(d\omega)\end{aligned}$$

and

$$\begin{aligned}\mu(F \times B) &= \int_F \mu_\omega(B)\,\mathbb{P}(d\omega) = \int_F \mu_\omega(B)\,(\theta(t)\mathbb{P})(d\omega) \\ &= \int_{\theta(t)^{-1}F} \mu_{\theta(t)\omega}(B)\,\mathbb{P}(d\omega).\end{aligned}$$

Thus if μ is invariant, then for each fixed B and t, the $\theta(t)^{-1}\mathcal{F}$-measurable function $\mu_{\theta(t)\cdot}(B)$ is a version of the conditional expectation $\mathbb{E}(\varphi(t,\cdot)\mu.(B)|\theta(t)^{-1}\mathcal{F})$ of the $\mathcal{F}$-measurable function $(\varphi(t,\cdot)\mu.)(B)$ (remember that $\theta(t)^{-1}\mathcal{F} \subset \mathcal{F}$).

The exceptional set here can depend on B and t. Since $\mathcal{B}$ is countably generated we can find a universal exceptional set, first for a countable generating algebra, then by the extension theorem for all of $\mathcal{B}$ (in continuous time still depending on t) outside of which (1.4.3) holds.

Conversely, if we start with (1.4.3) we arrive at the invariance of μ.

(ii) In the case that $\theta(t)$ is measurably invertible we have $\theta(t)^{-1}\mathcal{F} = \mathcal{F}$ and thus
$$(\varphi(t,\omega)\mu_\omega)(B) = \mu_{\theta(t)\omega}(B) \quad \text{P-a.s.},$$
form which we continue as in (i). □

1.4.6 Remark. (i) If property (1.4.4) is satisfied μ_ω is also called *equivariant* (with respect to φ over θ).

(ii) Of course (1.4.4) is sufficient for (1.4.3) and thus for the invariance of μ also in the case where θ is not necessarily invertible.

(iii) Let $\mu \in \mathcal{I}_\mathbb{P}(\varphi)$ with marginal $\rho = \pi_X \mu = \mathbb{E}\,\mu.$ on X. Then (1.4.3) and the θ-invariance of $\mathbb{P}$ imply that
$$\rho_t := \mathbb{E}\,\varphi(t,\cdot)\mu. = \rho \quad \text{for all } t \in \mathbb{T}.$$

(iv) A φ-invariant measure μ is called a *random Dirac measure* if there exists a random variable $x_0 : \Omega \to X$ with $\mu_\omega = \delta_{x_0(\omega)}$ P-a.s. Invariance in the case of two-sided time then reads as
$$\varphi(t,\omega)x_0(\omega) = x_0(\theta(t)\omega) \quad \text{P-a.s.} \quad \text{for all } t \in \mathbb{T}, \tag{1.4.5}$$
i.e. the orbit of the cocycle starting at the random initial value $x_0(\omega)$ is a stationary stochastic process in X. Relation (1.4.5) is also sufficient for the invariance of μ for one-sided time.

This situation in which μ is supported by the graph of a random variable will be encountered quite frequently. See Sect. 5.6 for affine RDS (and Remark 5.6.2 for nonlinear RDS) and Sect. 7.6 and Chap. 9 for various examples. ∎

1.4.7 Example (Invariant Product Measures). When is a product measure $\mu = \mathbb{P} \times \rho$ φ-invariant?

(i) In the case that θ is measurably invertible (in particular, if $\mathbb{T}$ is two-sided) $\mu_\omega = \rho$ P-a.s. if and only if $\varphi(t,\omega)\rho = \rho$ P-a.s. for all $t \in \mathbb{T}$, i.e. almost all mappings $\varphi(t,\omega)$ leave the measure ρ invariant – which will be a rare case.

(ii) If θ is not necessarily invertible, $\mu = \mathbb{P} \times \rho$ is invariant if and only if
$$\mathbb{E}(\varphi(t,\cdot)\rho|\theta(t)^{-1}\mathcal{F}) = \rho \quad \text{P-a.s.}$$
for all $t \in \mathbb{T}$. Suppose $\mathbb{T} = \mathbb{Z}^+$ or $\mathbb{R}^+$ and $\varphi(t,\cdot)$ and $\theta(t)^{-1}\mathcal{F}$ are independent for each $t \in \mathbb{T}$. Since then $(\varphi(t,\omega)\rho)(B) = \int_X 1_B(\varphi(t,\omega)x)\,\rho(dx)$ and $\theta(t)^{-1}\mathcal{F}$ are independent, the last condition becomes in this case
$$\mathbb{E}(\varphi(t,\cdot)\rho)(B) = \int_X \mathbb{P}\{\omega : \varphi(t,\omega)x \in B\}\,\rho(dx) = \rho(B), \tag{1.4.6}$$
in short
$$\mathbb{E}\,\varphi(t,\cdot)\rho = \rho,$$

meaning that ρ is *invariant on the average* with respect to φ. For RDS which are either iterations of i.i.d. mappings (see Sect. 2.1) or solutions of (classical) stochastic differential equations (see Sect. 2.3), the above-mentioned independence condition can be satisfied by a proper choice of the set-up. Moreover, the one-point motions are a Markov family with transition probability

$$P(t,x,B) = \mathbb{P}\{\omega : \varphi(t,\omega)x \in B\}.$$

Then (1.4.6) reads as

$$\int_X P(t,x,B)\,\rho(dx) = \rho(B).$$

Hence in these cases $\mu = \mathbb{P} \times \rho$ is φ-invariant if and only if ρ is φ-invariant on the average if and only if ρ is a stationary measure of the Markov transition probability P of the one-point motion of φ. ∎

The next lemma shows that if we are willing to accept an extension of the underlying DS θ (and if we only deal with one invariant probability μ) we can assume without loss of generality that μ is a random Dirac measure.

1.4.8 Lemma (Every Invariant Measure is Dirac on Extension).
Let $\mu \in \mathcal{I}_{\mathbb{P}}(\varphi)$ for a measurable/continuous/C^k RDS φ. If we extend the RDS by

$$(\bar{\Omega}, \bar{\mathcal{F}}, \bar{\mathbb{P}}, (\bar{\theta}(t))_{t\in\mathbb{T}}) := (\Omega \times X, \mathcal{F} \otimes \mathcal{B}, \mu, (\Theta(t))_{t\in\mathbb{T}}),$$

$$\bar{\varphi}(t, \bar{\omega}) := \varphi(t, \omega),$$

then $\bar{\varphi}$ is a measurable/continuous/C^k RDS on X over $\bar{\theta}$, and the measure $\bar{\mu}$ defined through its factorization $\bar{\mu}_{\bar{\omega}}(dx) := \delta_{x_0(\bar{\omega})}(dx)$, where $x_0(\bar{\omega}) = x_0(\omega, x) := x$, is $\bar{\varphi}$-invariant, $\bar{\mu} \in \mathcal{I}_{\bar{\mathbb{P}}}(\bar{\varphi})$.

Proof. $\bar{\mu}$ is invariant if and only if $\bar{\Theta}(t)\bar{\mu}(\bar{f}) = \bar{\mu}(\bar{f})$ for all measurable bounded functions $\bar{f}$ on $\bar{\Omega} \times X$, where $\bar{\Theta}(t)(\bar{\omega}, y) = (\Theta(t)(\omega, x), \varphi(t,\omega)y)$. But, putting $f(\omega, x) := \bar{f}(\omega, x, x)$,

$$\bar{\Theta}(t)\bar{\mu}(\bar{f}) = \Theta(t)\mu(f) = \mu(f) = \bar{\mu}(\bar{f}).$$

□

1.5 Invariant Measures for Continuous RDS

1.5.1 Polish State Space

We can sometimes construct, or at least can assure existence of, invariant measures for continuous RDS.

1.5 Invariant Measures for Continuous RDS

Let in this whole subsection (unless stated otherwise) X be a Polish space (covering, in particular, any locally compact Hausdorff space with a countable base (see e.g. Bauer [51: 29.13]) such as $\mathbb{R}^d$, a manifold, or a compact metrizable Hausdorff space).

By Proposition 1.4.3 any probability measure $\mu \in \mathcal{P}_\mathbb{P}(\Omega \times X)$ has thus a $\mathbb{P}$-a.s. unique factorization

$$\mu(d\omega, dx) = \mu_\omega(dx)\mathbb{P}(d\omega),$$

and μ can be naturally identified with this factorization. Let $\mathcal{C}_b(X)$ be the Banach space of real-valued bounded continuous functions on X, with sup norm $\|f\|_b := \sup_{x \in X} |f(x)|$.

Call a function $f : \Omega \to \mathcal{C}_b(X)$ *measurable* if $(\omega, x) \mapsto f(\omega, x)$ is measurable, and define

$$L^1_\mathbb{P}(\Omega, \mathcal{C}_b(X)) = \{f : \Omega \to \mathcal{C}_b(X) \text{ measurable, } \|f\| := \int_\Omega \|f(\omega, \cdot)\|_b d\mathbb{P} < \infty\},$$

where, as usual, f and g are identified if $\|f - g\| = 0$.

1.5.1 Remark. (i) Since X is separable, $\omega \mapsto \|f(\omega, \cdot)\|_b$ is measurable.

(ii) The joint measurability of f could be rewritten using the following well-known lemma (see Castaing and Valadier [96: Lemma 3.14]).

1.5.2 Lemma. *Suppose $f : \Omega \times X \to Z$, where X is separable metric, Z is metric, $\omega \mapsto f(\omega, x)$ is measurable for each $x \in X$, and $x \mapsto f(\omega, x)$ is continuous for each $\omega \in \Omega$. Then $(\omega, x) \mapsto f(\omega, x)$ is measurable.*

(iii) For each $f \in L^1_\mathbb{P}(\Omega, \mathcal{C}_b(X))$ and $\mu \in \mathcal{P}_\mathbb{P}(\Omega \times X)$ we have $f \in L^1(\mu)$, and, putting as usual $\mu(f) = \int f d\mu$,

$$|\mu(f)| \leq \|f\|.$$

Further, $f \mapsto \mu(f)$ is linear for each μ, and $\mu \mapsto \mu(f)$ is affine for each f. ∎

1.5.3 Definition (Topology of Weak Convergence in $\mathcal{P}_\mathbb{P}(\Omega \times X)$).
We call the smallest topology in $\mathcal{P}_\mathbb{P}(\Omega \times X)$ which makes $\mu \mapsto \mu(f)$ continuous for each $f \in L^1_\mathbb{P}(\Omega, \mathcal{C}_b(X))$ the *topology of weak convergence* on $\mathcal{P}_\mathbb{P}(\Omega \times X)$. A net $\{\mu^\alpha\}$ converges in this topology to μ if $\mu^\alpha(f) \to \mu(f)$ for each $f \in L^1_\mathbb{P}(\Omega, \mathcal{C}_b(X))$. ∎

Suppose $\mathcal{P}_\mathbb{P}(\Omega \times X)$ and $\mathcal{P}(X)$ are endowed with their respective topologies of weak convergence. Then the projection $\pi_X : \mathcal{P}_\mathbb{P}(\Omega \times X) \to \mathcal{P}(X)$ defined by assigning to each $\mu \in \mathcal{P}_\mathbb{P}(\Omega \times X)$ its marginal $\rho = \pi_X(\mu) = \mathbb{E}\,\mu.$ on X is continuous.

For a thorough study of the topology of weak convergence we refer to Crauel [114]. In particular, Crauel gives criteria for compactness ("Prokhorov theory") and proves that $\mathcal{P}_\mathbb{P}(\Omega \times X)$ endowed with the topology of weak convergence is itself a Polish space if and only if the probability space $(\Omega, \mathcal{F}, \mathbb{P})$ is countably generated (Theorem 5.6, p. 55).

It follows from the next lemma that the topology of weak convergence is Hausdorff.

1.5.4 Lemma. *Let X be a metric space and $\mu, \nu \in \mathcal{P}_\mathbb{P}(\Omega \times X)$. Then*

$$\mu = \nu \iff \int f \, d\mu = \int f \, d\nu \quad \text{for all} \quad f \in L^1_\mathbb{P}(\Omega, \mathcal{C}_b(X)).$$

Proof. We prove only the non-trivial direction $\Leftarrow$. It suffices to show that $\mu(A \times F) = \nu(A \times F)$ for all $A \in \mathcal{F}$ and all closed F. This is implied by Lebesgue's theorem if we can exhibit a sequence $f_n(\omega, x) = 1_A(\omega) g_n(x) \in L^1_\mathbb{P}(\Omega, \mathcal{C}_b(X))$ with $g_n \downarrow 1_F$. But $g_n(x) = (1 - n\, d(x, F))^+$, where d is the metric on X, is such a sequence since $d(x, F) = 0 \iff x \in F$. □

Let φ be a measurable RDS with corresponding skew-product $(\omega, x) \mapsto \Theta(t)(\omega, x) := (\theta(t)\omega, \varphi(t, \omega)x)$. $\Theta(t)$ acts on functions on $\Omega \times X$ via

$$(\Theta(t)f)(\omega, x) := f(\Theta(t)(\omega, x)).$$

It also acts on measures μ on $(\Omega \times X, \mathcal{F} \otimes \mathcal{B})$ via

$$(\Theta(t)\mu)(A) := \mu(\Theta(t)^{-1} A), \quad A \in \mathcal{F} \otimes \mathcal{B},$$

or

$$(\Theta(t)\mu)(f) := \mu(\Theta(t)f), \quad f \in L^1(\mu).$$

1.5.5 Proposition. *Let φ be a continuous RDS[4] on a Polish space X. Then:*

(i) The mappings $f \mapsto \Theta(t)f$, $t \in \mathbb{T}$, are a commuting family of continuous linear mappings of $L^1_\mathbb{P}(\Omega, \mathcal{C}_b(X))$ to itself (isometries if $\mathbb{T}$ is two-sided).

(ii) The mappings $\mu \mapsto \Theta(t)\mu$, $t \in \mathbb{T}$, are a commuting family of affine mappings of $\mathcal{P}_\mathbb{P}(\Omega \times X)$ to itself which are continuous with respect to the topology of weak convergence.

Proof. (i) For fixed t, $(\Theta(t)f)(\omega, x) = f(\theta(t)\omega, \varphi(t, \omega)x)$ is measurable and for fixed ω continuous with respect to x. Since $\|(\Theta(t)f)(\omega, \cdot)\|_b \leq \|f(\theta(t)\omega, \cdot)\|_b$ (and the last inequality is an equality for two-sided time since $\varphi(t, \omega)$ is bijective),

$$\|\Theta(t)f\| \leq \|f\|,$$

with equality for two-sided time. This implies that $\Theta(t)$ is bounded. Linearity is clear.

(ii) Clearly $\Theta(t)$ is affine since by definition $\Theta(t)(\alpha\mu + (1-\alpha)\nu) = \alpha\Theta(t)\mu + (1-\alpha)\Theta(t)\nu$ for all $\alpha \in [0, 1]$. We prove that $\Theta(t)$ is continuous. Take a net $\mu^\alpha \to \mu$, i.e. $\mu^\alpha(f) \to \mu(f)$ for each $f \in L^1_\mathbb{P}(\Omega, \mathcal{C}_b(X))$. Then also $(\Theta(t)\mu^\alpha)(f) \to (\Theta(t)\mu)(f)$ since by definition $(\Theta(t)\mu)(f) = \mu(\Theta(t)f)$, and $\Theta(t)f \in L^1_\mathbb{P}(\Omega, \mathcal{C}_b(X))$, by part (i) of the proposition. □

---

[4] All statements of Sects. 1.5, 1.6 and 1.7 remain true if the assumption that φ is a continuous RDS is replaced with the more general assumption that φ is a measurable RDS of continuous mappings.

1.5 Invariant Measures for Continuous RDS

1.5.6 Corollary. *Let φ be a continuous RDS on a Polish space X. Then $\mathcal{I}_\mathbb{P}(\varphi)$ is a (possibly empty) convex and closed subset of $\mathcal{P}_\mathbb{P}(\Omega \times X)$.*

Proof. Convexity and closedness follow from the fact that $\mu \mapsto \Theta(t)\mu$ is affine and continuous on $\mathcal{P}_\mathbb{P}(\Omega \times X)$, respectively. □

1.5.7 Remark. Suppose $\mathbb{P}$ is ergodic. Then the set of extreme elements of the convex set $\mathcal{I}_\mathbb{P}(\varphi)$ coincides with the set of Θ-ergodic elements (i. e. those for which there is no non-trivial invariant set for Θ, see Definition 1.6.8), and any two distinct extreme elements are mutually singular. See Crauel [114: Lemma 6.19]. ■

So far we do not know whether or not there are in fact elements in $\mathcal{I}_\mathbb{P}(\varphi)$. The following procedure sometimes produces such elements.

1.5.8 Theorem (Krylov-Bogolyubov Procedure for Continuous RDS).
Let φ be a continuous RDS on a Polish space X. Define for an arbitrary $\nu \in \mathcal{P}_\mathbb{P}(\Omega \times X)$ and $N \in \mathbb{T}$, $N > 0$,

$$\mu_N(\cdot) = \begin{cases} \frac{1}{N} \sum_{n=0}^{N-1} \Theta(n)\nu(\cdot), & \mathbb{T} \text{ discrete,} \\ \frac{1}{N} \int_0^N \Theta(t)\nu(\cdot)\,dt, & \mathbb{T} \text{ continuous} \end{cases} \quad (1.5.1)$$

(similarly for $N < 0$ if $\mathbb{T}$ is two-sided). Then every limit point of μ_N for $N \to \infty$ in the topology of weak convergence (or for $N \to -\infty$ for two-sided time) is in $\mathcal{I}_\mathbb{P}(\varphi)$, and any $\mu \in \mathcal{I}_\mathbb{P}(\varphi)$ arises in this way.

Proof. For the final assertion put $\nu = \mu$, which produces the constant sequence $\mu_N = \mu$.

We treat only the continuous time case. Assume $\mu_{N_k} \to \mu$. This implies that for each fixed $t_0 \in \mathbb{T}$, $\Theta(t_0)\mu_{N_k} \to \Theta(t_0)\mu$ since $\mu \mapsto \Theta(t_0)\mu$ is continuous.

Let now $f \in L^1_\mathbb{P}(\Omega, C_b(X))$, $t_0 > 0$ fixed and $N_k > t_0$. Since $t \mapsto \Theta(t)\nu(f)$ is measurable and bounded (by $\|f\|$), hence locally integrable, the integral in (1.5.1) is meaningful. We have

$$\begin{aligned}|\Theta(t_0)\mu_{N_k}(f) - \mu_{N_k}(f)| &= \frac{1}{N_k}\left|\left(\int_{N_k}^{N_k+t_0} - \int_0^{t_0}\right)\Theta(t)\nu(f)\,dt\right| \\ &\leq \frac{2t_0}{N_k}\|f\| \to 0 \ (k \to \infty),\end{aligned}$$

hence $\Theta(t_0)\mu(f) = \mu(f)$. By Lemma 1.5.4, $\Theta(t_0)\mu = \mu$. Since $t_0 > 0$ is arbitrary, μ is φ-invariant (this is true also for two-sided time, by Remark 1.4.2(i)). □

1.5.9 Remark. (i) If in Theorem 1.5.8 θ is measurably invertible (in particular, if time is two-sided) then the factorized form of μ_N in (1.5.1) is

$$\mu_{N,\omega} = \begin{cases} \frac{1}{N}\sum_{n=0}^{N-1} \varphi(n,\theta^{-n}\omega)\nu_{\theta^{-n}\omega} = \frac{1}{N}\sum_{n=0}^{N-1} \varphi(-n,\omega)^{-1}\nu_{\theta^{-n}\omega}, \\ \frac{1}{N}\int_0^N \varphi(t,\theta(t)^{-1}\omega)\nu_{\theta(t)^{-1}\omega}dt = \frac{1}{N}\int_0^N \varphi(-t,\omega)^{-1}\nu_{\theta(-t)\omega}\,dt, \end{cases}$$

where the second expression is valid for two-sided time. This follows from Lemma 1.4.4.

We can see that (for two-sided time) the random Krylov-Bogolyubov procedure builds up a measure $\mu_{N,\omega}$ in the ω-fiber $\{\omega\} \times X$ (i. e. at time $t = 0$) by first transporting for each fixed t the measure $\nu_{\theta(-t)\omega}$ from $\{\theta(-t)\omega\} \times X$ by means of the mapping $\varphi(-t,\omega)^{-1}$ to a measure $\varphi(-t,\omega)^{-1}\nu_{\theta(-t)\omega}$ in $\{\omega\} \times X$, and then by averaging all those measures in $\{\omega\} \times X$ with respect to $t \in [0, N]$ (see Fig. 1.5).

(ii) Particular case: If $\nu_\omega = \delta_{\kappa(\omega)}$, $\kappa : \Omega \to X$ a random variable, then $\mu_{N,\omega}(B)$ is the proportion of time from $[0, N]$ which the orbit $t \mapsto \varphi(-t,\omega)^{-1}\kappa(\theta(-t)\omega) \in \{\omega\} \times X$ spends in the subset $B \in \mathcal{B}$.

(iii) The Krylov-Bogolyubov procedure remains valid if the initial measure ν is replaced with a sequence ν_n, and if in (1.5.1) μ_N is formed with ν_n. ∎

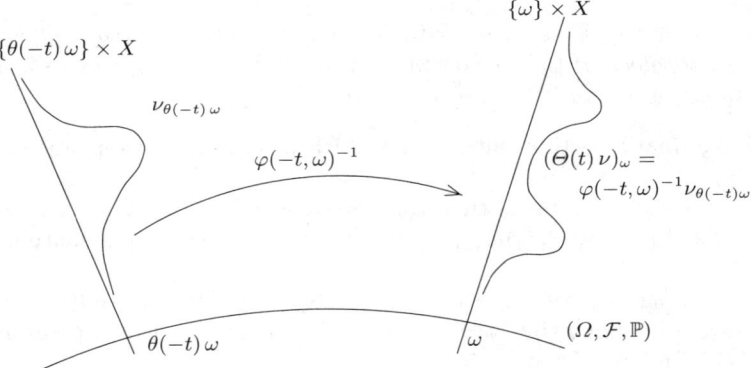

Fig. 1.5. The mechanism of the Krylov-Bogolyubov procedure

1.5.2 Compact Metric State Space

If the Polish space X is a compact metric space we can use classical duality theory.

For a compact metric space X the Riesz representation theorem asserts that $\mathcal{C}_b(X)^* = \mathcal{M}(X)$, where $\mathcal{M}(X)$ is the Banach space of signed measures on $(X, \mathcal{B})$ of finite total variation (this holds for any compact space). As a consequence,

$$L^1_{\mathbb{P}}(\Omega, \mathcal{C}_b(X))^* = L^\infty_{\mathbb{P}}(\Omega, \mathcal{M}(X)), \tag{1.5.2}$$

with the duality pairing given by

$$\langle f, \mu \rangle = \int_\Omega \int_X f(\omega, x) \mu_\omega(dx) \mathbb{P}(d\omega),$$

where $L^\infty_{\mathbb{P}}(\Omega, \mathcal{M}(X))$ is the space of $\mathbb{P}$-essentially bounded $\mathcal{M}(X)$-valued measurable functions. Here measurability is meant with respect to the Borel σ-algebra of the weak* topology on $\mathcal{M}(X)$ (see Ledrappier [230: p. 329] or Bourbaki [78: chapitre 6, § 2, N° 6]). We identify $\mathcal{P}_{\mathbb{P}}(\Omega \times X)$ with $L^\infty_{\mathbb{P}}(\Omega, \mathcal{P}(X))$ by identifying $\mu \in \mathcal{P}_{\mathbb{P}}(\Omega \times X)$ with its factorization $(\mu_\omega) \in L^\infty_{\mathbb{P}}(\Omega, \mathcal{P}(X))$,

$$\mathcal{P}_{\mathbb{P}}(\Omega \times X) \ni \mu \cong (\mu_\omega) \in L^\infty_{\mathbb{P}}(\Omega, \mathcal{P}(X)). \tag{1.5.3}$$

By the duality (1.5.2) and the identification (1.5.3) the topology of weak convergence of $\mathcal{P}_{\mathbb{P}}(\Omega \times X)$ as defined in Definition 1.5.3 coincides with the weak* topology in $L^\infty_{\mathbb{P}}(\Omega, \mathcal{P}(X))$.

1.5.10 Theorem (Existence of Invariant Measures). *Let X be a compact metric space. Then*

(i) $\mathcal{P}_{\mathbb{P}}(\Omega \times X) \cong L^\infty_{\mathbb{P}}(\Omega, \mathcal{P}(X))$ is a convex compact subset of $L^\infty_{\mathbb{P}}(\Omega, \mathcal{M}(X))$.

(ii) If φ is a continuous RDS on X, then the convex compact set of its invariant measures $\mathcal{I}_{\mathbb{P}}(\varphi)$ is non-void.

Proof. (i) Since by Alaoglu's theorem for each Banach space B the closed unit ball of B^* is compact in the B topology of B^*, the closed sets $L^\infty_{\mathbb{P}}(\Omega, \mathcal{P}(X))$ and $\mathcal{I}_{\mathbb{P}}(\varphi)$ are weak* compact in $B^* = L^\infty_{\mathbb{P}}(\Omega, \mathcal{M}(X))$.

(ii) By Proposition 1.5.5(ii), the mappings $\mu \mapsto \Theta(t)\mu$, $t \in \mathbb{T}$, are a commuting family of affine weak* continuous mappings of $L^\infty_{\mathbb{P}}(\Omega, \mathcal{P}(X)) \cong \mathcal{P}_{\mathbb{P}}(\Omega \times X)$ to itself. Hence by the Markov-Kakutani fixed point theorem (see Dunford and Schwartz [132: p. 456]) there is an element μ for which $\Theta(t)\mu = \mu$ for all $t \in \mathbb{T}$, i.e. $\mathcal{I}_{\mathbb{P}}(\varphi) \neq \emptyset$. □

1.5.11 Remark. (i) The Markov-Kakutani theorem can be applied to a not necessarily measurable cocycle of continuous maps.

(ii) By the Krein-Milman theorem (see Dunford and Schwartz [132: p. 440]) $\mathcal{I}_{\mathbb{P}}(\varphi)$ is equal to the closed convex hull of its extremal points. See Remark 1.5.7.

(iii) For a compact metric space X, $\mathcal{C}_b(X)$ is separable. Hence $L^1_{\mathbb{P}}(\Omega, \mathcal{C}_b(X))$ is separable if and only if the probability space $(\Omega, \mathcal{F}, \mathbb{P})$ is countably generated (see Appendix A.1). Assuming this, the topology of weak convergence of the compact Hausdorff space $\mathcal{P}_{\mathbb{P}}(\Omega \times X)$ will be even metrizable. Hence Choquet's theorem (see Phelps [277]) applies to $\mathcal{I}_{\mathbb{P}}(\varphi)$ and gives the following ergodic decomposition theorem: Suppose $\mathbb{P}$ is ergodic, then for each $\mu \in \mathcal{I}_{\mathbb{P}}(\varphi)$ there exists a unique Borel probability measure Q_μ such that $Q_\mu(\mathcal{E}) = 1$, $\mathcal{E}$ the extremal points of $\mathcal{I}_{\mathbb{P}}(\varphi)$, such that $\mu(f) = \int_\mathcal{E} e(f) \, dQ_\mu(e)$ for all $f \in L^1_{\mathbb{P}}(\Omega, \mathcal{C}_b(X))$. ∎

1.6 Invariant Measures on Random Sets

Again in this section let X be a Polish space.

It often happens that we want to construct a φ-invariant measure μ for which $\mu(A) = 1$ for some $A \in \mathcal{F} \otimes \mathcal{B}$, equivalently

$$\mu_\omega(A_\omega) = 1 \quad \mathbb{P}\text{-a.s.} ,$$

where, to recall, $A_\omega := \{x \in X : (\omega, x) \in A\} \in \mathcal{B}$ is the ω-section of A.

Let $A : \Omega \to \mathfrak{P}(\mathfrak{X})$, $\omega \mapsto A_\omega$, be a function whose values are subsets of X. Such a function is uniquely determined by its graph

$$\text{graph}(A_.) := \{(\omega, x) \in \Omega \times X : x \in A_\omega\} \subset \Omega \times X.$$

Conversely, every subset $A \subset \Omega \times X$ defines such a function via $\omega \mapsto A_\omega$.

Let for a metric space (X, d) the distance of x and B be defined by

$$d(x, B) := \inf\{d(x, y) : y \in B\}.$$

The following notion of a set-valued random variable will turn out to be crucial.

1.6.1 Definition (Random Set). The set-valued map $A : \Omega \to \mathfrak{P}(\mathfrak{X})$, $\omega \mapsto A_\omega$, where A_ω is closed (compact) for all $\omega \in \Omega$, is called a *random closed (compact) set* if for each $x \in X$ the map $\omega \mapsto d(x, A_\omega)$ is measurable. A *random open set* is a set-valued map U such that U^c is a random closed set. ∎

If A is a random closed set, then so is $\overline{A^c}$, the closure of A^c. If U is a random open set, then $\overline{U}$ is a random closed set. If A is a random closed set, then $\text{int}(A)$, the interior of A, is a random open set. If C_1 and C_2 are random compact sets, then so is $C_1 \cap C_2$ (see Crauel [114: Chap. 3] for all these statements).

The following important facts are quoted from Castaing and Valadier [96] (Theorem III.9, p. 67, and Theorem III.30, p. 80).

Recall that the σ-algebra $\mathcal{F}^u$ of *universally measurable sets* associated with the measurable space $(\Omega, \mathcal{F})$ is defined as

$$\mathcal{F}^u = \cap_Q \bar{\mathcal{F}}^Q,$$

where the intersection is taken over all probability measures Q on $(\Omega, \mathcal{F})$, and $\bar{\mathcal{F}}^Q$ denotes the completion of $\mathcal{F}$ with respect to Q.

1.6.2 Proposition. *Let the set-valued map $A : \Omega \to \mathfrak{P}(\mathfrak{X})$ take values in the subspace of closed subsets of a Polish space X. Then:*

(i) A is a random closed set if and only if for all open sets $U \subset X$ the set $\{\omega : A_\omega \cap U \neq \emptyset\}$ is measurable.

(ii) If A is a random closed set then

$$\text{graph}(A_\cdot) \in \mathcal{F} \otimes \mathcal{B}.$$

(iii) Conversely, if $\mathcal{F}$ contains the σ-algebra $\mathcal{F}^u$ of universally measurable sets (in particular, if $\mathcal{F}$ is $\mathbb{P}$-complete), then $\text{graph}(A_\cdot) \in \mathcal{F} \otimes \mathcal{B}$ implies that A is a random closed set.

The property of A being a random closed set is thus slightly stronger than $\text{graph}(A_\cdot)$ being measurable and A_ω being closed.

1.6.3 Proposition (Measurable Selection Theorem). *Let the set-valued map $A : \Omega \to \mathfrak{P}(\mathfrak{X})$ take values in the subspace of closed non-void subsets of a Polish space X. Then A is a random closed set if and only if there exists a sequence $(a_n)_{n \in \mathbb{N}}$ of measurable maps $a_n : \Omega \to X$ such that*

$$A_\omega = \text{closure } \{a_n(\omega) : n \in \mathbb{N}\} \quad \text{for all } \omega \in \Omega.$$

In particular, if A is a random closed set, then there exists a measurable selection, i.e. a measurable map $a : \Omega \to X$ such that $a(\omega) \in A_\omega$ for all $\omega \in \Omega$.

1.6.4 Definition (Support of a Measure). Let $\mu \in \mathcal{P}_\mathbb{P}(\Omega \times X)$.

(i) μ is said to be *supported* by the set $C \in \mathcal{F} \otimes \mathcal{B}$ (and C *supports* μ) if $\mu(C) = 1$, equivalently if

$$\mu_\omega(C_\omega) = 1 \quad \mathbb{P}\text{-a.s.}$$

Define
$$\mathcal{P}_\mathbb{P}(C) := \{\mu \in \mathcal{P}_\mathbb{P}(\Omega \times X) : \mu(C) = 1\}.$$

(ii) The *support* of μ is the set-valued map

$$\omega \mapsto C_\omega := \begin{cases} \text{supp}\mu_\omega, & \omega \in N_\mu^c, \\ X, & \omega \in N_\mu, \end{cases}$$

which by definition[5] takes its values in the subspace of closed non-void subsets of X. Here $N_\mu \in \mathcal{F}$ is a $\mathbb{P}$-null set outside of which μ_ω is a probability measure. ∎

1.6.5 Corollary. *The support $\omega \mapsto C_\omega$ of $\mu \in \mathcal{P}_\mathbb{P}(\Omega \times X)$ is a random closed set. In particular, $\text{graph}(C_\cdot) \in \mathcal{F} \otimes \mathcal{B}$.*

Proof. Since for any non-void open set $U \subset X$ and $\omega \in N_\mu^c$, $\mu_\omega(U) > 0$ if and only if $U \cap \text{supp}\mu_\omega \neq \emptyset$,

$$\begin{aligned} \{\omega : C_\omega \cap U \neq \emptyset\} &= N_\mu \cup \{\omega \in N_\mu^c : \text{supp}\mu_\omega \cap U \neq \emptyset\} \\ &= N_\mu \cup \{\omega \in N_\mu^c : \mu_\omega(U) > 0\} \end{aligned}$$

which is in $\mathcal{F}$ by the definition of μ_ω. Hence C is a random closed set (with $C_\omega \neq \emptyset$) by Proposition 1.6.2(i). □

---

[5] The *support* $\text{supp}\nu$ of a measure ν on $(X, \mathcal{B})$ is the intersection of all closed sets which support ν. We have $\nu(\text{supp}\nu) = 1$, which in fact holds for any topological space with a countable base.

1.6.6 Lemma. *Let μ^α be a net in $\mathcal{P}_\mathbb{P}(\Omega \times X)$ converging to μ in the topology of weak convergence. Then for each random closed set C*

$$\limsup_\alpha \mu^\alpha(C) \leq \mu(C),$$

and for each random open set U

$$\liminf_\alpha \mu^\alpha(U) \geq \mu(U).$$

Proof. Let C be a random closed set. Then $\omega \mapsto d(x, C_\omega)$ is measurable for each x and, clearly, $x \mapsto d(x, C_\omega)$ is continuous for each ω. Hence

$$f_n(\omega, x) := (1 - n\, d(x, C_\omega))^+ \in L^1_\mathbb{P}(\Omega, \mathcal{C}_b(X)),$$

$f_n \geq 1_C$, and $f_n \downarrow 1_C$. For each fixed n,

$$\limsup_\alpha \mu^\alpha(C) \leq \lim_\alpha \mu^\alpha(f_n) = \mu(f_n),$$

consequently

$$\limsup_\alpha \mu^\alpha(C) \leq \inf_n \mu(f_n) = \mu(C).$$

The last statement of the lemma follows by taking complements. $\square$

The final justification of the notion of a random set is given by the following consequence of the previous lemma.

1.6.7 Corollary. *(i) The set $\mathcal{P}_\mathbb{P}(C)$ of measures supported by a random closed set is convex and closed.*

(ii) Let φ be a continuous RDS and C be a random closed set. Then the set

$$\mathcal{I}_\mathbb{P}(\varphi|C) := \mathcal{I}_\mathbb{P}(\varphi) \cap \mathcal{P}_\mathbb{P}(C)$$

of φ-invariant measures supported by C is convex and closed.

Proof. (i) If $\mu^\alpha \to \mu$ and $\mu^\alpha(C) = 1$ then by Lemma 1.6.6 also $\mu(C) = 1$.
(ii) $\mathcal{I}_\mathbb{P}(\varphi|C)$ is the intersection of two convex and closed sets. $\square$

The key question is of course whether $\mathcal{I}_\mathbb{P}(\varphi|C)$ is non-void. To apply the random Krylov-Bogolyubov procedure or the Markov-Kakutani fixed point theorem we have to assure that $\mu \mapsto \Theta(t)\mu$ maps $\mathcal{P}_\mathbb{P}(C)$ into itself. This is guaranteed by an invariance property of C which we are going to introduce now.

1.6.8 Definition (Invariant Sets of RDS). Let φ be a measurable RDS and $C \subset \Omega \times X$ a set.
  (a) C is called *forward invariant* if for $t > 0$

$$C_\omega \subset \varphi(t, \omega)^{-1} C_{\theta(t)\omega} \quad \mathbb{P}\text{-a.s.},$$

equivalently

$$\varphi(t,\omega)C_\omega \subset C_{\theta(t)\omega} \quad \mathbb{P}\text{-a.s.}$$

(b) C is called *invariant* if for all $t \in \mathbb{T}$

$$C_\omega = \varphi(t,\omega)^{-1}C_{\theta(t)\omega} \quad \mathbb{P}\text{-a.s.},$$

for two-sided time equivalent to

$$\varphi(t,\omega)C_\omega = C_{\theta(t)\omega} \quad \mathbb{P}\text{-a.s.}$$

∎

1.6.9 Remark. (i) If there is no exceptional set in the definition of a (forward) invariant set we call it *strictly* (forward) invariant. A set C is strictly forward invariant if and only if $C \subset \Theta(t)^{-1}C$ for all $t > 0$, and strictly invariant if and only if $C = \Theta(t)^{-1}C$ for all $t \in \mathbb{T}$.

(ii) For two-sided time, it suffices to assume (strict) invariance for $t > 0$ only.

(iii) The definition of *(strict) backward invariance* is obtained from the forward definitions by replacing "$t > 0$" with "$t < 0$".

(iv) The term "positively invariant" is also used in place of "forward invariant".

∎

1.6.10 Exercise (Algebra of Invariant Sets). Prove the following random analogues of some properties of deterministic invariant sets (see Bhatia and Szegö [68: Sect. II.1]):

(A) Let φ be a measurable RDS. Then

1. arbitrary unions and intersections of strictly (forward) invariant sets are strictly (forward) invariant;
2. countable unions and intersections of (forward) invariant sets are (forward) invariant;
3. C is (strictly) forward invariant $\iff$ C^c is (strictly) backward invariant;
4. C is (strictly) invariant $\iff$ C^c is (strictly) invariant.

The (strictly) invariant measurable sets hence form a sub-σ-algebra of $\mathcal{F} \otimes \mathcal{B}$.

(B) Let φ be a continuous RDS. Then

1. If C is (strictly) forward invariant, then so is $\overline{C}$.
2. If time is two-sided, and if C is (strictly) (forward) invariant, then so is $\overline{C}$, $\text{int}(C)$, and ∂C.

The proof consists in applying the operations $\overline{(\cdot)}$, $\text{int}(\cdot)$ and $\partial(\cdot)$ on the relation defining the invariance of C.

∎

1.6.11 Proposition. Let φ be a continuous RDS and $\mu \in \mathcal{I}_\mathbb{P}(\varphi)$. Then the random closed set $C_\omega := \operatorname{supp}\mu_\omega$ is
   (i) forward invariant if θ is invertible,
   (ii) invariant if time is two-sided.

Proof. (i) For any continuous $f : X \to X$ and any Borel measure μ,
$$f(\operatorname{supp}\mu) \subset \operatorname{supp}(f\mu).$$
For invertible θ, μ is φ-invariant if and only if $\varphi(t,\omega)\mu_\omega = \mu_{\theta(t)\omega}$ $\mathbb{P}$-a.s. Hence
$$\varphi(t,\omega)C_\omega \subset C_{\theta(t)\omega} \quad \mathbb{P}\text{-a.s.}$$
(ii) For a homeomorphism f, $f(\operatorname{supp}\mu) = \operatorname{supp}(f\mu)$. □

1.6.12 Proposition. Let φ be a continuous RDS on a Polish space X.
   (i) Let $C \in \mathcal{F} \otimes \mathcal{B}$ be a forward invariant set. Then for each $t > 0$
$$\Theta(t)\mathcal{P}_\mathbb{P}(C) \subset \mathcal{P}_\mathbb{P}(C).$$

   (ii) If, in addition, C is a forward invariant random closed set, then any limit point of the random Krylov-Bogolyubov procedure with an arbitrary initial measure $\nu \in \mathcal{P}_\mathbb{P}(C)$ is in $\mathcal{I}_\mathbb{P}(\varphi|C)$.

Proof. (i) Take $\mu \in \mathcal{P}_\mathbb{P}(C)$. Since C is forward invariant,
$$1_{C_\omega}(x) \leq 1_{C_{\theta(t)\omega}}(\varphi(t,\omega)x) \quad \mathbb{P}\text{-a.s.},$$
thus
$$\begin{aligned}\Theta(t)\mu(C) &= \int_\Omega \int_X 1_C(\theta(t)\omega, \varphi(t,\omega)x)\, \mu_\omega(dx)\, \mathbb{P}(d\omega) \\ &= \int_\Omega \int_X 1_{C_{\theta(t)\omega}}(\varphi(t,\omega)x)\, \mu_\omega(dx)\, \mathbb{P}(d\omega) \\ &\geq \int_\Omega \mu_\omega(C_\omega)\, \mathbb{P}(d\omega) = \mu(C) = 1.\end{aligned}$$

(ii) By Corollary 1.6.7(i) $\mathcal{P}_\mathbb{P}(C)$ is closed. □

1.6.13 Theorem (Existence of Invariant Measures on Random Sets).
Let φ be a continuous RDS on a Polish space X. Let $K : \Omega \to \mathfrak{P}(\mathfrak{X})$, $\omega \mapsto K_\omega \neq \emptyset$, be a forward invariant random compact set. Then $\mathcal{P}_\mathbb{P}(K)$ is compact, and the compact subset $\mathcal{I}_\mathbb{P}(\varphi|K)$ of φ-invariant measures supported by K is non-void.

Proof. If X is compact metric then $\mathcal{P}_\mathbb{P}(K)$ is convex and compact in the topology of weak convergence, and $\mu \mapsto \Theta(t)\mu$, $t > 0$, is a commuting family of affine continuous mappings of $\mathcal{P}_\mathbb{P}(K)$ into itself, by the previous theorem. Hence the Markov-Kakutani theorem applies.

For the much more difficult case of a non-compact X we refer to Crauel [114: Corollary 6.13]. □

1.7 Markov Measures

This section is based on the work of Crauel [112, 113, 114].

1.7.1 Definition (Past, Future of an RDS). Let φ be a measurable RDS on a standard space $(X, \mathcal{B})$ with two-sided time. We call a sub-σ-algebra $\mathcal{F}^- \subset \mathcal{F}$ *past* of φ if it satisfies for all $t \geq 0$
  (1) $\varphi(-t, \cdot)$ is $\mathcal{F}^-$-measurable,
  (2) $\theta(-t)^{-1}\mathcal{F}^- \subset \mathcal{F}^-$.
The past $\mathcal{F}^-$ is called *exhaustive* if $\mathcal{F}_\infty^- := \sigma(\theta(t)^{-1}\mathcal{F}^- : t \geq 0) = \mathcal{F}$.
  Analogously, $\mathcal{F}^+ \subset \mathcal{F}$ is called *future* of φ if it satisfies for all $t \geq 0$
  (1) $\varphi(t, \cdot)$ is $\mathcal{F}^+$-measurable,
  (2) $\theta(t)^{-1}\mathcal{F}^+ \subset \mathcal{F}^+$.
The future $\mathcal{F}^+$ is called *exhaustive* if $\mathcal{F}_{-\infty}^+ := \sigma(\theta(-t)^{-1}\mathcal{F}^+ : t \geq 0) = \mathcal{F}$. ∎

The smallest possible (but in general not exhaustive) choice for $\mathcal{F}^\pm$ is of course the past

$$\mathcal{F}^- = \sigma\{\varphi(-t, \cdot)x : t \geq 0, \, x \in X\}$$

and the future

$$\mathcal{F}^+ = \sigma\{\varphi(t, \cdot)x : t \geq 0, \, x \in X\}$$

generated by φ.

The concept of past and future of an RDS with two-sided time allows one to restrict the RDS φ to RDS $\varphi^+ = \varphi$ and $\varphi^- = \varphi(-\cdot)$ with one-sided time $\mathbb{T}^+ = \mathbb{T} \cap \mathbb{R}^+$ over the DS $(\Omega, \mathcal{F}^\pm, \mathbb{P}|\mathcal{F}^\pm, (\theta(\pm t))_{t \in \mathbb{T}^+})$, where replacing $\mathcal{F}$ by $\mathcal{F}^\pm$ amounts to "throwing away information" in a controlled way.

1.7.2 Theorem (One-to-One Correspondence of Invariant Measures for Two-Sided and One-Sided Time). *(i) Let φ be a continuous RDS on a Polish space $(X, \mathcal{B})$ with two-sided time $\mathbb{T}$ with exhaustive future $\mathcal{F}^+ \subset \mathcal{F}$. Let φ^+ be the corresponding RDS with one-sided time $\mathbb{T}^+$ introduced above. Then there is a one-to-one correspondence between the set of φ-invariant measures $\mathcal{I}_\mathbb{P}(\varphi)$ and the set of φ^+-invariant measures $\mathcal{I}_\mathbb{P}(\varphi^+)$ given by*

$$\mathcal{I}_\mathbb{P}(\varphi) \ni \mu \mapsto \mu_\omega^+ := \mathbb{E}\left(\mu_\cdot | \mathcal{F}^+\right)_\omega \in \mathcal{I}_\mathbb{P}(\varphi^+), \tag{1.7.1}$$

and

$$\mathcal{I}_\mathbb{P}(\varphi^+) \ni \mu^+ \mapsto \mu_\omega := \lim_{t \to \infty} \varphi(t, \theta(-t)\omega)\mu_{\theta(-t)\omega}^+ \in \mathcal{I}_\mathbb{P}(\varphi). \tag{1.7.2}$$

The limit in (1.7.2) exists $\mathbb{P}$-a.s. and $\mathbb{E}(\mu | \mathcal{F}^+) = \mu^+$.
  In particular, $\mu \in \mathcal{I}_\mathbb{P}(\varphi)$ is ergodic if and only if $\mu^+ = \mathbb{E}(\mu|\mathcal{F}^+) \in \mathcal{I}_\mathbb{P}(\varphi^+)$ is ergodic.
  (ii) A completely analogous statement holds for the restriction φ^- to an exhaustive past $\mathcal{F}^-$.

Proof. We first prove (1.7.1) and assume that μ satisfies

$$\varphi(t,\omega)\mu_\omega = \mu_{\theta(t)\omega}, \quad t \in \mathbb{T}.$$

We now assume $t \geq 0$ and take the conditional expectation on both sides of the last equation with respect to $\theta(t)^{-1}\mathcal{F}^+$ which yields

$$\mathbb{E}(\varphi(t,\cdot)\mu.|\theta(t)^{-1}\mathcal{F}^+) = \mathbb{E}(\mu_{\theta(t)}.|\theta(t)^{-1}\mathcal{F}^+) = \mu_\cdot^+ \circ \theta(t).$$

Conditioning both sides on $\mathcal{F}^+$ leaves the right-hand side unchanged since $\theta(t)^{-1}\mathcal{F}^+ \subset \mathcal{F}^+$, while the left-hand side can be rewritten as

$$\mathbb{E}(\mathbb{E}(\varphi(t,\cdot)\mu.|\mathcal{F}^+)|\theta(t)^{-1}\mathcal{F}^+) = \mathbb{E}(\varphi(t,\cdot)\mu_\cdot^+|\theta(t)^{-1}\mathcal{F}^+),$$

where we used the fact that $\varphi(t,\cdot)$ is $\mathcal{F}^+$-measurable. Hence $\mu^+ \in \mathcal{I}_\mathbb{P}(\varphi^+)$.

We now prove (1.7.2). We show that (for any fixed $B \in \mathcal{B}$, but we omit the argument)

$$\nu_\cdot(t) := \varphi(t,\theta(-t)\cdot)\mu^+_{\theta(-t)}.$$

is a martingale with respect to the filtration $(\mathcal{G}_t)_{t\in\mathbb{T}^+}$, where $\mathcal{G}_t := \theta(-t)^{-1}\mathcal{F}^+$. Notice that $\mathcal{G}_t \uparrow \mathcal{G}_\infty = \mathcal{F}^+_{-\infty} = \mathcal{F}$.

Indeed, for $0 \leq s \leq t$

$$\begin{aligned}\mathbb{E}(\nu_\cdot(t)|\mathcal{G}_s) &= \mathbb{E}(\varphi(t,\cdot)\mu_\cdot^+ \circ \theta(-t)|\mathcal{G}_s) \\ &= \varphi(-s,\cdot)^{-1}\mathbb{E}(\varphi(t-s,\cdot)\mu_\cdot^+ \circ \theta(-t)|\mathcal{G}_s) \\ &= \varphi(-s,\cdot)^{-1}\mathbb{E}(\varphi(t-s,\cdot)\mu_\cdot^+|\theta(t-s)^{-1}\mathcal{F}^+) \circ \theta(-t) \\ &= \varphi(-s,\cdot)^{-1}\mu_\cdot^+ \circ \theta(t-s) \circ \theta(-t) \\ &= \nu_\cdot(s),\end{aligned}$$

where we have used the cocycle property and the fact that $\varphi(-s,\cdot)^{-1}$ is $\mathcal{G}_s$-measurable for the second equality sign, $\mathcal{G}_s = \theta(-t)^{-1}\theta(t-s)^{-1}\mathcal{F}^+$ for the third, and the assumed φ^+-invariance of μ^+ for the fourth.

Since $\nu_\omega(t)(B) \in [0,1]$ the martingale is uniformly integrable. It hence converges $\mathbb{P}$-a.s. and in $L^1(\mathbb{P})$,

$$\lim_{t\to\infty} \nu_\cdot(t) =: \nu_\cdot \quad \mathbb{P}\text{-a.s. and } L^1,$$

and $\nu_\cdot(t) = \mathbb{E}(\nu_\cdot|\mathcal{G}_t)$.

The limit is $\mathcal{F}^+_{-\infty}$-measurable by construction, hence $\mathcal{F}$-measurable since $\mathcal{F}^+$ is exhaustive.

Notice that the limit in (1.7.2) is the fiberwise expression of

$$\nu = \lim_{t\to\infty} \Theta(t)\mu^+.$$

Since $\Theta(s) : \mathcal{P}_\mathbb{P}(\Omega \times X) \to \mathcal{P}_\mathbb{P}(\Omega \times X)$ is continuous for all $s \in \mathbb{T}$ (here we use the assumption that X is Polish and $\varphi(t,\cdot)$ is continuous), the φ-invariance of the limit ν follows.

It remains to show that if we start with $\mu \in \mathcal{I}_\mathbb{P}(\varphi)$, go to $\mu^+ = \mathbb{E}(\mu|\mathcal{F}^+) \in \mathcal{I}_\mathbb{P}(\varphi^+)$ and then construct $\nu \in \mathcal{I}_\mathbb{P}(\varphi)$ by the above procedure, then $\nu = \mu$.

By the definition of $\nu(t)$, the $\mathcal{G}_t$-measurability of $\varphi(-t,\cdot)^{-1}$ and the φ-invariance of μ

$$\nu_\cdot(t) = \mathbb{E}(\mu_\cdot|\mathcal{G}_t), \quad t \geq 0.$$

While the left-hand side converges to ν the right-hand side converges to $\mathbb{E}(\mu|\mathcal{G}_\infty) = \mathbb{E}(\mu|\mathcal{F}) = \mu$ by the exhaustiveness of $\mathcal{F}^+$. ⌐

1.7.3 Remark. If in the last theorem $\mathcal{F}^+$ is not exhaustive the proof shows that the one-to-one correspondence is between those $\mu \in \mathcal{I}_\mathbb{P}(\varphi)$ which are $\mathcal{F}^+_{-\infty}$-measurable, and $\mathcal{I}_\mathbb{P}(\varphi^+)$. For a general $\mu \in \mathcal{I}_\mathbb{P}(\varphi)$ only $\mathbb{E}(\mu|\mathcal{F}^+_{-\infty})$ can be recovered from $\mu^+ = \mathbb{E}(\mu|\mathcal{F}^+)$ by the above procedure.

Similarly if $\mathcal{F}^-$ is not exhaustive. ■

1.7.4 Definition (Markov Measure). Let φ be a measurable RDS with two-sided time with past $\mathcal{F}^-$ and future $\mathcal{F}^+$.

A probability measure $\mu \in \mathcal{P}_\mathbb{P}(\Omega \times X)$ for which the factorization $\omega \mapsto \mu_\omega$ is $\mathcal{F}^-$-measurable or $\mathcal{F}^+$-measurable, i.e. $\mathbb{E}(\mu_\cdot|\mathcal{F}^\pm) = \mu_\cdot$ $\mathbb{P}$-a.s., is called a *Markov measure*. More specifically, an $\mathcal{F}^-/\mathcal{F}^+$-measurable μ is called a *forward/backward Markov measure*. ■

There is the following useful improvement of Theorem 1.6.13.

1.7.5 Theorem (Existence of Invariant Markov Measures). *Let φ be a continuous RDS on a Polish space X with two-sided time, past $\mathcal{F}^-$ and future $\mathcal{F}^+$. Let $\omega \mapsto K_\omega \neq \emptyset$ be a forward invariant $\mathcal{F}^-$-measurable random compact set. Then the set $\mathcal{I}_{\mathbb{P},\mathcal{F}^-}(\varphi|K)$ of φ-invariant forward Markov measures supported by K is compact and non-void.*

In particular, each limit point of the (forward) Krylov-Bogolyubov procedure with a forward Markov initial measure supported by K yields a φ-invariant forward Markov measure supported by K.

Similarly, if K is a backward invariant $\mathcal{F}^+$-measurable random compact set then the set $\mathcal{I}_{\mathbb{P},\mathcal{F}^+}(\varphi|K)$ of φ-invariant backward Markov measures supported by K is compact and non-void.

Proof. Let Γ be the set of forward Markov measure supported by K. Then Γ is non-void, convex and compact in the topology of weak convergence (Crauel [114: Theorem 4.43]). Further, by our assumptions,

$$\Theta(t)\Gamma \subset \Gamma,$$

which follows from the formula

$$(\Theta(t)\mu)_\omega = \varphi(t, \theta(-t)\omega)\mu_{\theta(-t)\omega}$$

(Lemma 1.4.4) and the definition of $\mathcal{F}^-$.

Similarly for the backward case. □

The reason for naming those measures Markov measures and for the attributes "forward" and "backward" is foreshadowed by the following corollary to Theorem 1.7.2 and will be discussed in Subsects. 2.1.3 (for $\mathbb{T} = \mathbb{Z}$) and 2.3.8 (for $\mathbb{T} = \mathbb{R}$).

1.7.6 Corollary (Past and Future Independent). *Let φ be a continuous RDS on a Polish space with past $\mathcal{F}^-$ and future $\mathcal{F}^+$. Suppose that $\mathcal{F}^-$ and $\mathcal{F}^+$ are independent. Then there is a one-to-one correspondence between the set $\mathcal{I}_{\mathbb{P},\mathcal{F}^-}(\varphi)$ of invariant forward Markov measures of φ and the subset of those elements of $\mathcal{I}_{\mathbb{P}}(\varphi^+)$ which are product measures, $\mu^+ = \mathbb{P} \times \rho^+$.*

Analogously for the invariant backward Markov measures.

1.8 Invariant Measures for Local RDS

Let φ be a local continuous RDS with two-sided time $\mathbb{T} = \mathbb{R}$ on a Polish[6] space X.

An invariant measure μ for a local RDS φ is formally defined as for a global one, e. g. by the equivariance property

$$\varphi(t,\omega)\mu_\omega = \mu_{\theta(t)\omega} \quad \mathbb{P}\text{-a.s. for all } t \in \mathbb{R}.$$

Can we have invariant measures in the presence of an explosion?

1.8.1 Definition (Set of Never Exploding Initial Values). Define

$$E(\omega) := \cap_{t \in \mathbb{R}} D(t, \omega)$$

to be the set of *never exploding initial values* (or initial values for which the trajectories "live from eternity to eternity"), i. e. those x for which $D(\omega, x) = \mathbb{R}$. ∎

Note that $E(\omega) \subset X$ is not necessarily connected and could be empty. We can write

$$E(\omega) = E^+(\omega) \cap E^-(\omega), \quad E^\pm(\omega) := \cap_{t \in \mathbb{R}^\pm} D(t, \omega),$$

where $E^\pm(\omega)$ are limits of a decreasing family of open sets.

1.8.2 Theorem. *Let φ be a local continuous RDS on a Polish space. Then:*
(i) $\operatorname{graph}(E(\cdot)) \in \mathcal{F} \otimes \mathcal{B}$, and $E(\omega)$ is a G_δ set for all $\omega \in \Omega$.
(ii) E is strictly invariant,

$$\Theta(t)^{-1}E = E \quad \text{for all } t \in \mathbb{R},$$

and any other strictly invariant set is contained in E.
(iii) E supports all φ-invariant measures, i. e.

$$\mathcal{I}_\mathbb{P}(\varphi) = \mathcal{I}_\mathbb{P}(\varphi|E).$$

---
[6] This is only needed here to assure existence and uniqueness of the factorization of $\mu \in \mathcal{P}_\mathbb{P}(\Omega \times X)$.

Proof. (i) We have in fact $E(\omega) = \cap_{t\in\mathbb{Z}} D(t,\omega)$, whence $E(\omega)$ is a G_δ set. Now

$$\begin{aligned}\mathrm{graph}(E(\cdot)) &= \{(\omega,x) : x \in E(\omega)\} \\ &= \cap_{t\in\mathbb{Z}}\{(\omega,x) : x \in D(t,\omega)\} \\ &= \cap_{t\in\mathbb{Z}} D(t) \in \mathcal{F} \otimes \mathcal{B}\end{aligned}$$

since $D(t)$, the t-section of the measurable set $D \subset \mathbb{R} \times \Omega \times X$, is measurable.

(ii) Let $t \in \mathbb{R}$ be fixed. We have $x \in E(\omega)$ if and only if $x \in D(s+t,\omega)$ for all $s \in \mathbb{R}$ if and only if $s+t \in D(\omega,x)$ for all $s \in \mathbb{R}$. By the local cocycle property, this is equivalent to $s \in D(\theta(t)\omega, \varphi(t,\omega)x)$ for all $s \in \mathbb{R}$, hence equivalent to $\varphi(t,\omega)x \in D(s,\theta(t)\omega)$ for all $s \in \mathbb{R}$, meaning that $\varphi(t,\omega)x \in E(\theta(t)\omega)$. This proves

$$\varphi(t,\omega)E(\omega) = E(\theta(t)\omega) \quad \text{for each } (\omega,t), \tag{1.8.1}$$

hence E is strictly invariant.

(iii) Since $\varphi(t,\omega) : D(t,\omega) \to D(-t,\theta(t)\omega)$, we necessarily have $\mu_\omega(D(t,\omega)) = 1$ for each $t \in \mathbb{R}$, $\mathbb{P}$-a.s. to render $\varphi(t,\omega)\mu_\omega$ meaningful $\mathbb{P}$-a.s. Since $E(\omega)$ is the intersection of countably many sets of full μ_ω-measure. This implies $\mu_\omega(E(\omega)) = 1$ $\mathbb{P}$-a.s. ⌐

1.8.3 Remark. (i) φ restricted to E is a nontrivial global "bundle RDS" in the sense of Sect. 1.9 with fibers $E(\omega) \subset X$. By (1.8.1), $E(\omega)$ and $E(\theta(t)\omega)$ are topologically equivalent via the homeomorphism $\varphi(t,\omega)$.

(ii) $\mathrm{int}(E)$ and $\partial E \cap E$ are also strictly invariant.

(iii) If there is a sequence $\ldots < t_{-1} < t_0 = 0 < t_1 < \ldots$ with $t_{|n|} \to \infty$ as $|n| \to \infty$ such that $\bar{D}(t_{n\pm 1},\omega) \subset D(t_n,\omega)$, then $E^+(\omega)$, $E^-(\omega)$ and hence $E(\omega) = E^+(\omega) \cap E^-(\omega)$ are closed.

(iv) See the appearance of E in the theory of random attractors (Subsect. 9.3.4). ∎

Local RDS with State Space $X = \mathbb{R}$

It is an elementary fact that every ergodic invariant measure of a local dynamical system generated by a C^1 vector field $\dot{x} = f(x)$ in $\mathbb{R}$ is a Dirac measure at an equilibrium point. This follows, for example, from Hale and Koçak [167: Lemma 7.1]. It basically carries over to local continuous RDS with state space $\mathbb{R}$.

1.8.4 Theorem (Invariant Measures in $X = \mathbb{R}$). *Let φ be a local continuous RDS with time $\mathbb{T} = \mathbb{R}$ and state space $X = \mathbb{R}$. Then*

(i) φ is monotone with respect to initial conditions, i.e. for each t, ω and x_1, $x_2 \in D(t,\omega)$ such that $x_1 < x_2$

$$\varphi(t,\omega)x_1 < \varphi(t,\omega)x_2.$$

(ii) For each t and ω, $D(t,\omega)$ is an open interval.
(iii) For each ω, $E(\omega)$ is a (possibly empty) interval.
(iv) Each ergodic φ-invariant measure μ_ω is a random Dirac measure, $\mu_\omega = \delta_{x_0(\omega)}$.

Proof. (i) Since $t \mapsto \varphi(t,\omega)x$ is continuous, $x_1 < x_2$ but $\varphi(t,\omega)x_1 \geq \varphi(t,\omega)x_2$ implies the existence of a t_0 for which $\varphi(t_0,\omega)x_1 = \varphi(t_0,\omega)x_2$. As a consequence, $\varphi(t_0,\omega) : D(t_0,\omega) \to R(t_0,\omega)$ is not injective which is a contradiction.

(ii) If the open set $D(t,\omega)$ is not an interval, there exist $x_1 < x < x_2$ with $x_1, x_2 \in D(t,\omega)$, but $x \notin D(t,\omega)$. This contradicts the injectivity of some $\varphi(t_0,\omega)$, as in the proof of (i).

(iii) The set $E(\omega) := \cap_{n\in\mathbb{Z}} D(n,\omega)$ is the intersection of open intervals, hence is an interval.

(iv) (Communicated by Hans Crauel) Suppose μ is an ergodic φ-invariant measure. Let $x_0(\omega)$ be the smallest median of μ_ω, i.e. the infimum of all points x for which

$$\mu_\omega([x,\infty)) \geq \frac{1}{2} \leq \mu_\omega((-\infty,x])$$

(the set of medians is a compact interval and $x_0(\cdot)$ is measurable).

Consider

$$C^-(\omega) := (-\infty, x_0(\omega)]$$

for which $\mu_\omega(C^-(\omega)) \geq \frac{1}{2}$ by definition. We claim that C^- is an invariant set.

Indeed, (i) implies that x is a median of μ_ω if and only if $\varphi(t,\omega)x$ is a median of $\varphi(t,\omega)\mu_\omega$. Since μ is invariant

$$x_0(\theta(t)\omega) = \varphi(t,\omega)x_0(\omega),$$

implying $\varphi(t,\omega)C^-(\omega) = C^-(\theta(t)\omega)$. Since μ is assumed to be ergodic $\mu_\omega(C^-(\omega)) = 1$ $\mathbb{P}$-a.s.

Using the same argument for $C^+(\omega) := [x_0(\omega),\infty)$ we obtain for $\{x_0(\omega)\} = C^-(\omega) \cap C^+(\omega)$

$$\mu_\omega(\{x_0(\omega)\}) = 1,$$

thus $\mu_\omega = \delta_{x_0(\omega)}$ $\mathbb{P}$-a.s. □

We will encounter many applications of this useful theorem (see Exercises 2.2.9, Subsect. 2.3.7 and Sect. 9.3).

1.9 Random Dynamical Systems on Subsets and Bundles. Isomorphisms

1.9.1 Bundle RDS

We will now introduce two useful extensions of our notion of an RDS which both have to do with replacing the trivial bundle $\Omega \times X$, in the first case with a subset of $\Omega \times X$ which is not a product set, in the second case with a nontrivial bundle. We will for convenience call both generalizations "bundle RDS".

RDS on Subsets

The restriction of an RDS onto a strictly forward invariant or strictly invariant subset of $\Omega \times X$ (see Sect. 1.6) creates an RDS on this subset. Here is a formal definition (see Bogenschütz [73]).

1.9.1 Definition (RDS on a Subset). Let $C \subset \Omega \times X$, $C \in \mathcal{F} \otimes \mathcal{B}$, be a subset with $C_\omega := \{x : (\omega, x) \in C\} \neq \emptyset$ for all $\omega \in \Omega$. An *RDS on the subset* C or a *bundle RDS* with fibers C_ω is a measurable mapping

$$\Theta : \mathbb{T} \times C \to C, \quad (t, \omega, x) \mapsto \Theta(t)(\omega, x) = (\theta(t)\omega, \varphi(t, \omega)x),$$

such that the mappings $\varphi(t, \omega) : C_\omega \to C_{\theta(t)\omega}$, $x \mapsto \varphi(t, \omega)x$, form a cocyle, i.e. satisfy identically

$$\varphi(0, \omega) = \mathrm{id}_{C_\omega},$$
$$\varphi(t+s, \omega) = \varphi(t, \theta(s)\omega) \circ \varphi(s, \omega).$$

■

If φ is an RDS on the subset C, then C is obviously strictly forward invariant for one-sided time, and strictly invariant for two-sided time, with $\varphi(t, \omega)^{-1} = \varphi(-t, \theta(t)\omega)$.

An invariant measure μ for an RDS φ on C has to satisfy $\mu(C) = 1$ (see Sect. 1.5). In particular, if X is a Polish space and C is a random closed set, then $\mathcal{P}_\mathbb{P}(C)$ is closed.

We also saw in Sect. 1.8 that to each local RDS there corresponds a natural global bundle RDS on the set E of the never exploding initial values.

RDS on Measurable Bundles

Nontrivial bundle DS arise naturally already in the deterministic case: A smooth DS φ on a manifold X generates through its linearization $T\varphi$ a DS on the tangent bundle TM (typically a non-trivial bundle). The chain rule is equivalent to $T\varphi$ being a skew product $(x, v) \mapsto (\varphi(t)x, T\varphi(t, x)v)$ which is linear on fibers, i.e. the mappings

$$T\varphi(t, x) : T_x X \to T_{\varphi(t)x} X$$

form a cocycle of linear mappings over φ. This motivates our definition of an RDS on a bundle, for which we need the following measurable version of the notion of a bundle.

1.9.2 Definition (Measurable Bundle). (i) A *measurable bundle* (Y, Ω, π) with *typical fiber* X consists of a measurable space $(Y, \mathcal{Y})$, a measurable space $(\Omega, \mathcal{F})$ whose one-point sets are measurable, a measurable space $(X, \mathcal{B})$, a measurable onto map $\pi : Y \to \Omega$, and a *global measurable trivialization*, i.e. a bimeasurable bijection $\Psi : Y \to \Omega \times X$ such that $\pi_\Omega \circ \Psi = \mathrm{id}_\Omega \circ \pi = \pi$ (equivalently: preserves fibers) (see Fig. 1.6). In particular, for any $\omega \in \Omega$

$$\psi(\omega) := \Psi|_{\pi^{-1}(\omega)} : \pi^{-1}(\omega) \to \{\omega\} \times X$$

is a bimeasurable bijection (in the respective trace σ-algebras).

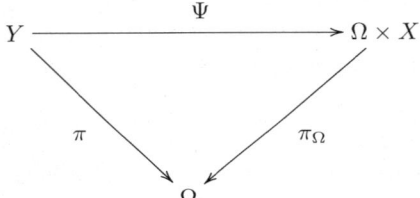

Fig. 1.6. Measurable bundle

(ii) If the typical fiber X has the additional structure of a d-dimensional vector space (topological space, smooth manifold) we assume the same structure for all fibers $\pi^{-1}(\omega)$, and $\psi(\omega)$ is assumed to preserve this structure (i.e. is a vector space isomorphism (homeomorphism, diffeomorphism)). We then speak of a *linear (continuous, smooth) measurable bundle*. Notice that there is no probability measure involved in this definition. ∎

1.9.3 Definition (Bundle RDS). (i) Let $(\Omega, \mathcal{F}, \mathbb{P}, (\theta(t))_{t \in T})$ be a metric DS and let (Y, Ω, π) be a measurable bundle with typical fiber X. A *(measurable) bundle RDS* over θ is a measurable DS Θ on Y which covers θ (equivalently: preserves fibers), i.e. it satisfies

$$\pi \circ \Theta(t) = \theta(t) \circ \pi \quad \text{for all} \quad t \in T. \tag{1.9.1}$$

(1.9.1) is equivalent (exercise) to the statement that the fiber mappings

$$\varphi(t, \omega) := \Theta(t)|_{\pi^{-1}(\omega)} : \pi^{-1}(\omega) \to \pi^{-1}(\theta(t)\omega)$$

constitute a cocycle, i.e. $\varphi(0, \omega) = \mathrm{id}_{\pi^{-1}(\omega)}$, $\varphi(t+s, \omega) = \varphi(t, \theta(s)\omega) \circ \varphi(s, \omega)$.

(ii) If (Y, Ω, π) is a linear (continuous, smooth) measurable bundle and if the fiber mappings

$$\varphi(t,\omega) : \pi^{-1}(\omega) \to \pi^{-1}(\theta(t)\omega)$$

respect this structure, i.e. are linear (continuous, C^k) – in the two-sided time case they are even linear isomorphisms (homeomorphisms, C^k diffeomorphisms), we speak of a *linear (continuous, C^k) bundle RDS*. ∎

The theory of invariant measures developed in Sects. 1.4 to 1.6 for RDS on the trivial bundle $\Omega \times X$ carries over in an obvious manner to RDS on measurable bundles.

Linear bundle RDS are extensively treated in Sect. 4.1.

1.9.4 Remark (RDS as DS with a Metric Factor θ). An even more general notion of an RDS would be a DS Θ with a metric factor θ, i.e. there is a measurable mapping π which satisfies

$$\pi \circ \Theta = \theta \circ \pi.$$

For Lebesgue spaces, Θ can be represented as a skew product over θ, see Sinai [318: p. 24] for details. ∎

1.9.2 Isomorphisms of RDS

One of the basic tasks of the theory of RDS is *classification*. For this task we need concepts enabling us to decide when two RDS can be considered basically identical or *isomorphic*. Our concepts for RDS are made to be consistent with and to extend those concepts already available in ergodic theory (metric isomorphism) and in the theory of topological or smooth dynamical systems (topological or smooth equivalence or conjugacy).

We will certainly not be willing to identify two RDS φ_1, φ_2 if the metric DS θ_1 and θ_2 over which they are defined are not isomorphic. Hence the first requirement is that (both systems have the same time $\mathbb{T}$ and) the metric factors $(\Omega_i, \mathcal{F}_i, \mathbb{P}_i, (\theta_i(t))_{t\in\mathbb{T}})$, $i=1,2$, are metrically isomorphic (see Appendix A.1). We thus can and will assume that for the purpose of classification all RDS are defined over the same DS $(\Omega, \mathcal{F}, \mathbb{P}, (\theta(t))_{t\in\mathbb{T}})$.

The second requirement depends on the category of RDS under consideration. For any category, we assume that the two RDS are measurably isomorphic or cohomologous in the sense of the following definition.

1.9.5 Definition (Measurable Isomorphism of Bundle RDS). Given two measurable bundle RDS Θ_1 on Y_1 and Θ_2 on Y_2 over θ. They are called *measurably isomorphic*, if there exists a measurable isomorphism Ψ of the corresponding bundles (i.e. a bimeasurable bijection $\Psi: Y_1 \to Y_2$ which covers the identity id_Ω, i.e. $\pi_2 \circ \Psi = \mathrm{id}_\Omega \circ \pi_1$) such that

$$\Theta_2(t) \circ \Psi = \Psi \circ \Theta_1(t) \quad \text{for all } t \in T.$$

(For RDS φ_i on subsets $C_i \subset \Omega \times X_i$ replace Y_i by C_i in the definition). "Measurably isomorphic" is an equivalence relation. ∎

46   Chapter 1. Basic Definitions. Invariant Measures

The definitions immediately yield (exercise) the following proposition.

1.9.6 Proposition (Cohomology of Cocycles). *Two measurable bundle RDS Θ_1 and Θ_2 are measurably isomorphic via Ψ if and only if the corresponding cocycles*

$$\varphi_i(t,\omega) := \Theta_i(t)|_{\pi_i^{-1}(\omega)} : \pi_i^{-1}(\omega) \to \pi_i^{-1}(\theta(t)\omega)$$

are cohomologous with cohomology

$$\psi(\omega) := \Psi|_{\pi_1^{-1}(\omega)} : \pi_1^{-1}(\omega) \to \pi_2^{-1}(\omega),$$

which means that

$$\varphi_2(t,\omega) \circ \psi(\omega) = \psi(\theta(t)\omega) \circ \varphi_1(t,\omega). \quad (1.9.2)$$

(For RDS φ_i on subsets $C_i \subset \Omega \times X_i$ replace $\pi_i^{-1}(\omega)$ by $C_{i,\omega}$.)

See Fig. 1.7 for the diagram of cohomology.

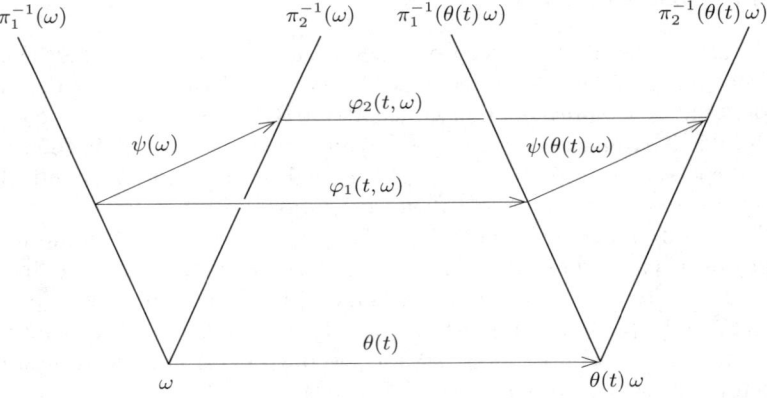

Fig. 1.7. Cohomology of cocycles

1.9.7 Remark (Origin of the Notions "Cocyle", etc.).
The notions *cocycle*, *cohomology* (and also *coboundary*, a cocycle which is cohomologous to the trivial cocycle $\varphi(t,\omega) = \mathrm{id}_X$) have their origin in homological algebra and are fundamental in algebraic ergodic theory, see e. g. Schmidt [312, 313] and Zimmer [353]. In the case of a group $\mathbb{T} = \mathbb{Z}$ or $\mathbb{R}$ and a cocycle φ taking values in a topological group H the cocycle property is exactly the condition that φ be a Borel 1-cocycle on $\mathbb{T}$ in the sense of Eilenberg-MacLane with values in the $\mathbb{T}$-module of measurable functions from Ω to H, where the $\mathbb{T}$-action is defined by $U_t f(\omega) := f(\theta(t)\omega)$. Similarly,

cohomology in the sense introduced in Proposition 1.9.6 is cohomology in the sense of Eilenberg-MacLane (using only Borel cochains). See Zimmer [353: pp. 66–67] for further references. ∎

If we have invariant measures for the RDS, or more structure on the fibers, we take these into account with the following more refined notion of "isomorphism".

1.9.8 Definition (Metric, Linear Isomorphism; Topological, C^k Equivalence).
(i) Two measurable bundle RDS Θ_1 and Θ_2 with invariant measures μ_1 and μ_2 are called *metrically isomorphic*, if there is a measurable isomorphism Ψ in the sense of Definition 1.9.5 which satisfies

$$\Psi(\mu_1) = \mu_2.$$

"Metrically isomorphic" is an equivalence relation.

(ii) Two linear (continuous, C^k) bundle RDS φ_1 and φ_2 on linear (continuous, smooth) bundles Y_1, Y_2 are called *linearly isomorphic (topologically, C^k equivalent)*, if there is a measurable isomorphism Ψ for which for all $\omega \in \Omega$

$$\psi(\omega) := \Psi|_{\pi_1^{-1}(\omega)} : \pi_1^{-1}(\omega) \to \pi_2^{-1}(\omega)$$

is a linear isomorphism (homeomorphism, C^k diffeomorphism). "Linearly isomorphic" ("topologically, C^k equivalent") is an equivalence relation. ∎

1.9.9 Exercise. Suppose two measurable bundle RDS φ_1 and φ_2 having invariant measures μ_1 and μ_2 are metrically isomorphic with isomorphism Ψ. Suppose further that the typical fibers are standard spaces so that the two invariant measures uniquely factorize as

$$\mu_i(d\omega, dx_i) = \mu_\omega^i(dx_i)\mathbb{P}(d\omega).$$

Then

$$\Psi(\mu_1) = \mu_2 \iff \psi(\omega)\mu_\omega^1 = \mu_\omega^2 \quad \mathbb{P}\text{-a.s.}$$

There has been great progress in the classification problem in recent years. For the metric isomorphism problem see Gundlach [164]. For the topological classification of RDS see Subsect. 9.2.1 and Nguyen Dinh Cong [261] for a complete treatment. For a random version of the Hartman-Grobman theorem see Sect. 7.4. For smooth equivalence (normal form problem) see Chap. 8. The linear classification problem is discussed in Sect. 4.1.

Chapter 2. Generation

Summary

In this foundational chapter we will associate (infinitesimal) generators with all reasonably regular RDS.

The task is easy for discrete time where all RDS have obvious generators, their time-one maps (Sect. 2.1). Products of i.i.d. random mappings and their associated Markov chains are studied in more detail in Subsect. 2.1.3.

In the continuous time case we ask for the stochastic equivalent of the deterministic one-to-one correspondence between dynamical systems and vector fields, symbolically written as

$$\text{dynamical system} = \exp(\text{vector field}).$$

There are two basically different answers:

(1) In Sect. 2.2 we obtain the following one-to-one correspondence: Every ω-wise random differential equation of the form $\dot{x}_t = f(\theta_t \omega, x_t)$ generates an RDS which is absolutely continuous with respect to time t (see Theorem 2.2.1 for the local case and Theorem 2.2.2 for the global case). Conversely, every RDS φ for which $t \mapsto \varphi(t, \omega)x$ is absolutely continuous is a solution of such a random differential equation (Theorem 2.2.13). All we need for the proof is a new look at classical results for ordinary differential equations.

(2) It seems surprising at first sight that there is a large and very important class of RDS for which $t \mapsto \varphi(t, \omega)x$ is, though continuous, not of bounded variation, and yet have generators. The key extra property is that $t \mapsto \varphi(t, \omega)x$ is a semimartingale. The generators are stochastic differential equations (Sect. 2.3).

As we are again aiming at a general one-to-one correspondence between those RDS and their generators we have to use Kunita's [224] general concepts, extended to two-sided time. Our main results are: Every stochastic differential equation driven by a semimartingale helix (additive cocycle) generates a semimartingale RDS (Theorem 2.3.26 for the global case, Theorem 2.3.29 for the local case). Conversely, every semimartingale RDS is generated by a stochastic differential equation driven by a semimartingale helix (Theorem 2.3.30). This one-to-one relation can be succinctly written as

$$\text{semimartingale cocycle} = \exp(\text{semimartingale helix}).$$

The proof relies crucially on our Perfection Theorem 1.3.2.

Criteria for when a classical (i.e. Brownian motion driven) stochastic differential equation generates a continuous or C^k RDS are given in Subsect. 2.3.6.

Subsect. 2.3.7 is devoted to the detailed treatment of a one-dimensional example.

In the final Subsect. 2.3.9 we sketch the route (historically and systematically) that lead from Markov processes via classical stochastic differential equations to random dynamical systems. As a climax, we characterize those invariant measures which are related to the solutions of the Fokker-Planck equation (Theorem 2.3.45).

Our hope is that this chapter can serve as a reference text for the generation problem.

2.1 Discrete Time: Products of Random Mappings

2.1.1 One-Sided Discrete Time

Let φ be a measurable/continuous/C^k RDS on X over θ with time $\mathbb{T} = \mathbb{Z}^+$ [1]. Introduce the *time-one mapping*

$$\psi(\omega) := \varphi(1,\omega) : X \to X. \qquad (2.1.1)$$

By repeatedly applying the cocycle property (1.1.2) we obtain

$$\varphi(n,\omega) = \begin{cases} \psi(\theta^{n-1}\omega) \circ \cdots \circ \psi(\omega), & n \geq 1, \\ \mathrm{id}_X, & n = 0. \end{cases} \qquad (2.1.2)$$

The RDS φ is measurable if and only if $(\omega, x) \mapsto \psi(\omega)x$ is measurable. It is continuous/C^k if and only if $x \mapsto \psi(\omega)x$ is, in addition, continuous/C^k.

Conversely, given a family of mappings $\psi(\omega) : X \to X$ such that $(\omega, x) \mapsto \psi(\omega)x$ is measurable, in addition $x \mapsto \psi(\omega)x$ is continuous/C^k. Then φ defined by (2.1.2) is a mesurable/continuous/C^k RDS. We say that φ is *generated* by ψ.

Hence every one-sided discrete time RDS has the form (2.1.2), i.e. is a *product of* (a stationary sequence of) *random mappings*, or an *iterated function system*, or a *system in a random environment* (see Fig. 2.1).

To emphasize the dynamical perspective, we can write the discrete time cocycle $\varphi(n,\omega)x$ as the "solution" of an initial value problem for a *random difference equation*

$$x_{n+1} = \psi(\theta^n \omega)x_n, \quad n \geq 0, \quad x_0 = x \in X. \qquad (2.1.3)$$

The sequence of random points $(\varphi(n,\omega)x)_{n \geq 0}$ in state space X is the *orbit* of the point x under the RDS φ.

---
[1] Recall that an RDS on $\mathbb{N}$ can be trivially extended to $\mathbb{Z}^+$ by putting $\varphi(0,\omega) = \mathrm{id}_X$.

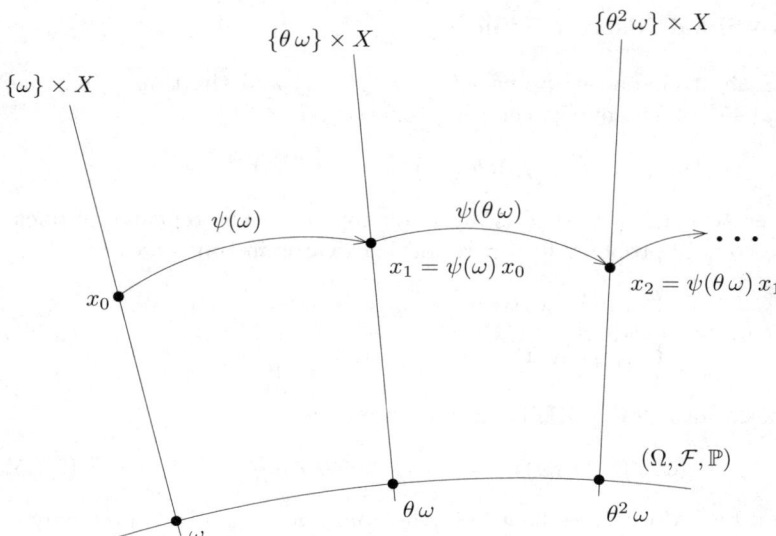

Fig. 2.1. Product of random mappings

2.1.1 Example (Linear and Affine RDS). Suppose a generator ψ takes values in some sub-semigroup $\mathcal{S}$ of the semigroup (with respect to composition $\circ$) of all self-mappings of X. Then the cocycle will stay there for all time. Cases of particular importance are for $X = \mathbb{R}^d$

(i) *Linear RDS, products of random matrices*: Let $\mathcal{S} = \mathbb{R}^{d \times d}$ be the semigroup of all $d \times d$ matrices, with matrix multiplication as composition. A linear RDS has thus the form

$$\Phi(n,\omega) = A_{n-1}(\omega) \cdots A_0(\omega), \quad A_k(\omega) := A(\theta^k \omega),$$

where $A : \Omega \to \mathbb{R}^{d \times d}$ is measurable. The theory of products of random matrices with the multiplicative ergodic theorem (see Chap. 3) is the core of the theory of RDS, with many fundamental papers such as Furstenberg and Kesten [154], Furstenberg [153], Oseledets [268], Ruelle [292], Guivar'ch and Raugi [163], Goldsheid and Margulis [161]. See also the books by Bougerol and Lacroix [77: Part A], and by Carmona and Lacroix [86: Chap. IV].

(ii) *Products of positive random matrices*: Let $\mathcal{S}$ be the semigroup of matrices $A = (a_{ij})_{d \times d}$ with nonnegative (or positive) entries. For a random Perron-Frobenius theory (and a survey of earlier results) see Arnold, Demetrius and Gundlach [15].

(iii) *Affine RDS*: Let $\mathcal{S}$ be the semigroup of all affine mappings of $\mathbb{R}^d$, hence $\psi(\omega)x = A(\omega)x + b(\omega)$ with $A : \Omega \to \mathbb{R}^{d \times d}$ and $b : \Omega \to \mathbb{R}^d$ measurable. Affine RDS are *iterated function systems* in the classical sense. They are important for encoding and visualizing fractals, see Hutchinson [177], Barnsley [49], and Sect. 5.6. ∎

2.1.2 Two-Sided Discrete Time

For $\mathbb{T} = \mathbb{Z}$ the time-one mapping $\psi(\omega) := \varphi(1, \omega)$ and the time-minus-one mapping $\varphi(-1, \omega)$ are, by the cocycle property, related by

$$\varphi(-1, \omega) = \varphi(1, \theta^{-1}\omega)^{-1} = \psi(\theta^{-1}\omega)^{-1},$$

so the generator $\psi(\omega) : X \to X$ is invertible for all ω. The repeated application of the cocycle property forwards and backwards in time gives

$$\varphi(n, \omega) = \begin{cases} \psi(\theta^{n-1}\omega) \circ \cdots \circ \psi(\omega), & n \geq 1, \\ \mathrm{id}_X, & n = 0, \\ \psi(\theta^n \omega)^{-1} \circ \cdots \circ \psi(\theta^{-1}\omega)^{-1}, & n \leq -1. \end{cases} \qquad (2.1.4)$$

This defines a measurable RDS φ if and only if

$$(\omega, x) \mapsto \psi(\omega)x \quad \text{and} \quad (\omega, x) \mapsto \psi(\omega)^{-1}x. \qquad (2.1.5)$$

are measurable. Moreover, the RDS φ is continuous or C^k if and only if $\psi(\omega) \in \mathrm{Homeo}(X)$ or $\mathrm{Diff}^k(X)$, respectively.

Conversely, given for each ω an invertible mapping $\psi(\omega) : X \to X$ such that the two mappings in (2.1.5) are measurable (in addition $\psi(\omega) \in \mathrm{Homeo}(X)$ or $\mathrm{Diff}^k(X)$). Then ψ defines via (2.1.4) a measurable (in addition continuous or C^k) RDS φ.

Consequently, all two-sided discrete time RDS have the form (2.1.4). The corresponding random difference equation

$$x_{n+1} = \psi(\theta^n \omega) x_n, \quad n \in \mathbb{Z}, \quad x_0 \in X,$$

can now be solved forwards and backwards in time.

Suppose that a generator ψ takes values in some subgroup of the group of invertible self-mappings of X. Then the cocycle will stay there for all time. The cases $X = \mathbb{R}^d$, $\psi(\omega) \in Gl(d, \mathbb{R})$, or $\psi(\omega)$ an invertible affine mapping, are of particular importance, see Example 2.1.1.

2.1.2 Exercise (Affine Equation). For $\mathbb{T} = \mathbb{Z}$ and $X = \mathbb{R}^d$ let $\psi(\omega)x = A(\omega)x + b(\omega)$ be the time-one mapping of the affine cocycle φ. We have

$$\varphi(1, \omega)x = A(\omega)x + b(\omega), \quad \varphi(-1, \omega)x = A(\theta^{-1}\omega)^{-1}(x - b(\theta^{-1}\omega))$$

and by induction

$$\varphi(n, \omega)x = \begin{cases} \Phi(n, \omega) \left(x + \sum_{j=0}^{n-1} \Phi(j+1, \omega)^{-1} b(\theta^j \omega) \right), & n \geq 1, \\ x, & n = 0, \\ \Phi(n, \omega) \left(x - \sum_{j=n}^{-1} \Phi(j+1, \omega)^{-1} b(\theta^j \omega) \right), & n \leq -1, \end{cases}$$

where Φ is the linear cocycle generated by A. Let now A be deterministic and have the form

$$A = \begin{pmatrix} A^- & 0 \\ 0 & A^+ \end{pmatrix}$$

with A^- having its spectrum inside and A^+ outside the unit disc in $\mathbb{C}$ and with $\log^+ |b| \in L^1(\mathbb{P})$. Put $b = \binom{b^-}{b^+}$. Then the unique φ-invariant measure is the random Dirac measure $\delta_{x_0(\omega)}$ with base point

$$x_0(\omega) = \begin{pmatrix} x_0(\omega)^- \\ x_0(\omega)^+ \end{pmatrix} = \begin{pmatrix} \sum_{n=1}^\infty (A^-)^{n-1} b^-(\theta^{-n}\omega) \\ -\sum_{n=0}^\infty (A^+)^{-n-1} b^+(\theta^n \omega) \end{pmatrix}.$$

Affine cocycles will be treated systematically in Sect. 5.6. ∎

2.1.3 Exercise. In the proofs of Lemma 7.3.6 and Proposition 7.5.2 we will make use of the following observation: Let

$$\psi(\omega)x := A(\omega)x + F(\omega, x), \quad (\omega, x) \in \Omega \times \mathbb{R}^d,$$

be measurable, and $A(\omega) \in Gl(d, \mathbb{R})$. Then ψ is the time-one mapping of a two-sided measurable RDS if and only if $x \mapsto h(\omega, x) := (I + A(\omega)^{-1} F(\omega, \cdot))x$ is (jointly) measurably invertible. It is, moreover, the time-one mapping of a two-sided continuous RDS if and only if $x \mapsto F(\omega, x)$ is continuous and $h^{-1}(\omega, \cdot)$ is continuous. ∎

2.1.3 RDS with Independent Increments

We now have a closer look at those RDS φ on X which are products of i.i.d. random mappings, equivalently, have independent increments (see Remark 1.1.9). It turns out that they have the extra property that their k-point motions $(\varphi(n)x_1, \ldots, \varphi(n)x_k)_{n \in \mathbb{T}}$ are a Markov family in X^k for each $k \in \mathbb{N}$ forwards in time (for $\mathbb{T} = \mathbb{Z}^+$), respectively forwards and backwards in time (for $\mathbb{T} = \mathbb{Z}$). We will in particular study the relation between invariant measures for the RDS φ and stationary measures for the one-point Markov transition operators $P^\pm$.

One-Sided Time

This case has been systematically dealt with by Kifer [207].

2.1.4 Theorem (Markov Chain Corresponding to RDS). *(i) Let φ be a measurable cocycle with time $\mathbb{T} = \mathbb{Z}^+$ which is a product of i.i.d. random mappings $\psi_n = \psi(\theta^n \cdot)$. Then for any fixed $x \in X$ (or a random variable x_0 independent of $(\psi_n)_{n \in \mathbb{Z}^+}$), the orbit (x_n^x) of $x_{n+1} = \psi_n x_n$, $x_0 = x$ (the one-point motion), is a homogeneous Markov chain with state space X and transition probability*

$$P(x, B) := \mathbb{P}\{\omega : \psi(\omega)x \in B\}. \tag{2.1.6}$$

More generally, the k-point motions are homogeneous Markov chains for any $k \in \mathbb{N}$.

(ii) *If φ is continuous and X is a metric space, then the Markov chain is Feller, i.e. the transition operator $f \mapsto P(f)$ defined by*

$$P(f)(x) := \int_X f(y)\, P(x, dy) = \mathbb{E} f(\psi(\cdot)x)$$

maps $\mathcal{C}_b(X)$ to itself.

Proof. (i) We only treat the case $k = 1$ and show first that P defined by (2.1.6) is a stochastic kernel.

(a) For each fixed x, $P(x, \cdot)$ is a probability on $(X, \mathcal{B})$ since it is the distribution of the random variable $\omega \mapsto \psi(\omega)x$.

(b) $P(\cdot, B)$ is measurable for each $B \in \mathcal{B}$. Indeed, denoting the mapping $(\omega, x) \mapsto \psi(\omega)x$ by Ψ, we have

$$P(x, B) = \mathbb{P}(A_x), \quad A = \Psi^{-1}B \in \mathcal{F} \otimes \mathcal{B}.$$

Let $\mathcal{A} = \{A \in \mathcal{F} \otimes \mathcal{B} : \mathbb{P}(A_x) \text{ is measurable in } x\}$. It can be easily seen that $\mathcal{A}$ contains product sets, is an algebra and a monotone system. Hence it is a σ-algebra, and $\mathcal{A} = \mathcal{F} \otimes \mathcal{B}$.

For the proof of the Markov property

$$\mathbb{P}(x^x_{n+1} \in B | \mathcal{F}_k) = \mathbb{P}(x^x_{n+1} \in B | x^x_k), \quad \mathcal{F}_k = \sigma(x^x_0, \ldots, x^x_k; x \in X), \quad k \le n,$$

we need the following well-known lemma.

2.1.5 Lemma. *Let $(\Omega, \mathcal{F}, \mathbb{P})$ be a probability space, $(X, \mathcal{B})$ a measurable space, $h : X \times \Omega \to \mathbb{R}$ bounded measurable, $\mathcal{C} \subset \mathcal{F}$ a sub-σ-algebra, $\xi : \Omega \to X$ $\mathcal{C}, \mathcal{B}$ measurable, and $h(x, \cdot)$ and $\mathcal{C}$ independent for each $x \in X$. Then*

$$\mathbb{E}(h(\xi(\cdot), \cdot) | \mathcal{C}) = \mathbb{E}(h(\xi(\cdot), \cdot) | \xi) = H \circ \xi, \quad H(x) := \mathbb{E}\, h(x, \cdot).$$

We apply the lemma with $h(x, \omega) = 1_B(\psi_n(\omega) \circ \cdots \circ \psi_k(\omega)x)$, $\mathcal{C} = \mathcal{F}_k$, $\xi = x^x_k$, which yields

$$\begin{aligned}\mathbb{E}(h(x^x_k, \cdot) | \mathcal{F}_k) &= \mathbb{P}(x^x_{n+1} \in B | \mathcal{F}_k) \\ &= \mathbb{P}(x^x_{n+1} \in B | x^x_k) \\ &= \mathbb{P}(\psi_n \circ \cdots \circ \psi_k z \in B)|_{z = x^x_k} \\ &= P_{k, n+1}(x^x_k, B).\end{aligned}$$

This proves the Markov property as well as a representation for the transition probability.

In particular, for $k = n$ we obtain

$$\mathbb{P}(x^x_{n+1} \in B | x^x_n = z) = \mathbb{P}(\psi_n z \in B) = \mathbb{P}\{\omega : \psi(\omega)z \in B\} = P(z, B),$$

i.e. the Markov chain is homogeneous due to the fact that the (ψ_n) are identically distributed.

In fact, $(x^x_n)_{x \in X}$ is a *Markov family* with common transition probability P.

(ii) See Kifer [207: pp. 16–17]. □

While the passing from a product of i.i.d. random mappings to a Markov chain is unique, the solution of the inverse problem of constructing a measurable/continuous/C^k RDS of i.i.d. random mappings with a prescribed transition probability is typically not unique (and so far largely unsolved). The general obvious reason for non-uniqueness is that a transition probability only determines the *statistics* of the *one*-point motion of a possible RDS, while the mapping $\psi(\omega)$ also describes the simultaneous motion of k points for arbitrary $k \in \mathbb{N}$.

However, there is always the following affirmative abstract answer to the inverse problem.

2.1.6 Theorem (RDS Corresponding to Markov Chain). *Let $(X, \mathcal{B})$ be a standard measurable space and $P(x, B)$ be a stochastic kernel. Then there exists a σ-algebra $\mathcal{H}$ on the semigroup H of measurable self-mappings of X such that $(f, x) \mapsto f(x)$ is $\mathcal{H} \otimes \mathcal{B}, \mathcal{B}$ measurable, and a probability m on $\mathcal{H}$ with $P(x, B) = m\{f : f(x) \in B\}$. Consequently, the coordinate mappings $\psi(\theta^n \omega) = \omega_n$ on $(\Omega, \mathcal{F}, \mathbb{P}, (\theta^n)_{n \in \mathbb{Z}^+})$, where $\Omega = H^{\mathbb{Z}^+}$, $\mathcal{F} = \mathcal{H}^{\mathbb{Z}^+}$, $\mathbb{P} = m^{\mathbb{Z}^+}$, θ the shift on Ω, generate an RDS on X of i.i.d. random mappings for which the transition probability is the prescribed one.*

For a proof and for more literature see e.g. Kifer [207: pp. 8–12]. Results on the representation of a Markov chain on a manifold by smooth maps were obtained by Quas [283], [284]. An up-to-date survey on the representation question is given by Dubischar [130].

If the state space X is compact metric, then every continuous RDS φ has at least one invariant measure μ (see Sect. 1.5) and every Feller transition probability P has at least one stationary measure ρ (see Appendix A.4). For an RDS which is a product of i.i.d. random mappings with one-sided time $\mathbb{Z}^+$ we have the following simple one-to-one correspondence (discovered by Ohno [266]) between φ-invariant product measures $\mu = \mathbb{P} \times \rho$ on $\Omega \times X$ and stationary measures ρ of the corresponding Markov chain on X.

2.1.7 Theorem (One-to-One Correspondence Between Invariant Product Measures and Stationary Measures). *Let φ be a measurable RDS of i.i.d. random mappings $\psi(\theta^n \omega)$ with discrete time $\mathbb{T} = \mathbb{Z}^+$ on a standard measurable space $(X, \mathcal{B})$ in its canonical representation, i.e. over the canonical metric DS $(\Omega, \mathcal{F}, \mathbb{P}, (\theta^n)_{n \in \mathbb{Z}^+})$ introduced in Theorem 2.1.6. Then*

(i) $\varphi(n, \cdot)$ and $\theta^{-n} \mathcal{F}$ are independent for all $n \in \mathbb{Z}^+$.

(ii) A measure $\mu = \mathbb{P} \times \rho \in \mathcal{P}_\mathbb{P}(\Omega \times X)$ is φ-invariant if and only if $\rho \in \mathcal{P}(X)$ is stationary for $P(x, B) = \mathbb{P}\{\omega : \psi(\omega)x \in B\}$.

(iii) A φ-invariant measure $\mu = \mathbb{P} \times \rho$ is ergodic if and only if ρ is ergodic.

Proof. (i) is obvious. The proof of (ii) uses the characterization of invariant measures given in Theorem 1.4.5 and relies on the independence of $\varphi(n, \cdot)\rho$ and $\theta^{-n} \mathcal{F}$. Details were already given in Example 1.4.7.

A proof of (iii) is given by Kifer [207: Theorem 2.1]. □

We stress that even for a product of i.i.d. random mappings in canonical respresentation there exist in general φ-invariant measures which are not product measures (see Exercise 2.1.2 and Subsect. 5.6.3).

The "colored noise" version where $\varphi(1, \theta^n \omega) = g(\xi_n)$ and (ξ_n) is a stationary Markov chain was treated by Arnold and Kliemann [24] and Crauel [112], among many others.

Two-Sided Time

Let φ be a product of i.i.d. random mappings $(\psi_n)_{n \in \mathbb{Z}}$ with two-sided time $\mathbb{T} = \mathbb{Z}$ on a standard space $(X, \mathcal{B})$. Assume that φ has a canonical representation, i.e. $\psi : \Omega \to G$ is measurable, where $(G, \mathcal{G})$ is a measurable group of self-mappings of X (i.e. $(f, g) \mapsto f \circ g$ and $f \mapsto f^{-1}$ are measurable) acting measurably on X (i.e. $(f, x) \mapsto f(x)$ is measurable), $\Omega = G^{\mathbb{Z}}$, $\mathcal{F} = \mathcal{G}^{\mathbb{Z}}$, $\mathbb{P} = m^{\mathbb{Z}}$ where $m = \mathcal{L}(\psi)$, θ is the two-sided shift and $\psi(\omega) = \omega_0$.

We will now utilize the theory of past and future of an RDS developed in Sect. 1.7.

For φ in a canonical representation a natural choice of past and future clearly is
$$\mathcal{F}^- = \sigma(\psi(\theta^n \cdot) : n \leq -1) = \sigma(\varphi(-n, \cdot) : n \in \mathbb{Z}^+)$$
and
$$\mathcal{F}^+ = \sigma(\psi(\theta^n \cdot) : n \geq 0) = \sigma(\varphi(n, \cdot) : n \in \mathbb{Z}^+).$$
Both $\mathcal{F}^-$ and $\mathcal{F}^+$ are exhaustive, $\mathcal{F}^- \cap \mathcal{F}^+ = \{\emptyset, \Omega\}$ and $\mathcal{F}^-$ and $\mathcal{F}^+$ are independent. Further $\varphi(-n, \cdot)$ and $\theta^n \mathcal{F}^-$ as well as $\varphi(n, \cdot)$ and $\theta^{-n} \mathcal{F}^+$ are independent for all $n \geq 0$.

We recall from Sect. 1.7 that $\mathcal{I}_{\mathbb{P}, \mathcal{F}^{\mp}}(\varphi)$ is the set of invariant forward/backward Markov measures of φ and that $\varphi^{\pm}$ is the restrictions of φ to one-sided time $\mathbb{Z}^+$ where $\mathcal{F}$ is replaced by $\mathcal{F}^{\pm}$.

The following theorem is an immediate consequence of Corollary 1.7.6.

2.1.8 Theorem (One-to-One Correspondence Between Markov Measures and Stationary Measures). *Let φ be a continuous RDS with two-sided time $\mathbb{T} = \mathbb{Z}$ on a Polish space which is a product of i.i.d. random mappings in canonical representation. Then*

(i) There is a one-to-one correspondence between the set $\mathcal{I}_{\mathbb{P}, \mathcal{F}^-}(\varphi)$ of invariant forward Markov measures of φ and the set
$$\{\mu^+ \in \mathcal{I}_{\mathbb{P}}(\varphi^+) : \mu^+ = \mathbb{P} \times \rho^+\} = \{\mathbb{P} \times \rho^+ : P^+(\rho^+) = \rho^+\},$$
i.e. the set of stationary measures of the forward one-point Markov transition probability $P^+(x, B) = \mathbb{P}\{\omega : \psi(\cdot)x \in B\}$. The correspondence is given by $\mu \mapsto \mathbb{E}\mu_{\cdot} = \rho^+$ and $\rho^+ \mapsto \mu_{\omega} = \lim_{n \to \infty} \varphi(n, \theta^{-n}\omega)\rho^+$.

In particular, a φ-invariant forward Markov measure μ is ergodic if and only if $\mathbb{E}\mu_{\cdot} = \rho^+$ is ergodic.

(ii) *Similarly, there is a one-to-one correspondence between the set* $\mathcal{I}_{\mathbb{P},\mathcal{F}^+}(\varphi)$ *of invariant backward Markov measures of φ and the set of stationary measures of the backward one-point Markov transition probability*

$$P^-(x, B) = \mathbb{P}\{\omega : \varphi(-1, \cdot)x \in B\} = \mathbb{P}\{\omega : \psi(\cdot)^{-1}x \in B\}.$$

The two-sided RDS φ typically has invariant measures which are not Markov measures (see Exercise 2.1.2). Since $\mathcal{I}_{\mathbb{P}}(\varphi)$ and $\mathcal{I}_{\mathbb{P}}(\varphi^\pm)$ are in a one-to-one correspondence by Theorem 1.7.2 the one-sided time restrictions $\varphi^\pm$ of φ also typically have more invariant measures than those corresponding to stationary measures of $P^\pm$.

The two transition probabilities P^+ and P^- are obviously not independent of each other, but their precise relation is still unclear. We have $\psi(\omega)x = y$ if and only if $\psi(\omega)^{-1}y = x$, which for a countable state space yields necessarily

$$P^-(y, \{x\}) = P^+(x, \{y\}) \quad \text{for all } x, y \in X, \tag{2.1.7}$$

i.e. the transition matrices $P^\pm$ are doubly-stochastic with $(P^+)^* = P^-$.

The inverse problem, "Is there an RDS of i.i.d. mappings on X which reproduces prescribed transition probabilities $P^\pm$?", is largely unsolved. For a countable state space condition (2.1.7) is also sufficient for the representation by bijections, see Dubischar [130].

2.2 Continuous Time 1: Random Differential Equations

Generators of deterministic continuous time dynamical systems are autonomous ordinary differential equations (or vector fields) (see Appendix B.4).

In case of an RDS, we should expect generators to be certain nonautonomous differential equations whose right-hand side depends on ω.

In this section we treat the (easy) *real noise case* in which the generator is indeed a certain family of ordinary differential equations with parameter ω, i.e. it can be solved "pathwise" for each fixed ω as a deterministic nonautonomous ordinary differential equation. All we need is a new way of looking at well-known classical results which we have, for convenience, collected in Appendix B.3.

2.2.1 RDS from Random Differential Equations

Let $\mathbb{T} = \mathbb{R}$, $X = \mathbb{R}^d$, and θ be a metric DS. We will now establish a basically one-to-one correspondence between (local) continuous/C^k RDS φ over θ which are absolutely continuous with respect to t and random differential equations $\dot{x}_t = f(\theta_t\omega, x_t)$ driven by θ. The correspondence is given by

58   Chapter 2. Generation

$$\varphi(t,\omega)x = x + \int_0^t f(\theta_s\omega, \varphi(s,\omega)x)\,ds, \qquad (2.2.1)$$

which is valid in the local case for all $t \in D(\omega, x)$, an open interval of $\mathbb{R}$ containing 0, and in the global case for all $t \in \mathbb{R}$. If (2.2.1) holds, we say that $t \mapsto \varphi(t,\omega)x$ solves or is a *solution* of the random differential equation[2] $\dot{x}_t = f(\theta_t\omega, x_t)$ (sometimes called *solution in the sense of Carathéodory*), or that the random differential equation *generates* φ. Note that (2.2.1) implies $\varphi(0,\omega)x = x$ for all ω and x.

If the solution is differentiable with respect to t and satisfies for $t \in D(\omega, x)$

$$\frac{d}{dt}\varphi(t,\omega)x = f(\theta_t\omega, \varphi(t,\omega)x), \quad \varphi(0,\omega)x = x, \qquad (2.2.2)$$

it is called a *classical* solution of $\dot{x}_t = f(\theta_t\omega, x_t)$. Non-classical solutions are important since they allow us to consider discontinuous noise.

2.2.1 Theorem (Local RDS from RDE). *Let $f : \Omega \times \mathbb{R}^d \to \mathbb{R}^d$ be measurable, consider the pathwise random differential equation*

$$\dot{x}_t = f(\theta_t\omega, x_t), \qquad (2.2.3)$$

and for fixed ω let $f_\omega(t,x) := f(\theta_t\omega, x)$.

(i) If $f_\omega \in L_{\mathrm{loc}}(\mathbb{R}, \mathcal{C}^{0,1})$[3] for all ω then (2.2.3) uniquely generates through its maximal solution a local continuous RDS φ over θ.

(ii) If $f_\omega \in L_{\mathrm{loc}}(\mathbb{R}, \mathcal{C}^{k,0})$ for all ω for some $k \geq 1$ then (2.2.3) uniquely generates a local C^k RDS φ over θ.

(iii) If $f_\omega \in \mathcal{C}^{0;0,1}$ for all ω then the continuous RDS φ from (i) is a classical solution, i.e. it satisfies (2.2.2). It is locally Lipschitz with respect to (t,x) and C^1 with respect to t.

(iv) If $f_\omega \in \mathcal{C}^{0;k,0}$ for all ω for some $k \geq 1$ then the C^k RDS φ from (ii) is a classical solution. It is C^1 with respect to (t,x).

Proof. The existence, uniqueness and regularity proofs are ω-wise adaptations of the deterministic proofs, see e.g. Amann [1: pp. 100 ff.].

The only non-standard assertion is the local cocycle property which we verify here, e.g. for the cases (i)/(ii). For this purpose fix (ω, x), let $D(\omega, x)$ be the interval of maximal existence of $\varphi(t,\omega)x$ and take $s \in D(\omega, x)$. We will show that

$$s + t \in D(\omega, x) \iff t \in D(\theta_s\omega, \varphi(s,\omega)x),$$

and in that case we have

$$\varphi(t+s, \omega)x = \varphi(t, \theta_s\omega)\varphi(s,\omega)x.$$

---

[2] "Random differential equation(s)" is henceforth abbreviated as "RDE".
[3] For the definition of the space $L_{\mathrm{loc}}(\mathbb{R}, \mathcal{C}^{0,1})$ etc. see Appendix B.1.2.

2.2 Continuous Time 1: Random Differential Eqs.

We distinguish between the four cases
1. $s > 0, t > 0$,
2. $s < 0, t < 0$ (analogous to 1.),
3. $s > 0, t < 0$,
4. $s < 0, t > 0$ (analogous to 3.).

Case 1, step 1: Let $t \in D(\theta_s\omega, \varphi(s,\omega)x)$. Hence

$$\begin{aligned}
\varphi(t,\theta_s\omega)\varphi(s,\omega)x &= \varphi(s,\omega)x + \int_0^t f(\theta_{u+s}\omega, \varphi(u,\theta_s\omega)\varphi(s,\omega)x)\,du \\
&= x + \int_0^s f(\theta_u\omega, \varphi(u,\omega)x)\,du \\
&\quad + \int_s^{s+t} f(\theta_u\omega, \varphi(u-s,\theta_s\omega)\varphi(s,\omega)x)\,du,
\end{aligned}$$

where we have used the fact that $s \in D(\omega, x)$ and made the substitution $u + s \to u$. Hence both integrals in the last lines are finite. We thus have found that the function

$$\psi(u,\omega)x := \begin{cases} \varphi(u,\omega)x, & 0 \le u \le s, \\ \varphi(u-s,\theta_s\omega)\varphi(s,\omega)x, & s \le u \le s+t, \end{cases}$$

satisfies

$$\psi(t+s,\omega)x = x + \int_0^{t+s} f(\theta_u\omega, \psi(u,\omega)x)\,du.$$

This proves that $t + s \in D(\omega, x)$, so by uniqueness of the solution

$$\varphi(t+s,\omega)x = \psi(t+s,\omega)x = \varphi(t,\theta_s\omega)\varphi(s,\omega)x.$$

Case 1, step 2: Let $t + s \in D(\omega, x)$. Then

$$\begin{aligned}
\varphi(t+s,\omega)x &= x + \int_0^{t+s} f(\theta_u\omega, \varphi(u,\omega)x)\,du \\
&= x + \int_0^s f(\theta_u\omega, \varphi(u,\omega)x)\,du + \int_s^{t+s} f(\theta_u\omega, \varphi(u,\omega)x)\,du \\
&= \varphi(s,\omega)x + \int_0^t f(\theta_u\theta_s\omega), \varphi(u+s,\omega)x)\,du,
\end{aligned}$$

where all integrals are finite. Putting $\bar{\omega} := \theta_s\omega$, $\bar{x} := \varphi(s,\omega)x$ and $\psi(u,\bar{\omega})\bar{x} := \varphi(u+s,\omega)x$, we have found that

$$\psi(t,\bar{\omega})\bar{x} = \bar{x} + \int_0^t f(\theta_u\bar{\omega}, \psi(u,\bar{\omega})\bar{x})\,du,$$

i.e. $t \in D(\bar{\omega}, \bar{x}) = D(\theta_s\omega, \varphi(s,\omega)x)$, so by uniqueness of solution

$$\varphi(t,\bar{\omega})\bar{x} = \varphi(t,\theta_s\omega)\varphi(s,\omega)x = \psi(t,\bar{\omega})\bar{x} = \varphi(t+s,\omega)x.$$

Case 3, step 1: Let $s > 0$, $t < 0$ and $t \in D(\theta_s\omega, \varphi(s,\omega)x)$.

Subcase (α): $s+t \geq 0$: Since $0 \leq s+t < s$, $s+t \in D(\omega, x)$ automatically. The local cocycle property then follows from the uniqueness of the solution.

Subcase (β): $s + t < 0$: Since $t < s + t < 0$, $s + t \in D(\theta_s\omega, \varphi(s,\omega)x)$ automatically, and the local cocycle property follows from the uniqueness of the solution.

Similarly for the remaining cases. □

As a particular case of the last theorem we obtain sufficient conditions for the generation of a *global* RDS by an RDE.

2.2.2 Theorem (Global RDS from RDE). *Assume the situation of Theorem 2.2.1.*

(i) If $f_\omega \in L_{\mathrm{loc}}(\mathbb{R}, \mathcal{C}_b^{0,1})$ for all ω then equation (2.2.3) uniquely generates a continuous RDS φ over θ.

(ii) If $f_\omega \in L_{\mathrm{loc}}(\mathbb{R}, \mathcal{C}_b^{k,0})$ for all ω and some $k \geq 1$ then (2.2.3) uniquely generates a C^k RDS φ over θ.

(iii) If $f_\omega \in \mathcal{C}_b^{0;0,1}$ for all ω then the continuous RDS from (i) is a classical solution of (2.2.3).

(iv) If $f_\omega \in \mathcal{C}_b^{0;k,0}$ for all ω and some $k \geq 1$ then the C^k RDS from (ii) is a classical solution of (2.2.3).

(v) In cases (ii) and (iv), consider the Jacobian of $\varphi(t,\omega)$ at x,

$$D\varphi(t,\omega,x) := \left(\frac{\partial (\varphi(t,\omega)x)_i}{\partial x_j} \right) \in Gl(\mathbb{R},d).$$

Then $(\varphi, D\varphi)$ is a C^{k-1} RDS on $\mathbb{R}^d \times \mathbb{R}^d$ over θ uniquely generated by $(\dot{x}_t, \dot{v}_t) = (f(\theta_t\omega, x_t), Df(\theta_t\omega, x_t)v_t)$. Hence $D\varphi$ uniquely solves the variational equation

$$D\varphi(t,\omega,x) = I + \int_0^t Df(\theta_s\omega, \varphi(s,\omega)x)\, D\varphi(s,\omega,x)\, ds,$$

thus is a matrix cocycle over the skew product $\Theta(t)(\omega, x) := (\theta_t\omega, \varphi(t,\omega)x)$. Finally, the determinant $\det D\varphi(t,\omega,x)$ satisfies Liouville's equation

$$\det D\varphi(t,\omega,x) = \exp \int_0^t (\mathrm{trace} Df)(\theta_s\omega, \varphi(s,\omega)x)\, ds,$$

and so is a scalar cocycle over Θ.

Proof. The proofs of these facts are ω-wise versions of their deterministic counterparts. The cocycle property of $(\varphi, D\varphi)$ over θ and of $D\varphi$ and $\det D\varphi$ over Θ follow again from the uniqueness of the ω-wise solution of the respective equation. □

2.2.3 Remark. (i) In both theorems we could have required that f_ω has the corresponding property for all ω in a θ-invariant set of full measure. Outside this set put $\varphi(t,\omega) = \mathrm{id}_{\mathbb{R}^d}$ for all $t \in \mathbb{R}$ to obtain a perfect cocycle.

(ii) Note that it does not in general suffice to have conditions on $f(\omega, \cdot)$ for each fixed ω.

(iii) The sufficient condition for a global continuous RDS in Theorem 2.2.2 (i) can be replaced by the following more general one: $f_\omega \in L_{\mathrm{loc}}(\mathbb{R}, \mathcal{C}^{0,1})$ and

$$\|f(\omega, x)\| \le \alpha(\omega)\|x\| + \beta(\omega),$$

where $t \mapsto \alpha(\theta_t \omega)$ and $t \mapsto \beta(\theta_t \omega)$ are locally integrable. ■

2.2.4 Remark (Variational Equation for Local RDS). The variational equation is valid also for the local C^k RDS of Theorem 2.2.1 (ii) and (iv) as long as $x \in D(t,\omega)$, the domain of $\varphi(t,\omega)$. It is globally valid for all $t \in \mathbb{R}$ provided $x \in E(\omega) = \cap_{t \in \mathbb{R}} D(t,\omega)$ (see Sects. 1.2 and 1.8 for details), because E is strictly φ-invariant, and since $D(t,\omega)$ is open, $\varphi(t,\omega)$ can be differentiated at $x \in E(\omega)$ for all t. ■

Sufficient Existence and Uniqueness Conditions

We will now present "static" sufficient conditions for $f_\omega \in L_{\mathrm{loc}}(\mathbb{R}, \mathcal{C}^{0,1})$ etc., the validity of which can often be read off from the right-hand side of $\dot{x}_t = f(\theta_t \omega, x_t)$.

2.2.5 Lemma. *Let $(\Omega, \mathcal{F}, \mathbb{P}, (\theta_t)_{t \in \mathbb{R}})$ be a metric DS and $g \in L^1(\Omega, \mathcal{F}, \mathbb{P})$. Then the measurable stationary stochastic process $t \mapsto g(\theta_t \omega)$ is locally integrable on an invariant ω set of full measure, and for all $a, b \in \mathbb{R}$*

$$\int_a^b g(\theta_t \cdot) dt \in L^1(\Omega, \mathcal{F}, \mathbb{P}).$$

Proof. The set of ω's for which $t \mapsto g(\theta_t \omega)$ is locally integrable is clearly measurable (exhaust $\mathbb{R}$ by countably many finite intervals) and θ-invariant due to the flow property of θ and the shift invariance of Lebesgue measure. We show that this set has full measure. By Fubini's theorem, utilizing $\mathbb{E}|g(\theta_t \cdot)| \equiv m < \infty$, for all $a, b \in \mathbb{R}$ with $a \le b$

$$\int_a^b \mathbb{E}|g(\theta_t \cdot)|\, dt = m(b-a) = \mathbb{E} \int_a^b |g(\theta_t \cdot)|\, dt < \infty,$$

implying that

$$\int_a^b |g(\theta_t \cdot)|\, dt < \infty \quad \mathbb{P}\text{-a. s.}$$

□

With this lemma we can deduce immediately the following theorem from Theorems 2.2.1 and 2.2.2.

2.2.6 Theorem (Sufficient Conditions for (Local) RDS). *Assume the situation of Theorem 2.2.1 and 2.2.2, with a measurable f.*

(i) If $f \in L^1_{\mathbb{P}}(\Omega, \mathcal{C}^{0,1})$ (or if $f \in L^1_{\mathbb{P}}(\Omega, \mathcal{C}^{k,0})$), then the set $\tilde{\Omega}$ of those ω's for which $f_\omega = f(\theta.\omega, \cdot) \in L_{\mathrm{loc}}(\mathbb{R}, \mathcal{C}^{0,1})$ (or $f_\omega \in L_{\mathrm{loc}}(\mathbb{R}, \mathcal{C}^{k,0})$) is θ-invariant and has full measure. After having redefined f outside $\tilde{\Omega}$ by $f(\omega, \cdot) :\equiv 0$ the random differential equation (2.2.3) uniquely generates a local continuous (or C^k) RDS.

(ii) If $f \in L^1_{\mathbb{P}}(\Omega, \mathcal{C}^{0,1}_b)$ (or if $f \in L^1_{\mathbb{P}}(\Omega, \mathcal{C}^{k,0}_b)$), then $f_\omega \in L_{\mathrm{loc}}(\mathbb{R}, \mathcal{C}^{0,1}_b)$ (or $f_\omega \in L_{\mathrm{loc}}(\mathbb{R}, \mathcal{C}^{k,0}_b)$) on an invariant set of full measure, and (after possibly redefining f outside this set as in (i)) the random differential equation (2.2.3) uniquely generates a continuous (or C^k) RDS. More generally, this is true if $f_\omega \in L_{\mathrm{loc}}(\mathbb{R}, \mathcal{C}^{0,1})$ (or if $f_\omega \in L_{\mathrm{loc}}(\mathbb{R}, \mathcal{C}^{k,0})$) and

$$\|f(\omega, x)\| \leq \alpha(\omega)\|x\| + \beta(\omega), \quad \alpha, \beta \in L^1(\Omega, \mathcal{F}, \mathbb{P}).$$

Proof. We use the definition of the spaces $L_{\mathrm{loc}}(\mathbb{R}, \mathcal{C}^{0,1})$ etc. and Lemma 2.2.5 for $g(\omega) = \|f(\omega, \cdot)\|_{0,1;K}$ etc. □

2.2.7 Remark. (i) The following conditions are equivalent to the condition $f \in L^1_{\mathbb{P}}(\Omega, \mathcal{C}^{0,1})$:

1. For some $x_0 \in \mathbb{R}^d$, $\mathbb{E}\|f(\cdot, x_0)\| < \infty$,
2. for all $R > 0$

$$\|f(\omega, x) - f(\omega, y)\| \leq L_R(\omega)\|x - y\|, \quad \|x\|, \|y\| \leq R,$$

and $\mathbb{E} L_R(\cdot) < \infty$.

(ii) The following conditions are equivalent to the condition $f \in L^1_{\mathbb{P}}(\Omega, \mathcal{C}^{0,1}_b)$:

1. For some $x_0 \in \mathbb{R}^d$, $\mathbb{E}\|f(\cdot, x_0)\| < \infty$,
2. for all $x, y \in \mathbb{R}^d$

$$\|f(\omega, x) - f(\omega, y)\| \leq L(\omega)\|x - y\|,$$

and $\mathbb{E} L(\cdot) < \infty$.

■

2.2.8 Example (Linear and Affine RDE). (i) Linear RDE: Let the measurable function $A : \Omega \to \mathbb{R}^{d \times d}$ satisfy $A \in L^1(\Omega, \mathcal{F}, \mathbb{P})$. Then $f_\omega(t, x) := A(\theta_t \omega)x$ satisfies $f_\omega \in L_{\mathrm{loc}}(\mathbb{R}, \mathcal{C}^\infty_b)$ $\mathbb{P}$-a. s. , hence $\dot{x}_t = A(\theta_t \omega)x_t$ generates a unique (up to indistinguishability) C^∞ (even linear) RDS Φ satisfying

$$\Phi(t, \omega) = I + \int_0^t A(\theta_s \omega) \Phi(s, \omega)\, ds$$

and

$$\det \Phi(t,\omega) = \exp \int_0^t \operatorname{trace} A(\theta_s \omega)\, ds.$$

Also, differentiating $\Phi(t,\omega)\Phi(t,\omega)^{-1} = I$ yields (see Sect. 6.1)

$$\Phi(t,\omega)^{-1} = I - \int_0^t \Phi(s,\omega)^{-1} A(\theta_s \omega)\, ds.$$

(ii) Affine RDE: The equation

$$\dot{x}_t = A(\theta_t \omega) x_t + b(\theta_t \omega), \quad A, b \in L^1(\Omega, \mathcal{F}, \mathbb{P}),$$

generates a unique (up to indistinguishability) C^∞ RDS. The variation of constants formula yields

$$\begin{aligned}\varphi(t,\omega)x &= \Phi(t,\omega)x + \int_0^t \Phi(t,\omega)\Phi(u,\omega)^{-1} b(\theta_u \omega)\, du \\ &= \Phi(t,\omega)x + \int_0^t \Phi(t-u, \theta_u \omega) b(\theta_u \omega)\, du,\end{aligned}$$

where Φ is the matrix cocycle generated by $\dot{x}_t = A(\theta_t \omega) x_t$. Consequently, the RDS φ consists of affine mappings (see Sect. 5.6). ∎

2.2.9 Exercise (RDE with Polynomial Right-Hand Side).
Let $f(\omega, x) = \sum_{j=0}^N a_j(\omega) x^j$ be a random polynomial with $N \geq 2$. If $a_j \in L^1(\Omega, \mathcal{F}, \mathbb{P})$ for all j, then the random differential equation

$$\dot{x}_t = \sum_{j=0}^N a_j(\theta_t \omega) x_t^j$$

uniquely generates a local C^∞ RDS in $\mathbb{R}^1$.

The case

$$\dot{x}_t = a(\theta_t \omega) x_t + b(\theta_t \omega) x_t^N$$

can be explicitly solved by transforming the equation to an affine equation via $y = x^{1-N}/(1-N)$. Write down explicitly all ingredients of the local RDS, in particular the domains $D(t,\omega)$ and ranges $R(t,\omega)$ of φ, the set of never exploding initial values $E(\omega)$, and the invariant measures. The cases $N = 2$ and $N = 3$ are treated by Xu [347] and in Subsect. 9.3.5. For the white noise case see Subsects. 2.3.7 and 9.3.1. ∎

2.2.2 The Memoryless Case

In most applications equation (2.2.3) has the particular form

$$\dot{x}_t = f(\theta_t \omega, x_t) = g(\xi_t(\omega), x_t), \qquad (2.2.4)$$

where $\xi_t(\omega)$ is a "nice" stationary stochastic process on some state space $(E, \mathcal{E})$. In other words, we assume that the right-hand side of (2.2.4) is *memoryless* in the sense that only the value of the perturbation *at* time t enters into the generator at t.

We assume for simplicity that $E = \mathbb{R}^m$ (everything remains valid if E is a manifold or a Polish space). We are interested in allowing discontinuous processes ξ such as jump Markov processes. A good class of processes to work with is the one whose paths are cadlag (i.e. are right continuous with left hand limits, see Appendix A.2 for details). This class is well-suited to deal with semimartingales and Markov processes, since, under mild conditions, those processes have cadlag versions.

We will now describe the canonical setting in which we study (2.2.4): We choose the measurable DS constructed in A.2 with sample space equal to the Polish space $\Omega := \mathcal{D}(\mathbb{R}, \mathbb{R}^m)$, with its Borel σ-algebra $\mathcal{F}$ and the classical shift $\theta_t \omega(\cdot) := \omega(t + \cdot)$. $\mathbb{P}$ can be any θ-invariant measure.

Define

$$\pi : \Omega \to \mathbb{R}^m, \quad \omega \mapsto \omega(0), \quad \pi(\mathbb{P}) = \rho.$$

Then the coordinate process

$$\xi_t(\omega) := \omega(t) = (\pi \circ \theta_t)(\omega) = \pi(\theta_t \omega)$$

is the canonical measurable stationary stochastic process with cadlag paths and prescribed distribution $\mathbb{P}$. Now let

$$g : \mathbb{R}^m \times \mathbb{R}^d \to \mathbb{R}^d$$

be continuous (this could be easily generalized), and put $f(\omega, x) := g(\omega(0), x)$. Since now $f(\theta_t \omega, x) = g(\omega(t), x) = g(\xi_t(\omega), x)$, we have an RDE of the form (2.2.4).

The function $(t, \omega, x) \mapsto g(\xi_t(\omega), x)$ is measurable, $t \mapsto g(\xi_t(\omega), x)$ is cadlag for all (ω, x) and $x \mapsto g(\xi_t(\omega), x)$ is continuous for all (t, ω).

We next give sufficient conditions on the function g assuring that (2.2.4) generates an RDS.

2.2.10 Theorem (Sufficient Conditions for Local RDS). *(i) Suppose ξ is cadlag and $g \in \mathcal{C}^{0;0,1}$. Then $g(\xi.(\omega), \cdot) \in L_{\mathrm{loc}}(\mathbb{R}, \mathcal{C}^{0,1})$ for all $\omega \in \Omega$ and the random differential equation $\dot{x}_t = g(\xi_t(\omega), x_t)$ has a unique maximal solution which is a local continuous RDS φ. The function $\varphi(\cdot, \omega)$ is differentiable with respect to t at those times at which $\xi.(\omega)$ is continuous.*

If $g \in \mathcal{C}^{0;k,0}$ then $g(\xi.(\omega), \cdot) \in L_{\mathrm{loc}}(\mathbb{R}, \mathcal{C}^{k,0})$, and the local RDS is C^k.

(ii) Suppose ξ is continuous and $g \in \mathcal{C}^{0;0,1}$. Then $g(\xi.(\omega),\cdot) \in \mathcal{C}^{0;0,1}$ for all $\omega \in \Omega$ and $\dot{x}_t = g(\xi_t(\omega), x_t)$ has a unique maximal classical solution which is a local continuous RDS φ.

If $g \in \mathcal{C}^{0;k,0}$ then $g(\xi.(\omega),\cdot) \in \mathcal{C}^{0;k,0}$ and the local RDS is C^k.

Proof. We just deal with the case (i) and show that $g \in \mathcal{C}^{0;0,1}$ implies $g(\xi.(\omega),\cdot) \in L_{\mathrm{loc}}(\mathbb{R},\mathcal{C}^{0,1})$. Indeed,

$$\sup_{(t,x)\in[a,b]\times K} \|g(\xi_t(\omega),x)\| =: C(\omega) < \infty$$

since g is continuous and ξ is cadlag, thus bounded on compact sets. Further, by definition $g \in \mathcal{C}^{0;0,1}$ means that g is locally Lipschitz with respect to x which is equivalent to being uniformly Lipschitz with respect to x on compact sets $K_1 \times K \subset \mathbb{R}^m \times \mathbb{R}^d$. But since ξ is cadlag, $\xi_{[a,b]}(\omega) \subset K_2(\omega)$, $K_2(\omega)$ a compact set in $\mathbb{R}^m$. Consequently, the local Lipschitz constant of $g(\xi.(\omega),\cdot)$ can be estimated as follows:

$$\sup_{t\in[a,b],\, x,y\in K,\, x\neq y} \frac{\|g(\xi_t(\omega),x) - g(\xi_t(\omega),y)\|}{\|x-y\|}$$

$$\leq \sup_{\xi\in K_2(\omega),\, x,y\in K,\, x\neq y} \frac{\|g(\xi,x) - g(\xi,y)\|}{\|x-y\|} =: L(\omega) < \infty.$$

As a result, $g(\xi.(\omega),\cdot) \in L_{\mathrm{loc}}(\mathbb{R},\mathcal{C}^{0,1})$. □

2.2.11 Theorem (Sufficient Conditions for Global RDS). *Suppose ξ is cadlag, $g \in \mathcal{C}^{0;0,1}$ (or $g \in \mathcal{C}^{0;k,0}$) and*

$$\|g(\xi,x)\| \leq \alpha(\xi)\|x\| + \beta(\xi),$$

where $t \mapsto \alpha(\xi_t(\omega))$, $t \mapsto \beta(\xi_t(\omega))$ are locally integrable (this will hold on an invariant set of full measure if $\alpha: \mathbb{R}^m \to \mathbb{R}$, $\beta: \mathbb{R}^m \to \mathbb{R}$ are continuous, or if $\alpha,\beta \in L^1(\mathbb{R}^m, \mathcal{B}^m, \rho)$, $\rho = \pi(\mathbb{P}) = \mathcal{L}(\xi_0)$), then the (if ξ is continuous: classical) solution of the random differential equation $\dot{x}_t = g(\xi_t(\omega), x_t)$ is a global continuous (or C^k) RDS.

2.2.12 Example (Noisy Parameters in ODE). Consider an ordinary differential equation $\dot{x}_t = g(\lambda, x_t)$ whose right-hand side depends on a parameter $\lambda \in \mathbb{R}^m$. In numerous applications it has to be assumed that such a parameter is "noisy". This amounts to replacing the fixed λ with a stationary stochastic process $\xi_t(\omega)$. We therefore arrive at $\dot{x}_t = g(\xi_t(\omega), x_t)$.

Assume, in particular, $g(\lambda, x) = A(\lambda)x + b(\lambda)$, with continuous functions $A: \mathbb{R}^m \to \mathbb{R}^{d\times d}$, $b: \mathbb{R}^m \to \mathbb{R}^d$. We have $g \in \mathcal{C}_b^{0;\infty}$. If $\xi_t(\omega)$ is cadlag then there is a global C^∞ RDS φ of affine mappings generated by

$$\dot{x}_t = A(\xi_t(\omega))x_t + b(\xi_t(\omega)),$$

and given by the formula of Example 2.2.8(ii). ■

2.2.3 Random Differential Equations from RDS

We now deal with the inverse problem of when for a given RDS φ on $\mathbb{R}^d$ over θ with time $\mathbb{T} = \mathbb{R}$ there exists a random differential equation $\dot{x}_t = F(t, \omega, x_t)$ which generates φ. We will also determine the only possible form in which F is coupled to θ, namely $F(t, \omega, x) = f(\theta_t \omega, x)$.

2.2.13 Theorem (RDE from RDS). *(i) Let φ be a local continuous RDS for which $t \mapsto \varphi(t, \omega)x$ is differentiable at $t = 0$ for all (ω, x). Put*

$$f(\omega, x) := \frac{d}{dt}\varphi(t, \omega)x|_{t=0}.$$

Then f is measurable, $t \mapsto \varphi(t, \omega)x$ is differentiable for all $t \in D(\omega, x)$ and

$$\frac{d}{dt}\varphi(t, \omega)x = f(\theta_t \omega, \varphi(t, \omega)x), \quad \varphi(0, \omega)x = x,$$

i.e. φ is a classical solution of $\dot{x}_t = f(\theta_t \omega, x_t)$ and f is uniquely determined by φ.

(ii) Let φ be a local continuous RDS for which $t \mapsto \varphi(t, \omega)x$ is absolutely continuous with respect to $t \in D(\omega, x)$ for all (ω, x). Then there exists a measurable function $f : \Omega \times \mathbb{R}^d \to \mathbb{R}^d$ for which for all (ω, x)

$$\varphi(t, \omega)x = x + \int_0^t f(\theta_s \omega, \varphi(s, \omega)x)ds, \quad t \in D(\omega, x),$$

i.e. φ is a solution of $\dot{x}_t = f(\theta_t \omega, x_t)$. The function f is unique in the sense that if $\bar{f}$ is another generator then for all (ω, x), $f(\theta_t \omega, \varphi(t, \omega)x) = \bar{f}(\theta_t \omega, \varphi(t, \omega)x)$ for Lebesgue-almost all $t \in D(\omega, x)$.

Proof. (i) We fix (ω, x) and choose an arbitrary $t \in D(\omega, x)$. Since $D(\omega, x)$ is open, we have $t + h \in D(\omega, x)$ for all h with $|h| \leq h_0(t, \omega, x)$. By the definition of a local RDS $t + h \in D(\omega, x) \iff h \in D(\theta_t \omega, \varphi(t, \omega)x)$, and

$$\varphi(t + h, \omega)x = \varphi(h, \theta_t \omega)\varphi(t, \omega)x.$$

Subtracting $\varphi(t, \omega)x$ from both sides, dividing by h, taking the limit for $h \to 0$ and taking into account that φ is differentiable at $t = 0$ yields

$$\frac{d}{dt}\varphi(t, \omega)x = \frac{d}{dh}\varphi(h, \theta_t \omega)\varphi(t, \omega)x|_{h=0} = f(\theta_t \omega, \varphi(t, \omega)x).$$

(ii) By assumption there is a g such that

$$\varphi(t, \omega)x - x = \int_0^t g(s, \omega, x)ds =: G(t, \omega, x).$$

g can be chosen to be measurable, in fact put for $s \in D(\omega, x)$

$$g(s,\omega,x) := \limsup_{n\to\infty} \frac{G(s+\frac{1}{n},\omega,x) - G(s,\omega,x)}{\frac{1}{n}}. \tag{2.2.5}$$

This function g is measurable, it is a version of the Radon-Nikodym density of G for all (ω,x), and the limit exists for Lebesgue-almost all $s \in D(\omega,x)$.

The local cocycle property of φ over θ translates into the local *helix property* of G over $\Theta = (\theta,\varphi)$: For $s \in D(\omega,x)$ we have $t+s \in D(\omega,x) \iff t \in D(\Theta(s)(\omega,x))$, and

$$G(t+s,\omega,x) = G(t,\Theta(s)(\omega,x)) + G(s,\omega,x).$$

Inserting this into (2.2.5) yields

$$g(s,\omega,x) = \limsup_{n\to\infty} \frac{G(\frac{1}{n},\Theta(s)(\omega,x))}{\frac{1}{n}} = g(0,\Theta(s)(\omega,x)). \tag{2.2.6}$$

Now define $f(\omega,x) := g(0,\omega,x)$ which is certainly measurable. By (2.2.6)

$$g(s,\omega,x) = f(\theta_s\omega, \varphi(s,\omega)x),$$

i.e. we have constructed an f for which for all (ω,x) and $t \in D(\omega,x)$

$$\varphi(t,\omega)x = x + \int_0^t f(\theta_s\omega, \varphi(s,\omega)x)ds,$$

meaning that φ has generator f, and f is unique in the sense claimed. □

The final result is that the following classes of objects are basically in one-to-one correspondence:

- solutions of RDE of the form $\dot{x}_t = f(\theta_t\omega, x_t)$, i.e. "driven" by the metric DS θ (note that $t \mapsto f(\theta_t\omega, \cdot)$ can also be viewed as a stationary stochastic process taking values in the space $\mathcal{C}^{0,1}$ etc. of vector fields),
- continuous/C^k RDS φ which are differentiable or absolutely continuous with respect to t.

2.2.4 The Manifold Case

If $X = M$ is a d-dimensional (C^∞) manifold, all statements made so far in this section (except the statements about the generation of global RDS) remain valid if properly interpreted (see Appendix B.5 for more details). For ease of reference, we collect some basic facts in the following theorem.

2.2.14 Theorem (Local RDS from RDE on Manifold). *Let $f : \Omega \to \mathfrak{X}^k(M)$, $1 \le k \le \infty$, be measurable, and consider the RDE*

$$\dot{x}_t = f(\theta_t\omega, x_t) \tag{2.2.7}$$

on M. Let for fixed ω, $t \mapsto f_\omega(t,\cdot) := f(\theta_t\omega,\cdot) \in \mathfrak{X}^k(M)$ satisfy

68   Chapter 2. Generation

$$f_\omega \in L_{\text{loc}}(\mathbb{R}, \mathcal{C}^{k,0}) \quad \text{for all } \omega \in \Omega \qquad (2.2.8)$$

(a sufficient condition for (2.2.8) is $f \in L^1_{\mathbb{P}}(\Omega, \mathcal{C}^{k,0})$).
Then

   (i) Equation (2.2.7) uniquely generates a local C^k RDS φ over θ. If $f_\omega \in C^{0;k,0}$ for all ω then φ is a classical solution of (2.2.7), and is C^1 with respect to (t,x).

   (ii) The derivative $T\varphi : TM \to TM$ is a local C^{k-1} RDS which uniquely solves

$$\dot{v}_t = Tf(\theta_t\omega, v_t), \qquad (2.2.9)$$

where Tf is the natural lift of f to TM.

   (iii) If M is a Riemannian manifold, and if we use the Riemannian connection on M to decompose $T_v TM$ into horizontal and vertical components, then $Tf(\omega, v)$ has horizontal component $f(\omega, x)$ and vertical component $\nabla f(\omega, x)v$, where ∇ denotes the covariant derivative, and equation (2.2.9) is equivalent to the coupled system of the original RDE (2.2.7) and the variational equation

$$v'_t = \nabla f(\theta_t\omega, \varphi(t,\omega)x)v_t.$$

Here the absolute derivative v'_t is taken along the integral curve $t \mapsto \varphi(t,\omega)x$ in the direction of the vector field $f(\theta_t\omega, \cdot)$ (see Appendix B.5).

   Furthermore, we have Liouville's equation for $\det T\varphi(t,\omega,x)$: For $t \in D(\omega,x)$

$$\begin{aligned}\det T\varphi(t,\omega,x) &= \exp\int_0^t \operatorname{div} f(\theta_s\omega, \varphi(s,\omega)x)\,ds \\ &= \exp\int_0^t \operatorname{trace}\nabla f(\theta_s\omega, \varphi(s,\omega)x)\,ds.\end{aligned}$$

   (iv) The RDS φ is global in case M is compact (but f still satisfies (2.2.8)), or, by embedding M into $\mathbb{R}^N$, $f(\omega, \cdot)$ is the restriction of a random vector field $\tilde f(\omega, \cdot)$ on $\mathbb{R}^N$ for which $\tilde f_\omega \in L_{\text{loc}}(\mathbb{R}, \mathcal{C}^{k,0}_b)$.

2.3 Continuous Time 2: Stochastic Differential Equations

2.3.1 Introduction. Two Cultures

We can go a big step beyond Sect. 2.2 and can assign generators to an important and large class of RDS which are just continuous but not absolutely continuous, in fact nowhere differentiable, and locally not of bounded variation with respect to t. Such generators cannot be pathwise differential equations,

2.3 Continuous Time 2: Stochastic Differential Eqs. 69

but will turn out to be so-called *stochastic differential equations*[4]. An SDE contains stochastic integrals, one of the central objects of stochastic analysis, which are defined only as limits in probability and not as ω-wise limits. In particular, ω-wise deterministic results cannot be utilized (as in Sect. 2.2) to establish the existence and uniqueness of an RDS generated by an SDE.

More specifically, the class of RDS which have an SDE as generator consists of those φ which have the additional statistical property of $t \mapsto \varphi(t,\omega)x$ being a semimartingale for each fixed x. The generating SDE will turn out to be driven by a semimartingale with stationary increments (called a "helix"). This establishes a basically one-to-one correspondence which can be succinctly expressed as

$$\text{semimartingale cocycle} = \exp(\text{semimartingale helix}),$$

where "exp" symbolizes an "engine" the implementation of which is the subject of this section.

For example, consider the simplest possible SDE,

$$dx_t = dW_t, \quad W = \text{Brownian motion (see Appendix A.3)}.$$

It generates the C^∞ RDS $\varphi(t,\omega)x = x + W_t(\omega)$ of random translations over the canonical DS θ describing Brownian motion. But $t \mapsto \varphi(t,\omega)x$ is known to be nowhere differentiable and of unbounded variation.

The general case is conceptually and technically very complicated and can only be fully appreciated with a good knowledge of stochastic analysis, which we assume for this section. The theory of SDE and stochastic analysis have been laid down in numerous excellent textbooks, of which we just mention a few more recent one's in alphabetic order: Belopolskaya and Dalecky [65], Elworthy [138], Gard [158], Gihman and Skorohod [160], Ikeda and Watanabe [178], Karatzas and Shreve [197], Malliavin [247], Protter [281], Revuz and Yor [286], Rogers and Williams [287], Stroock and Varadhan [326].

Two Cultures

In the theory of RDS, two well-established mathematical cultures meet, overlap, and sometimes collide:

1. *Dynamical systems:* the flow point of view. Typically $\mathbb{T} = \mathbb{R}$ or $\mathbb{Z}$ unless mappings are non-invertible which typically happens only for discrete time, or infinite-dimensional state space. On the measurable level, ergodic theory assumes an invariant measure, where invariance is defined with respect to the mappings of the flow. "Time" is purely algebraic and non-physical. There is no such thing as "past", "present", "future", or "evolution".

2. *Markov processes, stochastic analysis:* Here time is almost exclusively one-sided, i.e. $\mathbb{T} = \mathbb{R}^+$ or $\mathbb{Z}^+$ (or part of it). Markov processes are defined and studied via their transition semigroup forward in time. The necessity

---

[4] "Stochastic differential equation(s)" is henceforth abbreviated as "SDE".

to really construct stochastic processes with prescribed transition semigroup (their existence follows from Kolmogorov's fundamental theorem) motivated the creation of the theory of differential equations with white noise input, in other words: the theory of SDE, and finally stochastic analysis. Continuous time is $\mathbb{R}^+$, and a filtration $(\mathcal{F}_t)_{t\geq 0}$ (i.e. an increasing family of sub σ-algebras of (the completion of) $\mathcal{F}$) collects the information available at time t. Everything (at least until recently) has to be adapted, i.e. $\mathcal{F}_t$-measurable. "Invariant measure" now means that the measure is stationary with respect to the Markov transition semigroup. The filtration gives a direction to time and allows one to speak of the past etc. of t, thus getting closer to a physical concept of time.

In short, the key object of (continuous time $\mathbb{T} = \mathbb{R}$) dynamical systems is a flow $(\theta_t)_{t\in\mathbb{R}}$ of mappings, and (in ergodic theory) a measure $\mathbb{P}$ invariant with respect to it, while the chief ingredient of stochastic analysis is a filtration $(\mathcal{F}_t)_{t\geq 0}$ (on $\mathbb{T} = \mathbb{R}^+$) which allows to study evolution in time.

The first people who clearly spelled-out this gap between ergodic theory and stochastic analysis and made a first attempt to bridge it were de Sam Lazaro and Meyer [122], [123] with their theory of filtered flows $(\Omega, \mathcal{F}, \mathbb{P}, (\theta_t)_{t\geq 0}, (\mathcal{F}_t)_{t\geq 0})$ (for even earlier work see Krengel [220: p. 34]).

J. de Sam Lazaro and Meyer wrote on p. 2 in [123]: *"Mais la théorie ergodique...est plus proche, dans bien des cas, de la théorie des groupes que de celle des processus stochastiques, car il y manque l'idée probabiliste essentielle: celle d'une évolution dans le temps."* Such a concept of evolution in time is the filtration $\mathcal{F}_t$.

The work of de Sam Lazaro and Meyer was continued by Protter [280] who was able to characterize the class of semimartingales with stationary (but not necessarily independent) increments. These semimartingales qualify as integrators (driving noise) in SDE which generate RDS. The work of de Sam Lazaro, Meyer and Protter can be considered as a preparation for our systematic study of the relations between RDS and SDE presented below. We closely follow Arnold and Scheutzow [35] in our presentation which is based on the fundamental work of Baxendale [54], Le Jan and Watanabe [228] and Kunita [224].

In order not to overburden matters we first deal with the global case in $X = \mathbb{R}^d$, and discuss the local and manifold case later.

The gate from stochastic analysis to dynamical systems was really opened around 1980, when several people (Elworthy [137, 138], Baxendale [53], Bismut [70, 72, 71], Ikeda and Watanabe [178], Kunita [222, 223], Meyer [252] and others) realized and proved that writing down a classical SDE means much more than originally intended by the "fathers" of stochastic analysis: An SDE does not only generate a Markov family of solution processes indexed by the initial time and initial position, but it generates a two-parameter flow $\varphi_{s,t}(\omega)$ of random homeomorphisms or diffeomorphisms. However, this is still a static object and not yet an RDS. To prove that $\varphi_{0,t}(\omega)$ is in fact a cocycle over the DS by which the driving noise is modelled, more work is necessary.

There are certain "obstacles" in our way: While we can and will assume that our metric DS θ has two-sided time $\mathbb{T} = \mathbb{R}$, classical stochastic calculus, as already mentioned above, has been almost exclusively developed for one-sided time $\mathbb{T} = \mathbb{R}^+$ and a one-parameter filtration $(\mathcal{F}_t)_{t \in \mathbb{R}^+}$. Hence the first task we will have to accomplish (in Subsects. 2.3.2 and 2.3.3) is the extension of stochastic calculus to two-sided time $\mathbb{T} = \mathbb{R}$. Quite naturally we will work with a two-parameter filtration $(\mathcal{F}_s^t)_{s \leq t}$, which we will have to make consistent with θ.

Besides its aesthetic appeal, two-sided time is fundamental for the theory of smooth RDS as it allows a "better" multiplicative ergodic theorem (providing an invariant splitting of subspaces rather than only an invariant flag, hence making general invariant manifolds, normal forms, bifurcation theory etc. possible).

2.3.2 Semimartingales and Dynamical Systems: Stochastic Calculus for Two-Sided Time

Let $(\Omega, \mathcal{F}, \mathbb{P})$ denote from now on a complete probability space.

2.3.1 Definition (Two-Parameter Filtration). Assume $\mathcal{F}_s^t$, $s, t \in \mathbb{R}$, $s \leq t$, is a two-parameter family of sub-σ-algebras of $\mathcal{F}$ with the following properties:

1. $\mathcal{F}_s^t \subset \mathcal{F}_u^v$ for $u \leq s \leq t \leq v$,
2. $\mathcal{F}_s^{t+} := \cap_{u > t} \mathcal{F}_s^u = \mathcal{F}_s^t$, $\mathcal{F}_{s-}^t := \cap_{u < s} \mathcal{F}_u^t = \mathcal{F}_s^t$ for $s \leq t$,
3. $\mathcal{F}_s^t$ contains all $\mathbb{P}$-null sets of $\mathcal{F}$ for every $s \leq t$.

Then $\mathcal{F}_s^t$, $s \leq t$, is called a (*two-parameter*) *filtration* (on $(\Omega, \mathcal{F}, \mathbb{P})$). We define

$$\mathcal{F}_{-\infty}^t := \vee_{s \leq t} \mathcal{F}_s^t, \quad \mathcal{F}_s^\infty := \vee_{t \geq s} \mathcal{F}_s^t.$$

∎

As we want to solve an SDE from an initial time forwards *as well as* backwards in time, we follow Kunita [224] and introduce the following notions.

2.3.2 Definition (Forward/Backward Semimartingale). Let $\mathcal{F}_s^t$, $s \leq t$, be a filtration on $(\Omega, \mathcal{F}, \mathbb{P})$, and let $F: \mathbb{R} \times \mathbb{R} \times \Omega \to \mathbb{R}^d$, $(s, t, \omega) \mapsto F_s(t, \omega)$, be measurable and jointly continuous with respect to s, t for all $\omega \in \Omega$.

(a) F is called an $\mathcal{F}_s^t$-*forward semimartingale*, if for each $s \in \mathbb{R}$, $(F_s(s+t, \omega))_{t \geq 0}$ is an $(\mathcal{F}_s^{s+t})_{t \geq 0}$-semimartingale.
(b) F is called an $\mathcal{F}_s^t$-*backward semimartingale* if for each $s \in \mathbb{R}$, $(F_s(s-t, \omega))_{t \geq 0}$ is an $(\mathcal{F}_{s-t}^s)_{t \geq 0}$-semimartingale.
(c) F is called an $\mathcal{F}_s^t$-*semimartingale* (or just *semimartingale*) if F is both a forward and a backward semimartingale. ∎

2.3.3 Remark. (i) As in the classical case the class of $\mathcal{F}_s^t$-(forward, respectively backward) semimartingales remains invariant under an equivalent change of the underlying probability $\mathbb{P}$.

(ii) $\mathcal{F}_s^s$ is non-trivial in general. ∎

2.3.4 Definition (Filtered DS). Let $(\Omega, \mathcal{F}^0, \mathbb{P}, (\theta_t)_{t \in \mathbb{R}})$ be a metric DS, let $\mathcal{F}$ be the $\mathbb{P}$-completion of $\mathcal{F}^0$, and let $\mathcal{F}_s^t$, $s \leq t$, be a filtration on $(\Omega, \mathcal{F}, \mathbb{P})$. We call
$(\Omega, \mathcal{F}^0, \mathbb{P}, (\theta_t)_{t \in \mathbb{R}}, (\mathcal{F}_s^t)_{s \leq t})$ a *filtered DS*, if for all $s, t, u \in \mathbb{R}$, $s \leq t$, we have
$$\theta_u^{-1} \mathcal{F}_s^t = \mathcal{F}_{s+u}^{t+u}.$$
∎

2.3.5 Remark. One can construct a filtered DS from a metric DS in the following way in complete analogy with the construction of de Sam Lazaro and Meyer [123: p. 4] in the one-parameter case.

Let $(\Omega, \mathcal{F}^0, \mathbb{P}, (\theta_t)_{t \in \mathbb{R}})$ be a metric DS, $\mathcal{F}$ the $\mathbb{P}$-completion of $\mathcal{F}^0$, and let $\mathcal{F}^-$ and $\mathcal{F}^+$ be two sub-σ-algebras of $\mathcal{F}$ (representing "past" and "future", respectively; see Sect. 1.7) both containing all $\mathbb{P}$-null sets of $\mathcal{F}$ such that
$$\theta_t^{-1}(\mathcal{F}^-) \subset \mathcal{F}^- \quad \text{for all } t \leq 0$$
and
$$\theta_t^{-1}(\mathcal{F}^+) \subset \mathcal{F}^+ \quad \text{for all } t \geq 0.$$
Then define
$$\mathcal{F}^t := \theta_t^{-1}(\mathcal{F}^-), \quad \mathcal{F}_s := \theta_s^{-1}(\mathcal{F}^+), \quad \mathcal{F}_s^t := \mathcal{F}_s \cap \mathcal{F}^t \quad \text{for } s \leq t.$$

Clearly $\mathcal{F}^t$ is increasing in t and $\mathcal{F}_s$ is decreasing in s. Further $\mathcal{F}^t$ and $\mathcal{F}_s$ and hence $\mathcal{F}_s^t$ contain all $\mathbb{P}$-null sets of $\mathcal{F}$. By definition
$$\begin{aligned}\theta_u^{-1} \mathcal{F}_s^t &= \theta_u^{-1}(\mathcal{F}_s) \cap \theta_u^{-1}(\mathcal{F}^t) \\ &= \theta_u^{-1}\theta_s^{-1}(\mathcal{F}^+) \cap \theta_u^{-1}\theta_t^{-1}(\mathcal{F}^-) \\ &= \mathcal{F}_{s+u}^{t+u} \quad \text{for all } u \in \mathbb{R}, \quad s \leq t.\end{aligned}$$

It is proved in [123: p. 4] that $\mathcal{F}^t$, $t \geq 0$, is right-continuous. It follows immediately that $(\Omega, \mathcal{F}^0, \mathbb{P}, (\theta_t)_{t \in \mathbb{R}}, (\mathcal{F}_s^t)_{s \leq t})$ is a filtered DS which moreover satisfies
$$\mathcal{F}_s^t = \bigcap_{\varepsilon > 0} \mathcal{F}_{s-\varepsilon}^{t+\varepsilon} =: \mathcal{F}_{s-}^{t+} \quad \text{for } s \leq t.$$
∎

It will turn out that generators of cocycles are again cocycles – with composition of mappings replaced by addition in a vector space. Additive cocycles were called *helices* by de Sam Lazaro and Meyer [123] (the expression goes back to Kolmogorov [218]). Here is an abstract definition, already used in the Perfection Theorem 1.3.2

2.3.6 Definition (Helix). Let $(\Omega, \mathcal{F}^0, \mathbb{P}, (\theta_t)_{t\in\mathbb{R}})$ be a metric DS, and $(H, \circ)$ a group. $F : \mathbb{R} \times \Omega \to H$ is called a *(perfect) helix* or *(perfect) cocycle* if
$$F(t+s, \omega) = F(t, \theta_s\omega) \circ F(s, \omega) \qquad (2.3.1)$$
for all $s, t \in \mathbb{R}$ and all $\omega \in \Omega$.

F is called a *(very) crude helix* or a *(very) crude cocycle* if for every $s, t \in \mathbb{R}$, (2.3.1) only holds up to a $\mathbb{P}$-null set (which may depend on s (and t)). ∎

We will only use the term "helix" if H is Abelian, e.g. $H = (\mathbb{R}^d, +)$, whereas we will continue to use "cocycle" if H is a (non-Abelian) group of self-mappings of a space with respect to composition. Note that in contrast to Definition 1.1.1 an abstract cocycle/helix in the sense of Definition 2.3.6 automatically satisfies $F(0, \omega) = e_H$ for all $\omega \in \Omega$, e_H being the identity of the group $(H, \circ)$, since we have assumed that F takes values in a group.

2.3.7 Proposition. *Let $(H, \circ)$ be a group.*

(i) Assume $(\Omega, \mathcal{F}^0, \mathbb{P}, (\theta_t)_{t\in\mathbb{R}})$ is a metric DS, $\varepsilon > 0$ and $F : [0, \varepsilon] \times \Omega \to H$ satisfies
$$F(t+h, \omega) = F(h, \theta_t\omega) \circ F(t, \omega) \qquad (2.3.2)$$
for all $0 \leq t \leq t+h \leq \varepsilon$, $\omega \in \Omega$. Then F can be uniquely extended to a helix $\bar{F}$. If (2.3.2) holds only up to a $\mathbb{P}$-null set depending on h and t, then F can be extended to a very crude helix $\bar{F}$. If $(H, \circ)$ is a topological group and F is continuous/càd/càdlàg/càg/càglàd for all $\omega \in \Omega$ then the same is true for $\bar{F}$.

(ii) Assume $(\Omega, \mathcal{F}^0, \mathbb{P}, (\theta_t)_{t\in\mathbb{R}}, (\mathcal{F}_s^t)_{s\leq t})$ is a filtered DS, F is an H-valued very crude helix and $\mathcal{H}$ is some σ-algebra on H such that the composition map "$\circ$" is measurable. If for some $\varepsilon > 0$, $F(h, \cdot)$ is $\mathcal{F}_0^h, \mathcal{H}$-measurable for all $0 \leq h \leq \varepsilon$, then F is $\mathcal{F}_s^t$-adapted, i.e. $F(t, \cdot) \circ F(s, \cdot)^{-1}$ is $\mathcal{F}_s^t, \mathcal{H}$-measurable for all $-\infty < s \leq t < \infty$.

Proof. (i) Define $\bar{F}(s, \omega) := F(s, \omega)$ for $0 \leq s \leq \varepsilon$ and, by induction over k,

$$\bar{F}(k\varepsilon + s, \omega) := \bar{F}(s, \theta(k\varepsilon)\omega) \circ \bar{F}(0, \theta(k\varepsilon)\omega)^{-1} \circ \bar{F}(k\varepsilon, \omega) \quad \text{if } k \in \mathbb{N},$$

$$\bar{F}(k\varepsilon + s, \omega) := \bar{F}(s, \theta(k\varepsilon)\omega) \circ \bar{F}(\varepsilon, \theta(k\varepsilon)\omega)^{-1} \circ \bar{F}((k+1)\varepsilon, \omega) \quad \text{if } k \in -\mathbb{N},$$

for $0 \leq s \leq \varepsilon$. Note that $\bar{F}$ is well-defined. It is straightforward to check that $\bar{F}$ is a helix (resp., a very crude helix). If F satisfies equation (2.3.2) without an exceptional null set, then it is clear that $\bar{F}$ is the only function which extends F to a helix on $\mathbb{R}$. The last assertion is also clear.

(ii) For $0 \leq h \leq \varepsilon$ – using the assumption that $F(h, \cdot)$ is $\mathcal{F}_0^h, \mathcal{H}$-measurable and the definition of a filtered DS – we see that $F(h, \theta_t\cdot)$ is $\mathcal{F}_t^{t+h}, \mathcal{H}$-measurable for any $t \in \mathbb{R}$. By the very crude helix property and the completeness of $\mathcal{F}_t^{t+h}$ the same is true for $F(t+h, \cdot) \circ F(t, \cdot)^{-1}$. For $2\varepsilon \geq h \geq \varepsilon$ we have

$$F(t+h, \cdot) \circ F(t, \cdot)^{-1} = F(t+h, \cdot) \circ F(t+\varepsilon, \cdot)^{-1} \circ F(t+\varepsilon, \cdot) \circ F(t, \cdot)^{-1},$$

which is $\mathcal{F}_t^{t+h}, \mathcal{H}$-measurable. The assertion now follows by induction. ⊐

Next we study processes which are both semimartingales and helices at the same time.

2.3.8 Definition (Semimartingale Helix). Suppose we are given a filtered DS $(\Omega, \mathcal{F}^0, \mathbb{P}, (\theta_t)_{t \in \mathbb{R}}, (\mathcal{F}_s^t)_{s \leq t})$. An $(\mathbb{R}^d, +)$-valued helix F is called (*forward* resp., *backward* resp.) *semimartingale helix* if

$$F_s(t, \omega) := F(t, \omega) - F(s, \omega)$$

is an $\mathcal{F}_s^t$-(forward resp., backward resp.) semimartingale. ∎

The next proposition shows that there is a canonical way in which a semimartingale with stationary increments can be "ameliorated" to be a semimartingale helix over a filtered DS

2.3.9 Proposition. *Let E be a locally compact Hausdorff space with a countable base, let $\mathcal{C}(E, \mathbb{R}^d)$ be the space of continuous functions from E to $\mathbb{R}^d$ endowed with the (metrizable) topology of uniform convergence on compact sets, and let $\mathcal{C}_0(\mathbb{R}, \mathcal{C}(E, \mathbb{R}^d))$ be the space of continuous $\mathcal{C}(E, \mathbb{R}^d)$-valued functions on $\mathbb{R}$ which are zero at zero, also equipped with the topology of uniform convergence on compact sets.*

Let $\bar{\mathcal{F}}_s^t$, $s \leq t$, be a filtration on the complete probability space $(\bar{\Omega}, \bar{\mathcal{F}}, \bar{\mathbb{P}})$. Assume $\bar{F} : \mathbb{R} \times E \times \bar{\Omega} \to \mathbb{R}^d$, $(t, x, \bar{\omega}) \mapsto \bar{F}(t, x, \bar{\omega})$, is jointly continuous in (t, x) for all $\bar{\omega} \in \bar{\Omega}$, $\bar{F}(0, x, \bar{\omega}) = 0$ for all $x \in E$, $\bar{\omega} \in \bar{\Omega}$, and that

$$\bar{F}_s(t, x, \cdot) := \bar{F}(t, x, \cdot) - \bar{F}(s, x, \cdot)$$

is an $\bar{\mathcal{F}}_s^t$-(forward resp., backward resp.) semimartingale for every $x \in E$. Further assume that $\bar{F}$ has stationary increments in the sense that the law of $\{\bar{F}(t+h, x, \cdot) - \bar{F}(t, x, \cdot), x \in E, h \in \mathbb{R}\}$ on $\mathcal{C}_0(\mathbb{R}, \mathcal{C}(E, \mathbb{R}^d))$ does not depend on $t \in \mathbb{R}$.

Then there exists a filtered DS $(\Omega, \mathcal{F}^0, \mathbb{P}, (\theta_t)_{t \in \mathbb{R}}, (\mathcal{F}_s^t)_{s \leq t})$ with (forward resp., backward resp.) semimartingale helices $F(\cdot, x, \cdot)$, $x \in E$, which are also jointly continuous in (t, x) for every $\omega \in \Omega$ such that the laws of $\bar{F}$ and F on $\mathcal{C}_0(\mathbb{R}, \mathcal{C}(E, \mathbb{R}^d))$ coincide.

Proof. Define

$$\Omega := \mathcal{C}_0(\mathbb{R}, \mathcal{C}(E, \mathbb{R}^d)),$$

and let $\mathcal{F}^0$ be the Borel σ-algebra of Ω (which coincides with the σ-algebra generated by the evaluations). Note that the continuity and measurability assumptions imply that $\bar{F}$ is an $(\Omega, \mathcal{F}^0)$-valued random variable. Let $\mathbb{P}$ be its law. Define a shift θ by

$$(\theta_t \omega)(s) := \omega(t+s) - \omega(t), \quad \omega \in \Omega, \quad t, s \in \mathbb{R}.$$

Clearly $\theta_{s+t}\omega = \theta_t(\theta_s \omega)$ for all $\omega \in \Omega$, $s, t \in \mathbb{R}$. $\mathbb{P}$ is invariant under θ since $\bar{F}$ has stationary increments. Further, $\theta : \mathbb{R} \times \Omega \to \Omega$ is $\mathcal{B} \otimes \mathcal{F}^0$, $\mathcal{F}^0$

measurable since it is continuous and Ω and $\mathbb{R}$ are both topological spaces with a countable base, see Dudley [131: p. 90]. Define
$$F(t,\cdot,\omega):=\omega(t), \quad t\in\mathbb{R}, \quad \omega\in\Omega.$$
Then F is an $(H,\circ)=(\mathcal{C}(E,\mathbb{R}^d),+)$-valued helix since
$$F(t+h,\cdot,\omega)-F(t,\cdot,\omega)=\omega(t+h)-\omega(t)=\theta_t\omega(h)=F(h,\cdot,\theta_t\omega).$$
F and $\bar{F}$ have the same law by construction. Let $\mathcal{F}^-$ resp. $\mathcal{F}^+$ be the σ-algebra generated by the $\mathbb{P}$-null sets of $\mathcal{F}$ (the $\mathbb{P}$-completion of $\mathcal{F}^0$) and $\sigma(\omega(s),\, s\leq 0)$ resp. $\sigma(\omega(s),\, s\geq 0)$. Define $\mathcal{F}_s^t$ as in Remark 2.3.5, i.e. $\mathcal{F}_s^t$ is the completion of $\sigma(\omega(v)-\omega(u),\, s\leq u\leq v\leq t)$ by all $\mathcal{F}$-null sets. In particular, $\mathcal{F}_s^s$ is trivial for all $s\in\mathbb{R}$.

It remains to show that $F(\cdot,x,\cdot)$ is an $\mathcal{F}_s^t$-(forward resp., backward resp.) semimartingale for every $x\in E$. We only prove the forward statement – the backward one is analogous. The mapping
$$\bar{F}:(\bar{\Omega},\bar{\mathcal{F}},(\bar{\mathcal{F}}_s^t)_{s\leq t},\bar{\mathbb{P}})\to(\Omega,\mathcal{F},(\mathcal{F}_s^t)_{s\leq t},\mathbb{P})$$
is $\bar{\mathcal{F}},\mathcal{F}$-measurable, $\bar{F}\bar{\mathbb{P}}=\mathbb{P}$ and $\mathcal{G}_s^t:=\bar{F}^{-1}(\mathcal{F}_s^t)\subset\bar{\mathcal{F}}_s^t$, $t\geq s$. Fix $s\in\mathbb{R}$. Stricker's theorem (see Protter [281: p. 45]) implies that $\bar{F}(s+u,x,\cdot)-\bar{F}(s,x,\cdot)$, $u\geq 0$, is also a $(\mathcal{G}_s^{s+u})_{u\geq 0}$-semimartingale for every $x\in E$. Finally Theorem (10.37) in Jacod [185: p. 329] shows that $F(s+u,x,\cdot)-F(s,x,\cdot)$, $u\geq 0$, is an $(\mathcal{F}_s^{s+u})_{u\geq 0}$-semimartingale. This proves the proposition. □

The next proposition connects, for the helix case, our definition of a semimartingal with the usual one-parameter definition.

2.3.10 Proposition. *Given a filtered DS $(\Omega,\mathcal{F}^0,\mathbb{P},(\theta_t)_{t\in\mathbb{R}},(\mathcal{F}_s^t)_{s\leq t})$ and $F:\mathbb{R}\times\Omega\to\mathbb{R}^d$.*

F is a forward semimartingale helix if and only if F is a helix and $F|_{[0,\infty)}$ is an $(\mathcal{F}_0^t)_{t\geq 0}$-semimartingale in the usual sense.

Similarly, F is a backward semimartingale helix if and only if F is a helix and $F(-\cdot)|_{[0,\infty)}$ is an $(\mathcal{F}_{-t}^0)_{t\geq 0}$-semimartingale in the usual sense.

Finally, F is a semimartingale helix if and only if F is a helix, $F|_{[0,\infty)}$ is an $(\mathcal{F}_0^t)_{t\geq 0}$-semimartingale and $F(-\cdot)|_{[0,\infty)}$ is an $(\mathcal{F}_{-t}^0)_{t\geq 0}$-semimartingale, both in the usual sense.

Proof. We only prove the forward statement. By Proposition 2.3.7(ii) and the fact that $\theta(-s):(\Omega,\mathcal{F},(\mathcal{F}_0^t)_{t\geq 0},\mathbb{P})\to(\Omega,\mathcal{F},(\mathcal{F}_s^{s+t})_{t\geq 0},\mathbb{P})$ is an isomorphism by Definition 2.3.4, Jacod [185: Theorem (10.37), p. 329] and the helix property of F imply that if $F|_{[0,\infty)}$ is an $(\mathcal{F}_0^t)_{t\geq 0}$-semimartingale then $(F(s+u)-F(s))_{u\geq 0}$ is an $(\mathcal{F}_s^{s+u})_{u\geq 0}$-semimartingale. □

2.3.11 Example. A forward semimartingale helix need not be a backward semimartingale helix. As an example consider Brownian motion on $\mathbb{R}$ defined on the canonical space and choose for $s\leq t$ the filtration $\mathcal{F}_s^t:=\sigma(W(u)-W(v),\, -\infty<v<u\leq t)$ completed by null sets. Clearly W is a forward martingale helix but not a backward semimartingale helix. ∎

Next we introduce the forward and backward (local) characteristics of a (forward, backward) semimartingale helix.

2.3.12 Proposition. *Given a filtered DS* $(\Omega, \mathcal{F}^0, \mathbb{P}, (\theta_t)_{t \in \mathbb{R}}, (\mathcal{F}_s^t)_{s \leq t})$.
(i) Let F be an $\mathbb{R}^d$-valued forward semimartingale helix. Let

$$F(t) = M^+(t) + B^+(t), \quad t \geq 0,$$

be the canonical decomposition of the $(\mathcal{F}_0^t)_{t \geq 0}$-semimartingale $F|_{[0,\infty)}$ (cf. Theorem 2.3.10). Then there exist

1. *a strictly increasing continuous real-valued $\mathcal{F}_0^t$-adapted process $A^+(t)$, $t \geq 0$, with $A^+(0, \omega) = 0$,*
2. *an $\mathcal{F}_0^t$-predictable $\mathbb{R}^d$-valued process $b^+(t)$, $t \geq 0$,*
3. *an $\mathcal{F}_0^t$-predictable process $a^+(t)$, $t \geq 0$, taking values in the set of non-negative definite $d \times d$ matrices,*

such that for all $t \geq 0$, $\omega \in \Omega$ we have

$$B^+(t, \omega) = \int_0^t b^+(s, \omega) \, dA^+(s, \omega),$$

$$Q_{ij}^+(t, \omega) := \langle M_i^+(\cdot, \omega), M_j^+(\cdot, \omega) \rangle_t = \int_0^t a_{ij}^+(s, \omega) \, dA^+(s, \omega).$$

(ii) Analogously, let F be an $\mathbb{R}^d$-valued backward semimartingale helix, consider the canonical decomposition

$$F(t) = M^-(t) + B^-(t), \quad t \leq 0,$$

of the $(\mathcal{F}_t^0)_{t \leq 0}$-semimartingale $F|_{(-\infty, 0]}$ (backward direction). Then there exist

1. *a strictly increasing continuous real-valued $\mathcal{F}_t^0$-adapted process $A^-(t)$, $t \leq 0$, with $A^-(0, \omega) = 0$,*
2. *an $\mathcal{F}_t^0$-backward predictable $\mathbb{R}^d$-valued process $b^-(t)$, $t \leq 0$,*
3. *an $\mathcal{F}_t^0$-backward predictable process $a^-(t)$, $t \leq 0$, taking values in the set of non-negative definite $d \times d$ matrices,*

such that for all $t \leq 0$, $\omega \in \Omega$ we have

$$B^-(t, \omega) = \int_0^t b^-(s, \omega) \, dA^-(s, \omega),$$

$$Q_{ij}^-(t, \omega) := \langle M_i^-(\cdot, \omega), M_j^-(\cdot, \omega) \rangle_t = \int_0^t a_{ij}^-(s, \omega) \, dA^-(s, \omega).$$

Proof. (i) Take

$$A^+(t,\omega) := \sum_{i=1}^{d} \left(\int_0^t |dB_i^+(s,\omega)| + \langle M_i^+ \rangle_t \right) + t, \quad t \geq 0,$$

$$b_i^+(t,\omega) := \limsup_{s \uparrow t} \frac{B_i^+(t,\omega) - B_i^+(s,\omega)}{A^+(t,\omega) - A^+(s,\omega)}$$

and

$$a_{ij}^+(t,\omega) := \limsup_{s \uparrow t} \frac{Q_{ij}^+(t,\omega) - Q_{ij}^+(s,\omega)}{A^+(t,\omega) - A^+(s,\omega)}.$$

Then b_i^+ (resp. a_{ij}^+) is a version of the Radon-Nikodym derivative of B_i^+ (resp. Q_{ij}^+) with respect to A^+, according to Lebesgue's differentiation theorem (see e.g. Carmona and Nualart [87: pp. 69f.]).

All claims follow immediately.

(ii) Take the same definitions as in (i), with '−' quantities replacing all corresponding '+' quantities, and '↓' replacing '↑'. □

2.3.13 Definition (Local Characteristics of Semimartingale Helix).
Let $(\Omega, \mathcal{F}^0, \mathbb{P}, (\theta_t)_{t \in \mathbb{R}}, (\mathcal{F}_s^t)_{s \leq t})$ be a filtered DS, and let F be an $\mathbb{R}^d$-valued (forward resp., backward resp.) semimartingale helix. Then the quantities (a^+, b^+, A^+), resp. (a^-, b^-, A^-) are called the (*forward*, resp. *backward*) *local characteristics* of F.

In case of a semimartingale helix we can and will piece the '+' quantities and the '−' quantities together to yield the canonical decomposition

$$F(t) = M(t) + B(t), \quad t \in \mathbb{R},$$

and a triple (a, b, A) now defined on all of $\mathbb{R}$ and called the *local characteristics* of F. They satisfy for all $t \in \mathbb{R}$ and $\omega \in \Omega$

$$B(t,\omega) = \int_0^t b(s,\omega)\, dA(s,\omega),$$

$$Q_{ij}(t,\omega) := \langle M_i(\cdot,\omega), M_j(\cdot,\omega) \rangle_t = \int_0^t a_{ij}(s,\omega)\, dA(s,\omega).$$

■

2.3.14 Remark. (i) We defined the local characteristics of the forward semimartingale helix F as the local characteristics of the $(\mathcal{F}_0^t)$-semimartingale $F|_{[0,\infty)}$. Using Jacod's result we can argue as in the proof of Theorem 2.3.10 that for each $s \in \mathbb{R}$ the $(\mathcal{F}_s^{s+t})_{t \geq 0}$-semimartingale $F(s+t) - F(s)$, $t \geq 0$, has the canonical decomposition

$$F(s+t,\omega) - F(s,\omega) = M^{(s)}(t,\omega) + B^{(s)}(t,\omega),$$

where

$$M^{(s)}(t,\omega) = M^+(t,\theta_s\omega), \quad B^{(s)}(t,\omega) = B^+(t,\theta_s\omega)$$

are both $\mathcal{F}_s^{s+t}$-measurable, and the local characteristics $(a^{(s)}, b^{(s)}, A^{(s)})$ are related to (a^+, b^+, A^+) via

$$a^{(s)}(t,\omega) = a^+(t,\theta_s\omega), \quad b^{(s)}(t,\omega) = b^+(t,\theta_s\omega),$$

and

$$A^{(s)}(t,\omega) = A^+(t,\theta_s\omega).$$

Similarly for a (backward) semimartingale helix.

(ii) Given a semimartingale helix F, it is clear that the quadratic variation processes

$$Q(t,\omega) := \begin{cases} Q^-(t,\omega), & t \leq 0, \\ Q^+(t,\omega), & t \geq 0, \end{cases}$$

is a very crude helix and hence, by Theorem 1.3.2, has a version which is a helix. Further $Q(t) - Q(s)$ is $\mathcal{F}_s^t$-measurable. A and B however generally do not enjoy the (very crude) helix property, nor are $A(t) - A(s)$ and $B(t) - B(s)$ $\mathcal{F}_s^t$-measurable in general. Hence the canonical decomposition of a semimartingale helix does not in general consist of helices. One may ask whether we have chosen the "wrong" definition. As an alternative one could try to decompose the forward semimartingale helix F into $F = M + B$, where M and B are a forward local martingale helix and an $(\mathcal{F}_s^t)$-adapted helix of locally bounded variation, resp. Then $\langle M \rangle$ should also be an $(\mathcal{F}_s^t)$-adapted helix. Unfortunately, such a decomposition does not exist in general. ∎

2.3.3 Semimartingale Helices with Spatial Parameter

In this subsection we develop stochastic calculus for semimartingale helices depending on a parameter $x \in \mathbb{R}^d$. Our exposition is largely analogous to that of Kunita [224] (and similar to that of Carmona and Nualart [87]), but differs conceptually in two respects:

- We consistently consider two-sided time.
- We assume the helix property.

To save space we only formulate our statements for the (forward as well as backward) semimartingale case.

2.3.15 Definition (Semimartingale Helix with Spatial Parameters).
Let $(\Omega, \mathcal{F}^0, \mathbb{P}, (\theta_t)_{t\in\mathbb{R}}, (\mathcal{F}_s^t)_{s\leq t})$ be a filtered DS, $d \in \mathbb{N}$, $k \in \mathbb{Z}^+$, $0 \leq \delta \leq 1$. Assume that for each $x \in \mathbb{R}^d$, $F(x,t,\omega)$ is an $\mathbb{R}^d$-valued semimartingale helix and let

$$F(x,t,\omega) = M(x,t,\omega) + B(x,t,\omega), \quad t \in \mathbb{R},$$

be its canonical decomposition. F is called $\mathcal{C}_b^{k,\delta}$-*semimartingale helix* if for $\mathbb{P}$-almost all $\omega \in \Omega$

1. $M(\cdot, t, \omega)$ and $B(\cdot, t, \omega)$ are in $\mathcal{C}_b^{k,\delta}$ for all $t \in \mathbb{R}$,
2. for $|\alpha| \leq k$, the spatial derivative $D_x^\alpha M(x, t, \omega)$ is continuous with respect to (x, t) and for each x a local martingale (i.e. $M(x, t)$, $t \geq 0$, is an $(\mathcal{F}_0^t)_{t\geq 0}$-local martingale, and $M(x, -t)$, $t \geq 0$, is an $(\mathcal{F}_{-t}^0)_{t\geq 0}$-local martingale), and $D_x^\alpha B(x, t, \omega)$ is continuous with respect to (x, t) and for each x of locally bounded variation in t.

We will also use the corresponding notion with $\mathcal{C}_b^{k,\delta}$ replaced with $\mathcal{C}^{k,\delta}$. ∎

2.3.16 Remark. If the semimartingale $F(x, t, \omega)$ with spatial parameter x is only assumed to be a very crude helix, but if F is continuous with respect to (x, t) for $\mathbb{P}$-almost all $\omega \in \Omega$, then F is a very crude helix with values in the topological group $(\mathcal{C}^{0,0}, +)$ (which has a countable base). The reason is that for fixed $s, t \in \mathbb{R}$

$$F(x, t + s, \omega) = F(x, t, \theta_s \omega) + F(x, s, \omega)$$

holds for all $x \in \mathbb{R}^d$ on the complement of $\cup_{y \in \mathbb{Q}^d} N_{s,t,y}$, where $N_{s,t,y}$ is the set of measure zero where the helix property fails.

By Theorem 1.3.2, F can be perfected, i.e. there is a semimartingale helix $\tilde{F}$ indistinguishable from F. ∎

For a semimartingale helix with parameter $x \in \mathbb{R}^d$ the local characteristics (a, b, A) (see Definition 2.3.13) will of course also depend on x. However, in order to develop a reasonable theory of SDE driven by semimartingale helices we need to have more uniformity of the local characteristics with respect to x. We follow Kunita [224: pp. 79 and 84], and make the following assumption.

2.3.17. Assumption (Uniformly Controlled Local Characteristics). We assume that for the semimartingale helix $F(x, t, \omega)$ there exists a continuous adapted (without loss of generality strictly) increasing process $A(t, \omega)$ with $A(0, \omega) = 0$ such that

1. there exists a measurable function $b(x, t, \omega)$ which for every x is a predictable process, such that for all $x \in \mathbb{R}^d$, $t \in \mathbb{R}$, and $\omega \in \Omega$

$$B(x, t, \omega) = \int_0^t b(x, s, \omega) \, dA(s, \omega),$$

2. there exists a measurable function $a(x, y, t, \omega)$ which for every $x, y \in \mathbb{R}^d$ is a predictable process, such that for all $x, y \in \mathbb{R}^d$, $t \in \mathbb{R}$, $\omega \in \Omega$

$$Q(x, y, t, \omega) = \langle M(x, t, \omega), M(y, t, \omega) \rangle = \int_0^t a(x, y, s, \omega) \, dA(s, \omega).$$

The triple (a, b, A) is called the *local characteristics* of F. We will say that F is *uniformly controlled* by A. ∎

2.3.18 Remark. (i) If the local characteristics of F are (a, b, A) then by Remark 2.3.14 the local characteristics of

$$F(x, s+t, \omega) - F(x, s, \omega) = F(x, t, \theta_s\omega) = M(x, t, \theta_s\omega) + B(x, t, \theta_s\omega)$$

are $(a \circ \theta_s, b \circ \theta_s, A \circ \theta_s)$.

(ii) Carmona and Nualart [87: p. 101] introduce a similar assumption and say F is "uniformly controlled" by A, which we adopt in our case above.

(iii) Le Jan and Watanabe [228: p. 314] restrict themselves just to the case $A(t, \omega) \equiv t$.

(iv) Let F be a $\mathcal{C}^{0,0}$-semimartingale helix. The fact that $Q(x, y, t, \omega)$ is uniformly controlled holds without loss of generality and follows from Kunita [224] (Exercise 3.2.10(ii), p. 91). Kunita's argument needs a countably generated $\mathcal{F}^0$, which is the case for the canonical set-up of Theorem 2.3.9. By exploiting the joint continuity of $M(x, t)$, it is not hard to see that the result remains true without the hypothesis that $\mathcal{F}^0$ be countably generated.

(v) A sufficient condition for B to be uniformly controlled is that $t \mapsto B(\cdot, t, \omega)$ is locally of bounded variation as a function with values in the Fréchet space $\mathcal{C}^{0,0}$ (a similar condition is given by Carmona and Nualart [87: pp. 37 and 40]). ∎

To describe further regularity conditions for the local characteristics of a semimartingale helix we introduce the following spaces.

2.3.19 Definition. Let $k \in \mathbb{Z}^+$ and $0 \leq \delta \leq 1$. Define

$$\tilde{\mathcal{C}}_b^{k,\delta} := \{g \in \mathcal{C}(\mathbb{R}^d \times \mathbb{R}^d, \mathbb{R}^d) : \|g\|_{k,\delta}^{\sim} < \infty\}$$

where

$$\|g\|_{k,0}^{\sim} = \sup_{x,y} \frac{\|g(x,y)\|}{(1+\|x\|)(1+\|y\|)} + \sum_{1 \leq |\alpha| \leq k} \sup_{x,y} \|D_x^\alpha D_y^\alpha g(x,y)\|$$

and in case $\delta > 0$

$$\|g\|_{k,\delta}^{\sim} = \|g\|_{k,0}^{\sim} + \sup_{x \neq x', y \neq y'} \sum_{|\alpha|=k} \frac{\|g(x,y) - g(x',y) - g(x,y') + g(x',y')\|}{\|x-x'\|^\delta \|y-y'\|^\delta}.$$

Define $\tilde{\mathcal{C}}^{k,\delta}$ analogously. ∎

2.3.20 Definition. Let $k \in \mathbb{Z}^+$, $0 \leq \delta \leq 1$ and let $A : \mathbb{R} \to \mathbb{R}$ be a continuous increasing function with $A(0) = 0$. Define $L_{\text{loc}}(\mathbb{R}, dA, \mathcal{C}_b^{k,\delta})$ to be the set of measurable functions $f : \mathbb{R}^d \times \mathbb{R} \to \mathbb{R}^d$ for which

- $f(\cdot, t) \in \mathcal{C}_b^{k,\delta}$ for every $t \in \mathbb{R}$ ("for A-almost all t" would suffice),
- for every $t \in \mathbb{R}$

$$\left| \int_0^t \|f(\cdot, s)\|_{k,\delta} \, dA(s) \right| < \infty. \tag{2.3.3}$$

With the system of seminorms (2.3.3), $L_{\text{loc}}(\mathbb{R}, dA, \mathcal{C}_b^{k,\delta})$ is a Fréchet space. In addition, we have the continuous inclusions

$$L_{\text{loc}}(\mathbb{R}, dA, \mathcal{C}_b^{k,\delta}) \hookrightarrow L_{\text{loc}}(\mathbb{R}, dA, \mathcal{C}_b^{k,0}) \hookrightarrow L_{\text{loc}}(\mathbb{R}, dA, \mathcal{C}_b^{k-1,\varepsilon}).$$

The spaces L_{loc} for $\mathcal{C}^{k,\delta}$, $\tilde{\mathcal{C}}_b^{k,\delta}$ and $\tilde{\mathcal{C}}^{k,\delta}$ are analogously defined. ∎

Stratonovich Stochastic Integrals with Respect to a Semimartingale Helix[5]

Let now $F(x, t, \omega)$ be a $\mathcal{C}^{1,0}$-semimartingale helix with local characteristics (a, b, A) satisfying $a \in L_{\text{loc}}(\mathbb{R}, dA, \tilde{\mathcal{C}}^{2,\delta})$ for some $\delta > 0$ and $b \in L_{\text{loc}}(\mathbb{R}, dA, \mathcal{C}^{1,0})$ (meaning that it holds for all $\omega \in \Omega$). Let $f_s(t)$ be a semimartingale (in the sense of Definition 2.3.2(c)).

Forward Stratonovich Stochastic Integral. Let $s \leq t$. Then

$$\begin{aligned} I_s(t) &= \int_s^t F(f_s(u), \circ d^+ u) \\ &:= \lim_{\Delta \to 0} \text{ in pr. } \sum_{k=0}^{n-1} \frac{1}{2} \{F(f_s(t_{k+1}), t_{k+1}) + F(f_s(t_k), t_{k+1}) \\ &\quad - F(f_s(t_{k+1}), t_k) - F(f_s(t_k), t_k)\}, \end{aligned}$$

where the limit in probability runs through a sequence of partitions of $[s, t]$ for which the mesh $\Delta := \max_{0 \leq k \leq n-1}(t_{k+1} - t_k) \to 0$. This limit exists and is a forward semimartingale.

2.3.21 Remark. For $s \leq r \leq t$ the integral

$$\int_r^t F(f_s(u), \circ d^+ u)$$

also makes sense since $F(x, s+r+u) - F(x, s)$, $f_s(r+u)$, $u \geq 0$, are $(\mathcal{F}_s^{r+u})_{u \geq 0}$-semimartingales. ∎

Backward Stratonovich Stochastic Integral. Let $t \leq s$. Then

$$\begin{aligned} I_s(t) &= \int_t^s F(f_s(u), \circ d^- u) \\ &:= \lim_{\Delta \to 0} \text{ in pr. } \sum_{k=0}^{n-1} \frac{1}{2} \{F(f_s(t_{k+1}), t_{k+1}) + F(f_s(t_k), t_{k+1}) \\ &\quad - F(f_s(t_{k+1}), t_k) - F(f_s(t_k), t_k)\}, \end{aligned}$$

---

[5] The stochastic integrals can, of course, be defined with respect to semimartingales which are not necessarily helices (see Kunita [224: pp. 84–86]) – but this is not needed here.

where the limit in probability runs through a sequence of partitions of $[t,s]$ for which $\max_{0\leq k\leq n-1}(t_{k+1} - t_k) \to 0$. This limit exists and is a backward semimartingale.

We stress that there is no relation whatsoever between the forward and backward integrals.

Collecting both cases into one, we conclude that $(I_s(t))_{s,t\in\mathbb{R}}$ is a semimartingale.

The helix property of F allows us to reduce one endpoint of the integral to zero.

2.3.22 Proposition. *Let $s \in \mathbb{R}$ be fixed.*
(i) Let $t \geq 0$. Then

$$\int_s^{s+t} F(f_s(u), \circ d^+u) = \int_0^t (\theta_s F)(f_s(s+u), \circ d^+u) \quad \mathbb{P}\text{-a. s.},$$

where $\theta_s F(x,t,\omega) := F(x,t,\theta_s\omega)$. In case $f_s(s+t,\omega) = f_0(t,\theta_s\omega)$ for all $t \geq 0$ and $\omega \notin N_s$

$$\int_s^{s+t} F(f_s(u), \circ d^+u) = \theta_s \int_0^t F(f_0(u), \circ d^+u) \quad \mathbb{P}\text{-a. s.}$$

(ii) Let $t \leq 0$. Then

$$\int_{s+t}^s F(f_s(u), \circ d^-u) = \int_t^0 (\theta_s F)(f_s(s+u), \circ d^-u) \quad \mathbb{P}\text{-a. s.}$$

In case $f_s(s+t,\omega) = f_0(t,\theta_s\omega)$ for all $t \leq 0$ and $\omega \notin N_s$

$$\int_{s+t}^s F(f_s(u), \circ d^-u) = \theta_s \int_t^0 F(f_0(u), \circ d^-u) \quad \mathbb{P}\text{-a. s.}$$

This follows immediately from the definition of the corresponding integrals.

2.3.4 RDS from Stochastic Differential Equations

We will now establish the essentially one-to-one correspondence between semimartingale helices and semimartingale cocycles.

2.3.23 Definition (Semimartingale Cocycle). Let φ be a cocycle over a filtered DS θ for which

$$G_s(x,t,\omega) := \varphi(t,\omega)\varphi(s,\omega)^{-1}x - x$$

is a $\mathcal{C}^{k,\delta}$-semimartingale. Then φ is called a $\mathcal{C}^{k,\delta}$-*semimartingale cocycle*. Put

$$G(x,t,\omega) := G_0(x,t,\omega) = \varphi(t,\omega)x - x.$$

∎

2.3.24 Lemma. *Given a metric DS θ. The following statements are equivalent:*
  (i) φ is a cocycle over θ, i.e. $\varphi(t+s,\omega) = \varphi(t,\theta_s\omega) \circ \varphi(s,\omega)$.
  (ii) $G_s(x,t,\omega) = G(x, t-s, \theta_s\omega)$.
  (iii) G is a helix over the skew-product flow $\Theta = (\theta, \varphi)$, i.e.

$$G(x, t+s, \omega) = G(\varphi(s,\omega)x, t, \theta_s\omega) + G(x, s, \omega).$$

The proof follows from the definitions.

2.3.25 Corollary. *Given a cocycle φ over a filtered DS θ. φ is a $\mathcal{C}^{k,\tilde{\varepsilon}}$-semimartingale cocycle if and only if $G(\cdot, t, \omega)$, $t \geq 0$, is an $(\mathcal{F}_0^t)_{t\geq 0}$-$\mathcal{C}^{k,\tilde{\varepsilon}}$-semimartingale, and $G(\cdot, -t, \omega)$, $t \geq 0$, is an $(\mathcal{F}_{-t}^0)_{t\geq 0}$-$\mathcal{C}^{k,\delta}$-semimartingale. The canonical decomposition*

$$G_s(x,t,\omega) = M^{(s)}(x,t,\omega) + B^{(s)}(x,t,\omega)$$

satisfies for all $s, t \in \mathbb{R}$

$$M^{(s)}(x,t,\omega) = M(x, t-s, \theta_s\omega), \quad B^{(s)}(x,t,\omega) = B(x, t-s, \theta_s\omega),$$

where $G(x,t,\omega) = M(x,t,\omega) + B(x,t,\omega)$ is the canonical decomposition of G.

The proof is the same as for the helix case (see Remark 2.3.14).

In a first step we move from a semimartingale helix to a semimartingale cocycle via a Stratonovich SDE driven by the helix.

2.3.26 Theorem (Global RDS from Stratonovich SDE). *Given a filtered DS $(\Omega, \mathcal{F}^0, \mathbb{P}, (\theta_t)_{t\in\mathbb{R}}, (\mathcal{F}_s^t)_{s\leq t})$. Let F be a $\mathcal{C}_b^{k,\delta}$-semimartingale helix over θ, where $k \geq 1$ and $\delta > 0$. Assume that F has local characteristics (a_F, b_F, A_F) satisfying $a_F \in L_{\mathrm{loc}}(\mathbb{R}, dA_F, \tilde{\mathcal{C}}_b^{k+1,\delta})$, $b_F \in L_{\mathrm{loc}}(\mathbb{R}, dA_F, \mathcal{C}_b^{k,\delta})$ and $c_F \in L_{\mathrm{loc}}(\mathbb{R}, dA_F, \mathcal{C}_b^{k,\delta})$, where c_F is the Stratonovich-Itô correction term given in (2.3.5) below. Then there exists a unique (up to indistinguishability) global C^k RDS φ over θ which for any $\varepsilon < \delta$ is a $\mathcal{C}^{k,\varepsilon}$-semimartingale cocycle which solves the SDE*

$$dx_t = F(x_t, \circ dt),$$

i. e. more specifically, for each fixed $x \in \mathbb{R}^d$

$$\varphi(t)x - x = \begin{cases} \int_0^t F(\varphi(s)x, \circ d^+s), & 0 \leq t, \\ -\int_t^0 F(\varphi(s)x, \circ d^-s), & t \leq 0 \end{cases} \quad (2.3.4)$$

(this holds for fixed t and x outside a $\mathbb{P}$-null set $N_{t,x}$). The semimartingale cocycle φ has the local characteristics $A_\varphi = A_F$,

$$a_\varphi(x,y,t,\omega) = a_F(\varphi(t,\omega)x, \varphi(t,\omega)y, t, \omega)$$

and

$$b_\varphi(x,t,\omega) = b_F(\varphi(t,\omega)x, t, \omega) + c_F(\varphi(t,\omega)x, t, \omega),$$

where

$$c_F(x,t,\omega) := \frac{1}{2}\sum_{j=1}^{d} \frac{\partial a_F^j}{\partial x_j}(x,y,t,\omega)|_{y=x}. \qquad (2.3.5)$$

Proof. 1. Existence and uniqueness: We can use Kunita's results in [224] (we need the global versions of Theorems 4.7.3 and 4.7.4 for the Stratonovich case which Kunita did not state explicitly, but could be derived from Theorem 4.6.5 and Corollary 4.6.6 for the Itô case, taking into account the correction term c_F). The key point is that our driving process $F_s(x,t,\omega) := F(x,t,\omega) - F(x,s,\omega)$ is a forward as well as backward semimartingale whose local characteristics $(a_s, b_s, A_s) = (\theta_s a, \theta_s b, \theta_s A)$ satisfy the corresponding regularity conditions for each $s \in \mathbb{R}$ provided they do for $s = 0$, which we have assumed.

Consequently, there exists a $\mathcal{C}^{k,\varepsilon}$-semimartingale of C^k diffeomorphisms $(\varphi_{st}(\omega))_{s,t\in\mathbb{R}}$ which for each $x \in \mathbb{R}^d$ satisfies

$$\varphi_{st}x - x = \begin{cases} \int_s^t F(\varphi_{su}x, \circ d^+ u), & s \leq t, \\ -\int_t^s F(\varphi_{su}x, \circ d^- u), & t \leq s. \end{cases}$$

2. Two-parameter flow property: There is a version of φ which satisfies

$$\varphi_{ss}(\omega) = \mathrm{id}$$

and

$$\varphi_{st}(\omega) = \varphi_{ut}(\omega) \circ \varphi_{su}(\omega) \qquad (2.3.6)$$

identically. Indeed, since our mappings are invertible with $\varphi_{st}^{-1} = \varphi_{ts}$ it suffices to prove the flow property for the case $s < u < t$, which was done by Kunita [224: pp. 161f.].

3. Crude cocycle property: We claim that

$$\varphi(t,\omega) := \varphi_{0t}(\omega)$$

satisfies the crude cocycle property in the following form:

$$\varphi(t+s,\omega) = \varphi(t, \theta_s\omega) \circ \varphi(s,\omega) \quad \text{for all } t,s \in \mathbb{R}, \quad \omega \notin N_s, \qquad (2.3.7)$$

where N_s is a $\mathbb{P}$-null set which possibly depends on s (the dependence on t can be removed by utilizing the continuity with respect to t).

In view of (2.3.6), (2.3.7) is equivalent to

$$\varphi_{s,s+t}(\omega) = \varphi_{0t}(\theta_s\omega) \quad \text{for all } s,t \in \mathbb{R}, \quad \omega \notin N_s.$$

Consider first the case $t > 0$: We have for fixed $s \in \mathbb{R}$

2.3 Continuous Time 2: Stochastic Differential Eqs.

$$\varphi_{s,s+t}x - x = \int_s^{s+t} F(\varphi_{su}x, \circ d^+u)$$

$$= \int_0^t \theta_s F(\varphi_{s,s+u}x, \circ d^+u)$$

by Proposition 2.3.22(i), where $\theta_s F(x,t,\omega) := F(x,t,\theta_s\omega)$.

Now, obviously, $\varphi_{0t}x$ uniquely solves $dx = F(x, \circ d^+t)$ if and only if $\varphi_{0t}(\theta_s \cdot)x$ uniquely solves $dx = (\theta_s F)(x, \circ d^+t)$. Hence for each fixed $s \in \mathbb{R}$ and for all $t > 0$

$$\varphi_{s,s+t}(\omega) = \varphi_{0t}(\theta_s\omega) \quad \text{for } \omega \notin N_s.$$

The case $t < 0$ is analogous, using the backward integral and Proposition 2.3.22(ii).

4. Perfection: The topological group $\text{Diff}^k(\mathbb{R}^d)$ of C^k diffeomorphisms of $\mathbb{R}^d$ endowed with the usual complete metric is a Polish group (see Kunita [224: p. 115]) and hence has a countable base. Theorem 1.3.2 applies, providing us with an indistinguishable perfect cocycle.

5. Local characteristics: By definition the local characteristics of φ are those of the semimartingale

$$G(x,t) = \varphi(t)x - x = \begin{cases} \int_0^t F(\varphi(s)x, \circ d^+s), & 0 \leq t, \\ -\int_t^0 F(\varphi(s)x, \circ d^-s), & t \leq 0. \end{cases}$$

Decompose G as

$$G(x,t,\omega) = M_\varphi(x,t,\omega) + B_\varphi(x,t,\omega).$$

First consider the case $t \geq 0$: By Theorems 3.2.5, 3.2.4 and Lemma 3.2.2 of Kunita [224],

$$M_\varphi(x,t) = \int_0^t M_F(\varphi(s)x, d^+s)$$

with

$$Q_\varphi(x,y,t,\omega) = \int_0^t a_F(\varphi(s,\omega)x, \varphi(s,\omega)y, s, \omega)\, dA_F(s,\omega),$$

and

$$B_\varphi(x,t,\omega)$$
$$= \int_0^t b_F(\varphi(s,\omega)x, s, \omega)\, dA_F(s,\omega) + \int_0^t c_F(\varphi(s,\omega)x, s, \omega)\, dA_F(s,\omega),$$

with c_F given by (2.3.5) (Kunita [224: p. 132]).

For the case $t \leq 0$ we use formula (24) on p. 112 of [224] and obtain first

$$M_\varphi(x,t) = -\int_t^0 M_F(\varphi(s)x, d^-s),$$

and then (with our convention having $a(t) \geq 0$ and $\langle M \rangle_t$ increasing on all of $\mathbb{R}$)

$$Q_\varphi(x,y,t,\omega) = \int_0^t a_F(\varphi(s,\omega)x, \varphi(s,\omega)y, s, \omega) \, dA_F(s,\omega)$$

also for $t \leq 0$.

Further, with our convention $-\int_t^0 f\,dg =: \int_0^t f\,dg$ for $t < 0$,

$$B_\varphi(x,t,\omega)$$
$$= \int_0^t b_F(\varphi(s,\omega)x, s, \omega) \, dA_F(s,\omega) + \int_0^t c_F(\varphi(s,\omega)x, s, \omega) \, dA_F(s,\omega)$$

also for $t \leq 0$. □

2.3.27 Remark (Unified SDE for Two-Sided Time). So far, all of our stochastic integrals have lower limits less than or equal to upper limits. For a more symmetric formulation we introduce, as in the case of ordinary Lebesgue integral, the following *convention:* For each fixed $s \in \mathbb{R}$ and any semimartingale $f_s(t)$ put

$$\int_s^t F(f_s(u), \circ du) := \begin{cases} \int_s^t F(f_s(u), \circ d^+u), & s \leq t, \\ -\int_t^s F(f_s(u), \circ d^-u), & t \leq s. \end{cases}$$

Then (2.3.4) can be simply written as

$$\varphi(t)x - x = \int_0^t F(\varphi(s)x, \circ ds), \quad t \in \mathbb{R}, \tag{2.3.8}$$

in complete analogy to the deterministic case. ∎

2.3.28 Remark (SDE Based on Itô Integral). There is a similar, but less symmetric, theory based on the Itô (instead of Stratonovich) forward and backward stochastic integrals (see Kunita [224]). We will consider it only for the classical SDE, see Subsect. 2.3.6. ∎

Local RDS from Stochastic Differential Equations

2.3.29 Theorem (Local RDS from Stratonovich SDE). *Let $(\Omega, \mathcal{F}^0, \mathbb{P}, (\theta_t)_{t \in \mathbb{R}}, (\mathcal{F}_s^t)_{s \leq t})$ be a filtered DS and let F be a $\mathcal{C}^{k,\delta}$-semimartingale helix over θ, where $k \geq 1$ and $\delta > 0$. Assume that F has local characteristics (a_F, b_F, A_F) satisfying $a_F \in L_{\mathrm{loc}}(\mathbb{R}, dA_F, \tilde{\mathcal{C}}^{k+1,\delta})$ and $b_F \in L_{\mathrm{loc}}(\mathbb{R}, dA_F, \mathcal{C}^{k,\delta})$. Then there exists a unique (up to indistinguishability) local $\mathcal{C}^k$ RDS φ over θ which for any $\varepsilon < \delta$ is a local $\mathcal{C}^{k,\varepsilon}$-semimartingale[6] that is the unique maximal solution of the SDE*

$$dx_t = F(x_t, \circ dt).$$

---
[6] For the definition of a local $\mathcal{C}^{k,\varepsilon}$-semimartingale see Kunita [224: p. 91].

More precisely:

(i) (a) The forward explosion time $\tau^+(\omega, x)$ is an accessible stopping time in the following sense:

- *For each fixed x, $\tau^+(\cdot, x)$ is an $(\mathcal{F}_0^t)_{t \geq 0}$ stopping time,*
- *there exists a sequence $\tau_n^+(\omega, x) < \tau^+(\omega, x)$ of $(\mathcal{F}_0^t)_{t \geq 0}$ stopping times which are lower semicontinuous with respect to x and for which $\tau_n^+(\omega, x) \nearrow \tau^+(\omega, x)$ for all x, $\mathbb{P}$-a. s.*

(b) For each $n \in \mathbb{N}$ and $x \in \mathbb{R}^d$, the stopped process $t \mapsto \varphi(t \wedge \tau_n^+(\omega, x), \omega)x$ is the unique (up to indistinguishability) solution of the forward SDE

$$\varphi(t \wedge \tau_n^+(x))x - x = \int_0^t F(\varphi(s \wedge \tau_n^+(x))x, \circ d^+s), \quad 0 \leq t.$$

(ii) (a) The backward explosion time $\tau^-(\omega, x)$ is a backward accessible stopping time in the following sense:

- *For each fixed x, $\tau^-(\cdot, x)$ is an $(\mathcal{F}_t^0)_{t \leq 0}$ backward stopping time,*
- *there exists a sequence $\tau_n^-(\omega, x) > \tau^-(\omega, x)$ of $(\mathcal{F}_t^0)_{t \leq 0}$ backward stopping times which are upper semicontinuous with respect to x and for which $\tau_n^-(\omega, x) \downarrow \tau^-(\omega, x)$ for all x, $\mathbb{P}$-a. s.*

(b) For each $n \in \mathbb{N}$ and $x \in \mathbb{R}^d$, the stopped process $t \mapsto \varphi(t \vee \tau_n^-(\omega, x), \omega)x$ is the unique (up to indistinguishability) solution of the backward SDE

$$\varphi(t \vee \tau_n^-(x))x - x = -\int_t^0 F(\varphi(s \vee \tau_n^-(x))x, \circ d^-s), \quad 0 \geq t.$$

We refrain from giving details of the proof of this theorem as it can be derived from the previous proof combined with the usual truncation techniques, see Kunita [224: pp. 107–112].

2.3.5 Stochastic Differential Equations from RDS

We finally deal with the inverse problem: Given a cocycle φ, when is there a helix F which generates φ, i.e. such that $d\varphi_t = F(\varphi_t, \circ dt)$?

2.3.30 Theorem (Stratonovich SDE from RDS). *Given a filtered DS θ. Let φ be a C^k RDS over θ such that*

$$\begin{aligned}G_s(x, t, \omega) &:= \varphi(t, \omega)\varphi(s, \omega)^{-1}x - x = \varphi_{st}(\omega)x - x \\ &= M_\varphi^{(s)}(x, t, \omega) + B_\varphi^{(s)}(x, t, \omega)\end{aligned}$$

is a $C^{k,\delta}$-semimartingale with local characteristics $(a_\varphi, b_\varphi, A_\varphi)$ satisfying $a_\varphi \in L_{\text{loc}}(\mathbb{R}, dA_\varphi, \tilde{C}^{k,\delta})$, $b_\varphi \in L_{\text{loc}}(\mathbb{R}, dA_\varphi, C^{k,\delta})$ for some $k \geq 3$ and $\delta > 0$. Then there exists a unique (up to indistinguishability) stochastic process $F(x, t, \omega)$ which for some $\varepsilon \in (0, \delta)$ is a $C^{k-1,\varepsilon}$-semimartingale helix, such that φ is generated by F, i.e. φ and F satisfy

$$\varphi(t)x - x = \int_0^t F(\varphi(s)x, \circ\, ds), \quad t \in \mathbb{R}.$$

The local characteristics of F are obtained from those of φ as follows:

$$A_F = A_\varphi,$$
$$a_F(x, y, t, \omega) = a_\varphi(\varphi(t,\omega)^{-1}x, \varphi(t,\omega)^{-1}y, t, \omega),$$
$$b_F(x, t, \omega) = b_\varphi(\varphi(t,\omega)^{-1}x, t, \omega) - d_\varphi(x, t, \omega),$$

where

$$d_\varphi(x, t, \omega) := \frac{1}{2} \sum_{j=1}^d \frac{\partial}{\partial x_j} a_\varphi^{\cdot j}(\varphi(t,\omega)^{-1}x, \varphi(t,\omega)^{-1}y, t, \omega)|_{y=x}.$$

Proof. 1. Consider first the forward case and put for $s \le t$

$$F_s(x,t) := \int_s^t G_s(\varphi_{su}^{-1}x, \circ\, d^+u).$$

$G_s(x,t)$ is a $\mathcal{C}^{k,\delta}$-forward semimartingale by assumption and we have $a_\varphi \in L_{\text{loc}}(\mathbb{R}, dA_\varphi, \tilde{\mathcal{C}}^{k,\delta})$, $b_\varphi \in L_{\text{loc}}(\mathbb{R}, dA_\varphi, \mathcal{C}^{k,\delta})$. Further, $h_s(x,t) := \varphi_{st}^{-1}x$ is a $\mathcal{C}^{k-1,\varepsilon}$-forward semimartingale for some $\varepsilon \in (0,\delta)$ (Kunita [224], Theorem 4.4.2 and its proof) with local characteristics of corresponding regularity. Hence Theorem 3.3.4(i) of [224] applies and says that $F_s(x,t)$ is a $\mathcal{C}^{k-1,\rho}$-forward semimartingale for some $\rho \in (0,\varepsilon)$, with local characteristics of corresponding regularity.

We can thus use $F_s(x,t)$ as a forward integrator. Then Theorem 3.3.4(ii) of Kunita [224] yields

$$G_s(x,t) = \varphi_{st}x - x = \int_s^t F_s(\varphi_{su}x, \circ\, d^+u), \quad s \le t,$$

in particular

$$G(x,t) = \varphi(t)x - x = \int_0^t F_0(\varphi(s)x, \circ\, d^+s), \quad 0 \le t,$$

so $F_s(x,t)$ is a forward generator of φ_{st}, $s \le t$.

2. We now consider the backward case $t \le s$ and put

$$\hat{F}_s(x,t) := \int_t^s G_s(\varphi_{su}^{-1}x, \circ\, d^-u).$$

This is now a $\mathcal{C}^{k-1,\rho}$-backward semimartingale with local characteristics of corresponding regularity which satisfies

$$G_s(x,t) = \varphi_{st}x - x = -\int_t^s \hat{F}_s(\varphi_{su}x, \circ\, d^-u), \quad t \le s,$$

in particular

$$G(x,t) = \varphi(t)x - x = -\int_t^0 \hat{F}_0(\varphi(s)x, \circ d^-s), \quad t \leq 0,$$

so $\hat{F}_s(x,t)$ is a backward generator of φ_{st}, $t \leq s$.

3. We next show that in fact (up to indistinguishability)

$$F_s(x,t,\omega) = F(x,t,\omega) - F(x,s,\omega), \quad s \leq t,$$

where F is a (forward semimartingale) *helix*.

As a consequence of the flow property we have (see Kunita [224: p. 129])

$$F_s(x,u,\omega) = F_s(x,t,\omega) + F_t(x,u,\omega), \quad 0 \leq s \leq t \leq u, \quad \omega \in \Omega.$$

Putting

$$F(x,t,\omega) := F_0(x,t,\omega),$$

we obtain

$$F_s(x,t,\omega) = F(x,t,\omega) - F(x,s,\omega), \quad 0 \leq s \leq t, \, \omega \in \Omega.$$

Since by Lemma 2.3.24(ii) $G_s(x,t+s,\omega) = G(x,t,\theta_s\omega)$, we have

$$\begin{aligned} F_s(x,t+s) &= \int_s^{s+t} G_s(\varphi_{su}^{-1}x, \circ d^+u) \\ &= \int_s^{s+t} (\theta_s G)(\varphi_{su}^{-1}x, \circ d^+(u-s)) \\ &= \int_0^t (\theta_s G)(\varphi_{s,s+u}^{-1}x, \circ d^+u). \end{aligned}$$

Again the cocycle property $\varphi_{s,s+u}(\omega) = \varphi(u, \theta_s\omega)$ and Proposition 2.3.22(i) lead to

$$\begin{aligned} F_s(x, s+t) &= \theta_s \int_0^t G(\varphi(u)^{-1}x, \circ d^+u) \quad &(2.3.9) \\ &= \theta_s F(x,t), \, 0 \leq s \leq t, \, \omega \notin N_s. \end{aligned}$$

As a result,

$$F(x,s+t,\omega) - F(x,s,\omega) = F(x,t,\theta_s\omega), \quad s,t \in \mathbb{R}^+, \, \omega \notin N_s.$$

By Proposition 2.3.7(i), F can be extended to a (continuous) crude helix on $\mathbb{R}$, also called F.

Now Theorem 1.3.2 applies with $(H, \circ) = (C^{k-1,p}, +)$ and F can be perfected (the perfect version is again denoted by F).

By Proposition 2.3.10, F is a forward semimartingale. It satisfies for any $s \leq t$

$$F_s(x,t,\omega) = F(x,t,\omega) - F(x,s,\omega) = F(x,t-s,\theta_s\omega).$$

We have thus found a forward semimartingale helix F which generates the forward semimartingale cocycle φ_{st}, $s \leq t$. In particular,

$$\varphi(t)x - x = \int_0^t F(\varphi(s)x, \circ d^+s), \quad 0 \leq t.$$

4. In step 2 we have constructed the backward generator $\hat{F}_s(x,t)$, $t \leq s$. By the same arguments as in step 3 we prove that there is in fact a backward semimartingale helix $\hat{F}$ on $\mathbb{R}$ for which for each $t \leq s$

$$\hat{F}_s(x,t,\omega) = \hat{F}(x,t,\omega) - \hat{F}(x,s,\omega) = \hat{F}(x, t-s, \theta_s\omega).$$

In particular,

$$\varphi(t)x - x = -\int_t^0 \hat{F}(\varphi(s)x, \circ d^-s), \quad t \leq 0.$$

5. It remains to show that $\hat{F}$ and F are indistinguishable, i.e.

$$\hat{F}(x,t,\omega) = F(x,t,\omega) \quad \text{for all } x, t, \quad \mathbb{P}\text{-a.s.}$$

But this is exactly what Kunita [224] proves on pp. 150–151 (for the Itô case and hence involving a correction term which disappears here). We have thus found an F which is a $C^{k-1,\varepsilon}$-semimartingale helix for some $\varepsilon \in (0, \delta)$ and is the generator of our $C^{k,\delta}$-semimartingale cocycle.

6. Finally we calculate the local characteristics of F from those of φ. For $t \geq 0$ (to which we restrict ourselves)

$$\begin{aligned}
F(x,t) &= \int_0^t G(\varphi(s)^{-1}x, \circ d^+s) \\
&= \int_0^t M_\varphi(\varphi(s)^{-1}x, d^+s) + \int_0^t b_\varphi(\varphi(s)^{-1}x, s, \omega)\, dA_\varphi(s, \omega) \\
&\quad + \frac{1}{2}\sum_{j=1}^d \langle \int_0^t \frac{\partial M_\varphi}{\partial x_j}(\varphi(s)^{-1}x, d^+s), (\varphi(t)^{-1}x)_j \rangle.
\end{aligned}$$

Hence the canonical decomposition of $F(x,t) = M_F(x,t) + B_F(x,t)$ is given by

$$M_F(x,t) = \int_0^t M_\varphi(\varphi(s)^{-1}x, d^+s),$$

$$B_F(x,t) = \int_0^t b_\varphi(\varphi(s)^{-1}x, s)\, dA_\varphi(s) - \frac{1}{2}\sum_{j=1}^d$$
$$\left\langle \int_0^t \frac{\partial M_\varphi}{\partial x_j}(\varphi(s)^{-1}x, d^+s), \int_0^t (\partial\varphi(s)(\varphi(s)^{-1}x)^{-1} G(\varphi(s)^{-1}x, d^+s))_j \right\rangle,$$

where we have used formula (1) on p. 148 of Kunita [224].

Consequently

$$Q_F(x,y,t,\omega) = \int_0^t a_\varphi(\varphi(s,\omega)^{-1}x, \varphi(s,\omega)^{-1}y, s, \omega)\, dA_\varphi(s,\omega)$$

and

$$B_F(x,t,\omega) = \int_0^t \{b_\varphi(\varphi(s,\omega)^{-1}x, s, \omega)$$
$$- \frac{1}{2}\sum_{j=1}^d \frac{\partial}{\partial x_j}(a_\varphi^{\cdot j}(\varphi(s,\omega)^{-1}x, \varphi(s,\omega)^{-1}y, s, \omega))|_{y=x}\} dA_\varphi(s,\omega).$$

□

Summarizing our results, we have proved that the following classes of objects are basically the same:

- solutions of SDE $dx_t = F(x_t, \circ dt)$ driven by a semimartingale helix over a filtered DS θ,
- semimartingale cocycles (or RDS) over θ.

The "engine" "semimartingale cocycle = exp(semimartingale helix)" that we have implemented in this section converts vector-field valued additive helices into multiplicative helices with values in a group of diffeomorphisms, and vice versa. This mechanism is the most general one for the generation of RDS, as semimartingales are known to be the most general class of reasonable integrators.

2.3.6 White Noise

Let $(\Omega, \mathcal{F}^0, \mathbb{P}, (\theta_t)_{t\in\mathbb{R}})$ be the canonical metric DS describing $\mathbb{R}^m$-valued Brownian motion $W_t(\omega) = \omega(t)$ (see Appendix A.2.2). Let $\mathcal{F}$ be the $\mathbb{P}$-completion of the Borel σ-algebra $\mathcal{F}^0$.[7] Define

$$\mathcal{F}_s^t := \sigma(W_u - W_v : s \leq u, v \leq t) \vee \mathcal{N}, \quad s \leq t,$$

where $\mathcal{N}$ are the null sets of $\mathcal{F}$. In particular, $\mathcal{F}_s^s$ is trivial for each s.

2.3.31 Lemma. *(i) We have*

$$\mathcal{F}_s^{t+} = \mathcal{F}_s^t = \mathcal{F}_s^{t-} := \sigma(\cup_{s\leq u<t}\mathcal{F}_s^u),$$

$$\mathcal{F}_{s-}^t = \mathcal{F}_s^t = \mathcal{F}_{s+}^t := \sigma(\cup_{s<u\leq t}\mathcal{F}_u^t),$$

and $\theta_u^{-1}\mathcal{F}_s^t = \mathcal{F}_{s+u}^{t+u}$ for all $u \in \mathbb{R}$. Hence $(\Omega, \mathcal{F}^0, (\theta_t)_{t\in\mathbb{R}}, (\mathcal{F}_s^t)_{s\leq t})$ is a filtered DS.

---
[7] Recall that $(t,\omega) \mapsto \theta_t\omega$ is not $\mathcal{B}\otimes\mathcal{F}$, $\mathcal{F}$ measurable, but only $\mathcal{B}\otimes\mathcal{F}^0$, $\mathcal{F}^0$ measurable. This is why we have to keep $\mathcal{F}^0$ for the DS.

(ii) W_t is an $(\mathbb{R}^m, +)$-valued helix.

(iii) (a) For all $s \in \mathbb{R}$, $(W_{s+t} - W_s)_{t \geq 0}$ is an $(\mathcal{F}_s^{s+t})_{t \geq 0}$-Brownian motion, in particular an $\mathcal{F}_s^t$-forward martingale. (b) For all $s \in \mathbb{R}$, $(W_{s-t} - W_s)_{t \geq 0}$ is an $(\mathcal{F}_{s-t}^s)_{t \geq 0}$-Brownian motion, in particular an $\mathcal{F}_s^t$-backward martingale.

Hence W_t is a (forward as well as backward) $\mathcal{F}_s^t$-martingale helix with local characteristics $A \equiv t$, $b \equiv 0$ and $a \equiv 1$, thus

$$\langle W_t^i, W_t^j \rangle = tI, \quad I = m \times m \text{ unit matrix}.$$

Proof. (i) Since $W. - W_s$ is left-continuous at t and adapted, $W_t - W_s$ is $\mathcal{F}_s^{t-}$-measurable, so $\mathcal{F}_s^t \subset \mathcal{F}_s^{t-} \subset \mathcal{F}_s^t$. Similarly for the argument s for fixed t: Since $W_t - W.$ is right-continuous at s and adapted, $\mathcal{F}_s^t = \mathcal{F}_{s+}^t$.

As for the right-continuity of $\mathcal{F}_s^t$ with respect to t, we can appeal to a well-known theorem stating that for a Feller Markov process with right-continuous paths on a Polish state space the completed natural filtration is right-continuous, see e.g. Doob [129: p. 556].

(ii) Clear, by definition of θ.

(iii) (a) Fix $s \in \mathbb{R}$. By definition, $(W_{s+t} - W_s)_{t \geq 0}$ is an $(\mathcal{F}_s^{s+t})_{t \geq 0}$-Brownian motion if for each $t \geq 0$, $W_{s+t} - W_s$ is $\mathcal{F}_s^{s+t}$-measurable and for each $u \in [0, t]$, $\mathcal{F}_s^{s+u}$ and $W_{s+t} - W_{s+u}$ are independent, which is the case. In particular, $(W_{s+t} - W_s)_{t \geq 0}$ is an $(\mathcal{F}_s^{s+t})_{t \geq 0}$-martingale since for each $t \geq 0$ and $u \in [0, t]$

$$\mathbb{E}(W_{s+t} - W_s | \mathcal{F}_s^{s+u}) = \mathbb{E}(W_{s+t} - W_{s+u}) + W_{s+u} - W_s = W_{s+u} - W_s \quad \mathbb{P}\text{-a.s.}$$

Case (b) is analogous. □

Note that the canonical filtered DS describing Brownian motion could also have been constructed via Theorem 2.3.9.

RDS from Classical Stratonovich Stochastic Differential Equations

Let $f_0 \in \mathcal{C}_b^{k,\delta}$, $f_1, \ldots, f_m \in \mathcal{C}_b^{k+1,\delta}$, $k \geq 1$, $\delta > 0$, and let

$$F(x, t, \omega) = t f_0(x) + \sum_{j=1}^m W_t^j(\omega) f_j(x) = B(x, t, \omega) + M(x, t, \omega).$$

This is a $\mathcal{C}_b^{k,\delta}$-semimartingale helix with (non-random) local characteristics $A(t, \omega) \equiv t$,

$$a(x, y, t) \equiv \sum_{j=1}^m f_j(x) f_j^*(y), \quad b(x, t) \equiv f_0(x),$$

and Stratonovich-Itô correction term

$$c(x, t) \equiv \frac{1}{2} \sum_{j=1}^m \sum_{i=1}^d f_j^i(x) \frac{\partial}{\partial x_i} f_j(x).$$

If only $f_0 \in \mathcal{C}^{k,\delta}$, $f_1, \ldots, f_m \in \mathcal{C}^{k+1,\delta}$, then F is only a $\mathcal{C}^{k,\delta}$-semimartingale helix with the same local characteristics. Clearly $a \in L_{\text{loc}}(\mathbb{R}, dt, \tilde{\mathcal{C}}^{k+1,\delta}_{(b)})$ if and only if $a \in \tilde{\mathcal{C}}^{k+1,\delta}_{(b)}$, which is implied by our assumption $f_1, \ldots, f_m \in \mathcal{C}^{k+1,\delta}_{(b)}$. Further, $b \in L_{\text{loc}}(\mathbb{R}, dt, \mathcal{C}^{k,\delta}_{(b)})$ if and only if $b = f_0 \in \mathcal{C}^{k,\delta}_{(b)}$, and similarly for c. Hence we can apply Theorem 2.3.26 to obtain the following basic theorem for classical Stratonovich SDE.

2.3.32 Theorem (RDS from Classical Stratonovich SDE). *Let $f_0 \in \mathcal{C}^{k,\delta}_b$, $f_1, \ldots, f_m \in \mathcal{C}^{k+1,\delta}_b$, and $\sum_{j=1}^{m} \sum_{i=1}^{d} f_j^i \frac{\partial}{\partial x_i} f_j \in \mathcal{C}^{k,\delta}_b$ for some $k \geq 1$ and $\delta > 0$. Then:*

(i) The classical Stratonovich SDE

$$dx_t = f_0(x_t)dt + \sum_{j=1}^{m} f_j(x_t) \circ dW_t^j = \sum_{j=0}^{m} f_j(x_t) \circ dW_t^j, \quad t \in \mathbb{R}, \quad (2.3.10)$$

(with the convention $\circ dW_t^0 := dt$) generates a unique (up to indistinguishability) C^k RDS φ over the filtered DS describing Brownian motion. For any $\varepsilon \in (0, \delta)$, φ is a $\mathcal{C}^{k,\varepsilon}$-semimartingale cocycle and $(t, x) \mapsto \varphi(t, \omega)x$ belongs to $\mathcal{C}^{0,\beta;k,\varepsilon}$ for all $\beta < \frac{1}{2}$ and $\varepsilon < \delta$.

(ii) The RDS φ has stationary independent (multiplicative) increments, i.e. for all $t_0 \leq t_1 \leq \ldots \leq t_n$ the random variables

$$\varphi(t_1) \circ \varphi(t_0)^{-1}, \quad \varphi(t_2) \circ \varphi(t_1)^{-1}, \ldots, \quad \varphi(t_n) \circ \varphi(t_{n-1})^{-1}$$

are independent, and the law of $\varphi(t+h) \circ \varphi(t)^{-1}$ is independent of t (homogeneous Brownian motion in the group $\text{Diff}^k(\mathbb{R}^d)$).

(iii) If $D\varphi(t, \omega, x)$ denotes the Jacobian of $\varphi(t, \omega)$ at x then $(\varphi, D\varphi)$ is a C^{k-1} RDS uniquely generated by (2.3.10) together with

$$dv_t = \sum_{j=0}^{m} Df_j(x_t)v_t \circ dW_t^j, \quad t \in \mathbb{R}.$$

Hence $D\varphi$ uniquely solves the variational Stratonovich SDE on $\mathbb{R}$,

$$D\varphi(t, x) = I + \sum_{j=0}^{m} \int_0^t Df_j(\varphi(s)x)D\varphi(s, x) \circ dW_s^j, \quad t \in \mathbb{R},$$

and is thus a matrix cocycle over $\Theta = (\theta, \varphi)$.

Finally, the determinant $\det D\varphi(t, \omega, x)$ satisfies Liouville's equation on $\mathbb{R}$,

$$\det D\varphi(t, x) = \exp\left(\sum_{j=0}^{m} \int_0^t \text{trace} Df_j(\varphi(s)x) \circ dW_s^j\right), \quad t \in \mathbb{R},$$

and hence is a scalar cocycle over Θ.

Proof. Assertions (i) and (iii) are particular cases of Theorem 2.3.26. Assertion (ii) follows since the random variables in question are measurable with respect to $\mathcal{F}_{t_0}^{t_1}, \mathcal{F}_{t_1}^{t_2}, \ldots, \mathcal{F}_{t_{n-1}}^{t_n}$, hence independent. □

2.3.33 Remark. By applying the Stratonovich rule for calculating stochastic differentials to $(D\varphi(t,x))^{-1} D\varphi(t,x) = I$, we obtain a forward Stratonovich SDE for the inverse of the Jacobian,

$$(D\varphi(t,x))^{-1} = I - \sum_{j=0}^{m} \int_0^t (D\varphi(s,x))^{-1} Df_j(\varphi(s)x) \circ dW_s^j.$$

■

2.3.34 Example (Affine SDE). Consider on $\mathbb{T} = \mathbb{R}$ the affine SDE

$$dx_t = \sum_{j=0}^{m}(A_j x_t + b_j) \circ dW_t^j, \quad A_j \in \mathbb{R}^{d\times d}, \quad b_j \in \mathbb{R}^d.$$

Since $f_j(x) = A_j x + b_j \in \mathcal{C}_b^\infty$ and $(\sum_{j=1}^{m} A_j^2)x \in \mathcal{C}_b^\infty$, it uniquely generates a global C^∞ RDS, which consists of affine mappings given by the variation of constants formula

$$\varphi(t)x = \Phi(t)\left(x + \sum_{j=0}^{m}\int_0^t \Phi(s)^{-1} b_j \circ dW_s^j\right),$$

where Φ is the fundamental matrix of the corresponding linear SDE

$$dx_t = \sum_{j=0}^{m} A_j x_t \circ dW_t^j$$

which is a linear RDS over θ. ■

2.3.35 Remark (Cocycles with Independent Increments).
We can also recover the general results of Baxendale [54] and Kunita [224: Sect. 4.2]: Under assumptions on the local characteristics similar to ours above, a $\mathcal{C}_b^{k,\delta}$-valued helix F with independent increments (i. e. homogeneous Brownian motion in the linear space $\mathcal{C}_b^{k,\delta}$) generates a $\mathcal{C}^{k,\varepsilon}$-valued cocycle φ with independent increments (i. e. homogeneous Brownian motion in the group $\text{Diff}^k(\mathbb{R}^d)$). Conversely, every $\mathcal{C}^{k,\delta}$-valued cocycle with independent increments is generated by a $\mathcal{C}^{k,\varepsilon}$-valued helix with independent increments. It is exactly this converse that forces one to go beyond the classical model (2.3.10) of finitely many vector fields and consider more general vector field valued driving processes.

In short, the result can be symbolically written as

cocycle with indep. increments = exp(helix with indep. increments).

■

For completeness, we also state the local result.

2.3.36 Theorem (Local RDS from Classical Stratonovich SDE).
Let $f_0 \in \mathcal{C}^{k,\delta}$ and $f_1, \ldots, f_m \in \mathcal{C}^{k+1,\delta}$ for some $k \geq 1$ and $\delta > 0$. Then the classical Stratonovich SDE (2.3.10) uniquely generates a local C^k RDS φ over the filtered DS θ in the sense of Theorem 2.3.29. For any $\varepsilon \in (0, \delta)$, φ is a local $\mathcal{C}^{k,\varepsilon}$-semimartingale cocycle and $(t, x) \mapsto \varphi(t, \omega)x$ is in $\mathcal{C}^{0,\beta;k,\varepsilon}$ for all $\beta < \frac{1}{2}$ and $\varepsilon < \delta$.

Criteria for Global RDS from SDE

2.3.37 Definition (Strict Completeness of RDS). The local RDS φ of Theorem 2.3.36 is called *strictly forward complete* if

$$\mathbb{P}\{\omega : \tau^+(\omega, x) = \infty \text{ for all } x \in \mathbb{R}^d\} = 1,$$

and *strictly backward complete* if

$$\mathbb{P}\{\omega : \tau^-(\omega, x) = -\infty \text{ for all } x \in \mathbb{R}^d\} = 1.$$

It is called *strictly complete* if it is strictly forward complete and strictly backward complete. For these notions see Kunita [224: p. 182]. ∎

The local RDS φ of Theorem 2.3.36 is clearly indistinguishable from a global RDS if and only if it is strictly complete.

The only general conditions known for globality in the case of $X = \mathbb{R}^d$ are the global Lipschitz conditions for the local characteristics built into the assumptions of Theorem 2.3.32.

There is a weaker version of completeness for solutions of SDE related to the one-point motions of the local RDS φ: φ is called *forward complete* (or conservative, regular, non-explosive, see e. g. Khasminskii [206: Chap. III]) if $\mathbb{P}(\tau^+(\omega, x) = \infty) = 1$ for all x, and *backward complete* if $\mathbb{P}(\tau^-(\omega, x) = -\infty) = 1$ for all x. In contrast to the deterministic case, the fact that the local solution does not explode $\mathbb{P}$-a. s. for each fixed initial value in general does not imply that the local RDS is global. Examples are given by Léandre [229], and Carverhill and Elworthy [94]. Possibly, the simplest example in $\mathbb{R}^2$ is

$$dx_t = \begin{pmatrix} x_2^2 - x_1^2 \\ -2x_1 x_2 \end{pmatrix} \circ dW_t^1 + \begin{pmatrix} 2x_1 x_2 \\ x_2^2 - x_1^2 \end{pmatrix} \circ dW_t^2,$$

or $dz_t = -z_t^2 \circ dW_t$ in complex notation ($z = x_1 + ix_2$, $W = W^1 + iW^2$), with solution $\varphi(t, \omega, z) = z/(1 + zW_t(\omega))$.

Many important nonlinear physical and engineering systems (such as the noisy Duffing-van der Pol oscillator) are strictly forward complete, but not backward complete, see Schenk-Hoppé [303, 304] and Sect. 9.4.

While forward/backward completeness is determined by the transition probabilities of the corresponding Markov process, and therefore by the forward/backward generators $L^\pm$, strict completeness is not so determined. Indeed, Carverhill [88] constructed a pair of SDE with the same forward generator, but one generating a strictly forward complete RDS, and the other not doing so (this example is reproduced in Elworthy [139: p. 294]).

By collecting the exceptional sets for a countable dense set in $\mathbb{R}^d$, we can conclude from forward (backward) completeness that the open set $D(t,\omega)$ is dense in $\mathbb{R}^d$ $\mathbb{P}$-a.s., for $t > 0$ ($t < 0$).

For $X = \mathbb{R}$ and two-sided time $\mathbb{T} = \mathbb{R}$, forward/backward completeness of a local continuous RDS φ imply its strict forward/backward completeness and thus that it can be perfected to a global continuous RDS. Indeed, for each t, the open and $\mathbb{P}$-a.s. dense set $D(t,\omega)$ is an interval (see Theorem 1.8.4(ii)), and the only open and dense interval in $\mathbb{R}$ is $\mathbb{R}$ itself. Consequently, $D(t,\omega) = \mathbb{R}$ $\mathbb{P}$-a.s. for all $t \in \mathbb{R}$. Now the continuity of $(t,x) \mapsto \varphi(t,\omega)x$ implies that $D(t,\omega) = \mathbb{R}$ for all $t \in \mathbb{R}$, $\mathbb{P}$-a.s. (see Kunita [224: pp. 178–181]). Redefining φ on the exceptional ω-set as $\varphi(t,\omega) = \mathrm{id}_\mathbb{R}$ for all t gives a crude global cocycle which can be perfected.

The last remark and Feller's well-known necessary and sufficient non-explosion criteria for one-dimensional diffusions can be combined to give the following conditions which often can be verified in specific examples.

2.3.38 Theorem (Criterion for Global RDS from Scalar SDE). *Let*

$$dx_t = b(x_t)dt + \sigma(x_t) \circ dW_t = (b(x_t) \pm \frac{1}{2}\sigma(x_t)\sigma'(x_t))dt + \sigma(x_t)d^\pm W_t$$

be an SDE[8] *on $X = \mathbb{R}$ with time $\mathbb{T} = \mathbb{R}$. Assume $b \in \mathcal{C}^{k,\delta}$, $\sigma \in \mathcal{C}^{k+1,\delta}$, $k \geq 1$, $\delta > 0$, and $\sigma(x) > 0$ for all $x \in \mathbb{R}$. Then the local C^k RDS generated by the above SDE is global if and only if $\pm\infty$ are natural (i.e. non-exit and non-entrance) boundary points. The points $\pm\infty$ are non-exit boundary points (i.e. the RDS is forward complete and thus strictly forward complete) if and only if $K^+(\pm\infty) = \infty$, where*

$$K^+(x) = \int_0^x \frac{1}{\sigma(y)} \exp(-\int_0^y \frac{2b}{\sigma^2}dz) \left(\int_0^y \frac{1}{\sigma(z)} \exp(\int_0^z \frac{2b}{\sigma^2}du)dz \right) dy.$$

The points $\pm\infty$ are non-entrance boundary points (i.e. the RDS is backward complete and thus strictly backward complete) if and only if $K^-(\pm\infty) = \infty$, where

$$K^-(x) = \int_0^x \frac{1}{\sigma(y)} \exp(\int_0^y \frac{2b}{\sigma^2}dz) \left(\int_0^y \frac{1}{\sigma(z)} \exp(-\int_0^z \frac{2b}{\sigma^2}du)dz \right) dy.$$

---

[8] The Itô SDE is defined below.

For a proof see Kunita [224: pp. 181–184].

RDS from Itô Stochastic Differential Equations

For convenience we include the corresponding basic statements about the generation of RDS by classical Itô SDE.

Recall that for a continuous process $f_s(u)$, $s \leq u \leq t$, adapted to $\mathcal{F}_s^u$ and a scalar W_t the *forward Itô integral* is defined by

$$\int_s^t f_s(u)d^+W_u := \lim_{\Delta \to 0} \text{ in pr. } \sum_{k=0}^{n-1} f(t_k)(W_{t_{k+1}} - W_{t_k})$$

and that the forward Stratonovich integral (which needs a semimartingale integrand, see Subsect. 2.3.3) is related to the forward Itô integral by

$$\int_s^t f \circ d^+W = \int_s^t f d^+W + \frac{1}{2}(\langle f, W \rangle_t - \langle f, W \rangle_s),$$

where $\langle f, W \rangle_t$ denotes the quadratic covariation process of the semimartingales f and W. The *backward Itô integral* of a continuous process $f_s(u)$, $t \leq u \leq s$, adapted to the filtration $\mathcal{F}_u^s$ is defined by

$$\int_t^s f_s(u)d^-W_u := \lim_{\Delta \to 0} \text{ in pr. } \sum_{k=0}^{n-1} f(t_{k+1})(W_{t_{k+1}} - W_{t_k})$$

and the backward Stratonovich integral is related to it by

$$\int_t^s f \circ d^-W = \int_t^s f d^-W - \frac{1}{2}(\langle f, W \rangle_s^- - \langle f, W \rangle_t^-).$$

We stress that in each of the four stochastic integrals introduced above the lower integration limit is always less than or equal to the upper limit.

Let $f_0 \in \mathcal{C}_b^{k,\delta}$, $f_1, \ldots, f_m \in \mathcal{C}_b^{k+1,\delta}$ and $\sum_{j=1}^m \sum_{i=1}^d f_j^i \frac{\partial}{\partial x_i} f_j \in \mathcal{C}_b^{k,\delta}$ for some $k \geq 1$ and $\delta > 0$. Then the following forward/backward Stratonovich and Itô SDE are equivalent in the sense that they generate the same C^k RDS:

$$d^{\pm} x_t = \sum_{j=0}^m f_j(x_t) \circ d^{\pm} W_t^j, \quad d^{\pm} x_t = f_0^{\pm}(x_t)dt + \sum_{j=1}^m f_j(x_t) d^{\pm} W_t^j,$$

where

$$f_0^{\pm}(x) = f_0(x) \pm \frac{1}{2} \sum_{j=1}^m \sum_{i=1}^d f_j^i(x) \frac{\partial}{\partial x_i} f_j(x).$$

Here is the Itô counterpart of Theorem 2.3.32.

2.3.39 Theorem (RDS from Classical Itô SDE).
Let $f_0 \in \mathcal{C}_b^{k,\delta}$, $f_1,\ldots,f_m \in \mathcal{C}_b^{k+1,\delta}$ and $\sum_{j=1}^m \sum_{i=1}^d f_j^i \frac{\partial}{\partial x_i} f_j \in \mathcal{C}_b^{k,\delta}$ for some $k \geq 1$ and $\delta > 0$. Then there is a unique (up to indistinguishability) C^k RDS φ over the filtered DS describing Brownian motion which solves (again using the convention $dW_t^0 := dt$)

$$\varphi(t)x - x = \begin{cases} \sum_{j=0}^m \int_0^t f_j(\varphi(s)x) d^+W_s^j, & 0 \leq t, \\ -\sum_{j=0}^m \int_t^0 f_j^*(\varphi(s)x) d^-W_s^j, & t \leq 0, \end{cases} \quad (2.3.11)$$

where $f_j^* = f_j$ for $j = 1,\ldots,m$ and

$$f_0^*(x) = f_0(x) - \sum_{j=1}^m \sum_{i=1}^d f_j^i(x) \frac{\partial}{\partial x_i} f_j(x).$$

For any $\varepsilon \in (0,\delta)$, φ is a $\mathcal{C}^{k,\varepsilon}$-semimartingale cocycle and $(t,x) \mapsto \varphi(t,\omega)x$ is in $\mathcal{C}^{0,\beta;k,\varepsilon}$ for all $\beta < \frac{1}{2}$ and $\varepsilon < \delta$. The RDS φ has stationary independent increments.

Further, the Jacobian of φ and its determinant satisfy the same properties as in Theorem 2.3.32, where the variational Itô SDE is now obtained from linearizing (2.3.11).

We close with the following well-known conditions under which a *forward* Itô SDE generates a continuous or C^k RDS. These conditions are in general too weak to conclude the existence of a backward generator.

2.3.40 Theorem (RDS from Classical Forward Itô SDE). *(a) Let $f_j \in \mathcal{C}_b^{0,1}$ for $j = 0,\ldots,m$. Then there is a unique (up to indistinguishability) continuous RDS φ over the filtered DS describing Brownian motion which solves the forward Itô SDE*

$$\varphi(t)x - x = \sum_{j=0}^m \int_0^t f_j(\varphi(s)x) d^+W_s^j, \quad t \geq 0.$$

For any $\varepsilon < 1$, φ is a $\mathcal{C}^{0,\varepsilon}$-semimartingale cocycle and $(t,x) \mapsto \varphi(t,\omega)x$ is in $\mathcal{C}^{0,\beta;0,\varepsilon}$ for all $\beta < \frac{1}{2}$ and $\varepsilon < 1$. The RDS φ has stationary independent increments.

(b) If $f_0,\ldots,f_m \in \mathcal{C}_b^{k,\delta}$ for some $k \geq 1$ and $\delta > 0$, then φ is C^k.

2.3.7 An Example

If the parameters α and β in $\dot{x}_t = \alpha x_t - \beta x_t^N$, $N \geq 2$, are perturbed by white noise of intensities σ_1 and σ_2, respectively, we obtain

$$dx_t = (\alpha x_t - \beta x_t^N)dt + \sigma_1 x_t \circ dW_t^1 - \sigma_2 x_t^N \circ dW_t^2.$$

This equation can be explicitly solved by transforming it to an affine equation via $y = x^{1-N}/(1-N)$. We obtain the affine SDE

$$dy_t = (1-N)\alpha y_t\, dt - \beta dt + (1-N)\sigma_1 y_t \circ dW_t^1 + \sigma_2 \circ dW_t^2,$$

with solution

$$y(t;y_0) = e^{(1-N)(\alpha t + \sigma_1 W_t^1)}\left(y_0 + \int_0^t e^{(N-1)(\alpha s + \sigma_1 W_s^1)}(-\beta ds + \sigma_2 \circ dW_s^2)\right).$$

Transforming back via $x = ((1-N)y)^{\frac{1}{1-N}}$ gives the following explicit expression of the local C^∞ RDS φ for the original SDE:

$$\varphi(t,\cdot)x = \frac{xe^{\alpha t + \sigma_1 W_t^1}}{\left(1 + (N-1)x^{N-1}\int_0^t e^{(N-1)(\alpha s + \sigma_1 W_s^1)}(\beta\, ds - \sigma_2 \circ dW_s^2)\right)^{\frac{1}{N-1}}}. \tag{2.3.12}$$

We can read off from this expression the explosion time $\tau^\pm(\omega, x)$ as the first time at which the denominator (which is positive for small t) becomes zero, hence determining the domain $D(t,\omega)$ and range $R(t,\omega)$ of $\varphi(t,\omega)$: $D(t,\omega) \to R(t,\omega)$.

We first observe that for $\sigma_2 \neq 0$ the explosion time $\tau^\pm(\omega, x)$ is finite with positive probability for any $x \neq 0$, and $\tau^\pm(\omega, 0) = \pm\infty$. This implies that the set of never exploding initial values (which also carries all invariant measures) is trivial for these cases, $E(\omega) = \{0\}$. In particular, δ_0 is the only invariant measure (and the only solution of the Fokker-Planck equation).

It thus suffices to consider the case $\sigma_2 = 0$. We can also assume (possibly after time reversal) that $\beta > 0$, put for definiteness $\beta = 1$ (and $\sigma := \sigma_1$). Then the local C^∞ RDS for the SDE

$$dx_t = (\alpha x_t - x_t^N)dt + \sigma x_t \circ dW_t \tag{2.3.13}$$

is explicitly given by

$$x \mapsto \varphi_\alpha(t,\omega)x = \frac{xe^{\alpha t + \sigma W_t(\omega)}}{\left(1 + (N-1)x^{N-1}\int_0^t e^{(N-1)(\alpha s + \sigma W_s(\omega))}ds\right)^{\frac{1}{N-1}}}. \tag{2.3.14}$$

We now determine the domain $D_\alpha(t,\omega)$ and the range $R_\alpha(t,\omega)$ of $\varphi_\alpha(t,\omega)$: $D_\alpha(t,\omega) \to R_\alpha(t,\omega)$.

Case N Even:

We have

$$D_\alpha(t,\omega) = \begin{cases} (-d_\alpha(t,\omega), \infty), & t > 0, \\ \mathbb{R}, & t = 0, \\ (-\infty, d_\alpha(t,\omega)), & t < 0, \end{cases}$$

where

$$d_\alpha(t,\omega) = \frac{1}{\left((N-1)\left|\int_0^t e^{(N-1)(\alpha s+\sigma W_s(\omega))}ds\right|\right)^{\frac{1}{N-1}}} > 0,$$

and

$$R_\alpha(t,\omega) = D_\alpha(-t,\theta(t)\omega) = \begin{cases} (-\infty, r_\alpha(t,\omega)), & t > 0, \\ \mathbb{R}, & t = 0, \\ (-r_\alpha(t,\omega), \infty), & t < 0, \end{cases}$$

where

$$r_\alpha(t,\omega) = d_\alpha(-t,\theta(t)\omega) = \frac{e^{\alpha t+\sigma W_t(\omega)}}{\left((N-1)\left|\int_0^t e^{(N-1)(\alpha s+\sigma W_s(\omega))}ds\right|\right)^{\frac{1}{N-1}}} > 0.$$

We can now determine

$$E_\alpha(\omega) := \cap_{t\in\mathbb{R}} D_\alpha(t,\omega),$$

and obtain

$$E_\alpha(\omega) = \begin{cases} [0, d_\alpha^-(\omega)], & \alpha > 0, \\ \{0\}, & \alpha = 0, \\ [-d_\alpha^+(\omega), 0], & \alpha < 0, \end{cases}$$

where

$$0 < d_\alpha^\pm(\omega) = \frac{1}{\left((N-1)\left|\int_0^{\pm\infty} e^{(N-1)(\alpha s+\sigma W_s(\omega))}ds\right|\right)^{\frac{1}{N-1}}} < \infty.$$

The result for $\alpha = 0$ is a consequence of the following fact.

2.3.41 Lemma. *For any $\sigma \in \mathbb{R}$,*

$$\int_{-\infty}^0 e^{\sigma W_s(\omega)}ds = \int_0^\infty e^{\sigma W_s(\omega)}ds = \infty.$$

Proof. Put $\sigma > 0$ and $N(\omega) := \{t \geq 0 : W_t(\omega) \geq 0\}$. Then

$$\int_0^\infty e^{\sigma W_s(\omega)}ds \geq \int_0^\infty 1_{N(\omega)}(s)ds = \text{Leb } N(\omega).$$

By the arc sine law,

$$\limsup_{t\to\infty} \frac{\text{Leb}\{s \leq t : W_s(\omega) \geq 0\}}{t} = 1 \quad \mathbb{P}\text{-a. s.},$$

so that $\text{Leb } N(\omega) = \infty$, $\mathbb{P}$-a. s. □

By monotone convergence, $d_\alpha^- \downarrow 0$ for $\alpha \downarrow 0$, and $d_\alpha^+ \uparrow 0$ for $\alpha \uparrow 0$.

The **ergodic invariant measures** (which by Theorem 1.8.4 are random Dirac measures) are:

(i) for $\alpha > 0$: $\mu_{1,\omega}^\alpha = \delta_0$ and $\mu_{2,\omega}^\alpha = \delta_{d_\alpha^-(\omega)}$ which are both $\mathcal{F}_{-\infty}^0$-measurable (i.e. Markov measures; the density of $\mathbb{E}\,\mu_2^\alpha$ is determined by solving the Fokker-Planck equation in Example 4.2.15),

(ii) for $\alpha = 0$: $\mu_{1,\omega}^0 = \delta_0$,

(iii) for $\alpha < 0$: $\mu_{1,\omega}^\alpha = \delta_0$ and $\mu_{3,\omega}^\alpha = \delta_{-d_\alpha^+(\omega)}$, the latter being $\mathcal{F}_0^\infty$-measurable.

Besides these, there are no other ergodic invariant measures, since any other measure ν_ω would have to satisfy $\nu_\omega(\mathrm{int}E_\alpha(\omega)) = 1$, $\mathbb{P}$-a.s., which is contradicted by the fact that $\varphi_\alpha(t,\omega)x \to 0$ as $t \to \infty$ ($t \to -\infty$) for any $x \in \mathrm{int}E_\alpha(\omega)$ for $\alpha < 0$ ($\alpha > 0$).

Case N Odd:

Now
$$D_\alpha(t,\omega) = \begin{cases} \mathbb{R}, & t \geq 0, \\ (-d_\alpha(t,\omega), d_\alpha(t,\omega)), & t < 0, \end{cases}$$

$$R_\alpha(t,\omega) = D_\alpha(-t, \theta(t)\omega) = \begin{cases} (-r_\alpha(t,\omega), r_\alpha(t,\omega)), & t > 0, \\ \mathbb{R}, & t \leq 0, \end{cases}$$

$$E_\alpha(\omega) = \begin{cases} [-d_\alpha^-(\omega), d_\alpha^-(\omega)], & \alpha > 0, \\ \{0\}, & \alpha \leq 0. \end{cases}$$

The **ergodic invariant measures** are:

(i) for $\alpha > 0$: the three random Dirac measures $\mu_{1,\omega}^\alpha = \delta_0$, $\mu_{2,\omega}^\alpha = \delta_{-d_\alpha^-(\omega)}$ and $\mu_{3,\omega}^\alpha = \delta_{d_\alpha^-(\omega)}$, which are all Markov measures (for the corresponding solutions of the Fokker-Planck equation see Example 4.2.15),

(ii) for $\alpha = 0$: $\mu_{1,\omega}^0 = \delta_0$,

(iii) for $\alpha < 0$: $\mu_{1,\omega}^\alpha = \delta_0$.

That there are no other ergodic invariant measures is proved as in the case of an even N.

The Lyapunov exponent of φ for all invariant measures is calculated in Example 4.2.15. A study of the bifurcation behavior for the cases $N = 2$ and $N = 3$ is given in Subsect. 9.3.1. See Exercise 2.2.9 for the real noise case.

Other explicitly solvable SDE are given e.g. by Horsthemke and Lefever [175: pp. 139ff.], and Kloeden and Platen [214: pp. 117ff.].

2.3.8 The Manifold Case

There is a similar theory of semimartingale cocycles and their generation by semimartingale helices via SDE for the state space $X = M$, a manifold, which we refrain from developing in detail here (see Baxendale [54], Le Jan and Watanabe [228] and Kunita [224: Sect. 4.8] for more details).

Let us briefly discuss the case of a classical SDE. Given a standard Brownian motion W in $\mathbb{R}^m$ and vector fields f_j, $j = 0, \ldots, m$, the (forward as well as backward) Stratonovich SDE on the manifold M with time $\mathbb{T} = \mathbb{R}$ formally written (with $dW_t^0 := dt$) as

$$dx_t = \sum_{j=0}^{m} f_j(x_t) \circ dW_t^j \tag{2.3.15}$$

has the following coordinate-free meaning: A process $\varphi(t,\omega)x$ with values in M is called a solution of (2.3.15) if for each test function $h \in \mathcal{C}_K^\infty(M, \mathbb{R})$, the space of C^∞ functions on M with compact support,

$$h(\varphi(t)x) - h(x) = \begin{cases} \sum_{j=0}^{m} \int_0^t f_j(h)(\varphi(s)x) \circ d^+ W_s^j, & 0 \leq t, \\ -\sum_{j=0}^{m} \int_t^0 f_j(h)(\varphi(s)x) \circ d^- W_s^j, & t \leq 0. \end{cases}$$

For more details on SDE on manifolds see e. g. Belopolskaya and Dalecky [65], Emery [146], Elworthy [138, 139], Hackenbroch and Thalmaier [165], Ikeda and Watanabe [178], Kunita [224], and Rogers and Williams [287].

2.3.42 Theorem (Local RDS from SDE on Manifold). *Consider the Stratonovich SDE on M with time $\mathbb{T} = \mathbb{R}$*

$$dx_t = \sum_{j=0}^{m} f_j(x_t) \circ dW_t^j \tag{2.3.16}$$

and assume that $f_0 \in \mathcal{C}^{k,\delta}$ and $f_j \in \mathcal{C}^{k+1,\delta}$, $j = 1, \ldots, m$, $k \geq 1$, $0 < \delta \leq 1$. Then:

(i) There is a unique local C^k RDS φ over the canonical filtered DS corresponding to W which solves (2.3.16).

(ii) The derivative $T\varphi : TM \to TM$ is a local C^{k-1} RDS which is the unique solution of

$$dv_t = \sum_{j=0}^{m} Tf_j(v_t) \circ dW_t^j, \quad v_0 = v \in TM, \tag{2.3.17}$$

where Tf_j is the natural lift of the vector field f_j to a vector field on TM (in local coordinates, $Tf(x)v = (x, v, f(x), Df(x)v)$).

(iii) If M is a Riemannian manifold, and if we use the Riemannian connection on M to decompose $T_v TM$ into horizontal and vertical components, then $Tf_j(v)$ has horizontal component $f_j(x)$ and vertical component $\nabla f_j(x)v$, where $\nabla f_j(x)v$ denotes the covariant derivative of the vector field f_j in the direction of $v \in T_x M$, and equation (2.3.17) is equivalent to the two coupled SDE (2.3.16) and the variational equation

$$Dv_t = \sum_{j=0}^{m} \nabla f_j(x_t) v_t \circ dW_t^j, \quad v_0 = v \in T_x M.$$

Here Dv_t denotes the absolute stochastic differential.

Furthermore, we have Liouville's equation for $\det T\varphi(t,\omega,x)$: For $t \in D(\omega,x)$

$$\det T\varphi(t,x) = \exp \int_0^t \sum_{j=0}^m \mathrm{div}\, f_j(\varphi(s)x) \circ dW_s^j$$

$$= \exp \int_0^t \sum_{j=0}^m (\mathrm{trace}\,\nabla f_j)(\varphi(s)x) \circ dW_s^j.$$

(iv) The RDS φ is global in case M is compact, or if, by embedding M into an $\mathbb{R}^N$, the vector fields f_j are restrictions of vector fields $\tilde{f}_j$ in $\mathbb{R}^N$ for which $\tilde{f}_0 \in C_b^{k,\delta}$, $\tilde{f}_1, \ldots, \tilde{f}_m \in C_b^{k+1,\delta}$, and $\sum_{j=1}^m \sum_{i=1}^d \tilde{f}_j^i \frac{\partial}{\partial x_i} \tilde{f}_j \in C_b^{k,\delta}$.

Conditions for strict completeness of SDE on general non-compact manifolds (which is in general not implied by completeness if $\dim M \geq 2$, see Elworthy [137: p. 198]) are given by Li [235, 234].

2.3.9 RDS with Independent Increments

We now return to the situation and notation of Subsect. 2.3.6 and discuss the connection between SDE and Markov processes. This subject is classical and is treated in most of the books mentioned in Subsect. 2.3.1. For our brief survey (following Arnold [8]) we choose a more conceptual and descriptive style, often omitting technical conditions under which our statements hold.

The following problem is historically from the beginning of the theory of SDE: Suppose we are given an elliptic differential operator

$$L = \sum_{i=1}^d b^i(x)\frac{\partial}{\partial x_i} + \frac{1}{2}\sum_{k,l=1}^d a^{kl}(x)\frac{\partial^2}{\partial x_k \partial x_l} \qquad (2.3.18)$$

in $\mathbb{R}^d$ with smooth coefficients, where $a(x) = (a^{kl}(x))$ is for each $x \in \mathbb{R}^d$ a non-negative definite symmetric $d \times d$ matrix, we seek a differential equation

$$\dot{x}_t = f_0(x_t) + \sum_{j=1}^m \xi_t^j f_j(x_t), \quad x_0 = x \in \mathbb{R}^d, \qquad (2.3.19)$$

(also called *Langevin equation*) such that for each initial value $x_0^x = x$ its solution $(x_t^x)_{t \in \mathbb{R}^+}$ is a (homogeneous) Markov process for which the transition semigroup

$$T_t g(x) = \mathbb{E}\, g(x_t^x) = \int_{\mathbb{R}^d} g(y) P_t(x,dy) \qquad (2.3.20)$$

has the infinitesimal generator L and the vector fields $f_0, \ldots, f_m$ are determined by the coefficients b and a of L.

It turns out that $\xi_t = (\xi_t^1, \ldots, \xi_t^m)$ has to be Gaussian white noise (hence a generalized random process). Itô [182, 184, 183] gave a rigorous meaning to a "stochastic generator" of the form (2.3.19) by introducing stochastic integrals and writing (2.3.19) as an (Itô) SDE

$$dx_t = f_0(x_t)dt + \sum_{j=1}^m f_j(x_t)dW_t^j, \quad x_0 = x \in \mathbb{R}^d. \tag{2.3.21}$$

The f_j in (2.3.21) are derived from b and a as follows: Put $f_0 = b$. Take any factorization $a(x) = \sigma(x)\sigma(x)^*$, where $\sigma(x) = (\sigma_1(x), \ldots, \sigma_m(x))$ is a $d \times m$ matrix with columns $\sigma_j(x)$, and put $f_j = \sigma_j$, $j = 1, \ldots, m$.

The SDE (2.3.21) indeed generates a Markov family of processes $(x_\cdot^z)_{z \in \mathbb{R}^d}$ with time $\mathbb{R}^+$ indexed by their initial conditions, all with the same transition probability

$$P_t(x, B) = \mathbb{P}\{x_{t+s}^z \in B | x_s^z = x\} = \mathbb{P}\{\omega : x_t^x \in B\}. \tag{2.3.22}$$

The semigroup T_t given by (2.3.20) has generator L given by (2.3.18), and the coefficients b and a can be obtained from the solutions as *infinitesimal mean*

$$\lim_{t \downarrow 0} \frac{1}{t}\mathbb{E}(x_t^x - x) = b(x)$$

and *infinitesimal variance*

$$\lim_{t \downarrow 0} \frac{1}{t}\mathbb{E}(x_t^x - x)(x_t^x - x)^* = a(x).$$

Note that the semigroup T_t, hence the law $\mathbb{P}_x$ on $\mathcal{C}(\mathbb{R}^+, \mathbb{R}^d)$ of the solution $(x_t^x)_{t \in \mathbb{R}^+}$ of (2.3.21) is (in "nice" cases, e.g. under Oleinik's theorem, see Stroock and Varadhan [326: p. 77]) uniquely determined by L, hence by a and b, and is thus independent of the way we factorize $a(x)$. There are in general many different ways of factorization, thus many different SDE for the one and the same (in law) Markov family.

What we of course expect is that the RDS φ generated by the SDE (2.3.21) for different factorizations are different – which turns out to be the case. After all, the RDS describes the simultaneous motion of all initial points (and not just of one).

First of all, the RDS φ "remembers" all one-point motions since by the definition of φ, $t \mapsto \varphi(t, \cdot)x =: x_t^x$ is the solution of the SDE with initial value $x_0 = x$, but it also describes the simultaneous motion of n points, the *forward n-point motion*

$$\mathbb{R}^+ \ni t \mapsto \varphi(t, \omega)\mathbf{x}^{(n)} := (\varphi(t, \omega)x_1, \ldots, \varphi(t, \omega)x_n),$$

where $\mathbf{x}^{(n)} := (x_1, \ldots, x_n) \in (\mathbb{R}^d)^n$. The $(\mathbb{R}^d)^n$-valued stochastic process $\varphi(t, \cdot)\mathbf{x}^{(n)}$ is obtained by solving n copies of (2.3.21) with the same Wiener process, but with different initial conditions.

Similarly, the two-sided RDS φ also describes the *backward n-point motions* $\mathbb{R}^- \ni t \mapsto \varphi(t, \omega)\mathbf{x}^{(n)}$.

2.3.43 Theorem (Solution of SDE as Markovian RDS).
Assume the conditions of Theorem 2.3.39 for the SDE (2.3.21), so that it generates a two-sided C^k RDS φ with independent increments. Then φ is Markovian forwards and backwards in time in the following sense: For each $n \in \mathbb{N}$ and $\mathbf{x}^{(n)} \in (\mathbb{R}^d)^n$,

(i) the forward n-point motion $\mathbb{R}^+ \ni t \mapsto \varphi(t, \cdot)\mathbf{x}^{(n)}$ is a homogeneous Feller Markov process with respect to $\mathcal{F}_0^t$ with transition probability

$$P_t^{(n)+}(\mathbf{x}^{(n)}, B) = \mathbb{P}\{\omega : \varphi(t, \omega)\mathbf{x}^{(n)} \in B\}, \quad t \in \mathbb{R}^+,$$

(ii) the backward n-point motion $\mathbb{R}^- \ni t \mapsto \varphi(t, \cdot)\mathbf{x}^{(n)}$ is a homogeneous Feller Markov process with respect to $\mathcal{F}_t^0$ with transition probability

$$P_t^{(n)-}(\mathbf{x}^{(n)}, B) :=$$
$$\mathbb{P}\{\omega : \varphi(-t, \omega)\mathbf{x}^{(n)} \in B\} = \mathbb{P}\{\omega : \varphi(t, \omega)^{-1}\mathbf{x}^{(n)} \in B\}, \quad t \in \mathbb{R}^+.$$

The proof just uses the fact that φ has stationary independent increments, see Baxendale [54] or Kunita [224].

For each fixed $n \in \mathbb{N}$, the semigroup of the forward n-point motions

$$T_t^{(n)+}g(\mathbf{x}^{(n)}) := \mathbb{E}\, g(\varphi(t, \omega)\mathbf{x}^{(n)}), \quad t \in \mathbb{R}^+,$$

uniquely determines the law of the forward n-point motions. The family $(T_t^{(n)+})_{n=1,2,\ldots}$ is consistent in an obvious way and thus uniquely determines the distribution of the RDS $(\varphi(t))_{t \in \mathbb{R}^+}$. But the latter can be accomplished with much less information, which we will now explain.

The generator of $T_t^{(n)+}$,

$$L^{(n)+}g(\mathbf{x}^{(n)}) = \lim_{t \downarrow 0} \frac{1}{t}(T_t^{(n)+} - \mathrm{id})g(\mathbf{x}^{(n)}),$$

is (for $g \in \mathcal{C}^2$ with compact support)

$$L^{(n)+}g(\mathbf{x}^{(n)}) = \sum_{i=1}^d \sum_{p=1}^n b^i(x_p)\frac{\partial g(\mathbf{x}^{(n)})}{\partial x_p^i} + \frac{1}{2}\sum_{i,j=1}^d \sum_{p,q=1}^n a^{ij}(x_p, x_q)\frac{\partial^2 g(\mathbf{x}^{(n)})}{\partial x_p^i \partial x_q^j},$$

where $b(x) = f_0(x)$ and

$$a(x, y) := \lim_{t \downarrow 0} \frac{1}{t}\mathbb{E}(\varphi(t, \omega)x - x)(\varphi(t, \omega)y - y)^*.$$

The $d \times d$ matrix $a(x, y)$ is the *infinitesimal covariance* of the forward two-point motion $(\varphi(t, \omega)x, \varphi(t, \omega)y)$, which for the case of the SDE (2.3.21) is given by

$$a(x, y) = \sum_{j=1}^m f_j(x)f_j(y)^* = \sigma(x)\sigma(y)^*.$$

Now observe that the infinitesimal characteristics $a(x,y)$ and $b(x)$ can be read off from $L^{(2)+}$ (but not from $L = L^{(1)+}$), and if this is done the coefficients of all the $L^{(n)+}$ are determined.

In nice cases (see above) $T_t^{(n)+}$ is uniquely determined by $L^{(n)+}$. The final result is quite surprising: The law of the RDS φ as a stochastic process $\mathbb{R}^+ \ni t \mapsto \varphi(t,\omega) \in \mathrm{Diff}^k(\mathbb{R}^d)$ is uniquely determined by $a(x,y)$ and $b(x)$ and thus by the laws of the forward two-point motions (Baxendale [54]).

2.3.44 Remark. (i) The statement that the law of an RDS with stationary independent increments is uniquely determined by the laws of its forward two-point motions is essentially a consequence of the fact that W is Gaussian. The statement is thus generally wrong in the discrete time case.

(ii) With the above findings, we have also basically solved the *representation problem* for continuous time: Given a generator L of the form (2.3.18) with coefficients $a(x)$ and $b(x)$, is there an RDS φ whose forward one-point motions are homogeneous Markov processes with generator L?

We seek a representation by a Markovian RDS, in particular by one generated by an SDE. Then the law of φ is uniquely determined by the law of its two-point motions, hence by the functions $a(x,y)$ and $b(x)$. The one-point motions, however, only prescribe the diagonal values $a(x,x) = a(x)$.

Suppose we have a factorization $a(x) = \sigma(x)\sigma(x)^*$, where $\sigma(x) = (\sigma_1(x), \ldots, \sigma_m(x))$ is a $d \times m$ matrix with columns $\sigma_j(x)$, which is regular enough. Then
$$a(x,y) := \sigma(x)\sigma(y)^*$$
is an infinitesimal covariance, and L can be "realized" by the SDE (2.3.21) with $f_0(x) = b(x)$ and $f_j(x) = \sigma_j(x)$, $1 \le j \le m$.

The factorization, hence the representation, is in general not unique. ∎

Analogous facts hold for the backward n-point motions: The generator of the semigroup
$$T_t^{(n)-} g(\mathbf{x}^{(n)}) = \mathbb{E}\, g(\varphi(-t,\omega)\mathbf{x}^{(n)}), \quad t \in \mathbb{R}^+,$$
is
$$L^{(n)-} = \sum_{i=1}^d \sum_{p=1}^n b^{-,i}(x_p) \frac{\partial}{\partial x_p^i} + \frac{1}{2} \sum_{i,j=1}^d \sum_{p,q=1}^n a^{ij}(x_p, x_q) \frac{\partial^2}{\partial x_p^i \partial x_q^j},$$
where
$$b^-(x) = -f_0(x) + \sum_{i=1}^d \sum_{j=1}^m f_j^i(x) \frac{\partial f_j(x)}{\partial x^i}.$$

We stress that both $L^{(n)+}$ and $L^{(n)-}$ describe the n-point motions of the same two-sided RDS φ, but $L^{(n)+}$ forwards and $L^{(n)-}$ backwards in time. In general $L^{(n)+} \neq L^{(n)-}$, so the Fokker-Planck equations $(L^{(n)+})^* \nu_n^+ = 0$ and $(L^{(n)-})^* \nu_n^- = 0$ usually have different solutions.

Based on the concepts of past and future of an RDS developed in Sect. 1.7 it is now straightforward to carry over the statements on Markov measures formulated in Subsect. 2.1.3 for products of i.i.d. random mappings to the white noise case.

Assume the situation of Theorem 2.3.43 and denote by φ the C^k RDS with time $\mathbb{T} = \mathbb{R}$ over the canonical DS $(\Omega, \mathcal{F}, \mathbb{P}, (\theta_t)_{t \in \mathbb{R}})$ describing Brownian motion. We choose the past $\mathcal{F}^- = \sigma(\omega(t) : t \leq 0)$ and the future $\mathcal{F}^+ = \sigma(\omega(t) : t \geq 0)$ which are both exhaustive. Also, $\mathcal{F}^-$ and $\mathcal{F}^+$ are independent. Recall that those $\mu \in \mathcal{I}_{\mathbb{P}}(\varphi)$ which are $\mathcal{F}^-/\mathcal{F}^+$-measurable are called forward/backward Markov measures of φ and their sets are denoted by $\mathcal{I}_{\mathbb{P}, \mathcal{F}^\mp}(\varphi)$. The restriction of φ to one-sided time is denoted by $\varphi^\pm$. The justification of this name is given by the next theorem.

2.3.45 Theorem (One-to-One Correspondence Between Markov Measures and Stationary Measures). *Assume the conditions of Theorem 2.3.39 for the SDE (2.3.21), so that it generates a two-sided C^k RDS φ with independent increments. Then*

(i) There is a one-to-one correspondence between the set of invariant forward Markov measures $\mathcal{I}_{\mathbb{P}, \mathcal{F}^-}(\varphi)$, the set $\{\mu^+ \in \mathcal{I}_{\mathbb{P}}(\varphi^+) : \mu^+ = \mathbb{P} \times \rho^+\}$ and the set of stationary measures for L^+, the generator of the forward one-point motions, given by

$$\mu \mapsto \mathbb{E}\,\mu. = \rho^+ \mapsto \mu_\omega = \lim_{t \to \infty} \varphi(t, \theta_{-t}\omega)\rho^+.$$

In particular, $\mu \in \mathcal{I}_{\mathbb{P}, \mathcal{F}^-}(\varphi)$ is ergodic under φ if and only if ρ^+ is an ergodic stationary measure.

(ii) A completely analogous relation holds between the set of invariant backward Markov measures $\mathcal{I}_{\mathbb{P}, \mathcal{F}^+}(\varphi)$, the set $\{\mu^- \in \mathcal{I}_{\mathbb{P}}(\varphi^-) : \mu^- = \mathbb{P} \times \rho^-\}$ and the set of stationary measures for L^-, the generator of the backward one-point motions.

We stress that φ typically has more invariant measures than those which can be "seen" by the two Fokker-Planck equations $(L^+)^*\rho^+ = 0$ and $(L^-)^*\rho^- = 0$.

Partial statements of Theorem 2.3.45 were obtained by Carverhill [89]. The matter was systematically treated by Crauel [110, 112, 113].

If we start with the Stratonovich SDE $dx_t = \sum_{j=0}^{m} f_j(x_t) \circ dW_t^j$ instead of an Itô SDE, the forward/backward generators are more symmetric and have coefficients

$$b^\pm(x) = \pm f_0(x) + \frac{1}{2}\sum_{j=1}^{m} f_j^i(x)\frac{\partial f_j(x)}{\partial x^i}$$

and

$$a(x,y) = \sum_{j=1}^{m} f_j(x)f_j(y)^*.$$

In Hörmander form

$$L^\pm = \pm f_0 + \sum_{j=1}^{m} f_j^2.$$

Part II

Multiplicative Ergodic Theory

Chapter 3. The Multiplicative Ergodic Theorem in Euclidean Space

Summary

This chapter is devoted to the presentation of Oseledets's multiplicative ergodic theorem, the most fundamental theorem of the book. It provides a spectral theory for linear cocycles, and one can say without exaggeration that all what follows are just applications of this one theorem.

We hence have tried hard to give complete and precise formulations and clearly structured proofs of the various versions of the multiplicative ergodic theorem (Theorem 3.4.1 for one-sided time $\mathbb{T} = \mathbb{N}$ and $\mathbb{T} = \mathbb{R}^+$, Theorem 3.4.11 for two-sided time $\mathbb{T} = \mathbb{Z}$ and $\mathbb{T} = \mathbb{R}$). In particular, we carefully describe the invariance properties of the ω sets on which the statements hold.

As the proofs of the various versions of the multiplicative ergodic theorem are based on corresponding versions of the Furstenberg-Kesten theorem (Theorem 3.3.3 for one-sided time $\mathbb{T} = \mathbb{N}$ and $\mathbb{T} = \mathbb{R}^+$, Theorem 3.3.10 for two-sided time $\mathbb{T} = \mathbb{Z}$ and $\mathbb{T} = \mathbb{R}$), the same care has to be taken for the formulations and proofs of the latter.

A certain pedantry of our style is motivated by the aim of making this chapter a source of reference.

3.1 Introduction

It is well-known that the dynamics of the autonomous linear system $\dot{x}_t = Ax_t$ is completely described by linear algebra, more precisely, by the spectral theory of A.

The local theory of smooth nonlinear dynamical systems (such as stability theory, invariant manifold theory, local bifurcation theory, normal form theory) is now based on the following idea: Suppose the nonlinear vector field f has a singularity at x_0, $f(x_0) = 0$. We can regard the nonlinear system $\dot{x}_t = f(x_t)$ as a "small perturbation" of the linearized system $\dot{x}_t = Ax_t$. $A = \frac{\partial f}{\partial x}(x_0)$, in a neighborhood of x_0. We can then try to "lift" or "carry over" the dynamics (in particular, the stability properties of x_0) of the linearized system to the original nonlinear system.

This approach was adopted by Lyapunov in his celebrated classic *Problème général de la stabilité du mouvement* [240] (original Russian edition 1892, French translation 1907) also for nonautonomous systems. It can indeed be successfully carried over from a fixed point to a periodic orbit thanks to Floquet theory (spectral theory of $\dot{x}_t = A(t)x_t$, where $A(\cdot)$ is a continuous periodic matrix function) and then only partly further to quasi- and almost periodic orbits via the spectral theory of Sacker, Sell and Johnson [187, 298, 299, 316]. The key obstacle is the possibility of very complicated dynamical behavior of nonautonomous linear systems $\dot{x}_t = A(t)x_t$.

It might come as a surprise that an important class of nonautonomous linear systems, namely those driven by a metric DS (e. g. those generated by $\dot{x}_t = A(\theta_t \omega)x_t$ or $x_{n+1} = A(\theta^n \omega)x_n$) has a bona fide spectral theory, with probability one. This is the content of the celebrated *multiplicative ergodic theorem*[1] of Oseledets [268]. The MET provides us with exactly the right substitute for deterministic linear algebra, i. e. with the right kind of spectral objects (invariant subspaces, exponential growth rates = Lyapunov exponents) which allow the lift to nonlinear RDS and hence a local theory for nonlinear RDS. The MET thus occupies a pivotal position in this book.

Since the appearance of Oseledets' paper [268] in 1968 there have been numerous attempts to find alternative proofs of the MET[2]. The original proof relies on the triangularization of a linear cocycle and the classical ergodic theorem for the triangular cocycle. This technique was also used in the contemporaneous paper of Millionshchikov [254] who independently derived a portion of the MET, and was then taken up again by Palmer [269], Johnson, Palmer and Sell [192] (assuming a topological setting for the metric DS) and Margulis [251: Appendix A].

Another class of proofs uses the singular value decomposition of matrices in combination with Kingman's [209, 210, 211] *subadditive ergodic theorem* (see Raghunathan [285], Ruelle [292], Crauel [108], Ledrappier [230], Krengel [220], Cohen, Kesten and Newman [103], Goldsheid and Margulis [161], Berger [67]). The subadditive ergodic theorem allows a proof of the *Furstenberg-Kesten theorem* [154] in a few lines, and the Furstenberg-Kesten theorem is then a key ingredient of the proof of the MET.

Now begins the "hard work" (consisting of matrix calculations) in this class of proofs. It concerns the construction of invariant subspaces in which the smaller Lyapunov exponents are realized as limits. This "hard work" can be converted into the direct construction of complementary subspaces of lower growth rate (see Mañé [249]) or to the study of the action of the linear cocycle on projective space (also using the subadditive ergodic theorem) (see Walters [337]).

---

[1] "Multiplicative ergodic theorem" is henceforth abbreviated as "MET".
[2] In August 1997 I counted fifteen published proofs besides Oseledets'.

For extensions of Oseledets' theorem to semisimple Lie groups see Kaimanovich [195] and Zakharevich [350], and to local fields see Ragunathan [285].

For infinite-dimensional versions of the MET see e. g. Ruelle [293], Mañé [248], Thieullen [329], Mohammed [255, 256], Mohammed and Scheutzow [257], Flandoli and Schaumlöffel [150], Schaumlöffel [301], Lindemann [241, 242].

Here we will follow the "established" approach via the subadditive ergodic theorem, the singular value decomposition and exterior powers, and the Furstenberg-Kesten theorem, basically following Goldsheid and Margulis [161] in doing the "hard work".

3.2 Lyapunov Exponents, Singular Values, Exterior Powers

3.2.1 Deterministic Theory of Lyapunov Exponents

For an autonomous linear differential equation $\dot{x}_t = Ax_t$ or difference equation $x_{n+1} = Ax_n$ the origin $0 \in \mathbb{R}^d$ (and thus every point) is asymptotically stable if and only if it is exponentially stable if and only if all eigenvalues of A have negative real parts or absolute values less than 1, respectively. This can be readily seen by looking at the real Jordan canonical form J of A and by observing that if $A = P^{-1}JP$, $P \in Gl(d, \mathbb{R})$, then $\Phi(t) = e^{tA} = P^{-1}e^{tJ}P$ or $\Phi(n) = A^n = P^{-1}J^nP$.

For stability theory of nonautonomous linear systems $\dot{x}_t = A(t)x_t$ where $A(\cdot)$ is locally integrable, or $x_{n+1} = A_n x_n$ it turns out not to be the right thing to study the eigenvalues of $A(t)$ or A_n as they have little or nothing to do with the asymptotic properties of solutions.

We thus need to find a dynamical formulation of spectral theory of A, i. e. a formulation which describes spectral objects in terms of the long-term behavior of solutions.

Lyapunov [240] was aware of this problem when he introduced his *characteristic exponents* which today bear his name. We will give a brief review of his theory, following the profound monograph [85] of Bylov, Vinograd, Grobman and Nemytskii, in which proofs and many more facts on the deterministic theory of Lyapunov exponents for nonautonomous differential equations can be found. Other good sources of examples and counterexamples are Cesari [97: Chap. II] and Hahn [166: Chap. VIII].

As a preparation we quote a simple, but useful lemma.

3.2.1 Lemma (Lyapunov Index of a Function). *Let $\mathbb{T} = \mathbb{R}^+$ or $\mathbb{Z}^+$ or $\mathbb{N}$, $f : \mathbb{T} \to \mathbb{R}^d$, and call*

$$\lambda(f) := \limsup_{t \to \infty} \frac{1}{t} \log \|f(t)\| \in \mathbb{R} \cup \{-\infty, \infty\}$$

the Lyapunov index *of f*. Then
  (i) $\lambda(c) = 0$ if $c = \text{const} \neq 0$ (put $\lambda(0) := -\infty$),
  (ii) $\lambda(\alpha f) = \lambda(f)$ for all $\alpha \in \mathbb{R} \setminus \{0\}$,
  (iii) $\lambda(f+g) \leq \max(\lambda(f), \lambda(g))$ with equality if $\lambda(f) \neq \lambda(g)$.
  (iv) If $\mathbb{T} = \mathbb{R}^+$ and f is locally integrable, then

$$\lambda\left(\int_0^t f(s)\,ds\right) \leq \lambda(f) \quad \text{if} \quad \lambda(f) \geq 0,$$

$$\lambda\left(\int_t^\infty f(s)\,ds\right) \leq \lambda(f) \quad \text{if} \quad \lambda(f) < 0.$$

If f is measurable and locally bounded and g is locally integrable, then

$$\lambda\left(\int_0^t \langle f(s), g(s)\rangle\,ds\right) \leq \lambda(f) + \lambda\left(\int_0^t \|g(s)\|\,ds\right) \quad \text{if} \quad \lambda(f) \geq 0.$$

Similarly for $\mathbb{T} = \mathbb{Z}^+$ or $\mathbb{N}$ with integrals replaced by sums.
  (v)

$$\lambda(\langle f,g\rangle) \leq \lambda(f) + \lambda(g) \quad \text{(if the right-hand side makes sense)},$$

$$\lambda(\|f\|^\alpha) = \alpha \lambda(f) \quad for \quad \alpha \in \mathbb{R} \quad (put\ 0(\pm\infty) = 0).$$

The proof follows from the definition of $\lambda(f)$ and is left as an exercise.
  We now apply this to the function $f(t) = \Phi(t)x$.

3.2.2 Definition (Lyapunov Exponent). The *(forward) Lyapunov exponent* of the solution $\Phi(t)x$ of a non-autonomous linear differential equation $\dot{x}_t = A(t)x_t$ or difference equation $x_{n+1} = A_n x_n$ starting at time $t = 0$ at the state $x \in \mathbb{R}^d$ is defined to be the Lyapunov index of $\Phi(t)x$,

$$\lambda^+(x) = \lambda(x) := \limsup_{t \to \infty} \frac{1}{t} \log \|\Phi(t)x\|,$$

and for two-sided time the *backward Lyapunov exponent* of $\Phi(t)x$ is defined as the Lyapunov index of $\Phi(-t)x$,

$$\lambda^-(x) := \limsup_{t \to \infty} \frac{1}{t} \log \|\Phi(-t)x\| = \limsup_{t \to -\infty} \frac{1}{|t|} \log \|\Phi(t)x\|.$$

∎

Suppose $A(t)$ or (A_n) is bounded. Then $\lambda(\cdot)$ (and also $\lambda^-(\cdot)$) satisfies

1. $\lambda(x) \in \mathbb{R} \cup \{-\infty\}$ for all $x \in \mathbb{R}^d$, and $\lambda(0) = -\infty$,
2. $\lambda(\alpha x) = \lambda(x)$ for all $\alpha \in \mathbb{R} \setminus \{0\}$,
3. $\lambda(x+y) \leq \max(\lambda(x), \lambda(y))$ with equality if $\lambda(x) \neq \lambda(y)$.

A mapping $\lambda : \mathbb{R}^d \to \mathbb{R} \cup \{-\infty\}$ with these three properties is called a *characteristic exponent* (see e. g. Pesin [275]).

Further, the number p of distinct values that $\lambda(x)$ can take on for $x \neq 0$ is at most d. This can be seen by noting that vectors with different Lyapunov exponents are linearly independent. Write

$$-\infty \leq \lambda_p < \lambda_{p-1} < \ldots < \lambda_1 < \infty$$

for the different values of $\lambda(x)$ and call them the *Lyapunov exponents of the system*. Call λ_1 the *top Lyapunov exponent*.

The sets $V_\lambda := \{x : \lambda(x) \leq \lambda\}$ are linear subspaces of $\mathbb{R}^d$, $V_i := V_{\lambda_i}$ form a *filtration* (flag of subspaces)

$$\{0\} =: V_{p+1} \subset V_p \subset \ldots \subset V_1 = \mathbb{R}^d$$

(where all inclusions are proper), and

$$\lambda(x) = \lambda_i \iff x \in V_i \setminus V_{i+1}, \quad i = 1, \ldots, p.$$

The integer $d_i := \dim V_i - \dim V_{i+1}$ is the *multiplicity* of λ_i and the set $(\lambda_i, d_i)_{i=1,\ldots,p}$ is called the *(forward) Lyapunov spectrum*.

If $\lambda_1 < 0$, then $\dot{x}_t = A(t)x_t$ is exponentially stable. It is, however, in general not true that $\lambda_1 < 0$ implies the stability of the origin of the perturbed nonlinear system

$$\dot{x}_t = A(t)x_t + g(t, x_t), \quad |g(t,x)| \leq C|x|^{1+\alpha} \text{ for } |x| \leq h, \, C, \alpha > 0. \quad (3.2.1)$$

For this to hold true one needs a property of the linear system called *regularity*. The equation $\dot{x}_t = A(t)x_t$ or $x_{n+1} = A_n x_n$ is said to be *forward regular* if

$$\sum_{i=1}^p d_i \lambda_i = \liminf_{t \to \infty} \frac{1}{t} \log |\det \Phi(t)|.$$

For a forward regular system all lim sup's are in fact limits, and $\lambda_1 < 0$ implies exponential stability of (3.2.1).

For two-sided time, analogous facts hold for $\lambda^-(x)$: It has p^- different values

$$-\infty \leq \lambda_{p^-}^- < \ldots < \lambda_1^- < \infty$$

with multiplicities d_i^-, the filtration

$$\{0\} =: V_{p^-+1}^- \subset \ldots \subset V_1^- = \mathbb{R}^d,$$

and

$$\lambda^-(x) = \lambda_i^- \iff x \in V_i^- \setminus V_{i+1}^-.$$

The system $\dot{x}_t = A(t)x_t$ or $x_{n+1} = A_n x_n$ is called *backward regular* if

$$\sum_{i=1}^{p^-} d_i^- \lambda_i^- = \liminf_{t \to \infty} \frac{1}{t} \log |\det \Phi(-t)|.$$

In general the forward and backward spectrum and filtration are not related. We introduce such a relation as follows: The two-sided linear system is called *regular* if

1. it is forward regular and backward regular,
2. $p = p^-$, $d_i = d^-_{p+1-i}$, $\lambda_i = -\lambda^-_{p+1-i}$, $i = 1, \ldots, p$,
3. $V_{i+1} \cap V^-_{p+1-i} = \{0\}$, $i = 1, \ldots, p - 1$.

(2) and (3) imply that $\dim V_{i+1} + \dim V^-_{p+1-i} = d$ for $i = 1, \ldots, p-1$, and it makes sense for a regular system to define the subspaces

$$E_i := V_i \cap V^-_{p+1-i}, \quad i = 1, \ldots, p,$$

where $\dim E_i = d_i$, and which form a *splitting* of $\mathbb{R}^d$,

$$\mathbb{R}^d = E_1 \oplus \ldots \oplus E_p .$$

The filtrations can be recovered as $V_i = \oplus_{j=i}^p E_j$ and $V^-_{p+1-i} = \oplus_{j=1}^i E_j$. The splitting is *dynamically characterized* by

$$\lim_{t \to \pm\infty} \frac{1}{t} \log \|\Phi(t)x\| = \lambda_i \iff x \in E_i \setminus \{0\}. \quad (3.2.2)$$

3.2.3 Example (Autonomous Case). (i) Two-sided continuous time: Consider $\dot{x}_t = Ax_t$, $A \in \mathbb{R}^{d \times d}$, $t \in \mathbb{R}$. It follows immediately from the Jordan canonical form for A and $\Phi(t) = e^{tA}$ that $\dot{x}_t = Ax_t$ is regular with $-\infty < \lambda_p < \ldots < \lambda_1$ the different real parts of eigenvalues of A, E_i the sum of the generalized eigenspaces to eigenvalues with real parts equal to λ_i. We have $\Phi(t) \circ \pi_i = \pi_i \circ \Phi(t)$ where π_i is the projection onto E_i along $\oplus_{j \neq i} E_j$, equivalently $\Phi(t)E_i = E_i$ for all $t \in \mathbb{R}$, $i = 1, \ldots, p$. Also $\sum_1^p d_i \lambda_i = \text{trace} A$.

(ii) Two-sided discrete time: Consider $x_{n+1} = Ax_n$, $A \in Gl(d, \mathbb{R})$, $n \in \mathbb{Z}$. Then $\Phi(n) = A^n$, $n \in \mathbb{Z}$, and the system is regular with λ_i equal to the different values of $\log|\mu|$, μ an eigenvalue of A, and E_i the corresponding sum of generalized eigenspaces. Again $A \circ \pi_i = \pi_i \circ A$, equivalently $AE_i = E_i$, and $\sum_1^p d_i \lambda_i = \log|\det A|$.

(iii) One-sided discrete time: Consider $x_{n+1} = Ax_n$, $A \in \mathbb{R}^{d \times d}$, $n \in \mathbb{Z}^+$. The system is forward regular with λ_i as in (ii), where $\lambda_p = -\infty$ if $\mu = 0$ is an eigenvalue of A, and filtration

$$V_p \subset \ldots \subset V_1 = \mathbb{R}^d, \quad AV_i \subset V_i,$$

where V_i is the sum of the generalized eigenspaces for eigenvalues μ of A with $\log|\mu| \leq \lambda_i$. Note that the knowledge of $\lambda(x)$ only allows us to reconstruct the filtration $V_i = \{x : \lambda(x) \leq \lambda_i\}$, and not the eigenspace E_i corresponding to λ_i, as E_i is not dynamically characterizable for one-sided time (see (3.2.2)). ∎

3.2.4 Example (Periodic Case).
Let $A : \mathbb{R} \to \mathbb{R}^{d \times d}$ be continuous and periodic with period $\tau > 0$. Then $\dot{x}_t = A(t)x_t$ is regular. Indeed (see e.g. Coddington and Levinson [102: Sect. 3.5] or Farkas [148: Sect. 2.2]), write $\Phi(2\tau) = e^{2\tau R}$ with $R \in \mathbb{R}^{d \times d}$. The eigenvalues of R are called *Floquet characteristic exponents* and the fundamental matrix is represented by $\Phi(t) = P(t)e^{tR}$, where $P : \mathbb{R} \to Gl(d, \mathbb{R})$ is C^1 and 2τ-periodic, and P, P^{-1} and $\dot{P}$ are bounded on $\mathbb{R}$. The new moving coordinates $x = P(t)y$ transform $\dot{x}_t = A(t)x_t$ to $\dot{y}_t = Ry_t$. Now the Lyapunov exponents λ_i are the different values of $\frac{1}{2\tau}\log|\mu|$, where μ is an eigenvalue of $\Phi(2\tau)$, i.e. they are the different real parts of the Floquet characteristic exponents – which are uniquely determined by $A(\cdot)$. The E_i are the sums of the generalized eigenspaces of R corresponding to λ_i. Note that

$$\sum_{i=1}^{p} d_i \lambda_i = \frac{1}{\tau}\int_0^{\tau} \operatorname{trace} A(t)\, dt.$$

The invariance property of the splitting now reads $\Phi(t) \circ \pi_i(0) = \pi_i(t) \circ \Phi(t)$ where $\pi_i(t)$ is the projection onto $E_i(t) := P(t)E_i$ along $F_i(t) := \oplus_{j \neq i} E_j(t)$, so $\Phi(t)E_i(0) = E_i(t)$. This foreshadows the more general invariance property of the E_i's in the random case. For the embedding into an RDS see Example 4.1.12. ∎

Although it is hard if not impossible to verify regularity for a particular system, the surprise (and the key statement of the MET) is that linear RDS (including constant, periodic, quasi-periodic and almost-periodic $A(\cdot)$) are regular with probability one.

3.2.2 Singular Values

Let $\mathbb{R}^d$ be endowed with the standard scalar product, let (e_i) be the standard basis and let $O(d, \mathbb{R}) := \{U \in Gl(d, \mathbb{R}) : U^*U = I\}$ be the orthogonal group. We say for $A \in \mathbb{R}^{d \times d}$ that

$$A = VDU$$

is a *singular value decomposition* of A if $U, V \in O(d, \mathbb{R})$ and $D = \operatorname{diag}(\delta_1, \ldots, \delta_d)$ with $\delta_1 \geq \delta_2 \geq \ldots \geq \delta_d \geq 0$. The δ_i are called the *singular values* of A.

3.2.5 Lemma (Singular Value Decomposition).
Any $d \times d$ matrix A has a singular value decomposition. Moreover, the $\delta_1 \geq \ldots \geq \delta_d \geq 0$ are necessarily the eigenvalues of $\sqrt{A^*A}$ (also of $\sqrt{AA^*}$), and the columns of U^* are corresponding eigenvectors of $\sqrt{A^*A}$. In particular, $\|A\| = \delta_1$, where $\|\cdot\|$ is the operator norm associated with the standard Euclidean norm in $\mathbb{R}^d$, and $|\det A| = \delta_1 \ldots \delta_d$.

Proof. Any A may be written as $A = W\sqrt{A^*A}$ with $W \in O(d, \mathbb{R})$ (polar decomposition, see Gantmacher [157: IX.14]). Now $\sqrt{A^*A} \geq 0$, so it may be written as $\sqrt{A^*A} = U^{-1}DU$ with $D = \text{diag}(\delta_1, \ldots, \delta_d)$, where the δ_i are the eigenvalues of $\sqrt{A^*A}$. Together $A = WU^{-1}DU = VDU$ with $U, V \in O(d, \mathbb{R})$. Conversely, if $A = VDU$, then $A^*A = U^*D^2U$, thus $\sqrt{A^*A}U^* = U^*D$, and the δ_i, U^*e_i are the eigenvalues and eigenvectors of $\sqrt{A^*A}$, respectively. □

The geometric meaning of the singular value decomposition is as follows: δ_i is the length and U^*e_i the direction of the i'th principal axis of the ellipsoid $A(S^{d-1})$ obtained as the image of the unit sphere $S^{d-1} := \{x \in \mathbb{R}^d : \langle x, x \rangle = 1\}$ under the linear mapping A.

3.2.3 Exterior Powers

Let E be a real vector space of dimension d and for $1 \leq k \leq d$, let $\wedge^k E$, the *k-fold exterior power of E*, be the vector space of alternating k-linear forms on the dual space E^* (see e.g. Temam [328: Chap. V] or Kowalsky [219: §45]). Naturally $\wedge^1 E = E^{**} \cong E$, $\wedge^d E \cong \mathbb{R}$. The space $\wedge^k E$ can be identified with the set of formal expressions $\sum_{i=1}^m c_i(u_1^{(i)} \wedge \ldots \wedge u_k^{(i)})$ with $m \in \mathbb{N}$, $c_i \in \mathbb{R}$ and $u_j^{(i)} \in E$ if we do computations with the following conventions:

1. $u_1 \wedge \ldots \wedge (u_j + u_j') \wedge \ldots \wedge u_k =$
   $(u_1 \wedge \ldots \wedge u_j \wedge \ldots \wedge u_k) + (u_1 \wedge \ldots \wedge u_j' \wedge \ldots \wedge u_k)$,
2. $u_1 \wedge \ldots \wedge cu_j \wedge \ldots \wedge u_k = c(u_1 \wedge \ldots \wedge u_j \wedge \ldots \wedge u_k)$,
3. for any permutation π of $\{1, \ldots, k\}$
$$u_{\pi(1)} \wedge \ldots \wedge u_{\pi(k)} = \text{sign}(\pi)\, u_1 \wedge \ldots \wedge u_k.$$

The elements in $\wedge^k E$ of the form $u_1 \wedge \ldots \wedge u_k$ are called *decomposable k-vectors* and the set of decomposable k-vectors is denoted by $\wedge_0^k E$. Clearly $\wedge^k E = \text{span}(\wedge_0^k E)$.

The following facts can be readily checked.

3.2.6 Lemma (Exterior Powers of Spaces and Operators). *Let $\wedge^k E$ be the k-fold exterior power of E, where $1 \leq k \leq d = \dim E$.*

(i) If $e_1, \ldots, e_d$ is a basis of E, then $\{e_{i_1} \wedge \ldots \wedge e_{i_k} : 1 \leq i_1 < \ldots < i_k \leq d\}$ is a basis of $\wedge^k E$. In particular,
$$\dim \wedge^k E = \binom{d}{k}.$$

(ii) We have $u_1 \wedge \ldots \wedge u_k = 0$ if and only if $u_1, \ldots, u_k$ are linearly dependent.

(iii) Two sets $u_1, \ldots, u_k$ and $v_1, \ldots, v_k$ of k linearly independent vectors in E have the same linear span if and only if
$$u_1 \wedge \ldots \wedge u_k = \lambda(v_1 \wedge \ldots \wedge v_k) \quad \text{for some } \lambda \in \mathbb{R}.$$

Hence the image of the set $\wedge_0^k E$ of decomposable k-vectors in the projective space $P(\wedge^k E)$ can be identified with the Grassmannian $G_k(E)$ of k-dimensional subspaces of E.

(iv) If E has the scalar product $\langle \cdot, \cdot \rangle$, then the bilinear extension of

$$\langle u_1 \wedge \cdots \wedge u_k, v_1 \wedge \cdots \wedge v_k \rangle := \det(\langle u_i, v_j \rangle)_{k \times k}$$

defines a scalar product in $\wedge^k E$. In particular,

$$\|u_1 \wedge \cdots \wedge u_k\| = \sqrt{\det(\langle u_i, u_j \rangle)_{k \times k}}$$

is the volume of the k-dimensional parallelepiped spanned by $u_1, \ldots, u_k$. Further, the generalized Hadamard's inequality holds: For $1 \leq p < k$

$$\|u_1 \wedge \cdots \wedge u_k\| \leq \|u_1 \wedge \cdots \wedge u_p\| \|u_{p+1} \wedge \cdots \wedge u_k\|.$$

(v) If $A : E \to E$ is a linear operator then the linear extension of

$$\wedge^k A(u_1 \wedge \cdots \wedge u_k) := Au_1 \wedge \cdots \wedge Au_k$$

defines a linear operator $\wedge^k A$ on $\wedge^k E$, the k-fold exterior power of A, with $\wedge^1 A = A$, $\wedge^d A = \det A$ and $\wedge^k I = I$. $\wedge^k A$ has eigenvalues $\{\lambda_{i_1} \cdots \lambda_{i_k} : 1 \leq i_1 < \cdots < i_k \leq d\}$, where $\lambda_1, \ldots, \lambda_d$ are the eigenvalues of A. Further,

$$\wedge^k(AB) = (\wedge^k A)(\wedge^k B), \quad (\wedge^k A)^{-1} = \wedge^k A^{-1}, \quad \wedge^k(cA) = c^k \wedge^k A,$$

and

$$\det \wedge^k A = (\det A)^{\binom{d-1}{k-1}}.$$

If A is an upper/lower triangular matrix in some basis then so is $\wedge^k A$ in the corresponding basis of (i).

(vi) Let E have a scalar product and let $\wedge^k E$ be endowed with the scalar product introduced in part (iv). If U is orthogonal, then $\wedge^k U$ is orthogonal. Further,

$$\langle \wedge^k A(u_1 \wedge \ldots \wedge u_k), v_1 \wedge \ldots \wedge v_k \rangle = \det(\langle Au_i, v_j \rangle)_{k \times k},$$

hence, with $\| \wedge^k A \|$ denoting the corresponding operator norm,

$$\| \wedge^k (AB)\| \leq \| \wedge^k A\| \| \wedge^k B\|,$$

$$(\wedge^k A)^* = \wedge^k A^*.$$

(vii) Suppose A is a linear operator on E. Then

$$\wedge^k(e^{tA}) = e^{t\hat{A}^k},$$

where the linear operator $\hat{A}^k : \wedge^k E \to \wedge^k E$ is defined by the linear extension of

$$\hat{A}^k(u_1 \wedge \ldots \wedge u_k) := \sum_{i=1}^{k} u_1 \wedge \ldots \wedge u_{i-1} \wedge Au_i \wedge u_{i+1} \wedge \ldots \wedge u_k,$$

and satisfies $\widehat{(\alpha A + \beta B)}^k = \alpha \hat{A}^k + \beta \hat{B}^k$, $\alpha, \beta \in \mathbb{R}$, $\hat{A}^1 = A$, $\hat{A}^d = \operatorname{trace} A$, and $\hat{I}^k u = ku$. $\hat{A}^k$ has eigenvalues $\{\lambda_{i_1} + \ldots + \lambda_{i_k} : 1 \leq i_1 < \ldots < i_k \leq d\}$ if $\lambda_1, \ldots, \lambda_d$ are the eigenvalues of A. In particular,

$$\operatorname{trace} \hat{A}^k = \binom{d-1}{k-1} \operatorname{trace} A.$$

If E has a scalar product then $(\hat{A}^k)^* = \widehat{(A^*)}^k$, and

$$\langle \hat{A}^k(u_1 \wedge \ldots \wedge u_k), u_1 \wedge \ldots \wedge u_k \rangle = \operatorname{trace}(AP_k) \|u_1 \wedge \ldots \wedge u_k\|^2,$$

where P_k is the orthogonal projector onto the subspace $\operatorname{span}(u_1, \ldots, u_k)$ of E. In particular, $\|\hat{A}^k\| \leq k\|A\|$.

The importance of exterior powers in the study of linear systems stems mainly from the following two propositions.

3.2.7 Proposition (Singular Values of Exterior Power). Let A be a $d \times d$ matrix, let $A = VDU$ be a singular value decomposition and let $\delta_1 \geq \ldots \geq \delta_d \geq 0$ be the singular values of A. Then
 (i) $\wedge^k A = (\wedge^k V)(\wedge^k D)(\wedge^k U)$ is a singular value decomposition of $\wedge^k A$.
 (ii) $\wedge^k D = \operatorname{diag}(\delta_{i_1} \cdots \delta_{i_k} : 1 \leq i_1 < \ldots < i_k \leq d)$. In particular, the top singular value of $\wedge^k A$ is $\delta_1 \cdots \delta_k$, and the smallest is $\delta_{d-k+1} \cdots \delta_d$. Moreover, if A is positive-definite (semi-definite) then so is $\wedge^k A$.
 (iii) $\|\wedge^k A\| = \delta_1 \cdots \delta_k$, in particular $\|\wedge^d A\| = \delta_1 \cdots \delta_d = |\det A|$, and $\|\wedge^{k+m} A\| \leq \|\wedge^k A\| \|\wedge^m A\|$, $1 \leq k, m, k+m \leq d$, in particular $\|\wedge^k A\| \leq \|A\|^k$. Here $\|\cdot\|$ is the corresponding operator norm associated with the standard Euclidean norm in $\mathbb{R}^d$.

For later use we also note

3.2.8 Proposition (Metric on Projective Space). Let P^{d-1} be the projective space of one-dimensional subspaces of $\mathbb{R}^d$, and denote by $\bar{x}$ the class of $x \in \mathbb{R}^d \setminus \{0\}$ in P^{d-1}. Then
 (i)
$$\delta(\bar{x}, \bar{y}) := \left(1 - \left(\frac{\langle x, y \rangle}{\|x\| \|y\|}\right)^2\right)^{1/2}, \quad x, y \in \mathbb{R}^d \setminus \{0\},$$

is a metric in P^{d-1} which makes it a complete metric space.
 (ii) We have
$$\delta(\bar{x}, \bar{y}) = \frac{\|x \wedge y\|}{\|x\| \|y\|}.$$

Proof. (i) To prove the triangle inequality, note that
$$\delta(\bar{U}\cdot\bar{x},\bar{U}\cdot\bar{y}) = \delta(\bar{x},\bar{y})$$
for any $U \in O(d,\mathbb{R})$, where we have put $\bar{U}\cdot\bar{x} := \overline{Ux}$. It hence suffices to prove
$$\delta(\bar{x},\bar{y}) \leq \delta(\bar{x},\bar{z}) + \delta(\bar{z},\bar{y})$$
for unit vectors of the form $z = e_1$, $x = x_1 e_1 + x_2 e_2$, $y = y_1 e_1 + y_2 e_2 + y_3 e_3$, which is left as an exercise.

If α denotes the angle between the lines through x and y, $\langle x, y \rangle = \cos\alpha$, then $\delta(\bar{x},\bar{y}) = |\sin\alpha|$ (see Fig. 3.1). Thus δ has the same set of Cauchy sequences as the natural Riemannian metric (arc length) on P^{d-1} which is complete.

(ii) By definition
$$\|x \wedge y\|^2 = \det\begin{pmatrix} \langle x,x\rangle & \langle x,y\rangle \\ \langle y,x\rangle & \langle y,y\rangle \end{pmatrix} = \|x\|^2 \|y\|^2 - \langle x,y\rangle^2.$$

□

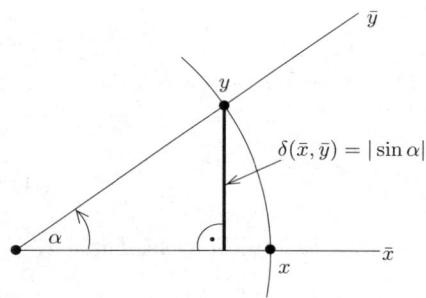

Fig. 3.1. The metric δ on projective space

3.3 The Furstenberg-Kesten Theorem

As a final preparation for the MET we prove a theorem of Furstenberg and Kesten [154] which now bears their names. As the MET (of which we present several versions) is directly based on the Furstenberg-Kesten theorem we also need several versions of the latter.

3.3.1 The Subadditive Ergodic Theorem

We first introduce a generalization of subadditivity to stochastic processes.

3.3.1 Definition (Subadditive Stochastic Process). Let $\mathbb{T} = \mathbb{N}$ or $\mathbb{R}^+$. A measurable stochastic process $(f_t(\omega))_{t \in \mathbb{T}}$ with values in $\mathbb{R} \cup \{-\infty\}$ is called (perfectly) *subadditive* over the metric DS $(\Omega, \mathcal{F}, \mathbb{P}, (\theta(t))_{t \in \mathbb{T}})$ if

$$f_{s+t}(\omega) \le f_s(\omega) + f_t(\theta(s)\omega)$$

holds identically. We call f superadditive if $-f$ is subadditive. ∎

As an example, let $f : \Omega \to \mathbb{R}$ be measurable and let for $\mathbb{T} = \mathbb{R}^+$, $t \mapsto f(\theta(\cdot)\omega)$ be locally integrable. Then $S_t(\omega) := \sum_{k=0}^{t-1} f(\theta(k)\omega)$ (for $\mathbb{T} = \mathbb{N}$) and $S_t(\omega) := \int_0^t f(\theta(s)\omega)\,ds$ (for $\mathbb{T} = \mathbb{R}^+$) are sub- as well as superadditive.

3.3.2 Theorem (Kingman's Subadditive Ergodic Theorem).
Let (f_n) be a subadditive sequence of random variables over the metric DS $(\Omega, \mathcal{F}, \mathbb{P}, (\theta^n)_{n \in \mathbb{N}})$.
Assume $f_1^+ \in L^1(\Omega, \mathcal{F}, \mathbb{P})$, where $a^+ := \max(0, a)$, and put

$$\gamma := \inf_{n \in \mathbb{N}} \frac{1}{n} \mathbb{E} f_n.$$

Then:
(a) There is a forward invariant set $\tilde{\Omega} \in \mathcal{F}$ of full measure and a measurable function $\tilde{f} : \Omega \to \mathbb{R} \cup \{-\infty\}$ with $\tilde{f}^+ \in L^1$ such that on $\tilde{\Omega}$

$$\lim_{n \to \infty} \frac{1}{n} f_n(\omega) = \tilde{f}(\omega), \qquad (3.3.1)$$

and $\tilde{f}(\theta(k)\omega) = \tilde{f}(\omega)$ for all $k \in \mathbb{N}$, $\tilde{f}(\omega) = \gamma$ in the ergodic case. Further,

$$\lim_{n \to \infty} \frac{1}{n} \mathbb{E} f_n = \gamma = \mathbb{E}\tilde{f} \in \mathbb{R} \cup \{-\infty\}.$$

(b) If, in addition, $f_n \in L^1$ for all n and if $\gamma > -\infty$, then assertion (a) holds with $\tilde{f} \in L^1$, and convergence in (3.3.1) also holds in L^1.

We decided to omit the proof as it is easily available in the literature, see e. g. Krengel [220] or Steele [324].

3.3.2 The Furstenberg-Kesten Theorem for One-Sided Time

Let $\Phi(n, \omega)$ be a linear RDS with time $T = \mathbb{N}$ (or $\mathbb{Z}^+$) over the metric DS $(\Omega, \mathcal{F}, \mathbb{P}, (\theta^n)_{n \in \mathbb{N}})$. Then

$$\Phi(n, \omega) = A(\theta^{n-1}\omega) \cdots A(\omega), \quad n \ge 1, \qquad (3.3.2)$$

where $A(\omega) := \Phi(1, \omega) \in \mathbb{R}^{d \times d}$ is the generator of Φ.

We want to study the asymptotic behavior of $\Phi(n,\omega)$, in particular of its singular values $\delta_k(\Phi(n,\omega))$, $k=1,\ldots,d$, for $n\to\infty$.

The cocycle property

$$\Phi(n+m,\omega) = \Phi(m,\theta^n\omega)\Phi(n,\omega)$$

lifts to $\wedge^k \mathbb{R}^d$, $1 \le k \le d$ (see Lemma 3.2.6, part (v)):

$$\wedge^k \Phi(n+m,\omega) = (\wedge^k \Phi)(m,\theta^n\omega)(\wedge^k \Phi)(n,\omega). \tag{3.3.3}$$

3.3.3 Theorem (Furstenberg-Kesten Theorem, One-Sided Time).
Let Φ be a linear cocycle with one-sided time $\mathbb{T} = \mathbb{N}$ or $\mathbb{R}^+$ over the metric DS $(\Omega, \mathcal{F}, \mathbb{P}, (\theta(t))_{t \in \mathbb{T}})$. Then the following statements hold:

(A) Non-invertible case $\mathbb{T} = \mathbb{N}$: If the generator $A : \Omega \to \mathbb{R}^{d\times d}$ satisfies

$$\log^+ \|A(\cdot)\| \in L^1(\Omega, \mathcal{F}, \mathbb{P}),$$

then
(i) For each $k = 1,\ldots, d$ the sequence

$$f_n^{(k)}(\omega) := \log \|\wedge^k \Phi(n,\omega)\|, \quad n \in \mathbb{N},$$

is subadditive and $f_1^{(k)+} \in L^1$.

(ii) There is a forward invariant set $\tilde{\Omega} \in \mathcal{F}$ of full measure and measurable functions $\gamma^{(k)} : \Omega \to \mathbb{R} \cup \{-\infty\}$ with $\gamma^{(k)+} \in L^1$ such that on $\tilde{\Omega}$

$$\lim_{n\to\infty} \frac{1}{n} \log \|\wedge^k \Phi(n,\omega)\| = \gamma^{(k)}(\omega)$$

and

$$\gamma^{(k)}(\theta\omega) = \gamma^{(k)}(\omega), \quad \gamma^{(k+m)}(\omega) \le \gamma^{(k)}(\omega) + \gamma^{(m)}(\omega),$$

$\gamma^{(k)}(\omega) = \mathbb{E}\,\gamma^{(k)}$ in the ergodic case. Further,

$$\lim_{n\to\infty} \frac{1}{n}\mathbb{E} \log \|\wedge^k \Phi(n,\cdot)\| = \mathbb{E}\,\gamma^{(k)} = \inf_{n\in\mathbb{N}} \frac{1}{n}\mathbb{E} \log \|\wedge^k \Phi(n,\cdot)\|.$$

(iii) The measurable functions Λ_k successively defined by

$$\Lambda_1(\omega) + \ldots + \Lambda_k(\omega) := \gamma^{(k)}(\omega),\ k = 1,\ldots, d,$$

(where we put $\Lambda_k(\omega) = -\infty$ if $\gamma^{(k)}(\omega) = -\infty$) have the following properties on $\tilde{\Omega}$:

$$\Lambda_k(\omega) = \lim_{n\to\infty} \frac{1}{n}\log \delta_k(\Phi(n,\omega)),$$

where $\delta_k(\Phi(n,\omega))$ are the singular values of $\Phi(n,\omega)$, and

$$\Lambda_k(\theta\omega) = \Lambda_k(\omega), \quad \Lambda_1(\omega) \ge \Lambda_2(\omega) \ge \ldots \ge \Lambda_d(\omega),$$

$\Lambda_k(\omega) = \mathbb{E}\,\Lambda_k$ in the ergodic case.
Further,

$$\lim_{n\to\infty} \frac{1}{n} \mathbb{E} \log \delta_k(\Phi(n,\cdot)) = \mathbb{E}\, \Lambda_k.$$

(B) *Invertible case* $\mathbb{T} = \mathbb{N}$: If $A : \Omega \to Gl(d,\mathbb{R})$ and

$$\log^+ \|A\| \in L^1(\Omega, \mathcal{F}, \mathbb{P}) \quad \text{and} \quad \log^+ \|A^{-1}\| \in L^1(\Omega, \mathcal{F}, \mathbb{P}), \quad (3.3.4)$$

then there exists an invariant set $\tilde{\Omega} \in \mathcal{F}$ of full measure on which the ω-wise statements of part (A) hold and all $\gamma^{(k)}$ (hence all Λ_k) are finite.

Moreover, all $\mathbb{E}\,\gamma^{(k)}$ (hence all $\mathbb{E}\,\Lambda_k$) are finite, and convergence also holds in L^1.

(C) *Invertible case* $\mathbb{T} = \mathbb{R}^+$: Let $\Phi(t,\omega) \in Gl(d,\mathbb{R})$, define

$$\alpha^+(\omega) := \sup_{0\le t\le 1} \log^+ \|\Phi(t,\omega)\|, \quad \alpha^-(\omega) := \sup_{0\le t\le 1} \log^+ \|\Phi(t,\omega)^{-1}\|$$

and assume that $\alpha^+ \in L^1$ and $\alpha^- \in L^1$. Then all statements of part (B) hold with n and θ replaced with t and $\theta(t)$, and $\tilde{\Omega} \in \mathcal{F}$ is now invariant with respect to $(\theta(t))_{t\in\mathbb{R}^+}$.

Proof. Part (A):

(i) The cocycle property (3.3.3) for $\wedge^k \Phi(n,\omega)$ and Lemma 3.2.6(vi) imply that $f_n^{(k)}(\omega) := \log \| \wedge^k \Phi(n,\omega)\|$ is subadditive. Furthermore, $f_1^{(1)+} \in L^1$ by assumption. This fact and $\| \wedge^k A\| \le \|A\|^k$ (Proposition 3.2.7(iii)) yield $f_1^{(k)+} \in L^1$.

(ii) This is an immediate consequence of Theorem 3.3.2. The forward invariant set $\tilde{\Omega}$ can be taken to be the intersection of the d forward invariant sets $\tilde{\Omega}_k$ for $k = 1,\ldots,d$. The property $\| \wedge^{k+m} \Phi\| \le \| \wedge^k \Phi\| \|\wedge^m \Phi\|$ (Proposition 3.2.7(iii)) implies $\gamma^{(k+m)}(\omega) \le \gamma^{(k)}(\omega) + \gamma^{(m)}(\omega)$. In particular, if $\gamma^{(k)}(\omega) = -\infty$ for some k, then $\gamma^{(m)}(\omega) = -\infty$ for all $m \ge k$.

(iii) By Proposition 3.2.7(iii) for $k = 1,\ldots,d$

$$\frac{1}{n} \log \| \wedge^k \Phi(n,\omega)\| = \frac{1}{n} \sum_{i=1}^k \log \delta_i(\Phi(n,\omega)).$$

We proceed successively as follows: Choose $\omega \in \tilde{\Omega}$. For $k = 1$

$$\gamma^{(1)} = \lim_{n\to\infty} \frac{1}{n} \log \delta_1(\Phi(n,\omega)) =: \Lambda_1(\omega)$$

exists. If $\gamma^{(1)}(\omega) = -\infty$, we are finished. For $\gamma^{(1)}(\omega) > -\infty$

$$\gamma^{(2)}(\omega) - \gamma^{(1)}(\omega) = \lim_{n\to\infty} \frac{1}{n} \log \delta_2(\Phi(n,\omega)) =: \Lambda_2(\omega)$$

exists, and so on. If finally $\gamma^{(d-1)}(\omega) > -\infty$,

$$\gamma^{(d)}(\omega) - \gamma^{(d-1)}(\omega) = \lim_{n\to\infty} \frac{1}{n} \log \delta_d(\Phi(n,\omega)) =: \Lambda_d(\omega)$$

exists. By this construction, $\gamma^{(k)}(\omega) = -\infty$ if and only if $\Lambda_k(\omega) = -\infty$. The inequalities $\Lambda_1(\omega) \geq \ldots \geq \Lambda_d(\omega)$ on $\tilde{\Omega}$ follow from $\delta_1 \geq \ldots \geq \delta_d$.

For the expectations, the limit

$$\mathbb{E}\,\gamma^{(1)} = \lim_{n\to\infty} \frac{1}{n}\mathbb{E}\log\delta_1(\Phi(n,\cdot)) = \mathbb{E}\,\Lambda_1$$

exists. If $\mathbb{E}\,\gamma^{(1)} = -\infty$ we are done as all other expectations $\mathbb{E}\,\gamma^{(k)}$ are also equal to $-\infty$. If not, the limit

$$\mathbb{E}\,\gamma^{(2)} - \mathbb{E}\,\gamma^{(1)} = \lim_{n\to\infty} \frac{1}{n}\mathbb{E}\log\delta_2(\Phi(n,\cdot)) = \mathbb{E}\,\Lambda_2$$

exists, and so on.

Part (B):

(a) We want to apply part (b) of Theorem 3.3.2, for which we have to verify that

$$f_n^{(k)}(\cdot) := \log \|\wedge^k \Phi(n,\cdot)\| \in L^1$$

for all n and k, and that $\mathbb{E}\,\gamma^{(k)} > -\infty$ for all k. Now

$$\|\wedge^k \Phi(n+1,\omega)\| \leq \|\wedge^k \Phi(n,\theta\omega)\| \|\wedge^k A(\omega)\|,$$

$$\|\wedge^k \Phi(n,\theta\omega)\| \leq \|\wedge^k \Phi(n+1,\omega)\| \|\wedge^k A(\omega)^{-1}\|,$$

and

$$\log \|\wedge^k A\| \leq k \log^+ \|A\|$$

result in

$$f_n^{(k)} \circ \theta - k \log^+ \|A^{-1}\| \leq f_{n+1}^{(k)} \leq f_n^{(k)} \circ \theta + k \log^+ \|A\|. \tag{3.3.5}$$

Since $\mathbb{P}$ is θ-invariant our assumptions (3.3.4) imply that $f_n^{(k)} \in L^1$ if $f_1^{(k)} = \log \|\wedge^k A\| \in L^1$.

(b) From $1 = \|AA^{-1}\| \leq \|A\|\|A^{-1}\|$ it follows that

$$-\log^+ \|A^{-1}\| \leq \log \|A\| \leq \log^+ \|A\|$$

and

$$|\log\|A\|| \leq \log^+ \|A\| + \log^+ \|A^{-1}\|.$$

Similarly,

$$|\log \|\wedge^k A\|| \leq k\,(\log^+ \|A\| + \log^+ \|A^{-1}\|).$$

Therefore $f_1^{(k)} \in L^1$ if $\log^+ \|A^{\pm 1}\| \in L^1$.

(c) We check that $\mathbb{E}\,\gamma^{(k)} > -\infty$. Indeed,

$$\begin{aligned}|\mathbb{E}\,\gamma^{(k)}| &= \lim_{n\to\infty} \frac{1}{n}|\mathbb{E}\log\|\wedge^k \Phi(n,\omega)\|| \\ &\leq k \lim_{n\to\infty} \frac{1}{n}\sum_{i=0}^{n-1}\mathbb{E}|\log\|A(\theta^i\omega)\|| = k\,\mathbb{E}|\log\|A\|| < \infty\end{aligned}$$

by step (b) of the proof.

(d) Dividing equation (3.3.5) by n we see that

$$\frac{f_{n+1}^{(k)}(\omega)}{n} \to \gamma^{(k)}(\omega) \iff \frac{f_n^{(k)}(\theta\omega)}{n} \to \gamma^{(k)}(\theta\omega),$$

and in that case $\gamma^{(k)}(\theta\omega) = \gamma^{(k)}(\omega)$. Hence the set $\tilde{\Omega}$ on which the limit exists and is finite for each k is invariant whenever $A(\omega) \in Gl(d, \mathbb{R})$ for all $\omega \in \Omega$. On this set, the limits are invariant. We know from Theorem 3.3.3(i) that $\tilde{\Omega}$ has full measure.

Part (C):

(a) Since $\log^+ \|\Phi(1,\omega)^{\pm 1}\| \in L^1$ part (B) applies. Let $\tilde{\Omega}_1$ be the $\theta(1)$-invariant set of full measure of those ω's for which the limits $\gamma^{(k)}(\omega)$ exist and are finite. These limits are $\theta(1)$-invariant on $\tilde{\Omega}_1$.

(b) By the continuity of Φ with respect to t, $\alpha^\pm$ is measurable. The integrability conditions $\alpha^\pm \in L^1$ and the ergodic theorem assure that the $\theta(1)$-invariant set $\tilde{\Omega}_2$ on which $\lim_{n\to\infty} \alpha^\pm(\theta(n)\omega)/n = 0$ has full measure. Hence $\tilde{\Omega} := \tilde{\Omega}_1 \cap \tilde{\Omega}_2$ is $\theta(1)$-invariant and has full measure.

(c) We next prove that $\tilde{\Omega}_2$ is $\theta(t)$-invariant. We have for $t \in [0,1]$

$$\begin{aligned}
\alpha^+(\theta(t)\omega) &= \sup_{0\le s\le 1} \log^+ \|\Phi(s+t,\omega)\Phi(t,\omega)^{-1}\| \\
&\le \sup_{0\le s\le 2} \log^+ \|\Phi(s,\omega)\| + \alpha^-(\omega) \\
&\le \alpha^+(\omega) + \alpha^-(\omega) + \sup_{0\le s\le 1} \log^+ \|\Phi(s,\theta(1)\omega)\Phi(1,\omega)\| \\
&\le 2\alpha^+(\omega) + \alpha^+(\theta(1)\omega) + \alpha^-(\omega),
\end{aligned}$$

similarly for α^-. Hence $\tilde{\Omega}_2$ is forward invariant.

For the converse, use for $s, t \in [0,1]$

$$\Phi(s+1-t, \theta(t)\omega) = \Phi(s, \theta(1)\omega)\Phi(1-t, \theta(t)\omega),$$

which gives

$$\alpha^+(\theta(1)\omega) \le 2\alpha^+(\theta(t)\omega) + \alpha^+(\theta(1)\theta(t)\omega) + \alpha^-(\theta(t)\omega),$$

similarly for α^-. Hence $\theta(t)\omega \in \tilde{\Omega}_2$ implies $\theta(1)\omega \in \tilde{\Omega}_2$ which holds if and only if $\omega \in \tilde{\Omega}_2$.

(d) We now prove that for each $\omega \in \tilde{\Omega}$ the limit

$$\lim_{t\to\infty} \frac{1}{t} \log \|\wedge^k \Phi(t,\omega)\| \tag{3.3.6}$$

exists and is equal to $\gamma^{(k)}(\omega)$.

The cocycle property gives for $t \in \mathbb{R}^+$ and $n = [t]$, the integer part of t,

$$\wedge^k \Phi(t,\omega) = \wedge^k \Phi(t-n, \theta(n)\omega) \wedge^k \Phi(n,\omega),$$

whence by the invertibility of Φ

$$\wedge^k \Phi(n,\omega) = \wedge^k \Phi(t-n,\theta(n)\omega)^{-1} \wedge^k \Phi(t,\omega).$$

Hence

$$\log \|\wedge^k \Phi(n,\omega)\| - k\alpha^-(\theta(n)\omega) \leq \log \|\wedge^k \Phi(t,\omega)\| \qquad (3.3.7)$$
$$\leq \log \|\wedge^k \Phi(n,\omega)\| + k\alpha^+(\theta(n)\omega).$$

We conclude from (3.3.7) that for $\omega \in \tilde{\Omega}$ the limits (3.3.6) exist and are equal to $\gamma^{(k)}(\omega)$.

(e) We finally prove that $\tilde{\Omega}$ is $\theta(t)$-invariant, i.e. $\theta(t)^{-1}\tilde{\Omega} = \tilde{\Omega}$, and that for $\omega \in \tilde{\Omega}$ and $t \in \mathbb{R}^+$, $\gamma^{(k)}(\theta(t)\omega) = \gamma^{(k)}(\omega)$. Consider the case $k = 1$ only, let $n = [t]$, and asssume that $s \geq n+1-t$. By the cocycle property,

$$\log \|\Phi(t+s-(n+1),\theta(n+1)\omega)\| - \log \|\Phi(n+1-t,\theta(t)\omega)^{-1}\|$$
$$\leq \log \|\Phi(s,\theta(t)\omega)\| \qquad (3.3.8)$$
$$\leq \log \|\Phi(t+s-(n+1),\theta(n+1)\omega)\| + \log \|\Phi(n+1-t,\theta(t)\omega)\|.$$

For $\omega \in \tilde{\Omega}$, the left-hand side and the right-hand side of (3.3.8) divided by s tend to $\gamma^{(1)}(\theta(n+1)\omega) = \gamma^{(1)}(\omega)$ for $s \to \infty$, by step (iv). Consequently, $\theta(t)\omega \in \tilde{\Omega}$, i.e. $\tilde{\Omega}$ is forward invariant. Moreover,

$$\lim_{s \to \infty} \frac{1}{s} \log \|\Phi(s,\theta(t)\omega)\|$$
$$= \lim_{n \to \infty} \frac{1}{n} \log \|\Phi(n,\theta(t)\omega)\| =: \gamma^{(1)}(\theta(t)\omega) = \gamma^{(1)}(\omega).$$

Rearranging the string of inequalities (3.3.8), we can also conclude that if $\theta(t)\omega \in \tilde{\Omega}$, then $\omega \in \tilde{\Omega}$, i.e. $\tilde{\Omega}$ is invariant. □

3.3.4 Corollary (Lyapunov Index of Determinant). *We have*

$$\lim_{n \to \infty} \frac{1}{n} \log |\det \Phi(n,\cdot)| = \gamma^{(d)} = \sum_{k=1}^d \Lambda_k = \mathbb{E}\left(\log |\det A(\cdot)|\,|\,\mathcal{I}\right) \quad \mathbb{P}\text{-a.s.},$$

where $\mathcal{I} \subset \mathcal{F}$ is the σ-algebra of θ-invariant sets, and

$$\mathbb{E}\gamma^{(d)} = \sum_{k=1}^d \mathbb{E}\Lambda_k = \mathbb{E}\log|\det A| \in \mathbb{R} \cup \{-\infty\},$$

where the expectations are finite under the assumptions (3.3.4).

Proof. By Proposition 3.2.7(iii)

$$\| \wedge^d \Phi(n,\omega) \| = |\det \Phi(n,\omega)| = \prod_{i=0}^{n-1} |\det A(\theta^i \omega)|.$$

We know that $\frac{1}{n}\log(\cdot)$ of the left-hand side converges on $\tilde{\Omega}$ to

$$\gamma^{(d)} = \sum_{k=1}^{d} \Lambda_k = \lim_{n\to\infty} \frac{1}{n} \log \| \wedge^d \Phi(n,\omega) \|.$$

The (generalized) ergodic theorem applies to the right-hand side since $\log^+ |\det A| = f_1^{(d)+} \in L^1$ (Theorem 3.3.3(i)) and yields

$$\lim_{n\to\infty} \frac{1}{n} \sum_{i=0}^{n-1} \log |\det A(\theta^i \omega)| = \mathbb{E}(\log |\det A(\cdot)| \,|\, \mathcal{I}) \in \mathbb{R} \cup \{-\infty\} \quad \mathbb{P}\text{-a.s.}$$

□

3.3.5 Remark (Furstenberg-Kesten Theorem for Other Norms.) We have worked with the standard scalar product in $\mathbb{R}^d$ to define "orthogonal", "singular value decomposition", the corresponding operator matrix norms in $\wedge^k \mathbb{R}^d$, etc. However, as all norms in $\mathbb{R}^{d\times d}$ are equivalent, the integrability condition (3.3.2) is satisfied or not for all norms simultaneously, and all limits are independent of the norm chosen.

The representation

$$\mathbb{E}\gamma^{(k)} = \inf_{n\in\mathbb{N}} \frac{1}{n} \mathbb{E} \log \| \wedge^k \Phi(n,\cdot) \|$$

is, however, only valid for so-called *matrix norms*, i.e. norms on $\mathbb{R}^{d\times d}$ with the additional property $\|AB\| \le \|A\| \|B\|$.

This remark equally applies to the other versions of the Furstenberg-Kesten theorem that we will present. ∎

3.3.6 Remark (Furstenberg-Kesten Theorem for Backward Cocycles). The Furstenberg-Kesten theorem also holds for $\mathbb{R}^{d\times d}$-valued backward cocycles $\Psi(n,\omega)$ over θ, satisfying

$$\Psi(n+m,\omega) = \Psi(n,\omega)\Psi(m,\theta^n \omega).$$

Indeed, also in this case for any matrix norm

$$\log \| \wedge^k \Psi(n+m,\omega) \| \le \log \| \wedge^k \Psi(n,\omega) \| + \log \| \wedge^k \Psi(m,\theta^n \omega) \|,$$

hence $\log \| \wedge^k \Psi(n,\omega) \|$ is subadditive. ∎

3.3.7 Remark. (i) Just as in step (c) of the last proof we can derive in any case
$$\mathbb{E}\,\gamma^{(k)} \leq k\,\mathbb{E}\log\|A\|,$$
so $\mathbb{E}\,\gamma^{(1)} > -\infty$ implies that $\mathbb{E}\log\|A\| > -\infty$.

(ii) Under the assumption $\log^+\|A\| \in L^1$ we have
$$\log^+\|A^{-1}\| \in L^1 \iff \log|\det A| \in L^1 \iff \text{all } \mathbb{E}\,\gamma^{(k)} \text{ are finite}.$$
By Theorem 3.3.3(B), $\log^+\|A^{\pm 1}\| \in L^1$ implies that $\mathbb{E}\,\gamma^{(d)} = \mathbb{E}\log|\det A|$ is finite. Conversely, since $\|A^{-1}\| = 1/\delta_d$, $|\det A|\,\|A^{-1}\| = \delta_1 \cdots \delta_{d-1} \leq \delta_1^{d-1} = \|A\|^{d-1}$ whence $\|A^{-1}\| \leq \|A\|^{d-1}/|\det A|$. This gives
$$\log^+\|A^{-1}\| \leq (d-1)\log^+\|A\| + |\log|\det A||.$$
This proves the first equivalence. The second one follows from Corollary 3.3.4. ∎

3.3.8 Definition (Lyapunov Spectrum). Suppose Φ is a linear cocycle over θ for which one of the versions of the Furstenberg-Kesten theorem holds. Denote by $\lambda_1(\omega) > \ldots > \lambda_{p(\omega)}(\omega)$ the *different* numbers in the sequence $\Lambda_1(\omega) \geq \ldots \geq \Lambda_d(\omega)$ (possibly $\lambda_{p(\omega)}(\omega) = -\infty$) and denote by $d_i(\omega)$ the frequency of appearance of $\lambda_i(\omega)$ in this sequence. On the corresponding forward invariant or invariant set $\tilde{\Omega}$ of full measure the functions $p : \tilde{\Omega} \to \{1,\ldots,d\}$, $d_i : \{\omega \in \tilde{\Omega} : p(\omega) \geq i\} \to \{1,\ldots,d\}$ and $\lambda_i : \{\omega \in \tilde{\Omega} : p(\omega) \geq i\} \to \mathbb{R} \cup \{-\infty\}$ are invariant, and constant in the ergodic case. The functions $\lambda_i(\cdot)$ are called the *Lyapunov exponents* of Φ, and the $d_i(\cdot)$ their *multiplicities*. The set
$$\mathcal{S}(\theta, \Phi) := \{(\lambda_i(\cdot), d_i(\cdot)) : i = 1,\ldots,p(\cdot)\}$$
is called the *Lyapunov spectrum* of Φ. We also write $\mathcal{S}(\theta, A)$ if Φ is generated by A. By Remark 3.3.5, the Lyapunov spectrum is independent of the choice of the norm in the space $\mathbb{R}^{d \times d}$. If θ is ergodic, we will agree that the Lyapunov spectrum consists of the corresponding set of constants (where possibly $\lambda_p = -\infty$). ∎

3.3.9 Example (Products of 2×2 Triangular Matrices). Let $A : \Omega \to Gl(2,\mathbb{R})$, where
$$A(\omega) = \begin{pmatrix} a(\omega) & c(\omega) \\ 0 & b(\omega) \end{pmatrix}, \quad a(\omega) \neq 0,\ b(\omega) \neq 0.$$
These matrices form a subgroup of $Gl(2,\mathbb{R})$ and the cocycle on $\mathbb{T} = \mathbb{N}$ over θ generated by A is
$$\Phi(n,\cdot) = A_{n-1} \cdots A_0 = \begin{pmatrix} a_{n-1} \cdots a_0 & \sum_{k=0}^{n-1} a_{n-1} \cdots a_{k+1} c_k b_{k-1} \cdots b_0 \\ 0 & b_{n-1} \cdots b_0 \end{pmatrix},$$
where $a_k(\omega) := a(\theta^k \omega)$, $b_k(\omega) := b(\theta^k \omega)$ and $c_k(\omega) := c(\theta^k \omega)$.

The following facts are readily verified (assume for simplicity θ to be ergodic):

(i) $\log^+ \|A^{\pm 1}\| \in L^1 \iff \log |a|, \log |b|$ and $\log^+ |c| \in L^1$, which we assume from now on.

(ii) By our assumptions,

$$\frac{1}{n} \sum_0^{n-1} \log |a_k| \to \mathbb{E} \log |a| =: \alpha,$$

$$\frac{1}{n} \sum_0^{n-1} \log |b_k| \to \mathbb{E} \log |b| =: \beta,$$

hence
$$\frac{1}{n} \log |\det \Phi(n, \cdot)| \to \gamma^{(2)} = \Lambda_1 + \Lambda_2 =: 2\lambda_\Sigma = \alpha + \beta.$$

(iii) The Lyapunov index of $\Phi(n, \omega)_{11}$ is α, that of $\Phi(n, \omega)_{22}$ is β, and that of $\Phi(n, \omega)_{12}$ is less than or equal to $\max(\alpha, \beta)$. Using the Euclidean norm in $\mathbb{R}^{2 \times 2}$ we obtain by Lemma 3.2.1

$$\frac{1}{n} \log \|\Phi(n, \omega)\| \to \gamma^{(1)} = \Lambda_1 = \max(\alpha, \beta),$$

hence for $\alpha \neq \beta$

$$\lambda_1 = \max(\alpha, \beta) > \lambda_\Sigma = \frac{1}{2}(\alpha + \beta) > \lambda_2 = \min(\alpha, \beta).$$

For $\alpha = \beta$, $\lambda_1 = \lambda_\Sigma = \alpha = \beta$ with multiplicity $d_1 = 2$.

(iv) Generalization: Suppose the generator of $\Phi(n, \omega)$ in $\mathbb{R}^d$ has the form

$$A = \begin{pmatrix} a_{11} & a_{12} & \cdots & a_{1d} \\ 0 & a_{22} & \cdots & a_{2d} \\ & & \ddots & \\ 0 & 0 & & a_{dd} \end{pmatrix}.$$

If $\log |a_{ij}| \in L^1$ for $1 \leq i \leq j \leq d$, then $\log^+ \|A^{\pm 1}\| \in L^1$. If the $\alpha_k := \mathbb{E} \log |a_{kk}|$ are arranged in decreasing order, $\alpha_{k_1} \geq \ldots \geq \alpha_{k_d}$, then $\Lambda_i = \alpha_{k_i}$, $i = 1, \ldots, d$. ∎

3.3.3 The Furstenberg-Kesten Theorem for Two-Sided Time

We now deal with the asymptotic behavior of the singular values of a cocycle with two-sided time for $t \to -\infty$ and its relation to the one for $t \to \infty$.

3.3.10 Theorem (Furstenberg-Kesten Theorem, Two-Sided Time).
Let Φ be a linear cocycle with two-sided time over the metric DS $(\Omega, \mathcal{F}, \mathbb{P}, (\theta(t))_{t\in\mathbb{T}})$.

(A) *Discrete time case* $\mathbb{T} = \mathbb{Z}$: Assume that the generator $A : \Omega \to Gl(d, \mathbb{R})$ of Φ satisfies

$$\log^+ \|A\| \in L^1(\Omega, \mathcal{F}, \mathbb{P}) \quad \text{and} \quad \log^+ \|A^{-1}\| \in L^1(\Omega, \mathcal{F}, \mathbb{P}). \quad (3.3.9)$$

Then $\Psi(n, \omega) := \Phi(-n, \omega)$ is a cocycle over θ^{-1} generated by $A^{-1} \circ \theta^{-1}$, and on an invariant set $\tilde{\Omega}$ of full measure we have for $k = 1, \ldots, d$

$$\gamma^{(k)-}(\omega) := \lim_{n\to\infty} \frac{1}{n} \log \| \wedge^k \Phi(-n, \omega)\| = \gamma^{(d-k)}(\omega) - \gamma^{(d)}(\omega)$$

and

$$\Lambda_k^-(\omega) := \lim_{n\to\infty} \frac{1}{n} \log \delta_k(\Phi(-n, \omega)) = -\Lambda_{d+1-k}(\omega),$$

where $\gamma^{(k)}$ and Λ_k are the corresponding limits for $\Phi(n, \omega)$. In particular, if $\mathcal{S}(\theta, \Phi) = \{(\lambda_i, d_i) : i = 1, \ldots, p\}$ is the Lyapunov spectrum of $\Phi(n, \omega)$ (as a cocycle over θ on $\mathbb{T} = \mathbb{N}$), then the Lyapunov spectrum of $\Phi(-n, \omega)$ (as a cocycle over θ^{-1} on $\mathbb{T} = \mathbb{N}$) is

$$\mathcal{S}(\theta^{-1}, \Phi(-\cdot)) = -\mathcal{S}(\theta, \Phi) := \{(-\lambda_i, d_i) : i = 1, \ldots, p\}.$$

(B) *Continuous time case* $\mathbb{T} = \mathbb{R}$: Assume $\alpha^+ \in L^1(\Omega, \mathcal{F}, \mathbb{P})$ and $\alpha^- \in L^1(\Omega, \mathcal{F}, \mathbb{P})$, where

$$\alpha^\pm(\omega) := \sup_{0\leq t \leq 1} \log^+ \|\Phi(t, \omega)^{\pm 1}\|.$$

Then all assertions of part (A) hold with n and $\mathbb{N}$ replaced with t and $\mathbb{R}^+$.

Proof. Part (A):
Clearly $\Psi = \Phi(-\cdot)$ is a cocycle over θ^{-1} with generator $A^{-1} \circ \theta^{-1}$. The generator satisfies the integrability conditions (3.3.9). The cocycle property relates positive and negative times via

$$\Phi(-n, \omega) = \Phi(n, \theta^{-n}\omega)^{-1}.$$

If $\delta_1 \geq \ldots \geq \delta_d > 0$ are the singular values of $A \in Gl(d, \mathbb{R})$, then $1/\delta_d \geq \ldots \geq 1/\delta_1 > 0$ are the singular values of A^{-1}. Hence

$$\delta_k(\Phi(-n, \omega)) = \delta_k(\Phi(n, \theta^{-n}\omega)^{-1}) = \delta_{d+1-k}^{-1}(\Phi(n, \theta^{-n}\omega)). \quad (3.3.10)$$

We now apply Theorem 3.3.3(B) first to $\Phi(n, \omega)$ which yields an invariant set $\tilde{\Omega}_1$ of full measure on which the corresponding statements hold, and then to $\Phi(-n, \omega)$ which yields an invariant set $\tilde{\Omega}_2$ of full measure on which for $k = 1, \ldots, d$

$$g_n^{(k)}(\omega) := \frac{1}{n} \log \delta_k(\Phi(-n, \omega)) \to \Lambda_k^-(\omega),$$

where Λ_k^- is invariant on $\tilde{\Omega}_2$, and convergence also holds in L^1. Now $g_n \to g$ in probability or L^1 and $g \circ \theta = g$ P-a.s. implies $g_n \circ \theta^n \to g$ in probability or L^1. This applied to equation (3.3.10) gives $\Lambda_k^- = -\Lambda_{d+1-k}$ P-a.s., consequently $\Lambda_k^-(\omega) = -\Lambda_{d+1-k}(\omega)$ for all ω in the invariant set of full measure

$$\tilde{\Omega} := \tilde{\Omega}_1 \cap \tilde{\Omega}_2 \cap \{\omega : \Lambda_k^-(\omega) = -\Lambda_{d+1-k}(\omega),\ k = 1, \ldots, d\}.$$

(In the ergodic case we could have just taken $\mathbb{E} \log(\cdot)$ in (3.3.10) to immediately arrive at the result.)

Part (B):

Theorem 3.3.3(C) can be applied to both cocycles. The integrability condition for $\Phi(-t, \omega)$ is satisfied, due to the following lemma.

3.3.11 Lemma. *The following three statements are equivalent:*

1. $\alpha^\pm(\omega) := \sup_{0 \leq t \leq 1} \|\Phi(t, \omega)^{\pm 1}\| \in L^1$,
2. $\beta^\pm(\omega) := \sup_{0 \leq t \leq 1} \|\Phi(-t, \omega)^{\pm 1}\| \in L^1$,
3. $\gamma^\pm(\omega) := \sup_{-1 \leq t \leq 1} \|\Phi(t, \omega)^{\pm 1}\| \in L^1$.

The proof is an exercise in the use of the cocycle property.

The remaining statements follow as in the proof of part (B). □

3.3.12 Remark (Average Lyapunov Exponent). Let Φ be a linear cocycle for which the assumptions of one of the versions of the Furstenberg-Kesten theorem are satisfied. Then we have

$$\lambda_p \leq \lambda_\Sigma := \frac{1}{d} \sum_{i=1}^p d_i \lambda_i = \frac{1}{d} \sum_{k=1}^d \Lambda_k = \frac{1}{d} \mathbb{E}\left(\log |\det A| \,|\, \mathcal{I}\right) \leq \lambda_1,$$

where $A = \Phi(1, \cdot)$ in case of $\mathbb{T} = \mathbb{R}^+$ and λ_Σ, the *average Lyapunov exponent*, can be calculated by using $|\det A|$ only (in contrast, the Lyapunov exponents are quantities about which explicit information is usually hard to obtain).

In the invertible (in particular: two-sided) case, we can without loss of generality assume that $\lambda_\Sigma = 0$ P-a.s. since

$$\mathcal{S}(\theta, |\det \Phi|^{-1/d} \Phi) = \mathcal{S}(\theta, \Phi) - \lambda_\Sigma.$$

■

3.3.13 Example (RDE with 2×2 Triangular Right-Hand Side.) Let $\dot{x}_t = A(\theta_t \omega) x_t$ in $\mathbb{R}^2$ with $A : \Omega \to \mathbb{R}^{2 \times 2}$ given by

$$A(\omega) = \begin{pmatrix} a(\omega) & c(\omega) \\ 0 & b(\omega) \end{pmatrix}.$$

Assume for simplicity that θ is ergodic.

(i) We have $A \in L^1$ if and only if $a, b, c \in L^1$, which we assume from now on. It implies that the RDE generates a linear cocycle $\Phi(t, \omega)$ with time

$\mathbb{T} = \mathbb{R}$ (see Example 2.2.8). It also implies the integrability conditions of the Furstenberg-Kesten theorem for $\mathbb{T} = \mathbb{R}$ (see Example 3.4.15).

(ii) Put $\alpha := \mathbb{E}\,a$, $\beta := \mathbb{E}\,b$,

$$\alpha_t(\omega) := \int_0^t a(\theta_s \omega)\,ds, \quad \beta_t(\omega) := \int_0^t b(\theta_s \omega)\,ds, \quad c_t(\omega) := c(\theta_t \omega).$$

The cocycle has the form

$$\Phi(t,\cdot) = \begin{pmatrix} e^{\alpha_t} & e^{\alpha_t}\int_0^t e^{\beta_s - \alpha_s} c_s\,ds \\ 0 & e^{\beta_t} \end{pmatrix}.$$

(iii) From $\det \Phi(t,\cdot) = \exp(\alpha_t + \beta_t)$ we obtain

$$\frac{1}{t}\log|\det \Phi(t,\omega)| \to \gamma^{(2)} = \Lambda_1 + \Lambda_2 = 2\lambda_\Sigma = \alpha + \beta.$$

Since the Lyapunov index of $\Phi(t,\omega)_{11}$ is α, that of $\Phi(t,\omega)_{22}$ is β and that of $\Phi(t,\omega)_{12}$ is less than or equal to $\max(\alpha,\beta)$, we obtain (using Lemma 3.2.1 and the Euclidean norm in $\mathbb{R}^{2\times 2}$)

$$\frac{1}{t}\log\|\Phi(t,\omega)\| \to \gamma^{(1)} = \Lambda_1 = \max(\alpha,\beta).$$

Hence for the case $\alpha \neq \beta$

$$\lambda_1 = \max(\alpha,\beta) > \lambda_\Sigma = \frac{1}{2}(\alpha + \beta) > \lambda_2 = \min(\alpha,\beta),$$

while for $\alpha = \beta$, $\lambda_1 = \lambda_\Sigma = \alpha = \beta$ with multiplicity $d_1 = 2$.

(iv) Generalization: Consider $\dot{x}_t = A(\theta_t \omega) x_t$ in $\mathbb{R}^d$ with the triangular matrix

$$A = \begin{pmatrix} a_{11} & a_{12} & \cdots & a_{1d} \\ 0 & a_{22} & \cdots & a_{2d} \\ & & \ddots & \\ 0 & 0 & & a_{dd} \end{pmatrix}.$$

Clearly $A \in L^1$ if and only if $a_{ij} \in L^1$ for all $1 \leq i \leq j \leq d$. Arranging the $\alpha_k := \mathbb{E}\,a_{kk}$ in decreasing order $\alpha_{k_1} \geq \ldots \geq \alpha_{k_d}$, we obtain $\Lambda_i = \alpha_{k_i}$, $i = 1,\ldots,d$. See Oseledets [268], Crauel [108] and Johnson, Palmer and Sell [192: Lemma 6.2]. ∎

3.4 The Multiplicative Ergodic Theorem

This section is devoted to the presentation, proof and discussion of the various versions of Oseledets's multiplicative ergodic theorem (MET), the most fundamental and most important theorem in this book. See Sect.3.1 for an overview of the scope and the various proofs of the MET.

We will first prove the MET for time $\mathbb{T} = \mathbb{N}$ (and more generally for one-sided time) following Goldsheid and Margulis [161] and then deal with two-sided time. We consider it worth the space to formulate and prove all versions in detail.

3.4.1 The MET for One-Sided Time

3.4.1 Theorem (MET for One-Sided Time). *Let Φ be a linear cocycle with one-sided time over the metric DS $(\Omega, \mathcal{F}, \mathbb{P}, (\theta(t))_{t \in \mathbb{T}})$. Then the following statements hold:*

(A) Non-invertible case $\mathbb{T} = \mathbb{N}$: If the generator $A : \Omega \to \mathbb{R}^{d \times d}$ satisfies

$$\log^+ \|A(\cdot)\| \in L^1(\Omega, \mathcal{F}, \mathbb{P}),$$

then there exists a forward invariant set $\tilde{\Omega} \in \mathcal{F}$ of full measure such that for each $\omega \in \tilde{\Omega}$

(i) The limit $\lim_{n \to \infty} (\Phi(n,\omega)^* \Phi(n,\omega))^{1/2n} =: \Psi(\omega) \geq 0$ exists.

(ii) Let $e^{\lambda_{p(\omega)}(\omega)} < \ldots < e^{\lambda_1(\omega)}$ be the different eigenvalues of $\Psi(\omega)$ (possibly $\lambda_{p(\omega)}(\omega) = -\infty$) and let $U_{p(\omega)}(\omega), \ldots, U_1(\omega)$ be the corresponding eigenspaces with multiplicities $d_i(\omega) := \dim U_i(\omega)$. Then

$$p(\theta\omega) = p(\omega),$$

$$\lambda_i(\theta\omega) = \lambda_i(\omega) \quad \text{for all } i \in \{1, \ldots, p(\omega)\},$$

$$d_i(\theta\omega) = d_i(\omega) \quad \text{for all } i \in \{1, \ldots, p(\omega)\}.$$

(iii) Put $V_{p(\omega)+1}(\omega) := \{0\}$ and for $i = 1, \ldots, p(\omega)$

$$V_i(\omega) := U_{p(\omega)} \oplus \ldots \oplus U_i(\omega),$$

so that

$$V_{p(\omega)}(\omega) \subset \ldots \subset V_i(\omega) \subset \ldots \subset V_1(\omega) = \mathbb{R}^d$$

defines a filtration of $\mathbb{R}^d$. Then for each $x \in \mathbb{R}^d \setminus \{0\}$ the Lyapunov exponent

$$\lambda(\omega, x) := \lim_{n \to \infty} \frac{1}{n} \log \|\Phi(n,\omega)x\|$$

exists as a limit and

$$\lambda(\omega, x) = \lambda_i(\omega) \iff x \in V_i(\omega) \setminus V_{i+1}(\omega),$$

equivalently

$$V_i(\omega) = \{x \in \mathbb{R}^d : \lambda(\omega, x) \leq \lambda_i(\omega)\}.$$

(iv) For all $x \in \mathbb{R}^d \setminus \{0\}$

$$\lambda(\theta\omega, A(\omega)x) = \lambda(\omega, x),$$

whence

$$A(\omega)V_i(\omega) \subset V_i(\theta\omega) \text{ for all } i \in \{1, \ldots, p(\omega)\}.$$

(v) If $(\Omega, \mathcal{F}, \mathbb{P}, (\theta^n)_{n \in \mathbb{N}})$ is ergodic, then the function $p(\cdot)$ is constant on $\tilde{\Omega}$, and the functions $\lambda_i(\cdot)$ and $d_i(\cdot)$ are constant on $\{\omega \in \tilde{\Omega} : p(\omega) \geq i\}$, $i = 1, \ldots, d$.

(B) Invertible case $\mathbb{T} = \mathbb{N}$: If $A : \Omega \to Gl(d, \mathbb{R})$ and

$$\log^+ \|A\| \in L^1(\Omega, \mathcal{F}, \mathbb{P}) \quad \text{and} \quad \log^+ \|A^{-1}\| \in L^1(\Omega, \mathcal{F}, \mathbb{P}),$$

then the set $\tilde{\Omega}$ of full measure on which (A) holds (and on which in the ergodic case $p(\cdot)$, $\lambda_i(\cdot)$, $d_i(\cdot)$ are constant) can be chosen to be invariant. Further, $\lambda_{p(\omega)}(\omega) > -\infty$ on $\tilde{\Omega}$, and

$$A(\omega)V_i(\omega) = V_i(\theta\omega) \text{ for all } i \in \{1, \ldots, p(\omega)\}.$$

(C) Invertible case $\mathbb{T} = \mathbb{R}^+$: Let $\Phi(t, \omega) \in Gl(d, \mathbb{R})$. Assume $\alpha^+ \in L^1$ and $\alpha^- \in L^1$, where

$$\alpha^\pm(\omega) := \sup_{0 \leq t \leq 1} \log^+ \|\Phi(t, \omega)^{\pm 1}\|.$$

Then all statements of part (B) hold with n, θ and $A(\omega)$ replaced with t, $\theta(t)$ and $\Phi(t, \omega)$, and the set $\tilde{\Omega} \in \mathcal{F}$ of full measure is now invariant with respect to $(\theta(t))_{t \in \mathbb{R}^+}$.

(D) Measurability: The function $\omega \mapsto p(\omega) \in \{1, \ldots, d\}$ (measurably extended from $\tilde{\Omega}$ to Ω) is measurable. The functions $\omega \mapsto \lambda_i(\omega) \in \mathbb{R} \cup \{-\infty\}$, $\omega \mapsto d_i(\omega) \in \{1, \ldots, d\}$, $\omega \mapsto U_i(\omega) \in \cup_{k=1}^d G_k(d)$ and $\omega \mapsto V_i(\omega) \in \cup_{k=1}^d G_k(d)$, $G_k(d)$ the Grassmann manifold of k-dimensional subspaces of $\mathbb{R}^d$ (measurably extended to $\{\omega : p(\omega) \geq i\} \in \mathcal{F}$) are measurable.

In particular, $\omega \mapsto V_i(\omega) \in \mathfrak{P}(\mathbb{R}^d)$ are random closed sets on their domain in the sense of Definition 1.6.1.

(E) Lyapunov spectrum: The collection $\{(\lambda_i(\cdot), d_i(\cdot))_{i=1,\ldots,p(\cdot)}\}$ of quantities from part (A) of the theorem coincides with the Lyapunov spectrum $\mathcal{S}(\theta, A)$ of the cocycle $\Phi(\cdot, \omega)$ (see Definition 3.3.8).

The proof of the MET follows in a few lines from the Furstenberg-Kesten theorem and the following deterministic proposition, the proof of which requires the "hard work" alluded to above.

3.4.2 Proposition (Deterministic MET). *Let $(A_n)_{n \in \mathbb{N}}$ be a sequence of $d \times d$ matrices which satisfies the following two conditions:*

1.
$$\limsup_{n \to \infty} \frac{1}{n} \log \|A_n\| \leq 0. \tag{3.4.1}$$

2. Putting $\Phi_n := A_n \cdots A_1$,
$$\lim_{n\to\infty} \frac{1}{n} \log \| \wedge^i \Phi_n \| = \gamma^{(i)} \in \mathbb{R} \cup \{-\infty\} \qquad (3.4.2)$$
exists for $i = 1, \ldots, d$.

Then:

(i) The limit $\lim_{n\to\infty}(\Phi_n^* \Phi_n)^{1/2n} =: \Psi \geq 0$ exists and the eigenvalues of Ψ coincide with (e^{Λ_i}), where the Λ_i are successively defined by $\Lambda_1 + \ldots + \Lambda_i := \gamma^{(i)}$, $i = 1, \ldots, d$ (put $\Lambda_i = -\infty$ if $\gamma^{(i)} = -\infty$) and satisfy
$$\Lambda_i = \lim_{n\to\infty} \frac{1}{n} \log \delta_i(\Phi_n), \quad i = 1, \ldots, d,$$
where $\delta_i(\Phi_n)$ is the i-th singular value of Φ_n.

(ii) Let $e^{\lambda_p} < \ldots < e^{\lambda_1}$ be the different eigenvalues of Ψ (possibly $\lambda_p = -\infty$) and let $U_p, \ldots, U_1$ be their corresponding eigenspaces with multiplicities $d_i = \dim U_i$. Write $V_{p+1} := \{0\}$,
$$V_i := U_p \oplus \ldots \oplus U_i, \quad i = 1, \ldots, p,$$
so that
$$V_p \subset \ldots \subset V_i \subset \ldots \subset V_1 = \mathbb{R}^d$$
defines a filtration of $\mathbb{R}^d$. Then for each $x \in \mathbb{R}^d \setminus \{0\}$ the Lyapunov exponent
$$\lambda(x) := \lim_{n\to\infty} \frac{1}{n} \log \|\Phi_n x\|$$
exists as a limit and
$$\lambda(x) = \lambda_i \iff x \in V_i \setminus V_{i+1},$$
equivalently
$$V_i = \{x \in \mathbb{R}^d : \lambda(x) \leq \lambda_i\}.$$

To derive the MET from this proposition, part of the following elementary lemma is needed.

3.4.3 Lemma (Growth of a Stationary Sequence). *Suppose we are given a metric DS $(\Omega, \mathcal{F}, \mathbb{P}, (\theta^n)_{n\in\mathbb{N}})$ and a random variable $f : \Omega \to \mathbb{R} \cup \{-\infty\}$. Then:*

(i) *If $f^+ \in L^1$, then the invariant set*
$$\Omega_1 := \{\omega : \limsup_{n\to\infty} \frac{1}{n} f(\theta^{n-1}\omega) \leq 0\}$$
has full measure.

(ii) *Let $f : \Omega \to \mathbb{R}$. Then the invariant set*
$$\Omega_2 := \{\omega : \liminf_{n\to\infty} \frac{1}{n} f(\theta^{n-1}\omega) \leq 0 \leq \limsup_{n\to\infty} \frac{1}{n} f(\theta^{n-1}\omega)\}$$

has full measure. If in addition $f^+ \in L^1$, then the invariant set
$$\Omega_3 := \{\omega : \limsup_{n\to\infty} \frac{1}{n} f(\theta^{n-1}\omega) = 0\}$$
has full measure.

If $f^- = \max(0,-f) \in L^1$, then the invariant set
$$\Omega_4 := \{\omega : \liminf_{n\to\infty} \frac{1}{n} f(\theta^{n-1}\omega) = 0\}$$
has full measure.

(iii) Let $f : \Omega \to \mathbb{R}$. If $f \in L^1$ (or if only $(f \circ \theta - f)^+ \in L^1$, or if $(f \circ \theta - f)^- \in L^1$, both implying $f \circ \theta - f \in L^1$ and $\mathbb{E}(f \circ \theta - f) = 0$), then the invariant set
$$\Omega_5 := \{\omega : \lim_{n\to\infty} \frac{1}{n} f(\theta^{n-1}\omega) = 0\}$$
has full measure.

Proof. The invariance of Ω_1, Ω_2, Ω_3 and Ω_4 follows from their definition.

(i) That Ω_1 has full measure is a consequence of the Borel-Cantelli lemma since
$$\sum_{n=1}^{\infty} \mathbb{P}(\frac{1}{n} f(\theta^{n-1}\cdot) > \varepsilon) = \sum_{n=1}^{\infty} \mathbb{P}(f > \varepsilon n) = \sum_{n=1}^{\infty} \mathbb{P}(f^+ > \varepsilon n) \leq \frac{1}{\varepsilon} \mathbb{E} f^+ < \infty,$$
where we have used the elementary fact that for $f \geq 0$ and any $\varepsilon > 0$
$$\varepsilon \sum_{n=1}^{\infty} \mathbb{P}(f > \varepsilon n) \leq \mathbb{E} f = \varepsilon \int_0^{\infty} \mathbb{P}(f > \varepsilon t)\, dt \leq \varepsilon \sum_{n=0}^{\infty} \mathbb{P}(f > \varepsilon n).$$

(ii) If f is finite-valued, then for any $\varepsilon > 0$
$$\lim_{n\to\infty} \mathbb{P}(|\frac{1}{n} f(\theta^{n-1}\omega)| > \epsilon) = \lim_{n\to\infty} \mathbb{P}(|f(\omega)| > \epsilon n) = 0,$$
whence $\frac{1}{n} f(\theta^{n-1}\cdot) \to 0$ in probability, from which the assertion follows. If $f^+ \in L^1$ combine (i) with what we have just proved.

(iii) If $(f \circ \theta - f)^+ \in L^1$, then the representation
$$f \circ \theta^n = f + \sum_{i=0}^{n-1} (f \circ \theta - f) \circ \theta^i$$
and the (extended version of the) ergodic theorem say that $\lim_{n\to\infty} \frac{1}{n} f \circ \theta^n =: z$ exists (and is possibly $-\infty$) with $\mathbb{E} z = \mathbb{E}(f \circ \theta - f)$ (possibly $-\infty$). By (ii) this limit is necessarily equal to 0 $\mathbb{P}$-a. s. Hence $\mathbb{E} z = \mathbb{E}(f \circ \theta - f) = 0$, in particular $f \circ \theta - f \in L^1$. Analogously for the case $(f \circ \theta - f)^- \in L^1$. □

Proof of Theorem 3.4.1

(A) Under the asssumption $\log^+ \|A\| \in L^1$, condition (3.4.1) of Proposition 3.4.2 is true for the sequence $A_n = A(\theta^{n-1}\omega)$ for ω in an invariant set Ω_1 of full measure, by Lemma 3.4.3(i) applied to $f(\omega) = \log \|A(\omega)\|$.

Under the same assumption, the Furstenberg-Kesten theorem 3.3.3(A), hence in particular (3.4.2), holds on a forward invariant set Ω_2 of full measure.

Consequently, (i), (ii), (iii) and (v) are true on $\tilde{\Omega} := \Omega_1 \cap \Omega_2$ which is forward invariant and full.

For (iv) just note that $\Phi(n, \theta\omega) A(\omega) = \Phi(n+1, \omega)$ which immediately implies $\lambda(\theta\omega, A(\omega)x) = \lambda(\omega, x)$.

Take $x \in V_i(\omega)$ which by (iii) is equivalent to $\lambda(\omega, x) \leq \lambda_i(\omega)$. Hence

$$\lambda(\theta\omega, A(\omega)x) = \lambda(\omega, x) \leq \lambda_i(\omega) = \lambda_i(\theta\omega),$$

entailing $A(\omega)x \in V_i(\theta\omega)$, or $A(\omega)V_i(\omega) \subset V_i(\theta\omega)$.

(B) follows from the fact that the Furstenberg-Kesten theorem 3.3.3(B) furnishes an invariant set of full measure on which all statements are true and all limits are finite (and constant in the ergodic case).

As for the invariance of $V_i(\omega)$, note that since $A(\omega)$ is non-singular, $\dim A(\omega)V_i(\omega) = \dim V_i(\omega)$. Hence (A)(iv) and $\dim V_i(\omega) = \dim V_i(\theta\omega)$ imply $A(\omega)V_i(\omega) = V_i(\theta\omega)$.

(C) The continuous time case can be reduced to the discrete time case as follows:

(i) The integrability conditions assure that the corresponding continuous time Furstenberg-Kesten theorem 3.3.3(C) is valid on a set $\tilde{\Omega}$ of full measure which is $\theta(t)$-invariant.

(ii) For fixed $\omega \in \tilde{\Omega}$ we introduce, as for $\mathbb{T} = \mathbb{N}$, the flag $F(t, \omega)$ corresponding to $(\Phi(t,\omega)^*\Phi(t,\omega))^{1/2}$ and show that

– $(F(t,\omega))_{t \in \mathbb{R}^+}$ converges in $F_\tau(d)$ to $F(\omega) := \lim_{n \to \infty} F(n, \omega)$ (the latter limit is known to exist on $\tilde{\Omega}$),
– the convergence is exponentially fast,

$$\limsup_{t \to \infty} \frac{1}{t} \log \delta(F(t,\omega), F(\omega)) \leq -h, \qquad (3.4.3)$$

from which all other steps of the proof follow.

(iii) The claims in (ii) would follow from

$$\limsup_{n \to \infty} \frac{1}{n} \log \delta(F(n,\omega), F(n+s,\omega)) \leq -h \text{ uniformly in } s \in [0,1], \qquad (3.4.4)$$

which replaces (3.4.12). Writing $n + s = t$, $n = [t]$ in (3.4.4) we obtain

$$\limsup_{t \to \infty} \frac{1}{t} \log \delta(F([t],\omega), F(t,\omega)) \leq -h,$$

in particular $F(t, \omega) \to F(\omega)$, and (3.4.3) follows, by (3.4.13) and the triangle inequality.

(iv) All that remains to be proved is (3.4.4). An inspection of the proof of Theorem 3.4.1 (step 3, case 1 and case 2(a)) shows that we have to assure that the right-hand side of

$$\|P_j(n+t,\omega)P_i(n,\omega)\| \leq \begin{cases} \|\Phi(t,\theta(n)\omega)\|\frac{\delta_i(\Phi(n,\omega))}{\delta_j(\Phi(n+t,\omega))}, & i > j, \\ \|\Phi(t,\theta(n)\omega)^{-1}\|\frac{\delta_j(\Phi(n+t,\omega))}{\delta_i(\Phi(n,\omega))}, & i < j, \end{cases}$$

decays to 0 with exponential speed $-|\lambda_i - \lambda_j|$ uniformly with respect to $t \in [0,1]$. This is the case provided

$$\limsup_{n\to\infty} \frac{1}{n} \sup_{0\leq t\leq 1} \log \|\Phi(t,\theta(n)\omega)^{\pm 1}\| = \limsup_{n\to\infty} \frac{1}{n}\alpha^{\pm}(\theta(n)\omega) \leq 0 \quad (3.4.5)$$

(note that since $\Phi(0,\omega) = I$, $\sup \log = \sup \log^+$). The integrability conditions $\alpha^{\pm} \in L^1$ and the ergodic theorem assure that in fact $\lim_{n\to\infty} \alpha^{\pm}(\theta(n)\omega)/n = 0$ on a set of full measure which is invariant with respect to $\theta(1)$. In step (iii) of the proof of part (C) of Theorem 3.3.3 we showed that this set is even $\theta(t)$-invariant, and its defining property was built into the $\theta(t)$-invariant set $\tilde{\Omega}$ on which we are working.

(D) The measurability of p, λ_i and d_i is a consequence of the measurability of the $\gamma^{(i)}$ which is assured by the Furstenberg-Kesten theorem. There remains to be considered the measurability of U_i and V_i. The following will become clear in the proof of Proposition 3.4.2: On the measurable set $\Omega_{i,k} := \{\omega \in \tilde{\Omega} : d_i(\omega) = k\}$, $U_i(\omega)$ will be the limit of a sequence $U_i(n,\omega)$ of measurable functions with values in the Grassmannian $G_k(d)$ and hence measurable itself, similarly for V_i. V_i is a random closed set by Proposition 1.6.2(i) and the fact that for an open set $U \subset \mathbb{R}^d$

$$\{\omega \in \Omega_{i,k} : V_i(\omega) \cap U \neq \emptyset\} = \{\omega \in \Omega_{i,k} : V_i(\omega) \in U_k\},$$

where $U_k = \{V \in G_k(d) : V \cap U \neq \emptyset\}$ is open in $G_k(d)$.

(E) is clear as we have constructed the objects in (A) by the Furstenberg-Kesten theorem on which Definition 3.3.8 is based. □

It remains to prove Proposition 3.4.2. For pedagogical reasons we first present the proof for dimension $d = 2$ and later for general d.

Proof [3] **of Proposition 3.4.2 for d = 2**
By assumption (3.4.2), the limits

$$\gamma^{(1)} = \lim_{n\to\infty} \frac{1}{n}\log\|\Phi_n\| = \lim_{n\to\infty} \frac{1}{n}\log\delta_1(\Phi_n) = \Lambda_1$$

and

---
[3] This proof was presented by Ilya Goldsheid in a lecture at the University of Bremen in June 1990.

140   Chapter 3. MET in Euclidean Space

$$\gamma^{(2)} = \lim_{n\to\infty} \frac{1}{n}\log\|\wedge^2 \Phi_n\| = \Lambda_1 + \Lambda_2$$

exist, and since $\|\wedge^2 \Phi_n\| = \delta_1(\Phi_n)\delta_2(\Phi_n)$ and if $\Lambda_1 > -\infty$,

$$\lim_{n\to\infty} \frac{1}{n}\log \delta_2(\Phi_n) = \gamma^{(2)} - \Lambda_1 = \Lambda_2$$

exists (if $\Lambda_1 = -\infty$, put $\Lambda_2 = -\infty$).

Let $\Phi_n = V_n D_n O_n$ be the singular value decomposition of Φ_n with $D_n = \text{diag}(\delta_1(\Phi_n), \delta_2(\Phi_n))$. Then

$$(\Phi_n^* \Phi_n)^{1/2n} = O_n^* D_n^{1/n} O_n$$

has eigenvalues $\delta_i(\Phi_n)^{1/n} \to e^{\Lambda_i}$ and eigenvectors equal to the columns $O_n^* e_1$, $O_n^* e_2$ of O_n^*. We will in general not be able to prove that O_n converges, due to the non-uniqueness of the singular value decomposition. However, it suffices to show that the spaces spanned by the eigenvectors converge.

3.4.4 Lemma. *Let S be a symmetric $d \times d$ matrix with spectral representation $S = \sum \lambda_k P_k$, where λ_k are the eigenvalues and P_k are the orthogonal projections onto the eigenspaces U_k corresponding to λ_k, uniquely given by*

$$P_k = \frac{1}{2\pi i} \int_{\Gamma_k} (zI - S)^{-1} dz, \qquad (3.4.6)$$

where Γ_k is a circle in the complex plane $\mathbb{C}$ around λ_k which excludes all other eigenvalues of S. Let $S_n = \sum \lambda_k(n) P_k(n)$ be a sequence such that for $n \to \infty$

1. $\lambda_k(n) \to \lambda_i$ *for* $k \in \Sigma_i \neq \emptyset$, *a group of indices,*
2. $\bar{P}_i(n) := \sum_{k \in \Sigma_i} P_k(n) \to P_i$.

Then $S_n \to S$.

Proof. By our assumptions,

$$S_n - S = \sum_i \left(\sum_{k \in \Sigma_i} (\lambda_k(n) - \lambda_i) P_k(n) + \lambda_i (\sum_{k \in \Sigma_i} P_k(n) - P_i) \right) \to 0$$

for $n \to \infty$. □

1. Case $\Lambda_1 = \Lambda_2 =: \lambda_1$

Here $D_n^{1/n} \to e^{\lambda_1} I$ and $\bar{P}_1(n) = P_1(n) + P_2(n) = I$, hence by Lemma 3.4.4 $(\Phi_n^* \Phi_n)^{1/2n} \to e^{\lambda_1} I$, which proves (i).

The filtration in (ii) is trivial, $V_1 = \mathbb{R}^2$. We need to prove that for each $x \in \mathbb{R}^2 \setminus \{0\}$

$$\lambda(x) = \lim_{n\to\infty} \frac{1}{n}\log \|\Phi_n x\| = \lambda_1.$$

But with $O_n x = x_n$

$$\|\Phi_n x\| = \|D_n x_n\| = \left(\delta_1(\Phi_n)^2 (x_n^1)^2 + \delta_2(\Phi_n)^2 (x_n^2)^2\right)^{1/2}.$$

If $\lambda_1 > -\infty$, then for every $\varepsilon > 0$ there is a $C_\varepsilon \in (0, \infty)$ such that for $i = 1, 2$ and all n

$$\frac{1}{C_\varepsilon} e^{n(\lambda_1 - \varepsilon)} \leq \delta_i(\Phi_n) \leq C_\varepsilon e^{n(\lambda_1 + \varepsilon)}.$$

Hence for every ε there is a $C_\varepsilon \in (0, \infty)$ independent of x [4] such that for all n and all $x \neq 0$

$$\|x\| \frac{1}{C_\varepsilon} e^{n(\lambda_1 - \varepsilon)} \leq \|\Phi_n x\| \leq \|x\| C_\varepsilon e^{n(\lambda_1 + \varepsilon)},$$

whence $\lambda(x) = \lambda_1$ for all $x \neq 0$. If $\lambda_1 = -\infty$, then the same estimates give $\lambda(x) = -\infty$ for all $x \neq 0$.

2. Case $\lambda_1 := \Lambda_1 > \Lambda_2 =: \lambda_2$

Now $D_n^{1/n} \to \mathrm{diag}(e^{\lambda_1}, e^{\lambda_2})$. The eigenvectors of $(\Phi_n^* \Phi_n)^{1/2n}$ are $u_1(n) = O_n^* e_1$ for $\delta_1(\Phi_n)^{1/n} \to e^{\lambda_1}$ and $u_2(n) = O_n^* e_2$ for $\delta_2(\Phi_n)^{1/n} \to e^{\lambda_2}$. We now prove that for the corresponding projections $P_1(n) \to P_1$ and $P_2(n) \to P_2$, hence by Lemma 3.4.4

$$(\Phi_n^* \Phi_n)^{1/2n} = \delta_1(\Phi_n) P_1(n) + \delta_2(\Phi_n) P_2(n) \to \Psi = e^{\lambda_1} P_1 + e^{\lambda_2} P_2.$$

This (in fact, a much sharper result) will follow from the next two lemmas.

3.4.5 Lemma. *Let P and Q be orthogonal projections in $\mathbb{R}^2$ with $\dim U = \dim V = 1$, where $U = \mathrm{im} P$ and $V = \mathrm{im} Q$. Then*

$$\|P - Q\| = \delta(U, V) = \|x \wedge y\|, \quad x \in U, y \in V, \|x\| = \|y\| = 1,$$

where δ denotes the distance in the projective space P^1 introduced in Proposition 3.2.8.

Proof. For elementary facts on projections see Kato [199: pp. 55–58].

(i) If P, Q are two orthogonal projections, then always $\|P - Q\| \leq 1$. We have $\|P - Q\| = 1$ if and only if $U \perp V$ if and only if $\delta(U, V) = 1$.

(ii) In general, if $x \in U$, $y \in V$, $\|x\| = \|y\| = 1$, then, using $x = \langle x, y \rangle y + \langle x, y^\perp \rangle y^\perp$,

$$\|(I - Q)P\| = |\langle x, y^\perp \rangle| = \|x \wedge y\| = \delta(U, V).$$

If $\|(I - Q)P\| < 1$, then $\|P - Q\| = \|(I - Q)P\|$ (Kato [199: Theorem 6.34]), and the result follows. □

---

[4] The fact that here and at later occasions C_ε is independent of x will be needed in the proof of a uniform convergence result for two-sided time, see Theorem 3.4.11(v).

3.4.6 Lemma. *Assume $d = 2$ and the conditions of Proposition 3.4.2 with $\lambda_1 > \lambda_2$. Let δ denote the distance in P^1 introduced in Proposition 3.2.8 and let $U_1(n)$ and $U_2(n)$ be the eigenspaces of $(\Phi_n^* \Phi_n)^{1/2n}$ corresponding to the eigenvalues $\delta_1(\Phi_n)^{1/n} > \delta_2(\Phi_n)^{1/n}$ (this inequality holds for all $n \geq n_0$). Then for $i = 1, 2$*

$$\limsup_{n \to \infty} \frac{1}{n} \log \delta(U_i(n), U_i(n+1)) \leq \lambda_2 - \lambda_1 \qquad (3.4.7)$$

(which is also true if $\lambda_2 = -\infty$). In particular, the sequence $(U_i(n))$ is a Cauchy sequence in P^1 and thus converges,

$$\lim_{n \to \infty} U_i(n) = U_i,$$

and the convergence is exponentially fast,

$$\limsup_{n \to \infty} \frac{1}{n} \log \delta(U_i(n), U_i) \leq \lambda_2 - \lambda_1. \qquad (3.4.8)$$

Proof. (i) It suffices to consider the case $i = 2$ since $U_1(n) = TU_2(n)$, $T \in O(2, \mathbb{R})$, thus $\delta(U_1(n), U_1(n+1)) = \delta(U_2(n), U_2(n+1))$.

(ii) First observe that we have $\Phi_n u_1(n) = \delta_1(\Phi_n) V_n e_1$, $\Phi_n u_2(n) = \delta_2(\Phi_n) V_n e_2$, thus $\|\Phi_n u_1(n)\| = \delta_1(\Phi_n)$, $\|\Phi_n u_2(n)\| = \delta_2(\Phi_n)$ and $\Phi_n u_1(n) \perp \Phi_n u_2(n)$. By the definition of Φ_{n+1},

$$\|\Phi_{n+1} u_2(n)\| = \|A_{n+1} \Phi_n u_2(n)\| \leq \|A_{n+1}\| \delta_2(\Phi_n).$$

Represent $u_2(n)$ as $u_2(n) = \alpha_n u_1(n+1) + \beta_n u_2(n+1)$ and note that

$$\begin{aligned}
\|\Phi_{n+1} u_2(n)\| &= \|\alpha_n \Phi_{n+1} u_1(n+1) + \beta_n \Phi_{n+1} u_2(n+1)\| \\
&= (\alpha_n^2 \delta_1(\Phi_{n+1})^2 + \beta_n^2 \delta_2(\Phi_{n+1})^2)^{1/2} \\
&\geq |\alpha_n| \delta_1(\Phi_{n+1}).
\end{aligned}$$

By Proposition 3.2.8

$$\delta(U_2(n), U_2(n+1)) = \|u_2(n) \wedge u_2(n+1)\| = |\alpha_n| \|u_1(n+1) \wedge u_2(n+1)\| = |\alpha_n|,$$

so

$$\delta(U_2(n), U_2(n+1)) \leq \|A_{n+1}\| \frac{\delta_2(\Phi_n)}{\delta_1(\Phi_{n+1})}$$

(note that $\delta_1(\Phi_{n+1}) > 0$, otherwise $\Lambda_1 = \Lambda_2 = -\infty$). The assumptions (3.4.1) and (3.4.2) of Proposition 3.4.2 and Lemma 3.2.1(v) yield (3.4.7). (3.4.8) follows from

$$\delta(U_2(n), U_2) \leq \sum_{i=n}^{\infty} \delta(U_2(i), U_2(i+1))$$

and the discrete time version of Lemma 3.2.1(iv). $\square$

We now complete the proof of Proposition 3.4.2 for $d = 2$. The result of Lemma 3.4.6 is that (both the eigenvalues and) the eigenspaces of $(\Phi_n^* \Phi_n)^{1/2n}$ converge to limits (e^{λ_i} and) U_i. By Lemma 3.4.5, this is equivalent to the convergence of the corresponding orthogonal projections, proving part (i) of the proposition.

To prove part (ii) take $x \in V_2 \setminus \{0\}$ where $V_2 = U_2 \subset V_1 = \mathbb{R}^2$, write $x = \|x\|v$, $v = \alpha_n u_1(n) + \beta_n u_2(n)$ with $\|v\|^2 = \alpha_n^2 + \beta_n^2 = 1$, thus

$$\|x\| |\beta_n| \delta_2(\Phi_n) \leq \|\Phi_n x\| = \|x\| \left(\alpha_n^2 \delta_1(\Phi_n)^2 + \beta_n^2 \delta_2(\Phi_n)^2 \right)^{1/2}. \quad (3.4.9)$$

We have proved in the above lemma that

$$\limsup_{n\to\infty} \frac{1}{n} \log |\alpha_n| = \limsup_{n\to\infty} \frac{1}{n} \log \delta(U_2(n), U_2) \leq \lambda_2 - \lambda_1,$$

thus $\beta_n^2 \to 1$. Lemma 3.2.1 (iii) and (v) applied to (3.4.9) gives

$$\lambda_2 \leq \liminf_{n\to\infty} \frac{1}{n} \log \|\Phi_n x\|$$
$$\leq \limsup_{n\to\infty} \frac{1}{n} \log \|\Phi_n x\| \leq \max(\lambda_2 - \lambda_1 + \lambda_1, \lambda_2) = \lambda_2.$$

More precisely, for each $\varepsilon > 0$ there is a $C_\varepsilon \in (0, \infty)$ independent of x such that for all n and $x \in V_2 \setminus \{0\}$

$$\|x\| \frac{1}{C_\varepsilon} e^{n(\lambda_2 - \varepsilon)} \leq \|\Phi_n x\| \leq \|x\| C_\varepsilon e^{n(\lambda_2 + \varepsilon)},$$

whence

$$\lambda(x) = \lim_{n\to\infty} \frac{1}{n} \log \|\Phi_n x\| = \lambda_2 \quad \text{if } x \in V_2 \setminus \{0\}.$$

Now take $x \in \mathbb{R}^2 \setminus V_2$ and write $x = \alpha u + \beta v$ with unit vectors $u \in U_1$, $v \in U_2 = V_2$, and $\alpha = \langle x, u \rangle \neq 0$. Write $v = \alpha_n u_1(n) + \beta_n u_2(n)$, $u = \gamma_n u_1(n) + \delta_n u_2(n)$. By the above lemma $\alpha_n \to 0$, $\delta_n \to 0$ with exponential speed $\leq \lambda_2 - \lambda_1$, so $\beta_n^2 \to 1$, $\gamma_n^2 \to 1$. We obtain the estimate

$$|\alpha| |\gamma_n| \delta_1(\Phi_n) \leq \|\Phi_n x\|$$
$$= \left((\alpha \gamma_n + \beta \alpha_n)^2 \delta_1(\Phi_n)^2 + (\alpha \delta_n + \beta \beta_n)^2 \delta_2(\Phi_n)^2 \right)^{1/2}.$$

More precisely, by Lemma 3.2.1(iii) and (v), for each $\varepsilon > 0$ there is a $C_\varepsilon \in (0, \infty)$ independent of x such that for all n and all $x \in \mathbb{R}^2 \setminus V_2$

$$\|x\| |\langle \frac{x}{\|x\|}, u \rangle| \frac{1}{C_\varepsilon} e^{n(\lambda_1 - \varepsilon)} \leq \|\Phi_n x\| \leq \|x\| C_\varepsilon e^{n(\lambda_1 + \varepsilon)},$$

whence

$$\lambda(x) = \lim_{n\to\infty} \frac{1}{n} \log \|\Phi_n x\| = \lambda_1 \quad \text{for } x \in \mathbb{R}^2 \setminus V_2.$$

This completes the proof of Proposition 3.4.2 for the case $d = 2$. □

3.4.7 Remark. The crucial fact in the proof of $\lambda(x) = \lambda_2$ for $x \in V_2 \setminus \{0\}$ is that $|\alpha_n| = \delta(U_2, U_2(n)) = |u_2 \wedge u_2(n)| = |\langle u_2^\perp, u_2(n)\rangle|$, $u_2 \in U_2$, decays with rate $\leq \lambda_2 - \lambda_1$, thus cancelling the rate λ_1 of $\delta_1(\Phi_n)$ in the term $|\alpha_n|\delta_1(\Phi_n)$ on the right-hand side of (3.4.9). It is therefore not just the convergence $U_2(n) \to U_2$ but its particular speed that makes the proof work. ∎

Proof of Proposition 3.4.2 for general d

The *general idea*, abstracted from the proof for $d = 2$, is as follows: Let $\Phi_n = V_n D_n O_n$, $D_n = \mathrm{diag}(\delta_1(\Phi_n), \ldots, \delta_d(\Phi_n))$, be the singular value decomposition of Φ_n. By assumption

$$\lim_{n \to \infty} D_n^{1/n} = \mathrm{diag}(e^{\Lambda_1}, \ldots, e^{\Lambda_d}).$$

Denote by $\lambda_1 > \ldots > \lambda_p$ the different numbers among the Λ_i. Let for $i = 1, \ldots, p$, $U_i(n)$ be spanned by the group Σ_i of those eigenvectors of $(\Phi_n^* \Phi_n)^{1/2n} = O_n^* D_n^{1/n} O_n$ corresponding to eigenvalues $\delta_{k(i)}(\Phi_n)^{1/n} \to e^{\lambda_i}$. Let $P_i(n)$ be the corresponding orthogonal projection. We prove that

1. $P_i(n) \to P_i$ $(n \to \infty)$ for $i = 1, \ldots, p$, which, by Lemma 3.4.4, yields $(\Phi_n^* \Phi_n)^{1/2n} \to \Psi = \sum e^{\lambda_i} P_i$ (assertion (i)),
2. the convergence in 1. has just the right speed to ensure assertion (ii).

Step 1: Construction of a sequence $(F(n))$ of flags

We can assume $d \geq 2$ since the case $d = 1$ is trivial. Let

$$\lambda_1 > \ldots > \lambda_p$$

be the different numbers in the sequence $\Lambda_1 \geq \ldots \geq \Lambda_d$, let d_i be the multiplicity of λ_i. If $p = 1$ we can use the proof for $d = 2$ (case $\Lambda_1 = \Lambda_2$) verbatim. Let now $p \geq 2$. Assume $\lambda_p > -\infty$. The modifications necessary if $\lambda_p = -\infty$ are either obvious or given in the course of the proof.

Put

$$\Delta_i := \lambda_i - \lambda_{i+1}, \ i = 1, \ldots, p-1, \quad \Delta := \min_{i=1,\ldots,p-1} \Delta_i > 0.$$

Choose an $\varepsilon \in (0, \Delta)$ and then an N_ε such that for all $i = 1, \ldots, p$

$$\left|\frac{1}{n} \log \delta_{k(i)}(\Phi_n) - \lambda_i\right| < \varepsilon \quad \text{for all } n \geq N_\varepsilon,$$

where $k(i)$ runs through the group Σ_i of d_i indices for which

$$\lim_{n \to \infty} \frac{1}{n} \log \delta_{k(i)}(\Phi_n) = \lambda_i.$$

Let

$$U_i(n) := \mathrm{span}\,(O_n^* e_{d_1 + \ldots + d_{i-1} + 1}, \ldots, O_n^* e_{d_1 + \ldots + d_i}), \ i = 1, \ldots, p,$$

and

$$V_i(n) := U_p(n) \oplus \ldots \oplus U_i(n), \ i = 1, \ldots, p.$$

The nested sequence of subspaces of $\mathbb{R}^d$ given by

$$F(n) = (V_p(n) \subset \ldots \subset V_i(n) \subset \ldots \subset V_1(n))$$

forms a filtration or *flag*, corresponding for all $\varepsilon \in (0, \Delta)$ and all $n \geq N_\varepsilon$ to the vector of dimensions

$$\tau = (d_p, d_p + d_{p-1}, \ldots, d_p + \ldots + d_1 = d)$$

which is independent of ε and n.

The set $F_\tau(d)$ of all flags corresponding to the vector of dimensions τ can be given the structure of a compact C^∞ manifold in a natural way (see Husemoller [176: Chap. 8]). The flag manifold $F_\tau(d)$ can be endowed with a complete metric as follows:

Take $F = (V_p \subset \ldots \subset V_1) \in F_\tau(d)$ and define $U_p = V_p$, U_i = orthogonal complement of V_{i+1} in V_i, $i = p-1, \ldots, 1$, so that

$$V_i = U_p \oplus \ldots \oplus U_i, \ i = 1, \ldots, p.$$

Choose $h = \frac{\Delta}{d-1}$ and define for any $F, \bar{F} \in F_\tau(d)$

$$\delta(F, \bar{F}) := \max_{i,j=1,\ldots,p;\, i \neq j} \max_{x \in U_i,\, y \in \bar{U}_j;\, \|x\|=\|y\|=1} |\langle x, y \rangle|^{h/|\lambda_i - \lambda_j|}. \quad (3.4.10)$$

If $\lambda_p = -\infty$, replace $h/|\lambda_p - \lambda_j|$ and $h/|\lambda_i - \lambda_p|$ by $\frac{1}{d-1}$. Then $\delta(\cdot, \cdot)$ is a complete metric on $F_\tau(d)$ compatible with the natural topology. For a proof (which mainly consists of verifying the triangle inequality) see Goldsheid and Margulis [161: pp. 17–19].

3.4.8 Remark. (i) It follows from the formula

$$\|A\| = \sup_{\|x\| \leq 1, \|y\| \leq 1} |\langle Ax, y \rangle|$$

that for two orthogonal projections P, Q in a Hilbert space

$$\|PQ\| = \|QP\| = \max_{x \in \text{im} P,\, y \in \text{im} Q;\, \|x\|=\|y\|=1} |\langle x, y \rangle|.$$

If for $F \in F_\tau(d)$, P_i denotes the orthogonal projection onto U_i, the metric $\delta(F, \bar{F})$ defined by (3.4.10) can be written as

$$\delta(F, \bar{F}) = \max_{i,j=1,\ldots,p;\, i \neq j} \|P_i \bar{P}_j\|^{h/|\lambda_i - \lambda_j|}. \quad (3.4.11)$$

(ii) For $d = 2$ we obtain $h/(\lambda_1 - \lambda_2) = 1$ which yields exactly the metric in $P^1 \cong F_{(1,2)}(2)$ used above in the proof for $d = 2$ (case $\lambda_1 > \lambda_2$). Indeed, for $F = (U \subset \mathbb{R}^2)$, $\bar{F} = (\bar{U} \subset \mathbb{R}^2)$

$$\delta(F, \bar{F}) = |\langle x, y^\perp \rangle| = |\langle y, x^\perp \rangle|,$$

where $x \in U$, $x^\perp \in U^\perp$, $y \in \bar{U}$, $y^\perp \in \bar{U}^\perp$ are unit vectors, and hence with $x = \langle x, y \rangle y + \langle x, y^\perp \rangle y^\perp$

$$\delta(U, \bar{U}) = \|x \wedge y\| = |\langle x, y^\perp \rangle| \, \|y^\perp \wedge y\| = \delta(F, \bar{F}).$$

∎

Step 2: Reduction to a lemma on the convergence of the sequence (F(n)) of flags

Proposition 3.4.2 is a consequence of the following lemma.

3.4.9 Lemma. *We have*

$$\limsup_{n \to \infty} \frac{1}{n} \log \delta(F(n), F(n+1)) \leq -h. \tag{3.4.12}$$

In particular, the sequence $(F(n))$ is a Cauchy sequence and thus converges in $F_\tau(d)$,

$$\lim_{n \to \infty} F(n) = F,$$

and this convergence is exponentially fast,

$$\limsup_{n \to \infty} \frac{1}{n} \log \delta(F(n), F) \leq -h. \tag{3.4.13}$$

We first derive Proposition 3.4.2 from this lemma. By (3.4.13) and the definition of the metric δ we have $F(n) = (V_p(n) \subset \ldots \subset V_1(n)) \to F = (V_p \subset \ldots \subset V_1)$ if and only if for each i, $U_i(n) \to U_i$ in $G_{d_i}(d)$, where we use the complete metric on $G_k(d)$ given by

$$\delta(U, \bar{U}) := \|(I - P)Q\| = \|(I - Q)P\|,$$

where P and Q are the orthogonal projections onto U and $\bar{U}$ in $G_k(d)$. Indeed, denoting by P_i and $P_i(n)$ the orthogonal projections onto U_i and $U_i(n)$, respectively, and using the identity $I = \sum_{j=1}^p P_j(n)$,

$$\delta(U_i, U_i(n)) = \|(I - P_i(n))P_i\| \leq \sum_{j=1,\ldots,p;\, j \neq i} \|P_j(n)\, P_i\| \to 0.$$

Recalling the form (3.4.11) of the metric on $F_\tau(d)$, $F(n) \to F$ in $F_\tau(d)$ implies $U_i(n) \to U_i$ in $G_{d_i}(d)$ for $i = 1, \ldots, p$.

Conversely, assume $\delta(U_i, U_i(n)) \to 0$ for all i. A result of Kato [199: p. 56] applies saying that if P and $\bar{P}$ are orthogonal projections with $\dim \operatorname{im} P = \dim \operatorname{im} \bar{P}$ and $\|(I - P)\bar{P}\| = c < 1$, then

$$\|P - \bar{P}\| = \|(I - P)\bar{P}\| = \|(I - \bar{P})P\| = c. \tag{3.4.14}$$

Hence $P_i(n) \to P_i$ for all i which implies $\delta(F(n), F) \to 0$.

In particular, $F(n) \to F$ implies $(\Phi_n^* \Phi_n)^{1/2n} \to \sum e^{\lambda_i} P_i$, i.e. part (i) of the proposition.

To derive part (ii) from the lemma let for fixed $i = 1, \ldots, p$, $x \in V_i \setminus V_{i+1}$, $\|x\| = 1$, and write

$$x = \sum_{j=1}^{p} P_j(n)\, x = \sum_{j=i}^{p} P_j\, x.$$

By assumption $P_i x \neq 0$. The vectors $(\Phi_n P_j(n) x)_{j=1,\ldots,p}$ are orthogonal and

$$\|P_j(n) x\|\, \underline{\delta}_j(\Phi_n) \leq \|\Phi_n P_j(n) x\| \leq \|P_j(n) x\|\, \bar{\delta}_j(\Phi_n),$$

where

$$\underline{\delta}_j(\Phi_n) := \min_{\Sigma_j} \delta_{k(j)}(\Phi_n), \quad \bar{\delta}_j(\Phi_n) := \max_{\Sigma_j} \delta_{k(j)}(\Phi_n)$$

with

$$\lambda(\underline{\delta}_j(\Phi_n)) = \lambda(\bar{\delta}_j(\Phi_n)) = \lambda_j.$$

Hence

$$\sum_{j=1}^{p} \|P_j(n) x\|^2\, \underline{\delta}_j(\Phi_n)^2 \leq \|\Phi_n x\|^2 \leq \sum_{j=1}^{p} \|P_j(n) x\|^2\, \bar{\delta}_j(\Phi_n)^2. \qquad (3.4.15)$$

Since $\|P_i(n) x\|\, \underline{\delta}_i(\Phi_n) \leq \|\Phi_n x\|$ and $\|P_i(n) x\| \to \|P_i x\| > 0$

$$\lambda_i \leq \liminf_{n \to \infty} \frac{1}{n} \log \|\Phi_n x\|.$$

We now estimate the summands in (3.4.15) from above. For $j = i, \ldots, p$ we use the trivial estimate $\|P_j(n) x\| \leq 1$ to obtain

$$\lambda(\|P_j(n) x\|\, \bar{\delta}_j(\Phi_n)) \leq \lambda_i.$$

For $j = 1, \ldots, i-1$ we use

$$\|P_j(n) x\| = \|P_j(n) \sum_{k=i}^{p} P_k x\| \leq \sum_{k=i}^{p} \|P_j(n) P_k\|$$

and (3.4.13) to obtain from (3.4.11) $\lambda(\|P_j(n) P_k\|^{h/|\lambda_j - \lambda_k|}) \leq -h$ and

$$\lambda(\|P_j(n) x\|\, \bar{\delta}_j(\Phi_n)) \leq \lambda_j + \max_{k=i,\ldots,p} \frac{-h|\lambda_j - \lambda_k|}{h} = \lambda_j + (\lambda_i - \lambda_j) = \lambda_i.$$

Hence $\limsup_{n \to \infty} \frac{1}{n} \log \|\Phi_n x\| \leq \lambda_i$ and altogether

$$\lim_{n \to \infty} \frac{1}{n} \log \|\Phi_n x\| = \lambda_i,$$

which is part (ii) of the proposition.

Step 3: Proof of Lemma 3.4.9

Since
$$\delta(F(n), F) \leq \sum_{k=n}^{\infty} \delta(F(k), F(k+1))$$

(3.4.13) follows from (3.4.12) by the discrete time version of Lemma 3.2.1(iv). It thus suffices to prove (3.4.12) which, introducing

$$\Delta_{ij}(n) := \|P_i(n)P_j(n+1)\| = \|P_j(n+1)P_i(n)\|,$$

is equivalent to

$$\lambda(\Delta_{ij}(n)) \leq -|\lambda_i - \lambda_j| \quad \text{for } i \neq j. \tag{3.4.16}$$

Case 1: $i > j$

Now $\lambda_i < \lambda_j$. For a unit vector $x \in U_i(n)$ and $y = P_j(n+1)x \in U_j(n+1)$

$$\|\Phi_{n+1}x\| = \|A_{n+1}\Phi_n x\| \leq \|A_{n+1}\| \, \bar{\delta}_i(\Phi_n)$$

and

$$\|\Phi_{n+1}x\| = (\|\Phi_{n+1}y\|^2 + \|\Phi_{n+1}(x-y)\|^2)^{1/2} \geq \|\Phi_{n+1}y\| = \underline{\delta}_j(\Phi_{n+1})\|y\|,$$

so

$$\|y\| \leq \Delta_{ij}(n) = \|P_j(n+1)P_i(n)\| \leq \|A_{n+1}\| \frac{\bar{\delta}_i(\Phi_n)}{\underline{\delta}_j(\Phi_{n+1})},$$

whence

$$\lambda(\Delta_{ij}(n)) \leq \lambda_i - \lambda_j = -|\lambda_i - \lambda_j|.$$

For $\lambda_p = -\infty$ we obtain $\lambda(\Delta_{pj}(n)) = -\infty$ for $j = 1, \ldots, p-1$.

Case 2: $i < j$

Now $\lambda_i > \lambda_j$.

(a) We first show that we can repeat the argument of case 1 if we have the additional assumption that $A_n \in Gl(d, \mathbb{R})$ and

$$\limsup_{n \to \infty} \frac{1}{n} \log \|A_n^{-1}\| \leq 0. \tag{3.4.17}$$

For a unit vector $x \in U_j(n+1)$ and $y = P_i(n)x \in U_i(n)$

$$\|\Phi_n x\| = \|A_{n+1}^{-1}\Phi_{n+1}x\| \leq \|A_{n+1}^{-1}\| \, \bar{\delta}_j(\Phi_{n+1})$$

and

$$\|\Phi_n x\| = (\|\Phi_n y\|^2 + \|\Phi_n(x-y)\|^2)^{1/2} \geq \|\Phi_n y\| = \underline{\delta}_i(\Phi_n)\|y\|,$$

so

$$\Delta_{ij}(n) = \|P_i(n)P_j(n+1)\| \leq \|A_{n+1}^{-1}\| \frac{\bar{\delta}_j(\Phi_{n+1})}{\underline{\delta}_i(\Phi_n)}$$

and hence

$$\lambda(\Delta_{ij}(n)) \leq \lambda_j - \lambda_i = -|\lambda_i - \lambda_j|.$$

(b) In the general situation we can reduce case 2 ($i < j$) to case 1 ($i > j$) as follows:

(i) Observe that $\lambda(\Delta_{ij}(n)) \leq -|\lambda_i - \lambda_j|$ for $i > j$ implies
$$\lim_{n \to \infty} \|(I - P_i(n+1))P_i(n)\| = 0, \quad i = 1, \ldots, p, \qquad (3.4.18)$$
hence by Kato's result (3.4.14)
$$\lim_{n \to \infty} \|P_i(n) - P_i(n+1)\| = 0, \quad i = 1, \ldots, p.$$

To prove (3.4.18), we proceed by induction. For $i = p$, using $I = \sum_{j=1}^{p} P_j(n+1)$,
$$\|(I - P_p(n+1))P_p(n)\| = \|\sum_{j=1}^{p-1} P_j(n+1)P_p(n)\| \leq \sum_{j=1}^{p-1} \Delta_{pj}(n) \to 0 \ (n \to \infty).$$

Suppose we have proved the claim up to the index $i \leq p$. Then it also holds for $i - 1 \geq 1$. For, with
$$I = \sum_{j=1}^{i-2} P_j(n+1) + P_{i-1}(n+1) + \sum_{j=i}^{p} P_j(n+1),$$

$$\|P_{i-1}(n)(I - P_{i-1}(n+1))\|$$
$$\leq \sum_{j=1}^{i-2} \|P_{i-1}(n)P_j(n+1)\| + \sum_{j=i}^{p} \|P_{i-1}(n)P_j(n+1)\|$$
$$\leq \sum_{j=1}^{i-2} \Delta_{i-1,j}(n) + \sum_{j=i}^{p} \|P_{i-1}(n)(I - P_j(n) + P_j(n))P_j(n+1)\|$$
$$\leq \sum_{j=1}^{i-2} \Delta_{i-1,j}(n) + \sum_{j=i}^{p} \|(I - P_j(n))P_j(n+1)\|$$
$$\to \ 0 \ (n \to \infty).$$

We can thus assume from now on that n is chosen so large that
$$\|P_i(n) - P_i(n+1)\| \leq \frac{1}{2}, \ i = 1, \ldots, p.$$

(ii) If A, A' and B are bounded operators in a Banach space with $A^2 = A$, $\|A - A'\| \leq 1/2$, then
$$\|AB\| \leq 2\|A'AB\|. \qquad (3.4.19)$$
For,
$$\|A'AB\| = \|A^2B - (A - A')AB\| \geq \|A^2B\| - \|(A - A')AB\|$$
$$\geq \|AB\| - \|A - A'\|\|AB\| \geq \frac{1}{2}\|AB\|.$$

(iii) For $i \neq j$

$$0 = P_i(n+1)P_j(n+1) = \sum_{k=1}^{p} P_i(n+1)P_k(n)P_j(n+1),$$

thus

$$P_i(n+1)P_i(n)P_j(n+1) = - \sum_{k=1,\ldots,p;\, k \neq i} P_i(n+1)P_k(n)P_j(n+1).$$

For n big enough by (3.4.19) and since $\|P_i(n) - P_i(n+1)\| \leq \frac{1}{2}$

$$\begin{aligned}
\Delta_{ij}(n) &\leq 2\|P_i(n+1)P_i(n)P_j(n+1)\| \\
&\leq 2 \sum_{k=1,\ldots,p;\, k \neq i} \|P_i(n+1)P_k(n)^2 P_j(n+1)\| \\
&\leq 2 \sum_{k=1,\ldots,p;\, k \neq i} \Delta_{ki}(n)\Delta_{kj}(n).
\end{aligned}$$

Since $1 \leq i < j \leq p$,

$$\Delta_{ij} \leq 2 \left(\sum_{k=1}^{i-1} \Delta_{ki}\Delta_{kj} + \sum_{k=i+1}^{j-1} \Delta_{ki}^0 \Delta_{kj} + \sum_{k=j}^{p} \Delta_{ki}^0 \Delta_{kj}^0 \right), \qquad (3.4.20)$$

where we have omitted the variable n and have marked those terms with a superscript 0 for which case 1 applies.

(iv) The Lyapunov index of the last term in (3.4.20) can be estimated as follows using the trivial bound $\Delta_{ij}(n) \leq 1$ and the result of case 1:

$$\begin{aligned}
\lambda \left(\sum_{k=j}^{p} \Delta_{ki}^0(n)\Delta_{kj}^0(n) \right) &\leq \lambda \left(\sum_{k=j}^{p} \Delta_{ki}^0(n) \right) \\
&= \max_{k=j,\ldots,p} (-|\lambda_k - \lambda_i|) = -|\lambda_j - \lambda_i|.
\end{aligned}$$

(v) We first derive the following crude estimate: For $1 \leq i < j \leq p$

$$\lambda(\Delta_{ij}(n)) \leq -\min(\Delta_1,\ldots,\Delta_i) \leq -\Delta < 0.$$

The proof proceeds by induction. First put $i = 1$ and use $\Delta_{kj}(n) \leq 1$. The first sum in (3.4.20) is not present and

$$\begin{aligned}
\lambda(\Delta_{1j}(n)) &\leq \max(\max_{k=2,\ldots,j-1}(-|\lambda_k - \lambda_1|), -|\lambda_j - \lambda_1|) \\
&= -\min_{k=2,\ldots,j} |\lambda_k - \lambda_1| = -\Delta_1.
\end{aligned}$$

Using this for the first term in (3.4.20),

$$\lambda(\Delta_{2j}(n)) \leq \max(-\Delta_1, \max_{k=3,\ldots,j-1}(-|\lambda_k - \lambda_2|), -|\lambda_j - \lambda_2|)$$
$$= \max(-\Delta_1, -\Delta_2) = -\min(\Delta_1, \Delta_2),$$

in general
$$\lambda(\Delta_{ij}(n)) \leq -\min(\Delta_1, \ldots, \Delta_i)$$

and finally
$$\lambda(\Delta_{p-1,p}(n)) \leq -\min(\Delta_1, \ldots, \Delta_{p-1}) = -\Delta.$$

(vi) We improve this crude estimate as follows: For the terms of the first sum in (3.4.20) we have $k < i < j$, and the crude estimate gives

$$\lambda(\Delta_{ki}(n)\Delta_{kj}(n)) \leq -2\Delta.$$

For the terms of the second sum we use the estimate of case 1 for Δ_{ki}^0 and the crude estimate for Δ_{kj} to obtain

$$\lambda(\Delta_{ki}^0(n)\Delta_{kj}(n)) \leq -|\lambda_k - \lambda_i| - \Delta.$$

Altogether
$$\lambda(\Delta_{ij}(n)) \leq \max(-2\Delta, \max_{k=i+1,\ldots,j-1}(-|\lambda_k - \lambda_i| - \Delta), -|\lambda_i - \lambda_j|)$$
$$= \max(-2\Delta, -\Delta_i - \Delta, -|\lambda_i - \lambda_j|)$$
$$= \max(-2\Delta, -|\lambda_i - \lambda_j|).$$

(vii) We have now two cases[5]:

— Either $-|\lambda_i - \lambda_j| \geq -2\Delta$. Then $\lambda(\Delta_{ij}(n)) \leq -|\lambda_i - \lambda_j|$, and we are done.
— Or $-|\lambda_i - \lambda_j| < -2\Delta$.

In the second case the new estimate $\lambda(\Delta_{ij}(n)) \leq -2\Delta$ is strictly better than the crude estimate.

The first term on the right-hand side of (3.4.20) is estimated as

$$\lambda\left(\sum_{k=1}^{i-1} \Delta_{ki}\Delta_{kj}\right) \leq \max_{k=1,\ldots,i-1} (\lambda(\Delta_{ki}) + \lambda(\Delta_{kj})).$$

Applying the estimate of step (vi) and taking into account the definition of Δ we arrive at

$$\lambda(\Delta_{ki}) \leq \max_{k=1,\ldots,i-1}(-2\Delta, -|\lambda_k - \lambda_i|) \leq -\Delta \quad \text{for each } k = 1, \ldots, i-1.$$

On the other hand

$$-|\lambda_k - \lambda_j| < -|\lambda_i - \lambda_j| < -2\Delta \quad \text{for each } k = 1, \ldots, i-1,$$

resulting in

---

[5] I am indebted to Anna Kwiecinska for suggesting an improvement in this step.

$$\lambda(\Delta_{kj}) \le \max_{k=1,\ldots,i-1}(-2\Delta, -|\lambda_k - \lambda_j|) = -2\Delta,$$

and finally

$$\lambda\left(\sum_{k=1}^{i-1}\Delta_{ki}\Delta_{kj}\right) \le -3\Delta.$$

Now the second term on the right-hand side of (3.4.20) is estimated as

$$\lambda\left(\sum_{k=i+1}^{j-1}\Delta^0_{ki}\Delta_{kj}\right) \le \max_{k=i+1,\ldots,j-1}(\lambda(\Delta^0_{ki}) + \lambda(\Delta_{kj})) = \lambda(\Delta^0_{k_0 i}) + \lambda(\Delta_{k_0 j})$$

for some $k_0 \in \{i+1,\ldots,j-1\}$. Applying the estimate of step (vi) to the right-hand side of the last estimate we obtain

$$\lambda(\Delta^0_{k_0 i}) + \lambda(\Delta_{k_0 j})$$
$$\le -|\lambda_{k_0} - \lambda_i| + \max(-2\Delta, -|\lambda_{k_0} - \lambda_j|)$$
$$= \begin{cases} -|\lambda_{k_0} - \lambda_i| - |\lambda_{k_0} - \lambda_j| = -|\lambda_i - \lambda_j|, & |\lambda_{k_0} - \lambda_j| \le 2\Delta, \\ -|\lambda_{k_0} - \lambda_i| - 2\Delta \le -3\Delta, & |\lambda_{k_0} - \lambda_j| > 2\Delta. \end{cases}$$

For the third term on the right-hand side of (3.4.20) we use the estimate of step (iv) and obtain

$$\lambda\left(\sum_{k=j}^{p}\Delta^0_{ki}\Delta^0_{kj}\right) \le -|\lambda_i - \lambda_j|.$$

Collecting the three estimates together we finally obtain

$$\lambda(\Delta_{ij}(n)) \le \max\left(\lambda(\sum_{k=1}^{i-1}\Delta_{ki}\Delta_{kj}), \lambda(\sum_{k=i+1}^{j-1}\Delta^0_{ki}\Delta_{kj}), \lambda(\sum_{k=j}^{p}\Delta^0_{ki}\Delta^0_{kj})\right)$$
$$\le \max(-3\Delta, -|\lambda_i - \lambda_j|). \quad (3.4.21)$$

We again have two cases:
− either $-|\lambda_i - \lambda_j| \ge -3\Delta$. Then $\lambda(\Delta_{ij}(n)) \le -|\lambda_i - \lambda_j|$, and we are done,
− or $-|\lambda_i - \lambda_j| < -3\Delta$.

In the second case we apply the estimate (3.4.21) to the right-hand side of (3.4.20) to obtain the new estimate

$$\lambda(\Delta_{ij}(n)) \le \max(-4\Delta, -|\lambda_i - \lambda_j|).$$

We continue this procedure until we obtain in finitely many steps

$$\lambda(\Delta_{ij}(n)) \le -|\lambda_i - \lambda_j|,$$

which was to be proved.

For $\lambda_p = -\infty$ the same procedure gives $\lambda(\Delta_{ij}(n)) = -\infty$ for all $i = 1,\ldots,p-1$.

This completes the proof of Lemma 3.4.9 and thus of Proposition 3.4.2. □

3.4.10 Remark. (i) The filtration and the spectrum of a linear cocycle are independent of the choice of the vector norm and a corresponding matrix norm in $\mathbb{R}^d$, see also Remark 3.3.2.

Moreover, let V be a real normed vector space of dimension d with norm $\|\cdot\|$ and let $\omega \mapsto T(\omega) \in \mathcal{L}(V)$ be a measurable random linear operator with $\log^+ \|T(\cdot)\| \in L^1$. Then the MET Theorem 3.4.1 holds for the linear cocycle $S(n,\omega) := T(\theta^{n-1}\omega) \circ \cdots \circ T(\omega)$ for any norm in V and the filtration and the spectrum are independent of that norm, similarly for the other versions of the MET that follow.

To see this, choose a basis F in V. The coordinate mapping $k_F : V \to \mathbb{R}^d$ becomes an isometric isomorphism if we choose the standard scalar product $\langle \cdot, \cdot \rangle_S$ in $\mathbb{R}^d$ and $\langle x, y \rangle_F := \langle k_F(x), k_F(y) \rangle_S$ in V. The norm $\|\cdot\|_F$ is equivalent to any other norm in V.

(ii) The MET is also valid in $\mathbb{C}^d$ (or any finite-dimensional normed $\mathbb{C}$-vector space V). Our proof, now based on the polar and singular value decomposition in a unitary space (see Gantmacher [157: IX.12]), holds verbatim.

(iii) Uniqueness of spectrum and flag: Suppose the linear cocycle $\Phi(n,\omega)$ satisfies the conditions of Theorem 3.4.1. Then its spectrum and flag are unique. This means that whenever there is another flag

$$\{0\} \subset W_q \subset \ldots \subset W_1 = \mathbb{R}^d$$

and numbers $\mu_q < \ldots < \mu_1$ with $\lambda(x) = \mu_k$ if and only if $x \in W_k \setminus W_{k+1}$, then $q = p$, $\mu_k = \lambda_k$ and $W_k = V_k$. The proof is left as an exercise. ∎

3.4.2 The MET for Two-Sided Time

3.4.11 Theorem (MET for Two-Sided Time). *Let Φ be a linear cocycle with two-sided time over the metric DS $(\Omega, \mathcal{F}, \mathbb{P}, (\theta(t))_{t \in \mathbb{T}})$.*

(A) Discrete time $\mathbb{T} = \mathbb{Z}$: Let

$$\Phi(n,\omega) = \begin{cases} A(\theta^{n-1}\omega) \cdots A(\omega), & n > 0, \\ I, & n = 0, \\ A^{-1}(\theta^n \omega) \cdots A^{-1}(\theta^{-1}\omega), & n < 0, \end{cases}$$

be generated by $A : \Omega \to Gl(d, \mathbb{R})$ and assume

$$\log^+ \|A(\cdot)\| \in L^1(\Omega, \mathcal{F}, \mathbb{P}) \quad \text{and} \quad \log^+ \|A^{-1}(\cdot)\| \in L^1(\Omega, \mathcal{F}, \mathbb{P}).$$

Then there exists an invariant set $\tilde{\Omega}$ of full measure on which the statements of the MET for $\mathbb{T} = \mathbb{N}$ (Theorem 3.4.1(B)) hold. Moreover, for each $\omega \in \tilde{\Omega}$ there exists a splitting

$$\mathbb{R}^d = E_1(\omega) \oplus \ldots \oplus E_{p(\omega)}(\omega)$$

of $\mathbb{R}^d$ into random subspaces $E_i(\omega)$ (so-called Oseledets *spaces) (depending measurably on ω, hence are random closed sets) with dimension $\dim E_i(\omega) = $*

$d_i(\omega)$ *(so-called Oseledets splitting) with the following properties: For $i \in \{1, \ldots, p(\omega)\}$*

(i) if $P_i(\omega) : \mathbb{R}^d \to E_i(\omega)$ denotes the projection onto $E_i(\omega)$ along $F_i(\omega) := \oplus_{j \neq i} E_j(\omega)$, then

$$A(\omega) P_i(\omega) = P_i(\theta\omega) A(\omega),$$

equivalently

$$A(\omega) E_i(\omega) = E_i(\theta\omega),$$

(ii) we have

$$\lim_{n \to \pm\infty} \frac{1}{n} \log \|\Phi(n,\omega)x\| = \lambda_i(\omega) \iff x \in E_i(\omega) \setminus \{0\},$$

(iii) convergence in (ii) is uniform with respect to $x \in E_i(\omega) \cap S^{d-1}$ for each fixed ω.

(B) Continuous time $\mathbb{T} = \mathbb{R}$: Assume that $\alpha^+ \in L^1$ and $\alpha^- \in L^1$, where

$$\alpha^{\pm}(\omega) := \sup_{0 \leq t \leq 1} \log^+ \|\Phi(t,\omega)^{\pm 1}\|.$$

Then all statements of part (A) hold with n, θ and $A(\omega)$ replaced with t, $\theta(t)$ and $\Phi(t,\omega)$, and the set $\tilde{\Omega} \in \mathcal{F}$ of full measure is now invariant with respect to $(\theta(t))_{t \in \mathbb{R}}$.

Proof. Part (A):

Theorem 3.4.1 (A) and (B) applied to the cocycle $\Phi(n,\omega)$ yields, on an invariant set of full measure, the finite spectrum

$$\mathcal{S}(\theta, \Phi)(\omega) = \{(\lambda_i(\omega), d_i(\omega))_{i=1,\ldots,p(\omega)}\}$$

and the forward filtration

$$V_{p(\omega)}(\omega) \subset \ldots \subset V_1(\omega) = \mathbb{R}^d, \quad A(\omega) V_i(\omega) = V_i(\theta\omega).$$

The same theorem applied to the cocycle $\Phi(-n,\omega)$ with time $n \in \mathbb{N}$, over θ^{-1} yields, on an invariant set of full measure, the finite spectrum

$$\mathcal{S}(\theta^{-1}, \Phi(-\cdot))(\omega) = \{(\lambda_i^-(\omega), d_i^-(\omega))_{i=1,\ldots,p^-(\omega)}\}$$

and the backward filtration

$$V_{p^-(\omega)}^-(\omega) \subset \ldots \subset V_1^-(\omega) = \mathbb{R}^d, \quad A^{-1}(\theta^{-1}\omega) V_i^-(\omega) = V_i^-(\theta^{-1}\omega).$$

The two-sided Furstenberg–Kesten theorem 3.3.10 assures that on an invariant set $\tilde{\Omega}$ of full measure all statements above are true and

$$\mathcal{S}(\theta, \Phi)(\omega) = -\mathcal{S}(\theta^{-1}, \Phi(-\cdot))(\omega),$$

more precisely, $p(\omega) = p^-(\omega)$ and for $i = 1, \ldots, p(\omega)$

$$d_i(\omega) = d^-_{p(\omega)+1-i}(\omega), \quad \lambda_i(\omega) = -\lambda^-_{p(\omega)+1-i}(\omega). \tag{3.4.22}$$

We now construct the spaces $E_i(\omega)$ as intersections of certain spaces from the forward and the backward filtration. We set

$$E_i(\omega) := V_i(\omega) \cap V^-_{p(\omega)+1-i}(\omega), \quad 1 \leq i \leq p(\omega)$$

(in particular, $E_1(\omega) := V^-_{p(\omega)}$, $E_{p(\omega)} := V_{p(\omega)}$). As a preparation we prove

3.4.12 Lemma. *On an invariant set $\tilde{\Omega}_1 \subset \tilde{\Omega}$ of full measure*
  (i) $V_{i+1}(\omega) \cap V^-_{p(\omega)+1-i}(\omega) = \{0\}$, $1 \leq i \leq p(\omega)$,
and
  (ii) $V_{i+1}(\omega) \oplus V^-_{p(\omega)+1-i}(\omega) = \mathbb{R}^d$, $1 \leq i \leq p(\omega)$.

Proof. We show that the statements of the lemma hold on a full set $\tilde{\Omega}_2 \subset \tilde{\Omega}$. Passing to $\tilde{\Omega}_1 = \cap_{n \in \mathbb{Z}} \theta^n \tilde{\Omega}_2$ we have an invariant set of full measure on which the lemma and the above statements hold. This set $\tilde{\Omega}_1$ also qualifies for the invariant set of full measure mentioned in the theorem.

(i) Select $p_0 \in \{2, \ldots, d\}$ arbitrary, but fixed. Define $\tilde{\Omega}_{p_0} := \{\omega \in \tilde{\Omega} : p(\omega) \geq p_0\}$. This is an invariant set. Assume $\mathbb{P}(\tilde{\Omega}_{p_0}) > 0$ and choose $i \in \{1, \ldots, p_0 - 1\}$. We prove that $\mathbb{P}(B) = 0$, where

$$B := \{\omega \in \tilde{\Omega}_{p_0} : V_{i+1}(\omega) \cap V^-_{p(\omega)+1-i}(\omega) \neq \{0\}\}.$$

The idea of the proof is as follows: We move for an $\omega \in B$ the vector $x \in V_{i+1}(\omega) \cap V^-_{p(\omega)+1-i}(\omega)$ from time 0 back to time $-N$ by $\Phi(-N, \omega)$ with rate at most $-\lambda_i(\omega)$. Then we move the same point from time $-N$ forward to time 0 by $\Phi(N, \theta^{-N}\omega)$ with rate at most $\lambda_{i+1}(\omega)$. Since $\lambda_{i+1}(\omega) < \lambda_i(\omega)$ the result will be a small quantity. But this would contradict the cocycle property which gives $x = \Phi(N, \theta^{-N}\omega)\Phi(-N, \omega)x$, unless $x = 0$.

Put $\delta(\omega) := \frac{1}{2}(\lambda_i(\omega) - \lambda_{i+1}(\omega))$. For each $\varepsilon > 0$ choose an integer N_ε so big that $\mathbb{P}(C_\varepsilon) > (1 - \frac{\varepsilon}{2})\mathbb{P}(\tilde{\Omega}_{p_0})$ and $\mathbb{P}(D_\varepsilon) = \mathbb{P}(\theta^{N_\varepsilon} D_\varepsilon) > (1 - \frac{\varepsilon}{2})\mathbb{P}(\tilde{\Omega}_{p_0})$, where

$$C_\varepsilon := \{\omega \in \tilde{\Omega}_{p_0} : \|\Phi(-N_\varepsilon, \omega)x\| < e^{N_\varepsilon(-\lambda_i(\omega)+\delta(\omega))}\|x\|$$
$$\text{for all } x \in V^-_{p(\omega)+1-i}(\omega) \setminus \{0\}\}$$

and

$$D_\varepsilon := \{\omega \in \tilde{\Omega}_{p_0} : \|\Phi(N_\varepsilon, \omega)x\| < e^{N_\varepsilon(\lambda_{i+1}(\omega)+\delta(\omega))}\|x\|$$
$$\text{for all } x \in V_{i+1}(\omega) \setminus \{0\}\}.$$

By the proof of Theorem 3.4.1, these estimates hold for all $n \geq N_0(\omega)$ independently of x, hence we can choose an integer N_ε such that

$$\mathbb{P}\{\omega \in \tilde{\Omega}_{p_0} : N_0(\omega) \leq N_\varepsilon\} > (1 - \frac{\varepsilon}{2})\mathbb{P}(\tilde{\Omega}_{p_0}).$$

Now

$$\mathbb{P}(B) = \mathbb{P}(B \cap C_\varepsilon \cap \theta^{N_\varepsilon} D_\varepsilon) + \mathbb{P}(B \cap (C_\varepsilon \cap \theta^{N_\varepsilon} D_\varepsilon)^c)$$
$$\leq \mathbb{P}(B \cap C_\varepsilon \cap \theta^{N_\varepsilon} D_\varepsilon) + \varepsilon.$$

We prove that $B \cap C_\varepsilon \cap \theta^{N_\varepsilon} D_\varepsilon = \emptyset$ by the above argument. Suppose $\omega \in B \cap C_\varepsilon \cap \theta^{N_\varepsilon} D_\varepsilon$.

(a) Since $\omega \in B$ there exists an $x \in V_{i+1}(\omega) \cap V^-_{p(\omega)+1-i}(\omega) \setminus \{0\}$ such that without loss of generality

$$1 = \|x\| = \|\Phi(N_\varepsilon, \theta^{-N_\varepsilon}\omega) \frac{\Phi(-N_\varepsilon, \omega)x}{\|\Phi(-N_\varepsilon, \omega)x\|}\| \, \|\Phi(-N_\varepsilon, \omega)x\|, \tag{3.4.23}$$

where we have used the cocycle property.

(b) Since $\omega \in C_\varepsilon$

$$\|\Phi(-N_\varepsilon, \omega)x\| < e^{N_\varepsilon(-\lambda_i(\omega)+\delta(\omega))}. \tag{3.4.24}$$

(c) Since $\omega \in \theta^{N_\varepsilon} D_\varepsilon$, for any $w \in V_{i+1}(\theta^{-N_\varepsilon}\omega) \setminus \{0\}$

$$\|\Phi(N_\varepsilon, \theta^{-N_\varepsilon}\omega)w\| < e^{N_\varepsilon(\lambda_{i+1}(\omega)+\delta(\omega))}\|w\|, \tag{3.4.25}$$

where we have used the invariance of λ_{i+1} and δ. Since $\Phi(-N_\varepsilon, \omega)V_{i+1}(\omega) = V_{i+1}(\theta^{-N_\varepsilon}\omega)$ on $\tilde{\Omega}$,

$$w = \frac{\Phi(-N_\varepsilon, \omega)x}{\|\Phi(-N_\varepsilon, \omega)x\|} \in V_{i+1}(\theta^{-N_\varepsilon}\omega).$$

Estimating (3.4.23) with (3.4.24) and (3.4.25) gives

$$1 < e^{N_\varepsilon(-\lambda_i(\omega)+\delta(\omega)+\lambda_{i+1}(\omega)+\delta(\omega))} = e^0 = 1,$$

which is a contradiction. Thus $\mathbb{P}(B) < \varepsilon$ with ε arbitrary, proving (i).

(ii) By (i) the sum in (ii) is direct. We have $\dim V_{i+1}(\omega) = \sum_{k=i+1}^{p(\omega)} d_k(\omega)$ and by (3.4.22)

$$\dim V^-_{p(\omega)+1-i}(\omega) = \sum_{k=1}^{i} d_k(\omega),$$

giving $\dim V_{i+1}(\omega) + \dim V^-_{p(\omega)+1-i}(\omega) = d$ and thus (ii). □

Continuing the proof of part (i) of Theorem 3.4.11(A) we use the elementary formula

$$\dim(U \cap V) = \dim U + \dim V - \dim(U + V), \quad U, V \subset \mathbb{R}^d,$$

and $V_i + V^-_{p+1-i} \supset V_{i+1} + V^-_{p+1-i} = \mathbb{R}^d$ (Lemma 3.4.12(ii)) to obtain

$$\begin{aligned}
\dim E_i(\omega) &= \dim V_i(\omega) + \dim V^-_{p(\omega)+1-i}(\omega) - \dim(V_i(\omega) + V^-_{p(\omega)+1-i}(\omega)) \\
&= \sum_{k=i}^{p(\omega)} d_k(\omega) + \sum_{k=1}^{i} d_k(\omega) - d \\
&= d_i(\omega).
\end{aligned}$$

We now prove that the sum of the E_i is direct. This is the case if and only if $E_i \cap F_i = \{0\}$ for all i, where $F_i := \sum_{j \neq i} E_j$. By definition

$$E_i \cap F_i = E_i \cap (U+V), \quad U = \sum_{j=1}^{i-1} E_j, \quad V = \sum_{j=i+1}^{p} E_j,$$

where for $i=1$, U is not present, and for $i=p$, V is not present. Note that

$$U = \sum_{j=1}^{i-1} (V_j \cap V^-_{p+1-j}) \subset V^-_{p+1-(i-1)}$$

and

$$V = \sum_{j=i+1}^{p} (V_j \cap V^-_{p+1-j}) \subset V_{i+1}.$$

Suppose $x \in E_i \cap F_i$. Then $x \in V_i$ and $x \in V^-_{p+1-i}$ and $x = u+v$, $u \in U$, $v \in V$.

Since $x \in V_i$ and $v \in V_{i+1} \subset V_i$, we have $x - v = u \in V_i \cap V^-_{p+1-(i-1)} = \{0\}$ (Lemma 3.4.12(i)), i.e. $u = 0$.

Since $x \in V^-_{p+1-i}$ and $x = v \in V_{i+1}$, we have $v \in V_{i+1} \cap V^-_{p+1-i} = \{0\}$ (Lemma 3.4.12(i)), i.e. $v = 0$.

Hence $\sum E_i$ is direct, in particular $\dim \sum E_i = \sum \dim E_i = d$.

(ii) Using the invariance of the forward and backward filtration (the latter one with ω replaced by $\theta\omega$), the invariance of $p(\cdot)$ and the elementary relation

$$A(U \cap V) = AU \cap AV \quad \text{for } A \in Gl(d, \mathbb{R})$$

we obtain

$$\begin{aligned}
A(\omega) E_i(\omega) &= A(\omega) V_i(\omega) \cap A(\omega) V^-_{p(\omega)+1-i}(\omega) \\
&= V_i(\theta\omega) \cap V^-_{p(\theta\omega)+1-i}(\theta\omega) \\
&= E_i(\theta\omega).
\end{aligned}$$

Since the complementary space $F_i(\omega) := \oplus_{j \neq i} E_j(\omega)$ is also invariant, i.e. satisfies $A(\omega) F_i(\omega) = F_i(\theta\omega)$ we even have $A(\omega) P_i(\omega) = P_i(\theta\omega) A(\omega)$.

(iii) By the two one-sided versions of the MET

$$\lim_{n \to \pm\infty} \frac{1}{n} \log \|\Phi(n,\omega) x\| = \lambda_i(\omega)$$

if and only if
$$x \in (V_i(\omega) \setminus V_{i+1}(\omega)) \cap (V^-_{p(\omega)+1-i}(\omega) \setminus V^-_{p(\omega)+1-(i-1)}(\omega)) =: M_i(\omega).$$
Now
$$M_i = V_i \cap V^c_{i+1} \cap V^-_{p+1-i} \cap (V^-_{p+1-(i-1)})^c.$$
Since $V_i \cap V^-_{p+1-(i-1)} = \{0\}$ (Lemma 3.4.12(i)), $V_i \cap (V^-_{p+1-(i-1)})^c = V_i \setminus \{0\}$, similarly $V^c_{i+1} \cap V^-_{p+1-i} = V^-_{p+1-i} \setminus \{0\}$. Hence
$$M_i = (V_i \setminus \{0\}) \cap (V^-_{p+1-i} \setminus \{0\}) = E_i \setminus \{0\}.$$

(iv) Uniform convergence in $E_i \cap S^{d-1}$: We consider the case $n \to \infty$ only and recall certain facts from the proof of Theorem 3.4.1 and Proposition 3.4.2. Since $V_i = V_{i+1} \oplus U_i$, $x \in E_i \subset V_i$ can be represented as $x = R_i x + Q_i x$, where R_i and Q_i are the corresponding orthogonal projections.

By the proof of Proposition 3.4.2 (and explicitly formulated in our proof for the case $d = 2$), for each $\varepsilon > 0$ there exists a random variable $C_\varepsilon : \tilde{\Omega} \to (0, \infty)$ which is independent of x such that for all n

- $\|\Phi(n,\omega)x\| \leq \|x\| e^{n(\lambda_i(\omega)+\varepsilon)} C_\varepsilon(\omega)$ in $V_i(\omega) \setminus V_{i+1}(\omega)$,
- $\|\Phi(n,\omega)R_i(\omega)x\| \leq \|R_i(\omega)x\| e^{n(\lambda_{i+1}(\omega)+\varepsilon)} C_\varepsilon(\omega)$ for all $R_i(\omega)x \in V_{i+1}(\omega)$,
- $\|Q_i(\omega)x\| e^{n(\lambda_i(\omega)-\varepsilon)} C_\varepsilon(\omega) \leq \|\Phi(n,\omega)Q_i(\omega)x\|$ for all $Q_i(\omega)x \in U_i(\omega) \subset V_i(\omega) \setminus V_{i+1}(\omega)$.

Since $\sup_{x \in E_i(\omega) \cap S^{d-1}} \|R_i(\omega)x\| \leq 1$ and
$$c_i(\omega) := \inf_{x \in E_i(\omega) \cap S^{d-1}} \|Q_i(\omega)x\| > 0$$
the above estimates yield for $x \in E_i(\omega) \cap S^{d-1}$
$$C_\varepsilon (c_i(\omega) e^{n(\lambda_i(\omega)-\varepsilon)} - e^{n(\lambda_{i+1}(\omega)+\varepsilon)}) \leq \|\Phi(n,\omega)x\| \leq e^{n(\lambda_i(\omega)+\varepsilon)} C_\varepsilon,$$
where the left-hand side and the right-hand side are independent of $x \in E_i(\omega) \cap S^{d-1}$.

Part (B):

The continuous time case can be reduced to the discrete time case as in the proof of Theorem 3.4.1(C). Note that the integrability conditions assure that the continuous time Furstenberg-Kesten theorem 3.3.10(B) is valid on a set $\tilde{\Omega}$ of full measure which is $\theta(t)$-invariant. □

3.4.13 Remark. (i) The uniformity statement of Theorem 3.4.11(iii) remains valid if S^{d-1} is replaced with a set in $\mathbb{R}^d$ bounded away from 0 and ∞.

(ii) For all $\omega \in \tilde{\Omega}$ the cocycle is (forward and backward) regular in the sense of Sect. 3.2.

(iii) The forward and backward filtration can be recovered from the Oseledets splitting via

$$V_i(\omega) = \oplus_{j=i}^{p(\omega)} E_j(\omega), \quad V^-_{p(\omega)+1-i}(\omega) = \oplus_{j=1}^{i} E_j(\omega).$$

(iv) Part (i) of the theorem gives for $\omega \in \tilde{\Omega}$

$$\Phi(n,\omega) P_i(\omega) = P_i(\theta^n \omega) \Phi(n,\omega) \quad \text{for all } n \in \mathbb{Z}.$$

This commutativity of the cocycle with the corresponding projections is a crucial advantage of two-sided time and will play a key rôle in the theory.

(v) The invariant splitting

$$\mathbb{R}^d = E_1(\omega) \oplus \ldots \oplus E_{p(\omega)}(\omega)$$

is in general not orthogonal (and the orthogonal splitting $\mathbb{R}^d = U_1(\omega) \oplus \ldots \oplus U_{p(\omega)}(\omega)$ appearing in Theorem 3.4.1 and its proof is generally not invariant).

(vi) Represent $x \neq 0$ as $x = \oplus_{i=1}^{p} P_i(\omega) x$. Then

$$\lambda(\omega, x) = \lambda_{k(\omega,x)},$$

where $k(\omega, x) := \min\{i : P_i(\omega) x \neq 0\}$.

(vii) Uniqueness of spectrum and splitting: For a linear cocycle Φ satisfying the conditions of Theorem 3.4.11 its spectrum and splitting are unique This means that if there is another splitting $F_q \oplus \ldots \oplus F_1 = \mathbb{R}^d$ with

$$\lim_{n \to \pm\infty} \frac{1}{n} \log \|\Phi(n, \cdot) x\| = \mu_i$$

if and only if $x \in F_i \setminus \{0\}$, then $q = p$, $\mu_i = \lambda_i$ and $F_i = E_i$. This follows from Remark 3.4.10(iii) applied to (V_i) and (V_i^-). ∎

3.4.14 Remark (Measurability Properties of Filtration). Let $\mathbb{T} = \mathbb{Z}$ or $\mathbb{R}$ and define $\mathcal{F}^+ := \sigma(\Phi(t,\cdot) : t \geq 0)$ (for discrete time we have $\mathcal{F}^+ = \sigma(A(\theta^n \cdot), n \geq 0))$ to be the future and $\mathcal{F}^- := \sigma(\Phi(t,\cdot) : t \leq 0)$ (for discrete time $\mathcal{F}^- = \sigma(A(\theta^n \cdot), n \leq -1))$ the past of the cocycle (see Sect. 1.7). By construction, the filtration (V_k) (in particular $E_p = V_p$) is $\mathcal{F}^+$ measurable, and (V_k^-) (in particular $E_1 = V_p^-$) is $\mathcal{F}^-$ measurable. If the sequence $(A(\theta^n \cdot))$ is i.i.d. or if Φ solves a linear stochastic differential equation, then $\mathcal{F}^+$ and $\mathcal{F}^-$, hence (V_k) and (V_k^-), in particular E_p and E_1 are independent. ∎

3.4.3 Examples

3.4.15 Example (Linear RDE in $\mathbb{R}^d$). By Example 2.2.8 the equation

$$\dot{x}_t = A(\theta_t \omega) x_t$$

generates a linear cocycle Φ on $\mathbb{T} = \mathbb{R}$ provided $A \in L^1$. We now prove that this condition also implies the integrability conditions $\alpha^\pm \in L^1$ of the MET for $\mathbb{T} = \mathbb{R}$. Take an ω in the invariant set for which $t \mapsto A(\theta_t \omega)$ is locally integrable. Then for $t \in \mathbb{R}^+$ (see Example 2.2.8)

160   Chapter 3. MET in Euclidean Space

$$\|\Phi(t,\omega)^{\pm 1}\| \leq 1 + \int_0^t \|A(\theta_s\omega)\| \, \|\Phi(s,\omega)^{\pm 1}\| \, ds.$$

By Gronwall's lemma B.3.4

$$\|\Phi(t,\omega)^{\pm 1}\| \leq \exp \int_0^t \|A(\theta_s\omega)\| \, ds \quad \text{for all} \quad t \in \mathbb{R}^+,$$

and finally

$$\sup_{0 \leq t \leq 1} \log^+ \|\Phi(t,\omega)^{\pm 1}\| \leq \int_0^1 \|A(\theta_s\omega)\| \, ds.$$

The right-hand side of the last equation is in L^1 by Lemma 2.2.5. The Furstenberg-Kesten theorem and the MET for $\mathbb{T} = \mathbb{R}$ are thus valid. In particular, the average Lyapunov exponent is, by Liouville's formula and the ergodic theorem (note that $|\text{trace} A| \leq c\|A\| \in L^1$)

$$\lambda_\Sigma = \frac{1}{d} \text{trace} \mathbb{E}\left(A | \mathcal{I}\right).$$

A prototypical and well-investigated example (see the articles in Arnold and Wihstutz [39] or Arnold, Crauel and Eckmann [14]) is the damped linear oscillator with a random restoring force,

$$\ddot{y}_t + 2\beta \dot{y}_t + (1 + \sigma f(\theta_t\omega))y_t = 0, \quad \beta, \sigma > 0, \, f \in L^1.$$

Putting $x_1 = y$, $x_2 = \dot{y}$,

$$\dot{x}_t = A(\theta_t\omega)x_t = \begin{pmatrix} 0 & 1 \\ -1 - \sigma f(\theta_t\omega) & -2\beta \end{pmatrix} x_t.$$

In this case $\lambda_\Sigma = -\beta$. ∎

3.4.16 Example (RDE with 2×2 Triangular Right-Hand Side, Continued). We continue Example 3.3.13 and explicitly calculate the Oseledets splittings (see also Crauel [108: Chap. 6]).

Case 1: $\alpha = \beta$

In this case, the filtrations and the splitting are trivial, $E_1 = V_1 = V_1^- = \mathbb{R}^2$. Assume for example that $a = b$, $c = 0$, $\alpha = \beta = 0$, and that the law of the iterated logarithm holds for α_t. Then the orbits $\Phi(t,\omega)x = \exp(\alpha_t(\omega))x$ typically undergo fluctuations of the order $\exp\sqrt{2t \log \log t}$. Note, in contrast, that the deterministic "cocycle" $\exp(tA)$ has at most polynomial growth inside an eigenspace corresponding to an eigenvalue with vanishing real part.

Case 2: $\lambda_1 = \alpha > \beta = \lambda_2$
$E_1 = V_2^- = \mathbb{R}\, e_1$ is deterministic, and

$$E_2(\omega) = \mathbb{R}\begin{pmatrix} u(\omega) \\ 1 \end{pmatrix}, \quad u(\omega) = -\int_0^\infty e^{\beta_t(\omega) - \alpha_t(\omega)} c_t(\omega) \, dt,$$

which is $\mathcal{F}^+$ measurable. The orbit of $\binom{u}{1}$ is

$$\Phi(t,\omega)\begin{pmatrix}u(\omega)\\1\end{pmatrix} = \begin{pmatrix}-e^{\alpha_t(\omega)}\int_t^\infty e^{\beta_s(\omega)-\alpha_s(\omega)}c_s(\omega)\,ds\\ e^{\beta_t(\omega)}\end{pmatrix} = e^{\beta_t(\omega)}\begin{pmatrix}u(\theta_t\omega)\\1\end{pmatrix}.$$

That u makes sense and the orbit of $\binom{u}{1}$ has Lyapunov exponent β is a consequence of the following simple lemma.

3.4.17 Lemma. *Let $u \in L^1$ and $v_t(\omega)$ be continuous with $v_t(\omega)/t \to \lambda < 0$ $\mathbb{P}$-a.s. Then*

$$\int_0^\infty e^{v_t(\omega)}u(\theta_t\omega)\,dt \quad \text{exists} \quad \mathbb{P}\text{-a.s.}$$

and

$$\limsup_{t\to\infty} \frac{1}{t}\log\left|\int_t^\infty e^{v_s(\omega)}u(\theta_s\omega)\,ds\right| \leq \lambda.$$

Proof. For each fixed t_0 and $c > 0$

$$\mathbb{E}\int_{t_0}^\infty e^{-ct}|u(\theta_t\omega)|\,dt < \infty.$$

Now for $\varepsilon_1 > 0$ with $-c = \lambda + \varepsilon_1 < 0$ and $\varepsilon_2 > 0$ choose a $t_0 = t_0(\varepsilon_1, \varepsilon_2)$ such that

$$v_t(\omega) \leq -ct \quad \text{for } t \geq t_0 \text{ at least with probability } 1 - \varepsilon_2.$$

Since

$$\int_0^{t_0} e^{v_t(\omega)}|u(\theta_t\omega)|\,dt < \infty$$

and ε_2 is arbitrary, the first assertion follows.

The second claim follows from the first by observing that

$$\int_t^\infty e^{v_s(\omega)}|u(\theta_s\omega)|\,ds \leq e^{(\lambda+2\varepsilon_1)t}\int_t^\infty e^{-\varepsilon_1 s}|u(\theta_s\omega)|\,ds$$

with probability at least $1 - \varepsilon_2$, where the second factor on the right-hand side is bounded. $\square$

Case 3: $\lambda_1 = \beta > \alpha = \lambda_2$
Now $E_2 = V_2 = \mathbb{R}\,e_1$ is deterministic, and

$$E_1(\omega) = \mathbb{R}\begin{pmatrix}u(\omega)\\1\end{pmatrix}, \quad u(\omega) = \int_{-\infty}^0 e^{\beta_t(\omega)-\alpha_t(\omega)}c_t(\omega)\,dt,$$

which is $\mathcal{F}^-$ measurable. ∎

3.4.18 Exercise (Products of 2×2 Triangular Matrices). Continue Example 3.3.9 by explicitly calculating the Oseledets splitting, analogous to the continuous time case of Example 3.4.16 (see also Berger [67: p. 155]). ∎

3.4.19 Example (Linear SDE in $\mathbb{R}^d$). The linear SDE in $\mathbb{R}^d$

$$dx_t = A_0 x_t dt + \sum_{j=1}^{m} A_j x_t \circ dW_t^j$$

generates a linear cocycle $\Phi(t,\omega)$ on $\mathbb{T} = \mathbb{R}$ over the canonical (ergodic) DS associated with Brownian motion. The integrability conditions of the MET are automatically satisfied, see Theorem 6.2.11. Liouville's formula (Theorem 2.3.32(iii)) yields the scalar cocycle

$$\det \Phi(t,\omega) = \exp((\operatorname{trace} A_0) t + \sum_{j=1}^{m} (\operatorname{trace} A_j) W_t^j(\omega)),$$

so

$$\lambda_\Sigma = \frac{1}{d} \operatorname{trace} A_0.$$

We have $\det \Phi(t,\omega) = 1$ for all $t \in \mathbb{R}$ $\mathbb{P}$-a.s. if and only if $\operatorname{trace} A_j = 0$ for all $j = 0, \ldots, m$. The SDE

$$dx_t = (A_0 - (\frac{1}{d} \operatorname{trace} A_0) I) x_t dt + \sum_{j=1}^{m} (A_j - (\frac{1}{d} \operatorname{trace} A_j) I) x_t \circ dW_t^j$$

generates the volume preserving cocycle $(\det \Phi(t,\omega))^{-1/d} \Phi(t,\omega) \in Sl(d,\mathbb{R})$, for which $\lambda_\Sigma = 0$, hence $\lambda_p \leq 0 \leq \lambda_1$. An important problem is whether actually $\lambda_p < 0 < \lambda_1$ holds. In dimension $d = 2$ we have either $\lambda_1 = 0$ with multiplicity $d_1 = 2$ or $\lambda_1 > 0 > \lambda_2 = -\lambda_1$.

For criteria for $\lambda_1 > 0$ or for the simplicity of the Lyapunov spectrum in the white and Markovian noise case see e.g. Guivarc'h and Raugi [163], Bougerol and Lacroix [77], Arnold and Kliemann [25], Bougerol [76, 75] and Goldsheid and Margulis [161]. For genericity results see Knill [215, 216] and Arnold and Nguyen [29]. ∎

Chapter 4. The Multiplicative Ergodic Theorem on Bundles and Manifolds

Summary

This chapter deals with extensions of the MET from matrix cocycles in $\mathbb{R}^d$ to linear cocycles on bundles, in particular to the linearization of a nonlinear RDS on the tangent bundle, by which we prepare the smooth ergodic theory of Part III.

In order to facilitate the transfer of data of the MET from one bundle to another we need a "tempered" version of coordinate change called Lyapunov cohomology (Sect. 4.1).

In Sect. 4.2 the MET for the linearization of a nonlinear smooth RDS on a manifold is derived (Theorem 4.2.6). In case the RDS is generated by an RDE or SDE, we also give criteria in terms of the vector fields and the invariant measure ensuring the validity of the integrability conditions of the MET (Theorem 4.2.10 for the RDE case, Theorems 4.2.12, 4.2.13 and 4.2.14 for the SDE case).

Sect. 4.3 is devoted to one of the most important techniques of smooth ergodic theory, namely the use of (random) norms which "eat up" the non-uniformity of the MET (Theorem 4.3.6). It is of course crucial that the random norms do not change the Lyapunov exponents (Corollary 4.3.10). We also obtain what is called the "strong version" of the MET (Theorem 4.3.12).

4.1 Tempered Random Variables and Lyapunov Cohomology

4.1.1 Tempered Random Variables

We first introduce a concept which is of fundamental importance for most parts of the theory of RDS and can even be considered one of its characteristic features.

In many estimates for RDS we will have random "constants" whose values have to be controlled along the orbits of the DS θ. For example, the MET for $\mathbb{T} = \mathbb{Z}$ gives for each $\varepsilon > 0$ a finite random variable C_ε such that

$$\|\Phi(n,\omega)\| \leq C_\varepsilon(\omega)e^{(\lambda_1+\varepsilon)n}, \quad n \in \mathbb{Z}^+. \tag{4.1.1}$$

There are reasons to consider the norm of $\Phi(-n,\omega)^{-1} = \Phi(n, \theta^{-n}\omega)$ for which we obtain from (4.1.1) that

$$\|\Phi(n, \theta^{-n}\omega)\| \leq C_\varepsilon(\theta^{-n}\omega)e^{(\lambda_1+\varepsilon)n}, \quad n \in \mathbb{Z}^+.$$

In order not to destroy this estimate we have to exclude exponential growth (at other occasions: exponential decay) of the sequence $C_\varepsilon(\theta^{-n}\omega)$.

4.1.1 Definition (Tempered Random Variable). (i) A random variable $R: \Omega \to (0,\infty)$ is called *tempered* with respect to the DS θ if for the associated stationary stochastic process $t \mapsto R(\theta(t)\cdot)$ the invariant set for which

$$\lim_{t\to\pm\infty} \frac{1}{t} \log R(\theta(t)\omega) = 0 \tag{4.1.2}$$

($t \to -\infty$ applies only to two-sided time) has full $\mathbb{P}$-measure.

(ii) $R: \Omega \to [0,\infty)$ is called *tempered from above* if

$$\lim_{t\to\pm\infty} \frac{1}{|t|} \log^+ R(\theta(t)\omega) = 0 \quad \mathbb{P}\text{-a.s.}, \tag{4.1.3}$$

while $R: \Omega \to (0,\infty]$ is called *tempered from below* if $\frac{1}{R}$ is tempered from above, equivalently, if, with $\log^- x := \max(0, -\log x)$,

$$\lim_{t\to\pm\infty} \frac{1}{|t|} \log^- R(\theta(t)\omega) = 0 \quad \mathbb{P}\text{-a.s.} \tag{4.1.4}$$

(iii) A random variable $f: \Omega \to \mathbb{R}^d$ is called *tempered* (from above or below) with respect to the DS θ if the stationary stochastic process $t \mapsto \|f(\theta(t)\cdot)\|$ is tempered (from above or below). ∎

Since

$$\lim_{t\to\pm\infty} \frac{1}{|t|} \log^+ R(\theta(t)\omega) \geq \limsup_{t\to\pm\infty} \frac{1}{|t|} \log R(\theta(t)\omega)$$

and

$$\lim_{t\to\pm\infty} \frac{1}{|t|} \log^- R(\theta(t)\omega) \geq -\liminf_{t\to\pm\infty} \frac{1}{|t|} \log R(\theta(t)\omega),$$

R is tempered if and only if it is tempered from above and from below.

The following lemma follows immediately from the properties of the function $x \mapsto \log^+ x$.

4.1.2 Lemma. *The set of real-valued tempered random variables with respect to a DS θ is a commutative ring with unit element.*

There is the following remarkable dichotomy found by Tanny [327] and O'Brien [264].

4.1.3 Proposition (Dichotomy for Linear Growth of Stationary Process). *Let $(\Omega, \mathcal{F}, \mathbb{P}, (\theta(t))_{t \in \mathbb{T}})$ be a metric DS and let $f : \Omega \to \mathbb{R}$ be measurable. Then*

$$\limsup_{t \to \infty} \frac{1}{t} f(\theta(t)\omega) = \limsup_{t \to -\infty} \frac{1}{|t|} f(\theta(t)\omega) \in \{0, \infty\} \quad \mathbb{P}\text{-}a.\,s.$$

and

$$\liminf_{t \to \infty} \frac{1}{t} f(\theta(t)\omega) = \liminf_{t \to -\infty} \frac{1}{|t|} f(\theta(t)\omega) \in \{-\infty, 0\} \quad \mathbb{P}\text{-}a.\,s.$$

($t \to -\infty$ applies only for two-sided time).
  (ii) For discrete time

$$f^+ \in L^1 \Rightarrow \limsup_{n \to \pm\infty} \frac{1}{|n|} f(\theta^n \omega) = 0 \quad \mathbb{P}\text{-}a.\,s.$$

($\Leftrightarrow$ in the i.i.d. case) and

$$f^- \in L^1 \Rightarrow \liminf_{n \to \pm\infty} \frac{1}{|n|} f(\theta^n \omega) = 0 \quad \mathbb{P}\text{-}a.\,s.$$

($\Leftrightarrow$ in the i.i.d. case).
  For continuous time

$$\sup_{0 \le t \le 1} f^+(\theta(t)\cdot) \in L^1 \Rightarrow \limsup_{t \to \pm\infty} \frac{1}{|t|} f(\theta(t)\omega) = 0 \quad \mathbb{P}\text{-}a.\,s.$$

and

$$\sup_{0 \le t \le 1} f^-(\theta(t)\cdot) \in L^1 \Rightarrow \liminf_{t \to \pm\infty} \frac{1}{|t|} f(\theta(t)\omega) = 0 \quad \mathbb{P}\text{-}a.\,s.$$

The $\mathbb{P}$-a.s. statements hold on an invariant set of full $\mathbb{P}$-measure.
  (iii) If θ is ergodic the above $\limsup$'s and $\liminf$'s are constant on an invariant set of full $\mathbb{P}$-measure.

Proof. (i) We follow O'Brien [264] but restrict ourselves to the case where θ is ergodic, time is discrete and to the consideration of the $\limsup_{n \to \infty}$.

By ergodicity the invariant random variable $\limsup f(\theta^n \cdot)/n$ is $\mathbb{P}$-a.s. constant. This constant c satisfies $c \in [0, \infty]$ since $f(\theta^n \cdot)/n \to 0$ in probability.

Assume $0 < c < \infty$. The event

$$U_n = \max_{k \in \mathbb{Z}^+}(\frac{1}{2c} f(\theta^{n+k}\cdot) - k)$$

is defined $\mathbb{P}$-a.s., and the sequence (U_n) is stationary ergodic. Also, $U_{n-1} = \max(f(\theta^{n-1}\cdot)/2c, U_n - 1)$ so that

$$\frac{1}{2c} f(\theta^n \cdot) \le U_n \le U_{n-1} + 1 \quad \mathbb{P}\text{-}a.\,s. \text{ for all } n. \tag{4.1.5}$$

For some $t \in \mathbb{R}$, $\mathbb{P}(U_n \leq t) > 0$ so that $\mathbb{P}(U_n \leq t \text{ i.o.}) = 1$ ("i.o." stands for "infinitely often"). Let $n_1, n_2, \ldots$ be the random positive epochs for which $U_n \leq t$, with $n_i < n_{i+1}$ for all i. Let $I_n = 1$ if $U_n \leq t$ and $I_n = 0$ otherwise. Applying the ergodic theorem to the stationary ergodic sequence (I_n), we see that
$$\lim_{i \to \infty} \frac{n_{i-1}}{n_i} = 1 \quad \mathbb{P}\text{-a.s.} \tag{4.1.6}$$
For $n_{i-1} \leq n < n_i$, we deduce from (4.1.5) and (4.1.6) that
$$\frac{1}{2cn} f(\theta^n \cdot) \leq \frac{U_n}{n} \leq \frac{U_{n_{i-1}} + n - n_{i-1}}{n} \leq \frac{t + n_i - n_{i-1}}{n_{i-1}} \to 0$$
as $i \to \infty$. This implies $\limsup f(\theta^n \cdot)/n = 0$ and hence contradicts our assumption.

(ii) All statements are easy consequences of the Borel-Cantelli lemma. □

Applied to $f = \log \|g\|$ for a measurable $g : \Omega \to \mathbb{R}^d$ the lemma yields a dichotomy for the Lyapunov index of the stationary process $t \mapsto g(\theta(t)\cdot)$. If this index is non-zero it is automatically equal to infinity. Temperedness hence prevents the case of superexponential growth of a stationary process from happening.

4.1.4 Remark (Existence of Non-Tempered Random Variables).
Let $(\Omega, \mathcal{F}, \mathbb{P})$ be a standard space and θ be aperiodic. Then there exists a random variable which is not tempered with respect to θ. The proof relies on the Rokhlin-Halmos lemma (see Arnold, Nguyen and Oseledets [31: Sect. 8]). ∎

4.1.2 Lyapunov Cohomology

We have stated the MET in Chap. 3 only for linear RDS on the trivial bundle $\Omega \times \mathbb{R}^d$. We need to extend it to nontrivial linear bundle RDS in the sense of Definition 1.9.3(ii) to cover e.g. the linearization $T\varphi$ of a C^1 RDS φ on a manifold M or the case of random norms $\|\cdot\|_\omega$ on $\mathbb{R}^d$.

The key notion to facilitate the transfer of spectral objects from one linear (bundle) RDS to another is the notion of *Lyapunov cohomology* which we will introduce first.

Let $(\Omega, \mathcal{F}, \mathbb{P}, (\theta(t))_{t \in \mathbb{T}})$ be a fixed metric DS. In this chapter we consider linear measurable bundles (Y, Ω, π) with typical fiber $\mathbb{R}^d$ (see Definition 1.9.2(ii)) and a linear (bundle) RDS $(\Theta(t))_{t \in \mathbb{T}}$ on it, i.e. the cocycle
$$\Phi(t, \omega) := \Theta(t) | \pi^{-1}(\omega) : \pi^{-1}(\omega) \to \pi^{-1}(\theta(t)\omega)$$
is linear.

Assume that two linear RDS $(\Theta_i(t))_{t \in \mathbb{T}}$, $i = 1, 2$, over θ on linear measurable bundles (Y_i, Ω, π_i) with typical fiber $\mathbb{R}^d$ are linearly isomorphic (see Definition 1.9.8(ii)), i.e. there is a linear measurable isomorphism Ψ of the corresponding bundles (Y_i, Ω, π_i) such that $\Theta_2(t) \circ \Psi = \Psi \circ \Theta_1(t)$ for all $t \in \mathbb{T}$. For the corresponding linear cocycles defined by

$$\Phi_i(t, \omega) := \Theta_i(t)|\pi_i^{-1}(\omega) : \pi_i^{-1}(\omega) \to \pi_i^{-1}(\theta(t)\omega)$$

this is equivalent to being cohomologous with cohomology $\psi = \Psi|_{\pi_1^{-1}(\cdot)}$, i.e. for all $t \in \mathbb{T}$

$$\Phi_2(t, \omega) \circ \psi(\omega) = \psi(\theta(t)\omega) \circ \Phi_1(t, \omega). \tag{4.1.7}$$

"Cohomology" is an equivalence relation in the set of linear RDS on measurable linear bundles with typical fiber $\mathbb{R}^d$.

4.1.5 Definition (Normed Linear Bundle). A linear measurable bundle (Y, Ω, π) is said to be *normed* if there is a measurable function on Y which is a norm $\|\cdot\|_\omega$ on each fiber $\pi^{-1}(\omega)$. The norm can in particular be induced by a scalar product $\langle \cdot, \cdot \rangle_\omega$, in which case we speak of a *Euclidean bundle*. ∎

For linear RDS on normed linear bundles we can define the Lyapunov exponent of a vector $x \in Y$ by

$$\lambda(x) := \limsup_{t \to \infty} \frac{1}{t} \log \|\Phi(t, \omega)x\|_{\theta(t)\omega}, \quad x \in \pi^{-1}(\omega).$$

We would like to carry "spectral theory" (i.e. the objects of the MET: Lyapunov exponents, filtrations and splittings) from one linear RDS to a cohomologous one. To make this work we have to impose a condition on the cohomology.

4.1.6 Definition (Lyapunov Cohomology). A cohomology

$$\psi(\omega) : (\pi_1^{-1}(\omega), \|\cdot\|_\omega^1) \to (\pi_2^{-1}(\omega), \|\cdot\|_\omega^2)$$

of two linear cocycles Φ_1 and Φ_2 over θ on normed linear bundles is called a *Lyapunov cohomology* if ψ and ψ^{-1} are tempered, i.e. if there is a θ-invariant set of full $\mathbb{P}$-measure on which

$$\lim_{t \to \pm\infty} \frac{1}{t} \log \|\psi(\theta(t)\omega)\|_{\theta(t)\omega} = 0, \quad \lim_{t \to \pm\infty} \frac{1}{t} \log \|\psi(\theta(t)\omega)^{-1}\|_{\theta(t)\omega} = 0 \tag{4.1.8}$$

($t \to -\infty$ applies to two-sided time). Here $\|\psi(\omega)\|_\omega$ and $\|\psi(\omega)^{-1}\|_\omega$ are the operator norms of $\psi(\omega) : (\pi_1^{-1}(\omega), \|\cdot\|_\omega^1) \to (\pi_2^{-1}(\omega), \|\cdot\|_\omega^2)$ and its inverse. ∎

"Lyapunov cohomology" is an equivalence relation in the set of linear RDS on normed linear bundles with typical fiber $\mathbb{R}^d$.

4.1.7 Proposition (Conditions for Lyapunov Cohomology). *A sufficient condition for a cohomology ψ between two linear RDS Φ_1 and Φ_2 on normed bundles to be a Lyapunov cohomology is that it satisfies*
  (i) for discrete time

$$\log^+ \|\psi(\cdot)\|_\cdot, \quad \log^+ \|\psi(\cdot)^{-1}\|_\cdot \in L^1(\mathbb{P}),$$

(ii) for continuous time

$$\sup_{0 \leq t \leq 1} \log^+ \|\psi(\theta(t)\cdot)\|_{\theta(t)\cdot}, \quad \sup_{0 \leq t \leq 1} \log^+ \|\psi(\theta(t)\cdot)^{-1}\|_{\theta(t)\cdot} \in L^1(\mathbb{P}).$$

Proof. This follows from Proposition 4.1.3(ii) for $f_1(\omega) = \log \|\psi(\omega)\|_\omega$ and $f_2(\omega) = \log \|\psi(\omega)^{-1}\|_\omega$, noting that

$$0 \leq \log^- \|\psi(\omega)\|_\omega = \log^+ \frac{1}{\|\psi(\omega)\|_\omega} \leq \log^+ \|\psi(\omega)^{-1}\|_\omega.$$

□

4.1.8 Definition (Spectral Theory of a Linear RDS). We say of a linear RDS Φ on a normed linear bundle that it *has spectral theory* if there is a forward invariant (for the invertible case: an invariant) set $\tilde{\Omega}$ of full $\mathbb{P}$ measure on which we have a *spectrum* $\mathcal{S}(\theta, \Phi) = \{(\lambda_i(\cdot), d_i(\cdot))_{i=1,\ldots,p(\cdot)}\}$ and a *filtration* $V_{p(\omega)}(\omega) \subset \ldots \subset V_1(\omega) = \pi^{-1}(\omega)$ (for two-sided time: a *splitting* $E_1(\omega) \oplus \ldots \oplus E_{p(\omega)}(\omega) = \pi^{-1}(\omega)$) with the properties listed in the corresponding MET (Theorem 3.4.1 for one-sided time and Theorem 3.4.11 for two-sided time). ∎

4.1.9 Proposition (Lyapunov Cohomology Preserves Spectral Theory). *Let Φ_1 and Φ_2 be two linear cocycles on normed bundles which are cohomologous via ψ.*

(i) If Φ_1 has spectral theory and ψ is a Lyapunov cohomology of Φ_1 and Φ_2, then also Φ_2 has spectral theory with

$$\mathcal{S}(\theta, \Phi_2) = \mathcal{S}(\theta, \Phi_1)$$

and for one-sided time the filtrations $V_i^{1,2}$ are related by

$$V_i^2(\omega) = \psi(\omega) V_i^1(\omega),$$

whereas for two-sided time the splittings $E_i^{1,2}$ are related by

$$E_i^2(\omega) = \psi(\omega) E_i^1(\omega).$$

(ii) Let $\mathbb{T} = \mathbb{Z}$ or $\mathbb{R}$. If both Φ_1 and Φ_2 have spectral theory, then ψ is a Lyapunov cohomology of Φ_1 and Φ_2.

Proof. (i) By assumption

$$\Phi_2(t, \omega) \circ \psi(\omega) = \psi(\theta(t)\omega) \circ \Phi_1(t, \omega), \qquad (4.1.9)$$

thus

$$\lim \frac{1}{t}\log \|\Phi_1(t,\omega)x\|^1_{\theta(t)\omega} - \lim \frac{1}{t}\log \|\psi(\theta(t)\omega)^{-1}\|_{\theta(t)\omega}$$
$$\leq \lim \frac{1}{t}\log \|\Phi_2(t,\omega)\psi(\omega)x\|^2_{\theta(t)\omega}$$
$$\leq \lim \frac{1}{t}\log \|\psi(\theta(t)\omega)\|_{\theta(t)\omega} + \lim \frac{1}{t}\log \|\Phi_1(t,\omega)x\|^1_{\theta(t)\omega}.$$

Since ψ is Lyapunov, on an invariant set of full measure

$$\lim_{t\to\infty(-\infty)} \frac{1}{t}\log \|\Phi_2(t,\omega)\psi(\omega)x\|^2_{\theta(t)\omega} = \lim_{t\to\infty(-\infty)} \frac{1}{t}\log \|\Phi_1(t,\omega)x\|^1_{\theta(t)\omega},$$

from which all statements of the MET for Φ_2 follow from those for Φ_1.

(ii) We stick to the case $\mathbb{T} = \mathbb{Z}$ and use the cohomology relation (4.1.9) in the form

$$\psi(\theta^n\omega) = \Phi_2(n,\omega) \circ \psi(\omega) \circ \Phi_1(n,\omega)^{-1}.$$

The assumptions and Lemma 3.4.3(ii) yield

$$0 \leq \limsup_{n\to\infty} \frac{1}{n}\log \|\psi(\theta^n\omega)\|_{\theta^n\omega}$$
$$\leq \lim_{n\to\infty} \frac{1}{n}\log \|\Phi_1(n,\omega)^{-1}\|^1_{\theta^n\omega,\omega} + \lim_{n\to\infty} \frac{1}{n}\log \|\Phi_2(n,\omega)\|^2_{\omega,\theta^n\omega} < \infty.$$

Hence by Proposition 4.1.3(i)

$$\limsup_{n\to\infty} \frac{1}{n}\log \|\psi(\theta^n\omega)\|_{\theta^n\omega} = 0.$$

The same reasoning yields

$$\liminf_{n\to\infty} \frac{1}{n}\log \|\psi(\theta^n\omega)\|_{\theta^n\omega} = 0.$$

Similarly for ψ^{-1} and $n \to -\infty$. Hence ψ is a Lyapunov cohomology. ⌐

4.1.10 Remark. (i) Proposition 4.1.9(i) says that each member of an equivalence class of Lyapunov cohomologous cocycles has spectral theory provided one member has and that the Lyapunov spectrum $\mathcal{S}(\theta,\Phi)$ is an invariant of a cocycle with respect to Lyapunov cohomology.

(ii) Proposition 4.1.9(ii) says that in the set of linear RDS having spectral theory cohomology and Lyapunov cohomology are identical. ■

There remains the basic problem of whether the integrability conditions of the MET[1] for a linear RDS Φ on a normed linear bundle imply that Φ has spectral theory, i.e. whether the MET holds. The answer is affirmative for a Euclidean bundle by the following theorem.

---

[1] "Integrability conditions of the MET" is henceforth abbreviated as "IC of the MET".

4.1.11 Theorem (MET on Euclidean Bundle). *(i) For a Euclidean bundle, the trivializing isomorphism Ψ can be chosen to be an isometry on fibers, i. e. if*

$$\psi(\omega) : (\pi^{-1}(\omega), \langle \cdot, \cdot \rangle_\omega) \to (\{\omega\} \times \mathbb{R}^d, \langle \cdot, \cdot \rangle_S)$$

then

$$\langle \psi(\omega)x, \psi(\omega)y \rangle_S = \langle x, y \rangle_\omega \quad \text{for all } x, y \in \pi^{-1}(\omega).$$

(ii) Assume that a linear RDS Φ on a Euclidean bundle satisfies the IC of the MET, namely α^+, $\alpha^- \in L^1(\mathbb{P})$, where for discrete time

$$\alpha^+(\omega) := \log^+ \|\Phi(1, \omega)\|_{\omega, \theta(1)\omega}, \quad \alpha^-(\omega) := \log^+ \|\Phi(1, \omega)^{-1}\|_{\theta(1)\omega, \omega}$$

(the second condition only applies in the invertible case), for continuous time

$$\alpha^+(\omega) := \sup_{0 \le t \le 1} \log^+ \|\Phi(t, \omega)\|_{\omega, \theta(t)\omega},$$

$$\alpha^-(\omega) := \sup_{0 \le t \le 1} \log^+ \|\Phi(t, \omega)^{-1}\|_{\theta(t)\omega, \omega}.$$

Then the MET holds.

Proof. (i) Define a positive-definite operator $A(\omega)$ of $\{\omega\} \times \mathbb{R}^d$ to itself by

$$\langle A(\omega)\psi(\omega)x, \psi(\omega)y \rangle_S = \langle \psi(\omega)x, A(\omega)\psi(\omega)y \rangle_S = \langle x, y \rangle_\omega, \quad x, y \in \pi^{-1}(\omega).$$

Define a new trivializing isomorphism by

$$\psi'(\omega) := A(\omega)^{1/2} \circ \psi(\omega),$$

which is an isometry since

$$\langle \psi'(\omega)x, \psi'(\omega)y \rangle_S = \langle A(\omega)\psi(\omega)x, \psi(\omega)y \rangle_S = \langle x, y \rangle_\omega.$$

(ii) The isometric trivialization constructed in (i) obviously satisfies the integrability conditions of Proposition 4.1.7, hence the cocycle

$$\tilde{\Phi}(t, \omega) := \psi(\theta(t)\omega) \circ \Phi(t, \omega) \circ \psi(\omega)^{-1}$$

on the trivial bundle $\Omega \times \mathbb{R}^d$ is Lyapunov cohomologous with Φ and satisfies the IC of the MET. Thus the MET holds for $\tilde{\Phi}$, hence also for Φ. □

4.1.12 Remark (Classification of Linear RDS). The basic classification problems for linear RDS on normed linear bundles are to decide whether two linear RDS are (Lyapunov) cohomologous, to characterize the equivalence classes by invariants, and to determine in each equivalence class the simplest possible element ("normal" or "canonical" form).

If we consider the classification problem with respect to Lyapunov cohomology for linear RDS on Euclidean bundles for which the IC of the MET hold, then the problem can be reduced by Theorem 4.1.11 to considering matrix cocycles on the trivial bundle $\Omega \times \mathbb{R}^d$ ($\mathbb{R}^d$ endowed with the standard scalar product) for which the MET holds.

Let $\Phi_1(t,\omega)$ be the matrix representation of a linear cocycle in $\mathbb{R}^d$ with respect to random bases $F(\omega) = (u_i(\omega))_{i=1,\ldots,d}$ and $F(\theta(t)\omega)$. If $\Phi_2(t,\omega)$ is the matrix representation in a second basis $G(\omega)$ and $G(\theta(t)\omega)$ then

$$\Phi_2(t,\omega) = P(\theta(t)\omega)\Phi_1(t,\omega)P(\omega)^{-1},$$

where $P(\omega) = P(\omega)_{G(\omega)}^{F(\omega)} \in Gl(d,\mathbb{R})$ is the basis transfer matrix from $F(\omega)$ to $G(\omega)$. If we want to "simplify" Φ_1 by choosing a more appropriate basis we have to make sure that P defines a Lyapunov cohomology.

Corollary 4.3.12 states that if the two-sided MET holds, then the change from the standard basis in $\mathbb{R}^d$ to a random basis which draws d_i orthonormal vectors from E_i defines a Lyapunov cohomology. Hence every cocycle Φ with spectrum $\{(\lambda_i, d_i)_{i=1,\ldots,p}\}$ is Lyapunov cohomologous to a block-diagonal cocycle

$$\tilde{\Phi}(t,\omega) = \begin{bmatrix} \tilde{\Phi}_1(t,\omega) & 0 & \cdots & 0 \\ 0 & \tilde{\Phi}_2(t,\omega) & \cdots & 0 \\ \vdots & \vdots & \ddots & \vdots \\ 0 & 0 & \cdots & \tilde{\Phi}_p(t,\omega) \end{bmatrix},$$

where $\tilde{\Phi}_i(t,\omega)$ is $d_i \times d_i$ and has just one Lyapunov exponent λ_i with multiplicity d_i. Classification of cocycles with the same spectrum is hence reduced to classifying the blocks, i.e. cocycles with one-point spectrum. By changing from $\tilde{\Phi}_i(t,\omega)$ to $|\det \tilde{\Phi}_i(t,\omega)|^{-1/d_i} \tilde{\Phi}_i(t,\omega)$ (see Remark 3.3.12) we can assume without loss of generality that the cocycle to be classified has one-point spectrum $\{0\}$. For a detailed study see Arnold, Nguyen Dinh Cong and Oseledets [31]. ∎

4.1.13 Example (Periodic Case). We continue Example 3.2.4 and reconsider $\dot{x}_t = A(t)x_t$, where $A : \mathbb{R} \to \mathbb{R}^{d \times d}$ is continuous and periodic with period τ. Let $\Phi(t) = P(t)e^{tR}$ be a real Floquet representation with Lyapunov exponents λ_i and E_i^0 the corresponding splitting.

We embed the equation into an RDS by associating the following (ergodic) metric DS to it: $\Omega := [0,\tau) \cong S^1$ ($\omega \mapsto \exp(2\pi i\omega/\tau)$), $\mathbb{P} :=$ normalized Lebesgue measure, $\theta(t)\omega := \omega + t \mod \tau$. The cocycle generated by $\dot{x} = A(\theta(t)\omega)x = A(t+\omega)x$ is

$$\Phi(t,\omega) = \Phi(t+\omega)\Phi(\omega)^{-1} = P(\theta(t)\omega)\,e^{tR}\,P(\omega)^{-1}.$$

Since P and P^{-1} are bounded, $\Phi(t,\omega)$ is Lyapunov cohomologous to e^{tR} with Lyapunov exponents λ_i and splitting $E_i(\omega) = P(\omega)E_i^0$. ∎

4.1.14 Example (Same Cocycle Under Different Norms). The use of appropriate random norms is a fundamental technique in the theory of continuous and smooth RDS, e.g. for the construction of invariant manifolds

(Chap. 7) and for normal form theory (Chap. 8). We will devote Sect. 4.3 to their construction.

Consider for example one linear cocycle Φ on a linear bundle under two different norms $\|\cdot\|_\omega^1$ and $\|\cdot\|_\omega^2$. Since all norms on a finite-dimensional vector space are equivalent, there are finite-valued random variables $c_i : \Omega \to (0, \infty)$ with
$$\frac{1}{c_2(\omega)}\|x\|_\omega^1 \leq \|x\|_\omega^2 \leq c_1(\omega)\|x\|_\omega^1 \quad \text{for } x \in \pi^{-1}(\omega).$$

The cohomology of Φ under $\|\cdot\|_\omega^1$ with Φ under $\|\cdot\|_\omega^2$ is given by the identity mapping $\psi(\omega) = \mathrm{id}_{\pi^{-1}(\omega)}$ which together with its inverse needs to be tempered in order to be Lyapunov. The integrability conditions of Proposition 4.1.7 assuring this will certainly be satisfied if the norms of $\psi(\omega)$ and $\psi(\omega)^{-1}$ are bounded away from $+\infty$ by a non-random constant (see the case of two non-random norms discussed in Remark 3.4.10 and the case of the tangent bundle of a compact Riemannian manifold in Remark 4.2.7). ∎

4.2 The Multiplicative Ergodic Theorem for RDS on Manifolds

4.2.1 Linearization of a C^1 RDS

The most important and historically oldest particular case of a nontrivial linear bundle RDS is the linearization $T\varphi$ of a C^1 RDS φ on a manifold, which will now be considered in detail.

First assume that φ is a (global, nonlinear) C^1 RDS on $\mathbb{R}^d$ with invariant measure μ. The *linearization*, or *derivative* of $\varphi(t, \omega)$ at $x \in \mathbb{R}^d$ (for fixed (ω, t)) is the Jacobian $d \times d$ matrix
$$D\varphi(t, \omega, x) = \frac{\partial \varphi(t, \omega)x}{\partial x} := \left(\frac{\partial(\varphi(t, \omega)y)_i}{\partial y_j}\right)\bigg|_{y=x},$$
which is a linear mapping of $\mathbb{R}^d$, $D\varphi(t, \omega, x) : \mathbb{R}^d \to \mathbb{R}^d$. The chain rule and the definition of a C^1 RDS immediately give the following statements.

4.2.1 Proposition. *Let φ be a C^1 RDS over θ on $\mathbb{R}^d$ with invariant measure μ. Then*

(i) $(\varphi, D\varphi)$, $(x, v) \mapsto (\varphi(t,\omega)x, D\varphi(t, \omega, x)v)$, is a continuous cocycle on $X = \mathbb{R}^d \times \mathbb{R}^d$ over θ.

(ii) $D\varphi$ is a linear cocycle on $X = \mathbb{R}^d$ over the metric dynamical system $(\Omega \times \mathbb{R}^d, \mathcal{F} \otimes \mathcal{B}, \mu, (\Theta(t))_{t \in \mathbb{T}})$, where $\Theta(t)(\omega, x) := (\theta(t)\omega, \varphi(t, \omega)x)$ is the skew product associated with φ.

4.2.2 Example. (i) If Ω is not present, $D\varphi$ is a linear cocycle over the deterministic flow φ with invariant measure μ on $\mathbb{R}^d$, a case much studied in smooth ergodic theory (mainly in its manifold version, see Remark 4.2.7(iv) below).

(ii) If the random variable x_0 generates a stationary orbit of φ, i.e. if $\varphi(t,\omega)x_0(\omega) = x_0(\theta(t)\omega)$, then the measure $\mu_\omega(dx) = \delta_{x_0(\omega)}(dx)$ is φ-invariant. In this case $\Phi(t,\omega) := D\varphi(t,\omega,x_0(\omega))$ is a matrix cocycle over θ. A fixed point $x_0 \in \mathbb{R}^d$ with $\varphi(t,\omega)x_0 = x_0$ is a particular case. ∎

We now regard $\mathbb{R}^d$ as a manifold and identify as usual the tangent bundle $T\mathbb{R}^d$ of $\mathbb{R}^d$ with its global trivialization $\mathbb{R}^d \times \mathbb{R}^d$,

$$T\mathbb{R}^d \cong \mathbb{R}^d \times \mathbb{R}^d, \quad T_x\mathbb{R}^d \cong \{x\} \times \mathbb{R}^d.$$

With this identification, $D\varphi(t,\omega,x)$ induces a linear mapping of $T_x\mathbb{R}^d \cong \{x\} \times \mathbb{R}^d$ to $T_{\varphi(t,\omega)x}\mathbb{R}^d \cong \{\varphi(t,\omega)x\} \times \mathbb{R}^d$ (i.e. we also keep track of the motion of the base point, $x \mapsto \varphi(t,\omega)x$) and as such we also denote it by

$$T\varphi(t,\omega,x): T_x\mathbb{R}^d \to T_{\varphi(t,\omega)x}\mathbb{R}^d.$$

Hence $T\varphi := (\varphi, D\varphi)$ is a cocycle over θ on $T\mathbb{R}^d$.

We now assume that we have a C^1 RDS φ with invariant measure μ on a manifold M. The *linearization* or *derivative* $T\varphi(t,\omega,x)$ of $\varphi(t,\omega)$ at $x \in M$ is the linear map

$$T\varphi(t,\omega,x): T_xM \to T_{\varphi(t,\omega)x}M, \quad v \mapsto w = T\varphi(t,\omega,x)v, \qquad (4.2.1)$$

(for its definition see Appendix B.5).

As usual, $T\varphi$ denotes the mapping on the tangent bundle (TM, M, π_M) which covers φ and is defined on fibers by (4.2.1).

4.2.3 Lemma. $(\Omega \times TM, \Omega \times M, \pi = (\mathrm{id}_\Omega, \pi_M))$ *is a linear measurable bundle with typical fiber $\mathbb{R}^d$.*

The proof consists in showing that there is a global measurable trivialization which is linear on fibers. This follows from part (i) of the next lemma.

4.2.4 Lemma (Global Trivialization of the Tangent Bundle). *Let M be a smooth manifold of dimension d.*

(i) The tangent bundle (TM, M, π_M) is a linear measurable bundle with base space M and typical fiber $\mathbb{R}^d$.

(ii) If M is a Riemannian manifold with Riemannian structure $\langle \cdot, \cdot \rangle_x$, then TM is a Euclidean measurable bundle and the global trivialization $\Psi: TM \to M \times \mathbb{R}^d$ in (i) can be chosen to be isometric on fibers, i.e.

$$\psi(x): (T_xM, \langle \cdot, \cdot \rangle_x) \to (\mathbb{R}^d, \langle \cdot, \cdot \rangle_S)$$

is an isometry. Here $\langle \cdot, \cdot \rangle_S$ denotes the standard scalar product of $\mathbb{R}^d$.

Proof. We just prove (ii) as it contains (i) since every smooth manifold possesses a smooth Riemannian metric.

The tangent bundle (TM, M, π_M) is a smooth vector bundle with typical fiber $\mathbb{R}^d$. By definition for each $x \in M$ there exists an open set U containing x and a diffeomorphism ψ covering id_U which trivializes the bundle locally, i.e. $\psi : \pi^{-1}(U) \to U \times \mathbb{R}^d$, and which for each $x \in U$ is a linear isomorphism as a mapping between fibers, $\pi^{-1}(x) = T_x M \to \{x\} \times \mathbb{R}^d$. By Theorem 1.8.20 of Klingenberg [213] this fiber mapping can be chosen to preserve the scalar product in fibers, i.e. is an isometry between $(T_x M, \langle \cdot, \cdot \rangle_x)$ and $(\mathbb{R}^d, \langle \cdot, \cdot \rangle_S)$.

We can now select a countable covering by bundle charts (U_i, ψ_i) trivializing TM locally and isometrically. "Disjointify" the U_i's by putting $B_1 := U_1$ and
$$B_n := U_n \setminus \cup_{i=1}^{n-1} B_i$$
to obtain a countable covering (B_n) of M by disjoint Borel sets. Finally put
$$\Psi : TM \to M \times \mathbb{R}^d, \quad v \mapsto \psi_n(v) \quad \text{for } v \in \pi^{-1}(B_n).$$

This map is a bimeasurable bijection covering id_M and is a linear isometric isomorphism on fibers. □

4.2.5 Proposition (Linearization of RDS is Linear Bundle RDS).
Let φ be a C^1 RDS on a d-dimensional manifold M over the metric DS $(\Omega, \mathcal{F}, \mathbb{P}, (\theta(t))_{t \in \mathbb{T}})$ with invariant measure μ. Let $T\varphi$ on TM be defined by $(x, v) \mapsto (\varphi(t, \omega)x, T\varphi(t, \omega, x)v)$. Then $T\varphi$ is
 (i) a continuous cocycle on TM over θ,
 (ii) a linear bundle RDS on $\Omega \times TM$ over the skew product $(\Omega \times M, \mathcal{F} \otimes \mathcal{B}, \mu, (\Theta(t))_{t \in \mathbb{T}})$, where $\Theta(t)(\omega, x) := (\theta(t)\omega, \varphi(t, \omega)x)$, with fiber mappings
$$T\varphi(t, \omega, x) : \pi^{-1}(\omega, x) = (\omega, T_x M)$$
$$\to \pi^{-1}(\Theta(t)(\omega, x)) = (\theta(t)\omega, T_{\varphi(t,\omega)x} M).$$

Proof. The cocycle property for $T\varphi(t, \omega, x)$ with respect to Θ follows by differentiating
$$\varphi(t + s, \omega)x = \varphi(t, \theta(s)\omega)\varphi(s, \omega)x$$
and the chain rule. □

4.2.2 The MET for RDS on Manifolds

We immediately state the two-sided time version.

4.2.6 Theorem (MET for RDS on Manifolds). *(A) Global case:* Let φ be a C^1 RDS on a Riemannian manifold M of dimension d over the metric DS $(\Omega, \mathcal{F}, \mathbb{P}, (\theta(t))_{t \in \mathbb{T}})$ with two-sided time and φ-invariant measure μ. Consider the linear bundle RDS $T\varphi$ on $\Omega \times TM$ over the metric DS

$(\Omega \times M, \mathcal{F} \otimes \mathcal{B}, \mu, (\Theta(t))_{t \in \mathbb{T}})$, where $\Theta(t)(\omega, x) = (\theta(t)\omega, \varphi(t, \omega)x)$ is the corresponding skew product flow. Assume that $T\varphi$ satisfies α^+, $\alpha^- \in L^1(\mathbb{P})$, where for $\mathbb{T} = \mathbb{Z}$

$$\alpha^+(\omega, x) := \log^+ \|T\varphi(1, \omega, x)\|_{x, \varphi(1,\omega)x},$$

$$\alpha^-(\omega, x) := \log^+ \|T\varphi(1, \omega, x)^{-1}\|_{\varphi(1,\omega)x, x}$$

and for $\mathbb{T} = \mathbb{R}$

$$\alpha^+(\omega, x) := \sup_{0 \leq t \leq 1} \log^+ \|T\varphi(t, \omega, x)\|_{x, \varphi(t,\omega)x}$$

and

$$\alpha^-(\omega, x) := \sup_{0 \leq t \leq 1} \log^+ \|T\varphi(t, \omega, x)^{-1}\|_{\varphi(t,\omega)x, x}.$$

Then there exists a Θ-invariant set $\Delta \subset \Omega \times M$ of full μ-measure on which the following statements hold:

There are $p(\omega, x)$, $1 \leq p(\omega, x) \leq d$, real numbers

$$-\infty < \lambda_{p(\omega,x)}(\omega, x) < \ldots < \lambda_1(\omega, x)$$

with multiplicities $d_i(\omega, x)$, $\sum_{i=1}^{p(\omega,x)} d_i(\omega, x) = d$, where p, λ_i and d_i are Θ-invariant random variables (which are constant if μ is ergodic), and a splitting

$$T_x M = E_1(\omega, x) \oplus \cdots \oplus E_{p(\omega,x)}(\omega, x)$$

of the tangent space $T_x M$ into random subspaces $E_i(\omega, x)$ (depending measurably on (ω, x), hence are random closed sets) with dimension $\dim E_i(\omega, x) = d_i(\omega, x)$ (the so-called Oseledets splitting) with the following properties: For $i \in \{1, \ldots, p(\omega, x)\}$

(i) if $P_i(\omega, x) : T_x M \to E_i(\omega, x)$ denotes the projection onto $E_i(\omega, x)$ along $F_i(\omega, x) := \oplus_{j \neq i} E_j(\omega, x)$ then

$$T\varphi(t, \omega, x) \circ P_i(\omega, x) = P_i(\Theta(t)(\omega, x)) \circ T\varphi(t, \omega, x),$$

equivalently

$$T\varphi(t, \omega, x) E_i(\omega, x) = E_i(\Theta(t)(\omega, x)),$$

(ii) we have

$$\lim_{t \to \pm\infty} \frac{1}{t} \log \|T\varphi(t, \omega, x)v\|_{\varphi(t,\omega)x} = \lambda_i(\omega, x) \iff v \in E_i(\omega, x) \setminus \{0\}.$$

(iii) Convergence in (ii) is uniform with respect to v in $E_i(\omega, x) \cap S_x^{d-1}$. The collection

$$\mathcal{S}(\theta, \varphi, \mu) := \{(\lambda_i(\cdot), d_i(\cdot))_{i=1,\ldots,p(\cdot)}\}$$

is called the **Lyapunov spectrum** of φ under μ.

(B) Local case: Let φ be a local C^1 RDS on a Riemannian manifold M with time $\mathbb{T} = \mathbb{R}$. Let μ be an invariant measure for φ which satisfies $\alpha^\pm \in L^1(\mu)$. Then all statements of Part (A) are true and the Θ-invariant set Δ of full μ-measure is a subset of $\mathrm{graph} E(\cdot)$.

Proof. Part (A): We use the isometric global trivialization $\Psi : TM \to M \times \mathbb{R}^d$ of Lemma 4.2.4. With $\psi(x) : T_xM \to \{x\} \times \mathbb{R}^d$

$$\Phi(n,\omega,x) := \psi(\varphi(n,\omega)x) \circ T\varphi(n,\omega,x) \circ \psi(x)^{-1}$$

is a linear cocycle on $\mathbb{R}^d$ over $(\Omega \times M, \mu, (\Theta(n))_{n\in\mathbb{Z}})$ which is trivially Lyapunov cohomologous to $T\varphi$ since $\|\psi(x)\|_{\langle\cdot,\cdot\rangle_x,\langle\cdot,\cdot\rangle_S} = 1$. Further,

$$\|\Phi(n,\omega,x)v\|_S = \|T\varphi(n,\omega,x)\psi(x)^{-1}v\|_{\varphi(n,\omega)x}, \quad v \in \mathbb{R}^d,$$

whence

$$\|\wedge^k \Phi(n,\omega,x)\|_S = \|\wedge^k T\varphi(n,\omega,x)\|_{x,\varphi(n,\omega)x}.$$

In particular, Φ satisfies the IC of the MET Theorem 3.4.11 which gives all statements for Φ and thus for $T\varphi$ by Proposition 4.1.9.

Part (B): We know from Theorem 1.8.2 that every invariant measure μ of a local C^1 RDS φ is supported by the strictly Θ-invariant set

$$E(\omega) := \cap_{t\in\mathbb{R}} D(t,\omega), \quad \varphi(t,\omega) : D(t,\omega) \to D(-t,\theta(t)\omega),$$

of initial values whose orbits do not explode,

$$\mu_\omega(E(\omega)) = 1 \quad \mathbb{P}\text{-a. s.}$$

Consequently, each $x \in E(\omega)$ satisfies $x \in D(t,\omega)$ for all $t \in \mathbb{R}$, hence $T\varphi(t,\omega,x) : T_xM \to T_{\varphi(t,\omega)x}M$ is well-defined for all $t \in \mathbb{R}$. The IC is as before, where $\alpha^\pm$ are well-defined measurable functions on $\mathrm{graph}E(\cdot) \in \mathcal{F} \otimes \mathcal{B}$. □

We saw in Chap. 2 that the conditions for generating a local RDS by means of a generator (random or stochastic differential equation) are much milder than those for a global RDS. Therefore, part (B) of the last theorem is a very useful extension.

4.2.7 Remark. (i) The MET remains valid and the spectrum is invariant if we change the Riemannian metric from $\langle\cdot,\cdot\rangle_x^1$ to $\langle\cdot,\cdot\rangle_x^2$ in a way such that the smooth functions $c_1, c_2 : M \to (0,\infty)$ fitting into

$$\frac{1}{c_2(x)}\|v\|_x^1 \leq \|v\|_x^2 \leq c_1(x)\|v\|_x^1$$

are tempered. In particular, if on a compact manifold the MET holds for one Riemannian metric, then it holds for all Riemannian metrics with the same spectrum and splitting.

We later need to change from a given Riemannian metric $\langle\cdot,\cdot\rangle_x$ to a random Riemannian metric $\langle\cdot,\cdot\rangle_{\omega,x}$ in T_xM which depends only measurably on (ω,x). The MET will hold under the new random norm if the $c_i : \Omega \times M \to (0,\infty)$ in

$$\frac{1}{c_2(\omega,x)}\|v\|_x^1 \leq \|v\|_{\omega,x}^2 \leq c_1(\omega,x)\|v\|_x^1,$$

are tempered, see Sect. 4.3.

(ii) Note that for a C^1 RDS with two-sided time $(t,x,v) \mapsto T\varphi(t,\omega,x)^{\pm 1}v$ is continuous for each ω (see Theorem 1.1.6(ii)). Hence the expressions appearing in the integrability conditions of the last theorem are finite measurable functions in (ω,x) which depend continuously on x. Therefore $\beta^{\pm} \in L^1(\mathbb{P})$, where

$$\beta(\omega)^{\pm} := \begin{cases} \sup_{x\in M} \log^+ \|T\varphi(1,\omega,x)^{\pm 1}\|, & \mathbb{T} = \mathbb{Z}, \\ \sup_{x\in M} \sup_{0\leq t\leq 1} \log^+ \|T\varphi(t,\omega,x)^{\pm 1}\|, & \mathbb{T} = \mathbb{R}, \end{cases}$$

would suffice for the validity of the MET for φ and for any invariant measure μ, see Subsect. 4.2.5.

A particular case which has been extensively studied is the deterministic case, where ω is not present (see Katok and Hasselblatt [200], Mañé [249], Pesin [274, 275], Ruelle [292]). Here φ is a C^1 DS on a Riemannian manifold M with invariant measure ρ on $(M, \mathcal{B})$. Then the MET holds with the dependence on ω ignored. If M is compact, the IC are always satisfied for any φ and any invariant measure ρ since $\|T\varphi(t,x)^{\pm 1}\|$ is continuous with respect to (t,x) and thus bounded on a compact set. ∎

4.2.8 Remark (Total Measure in the MET). The Θ-invariant set of full μ-measure $\Delta \subset \Omega \times M$ depends of course on φ. However, it has *full total measure* in the sense that $\mu(\Delta) = 1$ for all φ-invariant measures μ satisfying the IC. This eventually follows from the proof of the MET, in which first Δ is constructed as a Θ-forward invariant/invariant set and then we prove that it has full μ measure. ∎

4.2.9 Remark (Sacker-Sell Spectrum). Assume that μ is Θ-invariant and ergodic. The MET establishes a measurable splitting

$$\Omega \times TM = \bigoplus_{i=1}^{p} E_i \quad \text{(as a Whitney sum)}$$

into p linear measurable subbundles $(\omega, x) \mapsto \{(\omega, x)\} \times E_i(\omega, x)$ of constant dimension d_i and typical fiber $\mathbb{R}^{d_i}$, which are invariant with respect to Θ over the invariant set $\Delta \subset \Omega \times M$ of full μ-measure. The bundle E_i is characterized dynamically as containing exactly those tangent vectors which have Lyapunov exponent λ_i forwards and backwards in time.

If $(\theta(t))_{t\in\mathbb{T}}$ happens to consist of homeomorphisms of a compact and (dynamically) connected Hausdorff space Ω and $\varphi(t,\omega)$ is a C^1 cocycle on a compact manifold M, we can consider $T\varphi$ as a continuous linear skew product flow over Θ on $\Omega \times M$ (in which case we do not need an invariant measure). See Johnson, Palmer and Sell [192], Sacker and Sell [299] and Sell [316].

The *dichotomy spectrum* or *Sacker-Sell spectrum* $\mathcal{S}_{\text{dich}}$ of $T\varphi$ consists of all those $\lambda \in \mathbb{R}$ for which $\Phi_\lambda(t,\omega,x) := e^{-\lambda t} T\varphi(t,\omega,x)$ fails to have an exponential dichotomy. Recall that Φ_λ has an *exponential dichotomy* if there is a continuous projector $P(\omega,x)$ on $T_x M$ and constants $K \geq 0$, $\alpha > 0$ such that
$$\|\Phi_\lambda(t,\omega,x) P(\omega,x) \Phi_\lambda(s,\omega,x)^{-1}\| \leq K e^{-\alpha(t-s)}, \quad s \leq t,$$
$$\|\Phi_\lambda(t,\omega,x)(I - P(\omega,x)) \Phi_\lambda(s,\omega,x)^{-1}\| \leq K e^{-\alpha(s-t)}, \quad t \leq s$$
for all $(\omega,x) \in \Omega \times M$ and all $s,t \in \mathbb{T}$.

The spectral theorem of Sacker and Sell [299] states that
$$\mathcal{S}_{\text{dich}} = \cup_{j=1}^k [a_j, b_j],$$
the union of k, $1 \leq k \leq d$, non-overlapping compact intervals, and that to each interval there is a continuous invariant subbundle $F_j(\omega, x)$ such that we have a continuous invariant splitting
$$\Omega \times TM = \bigoplus_{j=1}^k F_j \quad \text{(as a Whitney sum)}.$$

If $v \in F_j(\omega, x) \setminus \{0\}$ then the four numbers
$$\lambda_s^\pm(\omega, x, v) := \limsup_{t \to \pm\infty} \frac{1}{t} \log \|T\varphi(t,\omega,x)v\|,$$
$$\lambda_i^\pm(\omega, x, v) := \liminf_{t \to \pm\infty} \frac{1}{t} \log \|T\varphi(t,\omega,x)v\|$$
lie in $[a_j, b_j]$ (Sacker and Sell [299: Theorem 3]).

Let us denote by $\mathcal{S}_\mu = \{\lambda_i(\mu) : i = 1, \ldots, p\}$ the Lyapunov exponents of $T\varphi$ with respect to the ergodic invariant measure μ, the notation emphasizing the dependence on μ. Johnson, Palmer and Sell [192] have proved that
$$\partial(\mathcal{S}_{\text{dich}}) \subset \bigcup \{\mathcal{S}_\mu : \mu \; \Theta\text{-invariant}\} \subset \mathcal{S}_{\text{dich}},$$
where $\partial(B)$ denotes the boundary of the set B.

More precisely, for each $a \in \partial(\mathcal{S}_{\text{dich}})$ there exists an ergodic invariant measure μ for which $a \in \mathcal{S}_\mu$ and for each $\lambda_i(\mu) \in \mathcal{S}_\mu$ there is exactly one spectral interval $[a_j, b_j]$ from $\mathcal{S}_{\text{dich}}$ with $\lambda_i(\mu) \in [a_j, b_j]$ and $E_i^\mu(\omega, x) \subset F_j(\omega, x)$ for all $(\omega, x) \in \Delta_\mu$, where
$$F_j(\omega, x) = \bigoplus_{i : \lambda_i(\mu) \in [a_j, b_j]} E_i^\mu(\omega, x).$$

For each invariant μ the measurable invariant splitting of the MET thus forms a refinement of the continuous invariant splitting of the Sacker-Sell theory. The "price" we have to pay for the refinement of the splitting is the loss of its continuity.

Johnson [189] constructed an example of an almost-periodic linear differential equation in $\mathbb{R}^3$ for which $\mathcal{S}_{\text{dich}} = [a_1, b_1]$, but $\mathcal{S}_\mu = \{a_1, a, b_1\}$ with $a_1 < a < b_1$.

We refer to Colonius and Kliemann [105] for an extensive treatment of the various spectral theories of linear nonautonomous systems. ∎

4.2.3 Random Differential Equations

We now search for conditions on the right-hand side of an RDE

$$\dot{x}_t = f(\theta_t \omega, x_t) \qquad (4.2.2)$$

on a Riemannian manifold M which suffice for the IC of the MET.

For $M = \mathbb{R}^d$ (4.2.2) generates a unique global C^1 RDS φ provided $f \in L^1_\mathbb{P}(\Omega, \mathcal{C}^{1,0}_b)$ (see Theorem 2.2.6). By definition and the mean value theorem, $f \in L^1_\mathbb{P}(\Omega, \mathcal{C}^{1,0}_b)$ if and only if $|f(\cdot, 0)\| \in L^1(\mathbb{P})$ and

$$\sup_{x \in \mathbb{R}^d} \|Df(\cdot, x)\| \in L^1(\mathbb{P}). \qquad (4.2.3)$$

Under this condition we have globally (on an invariant set $\tilde{\Omega}$ of full $\mathbb{P}$-measure)

$$\varphi(t, \omega)x = x + \int_0^t f(\theta_s \omega, \varphi(s, \omega)x) \, ds, \qquad (4.2.4)$$

$$D\varphi(t, \omega, x) = I + \int_0^t Df(\theta_s \omega, \varphi(s, \omega)x) D\varphi(s, \omega, x) \, ds, \qquad (4.2.5)$$

and

$$D\varphi(t, \omega, x)^{-1} = I - \int_0^t D\varphi(s, \omega, x)^{-1} Df(\theta_s \omega, \varphi(s, \omega)x) \, ds. \qquad (4.2.6)$$

From (4.2.5) and (4.2.6) by Gronwall's lemma,

$$\sup_{0 \le t \le 1} \log^+ \|D\varphi(t, \omega, x)^{\pm 1}\| \le \int_0^1 \|Df(\theta_s \omega, \varphi(s, \omega)x)\| \, ds. \qquad (4.2.7)$$

Suppose μ is a φ-invariant measure, and $\|Df(\cdot, \cdot)\| \in L^1(\mu)$ then the right-hand side, hence the left-hand side of (4.2.7) is in $L^1(\mu)$ and the IC of the MET are fulfilled. But by (4.2.3)

$$\|Df(\cdot, \cdot)\| \le \sup_{x \in \mathbb{R}^d} \|Df(\cdot, x)\| \in L^1(\mathbb{P}),$$

hence under our conditions for the existence of a global RDS, the IC of the MET are automatically satisfied for any μ. We summarize this as the part (A) of the following theorem.

4.2.10 Theorem (MET for RDE). *Consider the RDE (4.2.2).*
(A) Global case on $\mathbb{R}^d$: If $\|f(\cdot, 0)\| \in L^1(\mathbb{P})$ and $\sup_{x \in \mathbb{R}^d} \|Df(\cdot, x)\| \in L^1(\mathbb{P})$, then (4.2.2) uniquely generates a global C^1 RDS which satisfies the IC of the MET for any φ-invariant measure μ.
(B) Local case for Riemannian manifold M: Assume $f \in L^1_{\mathbb{P}}(\Omega, \mathcal{C}^{1,0})$. Then (4.2.2) uniquely generates a local C^1 RDS φ. Suppose μ is a φ-invariant measure which satisfies
$$\|\nabla f(\cdot, \cdot)\| \in L^1(\mu).$$
Then φ satisfies the IC of the MET under μ.

Proof. It remains to prove part (B).

The existence and uniqueness of φ is guaranteed by Theorem 2.2.14. Assume that $x \in E(\omega)$. Then $T\varphi$ solves $\dot{v}_t = Tf(\theta_t \omega, v_t)$, which is equivalent to the pair consisting of the original RDE and the variational equation

$$v'_t = \nabla f(\theta_t \omega, \varphi(t, \omega)x) v_t.$$

Using polar coordinates in TM (see Sect. 6.4) we obtain

$$\log \|T\varphi(t, \omega, x)v\| \qquad (4.2.8)$$
$$= \log \|v\| + \int_0^t \langle S\varphi(u, \omega, x)s, \nabla f(\theta_u \omega, \varphi(u, \omega)x) S\varphi(u, \omega, x)s \rangle\, du,$$

where $s = \frac{v}{\|v\|}$, and $S\varphi$ is the RDS on the unit sphere bundle SM induced by $T\varphi$. Using the elementary fact that for any $A \in \mathbb{R}^{d \times d}$ and any basis of unit vectors (v_i) of $\mathbb{R}^d$, $\|A\| \leq d \max_{1 \leq i \leq d} \|Av_i\|$, we obtain from (4.2.8)

$$\alpha^+(\omega, x) := \sup_{0 \leq t \leq 1} \log^+ \|T\varphi(t, \omega, x)\| \leq \log d + \int_0^t \|\nabla f(\theta_u \omega, \varphi(u, \omega)x)\|\, du.$$

The same estimate is obtained for α^- by using the same argument for $(T\varphi)^{*-1}$ which is generated by $v'_t = (-\nabla f)^* v_t$. □

For both the global and the local case, Liouville's formula yields (for ergodic μ)

$$\lambda_\Sigma = \frac{1}{d} \sum_{i=1}^p d_i \lambda_i = \frac{1}{d} \mathbb{E}_\mu \text{div} f = \frac{1}{d} \mathbb{E}_\mu \text{trace} \nabla f.$$

Recall that for $M = \mathbb{R}^d$, $\nabla f = Df$.

4.2.11 Exercise (Continuation of Exercise 2.2.9). The random differential equation in $\mathbb{R}$

$$\dot{x}_t = a(\theta_t \omega) x_t + b(\theta_t \omega) x_t^N, \quad N \geq 2, \quad a, b \in L^1(\mathbb{P}),$$

generates a local C^∞ RDS φ. Verify the validity of the IC for its invariant measures and determine the corresponding Lyapunov exponents. For the cases $N = 2$ and $N = 3$ see Subsect. 9.3.4. ∎

4.2.4 Stochastic Differential Equations

Now consider the SDE

$$dx_t = \sum_{j=0}^{m} f_j(x_t) \circ dW_t^j \qquad (4.2.9)$$

on a Riemannian manifold M. We assume that for some $0 < \delta \leq 1$

$$f_0 \in \mathcal{C}^{1,\delta}, \quad f_1, \ldots, f_m \in \mathcal{C}^{2,\delta}. \qquad (4.2.10)$$

Then (4.2.9) generates a local C^1 RDS φ over the canonical DS modeling Brownian motion (Theorem 2.2.14).

We would like to find conditions in terms of the vector fields f_j and the invariant measure μ which imply the IC of the MET. The task is complicated by the fact that $\omega \mapsto \mu_\omega$ generally depends on the whole history $\mathcal{F} = \mathcal{F}_{-\infty}^\infty$ of the Wiener process.

For example, if M is a compact manifold, (4.2.9) generates a global C^1 RDS φ (see Theorem 2.2.14(iv)). It was first proved by Carverhill [89: Appendix A.2] (for M a bounded open domain of $\mathbb{R}^d$) and Baxendale [59: Proposition 2.1] that φ automatically satisfies the IC of the MET for any μ. This follows from the stronger result

$$\mathbb{E}\left(\sup_{x \in M} \sup_{0 \leq t \leq 1} \|T\varphi(t, \cdot, x)^{\pm 1}\|\right) < \infty.$$

The latter result was considerably improved by Kifer [208] who showed that for any $p \geq 1$

$$\mathbb{E}\left(\sup_{0 \leq t \leq 1} \|\varphi(t, \cdot)\|_k + \sup_{0 \leq t \leq 1} \|\varphi(t, \cdot)^{-1}\|_k\right)^p < \infty,$$

where $\|\cdot\|_k$ is the norm in $\mathcal{C}^k(M, M)$, provided $f_0 \in \mathcal{C}^{k,\delta}$, $f_1, \ldots, f_m \in \mathcal{C}^{k+1,\delta}$ for some $k \geq 1$.

We hence have the following statement.

4.2.12 Theorem (MET for SDE on Compact Manifolds). *Let M be a compact Riemannian manifold and let the SDE (4.2.9) satisfy (4.2.10). Then the global C^1 RDS φ it generates satisfies the IC of the MET (Theorem 4.2.6) for any φ-invariant measure μ.*

Let now $M = \mathbb{R}^d$ and assume the stronger conditions

$$f_0 \in \mathcal{C}_b^{1,\delta}, \quad f_1, \ldots, f_m \in \mathcal{C}_b^{2,\delta}, \quad \sum_{j=1}^{m} D_{f_j} f_j \in \mathcal{C}_b^{1,\delta}, \qquad (4.2.11)$$

where $D_f f := \sum_{i=1}^{d} f^i \frac{\partial}{\partial x_i} f$. Then also the Itô-Stratonovich correction term

$$\hat{f}_0(x) := f_0(x) \pm \frac{1}{2} \sum_{j=1}^{m} D f_j f_j(x), \qquad (4.2.12)$$

where "+" holds for time $\mathbb{R}^+$ and "−" holds for $\mathbb{R}^-$, satisfies $\hat{f}_0 \in C_b^{1,\delta}$.

Under (4.2.11) the SDE (4.2.9) generates a global C^1 RDS φ. Further, the Jacobian $D\varphi(t,\omega,x)$ of $\varphi(t,\omega)$ at x solves

$$dv_t = \sum_{j=0}^{m} Df_j(x_t) v_t \circ dW_t^j = D\hat{f}_0(x_t) v_t dt + \sum_{j=1}^{m} Df_j(x_t) v_t \, dW_t^j. \qquad (4.2.13)$$

Using polar coordinates $r = |x| \in (0,\infty)$, $s = \frac{x}{|x|} \in S^{d-1}$ as in the linear case (see Sect. 6.2) we obtain that the SDE (4.2.13) is equivalent to the two SDE

$$dS_t = \sum_{j=0}^{m} h_j(S_t) \circ dW_t^j, \quad S_0 = s \in S^{d-1}, \qquad (4.2.14)$$

and

$$dR_t = \sum_{j=0}^{m} q_j(S_t) R_t \circ dW_t^j = Q(S_t) R_t dt + \sum_{j=1}^{m} q_j(S_t) R_t \, dW_t^j, \qquad (4.2.15)$$

$R_0 = r \in (0,\infty)$. Here $h_j := A_j s - \langle A_j s, s \rangle s$, $q_j := \langle A_j s, s \rangle$ and

$$Q := \langle A_0 s, s \rangle \pm \frac{1}{2} \sum_{j=1}^{m} \left(\langle B_j s, s \rangle + \langle A_j s, A_j s \rangle - 2 \langle A_j s, s \rangle^2 \right), \qquad (4.2.16)$$

where $A_j = Df_j(x)$ and

$$B_j = D(A_j(x) f_j(x)) = A_j(x)^2 + (D^2 f_j)(x) f_j(x)$$

is the Jacobian of $g_j(x) = A_j(x) f_j(x)$. Hence $|q_j(x,s)| \leq \|A_j(x)\|$ and

$$|Q(x,s)| \leq \|A_0(x)\| + \frac{1}{2} \sum_{j=1}^{m} (\|B_j(x)\| + 3\|A_j(x)\|^2) =: Q_0(x).$$

It suffices to deal with the first IC $\alpha^+ \in L^1(\mu)$. Integrating (4.2.15), proceeding as in the linear case (see Remark 6.2.12), denoting the marginal of μ on $\mathbb{R}^d$ by ρ and the integral with respect to a measure ν by $\mathbb{E}_\nu$,

$$\mathbb{E}_\mu \alpha^+ \leq \log d + \mathbb{E}_\rho Q_0 \qquad (4.2.17)$$

$$+ \max_i \sum_{j=1}^{m} \mathbb{E}_\mathbb{P} \mathbb{E}_{\mu_\omega} \sup_{0 \leq t \leq 1} \left| \int_0^t q_j(\varphi(u,\cdot)x, S_u(\cdot,x) e_i) \, dW_u^j \right|,$$

where (e_i) is the standard basis of $\mathbb{R}^d$. Due to our assumptions (4.2.11) $Q_0(x) \leq C$, hence $\mathbb{E}_\rho Q_0 < \infty$ for any μ. The expectation of the stochastic integrals in (4.2.17) can in general not be estimated by means of classical stochastic analysis.

There is, however, one important particular case which permits the use of Burkholder's inequality, namely the case where μ is a forward or backward Markov measure, i. e. where $\omega \mapsto \mu_\omega$ is measurable with respect to the past $\mathcal{F}^-$ or with respect to the future $\mathcal{F}^+$ of the Wiener process W (see Subsect. 2.9.3).

4.2.13 Theorem (MET for SDE, Invariant Markov Measure).
Consider the SDE (4.2.9) in $\mathbb{R}^d$.

(A) Global case: Let (4.2.9) satisfy conditions (4.2.11) so that it generates a global C^1 RDS φ and let μ be a φ-invariant Markov measure. Then the IC of the MET are automatically satisfied.

(B) Local case: Let (4.2.9) satisfy the conditions (4.2.10) so that it generates a local C^1 RDS φ and let μ be a φ-invariant Markov measure with marginal ρ. Then the IC of the MET are satisfied provided $\|Df_0\| \in L^1(\rho)$ and $\|D((Df_j)f_j)\|, \|Df_j\|^2 \in L^1(\rho)$ for $j = 1, \ldots, m$.

(C) The invariant random variables p, d_i and λ_i from the MET are independ of ω.

Proof. (A) Restricting ourselves to forward Markov measures, they are by Theorem 2.3.45 in a one-to-one correspondence with the stationary measures ρ of the one-point Markov process generated by (4.2.9) for time $\mathbb{R}^+$, equivalently, with the solutions of the Fokker-Planck equation $(L^+)^* \rho^+ = 0$, $L^+ := f_0 + \frac{1}{2}\sum_{j=1}^m (f_j)^2$. Further, restricting the RDS φ to one-sided time $\mathbb{T} = \mathbb{R}^+$ and to the σ-algebra $\mathcal{F}_0^\infty$, μ corresponds to the Θ-invariant product measure $\mu^+ = \mathbb{P} \times \rho^+$.

Making this restriction for the stochastic integrals in (4.2.17) Burkholder's inequality for any $p > 0$ and $s \in S^{d-1}$ and then Jensen's inequality for $p \geq 1$ yields

$$\mathbb{E}_\mu \left(\sup_{0 \leq t \leq 1} \left| \int_0^t q_j(\varphi(u,\cdot)x, S_u(\cdot,x)s) dW_u^j \right| \right)^{2p} \qquad (4.2.18)$$

$$\leq C_p \mathbb{E}_\rho \mathbb{E}_\mathbb{P} \left(\int_0^1 \|A_j(\varphi(u,\cdot)x)\|^2 du \right)^p \leq C_p \mathbb{E}_\rho \|A_j(\cdot)\|^{2p}.$$

By (4.2.11), $\|A_j(x)\| \leq C$, hence the right-hand side of (4.2.17) is finite for any $p \geq 1$ and any ρ. Consequently, even $\mathbb{E}_\mu(\alpha^+)^p < \infty$ for any $p \geq 1$.

(B) The estimates (4.2.17) and (4.2.18), using for example $p = 1$ in (4.2.18), show that the IC of the MET are met provided the conditions listed in the theorem are satisfied.

(C) If a random variable $f : \Omega \times \mathbb{R}^d \to \mathbb{R}$ is Θ-invariant and $\mathbb{P} \times \rho^+$ is an invariant measure for the RDS restricted to time $\mathbb{R}^+$, then $f(\omega, x) = \bar{f}(x)$

$\mathbb{P} \times \rho^+$-a.s. for some measurable $\bar f$. This is a consequence of the ergodicity of θ and the product structure of the invariant measure (for details see Kifer [207: §1.2]). □

Theorem 4.2.13(A) covers in particular the case of a linear SDE under the Markov measure δ_0 (see also Theorem 6.2.11 and Remark 6.2.12).

We are not able to extend these simple statements from Markov measures to arbitrary invariant measures. We have, however, the following quite satisfactory general criterion (see Arnold and Imkeller [22]).

4.2.14 Theorem (MET for SDE, General Invariant Measure).
Let the SDE (4.2.9) in $\mathbb{R}^d$ satisfy the conditions (4.2.11) so that it generates a global C^1 RDS φ. Assume further that $D\hat f_0$ is globally Lipschitz. Let μ be a φ-invariant measure with marginal ρ on $\mathbb{R}^d$ which satisfies

$$\mathbb{E} \int_{\mathbb{R}^d} \sqrt{\log^+ \|x\|}\, \mu.(dx) = \int_{\mathbb{R}^d} \sqrt{\log^+ \|x\|}\, \rho(dx) < \infty. \qquad (4.2.19)$$

Then φ under μ meets the IC of the MET.

We refrain from giving the proof here as it is based on techniques which are not used otherwise in this book.

4.2.15 Example (Continuation of Subsect. 2.3.7). In Subsect. 2.3.7 we determined all invariant measures (necessarily Dirac measures) of the local RDS φ generated by the SDE

$$dx_t = (\alpha x_t - x_t^N)dt + \sigma x_t \circ dW_t,$$

on the state space $\mathbb{R}$, where $N \geq 2$, $\alpha \in \mathbb{R}$ and $\sigma \geq 0$. We now calculate the Lyapunov exponent for each of these measures.

The linearized RDS $v_t = D\varphi(t,\omega,x)v$ satisfies the linearized SDE

$$dv_t = (\alpha - N(\varphi(t,\omega)x)^{N-1})v_t dt + \sigma v_t \circ dW_t,$$

hence

$$D\varphi(t,\omega,x)v = v \exp\left(\alpha t + \sigma W_t(\omega) - N \int_0^t (\varphi(s,\omega)x)^{N-1} ds\right).$$

Thus, if $\mu_\omega = \delta_{x_0(\omega)}$ is a φ-invariant measure, its Lyapunov exponent is

$$\begin{aligned}
\lambda(\mu) &= \lim_{t\to\infty} \frac{1}{t} \log \|D\varphi(t,\omega,x)v\| \\
&= \alpha - N \lim_{t\to\infty} \frac{1}{t} \int_0^t (\varphi(s,\omega)x)^{N-1} ds \\
&= \alpha - N\, \mathbb{E}\, x_0^{N-1},
\end{aligned}$$

provided the IC $x_0^{N-1} \in L^1(\mathbb{P})$ is satisfied.

Case N even:
(i) For $\alpha \in \mathbb{R}$, the IC for $\mu_{1,\omega}^\alpha = \delta_0$ is trivially satisfied and we obtain
$$\lambda(\mu_1^\alpha) = \alpha,$$
so μ_1^α is stable for $\alpha < 0$ and unstable for $\alpha > 0$.

(ii) For $\alpha > 0$, $\mu_{2,\omega}^\alpha = \delta_{d_-^\alpha(\omega)}$ is $\mathcal{F}_{-\infty}^0$ measurable, hence the density p^α of $\rho^\alpha = \mathbb{E}\,\mu_{2,\cdot}^\alpha$ satisfies the Fokker-Planck equation
$$L_\alpha^* p^\alpha = -((\alpha x + \frac{\sigma^2}{2}x - x^N)p^\alpha)' + \frac{\sigma^2}{2}(x^2 p^\alpha)'' = 0,$$
which has the unique probability density solution
$$p^\alpha(x) = N_\alpha x^{2\alpha/\sigma^2 - 1} \exp\left(-\frac{2x^{N-1}}{(N-1)\sigma^2}\right), \quad x > 0$$
(see Ariaratnam [4: Equation (15)]). Since
$$\mathbb{E}_{\mu_2^\alpha} x^{N-1} = \mathbb{E}\,(d_-^\alpha)^{N-1} = \int_0^\infty x^{N-1} p^\alpha(x)\,dx < \infty,$$
the IC is satisfied. The calculation of the Lyapunov exponent is accomplished by observing that
$$d_-^\alpha(\theta_t\omega)^{N-1} = \frac{e^{(N-1)(\alpha t + \sigma W_t(\omega))}}{(N-1)\int_{-\infty}^t e^{(N-1)(\alpha s + W_s(\omega))}\,ds} = \frac{\Psi(t)'}{(N-1)\Psi(t)},$$
where
$$\Psi(t) = \int_{-\infty}^t e^{(N-1)(\alpha s + W_s(\omega))}\,ds.$$
Hence by the ergodic theorem
$$\mathbb{E}\,(d_-^\alpha)^{N-1} = \frac{1}{N-1}\lim_{t\to\infty}\frac{1}{t}\log\Psi(t) = \alpha,$$
finally
$$\lambda(\mu_2^\alpha) = -(N-1)\alpha < 0.$$

(iii) For $\alpha < 0$, $\mu_{3,\omega}^\alpha = \delta_{-d_+^\alpha(\omega)}$ is $\mathcal{F}_0^\infty$ measurable. Since $\mathcal{L}(d_+^\alpha) = \mathcal{L}(d_-^{-\alpha})$ and N is even,
$$\mathbb{E}\,(-d_+^\alpha)^{N-1} = -\mathbb{E}\,(d_-^{-\alpha})^{N-1} = \alpha,$$
thus
$$\lambda(\mu_3^\alpha) = -(N-1)\alpha > 0.$$

Case N odd:
(i) For $\alpha \in \mathbb{R}$, $\mu_{1,\omega}^\alpha = \delta_0$, and
$$\lambda(\mu_1^\alpha) = \alpha.$$

(ii) For $\alpha > 0$, $\mu_{2,\omega}^\alpha = \delta_{-d_-^\alpha(\omega)}$ and $\mu_{3,\omega}^\alpha = \delta_{d_-^\alpha(\omega)}$ are $\mathcal{F}_{-\infty}^0$ measurable. The solution p_2^α of the Fokker-Planck equation corresonding to μ_2^α is given by the same expression as for the case N even and $p_3^\alpha(x) = p_2^\alpha(-x)$. The IC can be verified directly using the expression for p_2^α, and

$$\mathbb{E}\left(-d_-^\alpha\right)^{N-1} = \mathbb{E}\left(d_-^\alpha\right)^{N-1} = \alpha,$$

thus

$$\lambda(\mu_2^\alpha) = \lambda(\mu_3^\alpha) = -(N-1)\alpha < 0.$$

A detailed analysis of the bifurcation behavior of this RDS for the cases $N = 2$ and $N = 3$ is given in Subsect. 9.3.1. ∎

4.3 Random Lyapunov Metrics and Norms

One of the key technical steps in the construction of invariant manifolds of a deterministic diffeomorphism φ at the hyperbolic fixed point 0 is the change from the standard Euclidean norm in $\mathbb{R}^d$ to a norm under which $D\varphi(0)$ is contracting on its stable space and expanding on its unstable space. See e. g. Irwin [181: p. 155] or Ruelle [295: p. 19]. These new norms are called *adapted to* φ or *Lyapunov norms*.

This is also possible, in fact essential, in the random case. For the transfer of the results of the MET from the linearized RDS $\Phi(t,\omega) = D\varphi(t,\omega,0)$ to the original nonlinear RDS $\varphi(t,\omega)$ in a neighborhood of 0 we need control of the non-uniformity in the MET which is due to the fact that Lyapunov exponents are an asymptotic concept. Since the non-uniformity in $\Phi(t,\omega)$ that has to be suppressed by the new norm depends on ω, the new norm will inevitably depend on ω too.

The construction of the new random norms has to be done carefully since we want our linearized cocycle under the new norm to be Lyapunov cohomologous to itself under the original norm in order not to change the Lyapunov spectrum.

This section is based on the now classical work of Pesin [274], followed by that of Fathi, Herman and Yoccoz [149] and of Pugh and Shub [282], who all treat the case of the iteration of a deterministic diffeomorphism on a manifold M which leaves a measure invariant (smooth ergodic theory). We go beyond these papers and follow Boxler [79] and Dahlke [120] who develop random versions of the construction in which

- every subspace from the Oseledets splitting is equipped with a new scalar product,
- the dynamics in both time directions ($t \to -\infty$ and $t \to \infty$) is taken into account,
- time can also be continuous.

In this section we make the following **standing assumptions** on our underlying DS $(\Omega, \mathcal{F}, \mathbb{P}, (\theta(t))_{t \in \mathbb{T}})$:

- time is two-sided, $\mathbb{T} = \mathbb{Z}$ or $\mathbb{T} = \mathbb{R}$,
- θ is ergodic (this assumption is only made to simplify presentation).

4.3.1 The Control of Non-Uniformity in the MET

We treat the cases $\mathbb{T} = \mathbb{Z}$ and $\mathbb{T} = \mathbb{R}$ as one. Assume that a matrix cocycle $\Phi(t, \omega)$ is given which satisfies the IC of the MET Theorem 3.4.11.

We will now introduce an improved temperedness property.

4.3.1 Definition (Slowly Varying Random Variable). Given $\varepsilon \geq 0$ and a DS θ. A random variable $R : \Omega \to (0, \infty)$ is called ε-*slowly varying* with respect to θ if $\mathbb{P}$-a.s.

$$e^{-\varepsilon|t|}R(\omega) \leq R(\theta(t)\omega) \leq e^{\varepsilon|t|}R(\omega) \quad \text{for all} \quad t \in \mathbb{T}.$$

∎

4.3.2 Lemma. *(i) Each of the following conditions is equivalent with R_ε being ε-slowly varying:*

- $R_\varepsilon(\theta(t)\omega) \leq R_\varepsilon(\omega)e^{\varepsilon|t|}$ *for all* $t \in \mathbb{T}$, $\mathbb{P}$-*a. s.* ,
- $e^{-\varepsilon|t|}R_\varepsilon(\omega) \leq R_\varepsilon(\theta(t)\omega)$ *for all* $t \in \mathbb{T}$, $\mathbb{P}$-*a. s.* ,
- $\frac{1}{R_\varepsilon}$ *is ε-slowly varying.*

(ii) If R_ε is ε-slowly varying then it is δ-slowly varying for $\delta \geq \varepsilon$ and cR_ε^α is $|\alpha\varepsilon|$-slowly varying for each $c > 0$ and $\alpha \in \mathbb{R}$.

(iii) If $R_{\varepsilon,1}$ and $R_{\varepsilon,2}$ are ε-slowly varying, then so are $R_{\varepsilon,1} + R_{\varepsilon,2}$, $\max(R_{\varepsilon,1}, R_{\varepsilon,2})$ and $\min(R_{\varepsilon,1}, R_{\varepsilon,2})$. Further, $R_{\varepsilon,1} \cdot R_{\varepsilon,2}$ is 2ε-slowly varying. Also, the pointwise limit of ε-slowly varying random variables is ε-slowly varying.

The proof is elementary.

4.3.3 Proposition (Tempered Versus Slowly Varying). *(i) If R_ε is ε-slowly varying for some $\varepsilon \geq 0$ then it is tempered.*

(ii) Conversely, if $f : \Omega \to (0, \infty)$ is tempered and in case $\mathbb{T} = \mathbb{R}$ if $t \mapsto \log f(\theta(t)\omega)$ is continuous $\mathbb{P}$-a.s., then for any $\varepsilon > 0$ there is an ε-slowly varying random variable R_ε for which

$$\frac{1}{R_\varepsilon(\omega)} \leq f(\omega) \leq R_\varepsilon(\omega).$$

Proof. (i) We have

$$-\varepsilon \leq \liminf_{t \to \pm\infty} \frac{1}{t} \log R_\varepsilon(\theta(t)\omega) \leq \limsup_{t \to \pm\infty} \frac{1}{t} \log R_\varepsilon(\theta(t)\omega) \leq \varepsilon,$$

whence the result follows from Proposition 4.1.3(i).

(ii) By definition and utilizing the continuity of $t \mapsto \log f(\theta(t)\omega)$ in the case $\mathbb{T} = \mathbb{R}$, there is for each $\varepsilon > 0$ a C_ε (in general not ε-slowly varying) with
$$\frac{1}{C_\varepsilon(\omega)} e^{-\varepsilon|t|} \leq f(\theta(t)\omega) \leq C_\varepsilon(\omega) e^{\varepsilon|t|}$$
for all $t \in \mathbb{T}$. Part (i) of the proof of the next theorem shows that there is then a 2ε-slowly varying random variable $R_{2\varepsilon}$ for which
$$\frac{1}{R_{2\varepsilon}(\omega)} \leq f(\omega) \leq R_{2\varepsilon}(\omega).$$

$\square$

4.3.4 Theorem (Non-Uniformity of MET is Slowly Varying).
Assume that the linear cocycle Φ with two-sided time satisfies the IC of the MET. Then for each $\varepsilon > 0$ there exists an ε-slowly varying random variable $R_\varepsilon : \Omega \to [1, \infty)$ such that on the invariant set $\tilde{\Omega}$ of the MET the cocycle has the following properties:

(i) For each $i = 1, \ldots, p$, $x \in E_i(\omega)$, $t \in \mathbb{T}$
$$\frac{1}{R_\varepsilon(\omega)} e^{\lambda_i t - \varepsilon|t|} \|x\| \leq \|\Phi(t, \omega)x\| \leq R_\varepsilon(\omega) e^{\lambda_i t + \varepsilon|t|} \|x\|,$$

(ii) If M and N are disjoint subsets of $\mathbb{R}$ and if $\gamma(E_M(\omega), E_N(\omega)) \in [0, \frac{\pi}{2}]$ is the angle between $E_M(\omega) := \bigoplus_{\lambda_i \in M} E_i(\omega)$ and $E_N(\omega) := \bigoplus_{\lambda_i \in N} E_i(\omega)$ defined by
$$\cos \gamma(E_M(\omega), E_N(\omega)) := \max_{x \in E_M(\omega), y \in E_N(\omega), x, y \neq 0} \frac{|\langle x, y \rangle|}{\|x\| \|y\|},$$
then
$$\frac{1}{R_\varepsilon(\omega)} \leq \gamma(E_M(\omega), E_N(\omega)).$$

Proof. We work on the invariant set $\tilde{\Omega}$ throughout.

Assume we have found $R_{\varepsilon,1}$ for (i) and $R_{\varepsilon,2}$ for (ii) then $R_\varepsilon := \max(R_{\varepsilon,1}, R_{\varepsilon,2})$ will satisfy both (i) and (ii).

(i) Let $\varepsilon > 0$. The MET implies that there is a random variable $N(\varepsilon, \omega)$ such that for each $t \in \mathbb{T}$ with $|t| > N(\varepsilon, \omega)$, $i = 1, \cdots, p$, $x \in E_i(\omega)$
$$e^{\lambda_i t - \varepsilon|t|} \|x\| \leq \|\Phi(t, \omega)x\| \leq e^{\lambda_i t + \varepsilon|t|} \|x\|.$$

$N(\varepsilon, \omega)$ does not depend on x due to the uniform convergence of $\lim_{t \to \pm\infty} \frac{1}{t} \log \|\Phi(t, \omega)x\| = \lambda_i$ on the set $E_i(\omega) \cap S^{d-1}$. Define the random variables
$$A_{\varepsilon,1}(\omega) := \min_{i=1,\ldots,p} \min_{t \in \mathbb{T}} \min_{0 \neq x \in E_i(\omega)} \frac{\|\Phi(t, \omega)x\|}{e^{\lambda_i t - \varepsilon|t|} \|x\|},$$

$$A_{\varepsilon,2}(\omega) := \max_{i=1,\ldots,p} \max_{t\in\mathbb{T}} \max_{0\neq x\in E_i(\omega)} \frac{\|\Phi(t,\omega)x\|}{e^{\lambda_i t+\varepsilon|t|}\|x\|}.$$

We have $A_{\varepsilon,1}(\omega) > 0$ and $A_{\varepsilon,2}(\omega) < \infty$ since $\min_{t\in\mathbb{T}} \leq 1$ and $\max_{t\in\mathbb{T}} \geq 1$ outside of a finite interval of times and since for $\mathbb{T} = \mathbb{R}$ it is part of the definition of a linear cocycle that $t \mapsto \Phi(t,\omega)$ is continuous.

Now take $B_{\varepsilon,1} := \min(1, A_{\varepsilon,1}) \leq 1$, $B_{\varepsilon,2} := \max(1, A_{\varepsilon,2}) \geq 1$ and finally

$$C_\varepsilon(\omega) := \max\left(\frac{1}{B_{\varepsilon,1}(\omega)}, B_{\varepsilon,2}(\omega)\right) \in [1,\infty).$$

Then for all $i = 1,\ldots,p$, $x \in E_i(\omega)$, $t \in \mathbb{T}$

$$\frac{1}{C_\varepsilon(\omega)} e^{\lambda_i t - \varepsilon|t|} \|x\| \leq \|\Phi(t,\omega)x\| \leq C_\varepsilon(\omega) e^{\lambda_i t + \varepsilon|t|} \|x\|. \qquad (4.3.1)$$

C_ε is in general not yet ε-slowly varying. Rather, by the cocycle property for any $s, t \in \mathbb{T}$

$$\max_{0\neq x\in E_i(\theta(s)\omega)} \frac{\|\Phi(t,\theta(s)\omega)x\|}{e^{\lambda_i t+\varepsilon|t|}\|x\|}$$

$$= \max_{0\neq y\in E_i(\omega)} \frac{\|\Phi(t,\theta(s)\omega)\Phi(s,\omega)y\|}{e^{\lambda_i t+\varepsilon|t|}\|\Phi(s,\omega)y\|}$$

$$= \max_{0\neq y\in E_i(\omega)} \frac{\|\Phi(t+s,\omega)y\|}{e^{\lambda_i t+\varepsilon|t|}\|\Phi(s,\omega)y\|} \qquad (4.3.2)$$

$$= \max_{0\neq y\in E_i(\omega)} \frac{\|\Phi(t+s,\omega)y\|}{e^{\lambda_i(t+s)+\varepsilon|t+s|}\|y\|} \frac{e^{\lambda_i s-\varepsilon|s|}\|y\|}{\|\Phi(s,\omega)y\|} e^{\varepsilon(|s|+|t+s|-|t|)}$$

$$\leq e^{2\varepsilon|s|} \max_{0\neq y\in E_i(\omega)} \frac{\|\Phi(t+s,\omega)y\|}{e^{\lambda_i(t+s)+\varepsilon|t+s|}\|y\|} \left(\min_{0\neq y\in E_i(\omega)} \frac{\|\Phi(s,\omega)y\|}{e^{\lambda_i s-\varepsilon|s|}\|y\|}\right)^{-1},$$

whence

$$A_{\varepsilon,2}(\theta(s)\omega) \leq A_{\varepsilon,2}(\omega) \frac{1}{A_{\varepsilon,1}(\omega)} e^{2\varepsilon|s|}.$$

With a similar estimate for $A_{\varepsilon,1}$ we obtain

$$C_\varepsilon(\theta(t)\omega) \leq C_\varepsilon(\omega)^2 e^{2\varepsilon|t|}.$$

Working with $\theta(-t)\omega$ instead of ω we arrive at

$$\sqrt{C_\varepsilon(\omega)} e^{-2\varepsilon|t|} \leq C_\varepsilon(\theta(t)\omega) \leq C_\varepsilon(\omega)^2 e^{2\varepsilon|t|} \quad \text{for all} \quad t \in \mathbb{T}. \qquad (4.3.3)$$

It will turn out to be crucially important that for continuous time, $t \mapsto C_\varepsilon(\theta(t)\omega)$ is continuous (hence sup's and inf's are measurable – with no need of completion of $\mathcal{F}$), as follows from (4.3.2) and the continuity of $t \mapsto \Phi(t,\omega)$. We now prove that there exists a 3ε-slowly varying random variable $R_{\varepsilon,1}$ with

$$\frac{1}{R_{\varepsilon,1}(\omega)} \leq C_\varepsilon(\omega) \leq R_{\varepsilon,1}(\omega). \qquad (4.3.4)$$

Indeed, put
$$R'_{\varepsilon,1}(\omega) := \inf_{t\in\mathbb{T}}(1, C_\varepsilon(\theta(t)\omega)e^{3\varepsilon|t|}),$$
$$R''_{\varepsilon,1}(\omega) := \sup_{t\in\mathbb{T}}(1, C_\varepsilon(\theta(t)\omega)e^{-3\varepsilon|t|}).$$

$R'_{\varepsilon,1}$ and $R''_{\varepsilon,1}$ are measurable and $R'_{\varepsilon,1}(\omega) > 0$, $R''_{\varepsilon,1}(\omega) < \infty$ by (4.3.3). Hence

$$R'_{\varepsilon,1}(\omega)e^{-3\varepsilon|t|} \le C_\varepsilon(\theta(t)\omega) \le R''_{\varepsilon,1}(\omega)e^{3\varepsilon|t|} \quad \text{for all} \quad t \in \mathbb{T}. \tag{4.3.5}$$

Now for all $t, s \in \mathbb{T}$ by (4.3.5)

$$C_\varepsilon(\theta(t+s)\omega) \le R''_{\varepsilon,1}(\omega)e^{3\varepsilon|s|}e^{3\varepsilon|t|} \tag{4.3.6}$$

and also

$$C_\varepsilon(\theta(t+s)\omega) = C_\varepsilon(\theta(t)(\theta(s)\omega)) \le R''_{\varepsilon,1}(\theta(s)\omega)e^{3\varepsilon|t|}. \tag{4.3.7}$$

By definition $R''_{\varepsilon,1}$ is the smallest number satisfying the right-hand side inequality of (4.3.5) for all t and all $\omega \in \tilde\Omega$. Comparing (4.3.6) with (4.3.7),

$$R''_{\varepsilon,1}(\theta(s)\omega) \le R''_{\varepsilon,1}(\omega)e^{3\varepsilon|s|} \quad \text{for all} \quad s \in \mathbb{T},$$

i.e. $R''_{\varepsilon,1}$ is a 3ε-slowly varying random variable. Similarly

$$R'_{\varepsilon,1}(\omega)e^{-3\varepsilon|s|} \le R'_{\varepsilon,1}(\theta(s)\omega) \quad \text{for all} \quad s \in \mathbb{T}.$$

Finally,
$$R_{\varepsilon,1} := \max(R''_{\varepsilon,1}, \frac{1}{R'_{\varepsilon,1}})$$

is 3ε-slowly varying and satisfies (4.3.4) and thus, by (4.3.1), the assertion (i) of the theorem.

(ii) By Corollary 5.3.10, the angle γ is tempered on $\tilde\Omega$. For each $\varepsilon > 0$ there are thus random variables $M_{\varepsilon,1}$, $M_{\varepsilon,2}$ for which for all $t \in \mathbb{T}$

$$M_{\varepsilon,1}(\omega)e^{-\varepsilon|t|} \le \gamma(E_M(\theta(t)\omega), E_N(\theta(t)\omega)) \le M_{\varepsilon,2}(\omega)e^{\varepsilon|t|}.$$

By the same procedure as in the proof of (i) we construct now a 3ε-slowly varying random variable $R^{M,N}_{\varepsilon,2}$ satisfying for all $t \in \mathbb{T}$

$$\frac{1}{R^{M,N}_{\varepsilon,2}(\omega)}e^{-3\varepsilon|t|} \le \gamma(E_M(\theta(t)\omega), E_N(\theta(t)\omega)) \le R^{M,N}_{\varepsilon,2}(\omega)e^{3\varepsilon|t|}.$$

Taking finally
$$R_{\varepsilon,2} := \max\{R^{M,N}_{\varepsilon,2} : M \cap N = \emptyset\}$$

we have found a 3ε-slowly varying random variable which satisfies (ii). $\square$

4.3.5 Corollary. *Choose (and then fix) a constant $\kappa > 0$. Then for each $\varepsilon > 0$ there exists an ε-slowly varying random variable $G_\varepsilon : \Omega \to [1, \infty)$ such that on the invariant set $\tilde{\Omega}$ for all $i = 1, \ldots, p$, $x \in E_i(\omega)$, $t \in \mathbb{T}$*

$$\frac{1}{G_\varepsilon(\omega)} e^{\lambda_i t - \kappa |t|} \|x\| \leq \|\Phi(t, \omega) x\| \leq G_\varepsilon(\omega) e^{\lambda_i t + \kappa |t|} \|x\|. \tag{4.3.8}$$

Proof. Take $\delta(\varepsilon) = \min(\varepsilon, \kappa)$ and choose a δ-slowly varying random variable R_δ according to Theorem 4.3.4. Then $G_\varepsilon = R_{\delta(\varepsilon)}$ is ε-slowly varying since $\delta \leq \varepsilon$ and satisfies equation (4.3.8) since $\delta \leq \kappa$. □

4.3.2 Random Scalar Products

The following is the main theorem of Sect. 4.3.

4.3.6 Theorem (Random Scalar Product). *Assume that the linear cocycle Φ satisfies the IC of the MET with two-sided time. Choose (and then fix) a constant $\kappa > 0$. Introduce on the invariant set $\tilde{\Omega}$ of the MET and for any $x = \oplus_{i=1}^p x_i$ and $y = \oplus_{i=1}^p y_i$, with $x_i, y_i \in E_i(\omega)$*

$$\langle x, y \rangle_{\kappa, \omega} := \sum_{i=1}^p \langle x_i, y_i \rangle_{\kappa, \omega},$$

where for $u_i, v_i \in E_i(\omega)$

$$\langle u_i, v_i \rangle_{\kappa, \omega} := \begin{cases} \int_{-\infty}^{\infty} \frac{\langle \Phi(t, \omega) u_i, \Phi(t, \omega) v_i \rangle}{e^{2(\lambda_i t + \kappa |t|)}} dt, & \mathbb{T} = \mathbb{R}, \\ \sum_{n \in \mathbb{Z}} \frac{\langle \Phi(n, \omega) u_i, \Phi(n, \omega) v_i \rangle}{e^{2(\lambda_i n + \kappa |n|)}}, & \mathbb{T} = \mathbb{Z}. \end{cases} \tag{4.3.9}$$

Put $\langle x, y \rangle_{\kappa, \omega} := \langle x, y \rangle$ for $\omega \notin \tilde{\Omega}$. Then

(i) $\langle \cdot, \cdot \rangle_{\kappa, \omega}$ is a random scalar product in $\mathbb{R}^d$ which depends measurably on ω and under which the $E_i(\omega)$ are orthogonal.

(ii) For each $\varepsilon > 0$ there exists an ε-slowly varying random variable $B_\varepsilon : \Omega \to [1, \infty)$ such that

$$\frac{1}{B_\varepsilon(\omega)} \|\cdot\| \leq \|\cdot\|_{\kappa, \omega} \leq B_\varepsilon(\omega) \|\cdot\|,$$

where

$$\|x\|_{\kappa, \omega}^2 = \langle x, x \rangle_{\kappa, \omega} = \sum_{i=1}^p \|x_i\|_{\kappa, \omega}^2,$$

and

$$\|x_i\|_{\kappa, \omega}^2 = \begin{cases} \int_{-\infty}^{\infty} \frac{\|\Phi(t, \omega) x_i\|^2}{e^{2(\lambda_i t + \kappa |t|)}} dt, & \mathbb{T} = \mathbb{R}, \\ \sum_{n \in \mathbb{Z}} \frac{\|\Phi(n, \omega) x_i\|^2}{e^{2(\lambda_i n + \kappa |n|)}}, & \mathbb{T} = \mathbb{Z}, \end{cases}$$

defines the square of the random norm corresponding to $\langle \cdot, \cdot \rangle_{\kappa, \omega}$.

(iii) The function
$$(x, y, t) \mapsto \langle x, y \rangle_{\kappa, \theta(t)\omega}$$
is continuous with respect to the Euclidean metric in $\mathbb{R}^d \times \mathbb{R}^d \times \mathbb{T}$ and $\mathbb{R}$. In particular, $(x, y, t) \mapsto \langle \Phi(t, \omega)x, \Phi(t, \omega)y \rangle_{\kappa, \theta(t)\omega}$ and $(x, t) \mapsto \|\Phi(t, \omega)x\|_{\kappa, \theta(t)\omega}$ are continuous.

(iv) For all $i = 1, \ldots, p$, $x \in E_i(\omega)$, $t \in \mathbb{T}$
$$e^{\lambda_i t - \kappa|t|} \|x\|_\omega \leq \|\Phi(t, \omega)x\|_{\kappa, \theta(t)\omega} \leq e^{\lambda_i t + \kappa|t|} \|x\|_\omega. \tag{4.3.10}$$

Proof. (i) For $T = \mathbb{Z}$ note first that the series in (4.3.9) converges for each $\omega \in \tilde{\Omega}$ since
$$|\langle u, v \rangle_{\kappa, \omega}| \leq \sum_{n \in \mathbb{Z}} \frac{\|\Phi(n, \omega)u\|}{e^{\lambda_i n + \kappa|n|}} \frac{\|\Phi(n, \omega)v\|}{e^{\lambda_i n + \kappa|n|}}$$
$$\leq G_\varepsilon(\omega)^2 \left(\sum_{n \in \mathbb{Z}} e^{-\kappa|n|} \right) \|u\| \|v\|,$$

where we have applied Corollary 4.3.5 with κ replaced with $\frac{\kappa}{2}$ and the fact that G_ε is an ε-slowly varying random variable. Clearly $\langle x, y \rangle_{\kappa, \omega}$ is measurable with respect to ω as a series of measurable random variables. For each $\omega \in \Omega$ it is a scalar product in $\mathbb{R}^d$ and by definition $\langle x, y \rangle_{\kappa, \omega} = 0$ if $x \in E_i(\omega)$ and $y \in E_j(\omega)$, $i \neq j$. For $\mathbb{T} = \mathbb{R}$ the arguments are essentially the same.

(ii) First take $\mathbb{T} = \mathbb{Z}$. We first verify the estimate of $\|x\|_{\kappa, \omega}$ from below. The series defining $\langle x, y \rangle_{\kappa, \omega}$ in $E_i(\omega)$ contains in particular the summand $\langle x, y \rangle$ for $n = 0$, hence $\|x\| \leq \|x\|_{\kappa, \omega}$ in $E_i(\omega)$. For a general $x = \oplus_{i=1}^p x_i$ this yields
$$\|x\| \leq \sum_{i=1}^p \|x_i\| \leq \sum_{i=1}^p \|x_i\|_{\kappa, \omega} \leq \sqrt{p} \|x\|_{\kappa, \omega}.$$

The estimate of $\|x\|_{\kappa, \omega}$ from above is more complicated. Denote by
$$\pi_i(\omega) : \mathbb{R}^d \to E_i(\omega)$$
the projection onto $E_i(\omega)$ along $F_i(\omega) := \oplus_{j \neq i} E_j(\omega)$. Then for $x = \sum_{i=1}^p \pi_i(\omega)x$
$$\|x\|_{\kappa, \omega}^2 = \sum_{i=1}^p \|\pi_i(\omega)x\|_{\kappa, \omega}^2, \quad \|\pi_i(\omega)x\|_{\kappa, \omega}^2 = \sum_{n \in \mathbb{Z}} \frac{\|\Phi(n, \omega)\pi_i(\omega)x\|^2}{e^{2(\lambda_i n + \kappa|n|)}}.$$

Choose an $\frac{\varepsilon}{2}$-slowly varying random variable $G_{\varepsilon/2}$ corresponding to $\frac{\kappa}{2}$ according to Corollary 4.3.5. Then
$$\|\pi_i(\omega)x\|_{\kappa, \omega}^2 \leq c G_{\varepsilon/2}(\omega)^2 \|\pi_i(\omega)\|^2 \|x\|^2,$$
where $c = c(\kappa) = \sum_{n \in \mathbb{Z}} e^{-\kappa|n|} \in (1, \infty)$. Consequently

$$\|x\|_{\kappa,\omega} \leq \sqrt{c}G_{\varepsilon/2}(\omega)\|x\| \left(\sum_{i=1}^{p}\|\pi_i(\omega)\|^2\right)^{1/2}.$$

To estimate the norm of $\pi_i(\omega)$ (which is an orthogonal projection in the new norm, so $\|\pi_i(\omega)\|_{\kappa,\omega} = 1$) we use the following lemma.

4.3.7 Lemma (Norm of Projection). *Suppose that $(U, \langle \cdot, \cdot \rangle)$ is a vector space with scalar product and $\pi : U \to U$ is a projection, i.e. $\pi^2 = \pi$. Let $\gamma := \gamma(\mathrm{im}\,\pi, \ker \pi) \in [0, \frac{\pi}{2}]$ be the angle between $\mathrm{im}\,\pi$ and $\ker \pi$. Then*

$$\|\pi\| = \frac{1}{\sin \gamma}.$$

For a proof see Pugh and Shub [282: p. 9].

By Lemma 4.3.7

$$\|\pi_i(\omega)\| = \frac{1}{\sin \gamma(E_i(\omega), F_i(\omega))}.$$

By Theorem 4.3.4 there is an $\varepsilon/2$-slowly varying random variable $R_{\varepsilon/2}$ for which for all $i = 1,\ldots,p$, $\sin \gamma(E_i(\omega), F_i(\omega)) \geq \frac{1}{2}\gamma(E_i(\omega), F_i(\omega)) \geq \frac{1}{2R_{\varepsilon/2}(\omega)}$, thus

$$\|\pi_i(\omega)\| \leq 2R_{\varepsilon/2}(\omega).$$

Altogether

$$\|x\|_{\kappa,\omega} \leq 2\sqrt{cp}\,G_{\varepsilon/2}(\omega)R_{\varepsilon/2}(\omega)\|x\| =: B_\varepsilon(\omega)\|x\|,$$

where the random variable $B_\varepsilon : \Omega \to [1, \infty)$ is ε-slowly varying by Lemma 4.3.2.

We now treat the case $\mathbb{T} = \mathbb{R}$. The proof of the upper estimate is the same as for $\mathbb{T} = \mathbb{Z}$. Denote the resulting ε-slowly varying random variable by $B_{\varepsilon,1}$. For the lower estimate, take the ε-slowly varying random variable $G_\varepsilon(\omega)$ from Corollary 4.3.5 and use the lower estimate in (4.3.8) to obtain

$$\begin{aligned}\|\pi_i(\omega)x\|_{\kappa,\omega}^2 &= \int_{-\infty}^{\infty} \frac{\|\Phi(t,\omega)\pi_i(\omega)x\|^2}{e^{2(\lambda_i t + \kappa|t|)}}\,dt \\ &\geq \frac{1}{G_\varepsilon(\omega)^2}c(\kappa)\|\pi_i(\omega)x\|^2\end{aligned}$$

and hence

$$\begin{aligned}\|x\|_{\kappa,\omega} &= \left(\sum_{i=1}^p \|\pi_i(\omega)x\|_{\kappa,\omega}^2\right)^{1/2} \geq \frac{\sqrt{c(\kappa)}}{G_\varepsilon(\omega)}\left(\sum_{i=1}^p \|\pi_i(\omega)x\|^2\right)^{1/2} \\ &\geq \frac{\sqrt{c(\kappa)}}{G_\varepsilon(\omega)}\frac{1}{\sqrt{p}}\sum_{i=1}^p \|\pi_i(\omega)x\| \geq \frac{\sqrt{c(\kappa)}}{\sqrt{p}\,G_\varepsilon(\omega)}\|x\| \\ &=: \frac{1}{B_{\varepsilon,2}(\omega)}\|x\|.\end{aligned}$$

The final ε-slowly varying random variable is $B_\varepsilon = \max(B_{\varepsilon,1}, \frac{1}{B_{\varepsilon,2}})$.

(iii) This follows readily from the definition (4.3.9), the cocycle property, and the (crucial) fact that, by our definition of a linear cocycle, $(x,t) \mapsto \Phi(t,\omega)x$ is continuous. A detailed proof is given by Nguyen Dinh Cong [260: Theorem 3.7].

(iv) For $\mathbb{T} = \mathbb{Z}$, $x \in E_i(\omega)$ and $s \in \mathbb{T}$

$$\begin{aligned}
\|\Phi(s,\omega)x\|^2_{\kappa,\theta(s)\omega} &= \sum_{t \in \mathbb{Z}} \frac{\|\Phi(t,\theta(s)\omega)\Phi(s,\omega)x\|^2}{e^{2(\lambda_i t + \kappa|t|)}} \\
&= \sum_{t \in \mathbb{Z}} \frac{\|\Phi(t+s,\omega)x\|^2}{e^{2(\lambda_i(t+s) + \kappa|t+s|)}} e^{2(\lambda_i s + \kappa|t+s| - \kappa|t|)} \\
&\leq e^{2(\lambda_i s + \kappa|s|)} \|x\|^2_{\kappa,\omega},
\end{aligned}$$

analogously for the lower bound and for $\mathbb{T} = \mathbb{R}$. □

4.3.8 Remark (Lyapunov Metric, Lyapunov Norm). The quantities $\langle x,y \rangle_{\kappa,\omega}$ and $\|x\|_{\kappa,\omega}$ and their relatives on the tangent bundle are called the *Lyapunov metric* and the *Lyapunov norm* (e.g. by Pesin [274]). $\Omega \times \mathbb{R}^d$ with $\langle \cdot,\cdot \rangle_{\kappa,\omega}$ on the fiber $\{\omega\} \times \mathbb{R}^d$ is a Euclidean bundle in the sense of Definition 4.1.7. We stress that the Lyapunov metrics $\langle x,y \rangle_{\kappa,\omega}$ also depend on the fixed parameter $\kappa > 0$. ■

4.3.9 Remark (Uniformity Properties of Φ in the Random Norm).
(i) The key property of $\|\cdot\|_{\kappa,\omega}$ is that the random factors which are still present in equation (4.3.8) and which account for the non-uniformity of the cocycle in the original norm have disappeared in equation (4.3.10). An equivalent formulation of (4.3.10) is

$$e^{\lambda_i t - \kappa|t|} \leq \|\Phi(t,\omega)|_{E_i(\omega)}\|_{\kappa,\omega,\theta(t)\omega} \leq e^{\lambda_i t + \kappa|t|}.$$

For $T = \mathbb{Z}$ and $n = 1$ and $n = -1$ we e.g. obtain the uniform bounds

$$e^{-\lambda_i - \kappa} \leq \|A(\omega)|_{E_i(\omega)}\|_{\kappa,\omega,\theta\omega} \leq e^{\lambda_i + \kappa},$$

$$e^{-\lambda_i - \kappa} \leq \|A(\omega)^{-1}|_{E_i(\theta\omega)}\|_{\kappa,\theta\omega,\omega} \leq e^{\lambda_i + \kappa}.$$

If $\lambda_i < 0$ ($\lambda_i > 0$) and κ is so small that $\lambda_i + \kappa < 0$ ($\lambda_i - \kappa > 0$), then $\Phi(t,\omega)$ is contracting (expanding) on $E_i(\omega)$ for $t \geq 0$.

(ii) More generally, if $M \subset \mathbb{R}$ and $E_M(\omega) := \oplus_{\lambda_i \in M} E_i(\omega)$, then for each $x = \oplus_{\lambda_i \in M} x_i \in E_M(\omega)$, $\|\Phi(t,\omega)x\|_{\kappa,\theta(t)\omega}$ is continuous with respect to t by Theorem 4.3.6(iii) and

$$\sum_{\lambda_i \in M} e^{2(\lambda_i t - \kappa|t|)} \|x_i\|^2_{\kappa,\omega} \leq \|\Phi(t,\omega)x\|^2_{\kappa,\theta(t)\omega} \leq \sum_{\lambda_i \in M} e^{2(\lambda_i t + \kappa|t|)} \|x_i\|^2_{\kappa,\omega}.$$

With $\lambda_{\min} := \min_{\lambda_i \in M} \lambda_i$, $\lambda_{\max} := \max_{\lambda_i \in M} \lambda_i$ we obtain for $t \geq 0$: $\quad e^{(\lambda_{\min}-\kappa)t}\|x\|_{\kappa,\omega} \leq \|\Phi(t,\omega)x\|_{\kappa,\theta(t)\omega} \leq e^{(\lambda_{\max}+\kappa)t}\|x\|_{\kappa,\omega}$,

for $t \leq 0$: $\quad e^{(\lambda_{\max}+\kappa)t}\|x\|_{\kappa,\omega} \leq \|\Phi(t,\omega)x\|_{\kappa,\theta(t)\omega} \leq e^{(\lambda_{\min}-\kappa)t}\|x\|_{\kappa,\omega}$.

(iii) For $M = (-\infty, 0)$, $M = (0, \infty)$, $M = \{0\}$ we obtain the stable, unstable and center subspaces

$$E_s(\omega) := \oplus_{\lambda_i < 0} E_i(\omega), \quad E_u(\omega) := \oplus_{\lambda_i > 0} E_i(\omega), \quad E_c(\omega) := \oplus_{\lambda_i = 0} E_i(\omega).$$

For $\kappa > 0$ small enough the cocycle in the new norm is thus uniformly contracting on $E_s(\omega)$ and uniformly expanding on $E_u(\omega)$ (hence uniformly hyperbolic if $E_c(\omega)$ is not present).

More specifically, if β is any number in $(0, \min\{-\lambda_s - \kappa, \lambda_u - \kappa\})$, where

$$\lambda_s := \max_{\lambda_i < 0} \lambda_i < -\kappa < 0 < \kappa < \lambda_u := \min_{\lambda_i > 0} \lambda_i,$$

then

$$\|\Phi(t,\omega)|_{E_s(\omega)}\|_{\kappa,\omega,\theta(t)\omega} \leq e^{-\beta t} \quad \text{for all } t \geq 0, \tag{4.3.11}$$

$$\|\Phi(t,\omega)^{-1}|_{E_u(\theta(t)\omega)}\|_{\kappa,\theta(t)\omega,\omega} \leq e^{-\beta t} \quad \text{for all } t \geq 0, \tag{4.3.12}$$

and on $E_c(\omega)$ (if present at all)

$$e^{-\kappa|t|} \leq \|\Phi(t,\omega)|_{E_c(\omega)}\|_{\kappa,\omega,\theta(t)\omega} \leq e^{\kappa|t|} \quad \text{for all } t \in \mathbb{T}. \tag{4.3.13}$$

(iv) For $M = \mathbb{R}$

$$e^{(\lambda_p - \kappa)t} \leq \|\Phi(t,\omega)\|_{\kappa,\omega,\theta(t)\omega} \leq e^{(\lambda_1 + \kappa)t} \quad \text{for } t \geq 0,$$

$$e^{(\lambda_1 + \kappa)t} \leq \|\Phi(t,\omega)\|_{\kappa,\omega,\theta(t)\omega} \leq e^{(\lambda_p - \kappa)t} \quad \text{for } t \leq 0,$$

and so together, with $\Lambda := \max(\lambda_1, -\lambda_p)$

$$e^{-(\Lambda+\kappa)|t|} \leq \|\Phi(t,\omega)\|_{\kappa,\omega,\theta(t)\omega} \leq e^{(\Lambda+\kappa)|t|} \quad \text{for all } t \in \mathbb{T}, \tag{4.3.14}$$

$$e^{-(\Lambda+\kappa)|t|} \leq \|\Phi(t,\omega)^{-1}\|_{\kappa,\theta(t)\omega,\omega} \leq e^{(\Lambda+\kappa)|t|} \quad \text{for all } t \in \mathbb{T}, \tag{4.3.15}$$

i.e. the cocycle and its inverse are uniformly bounded in the random norm. ∎

4.3.10 Corollary. *The cocycle $\Phi(t,\omega)$ with respect to the Lyapunov norm $\|\cdot\|_{\kappa,\omega}$ (for any $\kappa > 0$ fixed) is Lyapunov cohomologous to itself with respect to any fixed norm in $\mathbb{R}^d$.*

Proof. By Remark 3.4.10 we can assume without loss of generality that the fixed norm is the standard Euclidean norm $\|\cdot\|_S$. By Example 4.1.14 we have to check whether we have (suppressing κ)

$$\lim_{t\to\pm\infty} \frac{1}{t} \log \|\mathrm{id}\|_{S,\theta(t)\omega} = 0, \quad \lim_{t\to\pm\infty} \frac{1}{t} \log \|\mathrm{id}\|_{\theta(t)\omega,S} = 0, \qquad (4.3.16)$$

where

$$\|\mathrm{id}\|_{S,\omega} := \sup_{x\neq 0} \frac{\|x\|_\omega}{\|x\|_S}, \quad \text{and} \quad \|\mathrm{id}\|_{\omega,S} := \sup_{x\neq 0} \frac{\|x\|_S}{\|x\|_\omega}.$$

By Theorem 4.3.6(ii) for each ε

$$\frac{1}{B_\varepsilon(\omega)} \leq \|\mathrm{id}\|_{S,\omega} \leq B_\varepsilon(\omega), \quad \frac{1}{B_\varepsilon(\omega)} \leq \|\mathrm{id}\|_{\omega,S} \leq B_\varepsilon(\omega),$$

which implies (4.3.16) by Proposition 4.3.3(i). □

4.3.11 Corollary (Projections to Oseledets Spaces). *Let M and N be disjoint subsets of $\mathbb{R}$ whose union is $\mathbb{R}$, and let $\pi_{M,N}(\omega)$ be the projection onto $E_M(\omega) := \oplus_{\lambda_i \in M} E_i(\omega)$ along $E_N(\omega) := \oplus_{\lambda_i \in N} E_i(\omega)$. Then for each $\varepsilon > 0$ there exists an ε-slowly varying random variable R_ε such that*

$$1 \leq \|\pi_{M,N}(\omega)\| \leq R_\varepsilon(\omega).$$

In particular, $\pi_{M,N}$ is tempered.

Proof. Combine Theorem 4.3.4(ii) with Lemma 4.3.7 to obtain the first assertion. Then use Proposition 4.3.3(i). □

4.3.12 Corollary ("Strong Version" of the MET). *Assume that the matrix cocycle $\Phi(t,\omega)$ is written in the standard basis $S = (u_i)$ of $\mathbb{R}^d$ endowed with the standard scalar product and its corresponding Euclidean norm. Construct a new random basis $G(\omega) = (v_i(\omega))$ adapted to the Oseledets splitting $\mathbb{R}^d = E_1(\omega) \oplus \cdots \oplus E_p(\omega)$ as follows: Pick measurable unit vectors (in the standard Euclidean norm) in such a way that the $v_1(\omega),\ldots,v_{d_1}(\omega)$ are orthogonal and taken from $E_1(\omega)$, the $v_{d_1+1}(\omega),\ldots,v_{d_1+d_2}(\omega)$ are orthogonal and taken from $E_2(\omega)$, ..., and the $v_{d_1+\cdots+d_{p-1}+1}(\omega),\ldots,v_d(\omega)$ are orthogonal and taken from $E_p(\omega)$. The matrix representation of the cocycle with respect to the bases $G(\omega)$ and $G(\theta(t)\omega)$ is*

$$\Psi(t,\omega) = P(\theta(t)\omega)\Phi(t,\omega)P(\omega)^{-1},$$

where $P(\omega) = P_{G(\omega)}^S$ is the basis transfer matrix from S to $G(\omega)$. Then the cocycle $\Psi(t,\omega)$ is block-diagonal,

$$\Psi(t,\omega) = \mathrm{diag}(\Psi_1(t,\omega),\ldots,\Psi_p(t,\omega)),$$

where the blocks $\Psi_i(t,\omega)$ are cocycles of size d_i and one-point spectrum $\{(\lambda_i,d_i)\}$. Moreover, the cocycles Φ and Ψ are Lyapunov cohomologous.

Proof. The existence of a measurable selection of a basis $G(\omega)$ is guaranteed by the Measurable Selection Theorem (Proposition 1.6.3) since by the MET the Oseledets spaces $E_i(\omega)$ are random closed sets (the procedure is carefully carried out by Crauel [108] and Walters [337]).

The columns of $Q(\omega) := P(\omega)^{-1}$ are the coordinates $k_S(v_j(\omega))$, where $q_{ij}(\omega)$ is obtained as the orthogonal projection of the unit vector $v_j(\omega)$ onto the subspace $\mathbb{R}u_i$, thus $|q_{ij}(\omega)| \leq 1$ and hence $\|P(\omega)^{-1}\| \leq C$.

Further, the columns of $P(\omega)$ are the coordinates $k_{G(\omega)}(u_j)$ obtained by first applying one of the projections $\pi_i(\omega) : \mathbb{R}^d \to E_i(\omega)$ and then by orthogonal projection onto one of the $v_k(\omega)$ in $E_i(\omega)$. Thus

$$|p_{ij}(\omega)| \leq \|\pi_i(\omega)\| = \frac{1}{\sin \gamma_i(\omega)},$$

where $\gamma_i(\omega)$ is the angle between $E_i(\omega)$ and $F_i(\omega) = \oplus_{j \neq i} E_j(\omega)$ (see Lemma 4.3.7). By Theorem 4.3.4(ii) there exists for each $\varepsilon > 0$ an ε-slowly varying random variable $R_\varepsilon(\omega) \geq 1$ for which $2 \sin \gamma_i(\omega) \geq \gamma_i(\omega) \geq 1/R_\varepsilon(\omega)$, thus

$$\frac{1}{C} \leq \|P(\omega)\| \leq 2C\, R_\varepsilon(\omega).$$

Analogously,

$$\frac{1}{2C\, R_\varepsilon(\omega)} \leq \|P(\omega)^{-1}\| \leq C.$$

Summarizing, we have the following lemma.

4.3.13 Lemma. *Let $P(\omega)$ be the basis transfer matrix from a fixed basis to a basis adapted to the Oseledets splitting. Then there exists a constant $C > 0$, and for each $\varepsilon > 0$ there exists an ε-slowly varying random variable $R_\varepsilon \geq 1$ such that*

$$\frac{1}{C} \leq \|P(\omega)\| \leq 2C\, R_\varepsilon(\omega), \quad \frac{1}{2C\, R_\varepsilon(\omega)} \leq \|P(\omega)^{-1}\| \leq C.$$

Continuing the proof of Corollary 4.3.12, the lemma implies in particular that P and P^{-1} are tempered, hence P is a Lyapunov cohomology between the two cocycles. □

4.3.3 Random Riemannian Metrics on Manifolds

Consider now a C^1 RDS $\varphi(t, \omega)$ on a Riemannian manifold $(M, \langle \cdot, \cdot \rangle_x)$ over an ergodic DS $(\Omega, \mathcal{F}, \mathbb{P}, (\theta(t))_{t \in \mathbb{T}})$ with two-sided time. Assume that μ is an ergodic φ-invariant measure. The linearized cocycle $T\varphi$ is a linear bundle RDS over the skew product $\Theta(t)(\omega, x) = (\theta(t)\omega, \varphi(t, \omega)x)$.

Assume that $T\varphi$ satisfies the IC of the MET Theorem 4.2.6 with respect to μ. Using the global trivialization $\Psi : TM \to M \times \mathbb{R}^d$ such that on fibers $\psi(x) : (T_xM, \langle \cdot, \cdot \rangle_x) \to (\mathbb{R}^d, \langle \cdot, \cdot \rangle_S)$ is an isometry (see Lemma 4.2.5), all results of the last subsection and their proofs carry over verbatim since they can be transferred from the cocycle $\Phi(t,\omega,x) = \psi(\varphi(t,\omega)x) \circ T\varphi(t,\omega,x) \circ \psi(x)^{-1}$ on $\mathbb{R}^d$ to $T\varphi(t,\omega,x)$, where

$$\langle T\varphi(t,\omega,x)u, T\varphi(t,\omega,x)v\rangle_{\varphi(t,\omega)x} = \langle \Phi(t,\omega,x)\psi(x)u, \Phi(t,\omega,x)\psi(x)v\rangle_S.$$

4.3.14 Theorem (Non-Uniformity of MET is Slowly Varying).
Let φ be a C^1 RDS on a Riemannian manifold $(M, \langle \cdot, \cdot \rangle_x)$. Let μ be an ergodic φ-invariant measure and assume that $T\varphi$ satisfies the IC of the MET with respect to μ (Theorem 4.2.6). Then for each $\varepsilon > 0$ there is an ε-slowly varying random variable $R_\varepsilon : \Omega \times M \to [1, \infty)$ over $\Theta(t)$ such that on the invariant set $\Delta \subset \Omega \times M$ of full μ-measure the cocycle $T\varphi(t,\omega,x)$ has the following properties:

(i) For each $i = 1, \ldots, p$, $u \in E_i(\omega, x)$, $t \in \mathbb{T}$

$$\frac{1}{R_\varepsilon(\omega,x)} e^{\lambda_i t - \varepsilon|t|} \|u\|_x \leq \|T\varphi(t,\omega,x)u\|_{\varphi(t,\omega)x} \leq R_\varepsilon(\omega,x) e^{\lambda_i t + \varepsilon|t|} \|u\|_x.$$

(ii) If M and N are disjoint subsets of $\mathbb{R}$ and if $\gamma(E_M(\omega,x), E_N(\omega,x)) \in [0, \frac{\pi}{2}]$ is the angle between $E_M(\omega,x) := \oplus_{\lambda_i \in M} E_i(\omega,x)$ and $E_N(\omega,x) := \oplus_{\lambda_i \in N} E_i(\omega,x)$ defined by

$$\cos\gamma(E_M(\omega,x), E_N(\omega,x)) := \max_{u \in E_M(\omega,x), v \in E_N(\omega,x), u,v \neq 0} \frac{|\langle u,v\rangle_x|}{\|u\|_x \|v\|_x},$$

then

$$\frac{1}{R_\varepsilon(\omega,x)} \leq \gamma(E_M(\omega,x), E_N(\omega,x)).$$

4.3.15 Corollary. *Choose (and then fix) a constant $\kappa > 0$. Then for each $\varepsilon > 0$ there exists an ε-slowly varying random variable $G_\varepsilon : \Omega \times M \to [1, \infty)$ such that on the invariant set $\Delta \subset \Omega \times M$ for all $i = 1, \ldots, p$, $u \in E_i(\omega,x)$, $t \in \mathbb{T}$*

$$\frac{1}{G_\varepsilon(\omega,x)} e^{\lambda_i t - \kappa|t|} \|u\|_x \leq \|T\varphi(t,\omega,x)u\|_{\varphi(t,\omega)x} \leq G_\varepsilon(\omega,x) e^{\lambda_i t + \kappa|t|} \|u\|_x.$$

4.3.16 Theorem (Random Riemannian Metric). *Let φ be a C^1 RDS on a Riemannian manifold $(M, \langle \cdot, \cdot \rangle_x)$. Let μ be an ergodic φ-invariant measure and assume that $T\varphi$ satisfies the IC of the MET with respect to μ. Choose (and then fix) a constant $\kappa > 0$ and introduce for any $(\omega,x) \in \Delta$ and $u = \oplus_{i=1}^p u_i$, $v = \oplus_{i=1}^p v_i \in T_x M$, $u_i, v_i \in E_i(\omega,x)$,*

$$\langle u, v\rangle_{\kappa,(\omega,x)} := \sum_{i=1}^p \langle u_i, v_i\rangle_{\kappa,(\omega,x)},$$

where for $u_i, v_i \in E_i(\omega,x)$

$$\langle u_i, v_i\rangle_{\kappa,(\omega,x)} := \begin{cases} \int_{-\infty}^\infty \frac{\langle T\varphi(t,\omega,x)u_i, T\varphi(t,\omega,x)v_i\rangle_{\varphi(t,\omega)x}}{e^{2(\lambda_i t + \kappa|t|)}} dt, & \mathbb{T} = \mathbb{R}, \\ \sum_{n \in \mathbb{Z}} \frac{\langle T\varphi(n,\omega,x)u_i, T\varphi(n,\omega,x)v_i\rangle_{\varphi(n,\omega)x}}{e^{2(\lambda_i n + \kappa|n|)}}, & T = \mathbb{Z}. \end{cases}$$

Put $\langle u, v \rangle_{\kappa,(\omega,x)} = \langle u, v \rangle_x$ for $(\omega, x) \notin \Delta$. Then

(i) $\langle \cdot, \cdot \rangle_{\kappa,(\omega,x)}$ is a random scalar product on TM which depends measurably on (ω, x) (i. e. is a measurable Riemannian metric) and under which the $E_i(\omega, x)$ are orthogonal.

(ii) For each $\varepsilon > 0$ there exists an ε-slowly varying random variable $B_\varepsilon : \Omega \times M \to [1, \infty)$ such that for the corresponding measurable Lyapunov norm $\| \cdot \|_{\kappa,(\omega,x)}$

$$\frac{1}{B_\varepsilon(\omega, x)} \| \cdot \|_x \leq \| \cdot \|_{\kappa,(\omega,x)} \leq B_\varepsilon(\omega, x) \| \cdot \|_x.$$

(iii) For all $i = 1, \ldots, p$, $u \in E_i(\omega, x)$, $t \in \mathbb{T}$

$$e^{\lambda_i t - \kappa |t|} \|u\|_{\kappa,(\omega,x)} \leq \|T\varphi(t,\omega,x)u\|_{\kappa,\Theta(t)(\omega,x)} \leq e^{\lambda_i t + \kappa |t|} \|u\|_{\kappa,(\omega,x)}.$$

4.3.17 Corollary. *The cocycle $T\varphi(t, \omega, x)$ with respect to the Lyapunov norm $\|u\|_{\kappa,(\omega,x)} = \langle u, u \rangle_{\kappa,(\omega,x)}^{1/2}$ is (for any $\kappa > 0$) Lyapunov cohomologous to itself with respect to the norm $\|u\|_x = \langle u, u \rangle_x^{1/2}$ associated with the Riemannian metric on M.*

4.3.18 Remark. (i) Remark 4.3.9 carries over to the manifold case in an obvious way. In particular, $\|T\varphi(t, \omega, x)\|_{\kappa,(\omega,x),\Theta(t)(\omega,x)}$ is uniformly bounded under the Lyapunov metric, and in case $0 \notin \mathcal{S}(\Theta, \mu)$ and for $\kappa > 0$ small enough $T\varphi(t, \omega, x)$ is uniformly hyperbolic.

(ii) Pesin [274], Fathi, Herman and Yoccoz [149] and Pugh and Shub [282] considered the case $\mathbb{T} = \mathbb{Z}$ and φ independent of ω, the classical case of smooth ergodic theory. See also Katok and Strelcyn [201] for an account and generalizations of Pesin's theory and Liu and Qian [244] for Pesin's theory for RDS. ∎

Chapter 5.
The MET for Related Linear and Affine RDS

Summary

Let V be a finite-dimensional vector space, and $A : V \to V$ be a linear operator. Then the standard constructions of linear algebra yield A^{-1} on V, A^* on the dual space V^*, $\wedge^k A$ on $\wedge^k V$ (the k-th exterior power of V where $1 \leq k \leq d$), in particular $\wedge^d A = \det A$ on $\wedge^d V$. If $U \subset V$ is A-invariant, A induces operators $A|_U$ on U, $A^\sim$ on the quotient space $U^\sim = V/U$ and $A^\perp$ on the orthogonal complement $U^\perp$. Further, if $B : W \to W$ is another linear operator, we have $A \oplus B$ on the direct sum $V \oplus W$ and $A \otimes B$ on the tensor product $V \otimes W$.

Assume now that Φ and Ψ are linear cocycles on V and W, respectively. In Sects. 5.1 to 5.4 we will systematically study the multiplicative ergodic theory of those cocycles obtained from the given ones by means of

(i) the above constructions, i.e. of Φ^{-1}, Φ^*, $\wedge^k \Phi$ (in particular, $\wedge^d \Phi = \det \Phi$), $\Phi^\sim$, $\Phi^\perp$, $\Phi \oplus \Psi$ and $\Phi \otimes \Psi$;

(ii) time reversal, i.e. by considering $\Phi(-\cdot)$;

(iii) by studying the above objects over θ and θ^{-1}.

The results of our study will be used in subsequent chapters of the book, e.g. exterior powers in the theory of rotation numbers (Sect. 6.5), tensor products and affine RDS in normal form theory (Chap. 8).

In Sect. 5.6 we apply multiplicative ergodic theory to the simplest possible nonlinear cocycles, namely those taking their values in the affine group, in the discrete time case also known as iterated function systems. The main result is Theorem 5.6.1 on the existence of a unique invariant measure for hyperbolic affine RDS.

5.1 Inverse and Adjoint

Suppose we are given a linear cocycle Φ on $\mathbb{R}^d$ over an ergodic[1] DS $(\Omega, \mathcal{F}, \mathbb{P}, (\theta(t))_{t \in \mathbb{T}})$ with two-sided time $\mathbb{T}$. For $\mathbb{T} = \mathbb{Z}$ (see Sect. 2.1) and with $A(\omega) := \Phi(1, \omega)$

---

[1] In the whole Chap. 5 we assume that the underlying DS is ergodic – but only to simplify presentation. All statements hold, with obvious modifications, in the non-ergodic case, too.

$$\Phi(n,\omega) = \begin{cases} A(\theta^{n-1}\omega)\ldots A(\omega), & n > 0, \\ I, & n = 0, \\ A^{-1}(\theta^n\omega)\ldots A^{-1}(\theta^{-1}\omega), & n < 0, \end{cases} \quad (5.1.1)$$

and the IC of the MET are $\log^+ \|A^{\pm 1}\| \in L^1(\mathbb{P})$. We say, as usual, that Φ is *generated* by (θ, A).

For $\mathbb{T} = \mathbb{R}$, the two ways of generating a linear cocycle are by solving a linear RDE (real noise case, see Sect. 2.2)

$$\dot{x}_t = A(\theta_t \omega) x_t, \quad A : \Omega \to \mathbb{R}^{d \times d}, \quad (5.1.2)$$

or a linear SDE (white noise case, see Sect. 2.3)

$$dx_t = A_0 x_t dt + \sum_{j=1}^m A_j x_t \circ dW_t^j, \quad A_0, A_1, \ldots, A_m \in \mathbb{R}^{d \times d}. \quad (5.1.3)$$

In case (5.1.2) the IC for Φ are implied by $A(\cdot) \in L^1(\mathbb{P})$ and we say that Φ is generated by (θ, A). In case (5.1.3) the canonical underlying DS, the shift on Wiener space, is ergodic and the IC of the MET are always satisfied (see Theorem 6.2.11). We say that Φ is generated by $(\theta, A_0, A_1, \ldots, A_m, W)$.

Once Φ is generated by one of the above mechanisms, we can consider it as a $Gl(d, \mathbb{R})$-valued cocycle (see Sect. 6.1), to which we can apply an arbitrary combination of

(i) the operation of time reversal $\Phi(\cdot) \mapsto \Phi(-\cdot)$,
(ii) matrix inversion $\Phi \mapsto \Phi^{-1}$,
(iii) matrix transposition $\Phi \mapsto \Phi^*$,

and by considering the result over θ or θ^{-1}. This yields 16 different objects (θ, Φ), $(\theta, \Phi(-\cdot))$, (θ, Φ^*),...,$(\theta^{-1}, \Phi(-\cdot)^{*-1})$, of which four turn out to be cocycles, namely (θ, Φ), (θ, Φ^{*-1}), $(\theta^{-1}, \Phi(-\cdot))$, $(\theta^{-1}, \Phi(-\cdot)^{*-1})$. Another four are $Gl(d, \mathbb{R})$-valued backward cocycles, namely (θ, Φ^{-1}), (θ, Φ^*), $(\theta^{-1}, \Phi(-\cdot)^{-1})$, $(\theta^{-1}, \Phi(-\cdot)^*)$, while the remaining eight objects $(\theta, \Phi(-\cdot))$ etc. are neither cocycles nor backward cocycles. By the canonical left action $(g, x) \mapsto gx$ of $Gl(d, \mathbb{R})$ on $\mathbb{R}^d$, the four cocycles induce linear cocycles, while the four backward cocycles induce linear backward cocycles on $\mathbb{R}^d$, see Remark 1.1.10.

We have the following facts about generators, spectrum and splitting of the four resulting linear cocycles.

5.1.1 Theorem (MET for Inverse and Adjoint). *Let (θ, Φ) be the linear cocycle generated by (θ, A) with $\log^+ \|A^{\pm 1}\| \in L^1(\mathbb{P})$ in case $\mathbb{T} = \mathbb{Z}$, by (θ, A) with $A \in L^1(\mathbb{P})$ in case of an RDE and by $(\theta, A_0, A_1, \ldots, A_m, W)$ in case of an SDE. Let $\Sigma := \mathcal{S}(\theta, \Phi)$ be its Lyapunov spectrum and $E_1(\omega), \ldots, E_p(\omega)$ its splitting.[2] Then the following holds:*

---

[2] In this chapter, all splittings are written in the order of decreasing Lyapunov exponents.

(i) The cocycle (θ, Φ^{*-1}) has generator (θ, A^{*-1}) in case $\mathbb{T} = \mathbb{Z}$, $(\theta, -A^*)$ in the RDE case and $(\theta, -A_0^*, -A_1^*, \ldots, -A_m^*, W)$ in the SDE case. Its spectrum is $-\Sigma$, and its splitting is

$$F_p(\omega)^\perp, \ldots, F_1(\omega)^\perp,$$

where $F_i(\omega) = \oplus_{j \neq i} E_j(\omega)$, $i = 1, \ldots, p$.

(ii) The cocycle $(\theta^{-1}, \Phi(-\cdot))$ has generator $(\theta^{-1}, A^{-1} \circ \theta^{-1})$ in case $\mathbb{T} = \mathbb{Z}$, $(\theta^{-1}, -A)$ in case of an RDE and $(\theta^{-1}, -A_0, A_1, \ldots, A_m, W(-\cdot))$ in the SDE case. Its spectrum is $-\Sigma$, and its splitting is $E_p(\omega), \ldots, E_1(\omega)$.

(iii) The cocycle $(\theta^{-1}, \Phi(-\cdot)^{*-1})$ has generator $(\theta^{-1}, A^* \circ \theta^{-1})$ in case $\mathbb{T} = \mathbb{Z}$, (θ^{-1}, A^*) in the RDE case and $(\theta^{-1}, A_0^*, -A_1^*, \ldots, -A_m^*, W(-\cdot))$ in the SDE case. Its spectrum is Σ and its splitting is $F_1(\omega)^\perp, \ldots, F_p(\omega)^\perp$.

Proof. (iii) follows from (i) applied to the data in (ii), so we only need to prove (i) and (ii).

(i) The generator of Φ^{*-1} is obtained for $\mathbb{T} = \mathbb{Z}$ by applying $(\cdot)^{*-1}$ to (5.1.1). For the RDE case we differentiate $\Phi \Phi^{-1} = I$ with respect to time t to obtain $\dot{\Phi}^{-1} = -\Phi^{-1}\dot{\Phi}\Phi^{-1}$, use $\dot{\Phi} = A\Phi$ and transposition to arrive at $\dot{\Phi}^{*-1} = -A^*\Phi^{*-1}$. Formally the same calculation goes through for a Stratonovich SDE.

The cocycle $\tilde{\Phi}(t, \omega) := \Phi(t, \omega)^{*-1}$ satisfies

$$(\tilde{\Phi}(t,\omega)^* \tilde{\Phi}(t,\omega))^{1/2t} \to \tilde{\Phi}(\omega)$$

by the MET, but

$$(\tilde{\Phi}(t,\omega)^* \tilde{\Phi}(t,\omega))^{1/2t} = \left((\Phi(t,\omega)^* \Phi(t,\omega))^{1/2t}\right)^{-1} \to \Phi(\omega)^{-1},$$

hence

$$\tilde{\Phi}(\omega) = \Phi(\omega)^{-1}.$$

In particular, following the proof of the MET 3.4.1, $\tilde{p} = p$, $\tilde{d}_k = d_{p+1-k}$, $\tilde{\lambda}_k = -\lambda_{p+1-k}$, and

$$\tilde{V}_k(\omega) = V_{p+2-k}(\omega)^\perp$$

(this is a one-sided result – we only need one-sided time and $\Phi(t, \omega) \in Gl(d, \mathbb{R})$ for this to hold). Now everything applied to the cocycle $\tilde{\Phi}(-t, \omega)$ over θ^{-1} (following now the proof of the two-sided MET 3.4.11) gives the spectrum $\tilde{p}^- = \tilde{p} = p$, $\tilde{d}_{p+1-k}^- = \tilde{d}_k = d_{p+1-k}$, $\tilde{\lambda}_{p+1-k}^- = -\tilde{\lambda}_k = \lambda_{p+1-k}$, and the backward filtration

$$\tilde{V}_{p+2-k}^-(\omega) = V_k^-(\omega)^\perp.$$

By definition for $1 \leq i \leq p$

$$\tilde{E}_i(\omega) := \tilde{V}_i(\omega) \cap \tilde{V}_{p+1-i}^-(\omega) = V_{p+2-i}(\omega)^\perp \cap V_{i+1}^-(\omega)^\perp.$$

Remembering that

$$V_{p+2-i}(\omega) = \oplus_{j=p+2-i}^{p} E_j(\omega), \quad V_{i+1}^{-}(\omega) = \oplus_{j=1}^{p-i} E_j(\omega),$$

and noting that for subspaces U, V we have $U^\perp \cap V^\perp = (U+V)^\perp$, we obtain

$$\tilde{E}_i(\omega) = (\oplus_{j \neq p+1-i} E_j(\omega))^\perp = F_{p+1-i}(\omega)^\perp.$$

(ii) The calculation of the generators of $\tilde{\Phi}(t, \omega) = \Phi(-t, \omega)$ for $\mathbb{T} = \mathbb{Z}$ and the RDE case are as above. In the SDE case for all $x \in \mathbb{R}^d$ and $t \in \mathbb{R}$

$$\tilde{\Phi}(t, \cdot)x = x + \int_0^{-t} A_0 \Phi(s, \cdot)x \, ds + \sum_{j=1}^m \int_0^{-t} A_j \Phi(s, \cdot)x \circ dW_s^j,$$

with the interpretation of the two-sided stochastic integral given in Sect. 2.3. By the calculus developed there,

$$\int_0^{-t} f(s) \circ dW_s = \int_0^t f(-s) \circ dW_{-s}.$$

Using this and inverting time in the Lebesgue integral gives

$$\tilde{\Phi}(t, \cdot)x = x - \int_0^t A_0 \tilde{\Phi}(s, \cdot)x \, ds + \sum_{j=1}^m \int_0^t A_j \tilde{\Phi}(s, \cdot)x \circ dW_{-s}^j,$$

i.e. $\tilde{\Phi}$ has generator $(\theta^{-1}, -A_0, A_1, \ldots, A_m, W(-\cdot))$.

The rest of the statement follows readily from

$$\lim_{t \to \pm\infty} \frac{1}{t} \log \|\tilde{\Phi}(t, \omega)x\| = -\lim_{t \to \mp\infty} \frac{1}{t} \log \|\Phi(t, \omega)x\| = -\lambda_i$$

if and only if $x \in E_i(\omega)$. □

The cocycle being generated by (θ^{-1}, A^*) means in more detail in the RDE case that it solves $\dot{x}_t = A^*(\theta(-t)\omega)x_t$, in particular $\dot{x}_t = A^*(\xi_{-t}(\omega))x_t$ if the original system has the form $\dot{x}_t = A(\xi_t(\omega))x_t$, etc.

5.1.2 Remark (More Cocycles). Inspecting the generators of the four linear cocycles above, e. g. for the RDE case, shows that some generators are "missing" in the list of Theorem 5.1.1, namely the family (θ, A^*), $(\theta, -A)$, $(\theta^{-1}, -A^*)$, (θ^{-1}, A). These are the generators of the four cocycles obtained from the four $Gl(d, \mathbb{R})$-valued backward cocycles through the right action $(g, x) \mapsto g^{-1}x$ of $Gl(d, \mathbb{R})$ on $\mathbb{R}^d$. Spectrum and splitting of the members of this family are again closely related with each other (the statement analogous to Theorem 5.1.1 is left as an exercise), but have in general nothing to do with the spectrum and splitting of the first family, hence there is no general relation between $\mathcal{S}(\theta, A)$ and $\mathcal{S}(\theta^{-1}, A) = \mathcal{S}(\theta, A^*)$. However, the following condition ensures equality. ∎

5.1.3 Proposition. *(i) Let for $\mathbb{T} = \mathbb{Z}$ or in the RDE case the generating process $A(\cdot)$ of a linear cocycle be time reversible, i.e.*

$$\mathcal{L}((A(\theta(t)))_{t\in\mathbb{T}}) = \mathcal{L}((A(\theta(-t)))_{t\in\mathbb{T}}),$$

where $\mathcal{L}(\xi)$ denotes the probability law of the random variable ξ. Then

$$\mathcal{L}(\Phi_{\theta,A}) = \mathcal{L}(\Phi_{\theta^{-1},A}), \tag{5.1.4}$$

hence $\mathcal{S}(\theta, A) = \mathcal{S}(\theta^{-1}, A)$.
(ii) In case of an SDE we always have

$$\mathcal{L}(\Phi_{(\theta,A_0,A_1,\ldots,A_m,W)}) = \mathcal{L}(\Phi_{(\theta^{-1},A_0,A_1,\ldots,A_m,-W(-\cdot))}),$$

hence $\mathcal{S}(\theta, A) = \mathcal{S}(\theta^{-1}, A)$.

Proof. (i) For $\mathbb{T} = \mathbb{Z}$ and $n > 0$

$$\begin{aligned}\mathcal{L}(\Phi_{\theta,A}(n,\omega)) &= \mathcal{L}(A(\theta^{n-1}\omega)\cdots A(\omega)) \\ &= \mathcal{L}(A((\theta^{-1})^{n-1}\omega)\cdots A(\omega)) \\ &= \mathcal{L}(\Phi_{\theta^{-1},A}(n,\omega)).\end{aligned}$$

Similarly for $n < 0$ and any finite-dimensional distribution of $\Phi_{\theta,A}(\cdot)$.

For $\mathbb{T} = \mathbb{R}$ the solution of $\dot{\Phi}(t) = A(t)\Phi(t)$, $\Phi(0) = I$, is a certain measurable function of A, $\Phi = f(A)$. If $\mathcal{L}(A) = \mathcal{L}(B)$ then $\mathcal{L}(f(A)) = \mathcal{L}(f(B))$.

Relation (5.1.4) implies

$$\mathcal{L}(\lim_{t\to\infty}(\Phi_{\theta,A}(t,\omega)^*\Phi_{\theta,A}(t,\omega))^{1/2t}) = \mathcal{L}(\lim_{t\to\infty}(\Phi_{\theta^{-1},A}(t,\omega)^*\Phi_{\theta^{-1},A}(t,\omega))^{1/2t})$$

and hence $\mathcal{L}(\mathcal{S}(\theta,A)) = \mathcal{L}(\mathcal{S}(\theta^{-1},A))$, in the ergodic case $\mathcal{S}(\theta,A) = \mathcal{S}(\theta^{-1},A)$.

(ii) The solution of $dx_t = A_0 x_t dt + \sum A_j x_t \circ dW^j$ and $dy_t = A_0 y_t dt + \sum A_j y_t \circ d\tilde{W}^j$ with $\tilde{W} := -W(-\cdot)$ have the same law since $\mathcal{L}(\tilde{W}) = \mathcal{L}(W)$. $\square$

The hypothesis of time reversibility cannot be dropped (for an example see Key [203]).

For $\mathbb{T} = \mathbb{Z}$ time reversibility is implied by exchangeability and exchangeability is implied by $(A(\theta^n \cdot))_{n\in\mathbb{Z}}$ being i.i.d.

For the RDE case assume that $\dot{x}_t = A(\xi_t(\omega))x_t$, where $(\xi_t)_{t\in\mathbb{R}}$ is a Feller Markov process on a locally compact state space E with a countable base (see Appendix A.2.3). Then time reversibility follows from reversibility (or "detailed balance") of the transition probability of ξ which in turn is equivalent to the generator being self-adjoint in $L^2(E,\rho)$, where $\rho = \mathcal{L}(\xi_0)$ (see e. g. Ikeda and Watanabe [178], p. 280).

5.2 The MET on Linear Subbundles and Quotient Spaces

Let in this subsection Φ be again a linear cocycle in $\mathbb{R}^d$ over an ergodic DS θ with two-sided time.

Recall from Sect. 1.6 that a mapping $\omega \mapsto U(\omega)$ taking values in the set of linear subspaces of $\mathbb{R}^d$ is called a *random linear subspace* if $\omega \mapsto d(x, U(\omega))$ is measurable for each $x \in \mathbb{R}^d$, where $d(x, C) := \inf\{|x-y| : y \in C\}$.

5.2.1 Lemma. *If U and V are random linear subspaces, then so are $U^\perp$, $U \cap V$ and $U + V$. Further, $\omega \mapsto \dim U(\omega)$ is measurable.*

The proof is left as an exercise (note e. g. that $U + V = (U^\perp \cap V^\perp)^\perp$).

A random linear subspace $U(\omega) \subset \mathbb{R}^d$ is called *invariant* with respect to Φ if
$$\Phi(t, \omega) U(\omega) = U(\theta(t)\omega) \quad \text{for all } t, \omega.$$

Since $m(\omega) := \dim U(\omega)$ is a θ-invariant random variable and thus a constant m, $U(\cdot)$ is a measurable linear bundle over $(\Omega, \mathcal{F}, \mathbb{P}, (\theta(t))_{t \in \mathbb{T}})$ with typical fiber $\mathbb{R}^m$. An important particular case is a deterministic invariant subspace $U \subset \mathbb{R}^d$.

The restriction of Φ on $U(\cdot)$,
$$\Phi_U(t, \omega) := \Phi(t, \omega)|_{U(\omega)} : U(\omega) \to U(\theta(t)\omega)$$
is a linear cocycle on the bundle $U(\cdot)$.

We define a second linear cocycle $\Phi^\sim$ on the quotient space $U(\omega)^\sim := \mathbb{R}^d / U(\omega)$ of equivalence classes (with respect to the equivalence relation $x \sim y \iff x - y \in U(\omega)$) by
$$U(\omega)^\sim \ni [x]_\omega \mapsto \Phi^\sim(t, \omega)[x]_\omega := [\Phi(t, \omega)x]_{\theta(t)\omega} \in U(\theta(t)\omega)^\sim,$$
where $[x]_\omega$ is the equivalence class containing $x \in \mathbb{R}^d$ over the fiber ω.

Finally, let $U(\omega)^\perp$ be the orthogonal complement of $U(\omega)$ in $\mathbb{R}^d$ (with respect to the standard scalar product)[3] and let the cocycle $\Phi^\perp$ on the linear bundle $U(\cdot)^\perp$ be defined by
$$U(\omega)^\perp \ni x \mapsto \Phi^\perp(t, \omega)x := (\Phi(t, \omega)x)^\perp_{\theta(t)\omega} \in U(\theta(t)\omega)^\perp,$$
where $y \mapsto y^\perp_\omega = \pi^\perp(\omega)(y)$ is the orthogonal projection onto $U(\omega)^\perp$ over ω.

The restriction of the canonical projection $\pi(\omega) : \mathbb{R}^d \to U(\omega)^\sim$, $x \mapsto [x]_\omega$, to $U(\omega)^\perp$ defines an isomorphism of $U(\omega)^\perp$ and $U(\omega)^\sim$, which is an isometry if we define a scalar product on $U(\omega)^\sim$ by $\langle [x]_\omega, [y]_\omega \rangle := \langle x^\perp_\omega, y^\perp_\omega \rangle$. With this done we can identify the cocycles $\Phi^\perp$ on $U(\cdot)^\perp$ and $\Phi^\sim$ on $U(\cdot)^\sim$ for our purposes.

---
[3] Note that $U(\omega)^\perp$ is in general not invariant with respect to Φ.

Choose an adapted measurable random basis

$$F(\omega) = (v_1(\omega), \ldots, v_m(\omega), v_{m+1}(\omega), \ldots, v_d(\omega))$$

in $\mathbb{R}^d$, i.e. such that $F_1(\omega) = (v_1(\omega), \ldots, v_m(\omega))$ is an orthogonal basis of $U(\omega)$ and $F_2(\omega) = (v_{m+1}(\omega), \ldots, v_d(\omega))$ is an orthogonal basis of $U(\omega)^\perp$. Then Φ represented in the bases $F(\omega)$, $F(\theta(t)\omega)$ takes the Lyapunov cohomologous (exercise) form

$$\tilde{\Phi}(t,\omega) = \begin{pmatrix} \Phi_{11}(t,\omega) & \Phi_{12}(t,\omega) \\ 0 & \Phi_{22}(t,\omega) \end{pmatrix},$$

where the cocycle $\Phi_{11}(t,\omega)$ on $\mathbb{R}^m$ represents $\Phi_U(t,\omega)$ in the bases $F_1(\omega)$ in $U(\omega)$ and $F_1(\theta(t)\omega)$ in $U(\theta(t)\omega)$, while the cocycle $\Phi_{22}(t,\omega)$ on $\mathbb{R}^{d-m}$ represents $\Phi^\sim(t,\omega)$ in the bases $[F_2(\omega)] := ([v_{m+1}(\omega)], \ldots, [v_d(\omega)])$ of $U(\omega)^\sim$ and $[F_2(\theta(t)\omega)]$ of $U(\theta(t)\omega)^\sim$, and represents $\Phi^\perp(t,\omega)$ in the bases $F_2(\omega) := (v_{m+1}(\omega), \ldots, v_d(\omega))$ of $U(\omega)^\perp$ and $F_2(\theta(t)\omega)$ of $U(\theta(t)\omega)^\perp$.

We will now determine the spectrum and splitting of Φ_U and $\Phi^\sim$ and their relation to the spectrum and splitting of Φ. As far as the spectrum is concerned we can basically follow Crauel [109] (Appendix A.1), who did the case of a deterministic invariant subspace and who generalized earlier work by Furstenberg and Kifer [155] (Lemma 3.6), Hennion [169] (Proposition 1) and Key [203] (Theorem 5).

5.2.2 Theorem (MET for Subbundle and Quotient). *Let Φ be a linear cocycle in $\mathbb{R}^d$ over the ergodic DS $(\Omega, \mathcal{F}, \mathbb{P}, (\theta(t))_{t \in \mathbb{T}})$ with two-sided time which satisfies the IC of the MET. Let Φ have spectrum $\mathcal{S}(\theta, \Phi) = \{(\lambda_i, d_i)_{i=1,\ldots,p}\}$ and splitting $E_1, \ldots, E_p$. Assume that the random linear subspace $U(\omega) \subset \mathbb{R}^d$ is non-trivial and Φ-invariant. Then:*

(i) The cocycle $\Phi_U = \Phi|_{U(\cdot)}$ satisfies the integrability conditions of the MET and has spectrum

$$\mathcal{S}(\theta, \Phi_U) = \{(\lambda_{k_i}, d_{k_i}^U)_{i=1,\ldots,r}\}$$

and splitting

$$E_{k_1} \cap U, \ldots, E_{k_r} \cap U,$$

where $1 \le k_1 < k_2 < \ldots < k_r \le p$ are the indices for which $d_k^U := \dim(E_k \cap U) > 0$. In particular, $U(\omega)$ has the form

$$U(\omega) = \oplus_{j=1}^r (U(\omega) \cap E_{k_j}(\omega)) = \oplus_{i=1}^p (U(\omega) \cap E_i(\omega)).$$

(ii) The cocycle $\Phi^\sim$ on $U^\sim$ (equivalently, the cocycle $\Phi^\perp$ on $U^\perp$) satisfies the IC of the MET and has spectrum

$$\mathcal{S}(\theta, \Phi^\sim) = \mathcal{S}(\theta, \Phi^\perp) = \{(\lambda_{m_i}, d_{m_i}^\sim)_{i=1,\ldots,q}\}$$

and splitting

$$E'_{m_1} \cap U^\perp, \ldots, E'_{m_q} \cap U^\perp,$$

where

$$E'_{m_i} = \left(\oplus_{k \neq i}(F_k^\perp \cap U^\perp)\right)^\perp$$

and $1 \leq m_1 < \ldots < m_q \leq p$ are the indices for which $d_k^\sim := \dim(F_k^\perp \cap U^\perp) = \dim(E'_k \cap U^\perp) > 0$.

(iii) We have
$$\mathcal{S}(\theta, \Phi) = \mathcal{S}(\theta, \Phi_U) \cup \mathcal{S}(\theta, \Phi^\sim),$$
where the multiplicities of λ_i with respect to Φ_U and $\Phi^\sim$ add up to d_i, $d_i = d_i^U + d_i^\sim$. In particular, denoting by $\lambda_1(\Phi)$ the top Lyapunov exponent of a cocyle Φ,
$$\lambda_1(\Phi) = \max(\lambda_1(\Phi_U), \lambda_1(\Phi^\sim)).$$

Proof. (i) The IC for Φ_U follow from
$$\begin{aligned}\|\Phi_U(t,\omega)^{\pm 1}\|_{\omega,\theta(t)\omega} &= \|(\Phi(t,\omega)|_{U(\omega)})^{\pm 1}\|_{\omega,\theta(t)\omega} \\ &= \sup_{x \in U(\omega), \|x\|=1} \|\Phi(t,\omega)^{\pm 1}x\| \leq \|\Phi(t,\omega)^{\pm 1}\|.\end{aligned}$$

The bundle $U(\omega)$ can be isometrically trivialized by picking a measurable orthonormal basis $F(\omega)$ in $(U(\omega), \langle \cdot, \cdot \rangle_S)$ and taking the coordinate mapping $\psi(\omega) = k_{F(\omega)} : (U(\omega), \langle \cdot, \cdot \rangle_S) \to (\mathbb{R}^m, \langle \cdot, \cdot \rangle_S)$, implying $\|\psi(\omega)^{\pm 1}\| = 1$. By Proposition 4.1.9 the MET holds for Φ_U, with spectrum $\mathcal{S}(\theta, \Phi_U) = \{(\bar{\lambda}_j, \bar{d}_j)_{j=1,\ldots,q}\}$, splitting $\bar{E}_1(\omega), \ldots, \bar{E}_q(\omega)$ and
$$U(\omega) = \oplus_{j=1}^q \bar{E}_j(\omega).$$

Let $1 \leq k_1 < \ldots < k_r \leq p$ be those indices k for which
$$d_k^U := \dim(U(\omega) \cap E_k(\omega)) > 0$$
(note that d_k^U is θ-invariant and thus constant).

Now $\bar{E}_1(\omega) \subset U(\omega) \cap E_{k_1}(\omega)$, $\bar{\lambda}_1 = \lambda_{k_1}$ and $\bar{d}_1 \leq d_{k_1}^U, \ldots, \bar{E}_q(\omega) \subset U(\omega) \cap E_{k_q}(\omega)$, $\bar{\lambda}_q = \lambda_{k_q}$, and $\bar{d}_q \leq d_{k_q}^U$. Hence $q \leq r$ and
$$U(\omega) = \oplus_{j=1}^q \bar{E}_j(\omega) \subset \oplus_{i=1}^p (U(\omega) \cap E_i(\omega)) \subset U(\omega),$$
consequently
$$\dim U = \sum_{i=1}^q \bar{d}_i \leq \sum_{i=1}^p d_i^U = \sum_{i=1}^r d_{k_i}^U = \dim U,$$
and hence $q = r$, $\bar{d}_i = d_{k_i}^U$, $\bar{E}_i(\omega) = U(\omega) \cap E_i(\omega)$ and
$$U(\omega) = \oplus_{i=1}^p U(\omega) \cap E_i(\omega).$$

(ii) It suffices to study the cocycle $\Phi^\perp$ on $U^\perp$. The integrability conditions for $\Phi^\perp$ follow from $(\Phi^\perp)^{-1} = (\Phi^{-1})^\perp$, $\|\pi^\perp(\omega)\| = 1$ and
$$\|(\Phi^\perp)^{\pm 1}\| = \sup_{x \in U^\perp, \|x\|=1} \|(\Phi^\perp)^{\pm 1}x\| \leq \|\Phi^{\pm 1}\|.$$

5.2 MET on Linear Subbundles

The IC of the MET then follow as in part (i).

Further, one readily checks that U is invariant for the cocycle Φ over θ if and only if $U^\perp$ is invariant for the cocycle Φ^{*-1} over θ. Since also $(\Phi^\perp)^* = (\Phi^*)^\perp$,

$$(\Phi^\perp)^{*-1} = (\Phi^{*-1})^\perp = \Phi^{*-1}|_{U^\perp}.$$

The cocycle Φ^{*-1} over θ has, by Theorem 5.1.1(i), the spectrum $\mathcal{S}(\theta, \Phi^{*-1}) = \{(\lambda_i^{(3)}, d_i^{(3)})_{i=1,\ldots,p}\}$ and filtration $E_1^{(3)}, \ldots, E_p^{(3)}$, where $\lambda_i^{(3)} = -\lambda_{p+1-i}$, $d_i^{(3)} = d_{p+1-i}$ and $E_i^{(3)} = F_{p+1-i}^\perp$, where $F_i := \oplus_{j\neq i} E_j$, $i = 1, \ldots, p$.

The cocycle $(\Phi^\perp)^{*-1} = \Phi^{*-1}|_{U^\perp}$ has, by part (i) of this theorem, spectrum

$$\mathcal{S}(\theta, (\Phi^\perp)^{*-1}) = \{(\lambda_i^{(4)}, d_i^{(4)})_{i=1,\ldots,q}\},$$

where $\lambda_i^{(4)} = \lambda_{l_i}^{(3)}$, $d_i^{(4)} = d_{l_i}^{(3)}$ and the indices $1 \leq l_1 < \ldots < l_q \leq p$ are those for which

$$d_k^{(3)} := \dim(E_k^{(3)} \cap U^\perp) = \dim(F_{p+1-k}^\perp \cap U^\perp) > 0,$$

while the filtration is

$$E_i^{(4)} = E_{l_i}^{(3)} \cap U^\perp = F_{p+1-l_i}^\perp \cap U^\perp.$$

Now the original cocycle $\Phi^\perp$ on $U^\perp$ is obtained as

$$\Phi^\perp = ((\Phi^\perp)^{*-1})^{*-1} = (\Phi^{*-1}|_{U^\perp})^{*-1}|_{U^\perp}.$$

By applying Theorem 5.1.1(i) a second time, its spectrum is

$$\mathcal{S}(\theta, \Phi^\perp) = -\mathcal{S}(\theta, (\Phi^\perp)^{*-1}) = \{(\lambda_i^{(5)}, d_i^{(5)})_{i=1,\ldots,q}\},$$

where

$$\lambda_i^{(5)} = -\lambda_{q+1-i}^{(4)} = -\lambda_{l_{q+1-i}}^{(3)} = \lambda_{p+1-l_{q+1-i}}$$

and

$$d_i^{(5)} = d_{q+1-i}^{(4)} = d_{l_{q+1-i}}^{(3)} = \dim(F_{p+1-l_{q+1-i}}^\perp \cap U^\perp) > 0,$$

and its filtration is (orthogonal complements to be taken in $U^\perp$)

$$E_i^{(5)} = G_{q+1-i}^\perp \cap U^\perp, \quad G_i = \oplus_{j\neq i} E_j^{(4)} = \oplus_{j\neq i}(F_{p+1-l_j}^\perp \cap U^\perp).$$

We now introduce the new index

$$m_i := p + 1 - l_{q+1-i}, \quad i = 1, \ldots, q,$$

with $1 \leq m_1 < \ldots < m_q \leq p$. Then

$$\lambda_i^{(5)} = \lambda_{m_i}, \quad d_i^{(5)} = \dim(F_{m_i}^\perp \cap U^\perp) =: \tilde{d}_{m_i} > 0$$

and

$$E_i^{(5)} = G_{q+1-i}^\perp \cap U^\perp = (\oplus_{k\neq i}(F_{m_k}^\perp \cap U^\perp))^\perp \cap U^\perp =: E'_{m_i} \cap U^\perp.$$

Note that
$$\begin{aligned}\dim(E'_{m_i} \cap U^\perp) &= \dim(U^\perp) - \dim(\oplus_{k \neq i}(F_k^\perp \cap U^\perp)) \\ &= \dim(U^\perp) - (\dim(U^\perp) - \dim(F_{m_i}^\perp \cap U^\perp)) \\ &= d_{\widetilde{m_i}},\end{aligned}$$

as it should be.

(iii) We now prove that
$$d_k = \dim E_k = d_k^U + d_k^{\sim},$$
where $d_k^U = \dim(E_k \cap U)$ and $d_k^{\sim} = \dim(F_k^\perp \cap U^\perp)$. Since
$$\begin{aligned}F_k^\perp \cap U^\perp &= (F_k + U)^\perp = (\oplus_{j \neq k} E_j + \oplus_j (E_j \cap U))^\perp \\ &= ((E_k \cap U) \oplus \oplus_{j \neq k} E_j)^\perp\end{aligned}$$

(having used relation (5.2.1) below), we have
$$d_k^{\sim} = d - (\dim(E_k \cap U) + d - d_k) = d_k - d_k^U.$$

$\square$

5.2.3 Corollary (Structure of Invariant Subbundle). *Every Φ-invariant measurable linear bundle $U(\omega)$ is subordinate to the Oseledets splitting, i. e. has the form*
$$U(\omega) = \oplus_{i=1}^p (U(\omega) \cap E_i(\omega)) \subset \oplus_{i=1}^p E_i(\omega). \quad (5.2.1)$$

Further, if $U = \oplus_{i=1}^p (U \cap E_i)$ and $V = \oplus_{i=1}^p (V \cap E_i)$ are Φ-invariant bundles, then so are
$$U + V = \oplus_{i=1}^p ((U \cap E_i) + (V \cap E_i))$$

and
$$U \cap V = \oplus_{i=1}^p (U \cap V \cap E_i).$$

One can repeatedly apply Theorem 5.2.2 to an invariant subbundle V of U, etc. See the "non-random" MET of Kifer [207: Chap. III] and Carverhill [91].

5.2.4 Exercise (Triangular 2×2 Matrices). Reconsider Example 3.4.16 and Exercise 3.4.18 in the light of this subsection. ∎

5.3 Exterior Powers, Volume, Angle

5.3.1 Exterior Powers

The k-th exterior power $\wedge^k \Phi(t,\omega)$ in $\wedge^k \mathbb{R}^d$, $1 \leq k \leq d$, of a cocycle Φ on $\mathbb{R}^d$ was already used in our proof of the MET (see Sect. 3.4; for basic facts on exterior powers see Lemma 3.2.6). We now determine the spectrum and splitting of $\wedge^k \Phi(t,\omega)$.

5.3.1 Theorem (MET for Exterior Power). *Let Φ be a linear cocycle in $\mathbb{R}^d$ over the ergodic DS $(\Omega, \mathcal{F}, \mathbb{P}, (\theta(t))_{t \in \mathbb{T}})$ with one- or two-sided time satisfying the IC of the respective MET. Let $\tilde{\Omega} \subset \Omega$ be the forward invariant (for two-sided time: invariant) set of full measure on which all statements of the MET for Φ hold, where the Lyapunov spectrum $\mathcal{S}(\theta, \Phi) = \{(\lambda_i, d_i)_{i=1,\ldots,p}\}$ is constant and*

$$\lim_{t \to \infty} (\Phi(t,\omega)^* \Phi(t,\omega))^{1/2t} = \Phi(\omega)$$

exists. Let

$$\Lambda_1 \geq \Lambda_2 \geq \ldots \geq \Lambda_d$$

be the list of the λ_i where λ_i appears d_i times. Then:

(i) The cocycle $\wedge^k \Phi(t,\omega)$ on $\wedge^k \mathbb{R}^d$, $1 \leq k \leq d$, satisfies the IC of the respective MET and the MET holds on the same forward invariant (resp. invariant) set $\tilde{\Omega}$. In particular,

$$\lim_{t \to \infty} ((\wedge^k \Phi(t,\omega))^* (\wedge^k \Phi(t,\omega)))^{1/2t} = \wedge^k \Phi(\omega)$$

and the spectrum $\mathcal{S}(\theta, \wedge^k \Phi) = \{(\lambda_i^{(k)}, d_i^{(k)})_{i=1,\ldots,p(k)}\}$ is determined as follows: The $\lambda_i^{(k)}$ are the different numbers in the list of $\binom{d}{k}$ numbers

$$\Lambda_{i_1} + \Lambda_{i_2} + \cdots + \Lambda_{i_k}, \quad 1 \leq i_1 < \ldots < i_k \leq d,$$

where this list starts with $\lambda_1^{(k)} = \Lambda_1 + \cdots + \Lambda_k$ and ends with $\lambda_{p(k)}^{(k)} = \Lambda_{d+1-k} + \cdots + \Lambda_d$, and $d_i^{(k)}$ is the number of times the value $\lambda_i^{(k)}$ appears in this list.

(ii) The flag $\{0\} \subset V_{p(k)}^{(k)}(\omega) \subset \ldots \subset V_1^{(k)}(\omega) = \wedge^k \mathbb{R}^d$ of $\wedge^k \Phi$ can be obtained from the flag $\{0\} \subset V_p(\omega) \subset \ldots \subset V_1(\omega) = \mathbb{R}^d$ of Φ as follows: Extract a measurable basis $F(\omega) = (v_1(\omega), \ldots, v_d(\omega))$ of $\mathbb{R}^d$ which is adapted to the flag of Φ, i.e. such that the first d_1 vectors are taken from $V_1(\omega) \setminus V_2(\omega), \ldots,$ the last d_p vectors are taken from $V_p(\omega) \setminus V_{p+1}(\omega)$. Then

$$V_i^{(k)} = \mathrm{span}\{v_{i_1} \wedge \ldots \wedge v_{i_k} : \Lambda_{i_1} + \cdots + \Lambda_{i_k} \leq \lambda_i^{(k)}, 1 \leq i_1 < \ldots < i_k \leq d\}. \tag{5.3.1}$$

(iii) If time is two-sided, then the splitting $E_i^{(k)}(\omega)$, $1 \leq i \leq p(k)$, of $\wedge^k \Phi(t,\omega)$ can be obtained from the splitting $E_i(\omega)$ of $\Phi(t,\omega)$ as follows:

Choose the measurable basis $F(\omega) = \{v_1(\omega), \ldots, v_d(\omega)\}$ adapted to the splitting of Φ, i.e. such that the first d_1 vectors are in $E_1(\omega)$, ..., the last d_p vectors are in $E_p(\omega)$. Then

$$E_i^{(k)} = \text{span}\{v_{i_1} \wedge \ldots \wedge v_{i_k} : \Lambda_{i_1} + \cdots + \Lambda_{i_k} = \lambda_i^{(k)}, 1 \leq i_1 < \ldots < i_k \leq d\}. \tag{5.3.2}$$

(iv) If Φ has a generator, then so has $\wedge^k \Phi$. In the discrete time case the generator of $\wedge^k \Phi$ is $(\theta, \wedge^k A)$. In the continuous time case $\mathbb{T} = \mathbb{R}$, the generators are $(\theta, \hat{A}^k)$ for the RDE case (and $\hat{A}^k \in L^1$ if $A \in L^1$) and $(\theta, \hat{A}_0^k, \ldots, \hat{A}_m^k, W)$ for the SDE case, where

$$\hat{A}^k(u_1 \wedge \ldots \wedge u_k) := \sum_{i=1}^k u_1 \wedge \ldots \wedge u_{i-1} \wedge Au_i \wedge u_{i+1} \wedge \ldots \wedge u_k.$$

Proof. (i) We assume knowledge of Lemma 3.2.6 and of the proofs of the respective MET's in Sect. 3.4. The IC for $\wedge^k \Phi(t, \omega)$ follow from

$$\|(\wedge^k \Phi)^{\pm 1}(t, \omega)\| = \|\wedge^k \Phi(t, \omega)^{\pm 1}\| \leq \|\Phi(t, \omega)^{\pm 1}\|^k.$$

Elementary manipulations with exterior powers yield

$$((\wedge^k \Phi(t, \omega))^* (\wedge^k \Phi(t, \omega)))^{1/2t} = \wedge^k (\Phi(t, \omega)^* \Phi(t, \omega))^{1/2t} \to \wedge^k \Phi(\omega) \geq 0,$$

where the eigenvalues of $\wedge^k \Phi(\omega)$ are $\exp(\Lambda_{i_1} + \cdots + \Lambda_{i_k})$, $\exp \Lambda_i$ being the eigenvalues of $\Phi(\omega)$. This proves the claim about the spectrum of $\wedge^k \Phi(t, \omega)$.

(ii) For the flag $V_i^{(k)}$ remember that

$$V_i^{(k)}(\omega) := U_{p(k)}^{(k)}(\omega) \oplus \cdots \oplus U_i^{(k)}(\omega),$$

where $U_i^{(k)}(\omega)$ is the eigenspace of $\wedge^k \Phi(\omega)$ corresponding to the eigenvalue $\exp \lambda_i^{(k)}$, with $\dim U_i^{(k)}(\omega) = d_i^{(k)}$. Now the right-hand side of the last equation is equal to the right-hand side of equation (5.3.1).

(iii) If time is two-sided, the splitting of $\wedge^k \Phi(t, \omega)$ is defined by

$$E_i^{(k)}(\omega) := V_i^{(k)}(\omega) \cap V_{p(k)+1-i}^{(k)}(\omega), \quad 1 \leq i \leq p(k).$$

But again the right-hand side of the last equation is equal to the right-hand side of equation (5.3.2).

(iv) This follows immediately from Lemma 3.2.6(vii). □

5.3.2 Remark. (i) The bases $F(\omega)$ appearing in part (ii) and (iii) of the theorem are so-called *normal bases* in the sense of Lyapunov (i.e. bases (v_i) for which $\sum \lambda(v_i)$ is minimal).

(ii) The order of the exponents $\lambda_i^{(k)}$ is in general not determined by the order of the λ_i. It is only clear that the top exponent of $\wedge^k \Phi(t, \omega)$ is $\lambda_1^{(k)} = \Lambda_1 + \cdots + \Lambda_k$ and the smallest exponent is $\lambda_{p(k)}^{(k)} = \Lambda_{d+1-k} + \cdots + \Lambda_d$.

(iii) Extending Remark 3.4.14, the subspace $E_1^{(k)}$ is $\mathcal{F}^-$ measurable, while the subspace $E_{p(k)}^{(k)}$ is $\mathcal{F}^+$ measurable, for each $k = 1, \ldots, d$. ∎

Theorem 5.3.1 has several important applications to the distorsion of volumes and the dynamics of angles under a cocycle (see in particular the theory of rotation numbers in Sect. 6.5).

5.3.2 Volume and Determinant

Recall that if $(V, \langle \cdot, \cdot \rangle)$ is a Euclidean vector space then

$$\|u_1 \wedge \ldots \wedge u_k\| = \text{vol}(u_1, \ldots, u_k) = (\det \langle u_i, u_j \rangle_{k \times k})^{1/2}$$

is the volume of the k-dimensional parallelepiped

$$\Pi(u_1, \ldots, u_k) := \{x = \alpha_1 u_1 + \cdots + \alpha_k u_k, \ 0 \leq \alpha_i \leq 1, \ i = 1, \ldots, k\} \subset V$$

spanned by $u_1, \ldots, u_k$. Since

$$\text{vol}(\Phi(t, \omega) u_1, \ldots, \Phi(t, \omega) u_k) = \|\wedge^k \Phi(t, \omega)(u_1 \wedge \ldots \wedge u_k)\|$$

Theorem 5.3.1 immediately gives the following result.

5.3.3 Corollary (Lyapunov Index of Volume). *Assume the situation of Theorem 5.3.1. Then on $\tilde{\Omega}$ for all $u_1, \ldots, u_k \in \mathbb{R}^d \setminus \{0\}$, $1 \leq k \leq d$,*

$$\lim_{t \to \infty} \frac{1}{t} \log \text{vol}(\Phi(t, \omega) u_1, \ldots, \Phi(t, \omega) u_k) = \lambda_{i_0}^{(k)},$$

where $i_0 = i_0(\omega) := \min\{j : u_1 \wedge \ldots \wedge u_k \in V_j^{(k)}(\omega) \setminus V_{j+1}^{(k)}(\omega)\}$.

5.3.4 Definition (Determinant of a Linear Map on Subspace). Let $T : (V, \langle \cdot, \cdot \rangle_V) \to (W, \langle \cdot, \cdot \rangle_W)$ be a linear map of Euclidean spaces and let $U \subset V$ be a k-dimensional subspace ($k > 0$) with basis $u_1, \ldots, u_k$. We define the *determinant* of $T|_U$ by

$$\det T|_U := \frac{\text{vol}_W(Tu_1, \ldots, Tu_k)}{\text{vol}_V(u_1, \ldots, u_k)} = \frac{\|Tu_1 \wedge \ldots \wedge Tu_k\|_W}{\|u_1 \wedge \ldots \wedge u_k\|_V}.$$

Equivalently,

$$\det T|_U = \frac{\|(\wedge^k T)(u_1 \wedge \ldots \wedge u_k)\|_W}{\|u_1 \wedge \ldots \wedge u_k\|_V} = \|(\wedge^k T)|_{\wedge^k U}\| = \|\wedge^k (T|_U)\|$$

and $\det T|_U$ is well-defined since $\wedge^k U = \mathbb{R}(u_1 \wedge \ldots \wedge u_k)$ is one-dimensional. $\det T|_U$ thus measures the distortion of volumes in U under T. We have the usual properties of the determinant:

- $\det T|_U$ is independent of the basis of U, and if A_U is a matrix representation of $T|_U$ with respect to orthonormal bases of U and $T(U)$, then $\det T|_U = |\det A_U|$,
- $\det T|_U \geq 0$ and $\det T|_U > 0 \iff T : U \to T(U)$ is bijective,

– if $T: V \to W$ and $S: W \to G$, then

$$\det(S \circ T)|_U = (\det S|_{T(U)})(\det T|_U).$$

For $U = V = W$ we recover the (absolute value of the) classical determinant $\det T = \|\wedge^d T\|$. ∎

5.3.5 Corollary (Lyapunov Index of Determinant on Invariant Subspace). *Assume the situation of Theorem 5.3.1 with two-sided time. Let*

$$U(\omega) = \oplus_{i=1}^{p} U(\omega) \cap E_i(\omega), \quad d_i^U := \dim(U(\omega) \cap E_i(\omega)) \geq 0,$$

be a non-trivial Φ-invariant bundle. Then $\det \Phi(t, \omega)|_{U(\omega)}$ is a positive scalar cocycle satisfying the IC of the MET and

$$\lim_{t \to \pm\infty} \frac{1}{t} \log \det \Phi(t, \omega)|_{U(\omega)} = \sum_{i=1}^{p} d_i^U \lambda_i.$$

For discrete time $\mathbb{T} = \mathbb{Z}$ and $A(\omega) = \Phi(1, \omega)$

$$\lim_{n \to \pm\infty} \frac{1}{n} \log \det \Phi(n, \omega)|_{U(\omega)} = \mathbb{E} \log \det A(\cdot)|_{U(\cdot)}.$$

Proof. By Theorem 5.2.2(i) the cocycle $\Phi(t, \omega)|_{U(\omega)}$ has spectrum $\{(\lambda_{k_i}, d_{k_i}^U)_{i=1,\ldots,r}\}$, where the indices $1 \leq k_1 < \ldots < k_r \leq p$ are those for which $d_i^U > 0$. The cocycle $\wedge^m \Phi(t, \omega)|_{\wedge^m U(\omega)}$, $m := \dim U(\omega) = \sum_{i=1}^{p} d_i^U > 0$, is a scalar cocycle on the one-dimensional $\wedge^m \Phi$-invariant subbundle $\wedge^m U(\omega) \subset \wedge^m \mathbb{R}^d$ and by Theorem 5.3.1 has the single the Lyapunov exponent $\sum_{j=1}^{r} d_{k_j}^U \lambda_{k_j} = \sum_{i=1}^{p} d_i^U \lambda_i$. The result follows from $\det \Phi(t, \omega)|_{U(\omega)} = \|\wedge^m \Phi(t, \omega)|_{\wedge^m U(\omega)}\|$.

For discrete time, e.g. for $n > 0$,

$$\log \det \Phi(n, \omega)|_{U(\omega)} = \sum_{k=0}^{n-1} \log \det A(\theta^k \omega)|_{U(\theta^k \omega)},$$

so that the ergodic theorem can be applied. □

For continuous time formulas for the Lyapunov index of volumes and determinants see Sect. 6.3.

We list the following important particular cases.

5.3.6 Example (Lyapunov Index of Determinant on Oseledets Space). *Assume the situation of Theorem 5.3.1 with two-sided time. Then:*

(i) *If $E_i(\omega)$ is an Oseledets space, then*

$$\lim_{t \to \pm\infty} \frac{1}{t} \log \det \Phi(t, \omega)|_{E_i(\omega)} = d_i \lambda_i.$$

(ii) Let $E_s(\omega) := \oplus_{\lambda_i<0} E_i(\omega)$, $E_c(\omega) := \oplus_{\lambda_i=0} E_i(\omega)$, $E_u(\omega) := \oplus_{\lambda_i>0} E_i(\omega)$ be the stable, center and unstable Oseledets space, respectively. Then

$$\lim_{t\to\pm\infty} \frac{1}{t} \log \det \Phi(t,\omega)|_{E_\alpha(\omega)} = \begin{cases} \sum_{\lambda_i<0} d_i\lambda_i, & \alpha = s, \\ 0, & \alpha = c, \\ \sum_{\lambda_i>0} d_i\lambda_i, & \alpha = u, \end{cases}$$

if E_α is non-trivial. ∎

5.3.3 Angles

Recall that the angle $\gamma(x,y) \in [0,\pi/2]$ between two vectors $x, y \in \mathbb{R}^d \setminus \{0\}$ is defined by

$$\sin \gamma(x,y) := \frac{(\|x\|^2 \|y\|^2 - \langle x,y\rangle^2)^{1/2}}{\|x\| \|y\|} = \frac{\|x \wedge y\|}{\|x\| \|y\|}.$$

5.3.7 Corollary (Lyapunov Index of Angle Between Vectors). *Assume the situation of Theorem 5.3.1 with two-sided time. Then for all $\omega \in \tilde{\Omega}$:*
(i) For all linearly independent $x, y \in \mathbb{R}^d$

$$\lim_{t\to\infty} \frac{1}{t} \log \sin\gamma(\Phi(t,\omega)x, \Phi(t,\omega)y) = \lim_{t\to\infty} \frac{1}{t} \log \gamma(\Phi(t,\omega)x, \Phi(t,\omega)y)$$
$$= \lambda^{(2)}(x \wedge y) - \lambda(x) - \lambda(y) \le 0.$$

(ii) Specifically, let $x = \oplus_{i=1}^p x_i(\omega)$, $y = \oplus_{i=1}^p y_i(\omega)$ be linearly independent and $i_0(\omega,x) := \min\{i : x_i(\omega) \ne 0\}$. Then

$$\lim_{t\to\infty} \frac{1}{t} \log \gamma(\Phi(t,\omega)x, \Phi(t,\omega)y) < 0$$

if and only if $x_{i_0(\omega,x)} \wedge y_{i_0(\omega,y)} = 0$ (i.e. $x_{i_0(\omega,x)}$ and $y_{i_0(\omega,y)}$ are linearly dependent). Otherwise

$$\lim_{t\to\infty} \frac{1}{t} \log \gamma(\Phi(t,\omega)x, \Phi(t,\omega)y) = 0.$$

Proof. (i) Equality of the two limits follows from $\gamma/2 \le \sin\gamma \le \gamma$, while existence and the formula follow from Theorem 5.3.1 since

$$\sin\gamma(\Phi(t,\omega)x, \Phi(t,\omega)y) = \frac{\|\Phi(t,\omega)x \wedge \Phi(t,\omega)y\|}{\|\Phi(t,\omega)x\| \|\Phi(t,\omega)y\|} = \frac{\|\wedge^2 \Phi(t,\omega)(x\wedge y)\|}{\|\Phi(t,\omega)x\| \|\Phi(t,\omega)y\|}.$$

(ii) With the representations $x = \oplus_{i=1}^p x_i$ and $y = \oplus_{j=1}^p y_j$ we have

$$\lambda^{(2)}(x \wedge y) = \max_{x_i \wedge y_j \ne 0}(\lambda_i + \lambda_j),$$

$$\lambda(x) = \max_{x_i \ne 0} \lambda_i = \lambda_{i_0(x)}, \quad \lambda(y) = \max_{y_j \ne 0} \lambda_j = \lambda_{i_0(y)}.$$

In case $i_0(x) \neq i_0(y)$, we always have $x_{i_0(x)} \wedge y_{i_0(y)} \neq 0$ and
$$\lambda^{(2)}(x \wedge y) - \lambda(x) - \lambda(y) = \lambda_{i_0(x)} + \lambda_{i_0(y)} - \lambda_{i_0(x)} - \lambda_{i_0(y)} = 0.$$
Now assume $i_0 := i_0(x) = i_0(y)$. Then either $x_{i_0} \wedge y_{i_0} \neq 0$ (implying $d_{i_0} > 1$), hence
$$\lambda^{(2)}(x \wedge y) - \lambda(x) - \lambda(y) = 2\lambda_{i_0} - \lambda_{i_0} - \lambda_{i_0} = 0,$$
or $x_{i_0} \wedge y_{i_0} = 0$. Then by linear independence of x and y, $p \geq 2$ and
$$i_1(\omega, x, y) := \min\{i > i_0 : x_i \neq 0 \text{ or } y_i \neq 0\} \in \{2, \ldots, p\}$$
is well-defined, whence
$$\lambda^{(2)}(x \wedge y) - \lambda(x) - \lambda(y) = \lambda_{i_1} + \lambda_{i_0} - \lambda_{i_0} - \lambda_{i_0} = \lambda_{i_1} - \lambda_{i_0} < 0.$$
□

5.3.8 Example. If either (i) $x \in E_i(\omega) \setminus \{0\}$ and $y \in E_j(\omega) \setminus \{0\}$, $i \neq j$, or (ii) $x, y \in E_i(\omega)$ are linearly independent, then
$$\lim_{t \to \pm\infty} \frac{1}{t} \log \gamma(\Phi(t,\omega)x, \Phi(t,\omega)y) = 0.$$
■

5.3.9 Remark. (i) Since by Proposition 3.2.8 $\delta(\bar{x}, \bar{y}) = \sin \gamma(x,y)$ is a metric on projective space P^{d-1}, Corollary 5.3.7 makes a statement about (exponentially fast) clustering of orbits of the cocycle induced by Φ on P^{d-1} (see also Sect. 6.2).

(ii) In particular, inside an Oseledets space there is never exponentially fast clustering. The clustering is exponentially fast if and only if the "dominating components" of x and y are linearly dependent. This is e.g. the case with probability 1 if the top Lyapunov exponent is simple and $i_0(\omega, x) = 1$ P-a. s. for any $x \neq 0$. The latter holds in the $\mathbb{T} = \mathbb{Z}$ (i.i.d.) case if the distribution of $A_n \in Gl(d, \mathbb{R})$ satisfies certain conditions (Furstenberg [153: Theorem 8.3], Guivarc'h and Raugi [163], Bougerol and Lacroix [77: Sect. VI.3]), and in the case $\mathbb{T} = \mathbb{R}$ for SDE and RDE with Markovian noise under a Lie algebra condition on P^{d-1} (Arnold et al. [27, 32], Imkeller [179]). ■

Recall that the angle between subspaces $U, V \subset \mathbb{R}^d$ is defined by
$$\sin \gamma(U,V) := \inf_{x \in U, y \in V, x,y \neq 0} \sin \gamma(x,y) = \inf_{x \in U, y \in V, x,y \neq 0} \frac{\|x \wedge y\|}{\|x\| \|y\|}.$$

5.3.10 Corollary (Angle Between Oseledets Spaces is Tempered). *Assume the situation of Theorem 5.3.1 with two-sided time. Then for all $\omega \in \tilde{\Omega}$*

(i) For $i \neq j$

$$\lim_{t\to\pm\infty} \frac{1}{t} \log \gamma(E_i(\theta(t)\omega), E_j(\theta(t)\omega)) = 0,$$

i. e. the angle between Oseledets spaces is tempered.

(ii) *More generally, if M and N are disjoint subsets of $\mathbb{R}$ and*

$$E_M(\omega) := \oplus_{\lambda_i \in M} E_i(\omega), \quad E_N(\omega) := \oplus_{\lambda_i \in N} E_i(\omega),$$

then

$$\lim_{t\to\pm\infty} \frac{1}{t} \log \gamma(E_M(\theta(t)\omega), E_N(\theta(t)\omega)) = 0.$$

Proof. (i) It suffices again to prove the statement for γ replaced by $\sin \gamma$. With $\Phi(t,\omega) E_i(\omega) = E_i(\theta(t)\omega)$,

$$\sin \gamma(E_i(\theta(t)\omega), E_j(\theta(t)\omega))$$
$$= \inf_{x \in E_i(\omega), y \in E_j(\omega), \|x\|=\|y\|=1} \frac{\|\wedge^2 \Phi(t,\omega)(x \wedge y)\|}{\|\Phi(t,\omega)x\| \|\Phi(t,\omega)y\|}.$$

Now $\lambda^{(2)}(x \wedge y) = \lambda_i + \lambda_j$, $\lambda(x) = \lambda_i$, $\lambda(y) = \lambda_j$, and $x \wedge y \in E^{(2)}(\omega)$, where $E^{(2)}(\omega)$ is the space in the splitting of $\wedge^2 \Phi$ corresponding to $\lambda^{(2)} = \lambda_i + \lambda_j$. By the uniformity statement of the two-sided MET 3.4.11, the convergence in

$$\lim_{t\to\infty} \frac{1}{t} \log \|\Phi(t,\omega)x\| = \lambda_i$$

is uniform with respect to $x \in E_i(\omega) \cap S^{d-1}$ and convergence in

$$\lim_{t\to\infty} \frac{1}{t} \log \| \wedge^2 \Phi(t,\omega)(x \wedge y)\| = \lambda_i + \lambda_j$$

is uniform with respect to $x \wedge y$ such that $x \in E_i(\omega) \cap S^{d-1}$ and $y \in E_j(\omega) \cap S^{d-1}$ (these $x \wedge y$ form a compact subset of $E^{(2)}(\omega) \setminus \{0\}$), so that for each $\varepsilon > 0$ there is a $T(\omega, \varepsilon)$ such that for all $t \geq T(\omega, \varepsilon)$

$$1 \geq \sin \gamma(E_i(\theta(t)\omega), E_j(\theta(t)\omega)) \geq e^{-\varepsilon t},$$

similarly for $t \to -\infty$.

(ii) For $x = \oplus_{\lambda_i \in M} x_i \in E_M$ and $y = \oplus_{\lambda_j \in N} y_j \in E_N$, $x \wedge y = \sum x_i \wedge y_j$, where $x_i \wedge y_j = 0$ if and only if $x_i = 0$ or $y_j = 0$. Therefore

$$\lambda^{(2)}(x \wedge y) = \max_{x_i \wedge y_j \neq 0}(\lambda_i + \lambda_j) = \max_{x_i \neq 0} \lambda_i + \max_{y_j \neq 0} \lambda_j = \lambda(x) + \lambda(y).$$

Similarly for $t \to -\infty$ with max replaced by min. The result follows again by uniformity of convergence in the MET and the fact that $\Phi(t,\omega) E_{M,N}(\omega) = E_{M,N}(\theta(t)\omega)$. □

5.4 Tensor Product

Let $V_1, \ldots, V_k$ be finite-dimensional real vector spaces of dimensions $\dim V_i = n_i$, $i = 1, \ldots, k$, and let $V = V_1 \otimes \cdots \otimes V_k$ be their tensor product, a vector space of dimension $n = n_1 n_2 \cdots n_k$. Linear cocycles Φ_i on V_i uniquely induce a linear cocycle $\Phi_1 \otimes \cdots \otimes \Phi_k$, the tensor product of $\Phi_1, \ldots, \Phi_k$, on V, whose spectrum, splitting and generator can be obtained from the spectra, splittings and generators of the Φ_i.

As we will not need the full machinery, and since the extension to general k is obvious, we will restrict ourselves to the case $k = 2$. The matrix version of $\Phi_1 \otimes \Phi_2$ in $\mathbb{R}^{n_1} \otimes \mathbb{R}^{n_2} \cong \mathbb{R}^{n_1 n_2}$ is called *Kronecker product* and will turn out to be basic for normal form theory of RDS (see Chap. 8).

The tensor product $V = V_1 \otimes V_2$ of V_1 and V_2 is defined to be the vector space $L(V_1^* \times V_2^*, \mathbb{R})$ of bilinear forms on $V_1^* \times V_2^*$ (2-contravariant tensors). Define for $x \in V_1$, $y \in V_2$ the element $x \otimes y \in V_1 \otimes V_2$ by

$$(x \otimes y)(\gamma_1, \gamma_2) := \gamma_1(x)\gamma_2(y), \quad \gamma_i \in V_i^* = L(V_i, \mathbb{R}).$$

Then $V_1 \otimes V_2$ can be identified with the set of linear combinations $\sum_{i=1}^m c_i (x_i \otimes y_i)$, $m \in \mathbb{N}$, $c_i \in \mathbb{R}$, $x_i \in V_1$, $y_i \in V_2$, where the following rules of computation hold:
1. $(x_1 + x_2) \otimes y = x_1 \otimes y + x_2 \otimes y$, $\quad x \otimes (y_1 + y_2) = x \otimes y_1 + x \otimes y_2$,
2. $(\alpha x) \otimes y = x \otimes (\alpha y) = \alpha(x \otimes y)$.

The following facts (which we also need in Chap. 8) can be found in many textbooks on (multi-)linear algebra (see e.g. Kowalsky [219: Chap. 11]) or readily obtained from the definitions.

5.4.1 Lemma (Tensor Product of Spaces and Operators). *Let $V = V_1 \otimes V_2$ be the tensor product of the finite-dimensional vector spaces V_1 and V_2 with dimensions $n_1 = \dim V_1$, $n_2 = \dim V_2$. Then we have the following properties:*

(i) If (e_i) is a basis of V_1 and (f_j) is a basis of V_2, then $(e_i \otimes f_j)$ is a basis of $V_1 \otimes V_2$. In particular,

$$\dim V_1 \otimes V_2 = (\dim V_1)(\dim V_2) = n_1 n_2.$$

(ii) The splittings $V_1 = \oplus_{i=1}^{p_1} V_i^{(1)}$, $V_2 = \oplus_{j=1}^{p_2} V_j^{(2)}$ induce the splitting

$$V_1 \otimes V_2 = \oplus_{(i,j)=(1,1)}^{(p_1, p_2)} V_i^{(1)} \otimes V_j^{(2)}.$$

(iii) If $T_i : V_i \to V_i$ are linear operators, then the linear extension of

$$(T_1 \otimes T_2)(x \otimes y) := T_1 x \otimes T_2 y$$

defines a linear operator $T_1 \otimes T_2$ on $V_1 \otimes V_2$, the tensor product *of T_1 and T_2. We have*

$$T_1 \otimes (aT_2) = (aT_1) \otimes T_2 = a(T_1 \otimes T_2),$$

$$(T_1 + S_1) \otimes T_2 = T_1 \otimes T_2 + S_1 \otimes T_2,$$
$$T_1 \otimes (T_2 + S_2) = T_1 \otimes T_2 + T_1 \otimes S_2,$$
$$(T_1 \otimes T_2) \circ (S_1 \otimes S_2) = (T_1 \circ S_1) \otimes (T_2 \circ S_2). \tag{5.4.1}$$

In particular,
$$(T_1 \otimes I_2) \circ (I_1 \otimes T_2) = T_1 \otimes T_2 = (I_1 \otimes T_2) \circ (T_1 \otimes I_2), \tag{5.4.2}$$

where I_i are the identity operators on V_i.

If T_1 and T_2 are invertible, then so is $T_1 \otimes T_2$ and
$$(T_1 \otimes T_2)^{-1} = T_1^{-1} \otimes T_2^{-1}. \tag{5.4.3}$$

(iv) If T_1 has eigenvalues (α_i) and eigenvectors (x_i) and if T_2 has eigenvalues (β_j) and eigenvectors (y_j), then $T_1 \otimes T_2$ has eigenvalues $(\alpha_i \beta_j)$ and eigenvectors $(x_i \otimes y_j)$ (with corresponding multiplicities). In particular,
$$\det T_1 \otimes T_2 = (\det T_1)^{n_2} (\det T_2)^{n_1}, \quad \operatorname{trace} T_1 \otimes T_2 = (\operatorname{trace} T_1)(\operatorname{trace} T_2).$$

(v) Let in the bases of part (i) the matrix representations of T_1, T_2 and $T_1 \otimes T_2$ be A, B and $A \otimes B$, respectively. Then
$$A \otimes B = \begin{bmatrix} a_{11}B & \cdots & a_{1,n_1}B \\ \vdots & \ddots & \vdots \\ a_{n_1,1}B & \cdots & a_{n_1,n_1}B \end{bmatrix},$$

i.e. the $n_1 n_2 \times n_1 n_2$ matrix $A \otimes B$ is the **Kronecker product** of the $n_1 \times n_1$ matrix A and the $n_2 \times n_2$ matrix B.

(vi) We have
$$e^{tT_1} \otimes e^{tT_2} = e^{t(T_1 \otimes I_2 + I_1 \otimes T_2)}.$$

(vii) If the spaces V_i have scalar products $\langle \cdot, \cdot \rangle_i$, then the bilinear extension of
$$\langle x_1 \otimes y_1, x_2 \otimes y_2 \rangle := \langle x_1, x_2 \rangle_1 \langle y_1, y_2 \rangle_2$$

defines a scalar product in $V_1 \otimes V_2$. In particular,
(a) $\|x \otimes y\| = \|x\|_1 \|y\|_2$,
(b) $\|T_1 \otimes T_2\| = \|T_1\|_1 \|T_2\|_2$,
(c) $(T_1 \otimes T_2)^* = T_1^* \otimes T_2^*$,
(d) If U_i are orthogonal, then so is $U_1 \otimes U_2$, and $(U_1 \otimes U_2)^* = U_1^* \otimes U_2^*$.

The following theorem is a more or less immediate consequence of the lemma.

5.4.2 Theorem (MET for Tensor Product). *Let Φ_1 in $\mathbb{R}^{n_1}$ and Φ_2 in $\mathbb{R}^{n_2}$ be linear cocycles over the ergodic DS $(\Omega, \mathcal{F}, \mathbb{P}, (\theta(t))_{t \in \mathbb{T}})$ with two-sided time satisfying the IC of the MET. Let $\tilde{\Omega} = \tilde{\Omega}_1 \cap \tilde{\Omega}_2$ be the invariant set of full measure on which the statements of the MET for Φ_1 and Φ_2 hold simultaneously. Let $\mathcal{S}(\theta, \Phi_i) = \{(\lambda_j^{(i)}, d_j^{(i)}) : j = 1, \ldots, p_i\}$ be the spectrum and $(E_j^{(i)})_{j=1,\ldots,p_i}$ the splitting of Φ_i, respectively. Then:*

(i) $\Phi(t, \omega) := \Phi_1(t, \omega) \otimes \Phi_2(t, \omega)$ is a linear cocycle on $\mathbb{R}^{n_1} \otimes \mathbb{R}^{n_2}$ which satisfies the IC of the MET.

(ii) The MET holds on $\tilde{\Omega}$ and

$$\lim_{t \to \infty} (\Phi(t, \omega)^* \Phi(t, \omega))^{1/2t} = \Phi(\omega) = \Phi_1(\omega) \otimes \Phi_2(\omega) > 0$$

where $\Phi_i(\omega) := \lim_{t \to \infty} (\Phi_i(t, \omega)^* \Phi_i(t, \omega))^{1/2t} > 0$. The spectrum of $\Phi_1 \otimes \Phi_2$ is

$$\begin{aligned} \mathcal{S}(\theta, \Phi_1 \otimes \Phi_2) &= \mathcal{S}(\theta, \Phi_1) + \mathcal{S}(\theta, \Phi_2) \\ &= \{\lambda = \lambda_i^{(1)} + \lambda_j^{(2)} : \lambda_i^{(1)} \in \mathcal{S}(\theta, \Phi_1), \lambda_j^{(2)} \in \mathcal{S}(\theta, \Phi_2)\} \end{aligned}$$

and the multiplicity of $\lambda \in \mathcal{S}(\theta, \Phi_1 \otimes \Phi_2)$ is

$$d_\lambda = \sum_{\lambda_i^{(1)} + \lambda_j^{(2)} = \lambda} d_i^{(1)} d_j^{(2)}.$$

The Oseledets spaces of $\Phi_1 \otimes \Phi_2$ are

$$E_\lambda = \oplus_{\lambda_i^{(1)} + \lambda_j^{(2)} = \lambda} E_i^{(1)} \otimes E_j^{(2)}, \quad \lambda \in \mathcal{S}(\theta, \Phi_1 \otimes \Phi_2),$$

and $\dim E_\lambda = d_\lambda$.

(iii) If the Φ_i have generators then so does $\Phi_1 \otimes \Phi_2$. In the discrete time case $\mathbb{T} = \mathbb{Z}$, $(\theta, A_1 \otimes A_2)$ is the generator of $\Phi_1 \otimes \Phi_2$. For $\mathbb{T} = \mathbb{R}$ and the RDE case $\dot{\Phi}_i(t, \omega) = A_i(\theta_t \omega) \Phi_i(t, \omega)$,

$$\frac{d}{dt}(\Phi_1 \otimes \Phi_2)(t, \omega) = (A_1(\theta_t \omega) \otimes I_2 + I_1 \otimes A_2(\theta_t \omega))(\Phi_1 \otimes \Phi_2)(t, \omega).$$

For $\mathbb{T} = \mathbb{R}$ and the SDE case $d\Phi_i(t) = \sum_{j=0}^m A_j^{(i)} \Phi_i(t) \circ dW_t^j$,

$$d(\Phi_1 \otimes \Phi_2)(t) = \sum_{j=0}^m (A_j^{(1)} \otimes I_2 + I_1 \otimes A_j^{(2)})(\Phi_1 \otimes \Phi_2)(t) \circ dW_t^j.$$

Proof. (i) The cocycle property follows from equation (5.4.1) of Lemma 5.4.1, the integrability condition follows from part (vii)(b) of the same lemma.

(ii) The claim about the matrix $\Phi(\omega)$ follows from

$$\begin{aligned} &(\Phi_1(t, \omega)^* \Phi_1(t, \omega) \otimes \Phi_2(t, \omega)^* \Phi_2(t, \omega))^{1/2t} \\ &= (\Phi_1(t, \omega)^* \Phi_1(t, \omega))^{1/2t} \otimes (\Phi_2(t, \omega)^* \Phi_2(t, \omega))^{1/2t}. \end{aligned}$$

The assertion about spectrum and splitting follows from

$$\|(\Phi_1(t,\omega) \otimes \Phi_2(t,\omega))(x \otimes y)\| = \|\Phi_1(t,\omega)x\|_1 \|\Phi_2(t,\omega)y\|_2$$

and the dynamical characterization of the splitting.

(iii) If $\dot{\Phi}_i = A_i(\theta_t\omega)\Phi_i$, then

$$\frac{d}{dt}\Phi_1(t,\omega) \otimes \Phi_2(t,\omega)$$
$$= A_1(\theta_t\omega)\Phi_1(t,\omega) \otimes \Phi_2(t,\omega) + \Phi_1(t,\omega) \otimes A_2(\theta_t\omega)\Phi_2(t,\omega)$$
$$= (A_1(\theta_t\omega) \otimes I_2 + I_1 \otimes A_2(\theta_t\omega))(\Phi_1(t,\omega) \otimes \Phi_2(t,\omega)).$$

Analogously for the Stratonovich SDE case. □

5.5 Manifold Versions

All standard constructions of linear algebra by which we derived the cocycles investigated in sections 5.1 to 5.4 also make sense on the level of vector bundles. Hence if Φ is a linear bundle RDS on a Euclidean bundle for which the MET holds (see Theorem 4.1.11) we can derive similar statements, which we refrain from formulating as they should be obvious for the reader by now.

We briefly recall the basic situation for the tangent bundle of a Riemannian manifold: If φ is a C^1 RDS on a Riemannian manifold $(M, \langle \cdot, \cdot \rangle_x)$ with invariant measure μ, then $T\varphi(t,\omega,x) : T_xM \to T_{\varphi(t,\omega)x}M$ is a linear bundle RDS on $\Omega \times TM$ over the DS $(\Omega \times M, \mathcal{F} \otimes \mathcal{B}, \mu, (\Theta(t))_{t \in \mathbb{T}})$, where $\Theta(t)(\omega, x) := (\theta(t)\omega, \varphi(t,\omega)x)$.

Now if the MET holds for $T\varphi$ with spectrum Σ and splitting $E_i(\omega, x)$ then, for example, $T\varphi(t,\omega,x)^{*-1}$ is a linear bundle RDS over Θ with spectrum $-\Sigma$ and splitting $(\oplus_{j \neq p+1-i} E_j(\omega, x))^\perp$, etc. Here $(T_xM)^*$ is, as usual, canonically identified with T_xM.

In Chap. 6, we will use the manifold versions of the MET for $\wedge^k T\varphi$ on $\wedge^k TM$ and its offsprings for the volume, the determinant and the angle (whose definitions in $(T_xM, \langle \cdot, \cdot \rangle_x)$ remain unchanged).

5.6 Affine RDS

5.6.1 Representation

The group $A(d, \mathbb{R})$ of invertible affine transformations of $\mathbb{R}^d$ has the structure of a semidirect product of $Gl(d, \mathbb{R})$ and $\mathbb{R}^d$. More specifically, every element $g \in A(d, \mathbb{R})$ is represented by $g = (A, b)$, where $g(x) = Ax + b$, with $A \in Gl(d, \mathbb{R})$ and $b \in \mathbb{R}^d$, and

$$g_1 g_2 = (A_1 A_2, A_1 b_2 + b_1), \quad g_i = (A_i, b_i),$$

$$g^{-1} = (A^{-1}, -A^{-1}b), \quad e = (I, 0).$$

$A(d, \mathbb{R})$ is a Lie group of diffeomorphisms of $\mathbb{R}^d$ with dimension $d^2 + d$ and with Lie algebra $\mathfrak{a}(d, \mathbb{R}) = \mathfrak{gl}(d, \mathbb{R}) \dotplus \mathbb{R}^d$ (semidirect sum).

Let $\mathbb{T}$ be two-sided. A mapping $\varphi : \mathbb{T} \times \Omega \to A(d, \mathbb{R})$, $(t, \omega) \mapsto \varphi(t, \omega) = (\Phi(t, \omega), \psi(t, \omega))$ which is measurable and for which $t \mapsto \varphi(t, \omega)$ is continuous is called an $A(d, \mathbb{R})$-valued cocycle if

$$\varphi(t + s, \omega) = \varphi(t, \theta(s)\omega) \cdot \varphi(s, \omega). \tag{5.6.1}$$

It follows that $\varphi(0, \omega) = e$.

(5.6.1) is equivalent to

$$\Phi(t + s, \omega) = \Phi(t, \theta(s)\omega)\Phi(s, \omega) \tag{5.6.2}$$

(i. e. Φ is a $Gl(d, \mathbb{R})$-valued cocycle) and

$$\psi(t + s, \omega) = \Phi(t, \theta(s)\omega)\psi(s, \omega) + \psi(t, \theta(s)\omega), \quad \psi(0, \omega) = 0. \tag{5.6.3}$$

Every $A(d, \mathbb{R})$-cocycle $\varphi(t, \omega)$ induces by the left action of $A(d, \mathbb{R})$ on $\mathbb{R}^d$, $(g, x) \mapsto g(x) := Ax + b$, an affine cocycle $\varphi(t, \omega)x = \Phi(t, \omega)x + \psi(t, \omega)$ in $\mathbb{R}^d$. We now determine its generator and a more explicit form of the cocycle by means of the variation of constants formula.

(i) $\mathbb{T} = \mathbb{Z}$: The time-one mapping is $\varphi(\omega)x = A(\omega)x + b(\omega)$ and the cocycle φ is generated by the affine difference equation

$$x_{n+1} = \varphi(\theta^n\omega)x_n = A(\theta^n\omega)x_n + b(\theta^n\omega), \quad x_0 = x \in \mathbb{R}^d,$$

the solution of which is (see exercise 2.1.2)

$$\varphi(n, \omega)x = \begin{cases} \Phi(n, \omega)\left(x + \sum_{j=0}^{n-1} \Phi(j+1, \omega)^{-1}b(\theta^j\omega)\right), & n \geq 1, \\ x, & n = 0, \\ \Phi(n, \omega)\left(x - \sum_{j=n}^{-1} \Phi(j+1, \omega)^{-1}b(\theta^j\omega)\right), & n \leq -1, \end{cases} \tag{5.6.4}$$

where Φ is the linear cocycle generated by $x_{n+1} = A(\theta^n\omega)x_n$.

(ii) $\mathbb{T} = \mathbb{R}$ (RDE case): Every $A(d, \mathbb{R})$-valued cocycle which is absolutely continuous with respect to t induces an affine cocycle in $\mathbb{R}^d$ which is generated by an affine RDE

$$\dot{x}_t = A(\theta_t\omega)x_t + b(\theta_t\omega) .$$

If $A, b \in L^1(\mathbb{P})$ the solution is (see Example 2.2.8)

$$\varphi(t, \omega)x = \Phi(t, \omega)\left(x + \int_0^t \Phi(s, \omega)^{-1}b(\theta_s\omega)ds\right) = \int_0^t \Phi(t - s, \theta_s\omega)b(\theta_s\omega)ds, \tag{5.6.5}$$

where Φ is the linear cocycle generated by the RDE $\dot{x}_t = A(\theta_t\omega)x_t$.

(iii) $\mathbb{T} = \mathbb{R}$ (SDE case): Here the cocycle is generated by the affine SDE

$$dx_t = \sum_{j=0}^{m}(A_j x_t + b_j) \circ dW_t^j .$$

Its solution is (see Example 2.3.34)

$$\varphi(t,\cdot)x = \Phi(t,\cdot)\left(x + \sum_{j=0}^{m}\int_0^t \Phi(s,\cdot)^{-1}b_j \circ dW_s^j\right) .$$

5.6.2 Invariant Measure in the Hyperbolic Case

If a linear RDS is perturbed by additive noise, then the fixed point $x = 0$ should survive as a stationary solution. We will prove that this is indeed the case, provided the fixed point is hyperbolic.

This subsection is mainly based on Arnold and Crauel [13], Arnold and Xu [42] and for the white noise case on Arnold and Imkeller [19]. We first prove a general theorem for affine cocycles in $\mathbb{R}^d$ which do not necessarily have a generator. The notion of temperedness (see Definition 4.1.1) will prove to be crucial here.

5.6.1 Theorem (Invariant Measure for Hyperbolic Affine RDS).
(i) Let $\varphi(t,\omega)x = \Phi(t,\omega)x + \psi(t,\omega)$ be an affine RDS in $\mathbb{R}^d$ with two-sided time such that the linear part Φ satisfies the IC of the MET. Assume that Φ is hyperbolic, i.e. that all Lyapunov exponents are non-zero. Further assume that

$$\Psi^s(\omega) := \sup_{t\in\mathbb{T}^+}\|\pi^s(\omega)\psi(t,\theta(-t)\omega)\| < \infty \quad \mathbb{P}\text{-a. s.} \tag{5.6.6}$$

and

$$\Psi^u(\omega) := \sup_{t\in\mathbb{T}^-}\|\pi^u(\omega)\psi(t,\theta(-t)\omega)\| < \infty \quad \mathbb{P}\text{-a. s.} , \tag{5.6.7}$$

where $\pi^{s,u}(\omega) : \mathbb{R}^d \to E^{s,u}(\omega)$ is the projection onto $E^{s,u}(\omega)$ along $E^{u,s}(\omega)$, $E^{s,u}$ being the stable/unstable space of Φ.

Then there exists a unique φ-invariant measure μ. This measure is a random Dirac measure, $\mu_\omega = \delta_{\xi(\omega)}$, where $\xi(\omega)$ is given by

$$\xi(\omega) = \begin{pmatrix} \xi^s(\omega) \\ \xi^u(\omega) \end{pmatrix} = \begin{pmatrix} \lim_{t\to\infty} \text{ in pr } \pi^s(\omega)\psi(t,\theta(-t)\omega) \\ \lim_{t\to-\infty} \text{ in pr } \pi^u(\omega)\psi(t,\theta(-t)\omega) \end{pmatrix}$$

and $\xi^{s,u}(\omega) := \pi^{s,u}(\omega)\xi(\omega)$.
We have for all $x \in \mathbb{R}^d$

$$\xi^s(\omega) = \lim_{t\to\infty} \text{ in pr } \pi^s(\omega)\varphi(-t,\omega)^{-1}x \tag{5.6.8}$$

and

$$\xi^u(\omega) = \lim_{t\to\infty} \text{ in pr } \pi^u(\omega)\varphi(t,\omega)^{-1}x. \tag{5.6.9}$$

If the limits in the definition of ξ are $\mathbb{P}$-a.s., then the limits in (5.6.8) and (5.6.9) are also $\mathbb{P}$-a.s.

(ii) If, moreover, the random variables $\Psi^{s,u}$ are tempered from above[4], the limit in probability in the definition of ξ is a $\mathbb{P}$-a.s. limit and ξ is also tempered. Further, the stationary stochastic process $\xi(\theta(t)\omega)$ has continuous trajectories.

Proof. (i) (a) We first prove that the expressions in the definition of ξ make sense by proving that the sequences in question are Cauchy in probability.

We use the random norms of Sect. 4.3 and recall that for all $\varepsilon > 0$ there exists an ε-slowly varying function $B_\varepsilon : \Omega \to [1,\infty)$, i.e. satisfying

$$e^{-\varepsilon|t|}B_\varepsilon(\omega) \leq B_\varepsilon(\theta(t)\omega) \leq e^{\varepsilon|t|}B_\varepsilon(\omega)$$

such that

$$\frac{1}{B_\varepsilon(\omega)}\|\cdot\| \leq \|\cdot\|_{\kappa,\omega} \leq B_\varepsilon(\omega)\|\cdot\|.$$

Choose κ so that

$$\lambda_s := \max_{\lambda_i < 0} \lambda_i < -\kappa < 0 < \kappa < \min_{\lambda_i > 0} \lambda_i =: \lambda_u,$$

then choose $\beta \in (0, \min(-\lambda_s - \kappa, \lambda_u - \kappa))$, and finally choose $\varepsilon > 0$ such that $\beta - 2\varepsilon > 0$.

By (5.6.3),

$$\psi(t+\tau, \theta(-t-\tau)\omega) = \psi(t, \theta(-t)\omega) + \Phi(t, \theta(-t)\omega)\psi(\tau, \theta(-t-\tau)\omega),$$

hence

$$\pi^{s,u}(\omega)\psi(t+\tau, \theta(-t-\tau)\omega)$$
$$= \pi^{s,u}(\omega)\psi(t, \theta(-t)\omega) + \Phi(t, \theta(-t)\omega)\pi^{s,u}(\theta(-t)\omega)\psi(\tau, \theta(-t-\tau)\omega).$$

Thus for $t, \tau > 0$ and the stable part

$$\begin{aligned}
&\|\pi^s(\omega)\psi(t+\tau, \theta(-t-\tau)\omega) - \pi^s(\omega)\psi(t, \theta(-t)\omega)\| \\
&\leq \|\Phi(t, \theta(-t)\omega)\pi^s(\theta(-t)\omega)\psi(\tau, \theta(-t-\tau)\omega)\| \\
&\leq \|\Phi(t, \theta(-t)\omega)|_{E^s(\theta(-t)\omega)}\| \cdot \|\pi^s(\theta(-t)\omega)\psi(\tau, \theta(-t-\tau)\omega)\| \\
&\leq B_\varepsilon(\theta(-t)\omega)^2\|\Phi(t, \theta(-t)\omega)|_{E^s(\theta(-t)\omega)}\|_{\kappa,\theta(-t)\omega}\Psi^s(\theta(-t)\omega) \\
&\leq B_\varepsilon(\theta(-t)\omega)^2 e^{-\beta t}\Psi^s(\theta(-t)\omega) \\
&\leq e^{(-\beta+2\varepsilon)t}B_\varepsilon(\omega)^2\Psi^s(\theta(-t)\omega) \\
&\to 0 \text{ in probability as } t \to \infty,
\end{aligned}$$

where we have used that $\|\Phi(t,\omega)|_{E^s(\omega)}\|_{\kappa,\omega,\theta(t)\omega} \leq e^{-\beta t}$.

---

[4] We henceforth write "tempered" for "tempered from above".

Consequently, $\pi^s(\omega)\psi(t,\theta(-t)\omega)$ is Cauchy in probability, hence converges in probability as $t \to \infty$.

Similarly for the unstable component.

(b) Invariance: We now show that $\mu_\omega = \delta_{\xi(\omega)}$ is invariant, i.e. that for all $t \in \mathbb{T}$

$$\pi^{s,u}(\theta(t)\omega)\varphi(t,\omega)\xi(\omega) = \pi^{s,u}(\theta(t)\omega)\xi(\theta(t)\omega).$$

For the stable part

$$\begin{aligned}
&\pi^s(\theta(t)\omega)\varphi(t,\omega)\psi(\tau,\theta(-\tau)\omega) \\
&= \Phi(t,\omega)\pi^s(\omega)\psi(\tau,\theta(-\tau)\omega) + \pi^s(\theta(t)\omega)\psi(t,\omega) \\
&\to \Phi(t,\omega)\pi^s(\omega)\xi(\omega) + \pi^s(\theta(t)\omega)\psi(t,\omega) \text{ in probability as } \tau \to \infty \\
&= \pi^s(\theta(t)\omega)\varphi(t,\omega)\xi(\omega).
\end{aligned}$$

On the other hand, putting $\bar{\omega} = \theta(-\tau)\omega$,

$$\begin{aligned}
&\pi^s(\theta(t)\omega)\varphi(t,\omega)\psi(\tau,\theta(-\tau)\omega) \\
&= \Phi(t,\theta(\tau)\bar{\omega})\pi^s(\theta(\tau)\bar{\omega})\psi(\tau,\bar{\omega}) + \pi^s(\theta(t+\tau)\bar{\omega})\psi(t,\theta(\tau)\bar{\omega}) \\
&= \pi^s(\theta(t+\tau)\bar{\omega})\psi(t+\tau,\bar{\omega}) \\
&= \pi^s(\theta(t)\omega)\psi(t+\tau,\theta(-t-\tau)(\theta(t)\omega)) \\
&\to \pi^s(\theta(t)\omega)\xi(\theta(t)\omega) \text{ in probability as } \tau \to \infty.
\end{aligned}$$

Hence

$$\pi^s(\theta(t)\omega)\varphi(t,\omega)\xi(\omega) = \pi^s(\theta(t)\omega)\xi(\theta(t)\omega),$$

similarly for the unstable part.

(c) Uniqueness: We show that if μ is a φ-invariant measure then necessarily $\mu_\omega = \delta_{\xi(\omega)}$, $\mathbb{P}$-a.s.

Since a measure on a product space is a Dirac measure if and only if its marginals are Dirac measures it suffices to prove that

$$\pi^{s,u}(\omega)\mu_\omega = \delta_{\xi^{s,u}(\omega)}.$$

Let μ be φ-invariant and let $f \in \mathcal{C}_b(\mathbb{R}^d)$. Then

$$(\pi^{s,u}(\theta(t)\omega) \circ \varphi(t,\omega))\mu_\omega(f) = \pi^{s,u}(\theta(t)\omega)\mu_{\theta(t)\omega}(f) =: \zeta_f^{s,u}(\theta(t)\omega)$$

and

$$(\pi^{s,u}(\theta(t)\omega) \circ \varphi(t,\omega))\delta_{\xi(\omega)}(f) = \pi^{s,u}(\theta(t)\omega)\delta_{\xi(\theta(t)\omega)}(f) =: \eta_f^{s,u}(\theta(t)\omega).$$

Consequently,

$$\eta_f^{s,u}(\theta(t)\omega) - \zeta_f^{s,u}(\theta(t)\omega) \quad (5.6.10)$$

$$\begin{aligned}
&= f(\pi^{s,u}(\theta(t)\omega)\varphi(t,\omega)\xi(\omega)) - \int f(\pi^{s,u}(\theta(t)\omega)\varphi(t,\omega)x)d\mu_\omega(x) \\
&= \int (f(\pi^{s,u}(\theta(t)\omega)\varphi(t,\omega)\xi(\omega)) - f(\pi^{s,u}(\theta(t)\omega)\varphi(t,\omega)x))d\mu_\omega(x).
\end{aligned}$$

Due to the affine structure of φ,

$$\pi^{s,u}(\theta(t)\omega)\varphi(t,\omega)\xi(\omega) - \pi^{s,u}(\theta(t)\omega)\varphi(t,\omega)x = \Phi(t,\omega)\pi^{s,u}(\omega)(\xi(\omega) - x).$$

Since Φ is hyperbolic, the stable/unstable part of the right-hand side of the last equation goes to zero for all $x \in \mathbb{R}^d$ for $t \to \infty/t \to -\infty$ exponentially fast, $\mathbb{P}$-a. s.

Suppose f is uniformly continuous. Then, by the dominated convergence theorem, the last line of (5.6.10) converges to zero as $t \to \infty/t \to -\infty$, while the first line is a stationary stochastic process. Thus necessarily

$$\eta_f^{s,u}(\omega) - \zeta_f^{s,u}(\omega) = 0 \quad \mathbb{P}\text{-a. s.}$$

Since uniformly continuous $f \in \mathcal{C}_b(\mathbb{R}^d)$ suffice to identify a measure (see Gänssler and Stute [156: p. 66]), we obtain $\mu_\omega = \delta_{\xi(\omega)}$ $\mathbb{P}$-a. s.

(d) We now prove (5.6.8). Observe that

$$\begin{aligned}
\pi^s(\omega)\varphi(-t,\omega)^{-1}x &= \pi^s(\omega)\Phi(t,\theta(-t)\omega)(x - \psi(-t,\omega)) \\
&= \Phi(t,\theta(-t)\omega)\pi^s(\theta(-t)\omega)x + \pi^s(\omega)\psi(t,\theta(-t)\omega) \\
&= \Phi(t,\theta(-t)\omega)\pi^s(\theta(-t)\omega)x + \xi^s(\omega) \\
&\quad + (\pi^s(\omega)\psi(t,\theta(-t)\omega) - \xi^s(\omega)).
\end{aligned}$$

Now, using random norms and the fact that $\|\pi^s(\omega)\|_{\kappa,\omega,\omega} = 1$ and $\|\Phi(t,\omega)|_{E^s(\omega)}\|_{\kappa,\omega,\theta(t)\omega} \leq \exp(-\beta t)$,

$$\begin{aligned}
\|\Phi(t,\theta(-t)\omega)\pi^s(\theta(-t)\omega)x\| &\leq B_\varepsilon(\omega)e^{-\beta t}\|\pi^s(\theta(-t)\omega)x\|_{\kappa,\theta(-t)\omega} \\
&\leq B_\varepsilon(\omega)e^{-\beta t}B_\varepsilon(\theta(-t)\omega)\|x\| \\
&\leq B_\varepsilon(\omega)^2 e^{(-\beta+\varepsilon)t}\|x\| \\
&\to 0 \text{ exponentially fast as } t \to \infty, \mathbb{P}\text{-a. s.}
\end{aligned}$$

Hence (5.6.8) follows, with mode of convergence the same as in the definition of ξ.

Similarly for the unstable component.

(ii) Step (i)(a) of the proof shows that, under our assumptions, $\pi^{s,u}(\omega)\psi(t,\theta(-t)\omega)$ is Cauchy $\mathbb{P}$-a. s. exponentially fast, hence converges $\mathbb{P}$-a. s. exponentially fast. Moreover, for each $\tau > 0$ and the stable part,

$$\begin{aligned}
\|\xi^s(\omega)\| &\leq \|\pi^s(\omega)\psi(\tau,\theta(-\tau)\omega) - \xi^s(\omega)\| + \|\pi^s(\omega)\psi(\tau,\theta(-\tau)\omega)\| \\
&\leq e^{(-\beta+2\varepsilon)\tau}B_\varepsilon(\omega)^2\Psi^s(\theta(-\tau)\omega) + \Psi^s(\omega).
\end{aligned}$$

Hence ξ^s, similarly ξ^u, and thus ξ is tempered.

Finally, $\xi(\theta(t)\omega) = \varphi(t,\omega)\xi(\omega)$ has continuous trajectories since $t \mapsto \varphi(t,\omega)x$ is continuous by definition. □

5.6.2 Remark (Invariant Measures for Nonlinear RDS).
Theorem 5.6.1 in combination with the Banach fixed point theorem can be used to establish the existence and uniqueness of an invariant measure $\mu_\omega = \delta_{\xi(\omega)}$ for the RDS
$$\varphi(t,\omega)x = \Phi(t,\omega)x + \psi(t,\omega) + N(t,\omega,x),$$
where Φ is hyperbolic, ψ satisfies assumptions similar to (5.6.6) and (5.6.7), but now with respect to the random norm, and the nonlinear term N fulfills a Lipschitz condition
$$\sup_{-1\leq t\leq 1} \|N(t,\omega,x) - N(t,\omega,y)\|_{\kappa,\theta(t)\omega} \leq L\|x-y\|_{\kappa,\omega},$$
with L small enough, see Arnold and Xu [42: Theorem 5.1] and (for a local version) Arnold and Boxler [11]. ∎

5.6.3 Corollary (Invariant Measure for Linear RDS).
Let Φ be a linear cocycle with two-sided time satisfying the IC of the MET. If Φ is hyperbolic, then its unique invariant measure is δ_0, the Dirac measure at 0.

5.6.4 Corollary (Spectrum, Splitting and Invariant Manifolds).
The splitting and spectrum of $T\varphi(t,\omega)$ under μ_ω are the same as for Φ, hence $\mathbb{P}$-a. s.
$$E^i(\omega, \xi(\omega)) = (\xi(\omega), E^i(\omega)).$$
Thus for the stable and unstable manifolds of φ under μ_ω
$$M^{s,u}(\omega, \xi(\omega)) = \xi(\omega) + E^{s,u}(\omega)$$
and the invariant foliation of $\mathbb{R}^d$ is $M^{s,u}(\omega, x) = x + E^{s,u}(\omega)$.

In particular, for $\lambda_1 < 0$
$$M^s(\omega, \xi(\omega)) = \mathbb{R}^d, \quad M^u(\omega, \xi(\omega)) = \{\xi(\omega)\}.$$

Proof. For $\varphi(t,\omega)x = \Phi(t,\omega)x + \psi(t,\omega)$,
$$T\varphi(t,\omega,x)v = \Phi(t,\omega)v,$$
hence spectrum and splitting of $T\varphi$ under μ are the same as for Φ, so for μ_ω-almost all $x \in \mathbb{R}^d$, i.e. for $x = \xi(\omega)$,
$$E^i(\omega, x) = (x, E^i(\omega)).$$
If we define for any $x \in \mathbb{R}^d$
$$M^{s,u}(\omega, x) := \{y \in \mathbb{R}^d : \|\varphi(t,\omega)y - \varphi(t,\omega)x\| \to 0 \text{ as } t \to \pm\infty\},$$
then $y \in M^{s,u}(\omega, x)$ if and only if $\varphi(t,\omega)y - \varphi(t,\omega)x = \Phi(t,\omega)(y-x) \to 0$ ($t \to \pm\infty$) if and only if $y - x \in E^{s,u}(\omega)$ or $y \in x + E^{s,u}(\omega)$. For $x = \xi(\omega)$ we obtain the stable/unstable manifold of φ under μ and for arbitrary x we obtain the invariant foliation of the state space (see subsection 7.4.1). □

We now deal with the case where φ has a generator.

5.6.5 Theorem (Invariant Measure for Affine RDS with Generator). *(i)* $\mathbb{T} = \mathbb{Z}$: *Suppose φ is generated by the affine random difference equation*
$$x_{n+1} = A(\theta^n \omega) x_n + b(\theta^n \omega)$$
in $\mathbb{R}^d$. Assume
$$\log^+ \|A^{\pm 1}\| \in L^1(\mathbb{P}), \quad \log^+ \|b\| \in L^1(\mathbb{P})$$
(the integrability condition for b can be replaced by the assumption that b is tempered). If A generates a hyperbolic linear cocycle Φ, then the unique φ-invariant measure is concentrated at
$$\xi(\omega) = \begin{pmatrix} \xi^s(\omega) \\ \xi^u(\omega) \end{pmatrix} = \begin{pmatrix} \sum_{n=-\infty}^{-1} \Phi(n+1,\omega)^{-1} \pi^s(\theta^{n+1}\omega) b(\theta^n \omega) \\ -\sum_{n=0}^{\infty} \Phi(n+1,\omega)^{-1} \pi^u(\theta^{n+1}\omega) b(\theta^n \omega) \end{pmatrix}.$$

Moreover, ξ is tempered.

(ii) $\mathbb{T} = \mathbb{R}$, *real noise case: Suppose φ is generated by the affine RDE*
$$\dot{x}_t = A(\theta_t \omega) x_t + b(\theta_t \omega)$$
in $\mathbb{R}^d$, where $A, b \in L^1(\mathbb{P})$ (the integrability condition for b can be replaced by the assumption that b is tempered). If A generates a hyperbolic linear cocycle Φ, then the unique φ-invariant measure is concentrated at
$$\xi(\omega) = \begin{pmatrix} \xi^s(\omega) \\ \xi^u(\omega) \end{pmatrix} = \begin{pmatrix} \int_{-\infty}^{0} \Phi(t,\omega)^{-1} \pi^s(\theta_t \omega) b(\theta_t \omega) dt \\ -\int_{0}^{\infty} \Phi(t,\omega)^{-1} \pi^u(\theta_t \omega) b(\theta_t \omega) dt \end{pmatrix}.$$

Moreover, ξ is tempered and the stationary stochastic process $\xi(\theta_t \omega)$ has absolutely continuous trajectories.

(iii) $\mathbb{T} = \mathbb{R}$, *white noise case: Suppose φ is generated by the affine SDE*
$$dx_t = \sum_{j=0}^{m} (A_j x_t + b_j) \circ dW_t^j$$
in $\mathbb{R}^d$ (no IC needed). If $dx_t = \sum_{j=0}^{m} A_j x_t \circ dW^j$ generates a hyperbolic linear cocycle Φ, then the unique φ-invariant measure is concentrated at
$$\xi(\cdot) = \begin{pmatrix} \xi^s(\cdot) \\ \xi^u(\cdot) \end{pmatrix} = \begin{pmatrix} \sum_{j=0}^{m} \int_{-\infty}^{0} \Phi(t,\cdot)^{-1} (\pi^s(\theta_t \cdot) b_j) \circ dW_t^j \\ -\sum_{j=0}^{m} \int_{0}^{\infty} \Phi(t,\cdot)^{-1} (\pi^u(\theta_t \cdot) b_j) \circ dW_t^j \end{pmatrix}.$$

The integrals exist as limits $\mathbb{P}$-a. s. of the non-adapted Stratonovich integrals $\int_{-T}^{0}$ and $\int_{0}^{T}$ as $T \to \infty$. Moreover, ξ is tempered and the stationary stochastic process $\xi(\theta_t \omega)$ has continuous trajectories.

5.6 Affine RDS

Proof. (i) $\mathbb{T} = \mathbb{Z}$: By (5.6.4) and the cocycle property, we obtain for $n \geq 1$

$$\pi^s(\omega)\psi(n,\theta^{-n}\omega) = \sum_{k=0}^{n-1} \Phi(k,\theta^{-k}\omega)\pi^s(\theta^{-k}\omega)b(\theta^{-k-1}\omega).$$

Using random norms as in the proof of Theorem 5.6.1, we obtain

$$\Psi^s(\omega) = \sup_{n \geq 1} \|\pi^s(\omega)\psi(n,\theta^{-n}\omega)\|$$

$$\leq \sum_{k=0}^{\infty} \|\Phi(k,\theta^{-k}\omega)\pi^s(\theta^{-k}\omega)b(\theta^{-k-1}\omega)\|$$

$$\leq B_\varepsilon(\omega)^2 \sum_{k=0}^{\infty} e^{(-\beta+\varepsilon)k} \|b(\theta^{-k-1}\omega)\| < \infty$$

since b is tempered, or by Lemma 3.4.3(i) if $\log^+ \|b\| \in L^1(\mathbb{P})$. Hence the expression for ξ^s makes sense. Further, Ψ^s is tempered, hence ξ^s is also tempered and

$$\pi^s(\omega)\psi(n,\theta^{-n}\omega) \to \xi^s(\omega) \quad \mathbb{P}\text{-a.s. exponentially fast}$$

as $n \to \infty$.

Similarly for Ψ^u and ξ^u.

(ii) $\mathbb{T} = \mathbb{R}$, real noise case: By (5.6.5),

$$\psi(t,\theta_{-t}\omega) = \int_0^t \Phi(u,\theta_{-u}\omega)b(\theta_{-u}\omega)\,du.$$

Using again random norms,

$$\Psi^s(\omega) = \sup_{t \geq 0} \|\pi^s(\omega)\psi(t,\theta_{-t}\omega)\|$$

$$\leq \int_0^\infty \|\Phi(t,\theta_{-t}\omega)\pi(\theta_{-t}\omega)b(\theta_{-t}\omega)\|dt$$

$$\leq B_\varepsilon(\omega)^2 \int_0^\infty e^{(-\beta+\varepsilon)t} \|b(\theta_{-t}\omega)\|dt < \infty$$

since b is tempered, or by Lemma 3.4.17 if $b \in L^1(\mathbb{P})$. Hence the expression for ξ^s makes sense. Further, Ψ^s is tempered since

$$\Psi^s(\theta_t\omega) \leq B_\varepsilon(\omega)^2 e^{(\beta+\varepsilon)t} \int_t^\infty e^{(-\beta+\varepsilon)u} \|b(\theta_{-u}\omega)\|\,du$$

and b is tempered, or again by Lemma 3.4.17 if $b \in L^1$. Thus ξ^s is tempered and

$$\pi^s(\omega)\psi(t,\theta_{-t}\omega) \to \xi^s(\omega) \quad \mathbb{P}\text{-a.s. exponentially fast}$$

as $t \to \infty$.

Similarly for Ψ^u and ξ^u.

(iii) For the SDE case the problems steming from the non-adaptedness of ξ require new techniques which will be developed in Sect. 8.5. See Theorem 8.5.15. □

5.6.6 Corollary (Stable Case). *If $\lambda_1 < 0$ then*
(i)
$$\xi(\omega) = \begin{cases} \sum_{n=-\infty}^{-1} \Phi(n+1,\omega)^{-1} b(\theta^n \omega), & \mathbb{T} = \mathbb{Z}, \\ \int_{-\infty}^{0} \Phi(t,\omega)^{-1} b(\theta_t \omega)\, dt, & \mathbb{T} = \mathbb{R} \ (RDE), \\ \sum_{j=0}^{m} \int_{-\infty}^{0} \Phi(t,\omega)^{-1} b_j \circ dW_t^j(\omega), & \mathbb{T} = \mathbb{R} \ (SDE). \end{cases}$$

In particular, $\mu_\omega = \delta_{\xi(\omega)}$ is a Markov measure.
(ii) For any $x \in \mathbb{R}^d$ and $t \to \infty$

$$\varphi(-t,\omega)^{-1} x \to \xi(\omega), \quad \varphi(t,\omega)x - \xi(\theta(t)\omega) \to 0 \quad \mathbb{P}\text{-a. s.} ,$$

both exponentially fast.
(iii) For any probability measure ρ on $\mathbb{R}^d$

$$\lim_{t \to \infty} \varphi(-t,\omega)^{-1} \rho = \delta_{\xi(\omega)} \quad \text{weakly} \quad \mathbb{P}\text{-a. s.}$$

and for all $x \in \mathbb{R}^d$

$$\lim_{t \to \infty} \mathcal{L}(\varphi(t,\cdot)x) = \mathcal{L}(\xi),$$

where $\mathcal{L}(\xi)$ denotes the law of the random variable ξ.

Proof. As (i) and (ii) are clear, we only prove (iii). For $f \in \mathcal{C}_b(\mathbb{R}^d)$ and by the fact that $\varphi(-t,\omega)^{-1} x \to \xi(\omega)$ $\mathbb{P}$-a. s.

$$\begin{aligned} \varphi(-t,\omega)^{-1} \rho(f) &= \int f(\varphi(-t,\omega)^{-1} x) \rho(dx) \\ &\to \int f(\xi(\omega)) \rho(dx) = \delta_{\xi(\omega)}(f). \end{aligned}$$

Furthermore,

$$\mathcal{L}(\varphi(t,\cdot)x) = \mathcal{L}(\varphi(t,\theta(-t)\cdot)x) = \mathcal{L}(\varphi(-t,\cdot)^{-1} x) \to \mathcal{L}(\xi),$$

by the above for $\rho = \delta_x$. □

5.6.7 Remark (Invariant Measures in the Non-Hyperbolic Case: Cohomological Equations, and Coboundaries). If an affine cocycle φ has a linear part Φ which is not hyperbolic, then φ has an invariant measure only under rare circumstances. For a complete treatment of the case of a non-random A for $\mathbb{T} = \mathbb{Z}$ and $\mathbb{T} = \mathbb{R}$ (RDE) see Arnold and Wihstutz [38].

To find a φ-invariant Dirac measure $\mu_\omega = \delta_{\xi(\omega)}$ amounts to looking for a random variable ξ for which $\varphi(t,\omega)\xi(\omega) = \xi(\theta(t)\omega)$, or solving the equation

$$\xi(\theta(t)\omega) - \Phi(t,\omega)\xi(\omega) = \psi(t,\omega). \tag{5.6.11}$$

ψ is then called a *coboundary* with respect to the cocycle Φ (or *twisted* by Φ, see Katok and Hasselblatt [200: p. 100]) if there exists a random variable ξ such that the *cohomological equation* (5.6.11) is satisfied.

For $\Phi \equiv I$, we obtain the definition of a classical additive coboundary, i.e. $\psi(t,\omega)$ is a helix which is cohomologous to the trivial helix 0, and two solutions differ only by a constant if θ is ergodic.

For the generator cases, the cohomological equation can be equivalently written in the "differential form"

$$L\xi = b, \tag{5.6.12}$$

where L is the *cohomological operator*. More precisely,
(i) for $\mathbb{T} = \mathbb{Z}$, $L\xi(\omega) = \xi(\theta\omega) - A(\omega)\xi(\omega) = b(\omega)$, i.e. with $Uf(\omega) := f(\theta\omega)$, $L = U - A$,
(ii) for $\mathbb{T} = \mathbb{R}$ (RDE), $L\xi(\theta_t\omega) = \frac{d}{dt}\xi(\theta_t\omega) - A(\theta_t\omega)\xi(\theta_t\omega) = b(\theta_t\omega)$, i.e. $L = \frac{d}{dt} - A(\theta_t\omega)$,
(iii) for $\mathbb{T} = \mathbb{R}$ (SDE),

$$dL(\xi) = d\xi - \sum_{j=0}^{m} A_j \xi \circ dW^j = \sum_{j=0}^{m} b_j \circ dW^j,$$

i.e.

$$dL(\cdot) = d(\cdot) - \sum_{j=0}^{m} A_j(\cdot) \circ dW^j.$$

We will make use of this in Chap. 8. ∎

5.6.3 Time Reversibility and Iterated Function Systems

We now compare our results on invariant measures for affine cocycles with those of Barnsley and Elton [50] and Elton [136] on stationary measures for iterated function systems.

Let φ be an affine RDS with discrete time $\mathbb{T} = \mathbb{Z}$ and time-one mapping $\varphi(\omega)x := \varphi(1,\omega)x = A(\omega)x + b(\omega)$.

Under the assumptions $\log^+ \|A^{\pm 1}\| \in L^1(\mathbb{P})$, $\log^+ \|b\| \in L^1(\mathbb{P})$ and $\lambda_1 < 0$ we have constructed in Corollary 5.6.6 the unique invariant (Markov) measure $\delta_{\xi(\omega)}$, and for any $x \in \mathbb{R}^d$, $\mathbb{P}$-a.s.

$$\xi(\omega) = \lim_{n \to \infty} \varphi(-n,\omega)^{-1} x = \lim_{n \to \infty} \varphi(\theta^{-1}\omega) \circ \cdots \circ \varphi(\theta^{-n}\omega)x,$$

i.e. we recover ξ by moving any x from time $-n$ forward to time 0 and letting $n \to \infty$.

We also have

$$\rho := \mathcal{L}(\xi) = \lim_{n \to \infty} \mathcal{L}(\varphi(n,\cdot)x) = \lim_{n \to \infty} \mathcal{L}(\varphi(\theta^{n-1}\cdot) \circ \cdots \circ \varphi(\cdot)x),$$

i.e. we recover the law of ξ by moving any x from time 0 forward to time n and letting $n \to \infty$. The orbit $\varphi(n,\omega)x$ itself does not converge, but approaches the stationary process $\xi(\theta^n\omega)$ exponentially fast.

Barnsley and Elton [50] constructed ρ in the i.i.d. case and for $\lambda_1 < 0$ by showing that $\rho = \mathcal{L}(\bar\xi)$, where for all $x \in \mathbb{R}^d$ and $\mathbb{P}$-a. s.

$$\bar\xi(\omega) := \lim_{n \to \infty} \varphi(\omega) \circ \cdots \circ \varphi(\theta^{n-1}\omega)x,$$

i.e. by somehow going from time $n > 0$ backward to time 0 and letting $n \to \infty$.

Since obviously $\xi \neq \bar\xi$ (ξ is measurable with respect to the past, while $\bar\xi$ is measurable with respect to the future of φ – in the i.i.d. case ξ and $\bar\xi$ are independent), the question arises why $\mathcal{L}(\xi) = \mathcal{L}(\bar\xi)$.

The general answer is given in the next theorem. For the discrete time case and the continuous time (RDE) case we write $\varphi_{\theta,A,b}$ for the affine cocycle generated by θ, A and b.

5.6.8 Theorem (Invariant Measure for Time Reversible Case).
Assume the situation of Theorem 5.6.5 and let for $\mathbb{T} = \mathbb{Z}$ or $\mathbb{T} = \mathbb{R}$ (RDE case) $(A(\theta(t)\cdot), b(\theta(t)\cdot))_{t \in \mathbb{T}}$ be time reversible, i. e.

$$\mathcal{L}((A(\theta(t)\cdot), b(\theta(t)\cdot))_{t \in \mathbb{T}}) = \mathcal{L}((A(\theta(-t)\cdot), b(\theta(-t)\cdot))_{t \in \mathbb{T}}).$$

Then

$$\mathcal{L}\left((\varphi_{\theta,A,b}(-t,\cdot)^{-1})_{t \geq 0}\right) = \mathcal{L}\left((\varphi_{\theta^{-1},A,b}(t,\theta(t)\cdot))_{t \geq 0}\right). \tag{5.6.13}$$

For $\mathbb{T} = \mathbb{R}$ (SDE case) the analogue of (5.6.13) always holds.

In particular, the $\mathbb{P}$-a. s. limit

$$\xi(\omega) := \lim_{t \to \infty} \varphi_{\theta,A,b}(-t,\omega)^{-1}x$$

exists if and only if the $\mathbb{P}$-a. s. limit

$$\bar\xi(\omega) := \lim_{t \to \infty} \varphi_{\theta^{-1},A,b}(t,\theta(t)\omega)x$$

exists, in case of which $\mathcal{L}(\xi) = \mathcal{L}(\bar\xi)$.

If $\lambda_1 < 0$ then both ξ and $\bar\xi$ exist, and are independent of x.

Proof. We may – and will – restrict ourselves to the case $\mathbb{T} = \mathbb{Z}$.

We have for $n > 0$

$$\varphi_{\theta,A,b}(-n,\omega)^{-1} = \varphi_{\theta,A,b}(n,\theta^{-n}\omega) = \varphi(\theta^{-1}\omega) \circ \cdots \circ \varphi(\theta^{-n}\omega).$$

Time reversibility implies

$$\mathcal{L}((\varphi_{\theta,A,b}(-n,\cdot)^{-1})_{n \geq 0}) = \mathcal{L}((\varphi(\theta\cdot) \circ \cdots \circ \varphi(\theta^n\cdot))_{n \geq 0}).$$

Now

$$\varphi(\theta\omega) \circ \cdots \circ \varphi(\theta^n\omega) = \varphi_{\theta^{-1},A,b}(n,\theta^n\omega),$$

from which (5.6.13) follows. In the SDE case note that $\mathcal{L}(-W(-\cdot)) = \mathcal{L}(W)$.

If all finite-dimensional distributions of two stochastic processes are the same, then one of them converges $\mathbb{P}$-a.s. if and only if the other one does so. Further, the laws of the limiting random variables coincide. □

5.6.9 Remark. (i) Note that if $(A(\theta(t)\cdot), b(\theta(t)\cdot))_{t \in \mathbb{T}}$ is time-reversible, then so is $(A(\theta(t)\cdot))_{t \in \mathbb{T}}$, hence $\mathcal{S}(\theta, A) = \mathcal{S}(\theta^{-1}, A)$ by Proposition 5.1.3.

The discrete time i.i.d. case is always time-reversible.

(ii) Assume $\lambda_1 < 0$. For the case $\mathbb{T} = \mathbb{Z}$ (i.i.d.), or for the case $\mathbb{T} = \mathbb{R}$ (SDE), $\mathcal{L}(\xi) = \mathbb{E}\,\delta_{\xi(\cdot)} = \rho^+$ is the unique stationary measure of the corresponding one-point motions $(\varphi(t,\omega)x)_{t \geq 0}$ (see Subsect. 2.3.1 for discrete time and Subsect. 2.3.9 for continuous time), i.e. the unique solution of

$$\rho^+(\cdot) = \int_{\mathbb{R}^d} P_t^+(x,\cdot)\rho^+(dx), \quad P_t^+(x,B) := \mathbb{P}\{\omega : \varphi(t,\omega)x \in B\}.$$

For $\mathbb{T} = \mathbb{R}$, this is equivalent to ρ^+ being the unique solution of the Fokker-Planck equation $(L^+)^*\rho^+ = 0$, L^+ being the generator of $\varphi(t,\omega)x$.

For $\lambda_p > 0$ we have a similar situation, with ρ^- now being the unique stationary measure of the one-point motions $(\varphi(-t,\omega)x)_{t \geq 0}$, i.e. the unique solution of

$$\rho^-(\cdot) = \int_{\mathbb{R}^d} P_t^-(x,\cdot)\rho^-(dx),$$

where

$$P_t^-(x,B) := \mathbb{P}\{\varphi(-t,\omega)x \in B\} = \mathbb{P}\{\varphi(t,\omega)^{-1}x \in B\}.$$

For $\mathbb{T} = \mathbb{R}$, this is equivalent to ρ^- being the unique solution of the Fokker-Planck equation $(L^-)^*\rho^- = 0$, L^- being the generator of $\varphi(-t,\omega)x$. ∎

Chapter 6. RDS on Homogeneous Spaces of the General Linear Group

Summary

Following the pioneering work of Furstenberg [153], the study of the asymptotic behavior (law of large numbers, positivity of top Lyapunov exponent alias Furstenberg constant, simplicity of the Lyapunov spectrum, central limit theorem, large deviations principle, etc.) of products of i.i.d. random matrices (more generally, products of i.i.d. random variables in a Lie group) became an important research area. This development was highlighted in 1985 by Guivarc'h and Raugi's profound paper [163] and the book [77] by Bougerol and Lacroix.

Furstenberg made the following basic observation: In order to determine the asymptotic behavior of the product $\Phi_n = A_n \cdots A_1$ of group-valued i.i.d. random variables (A_n), one has to let the group act on certain manifolds M (Furstenberg boundaries – in the matrix case Grassmannian and flag manifolds), $(g, x) \mapsto gx$, and study e. g. the stationary measures of the transition probability $P(x, B) = \mathbb{P}\{g : gx \in B\}$ of the Markov chain $x_n = \Phi_n x_0$ on M. Furstenberg's approach proved to be extremely fertile and practically every author investigating products of random matrices has made use of it.

One of the lasting outcomes was Furstenberg's formula for the top Lyapunov exponent as an integral over projective space [153: Formula (7.5)]. This formula was independently found by Khasminskii [205, 206] for linear SDE.

As explained in the Preface, we decided to omit the subject – except for the study of the action of linear cocycles on certain homogeneous spaces, thus permitting us in particular to derive Furstenberg-Khasminskii formulas for all Lyapunov exponents.

More specifically, in this chapter we will study some nonlinear RDS which are induced by a $Gl(d, \mathbb{R})$-valued cocycle Φ on homogeneous spaces of $Gl(d, \mathbb{R})$.

The most interesting of these homogeneous spaces for us are: the unit sphere S^{d-1}, the projective space P^{d-1}, the Grassmannian manifold $G_k(d)$ of k-planes, the Grassmannian manifold $G_k^+(d)$ of oriented k-planes, the Stiefel manifold $St_k(d)$ of orthonormal k-frames (the action of $Gl(d, \mathbb{R})$ is followed by orthonormalization) and the flag manifolds $F_\tau(d)$, where $\tau = (d_1, \ldots, d_r)$, $1 \leq d_1 < \ldots < d_r \leq d$, is a multi-index of dimensions d_i of subspaces V_i and $f = (V_1 \subset V_2 \subset \ldots \subset V_r)$ is a generic element of $F_\tau(d)$.

We also need to study RDS on a principal bundle and investigate under which conditions it induces an RDS on the base manifold. In case that the original RDS has a generator, we determine the generator of the induced system. The cases of particular interest for us are the principal bundles $St_2(d) \stackrel{\mathbb{Z}_2^2}{\to} F_{1,2}(d)$, $St_k(d) \stackrel{O(k,\mathbb{R})}{\to} G_k(d)$ and $St_k(d) \stackrel{SO(k,\mathbb{R})}{\to} G_k^+(d)$ which are the basis of our theory of rotation numbers in Sect. 6.5.

In Sect. 6.1 we collect some "abstract nonsense" about group-valued cocycles and cocycles on principal bundles.

Sect. 6.2 is devoted to the dynamics and ergodic theory of the RDS induced on S^{d-1} and P^{d-1} by a linear RDS Φ. We determine all invariant measures (Theorem 6.2.3) and their spectrum and splitting (Theorem 6.2.20). On this basis we obtain Furstenberg-Khasminskii formulas (i.e. phase averages over S^{d-1}) for all Lyapunov exponents of Φ (Theorem 6.2.8 for the real noise case and Theorem 6.2.14 for the white noise case). The white noise formulas are derived by anticipative calculus.

In order to obtain Furstenberg-Khasminskii formulas for sums of Lyapunov exponents (Theorem 6.3.3) we have to study the RDS induced by Φ on Grassmannian manifolds (see Sect. 6.3).

Manifold versions of the above are treated in Sect. 6.4.

Finally, Sect. 6.5 is devoted to our concept of rotation numbers, a simultaneous generalization of the two-dimensional concept and the notion of "imaginary part of eigenvalue of a matrix" to d dimensions and to nonlinear RDS. We prove an MET for rotation numbers which states that if the Lyapunov spectrum is simple, then every two-plane has a rotation number ρ, and this ρ is taken from a finite collection of basic rotation numbers ρ_{ij} realized in the canonical planes $E_i \wedge E_j$ (Theorem 6.5.14 for the real noise case and Theorem 6.5.16 for the white noise case).

6.1 Cocycles on Lie Groups and Their Homogeneous Spaces

A continuous or C^k cocycle with two-sided time on a state space X defines a group-valued cocycle on $(\text{Homeo}(X), \circ)$ or $(\text{Diff}^k(X), \circ)$, where "$\circ$" means composition and where those groups act naturally on X on the left (see Sect. 1.1).

Conversely, if G is a group, then a G-valued cocycle induces a cocycle on any space X on which G acts on the left. We are particularly interested in the case where G is a Lie group and $X = G/H$ is a homogeneous space, for which case we will determine the generators of the induced cocycles.

6.1.1 Group-Valued Cocycles and Their Generators

6.1.1 Definition (Group-Valued Cocycle). Let $(\Omega, \mathcal{F}, \mathbb{P}, (\theta(t))_{t \in \mathbb{T}})$ be a metric DS with two-sided time. Let $(G, \cdot)$ be a measurable group. A measurable function
$$\varphi : \mathbb{T} \times \Omega \to G$$
is called a *G-valued cocycle* if it satisfies[1]
$$\varphi(t + s, \omega) = \varphi(t, \theta(s)\omega) \cdot \varphi(s, \omega). \tag{6.1.1}$$
φ is called *continuous* if G is a topological group and $t \mapsto \varphi(t, \omega)$ is continuous. ∎

It follows that $\varphi(0, \omega) = e$, the unit element of G, and $\varphi(t, \omega)^{-1} = \varphi(-t, \theta(t)\omega)$.

The study of θ through its G-valued cocycles and their cohomology invariants has become known as *Mackey's program* in algebraic ergodic theory (see Schmidt [312, 313]).

Generators

As all G-valued cocycles with discrete time are obviously generated by their time-one mapping, we consider the case $\mathbb{T} = \mathbb{R}$.

Let G be a Lie group and let $\mathfrak{g}$ be its Lie algebra. A vector field X on G is called *right-invariant* if $R_{g*}X = X$ (for R_g see Definition 6.1.4).

The following reasoning is completely analogous to the one in Sect. 2.2: Let
$$A : \Omega \to T_e G \cong \mathfrak{g}.$$
be a $\mathfrak{g}$-valued random variable. Then there exists a unique right-invariant random vector field $X(\omega)$ defined by
$$X(\omega, g) := R_{g*}A(\omega) \in T_g G,$$
giving the corresponding right-invariant RDE on G
$$\dot{\varphi}(t, \omega) = X(\theta_t \omega, \varphi(t, \omega)) = R_{\varphi(t, \omega)*}A(\theta_t \omega), \quad \varphi(0, \omega) = e. \tag{6.1.2}$$
If $A(\theta.\omega) \in L_{\text{loc}}(\mathbb{R}, \mathfrak{g})$, then the RDE (6.1.2) uniquely generates a global G-valued cocycle (use for example Kobayashi and Nomizu [217: p. 69]).

Conversely, for a G-valued cocycle which is absolutely continuous with respect to t there exists $A(\omega) := \dot{\varphi}(0, \omega)$ (if φ is C^1) such that φ solves the right-invariant RDE (6.1.2).

We have thus obtained the following statements.

---
[1] On the group level it would be equally natural to write the factors in (6.1.1) the other way around. Only when G acts on a space X on the left, or on the right, is the symmetry broken. In the deterministic case (no ω) and in the random case if G is Abelian, the two notions coincide.

6.1.2 Theorem (Cocycles Through Right-Invariant RDE). *There is a basically one-to-one correspondence between*
  (i) G-valued cocycles which are absolutely continuous with respect to t,
  (ii) right-invariant RDE on G of the form (6.1.2).

6.1.3 Example. (i) $G = Gl(d, \mathbb{R})$, $\mathfrak{g} = \mathbb{R}^{d \times d}$. Here $R_{g*}A = Ag$, hence G-valued cocycles are generated by right-invariant RDE

$$\dot{\Phi}(t, \omega) = A(\theta_t \omega)\Phi(t, \omega), \quad \Phi(0, \omega) = I.$$

(ii) $G = \mathbb{R}^d$, $\mathfrak{g} = \mathbb{R}^d$. G is Abelian. A G-cocycle (also called a *helix* in the additive case) satisfies

$$\varphi(t + s, \omega) = \varphi(t, \theta(s)\omega) + \varphi(s, \omega).$$

A right-invariant RDE on $\mathbb{R}^d$ has the form

$$\dot{\varphi}(t, \omega) = \varphi(t, \omega) + b(\theta_t \omega), \quad \varphi(0, \omega) = 0,$$

where $b : \Omega \to \mathbb{R}^d$ satisfies $b(\theta.\omega) \in L_{\text{loc}}(\mathbb{R}, \mathbb{R}^d)$, with solution

$$\varphi(t, \omega) = \int_0^t b(\theta_s \omega) ds.$$

This is the most general absolutely continuous $\mathbb{R}^d$-valued cocycle.

(iii) $G = A(d, \mathbb{R})$, the group of affine transformations of $\mathbb{R}^d$. This case is treated in detail in Sect. 5.6. Examples (i) and (ii) above are particular cases of (iii). ∎

In the white noise case choose $A_0, \ldots, A_m \in \mathfrak{g}$. Consider the helix

$$F(t, \omega) = tA_0 + \sum_{j=1}^{m} W_t^j(\omega) A_j,$$

which is Brownian motion in $\mathfrak{g}$. Then the right-invariant SDE generating a G-valued cocycle reads

$$d\varphi(t) = R_{\varphi(t)*} \circ dF(t) = \sum_{j=0}^{m} R_{\varphi(t)*} A_j \circ dW_t^j, \quad \varphi(0) = e. \quad (6.1.3)$$

6.1.2 Cocycles Induced by Actions

6.1.4 Definition (Action of a Group on a Space). Let $(G, \cdot)$ be a measurable group, and $(X, \mathcal{B})$ be a measurable space. A measurable mapping $\gamma : G \times X \to X$, $(g, x) \mapsto \gamma_g(x)$ such that $x \mapsto \gamma_g(x)$ is a bimeasurable bijection of X is called a *left action* (G acts on X on the left) if

$$\gamma_{g_1 \cdot g_2} = \gamma_{g_1} \circ \gamma_{g_2},$$

and a *right action* (G acts on X on the right) if

$$\gamma_{g_1 \cdot g_2} = \gamma_{g_2} \circ \gamma_{g_1}.$$

∎

The group G acts naturally on itself, i.e. $X = G$,
  (i) on the left by $(g, x) \mapsto \gamma_g(x) = g \cdot x =: L_g(x)$, since $L_{g_1 \cdot g_2} = L_{g_1} \circ L_{g_2}$,
  (ii) on the right by $(g, x) \mapsto \gamma_g(x) = x \cdot g =: R_g(x)$, since $R_{g_1 \cdot g_2} = R_{g_2} \circ R_{g_1}$.

We write $X = G_L$ or G_R for the state space $X = G$ with the left or right action of G.

Combining a cocycle with a left or right action, the symmetry is broken as follows.

6.1.5 Lemma. *Let φ be a G-valued cocycle and let γ be a left or right action of G on X. Define*

$$\tilde\varphi(t, \omega, x) := \gamma_{\varphi(t,\omega)}(x). \tag{6.1.4}$$

Then $\tilde\varphi$ is measurable and is a cocycle or backward cocycle on X, respectively.

The proof follows immediately from the definitions.

We now investigate the question whether a cocycle $\tilde\varphi$ induced by the left action γ of the Lie group G on a manifold M by a cocycle φ on G via (6.1.4) has a generator if φ does. The answer is given by the non-autonomous version of a well-known fact from differential geometry (see Kobayashi and Nomizu [217: p. 40]).

6.1.6 Theorem (Generator of Cocycle Induced by Action). *Let φ be a G-valued cocycle generated by the right-invariant RDE $\dot\varphi(t, \omega) = R_{\varphi(t,\omega)*} A(\theta_t \omega)$, $\varphi(0, \omega) = e$, and let γ be a left action of G on the manifold M. Then the cocycle $\tilde\varphi$ defined by (6.1.4) is generated by the RDE*

$$\dot x_t = \tilde A(\theta_t\omega, x_t), \quad x_0 = x \in M, \tag{6.1.5}$$

where the random vector field $\tilde A(\omega, \cdot) \in \mathfrak{X}(M)$ is given by

$$\tilde A(\omega, x) := \frac{d}{dt}\gamma_{\varphi(t,\omega)}(x)|_{t=0}. \tag{6.1.6}$$

6.1.7 Example. Let $G = Gl(d, \mathbb{R})$ and $M = \mathbb{R}^d$. Then $\dot\Phi(t, \omega) = A(\theta_t\omega)\Phi(t, \omega)$ and $\gamma_g(x) = gx$ induce $\tilde\varphi(t, \omega, x) = \Phi(t, \omega)x$, a linear cocycle in $\mathbb{R}^d$, which solves $\dot x_t = A(\theta_t\omega)x_t$. ∎

In the white noise case, if φ solves (6.1.3) and γ is a left action on a manifold M, then the induced cocycle $\tilde\varphi$ defined by (6.1.4) solves the SDE

$$dx_t = \sum_{j=0}^{m} \tilde{A}_j(x_t) \circ dW_t^j, \qquad (6.1.7)$$

where $\tilde{A}_j \in \mathfrak{X}(M)$ is the vector field corresponding to the flow $\tilde{\varphi}_j(t,x) = \gamma_{\exp(tA_j)}(x)$ on M.

Theorem 6.1.6 becomes even nicer if M is a homogeneous space of G, i. e. if G acts transitively on M (meaning that $\{\gamma_g(x) : g \in G\} = M$ for all $x \in M$). Without loss of generality we can assume that $M = G/H = (gH)_{g \in G}$ is the set of left-cosets of a closed subgroup $H \subset G$.

There is a natural left action of G on G/H defined by

$$\gamma_g(xH) := gxH.$$

If $\pi : G \to G/H$, $g \mapsto gH$, is the canonical projection then

$$\pi \circ L_g = \gamma_g \circ \pi.$$

6.1.8 Corollary (Generator on Homogeneous Space). *Let φ be a G-valued cocycle with generator $\dot{\varphi}(t,\omega) = R_{\varphi(t,\omega)*}A(\theta_t\omega)$. Let $M = G/H$ be any homogeneous space of G with the natural left action γ of G. Then the cocycle $\tilde{\varphi} = \gamma_\varphi = \pi \circ L_\varphi$ has a generator $\dot{x}_t = \tilde{A}(\theta_t\omega, x_t)$ on M, where*

$$\tilde{A}(\omega, \cdot) = \pi_* A(\omega, \cdot).$$

Proof. The existence of a generator follows from Theorem 6.1.6. For $x \in G$,

$$\begin{aligned}
\tilde{A}(\omega, \pi(x)) &= \frac{d}{dt}\tilde{\varphi}(t,\omega,\pi(x))|_{t=0} = \frac{d}{dt}\gamma_{\varphi(t,\omega)} \circ \pi(x)|_{t=0} \\
&= \frac{d}{dt}\pi(\varphi(t,\omega) \cdot x)|_{t=0} \\
&= \pi_*(\varphi(t,\omega) \cdot x)R_{(\varphi(t,\omega) \cdot x)*}A(\theta_t\omega)|_{t=0} \\
&= \pi_*(x)R_{x*}A(\omega).
\end{aligned}$$

$\square$

6.1.9 Corollary (Generator on Homogeneous Space of $Gl(d,\mathbb{R})$). *Let Φ be a $Gl(d,\mathbb{R})$-valued cocycle generated by $\dot{\Phi}(t,\omega) = A(\theta_t\omega)\Phi(t,\omega)$, $\Phi(0,\omega) = I$. Then on any homogeneous space M of $Gl(d,\mathbb{R})$, Φ induces a cocycle $\tilde{\Phi} = \gamma_\Phi = \pi \circ L_\Phi$ with generator $\tilde{A} = \pi_* A$. Here γ is the natural left action of $Gl(d,\mathbb{R})$ on M and $\pi : Gl(d,\mathbb{R}) \to M$ is the canonical projection.*

For the white noise case, the right-invariant SDE (6.1.3) on G induces an SDE (6.1.7) on M which for a homogeneous space $M = G/H$ becomes

$$dx_t = \sum_{j=0}^{m} (\pi_* A_j)(x_t) \circ dW_t^j. \qquad (6.1.8)$$

Cocycles on Principal Bundles

Let $P(M, G)$ be a principal fiber bundle, i.e. P (the total space) is a manifold, G is a Lie group (structure group) acting freely[2] on the right on P (written $(p, g) \mapsto \gamma_g(p) = pg$), $M = P/G$ is the orbit space (base space), $\pi : P \to M$ the canonical projection (with $\pi \circ \gamma_g = \pi$ for all $g \in G$) such that P is locally trivial, i.e. for each $x \in M$ there exists a neighborhood U such that $\pi^{-1}(U) \cong U \times G$ by a diffeomorphism $\psi = (\pi, \chi)$ which intertwines the right action γ_g of G on P and the right action R_g of G on itself, $\chi \circ \gamma_g = R_g \circ \chi$.

It follows that all fibers $\pi^{-1}(x)$ are closed submanifolds of P on which the right action of G is transitive and free, hence $\pi^{-1}(x) \cong G$.

6.1.10 Theorem (Cocycle on Principal Bundle). *Let φ be a cocycle on the principal bundle P. Then the following is equivalent:*

(i) φ projects to a cocycle $\tilde{\varphi}$ on M via π, i.e. $\tilde{\varphi} \circ \pi = \pi \circ \varphi$,

(ii) φ is right-invariant with respect to G, i.e.

$$\gamma_g \circ \varphi(t, \omega, \cdot) = \varphi(t, \omega, \cdot) \circ \gamma_g \quad \text{for all } g \in G.$$

In case the cocycle φ has a generator $\dot{x}_t = X(\theta_t \omega, x_t)$ on P, (ii) is equivalent to either of the following statements:

(iii) $X(\omega)$ is a right-invariant generator on P, i.e.

$$\gamma_{g*} X(\omega) = X(\omega).$$

(iv) $X(\omega)$ can be projected to M, i.e. $\tilde{X}(\omega) = \pi_* X(\omega)$ is a random vector field on M that is π-related to $X(\omega)$, in which case $\tilde{X}(\omega)$ is the generator of $\tilde{\varphi}$.

The proof follows again immediately from well-known facts of differential geometry.

6.2 RDS Induced on S^{d-1} and P^{d-1}

We now follow Furstenberg's program [153] and study the action of a linear cocycle Φ in $\mathbb{R}^d$ induced on the manifolds S^{d-1} and P^{d-1} (more precisely, the left action of a $Gl(d, \mathbb{R})$-valued cocycle on the above homogeneous spaces, see Sect. 6.1).

While working on S^{d-1} is very convenient for calculations, the situation on P^{d-1} is "less redundant" since P^{d-1} is "smaller" than S^{d-1}, hence certain trivial ambiguities are removed.

Only if necessary we will distinguish between a quantity $(\hat{\cdot})$ associated with S^{d-1} and $(\bar{\cdot})$ associated with P^{d-1}. We will, however, mainly elaborate the case S^{d-1} (and then omit the bar in $(\hat{\cdot})$) and leave the formulation of the completely parallel statements for P^{d-1} to the reader.

---

[2] γ is a free action of G on P if $\gamma_g(x) = x$ for some $x \in P$ implies $g = e$.

We assume throughout that θ is **ergodic** and that time $\mathbb{T}$ is two-sided.

6.2.1 Invariant Measures

The Manifolds S^{d-1} and P^{d-1}

Although both S^{d-1} and P^{d-1} are homogeneous spaces of $Gl(d, \mathbb{R})$ we give a more concrete and elementary description here.

The *unit sphere* of $(\mathbb{R}^d, \langle \cdot, \cdot \rangle)$ is the manifold

$$S^{d-1} := \{x \in \mathbb{R}^d : \langle x, x \rangle = 1\} \subset \mathbb{R}^d.$$

It inherits its natural Riemannian structure from $\mathbb{R}^d$ by the inclusion mapping $i : S^{d-1} \to \mathbb{R}^d$, which is an isometric imbedding.

Denote by

$$\bar{\pi} : \mathbb{R}^d \setminus \{0\} \to S^{d-1}, \quad x \mapsto \bar{\pi}(x) = \frac{x}{\|x\|} = s \in S^{d-1},$$

the canonical projection onto S^{d-1}. Note that $\mathbb{R}^d \setminus \{0\} \cong S^{d-1} \times (0, \infty)$ is a flat (principal) bundle over S^{d-1} with fibers (structure group) $\bar{\pi}^{-1}(s) = (0, \infty)$. The coordinates $s = \frac{x}{\|x\|}$, $r = \|x\|$ of a point $x = rs \in \mathbb{R}^d \setminus \{0\}$ in this bundle are called *polar coordinates*.

The tangent bundle $T(\mathbb{R}^d \setminus \{0\})$ of $\mathbb{R}^d \setminus \{0\}$ is globally trivial and is identified with $(\mathbb{R}^d \setminus \{0\}) \times \mathbb{R}^d$. We also identify the tangent bundle TS^{d-1} of S^{d-1} with the subbundle of $S^{d-1} \times \mathbb{R}^d$ defined by identifying $(s, T_s S^{d-1})$ with $(s, \{v \in \mathbb{R}^d : \langle v, s \rangle = 0\}) = (s, s^\perp)$,

$$TS^{d-1} = \{(s, v) : s \in S^{d-1}, \langle v, s \rangle = 0\} \subset S^{d-1} \times \mathbb{R}^d \subset \mathbb{R}^d \times \mathbb{R}^d$$

(we can also think of $T_s S^{d-1}$ as the affine space $s + s^\perp$ in the space $\mathbb{R}^d$ in which S^{d-1} is imbedded).

Now $T\bar{\pi} : T(\mathbb{R}^d \setminus \{0\}) \to TS^{d-1}$ sends (x, v), $x \in \mathbb{R}^d \setminus \{0\}$, $v \in T_x(\mathbb{R}^d \setminus \{0\}) \cong \mathbb{R}^d$ to $(\bar{\pi}(x), T_x \bar{\pi}(v))$, where

$$T_x \bar{\pi}(v) = \frac{1}{\|x\|} (v - \langle v, \bar{\pi}(x) \rangle \bar{\pi}(x)) \in \bar{\pi}(x)^\perp = T_{\bar{\pi}(x)} S^{d-1}, \qquad (6.2.1)$$

and $\|x\| T_x \bar{\pi}$ is the orthogonal projection of $\mathbb{R}^d$ onto $\bar{\pi}(x)^\perp$.

A C^k diffeomorphism f of $\mathbb{R}^d \setminus \{0\}$ induces a C^k diffeomorphism $\bar{f}$ on S^{d-1} (i.e. $\bar{f} \circ \bar{\pi} = \bar{\pi} \circ f$) if f is positively homogeneous (or just homogeneous or even linear). In this case the unique $\bar{f}$ is

$$\bar{f}(s) = \frac{f(s)}{\|f(s)\|}, \quad s \in S^{d-1}.$$

As a consequence, a linear mapping $A \in Gl(d, \mathbb{R})$ induces naturally a smooth diffeomorphism $\bar{A}$ on S^{d-1} via

$$S^{d-1} \ni s \mapsto \frac{As}{\|As\|} =: \bar{A}s \in S^{d-1},$$

which satisfies $\bar{A} \circ \bar{\pi} = \bar{\pi} \circ A$, hence

$$\overline{AB} = \bar{A} \circ \bar{B}, \quad A, B \in Gl(d, \mathbb{R}).$$

Further, by homogeneity, $\bar{A} \circ g = g \circ \bar{A}$ for $g \in G = \{\pm \mathrm{id}_{S^{d-1}}\}$.

Every linear mapping $x \mapsto Ax$ can be described equivalently in polar coordinates by $(s, r) \mapsto (\bar{A}s, \|As\|r)$.

A C^k vector field X on $\mathbb{R}^d \setminus \{0\}$ induces a C^k vector field $\bar{X} = T\bar{\pi}(X)$ on S^{d-1} if X is positively homogeneous, (or just homogeneous or even linear). In this case the unique $\bar{X}$ is

$$\bar{X}(s) = X(s) - \langle X(s), s \rangle s, \quad s \in S^{d-1},$$

and the corresponding flows φ and $\bar{\varphi}$ are related by

$$\bar{\pi} \circ \varphi(t) = \bar{\varphi}(t) \circ \bar{\pi}.$$

To see this note that X induces the vector field $\bar{X}$ if and only if X and $\bar{X}$ are $\bar{\pi}$-related in the sense that for all $x, y \in \bar{\pi}^{-1}(s)$, $T\bar{\pi}(X)(x) = T\bar{\pi}(X)(y)$ (Boothby [74: p. 119]). Now use (6.2.1).

Hence every linear vector field $X_A(x) = Ax$, $A \in \mathbb{R}^{d \times d}$, induces a smooth vector field

$$\bar{X}_A(s) = As - \langle As, s \rangle s$$

on S^{d-1}.

The *real projective space* P^{d-1} of one-dimensional subspaces of $\mathbb{R}^d$ is the quotient space $\mathbb{R}^d \setminus \{0\} / \sim$, where the equivalence relation $\sim$ is defined by $x \sim y$ if and only if there exists a $\lambda \in \mathbb{R}$ with $x = \lambda y$. Let

$$\hat{\pi} : \mathbb{R}^d \setminus \{0\} \to P^{d-1}, \quad x \mapsto \hat{\pi}(x) = \hat{x},$$

be the canonical projection. $\mathbb{R}^d \setminus \{0\}$ is thus a (non-flat) (principal) bundle over P^{d-1} with typical fiber (structure group) $\mathbb{R}^* := \mathbb{R} \setminus \{0\}$.

There is yet another way of looking at S^{d-1} and P^{d-1}: Note that the group

$$G = \{\pm \mathrm{id}_{S^{d-1}}\} \cong \mathbb{Z}_2$$

is the cyclic group of order 2 with generator $g = -\mathrm{id}_{S^{d-1}}$ and that

$$P^{d-1} = S^{d-1}/G,$$

i.e. P^{d-1} is obtained from S^{d-1} by identifying antipodal points $\pm s$. S^{d-1} is thus a two-fold covering[3] of P^{d-1}, with covering projection

---

[3] We have $S^1 \cong P^1$.

$$\pi : S^{d-1} \to P^{d-1}$$

satisfying $\pi \circ g = \pi$ for $g \in G$. It is also a principal bundle with typical fiber G. We endow P^{d-1} with the Riemannian metric which makes π a local isometry. In particular,

$$\langle \cdot, \cdot \rangle_s = g_* \langle \cdot, \cdot \rangle_{gs} = \langle \cdot, \cdot \rangle_{\pi(s)},$$

where for $g = -\mathrm{id}_{S^{d-1}}$, $g_* : TS^{d-1} \to TS^{d-1}$ sends $v \in T_s S^{d-1}$ to $g_*(v) = -v \in T_{-s} S^{d-1}$ and $T\pi : TS^{d-1} \to TP^{d-1}$ identifies $v \in T_s S^{d-1}$ and $-v \in T_{-s} S^{d-1}$. Further,

$$\pi \circ \bar{\pi} = \hat{\pi} \quad \text{on } \mathbb{R}^d \setminus \{0\}.$$

A C^k diffeomorphism f on $\mathbb{R}^d \setminus \{0\}$ induces a C^k diffeomorphism $\hat{f}$ on P^{d-1} (i.e. $\hat{f} \circ \hat{\pi} = \hat{\pi} \circ f$) if and only if f induces a diffeomorphism $\bar{f}$ on S^{d-1} with $\bar{f}(-s) = -\bar{f}(s)$.

If this holds, the unique $\hat{f}$ is given by

$$\hat{f}(\hat{x}) = \widehat{f(x)} = \pi(\bar{f}(s)).$$

The conditions are in particular satisfied if f is homogeneous or even linear.

As a consequence, a linear mapping $x \mapsto Ax$, $A \in Gl(d, \mathbb{R})$, induces naturally a smooth diffeomorphism $\hat{A} \in PGl(d, \mathbb{R}) := Gl(d, \mathbb{R})/\mathbb{R}^* I$, the projective general linear group, on P^{d-1} via

$$P^{d-1} \ni \hat{x} \mapsto \widehat{Ax} =: \hat{A}\hat{x} \in P^{d-1}$$

which satisfies $\hat{A} \circ \hat{\pi} = \hat{\pi} \circ A$ on $\mathbb{R}^d \setminus \{0\}$ and $\pi \circ \bar{A} = \hat{A} \circ \pi$ on S^{d-1}, hence

$$\widehat{AB} = \hat{A} \circ \hat{B}, \quad A, B \in Gl(d, \mathbb{R}).$$

Further, a C^k vector field X on $\mathbb{R}^d \setminus \{0\}$ induces a C^k vector field $\hat{X} = T\hat{\pi}(X)$ on P^{d-1} if and only if X induces a vector field $\bar{X}$ on S^{d-1} with $\bar{X}(-s) = -\bar{X}(s)$.

If this holds, the unique $\hat{X}$ is given by

$$\hat{X} = T\hat{\pi}(X) = T\pi(\bar{X}) = (T\pi \circ T\bar{\pi})(X)$$

and the corresponding flows φ, $\bar{\varphi}$ and $\hat{\varphi}$ are related by

$$\hat{\varphi}(t) \circ \hat{\pi} = \hat{\pi} \circ \varphi(t), \quad \bar{\varphi}(t) \circ \bar{\pi} = \bar{\pi} \circ \varphi(t), \quad \hat{\varphi}(t) \circ \pi = \pi \circ \bar{\varphi}(t).$$

The conditions are particularly satisfied if X is homogeneous or even linear.

Hence every linear vector field $X_A(x) = Ax$, $A \in \mathbb{R}^{d \times d}$, induces a smooth vector field on P^{d-1} given by

$$\hat{X}_A(\hat{x}) = T\pi(\bar{X}_A)(s) = T\pi(As - \langle As, s \rangle s).$$

All of the above relations are summarized for the mapping case in the following commutative diagram of Fig. 6.1.

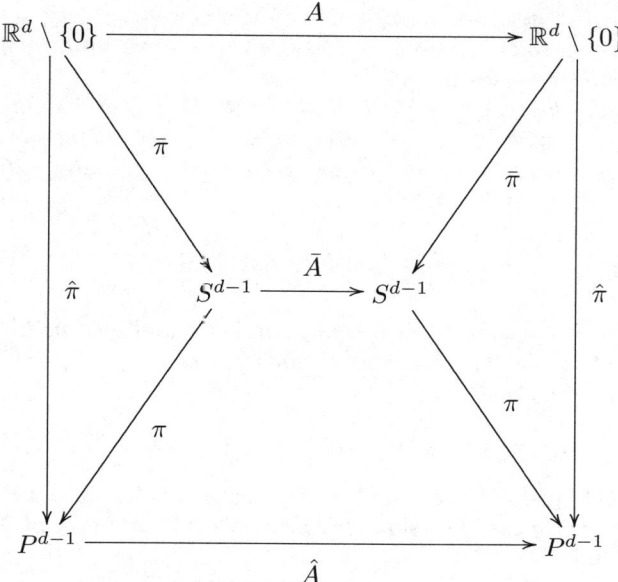

Fig. 6.1. Actions induced by a matrix

Invariant Measures for the Induced RDS on S^{d-1} and P^{d-1}

With the help of the MET we will determine the invariant measures of the nonlinear RDS $\varphi := \bar{\Phi}$ on S^{d-1} and $\hat{\Phi}$ on P^{d-1} induced by a linear RDS Φ in $\mathbb{R}^d$ (but only spell out the results for S^{d-1}).

6.2.1 Proposition. *Let Φ be a linear RDS in $\mathbb{R}^d$ over the ergodic metric DS θ with two-sided time and let*

$$\varphi(t,\omega)s = \bar{\Phi}(t,\omega)s := \frac{\Phi(t,\omega)s}{\|\Phi(t,\omega)s\|}, \quad s \in S^{d-1}.$$

Then:

(i) φ is a C^∞ RDS on S^{d-1} over θ satisfying

$$\varphi(t,\omega) \circ \bar{\pi} = \bar{\pi} \circ \Phi(t,\omega).$$

(ii) The function
$$q(t,\omega,s) := \log \|\Phi(t,\omega)s\| \qquad (6.2.2)$$

is a helix over $\Theta(t)(\omega,s) := (\theta(t)\omega, \varphi(t,\omega)s)$, i. e.

$$q(t+u,\omega,s) = q(t,\omega,s) + q(u, \Theta(t)(\omega,s))$$

identically on $\mathbb{T} \times \mathbb{T} \times \Omega \times S^{d-1}$.

(iii) The set $\mathcal{I}_\mathbb{P}(\varphi)$ of φ-invariant measures is nonvoid convex compact and is invariant with respect to $G = \{\pm \mathrm{id}_{S^{d-1}}\}$ (i. e. if μ is φ-invariant, then so is $g\mu$, the measure with factorization $g\mu_\omega$).

(iv) Suppose Φ satisfies the IC of the MET, let λ_1 be the largest and λ_p the smallest Lyapunov exponent of Φ. Then the helix q defined by (6.2.2) satisfies the IC of the subadditive ergodic theorem for any φ-invariant measure μ. If μ is such a measure,

$$\lambda_p \leq \gamma(\mu)(\omega, s) := \lim_{t \to \pm\infty} \frac{1}{t} \log \|\Phi(t,\omega)s\| \leq \lambda_1 \qquad (6.2.3)$$

where the limits exist on an invariant set of full μ measure and also in L^1, and on this invariant set $\gamma(\mu)$ is an invariant random variable.

In case μ is ergodic,

$$\gamma(\mu) = \mathbb{E}_\mu \log \|\Phi(1,\omega)s\|.$$

Proof. (i)(ii) The cocycle property of φ follows from that of Φ and $\overline{AB} = \bar{A} \circ \bar{B}$. The helix property of q over Θ follows from the cocycle property of Φ and $\|ABs\| = \|A\bar{B}s\| \|Bs\|$.

(iii) Since S^{d-1} is compact metric, $\mathcal{P}_\mathbb{P}(\Omega \times S^{d-1})$ and $\mathcal{I}_\mathbb{P}(\varphi)$ are nonvoid compact convex sets (see Theorem 1.5.10). If $\mu \in \mathcal{I}_\mathbb{P}(\varphi)$, then $g\mu \in \mathcal{I}_\mathbb{P}(\varphi)$ since $\bar{A} \circ g = g \circ \bar{A}$.

(iv) We have

$$-\log^+ \|\Phi(t,\omega)^{-1}\| \leq -\log \|\Phi(t,\omega)^{-1}\|$$
$$\leq \log \|\Phi(t,\omega)s\| \leq \log \|\Phi(t,\omega)\| \leq \log^+ \|\Phi(t,\omega)\|.$$

Thus the helix $q(t,\omega,s) = \log \|\Phi(t,\omega)s\|$ over Θ is bounded above by $\log \|\Phi(t,\omega)\|$, which is subadditive over θ, and bounded below by $-\log \|\Phi(t,\omega)^{-1}\|$, which is superadditive over θ, both satisfying the IC of the Furstenberg-Kesten theorem (Theorem 3.3.3, invertible cases). Hence for any Θ-invariant μ, the subadditive ergodic theorem holds. First assume $\mathbb{T} = \mathbb{Z}$. By additivity and the technique used to prove Theorem 3.3.3(B),

$$\lim_{n \to \infty} \frac{1}{n} \log \|\Phi(n,\omega)s\| =: \gamma(\mu)(\omega, s)$$

exists on an invariant set of full μ measure and also in $L^1(\mu)$, and the limit is invariant on this set. Further,

$$\lambda_p \leq \gamma(\mu)(\omega, s) \leq \lambda_1$$

(for the left-hand side inequality note that $\Phi(n,\omega)^{-1}$ is a backward cocycle over θ with spectrum $-\Sigma$). For $\mathbb{T} = \mathbb{R}$, we use additivity, the discrete time result and the technique used to prove Theorem 3.3.3(C) to obtain the same statements.

Now $q(-t, \omega, s)$ has analogous properties with respect to θ^{-1} and Θ^{-1}, hence
$$-\lambda_1 \leq \gamma^-(\mu)(\omega, s) =: \lim_{t \to \infty} \frac{1}{t} \log \|\Phi(-t, \omega)s\| \leq -\lambda_p.$$
Since by additivity $q(-t, \omega, s) = -q(t, \Theta(-t)(\omega, s))$, the reasoning used in the proof of Theorem 3.3.10 yields $\gamma^-(\mu) = -\gamma(\mu)$.

Since additivity implies $\mathbb{E}_\mu q(t, \cdot) = t\, \mathbb{E}_\mu q(1, \cdot)$,
$$\gamma(\mu) = \mathbb{E}_\mu q(1, \cdot)$$
in the ergodic case. □

6.2.2 Remark (Lift of Measure from P^{d-1} to S^{d-1}). In this remark we write $\bar\Phi$ and $\bar\mu$ for φ and μ.

The set $\mathcal{I}_\mathbb{P}(\hat\Phi)$ of $\hat\Phi$-invariant measures is also nonvoid convex and compact. Further, the mapping
$$\pi : \mathcal{I}_\mathbb{P}(\bar\Phi) \to \mathcal{I}_\mathbb{P}(\hat\Phi), \quad \bar\mu \mapsto \hat\mu := \pi\bar\mu = \pi g\bar\mu,$$
is continuous, affine and surjective (the latter meaning that if $\hat\mu$ is $\hat\Phi$-invariant on P^{d-1}, then there exists a $\bar\Phi$-invariant $\bar\mu$ on S^{d-1} which is a *lift* of $\hat\mu$, i.e. $\pi\bar\mu = \hat\mu$; in this case $g\bar\mu$, $g \in G$, is also a lift of $\hat\mu$).

This can be seen as follows: The continuous affine mapping
$$\pi : \mathcal{P}_\mathbb{P}(\Omega \times S^{d-1}) \to \mathcal{P}_\mathbb{P}(\Omega \times P^{d-1}), \quad \bar\mu \mapsto \pi\bar\mu, \qquad (6.2.4)$$
satisfies $\pi(\mathcal{I}_\mathbb{P}(\bar\Phi)) \subset \mathcal{I}_\mathbb{P}(\hat\Phi)$ since $\hat A \circ \pi = \pi \circ \bar A$ and $\pi(\bar\mu) = \pi(g\bar\mu)$ since $\pi \circ g = \pi$.

The existence of a lift $\bar\mu$ of $\hat\mu \in \mathcal{I}_\mathbb{P}(\hat\Phi)$ is equivalent to $\pi : \mathcal{I}_\mathbb{P}(\bar\Phi) \to \mathcal{I}_\mathbb{P}(\hat\Phi)$ being surjective. This follows from the surjectivity of (6.2.4) as follows: Assume the compact convex set $\pi^{-1}(\hat\mu) \subset \mathcal{P}_\mathbb{P}(\Omega \times S^{d-1})$ is nonvoid. For $\hat\mu \in \mathcal{I}_\mathbb{P}(\hat\Phi)$,
$$\Theta(t)\pi^{-1}(\hat\mu) = \pi^{-1}(\hat\mu),$$
implying that $\pi^{-1}(\hat\mu) \cap \mathcal{I}_\mathbb{P}(\bar\Phi) \neq \emptyset$ (see Corollary 1.6.13).

It remains to be proved that π is surjective. By the Krein-Milman theorem, $\mathcal{P}_\mathbb{P}(\Omega \times P^{d-1})$ is the closed convex hull of its extremal points, which are the random Dirac measures $\hat\mu_\omega = \delta_{\hat x(\omega)}$. If $\lambda : P^{d-1} \to S^{d-1}$ is a measurable section of $\pi : S^{d-1} \to P^{d-1}$, i.e. $\pi \circ \lambda = \mathrm{id}_{P^{d-1}}$, then $\bar\mu_\omega := \delta_{\lambda(\hat x(\omega))} = \delta_{s(\omega)}$ is an extremal point of $\mathcal{P}_\mathbb{P}(\Omega \times S^{d-1})$ with $\pi\bar\mu = \hat\mu$. Now let $\hat\mu \in \mathcal{P}_\mathbb{P}(\Omega \times P^{d-1})$ be arbitrary and assume $\hat\mu_n \to \hat\mu$, $\hat\mu_n = \sum p_{k,n} \delta_{\hat x_{k,n}(\omega)}$. Then $\bar\mu_n = \sum p_{k,n} \delta_{\lambda(\hat x_{k,n}(\omega))} = \sum p_{k,n} \delta_{s_{k,n}(\omega)}$ has the property $\pi\bar\mu_n = \hat\mu_n$ since π is affine.

Now pick a converging subsequence of $(\bar\mu_n)$, $\bar\mu_{n'} \to \bar\mu$ (which is possible by compactness). By the continuity of π,
$$\pi\bar\mu = \pi(\lim \bar\mu_{n'}) = \lim \pi\bar\mu_{n'} = \lim \hat\mu_{n'} = \hat\mu,$$
thus $\bar\mu \in \pi^{-1}(\hat\mu)$. ■

6.2.3 Theorem (Invariant Measures on S^{d-1}). *Let Φ be a linear RDS in $\mathbb{R}^d$ over the ergodic metric DS θ with two-sided time satisfying the IC of the MET. Let $\mathcal{S}(\theta, \Phi) = \{(\lambda_i, d_i)_{i=1,\ldots,p}\}$ be its Lyapunov spectrum and $\mathbb{R}^d = \oplus_{i=1}^p E_i(\omega)$ its Oseledets splitting. Then:*

(i) On the invariant set $\tilde{\Omega}$ of full $\mathbb{P}$ measure of the MET,

$$\bar{E}_i(\omega) := \bar{\pi}(E_i(\omega)) = E_i(\omega) \cap S^{d-1}, \quad i = 1, \ldots, p,$$

are disjoint strictly φ-invariant random compact submanifolds of S^{d-1} of dimension $d_i - 1$.

(ii) For each ergodic φ-invariant measure μ on S^{d-1} there exists a unique $i \in \{1, \ldots, p\}$ such that μ realizes λ_i in the sense that

$$\lim_{t \to \pm\infty} \frac{1}{t} \log \|\Phi(t, \omega)s\| = \lambda_i \quad \mu\text{-a.s.} \tag{6.2.5}$$

Equivalently, μ is supported by the random compact set $\bar{E}_i$, i.e.

$$\mu_\omega(\bar{E}_i(\omega)) = 1 \quad \mathbb{P}\text{-a. s.}$$

Conversely, for each $i \in \{1, \ldots, p\}$ there is a φ-invariant measure on S^{d-1} which realizes λ_i.

(iii) Each φ-invariant measure μ on S^{d-1} has the form

$$\mu = \sum_{i=1}^p \alpha_i \mu_i, \quad \alpha_i \geq 0, \quad \sum_{i=1}^p \alpha_i = 1, \quad \mu_i \text{ realizes } \lambda_i.$$

Proof. (i) The strict φ-invariance of $\bar{E}_i$ follows from the strict Φ-invariance of E_i and $\bar{A} \circ \bar{\pi} = \bar{\pi} \circ A$.

(ii) Since μ is ergodic both limits in (6.2.3) are equal to the same constant $\gamma(\mu)$. But the MET for Φ and the characterization of the $E_i(\omega)$ assert that for $\omega \in \tilde{\Omega}$ the two limits

$$\lim_{t \to \pm\infty} \frac{1}{t} \log \|\Phi(t, \omega)s\| = \lambda^\pm(\omega, s)$$

are equal if and only if $\lambda^\pm(\omega, s) = \lambda_i$ for some i, equivalently if and only if $s \in \bar{E}_i(\omega)$ for some i. Since $\lambda^\pm(\omega, s) = \gamma(\mu)$ μ-a.s. this i is independent of (ω, s), μ-a.s., and $\mu_\omega(\bar{E}_i(\omega)) = 1$, $\mathbb{P}$-a. s.

Conversely, let $\mathcal{I}_\mathbb{P}(\varphi|\bar{E}_i)$ be the weakly compact set of φ-invariant measures supported by the strictly invariant (hence forward invariant) random compact set $\bar{E}_i$. By Theorem 1.6.13, $\mathcal{I}_\mathbb{P}(\varphi|\bar{E}_i)$ is nonvoid and equal to the closed convex hull of its ergodic elements (see Remark 1.5.11). Note that any $\mu \in \mathcal{I}_\mathbb{P}(\varphi|\bar{E}_i)$ (and not just the ergodic ones) fulfills (6.2.5).

(iv) Assume that μ is invariant. By Proposition 6.2.1 (iv)

$$\lim_{t \to \pm\infty} \frac{1}{t} \log \|\Phi(t, \omega)s\| = \gamma(\mu)(\omega, s) \quad \mu\text{-a.s.}$$

By the MET the possible values of the invariant random variable $\gamma(\mu)$ are the λ_i, with

$$\mu\{(\omega,s) : \gamma(\mu)(\omega,s) = \lambda_i\} = \mu\{(\omega,s) : s \in \bar{E}_i(\omega)\} =: \alpha_i \geq 0$$

and
$$\mu\{(\omega,s) : s \in \cup_{i=1}^p \bar{E}_i(\omega)\} = \sum_{i=1}^p \alpha_i = 1.$$

Define for $\alpha_i > 0$
$$\mu_{i,\omega}(\cdot) := \frac{1}{\alpha_i}\mu_\omega(\cdot \cap \bar{E}_i(\omega))$$

and for $\alpha_i = 0$ choose any μ_i realizing λ_i (see (ii)). Then μ_i is invariant (though not necessarily ergodic) and realizes λ_i, and we have

$$\mu_\omega = \sum_{i=1}^p \alpha_i \mu_{i,\omega}.$$

∎

6.2.4 Remark (Markov Measures on S^{d-1}). The Oseledets space E_1 is $\mathcal{F}^-$-measurable and E_p is $\mathcal{F}^+$-measurable (see Remark 3.4.14). By Theorem 1.7.5 the sets $\mathcal{I}_{\mathbb{P},\mathcal{F}^-}(\varphi|\bar{E}_1)$ and $\mathcal{I}_{\mathbb{P},\mathcal{F}^+}(\varphi|\bar{E}_p)$ are nonvoid compact convex sets. In particular, there is always an (ergodic) invariant forward Markov measure realizing λ_1 and an (ergodic) invariant backward Markov measure realizing λ_p. ■

Random Eigenvectors

The block-diagonalization of a linear cocycle Φ with two-sided time satisfying the IC of the MET can be easily accomplished as follows (see Corollary 4.3.12): We choose a random basis $G(\omega) = (v_i(\omega))$ which is adapted to the Oseledets splitting. If $P(\omega) = P^S_{G(\omega)}$ is the basis transfer matrix from the standard basis S to $G(\omega)$, then

$$\Psi(t,\omega) = P(\theta(t)\omega)\Phi(t,\omega)P(\omega)^{-1} = \mathrm{diag}(\Psi_1(t,\omega),\ldots,\Psi_p(t,\omega)) \quad (6.2.6)$$

is block-diagonal and Lyapunov cohomologous to Φ. For the proof one uses the fact that the property $\Phi(t,\omega)E_i(\omega) = E_i(\theta(t)\omega)$ implies that if $v(\omega) \in E_i(\omega)$ then both $v(\theta(t)\omega) \in E_i(\theta(t)\omega)$ and $\Phi(t,\omega)v(\omega) \in E_i(\theta(t)\omega)$. However, in general $\varphi(t,\omega)v(\omega) \neq v(\theta(t)\omega)$. Can we find random vectors for which equality holds, i.e. which are φ-invariant?

6.2.5 Definition (Random Eigenvector, Eigenvalue). Let Φ be a linear RDS in $\mathbb{R}^d$. A random variable $u : \Omega \to S^{d-1}$ is called *random eigenvector* of Φ if it satisfies
$$\varphi(t,\omega)u(\omega) = u(\theta(t)\omega),$$
equivalently
$$\Phi(t,\omega)u(\omega) = \|\Phi(t,\omega)u(\omega)\|\,u(\theta(t)\omega) =: Q(t,\omega)u(\theta(t)\omega),$$

and the cocycle $Q(t,\omega) = \|\Phi(t,\omega)u(\omega)\| > 0$ is called the corresponding *random eigenvalue* of Φ. ■

$u(\cdot)$ being a random eigenvector is equivalent to $\mu_\omega = \delta_{u(\omega)}$ being φ-invariant (and necessarily ergodic).

For example, if $d_i = 1$ then $\hat{E}_i(\omega) := \hat{\pi}(E_i(\omega)) = \pi(\bar{E}_i(\omega))$ is one-point and there is exactly one ergodic $\hat{\Phi}$-invariant measure realizing λ_i, namely $\hat{\mu}_\omega = \delta_{\hat{E}_i(\omega)}$. Now $\bar{E}_i(\omega)$ consists of two points. $\hat{\mu}$ can be either lifted to the two ergodic φ-invariant Dirac measures $\mu_\omega = \delta_{s_i(\omega)}$ and $g\mu_\omega = \delta_{-s_i(\omega)}$, $g = -\mathrm{id}_{S^{d-1}}$ (i.e. $g\mu \neq \mu$) or to the single ergodic φ-invariant measure

$$\mu_\omega = \frac{1}{2}(\delta_{s_i(\omega)} + \delta_{-s_i(\omega)}) \tag{6.2.7}$$

(the weight $\frac{1}{2}$ coming from $g\mu = \mu$). In the second case there is no (measurable) way of distinguishing between the two points in $\bar{E}_i(\omega)$.

However, if we are willing to accept a possible extension of the underlying DS θ, we can have the following random analogue of the deterministic diagonalization of a matrix by means of a basis of eigenvectors.

6.2.6 Theorem (Existence of Random Eigenvectors). *Let Φ be a linear RDS with two-sided time over the ergodic DS θ satisfying the IC of the MET. Then the following holds, possibly over an extension $\bar{\theta}$ of θ:*

(i) In each Oseledets space $E_i(\omega)$ there exists at least one random eigenvector $u_i(\omega)$ and the corresponding eigenvalue is $Q_i(t,\omega) = \|\Phi(t,\omega)u_i(\omega)\|$, with

$$\lambda_i = \mathbb{E} \log \|\Phi(1,\cdot)u_i(\cdot)\|.$$

(ii) If the spectrum of Φ is simple, then there exists a basis of random eigenvectors

$$G(\omega) = (u_1(\omega), \ldots, u_d(\omega)),$$

where $u_i(\omega) \in E_i(\omega)$ $\mathbb{P}$-a.s., with random eigenvalues $Q_i(t,\omega) = \|\Phi(t,\omega)u_i(\omega)\|$. Thus $\Psi(t,\omega)$, the cocycle Φ written in the bases $G(\omega)$ and $G(\theta(t)\omega)$, is Lyapunov cohomologous to Φ and diagonal,

$$\Psi(t,\omega) = \mathrm{diag}(\|\Phi(t,\omega)u_1(\omega)\|, \ldots, \|\Phi(t,\omega)u_d(\omega)\|), \tag{6.2.8}$$

with $\mathbb{E} \log \|\Phi(1,\cdot)u_i(\cdot)\| = \lambda_i$ for all $i = 1, \ldots, d$.

Proof. We follow Arnold and Xu [40] (Appendix B).

(i) Assume without loss of generality that the MET holds without an exceptional set. We consider the random compact set

$$C_\omega := \bar{E}_1(\omega) \times \cdots \times \bar{E}_p(\omega) \subset (S^{d-1})^p,$$

which is strictly invariant with respect to φ. Hence $\mathcal{I}_\mathbb{P}(\varphi|C)$ is nonvoid compact convex, thus there exists an ergodic φ-invariant measure μ_ω supported by C. If μ_ω happens to be a Dirac measure $\delta_{u_1(\omega),\ldots,u_p(\omega)}$ we are done.

Otherwise, we consider the extended DS $(\bar{\Omega}, \bar{\mathbb{P}}, \bar{\theta})$, where $\bar{\Omega} = C$, $\bar{\mathbb{P}} = \mu$, $\bar{\theta} = (\theta, \varphi)$, with $\varphi(t, \bar{\omega}) = \varphi(t, \omega)$ trivially extended and define random vectors by

$$u_i(\bar{\omega}) = u_i(\omega, s_1, \ldots, s_p) := s_i, \quad i = 1, \ldots, p.$$

Then

$$\varphi(t,\bar{\omega})u_i(\bar{\omega}) = \varphi(t,\omega)s_i = u_i(\bar{\theta}(t)\bar{\omega}).$$

(ii) This follows from (i) for $p = d$. □

If $d_i > 1$ the random eigenvector in $E_i(\omega)$ just constructed might be the only possible one (see the deterministic case). For more examples see Crauel [108].

Note that all diagonal elements in (6.2.8) are positive, meaning that any "flip behavior" is built into the basis of random eigenvectors.

6.2.2 Furstenberg-Khasminskii Formulas

The study of the action φ induced by Φ on S^{d-1} in Subsect. 6.2.1 gave rise to the Θ-helix

$$q(t,\omega,s) := \log\|\Phi(t,\omega)s\|$$

(Proposition 6.2.1) and for each λ_i there is a Θ-invariant measure μ_i realizing λ_i,

$$\lim_{t\to\pm\infty}\frac{1}{t}q(t,\omega,s) = \lambda_i \quad \mu_i\text{-a.s.} \tag{6.2.9}$$

(Theorem 6.2.3).

When Φ has a generator, the helix q has a generator too, i.e. it can be represented as an integral (as a sum if $\mathbb{T} = \mathbb{Z}$) of a stationary stochastic process. Hence the left-hand side of (6.2.9) becomes a time average, so that the classical ergodic theorem applies and gives an expression for λ_i as a phase average (expectation) with respect to μ_i. We call these expressions for λ_i the *Furstenberg-Khasminskii formulas*. The prototypes of these formulas for the top exponent λ_1 were discovered independently by Furstenberg [153] for discrete time and by Khasminskii [205] for linear stochastic differential equations. This was well before Oseledets [268] proved the MET in 1968.

We only treat the continuous time case here and leave the case $\mathbb{T} = \mathbb{Z}$ as an exercise.

6.2.7 Exercise (Furstenberg-Khasminskii Formulas, $\mathbb{T} = \mathbb{Z}$). Let A with $\log^+\|A^{\pm 1}\| \in L^1(\mathbb{P})$ be the generator of a linear cocycle Φ and let φ be the corresponding cocycle on S^{d-1}. If μ_i is φ-invariant and realizes λ_i, then

$$\lambda_i = \int_{\Omega\times S^{d-1}} \log\|A(\omega)s\|\,d\mu_i(\omega,s)$$

∎

Real Noise Case

Let Φ be a linear cocycle in $\mathbb{R}^d$ generated by the RDE

$$\dot{x}_t = A(\theta_t\omega)x_t. \tag{6.2.10}$$

In polar coordinates $s = \frac{x}{\|x\|} \in S^{d-1}$, $r = \|x\| \in (0,\infty)$, this equation is equivalent to

$$\dot{s}_t = h(\theta_t\omega, s_t), \quad (6.2.11)$$

$$\dot{r}_t = Q(\theta_t\omega, s_t)r_t, \quad (6.2.12)$$

where

$$h(\omega, s) := A(\omega)s - \langle A(\omega)s, s\rangle s$$

is the projection of the linear vector field $A(\omega)x$ and

$$Q(\omega, s) := \langle A(\omega)s, s\rangle.$$

Equations (6.2.11) and (6.2.12) follow easily from (6.2.10) by differentiating $s = x/\langle x, x\rangle^{1/2}$ and $r = \langle x, x\rangle^{1/2}$ (exercise).

Equation (6.2.11) is decoupled from (6.2.12) and generates the cocycle φ, while (6.2.12) with φ inserted is equivalent to

$$\frac{d}{dt}\log r_t = Q(\Theta(t)(\omega, s))$$

and yields with initial condition $r_0 = 1$ the Θ-helix

$$q(t, \omega, s) = \log\|\Phi(t, \omega)s\| = \int_0^t Q(\Theta(u)(\omega, s))du.$$

Theorem 6.2.3 and the classical ergodic theorem give

6.2.8 Theorem (Furstenberg-Khasminskii Formulas, Real Noise Case). *Suppose $A \in L^1(\mathbb{P})$, so (6.2.10) generates a linear cocycle Φ for which the MET holds. Let φ be the corresponding cocycle on S^{d-1} generated by (6.2.11) and $Q(\omega, s) := \langle A(\omega)s, s\rangle$. Then*
  (i) For any φ-invariant measure μ, $Q \in L^1(\mu)$.
  (ii) If μ_i is φ-invariant and realizes λ_i, then

$$\lambda_i = \mathbb{E}_{\mu_i}Q.$$

Proof. The integrability of Q follows from that of A by

$$-\|A\| \leq \lambda_{\min}(\frac{A + A^*}{2}) \leq \langle As, s\rangle \leq \lambda_{\max}(\frac{A + A^*}{2}) \leq \|A\|. \quad (6.2.13)$$

□

A very satisfactory theory based on Theorem 6.2.8 can be developed in the Markovian context for the case where ξ_t is a "nice" ergodic diffusion process and $\dot{x}_t = A(\xi_t)x_t$. See Arnold, Kliemann and Oeljeklaus [27], Arnold and Kliemann [26] and Bougerol [76, 75].

We will finally use the MET to decouple a linear RDE. While in the deterministic case the (block-)diagonalization of a linear flow $\Phi(t) = \exp(tA)$ and of its generator A are the same problem since $Te^{tA}T^{-1} = e^{t(TAT^{-1})}$, this is not true in the random case: Suppose we (block-)diagonalize the linear cocycle Φ by means of a basis adapted to the Oseledets splitting as in (6.2.6).

Then the resulting (block-)diagonal cocycle $\Psi(t,\omega) = P(\theta_t\omega)\Phi(t,\omega)P(\omega)^{-1}$ is in general not absolutely continuous with respect to t, hence has no generator. This means that the basis which (block-)diagonalizes Φ in general does not (block-)diagonalize the RDE $\dot{x}_t = A(\theta_t\omega)x_t$. What we need to accomplish the latter is a basis of random eigenvectors (see Theorem 6.2.6 for a condition).

6.2.9 Corollary (Diagonalization of a Linear RDE). *Suppose the linear cocycle Φ generated by $\dot{x}_t = A(\theta_t\omega)x_t$ with $A \in L^1(\mathbb{P})$ has a basis $G(\omega) = (u_1(\omega), \ldots, u_d(\omega))$ of random eigenvectors. Then $\Psi(t,\omega)$, the cocycle $\Phi(t,\omega)$ written in the bases $G(\omega)$ and $G(\theta_t\omega)$, is diagonal and generated by the (d uncoupled scalar) RDE*

$$\dot{y}_t = \mathrm{diag}(Q_1(\theta_t\omega), \ldots, Q_d(\theta_t\omega))y_t, \quad (6.2.14)$$

where

$$Q_i(\omega) := Q(\omega, u_i(\omega)) = \langle A(\omega)u_i(\omega), u_i(\omega)\rangle$$

with $\lambda_i = \mathbb{E}\, Q_i$, $i = 1, \ldots, d$. Hence Ψ has the form

$$\Psi(t,\omega) = \mathrm{diag}(\exp\int_0^t Q_1(\theta_r\omega)dr, \ldots, \exp\int_0^t Q_d(\theta_r\omega)dr).$$

Proof. We start with equation (6.2.8). Since Φ has generator A,

$$\Psi_i(t,\omega) = \|\Phi(t,\omega)u_i(\omega)\| = \exp\int_0^t Q(\theta_r\omega, \varphi(r,\omega)u_i(\omega))dr.$$

Since $u_i(\omega)$ is a random eigenvector, $\varphi(r,\omega)u_i(\omega) = u_i(\theta_r\omega)$, hence

$$\dot{\Psi}_i(t,\omega) = Q_i(\theta_t\omega)\Psi_i(t,\omega).$$

$\square$

We will need this corollary in Sect. 8.2. We can write (6.2.14) as

$$\dot{y}_t = \mathrm{diag}(\lambda_i + \bar{Q}_i(\theta_t\omega))y_t,$$

where $\bar{Q}_i := Q_i - \lambda_i$ has mean 0. This is the closest one can in general get to an equation with constant coefficients by means of a linear random coordinate transformation.

White Noise Case

Now let the linear cocycle Φ in $\mathbb{R}^d$ be generated by the SDE

$$dx_t = A_0 x_t dt + \sum_{j=1}^m A_j x_t \circ dW_t^j = \sum_{j=0}^m A_j x_t \circ dW_t^j. \quad (6.2.15)$$

Since the real noise calculations remain formally valid for Stratonovich integrals, (6.2.15) written in polar coordinates is

254   Chapter 6. RDS on Homogeneous Spaces

$$ds_t = \sum_{j=0}^{m} h_j(s_t) \circ dW_t^j, \qquad (6.2.16)$$

$$dr_t = \sum_{j=0}^{m} q_j(s_t) r_t \circ dW_t^j, \qquad (6.2.17)$$

where for $A_j \in \mathbb{R}^{d \times d}$

$$h_j(s) := A_j s - \langle A_j s, s \rangle s, \quad q_j(s) := \langle A_j s, s \rangle.$$

Again equation (6.2.16) is decoupled from (6.2.17) and generates the cocycle φ on S^{d-1}, while (6.2.17) yields with initial condition $r_0 = 1$ the Θ-helix

$$q(t, \cdot, s) = \log \|\Phi(t, \cdot)s\| = \sum_{j=0}^{m} \int_0^t q_j(\varphi(u, \cdot)s) \circ dW_u^j. \qquad (6.2.18)$$

The following integrability condition assures the applicability of the MET to Φ and of the classical ergodic theorem to q in (6.2.18); it also assures the substitution of non-adapted initial values into the stochastic integrals of (6.2.18) in place of s.

6.2.10 Proposition. *For any $1 \leq j \leq m$ and $p \geq 1$*

$$\mathbb{E} \left(\sup_{s \in S^{d-1}} \sup_{0 \leq t \leq 1} \left| \int_0^t q_j(\varphi(u, \cdot)s) \circ d^+ W_u^j \right| \right)^p < \infty. \qquad (6.2.19)$$

The analogous statement holds for $\sup_{-1 \leq t \leq 0}$.

We omit the rather technical proof which consists of verifying the conditions of Proposition 3 of [19], the application of which yields (6.2.19).

Furstenberg-Khasminskii formulas in the white noise case were derived by Khasminskii [205] for the largest exponent (but under a rather restrictive condition), for the largest and smallest exponent by Arnold, Oeljeklaus and Pardoux [32] and for the general case by Arnold and Imkeller [18], the latter paper being based on a non-adapted substitution formula derived by Arnold and Imkeller in [19].

6.2.11 Proposition (Preliminary Furstenberg-Khasminskii Formulas, White Noise Case). *Let Φ be the solution of the SDE (6.2.15). Then:*
(i) $\alpha^\pm \in L^p(\mathbb{P})$ for any $p \geq 1$, where

$$\alpha^\pm(\omega) := \sup_{0 \leq t \leq 1} \log^+ \|\Phi(t, \omega)^{\pm 1}\|.$$

In particular, Φ always satisfies the IC of the MET.
(ii) For any φ-invariant measure μ_i realizing λ_i we have

$$\lambda_i = \sum_{j=0}^{m} \mathbb{E}_{\mu_i} \int_0^1 q_j(\varphi(t,\cdot)s) \circ d^+ W_t^j$$

$$= \mathbb{E}_{\mu_i} Q^+ + \sum_{j=1}^{m} \mathbb{E}_{\mu_i} \int_0^1 q_j(\varphi(t,\cdot)s) \, d^+ W_t^j$$

and

$$\lambda_i = \sum_{j=0}^{m} \mathbb{E}_{\mu_i} \int_{-1}^{0} q_j(\varphi(t,\cdot)s) \circ d^- W_t^j$$

$$= \mathbb{E}_{\mu_i} Q^- + \sum_{j=1}^{m} \mathbb{E}_{\mu_i} \int_{-1}^{0} q_j(\varphi(t,\cdot)s) \, d^- W_t^j,$$

where

$$Q^\pm(s) := q_0(s) \pm \frac{1}{2} \sum_{j=1}^{m} h_j(q_j)(s), \tag{6.2.20}$$

and

$$h_j(q_j)(s) = \langle (A_j + A_j^*)s, A_j s \rangle - 2\langle A_j s, s \rangle^2. \tag{6.2.21}$$

(iii) If μ_i is $\mathcal{F}^-$-measurable realizing λ_i with marginal ρ_i^+ on S^{d-1}, then

$$\lambda_i = \mathbb{E}_{\mu_i} Q^+ = \int_{S^{d-1}} Q^+(s) \rho_i^+(ds).$$

If μ_i is $\mathcal{F}^+$-measurable realizing λ_i with marginal ρ_i^- on S^{d-1}, then

$$\lambda_i = \mathbb{E}_{\mu_i} Q^- = \int_{S^{d-1}} Q^-(s) \rho_i^-(ds).$$

Proof. (i) Since log is increasing, $\|\Phi(0,\omega)\| = 1$ and $|Q_0(s)| \leq \|A_0\|$, relation (6.2.18) gives

$$\alpha^+(\cdot) = \sup_{0 \leq t \leq 1} \log^+ \|\Phi(t,\cdot)\| = \sup_{s \in S^{d-1}} \sup_{0 \leq t \leq 1} \log \|\Phi(t,\cdot)s\|$$

$$= \sup_{s \in S^{d-1}} \sup_{0 \leq t \leq 1} \sum_{j=0}^{m} \int_0^t q_j(\varphi(u,\cdot)s) \circ d^+ W_u^j$$

$$\leq \sum_{j=0}^{m} \sup_{s \in S^{d-1}} \sup_{0 \leq t \leq 1} \left| \int_0^t q_j(\varphi(u,\cdot)s) \circ d^+ W_u^j \right|$$

$$\leq \|A_0\| + \sum_{j=1}^{m} \sup_{s \in S^{d-1}} \sup_{0 \leq t \leq 1} \left| \int_0^t q_j(\varphi(u,\cdot)s) \circ d^+ W_u^j \right|.$$

By Proposition 6.2.10, $\alpha^+ \in L^p(\mathbb{P})$ for any $p \geq 1$.

As for $\alpha^- = \sup_{-1 \leq t \leq 0} \log^+ \|\Phi(t,\cdot)^{-1}\|$, recall that $\Psi(t,\omega) = \Phi(t,\omega)^{*-1}$ is generated by

$$dy_t = \sum_{j=0}^{m}(-A_j^*)y_t \circ dW_t^j$$

(see Theorem 5.1.1(i)) to which the above reasoning applies.

(ii) The first equality signs in the formulas for λ_i follow immediately from Theorem 6.2.3. The second equality signs are a consequence of the following form of the Stratonovich-Itô correction term. We have

$$\int_0^t q_j(\varphi(u,\cdot)s) \circ dW_u^j = \int_0^t q_j(\varphi(u,\cdot)s) \, dW_u^j + \frac{1}{2}\text{sign}(t) \int_0^t h_j(q_j)(\varphi(u,\cdot)s) \, du,$$

where $\text{sign}(t)$ takes care of the cases $t > 0$ (forward integral) and $t < 0$ (backward integral), and $h_j(q_j)$, the function obtained by applying the vector field h_j to the function q_j, is given by

$$h_j(q_j)(s) := \langle h_j(s), \text{grad}_{S^{d-1}} q_j(s) \rangle = \langle (A_j + A_j^*)s, A_j s \rangle - 2\langle A_j s, s \rangle^2$$

since

$$\text{grad}_{S^{d-1}} q_j(s) = (A_j + A_j^*)s - \langle (A_j + A_j^*)s, s \rangle s.$$

(iii) If μ_i is $\mathcal{F}^-$-measurable then

$$F(t,\omega) := \int_{S^{d-1}} q_j(\varphi(t,\omega)s)\mu_{i,\omega}(ds)$$

is adapted, hence using $g(s) := \frac{1}{2}h_j(q_j)(s)$ and recalling that the expectation of an Itô integral with respect to an adapted integrand vanishes,

$$\mathbb{E} \int_0^1 F(t,\cdot) \circ d^+W_t^j = \mathbb{E} \int_0^1 \int_{S^{d-1}} g(\varphi(t,\omega)s)\,\mu_{i,\omega}(ds)\,dt.$$

Since μ_i is φ-invariant,

$$\int_{S^{d-1}} g(\varphi(t,\omega)s)\mu_{i,\omega}(ds) = \int_{S^{d-1}} g(s)\mu_{i,\theta_t\omega}(ds) =: G(\theta_t\omega),$$

thus

$$\mathbb{E}_{\mu_i} \int_0^1 q_j(\varphi(t,\omega)s) \circ d^+W_t^j$$
$$= \mathbb{E} \int_0^1 \int_{S^{d-1}} q_j(\varphi(t,\omega)s)\mu_{i,\omega}(ds) \circ d^+W_t^j$$
$$= \mathbb{E}\,G(\cdot) = \int_{S^{d-1}} g(s)\mathbb{E}\mu.(ds) = \int_{S^{d-1}} g(s)\rho_i^+(ds).$$

Similarly for an $\mathcal{F}^+$-measurable μ_i. □

6.2.12 Remark (IC of the MET). The fact that $\alpha^\pm \in L^p(\mathbb{P})$ for any $p \geq 1$, hence the IC for Φ, can also be directly obtained as follows: First note that for any $A \in \mathbb{R}^{d \times d}$ and for any basis of unit vectors (u_i) in $\mathbb{R}^d$

$$\|A\| \leq d \max_{1 \leq i \leq d} \|Au_i\|.$$

Hence

$$\|\Phi(t, \cdot)\| \leq d \max_i \exp\left(\sum_{j=0}^{m} \int_0^t q_j(\varphi(u, \cdot)u_i) \circ d^+ W_u^j\right)$$

and

$$\alpha^+ \leq \log d + \|A_0\| + \max_i \sum_{j=1}^{m} \sup_{0 \leq t \leq 1} \left|\int_0^t q_j(\varphi(u, \cdot)u_i) \circ d^+ W_u^j\right|.$$

For proving $\alpha^+ \in L^p(\mathbb{P})$ for all $p \geq 1$ it thus suffices to prove that for each fixed $s \in S^{d-1}$

$$\sup_{0 \leq t \leq 1} \left|\int_0^t q_j(\varphi(u, \cdot)s) \circ d^+ W_u^j\right| \in L^p(\mathbb{P}).$$

Converting again the Stratonovich integral into an Itô integral it remains to show that $M_1^* := \sup_{0 \leq t \leq 1} |M_t| \in L^p(\mathbb{P})$, where

$$M_t := \int_0^t q_j(\varphi(u, \cdot)s) \, d^+ W_u^j$$

is a continuous square-integrable martingale.

By the Burkholder-Davis-Gundy inequality (Ikeda and Watanabe [178: p. 110]), for any $p \in (0, \infty)$

$$\begin{aligned}
\mathbb{E}(M_1^*)^{2p} &\leq C_p \mathbb{E}(\langle M, M \rangle_1)^p = C_p \mathbb{E} \left(\int_0^1 q_j(\varphi(u, \cdot)s)^2 du\right)^p \\
&\leq C_p \|A_j\|^{2p} < \infty.
\end{aligned}$$

For showing $\alpha^- \in L^p(\mathbb{P})$ consider the cocycle $\Psi(t, \omega) = \Phi(t, \omega)^{*-1}$ which solves $dy_t = \sum_{j=0}^{m} (-A_j^*) y_t \circ dW_t^j$. ∎

6.2.13 Remark (Stationary Measures for Extreme Exponents).
By Remark 6.2.4, there is always a φ-invariant forward Markov measure μ_1 realizing λ_1 and an invariant backward Markov measure μ_p realizing λ_p.

By Theorem 2.3.45 the φ-invariant forward Markov measures are in a one-to-one correspondence with the stationary measures of the diffusion process given by (6.2.15) for time $\mathbb{R}^+$, i.e. with the solutions of the Fokker-Planck equation

$$(L^+)^* \rho^+ = 0, \quad L^+ := h_0 + \frac{1}{2} \sum_{j=1}^{m} h_j^2.$$

Likewise, there is a one-to-one correspondence between φ-invariant backward Markov measures and solutions to the Fokker-Planck equation of the diffusion process given by (6.2.15) for time $\mathbb{R}^-$,

258   Chapter 6. RDS on Homogeneous Spaces

$$(L^-)^*\rho^- = 0, \quad L^- := -h_0 + \frac{1}{2}\sum_{j=1}^m h_j^2.$$

■

For obtaining true formulas for λ_i it remains to calculate the expectations of the stochastic integrals with respect to the measures μ_i for the general case as explicitly as possible, a highly nontrivial problem due to the non-adaptedness of these measures. We only elaborate the case $t \to \infty$.

6.2.14 Theorem (Furstenberg-Khasminskii Formulas, White Noise Case). *Let Φ be generated by the linear SDE (6.2.15) and assume that Φ has simple Lyapunov spectrum. Then for $1 \leq i \leq d$*

$$\lambda_i = \mathbb{E}\, Q^+(\bar{E}_i) - \sum_{j=1}^m \mathbb{E}\,\mathrm{trace}(A_j + A_j^*)N_i A_j R_i, \tag{6.2.22}$$

where Q^+ is given by (6.2.20), N_i is a random matrix defined by

$$N_i(\omega) := (I - S_i(\omega))^{-1} P_i^+(\omega) P_{d+1-i}^-(\omega)(I - P_i^+(\omega)),$$

$$S_i(\omega) := P_i^+(\omega) P_{d+1-i}^-(\omega) P_i^+(\omega) - R_i(\omega),$$

and $P_i^+(\omega)$, $P_{d+1-i}^-(\omega)$ and $R_i(\omega)$ are the orthogonal projections on $V_i^+(\omega)$, $V_{d+1-i}^-(\omega)$ and on the Oseledets spaces $E_i(\omega) := V_i^+(\omega) \cap V_{d+1-i}^-(\omega)$, respectively.

Proof. Under the integrability condition (6.2.19) the following *anticipative substitution formula* holds: For any random variable $X : \Omega \to S^{d-1}$

$$\int_0^1 q_j(\varphi(t,\cdot)s) \circ d^+W_t^j \bigg|_{s=X(\cdot)} = \int_0^1 q_j(\varphi(t,\cdot)X(\cdot)) \circ d^+W_t^j, \tag{6.2.23}$$

where both sides make sense (see Arnold and Imkeller [19: Corollary 3]).

Now suppose for simplicity that $\mu_{i,\omega} = \delta_{s_i(\omega)}$, $s_i(\omega) \in \bar{E}_i(\omega)$, realizes λ_i. By (6.2.23) and Proposition 6.2.11

$$\lambda_i = \mathbb{E}\int_0^1 q_0(s_i(\theta_t\cdot))dt + \sum_{j=1}^m \mathbb{E}\int_0^1 q_j(s_i(\theta_t\cdot)) \circ d^+W_t^j.$$

Since $\langle A_j s_i, s_i \rangle = \mathrm{trace}\, A_j R_i$

$$\mathbb{E}\int_0^1 q_0(s_i(\theta_t\cdot))dt = \mathbb{E}\, q_0(s_i) = \mathbb{E}\,\mathrm{trace}(A_0 R_i)$$

and

$$C_{ij} := \mathbb{E} \int_0^1 q_j(s_i(\theta_t \cdot)) \circ d^+ W_t^j = \mathbb{E} \int_0^1 \text{trace}(A_j R_i(\theta_t \cdot)) \circ d^+ W_t^j$$

for $1 \le i \le d$ and $1 \le j \le m$. "All" we have to calculate is C_{ij} which is a technically formidable task, but can be accomplished by using Malliavin calculus. For details see [18]. □

Particular cases:
 1. Case $i = 1$: Here $N_1 = 0$ so the correction term in (6.2.22) vanishes identically and we recover the classical Furstenberg-Khasminskii formula for λ_1,

$$\lambda_1 = \mathbb{E} Q^+(\bar{E}_1) = \int_{S^{d-1}} Q^+(s) \rho_1(ds), \tag{6.2.24}$$

where ρ_1 is the distribution of $\bar{E}_1$ which solves the Fokker-Planck equation $(L^+)^* \rho_1 = 0$ since $\bar{E}_1$ is $\mathcal{F}^-$-measurable.

 2. Case $i = d$: Here $N_d = I - R_d$, hence

$$\text{trace}(A_j + A_j^*) N_d A_j R_d = h_j(q_j)(\bar{E}_d),$$

thus

$$\lambda_d = \mathbb{E} Q^-(\bar{E}_d) = \int_{S^{d-1}} Q^-(s) \rho_d(ds),$$

where ρ_d is the distribution of $\bar{E}_d$ which solves the Fokker-Planck equation $(L^-)^* \rho_d = 0$ since $\bar{E}_d$ is $\mathcal{F}^+$-measurable.

6.2.15 Remark (Simple Lyapunov Spectrum). Assume that the subgroup G of $Gl(d, \mathbb{R})$ generated by the matrices $A_0, \ldots, A_m$ equals $Sl(d, \mathbb{R})$ or $Gl^+(d, \mathbb{R})$. Then the Lyapunov spectrum of Φ is simple (see Bougerol and Lacroix [77: Example 9.3]). We have $G = Gl^+(d, \mathbb{R})$ for an open and dense set of $(A_0, \ldots, A_m) \in (\mathbb{R}^{d \times d})^{m+1}$. Thus for matrices from this set (in particular: generically) the corresponding SDE $dx_t = \sum_{j=0}^m A_j x_t \circ dW_t^j$ has simple Lyapunov spectrum. ■

We would like to draw the reader's attention to a line of research with many applications (which is, due to our chosen restrictions, beyond the scope of this book) by quoting the following theorem.

6.2.16 Theorem. *Given the linear SDE (6.2.15). Assume that the vector fields of the corresponding SDE (6.2.16) (interpreted on P^{d-1}) satisfy the following hypoellipticity condition*

$$\dim \mathcal{LA}(h_0, \ldots, h_m)(\hat{x}) = d - 1 \quad \text{for all } \hat{x} \in P^{d-1}, \tag{6.2.25}$$

where $\mathcal{LA}(Z)$ denotes the Lie algebra generated by the set Z of vector fields. Then the following holds:

(i) The diffusion process on P^{d-1} generated by (6.2.16) with generator $L = h_0 + \frac{1}{2} \sum_{j=1}^m (h_j)^2$ admits a unique stationary measure ρ. The measure ρ has a C^∞ density with respect to Lebesgue measure on P^{d-1}.

(ii) We have
$$\lambda_1 = \int_{\mathbb{P}^{d-1}} Q^+(\hat{x})\rho(d\hat{x}). \qquad (6.2.26)$$
In particular, the invariant Markov measure μ corresponding to ρ is supported by $\hat{E}_1$, and is the only such Markov measure.

(iii) Law of large numbers: For each fixed $x \neq 0$ (more generally, for each random $x(\cdot) \in \mathcal{F}^-$)
$$\lambda(\omega, x) = \lim_{t \to \infty} \frac{1}{t} \log \|\Phi(t, \omega)x\| = \lambda_1 \quad \mathbb{P}\text{-a. s.} \qquad (6.2.27)$$
In other words, the projection of any fixed vector x onto the random subspace E_1 does not vanish $\mathbb{P}$-a. s.

The only hard part is to prove the uniqueness of ρ. Arnold, Oeljeklaus and Pardoux [32] use a transfer mechanism between diffusion processes and geometric control theory known as "support theorems" and the characterization of $C = \text{supp}\rho$ as an invariant control set by Kliemann [212, 26]. It turns out that under (6.2.25), C is unique and has nonvoid interior. For an up-to-date account see Colonius and Kliemann [105]. There is also a central limit theorem and large deviation theory associated with (6.2.27), etc. (see Arnold and Kliemann [26], Baxendale [58], Arnold and Khasminskii [23], and the references therein).

The formula (6.2.26) has been and still is the starting point of numerous asymptotic studies of λ_1 for small or large white noise, to prove stabilization or destabilization by noise, the existence of stochastic bifurcation (see Chap. 9), etc. From the vast literature we just mention Ariaratnam and Xie [7], Arnold, Eizenberg and Wihstutz [17], Kao and Wihstutz [196], Pardoux and Wihstutz [270, 271], Pinsky [278], and Pinsky and Wihstutz [279].

In Sect. 8.4 we need the following white noise analogue of Corollary 6.2.9.

6.2.17 Corollary (Diagonalization of Linear SDE). *Suppose the linear RDS generated by (6.2.15) has a basis $G(\omega) = (u_1(\omega), \ldots, u_d(\omega))$ of random eigenvectors. Then $\Psi(t, \omega)$, the cocycle $\Phi(t, \omega)$ written in the bases $G(\omega)$ and $G(\theta_t \omega)$, is diagonal and generated by the (d uncoupled scalar) SDE*
$$dy_t = \sum_{j=0}^{m} \text{diag}(q_j(u_1(\theta_t \cdot)), \ldots, q_j(u_d(\theta_t \cdot)))y_t \circ dW_t^j$$
with stationary anticipative coefficients.

See Arnold and Imkeller [19: Theorem 9].

6.2.3 Spectrum and Splitting

In Theorem 6.2.3 we have determined all invariant measures of φ, the C^∞ RDS on S^{d-1} induced by the linear RDS Φ on $\mathbb{R}^d$. We now apply the manifold version of the MET (Theorem 4.2.6) to $T\varphi$ and a φ-invariant measure μ and explicitly determine its Lyapunov spectrum and Oseledets splitting.

6.2.18 Lemma. Let $1 \leq k \leq d-1$, $v_1, \ldots, v_k \in \mathbb{R}^d$ and $s \in S^{d-1}$. Further, let
$$w_i = T_s\bar{\pi}(v_i) = v_i - \langle v_i, s\rangle s, \quad i = 1, \ldots, k,$$
be the orthogonal projection of v_i to $s^\perp$. Then
$$\|w_1 \wedge \ldots \wedge w_k\|_k = \|v_1 \wedge \ldots \wedge v_k \wedge s\|_{k+1}.$$

Proof. By definition (see Lemma 3.2.6(iv)), using $s \in S^{d-1}$ and $w_i \perp s$,
$$\begin{aligned}\|w_1 \wedge \ldots \wedge w_k \wedge s\|_{k+1}^2 &= \det\begin{pmatrix}(\langle w_i, w_j\rangle)_{k\times k} & 0 \\ 0 & 1\end{pmatrix} \\ &= \|w_1 \wedge \ldots \wedge w_k\|_k^2.\end{aligned}$$

Further, using the elementary rules of Subsect. 3.2.3,
$$\begin{aligned}w_1 \wedge \ldots \wedge w_k \wedge s &= (v_1 - \langle v_1, s\rangle s) \wedge \ldots \wedge (v_k - \langle v_k, s\rangle s) \wedge s \\ &= v_1 \wedge \ldots \wedge v_k \wedge s \\ &\quad - \sum_{i=1}^k \langle v_i, s\rangle w_1 \wedge \ldots \wedge w_{i-1} \wedge s \wedge w_{i+1} \wedge \ldots \wedge w_k \wedge s \\ &= v_1 \wedge \ldots \wedge v_k \wedge s.\end{aligned}$$

$\square$

6.2.19 Proposition. Let $A \in Gl(d, \mathbb{R})$, and $\bar{A}: S^{d-1} \to S^{d-1}$ be the diffeomorphism induced by A. Then for any $s \in S^{d-1}$ the linear operator
$$T\bar{A}(s): T_s S^{d-1} = s^\perp \to T_{\bar{A}s} S^{d-1} = (\bar{A}s)^\perp$$
is given by
$$T\bar{A}(s)(v) = \frac{1}{\|As\|}\left(Av - \frac{1}{\|As\|^2}\langle Av, As\rangle As\right), \quad v \in s^\perp \subset \mathbb{R}^d. \quad (6.2.28)$$
Moreover,
$$\|T\bar{A}(s)(v)\| = \frac{\|Av \wedge As\|}{\|As\|^2},$$
hence for all $s \in S^{d-1}$
$$\|T\bar{A}(s)\| = \frac{\sup_{\|v\|=1, v\perp s}\|Av \wedge As\|}{\|As\|^2} \leq \|A\|^2 \|A^{-1}\|^2.$$

Proof. Since $\bar{A} \circ \bar{\pi} = \bar{\pi} \circ A$,
$$T(\bar{A} \circ \bar{\pi}) = T(\bar{\pi} \circ A) = T\bar{\pi} \circ TA = T\bar{\pi} \circ A,$$
more precisely, since $\bar{\pi}(s) = s$,

$$T\bar{A}(s)(v) = T_{As}\bar{\pi}(Av),$$

so that (6.2.1) gives (6.2.28).

We now use Lemma 6.2.18 for $k=1$, $v_1 = Av$, $w_1 = Av - \langle Av, \bar{A}s\rangle \bar{A}s$ and $\bar{A}s$ instead of s, which gives

$$\|T\bar{A}(s)(v)\| = \frac{1}{\|As\|}\|Av - \langle Av, \bar{A}s\rangle \bar{A}s\|$$

$$= \frac{1}{\|As\|}\|Av \wedge \frac{As}{\|As\|}\| = \frac{\|\wedge^2 A(v \wedge s)\|}{\|As\|^2}.$$

Since by Proposition 3.2.7(iii)

$$\|\wedge^2 A(v \wedge s)\| \le \|\wedge^2 A\|\,\|v\| \le \|A\|^2\|v\|$$

and

$$\|A^{-1}\| = \sup_{\|s\|=1} \frac{1}{\|As\|},$$

we obtain

$$\|T\bar{A}(s)\| = \frac{\sup_{\|v\|=1, v\perp s}\|Av \wedge As\|}{\|As\|^2} \le \|A\|^2\,\|A^{-1}\|^2.$$

$\square$

6.2.20 Theorem (Spectrum and Splitting of Induced RDS on S^{d-1}). *Let Φ be a linear RDS on $\mathbb{R}^d$ over an ergodic DS θ with two-sided time satisfying the IC of the MET with spectrum $\mathcal{S}(\theta, \Phi) = \{(\lambda_i, d_i)_{i=1,\ldots,p}\}$ and splitting $\mathbb{R}^d = \oplus_{i=1}^p E_i(\omega)$. Denote by φ the RDS on S^{d-1} induced by Φ.*

Let $i_0 \in \{1, \ldots, p\}$, and let $\mu = \mu_{i_0}$ be a φ-invariant measure realizing λ_{i_0}. Then:

(i) $T\varphi$ and μ_{i_0} satisfy the IC of the MET.

(ii) The Lyapunov spectrum

$$\mathcal{S}(\theta, \varphi, \mu_{i_0}) = \{(\bar{\lambda}_j, \bar{d}_j)_{j=1,\ldots,p}\}$$

is given by

$$\bar{\lambda}_j = \begin{cases} \lambda_j - \lambda_{i_0}, & j \ne i_0, \\ 0, & j = i_0, \end{cases} \quad \bar{d}_j = \begin{cases} d_j, & j \ne i_0, \\ d_{i_0} - 1, & j = i_0, \end{cases}$$

where for $d_{i_0} = 1$ the case $j = i_0$ is not present, in which case we only have $p - 1$ Lyapunov exponents.

(iii) On the invariant set of full μ_{i_0}-measure (which is contained in $\bar{E}_{i_0}(\omega) = E_{i_0} \cap S^{d-1}$) the splitting $T_s S^{d-1} = \oplus_{j=1}^p \bar{E}_j(\omega, s)$ is

$$\bar{E}_j(\omega, s) = \mathrm{span}\{v - \langle v, s\rangle s : v \in E_j(\omega)\} \subset s^\perp = T_s S^{d-1}, \quad j = 1, \ldots, p$$

(for $d_{i_0} = 1$, $j \ne i_0$ only, and $\bar{E}_{i_0}(\omega, s)$ is not present).

Proof. (i) By Proposition 6.2.19,
$$\log^+ \|T\bar{A}(s)\| \leq 2(\log^+ \|A\| + \log^+ \|A^{-1}\|).$$

(ii) and (iii): We prove our assertion by exhibiting a basis of vectors which span $\bar{E}_j(\omega, s)$ and have growth rate $\lambda_j - \lambda_{i_0}$ for $t \to \pm\infty$. By uniqueness of spectrum and splitting, we are done.

Pick (ω, s) in the invariant set of full μ_{i_0}-measure (in particular, $s \in E_{i_0}(\omega)$). Now choose a random basis which is adapted to the splitting (E_i) (i.e. d_1 orthogonal unit vectors in E_1, d_2 in E_2, etc.) such that s is the first vector chosen in E_{i_0}. Denote the basis vectors without s by F.

The images $w = v - \langle v, s \rangle s$ of the vectors $v \in F$ under $T_s\bar{\pi}$, where $T_s\bar{\pi}$ is the orthogonal projection of $\mathbb{R}^d$ onto $s^\perp$ defined by $T_s\bar{\pi}(v) = v - \langle v, s \rangle s$, form a basis of $T_s S^{d-1}$, where $\bar{E}_j(\omega, s) = T_s\bar{\pi}(E_j(\omega))$ is spanned by the images of the basis vectors in $E_j(\omega)$.

We have for $w \in \bar{E}_j(\omega, s)$ by Proposition 6.2.19
$$\begin{aligned}
\|T\varphi(t, \omega, s)w\| &= \|\Phi(t,\omega)w \wedge \Phi(t,\omega)s\|/\|\Phi(t,\omega)s\|^2 \\
&= \|\Phi(t,\omega)v \wedge \Phi(t,\omega)s\|/\|\Phi(t,\omega)s\|^2 \\
&= \|\wedge^2 \Phi(t,\omega)(v \wedge s)\|/\|\Phi(t,\omega)s\|^2.
\end{aligned}$$

Since by construction $\Phi(t,\omega)s$ has growth rate λ_{i_0} and $\wedge^2 \Phi(t,\omega)(v \wedge s)$ has growth rate $\lambda_j + \lambda_{i_0}$ (Theorem 5.3.1),
$$\lim_{t \to \pm\infty} \frac{1}{t} \log \|T\varphi(t,\omega,s)w\| = \lambda_j + \lambda_{i_0} - 2\lambda_{i_0} = \lambda_j - \lambda_{i_0}.$$
□

The spectrum of φ under μ_{i_0} is thus obtained by subtracting λ_{i_0} from the exponents of Φ and by reducing the multiplicity of the exponent 0 by 1 (multiplicity 0 means removal).

If μ realizes a Lyapunov exponent which is simple, then φ is hyperbolic under μ.

If Φ has simple spectrum, then φ is hyperbolic under any φ-invariant measure. The only stable measures are the ones realizing λ_1; a measure realizing λ_d has exponents $\lambda_j - \lambda_d > 0$, $j = 1, \ldots, d-1$.

We refrain from treating the RDS $\hat{\Phi}$ induced on P^{d-1} for which completely analogous statements hold.

6.3 RDS Induced on Grassmannian Manifolds

We now aim at Furstenberg-Khasminskii formulas for sums of k Lyapunov exponents of Φ for which we have to study the RDS φ_k induced by Φ on the Grassmannian $G_k(d)$.

6.3.1 Invariant Measures

In this subsection we basically follow Crauel [109], [111].

The Manifolds $G_k(d)$ and $G_k^+(d)$

Let $1 \leq k \leq d$. The *Grassmannian manifold* $G_k(d)$ is the manifold of k-dimensional linear subspaces of $\mathbb{R}^d$. It is a compact smooth manifold of dimension $\dim G_k(d) = k(d-k)$, which can be easily seen by considering $G_k(d)$ as a homogeneous space G/H of $G = O(d, \mathbb{R})$, where

$$H = \begin{pmatrix} O(k, \mathbb{R}) & 0 \\ 0 & O(d-k, \mathbb{R}) \end{pmatrix}.$$

Note that $G_1(d) = P^{d-1}$, $G_{d-1}(d) \cong P^{d-1}$ and $G_d(d)$ is a one-point set.

However, we here prefer to represent $G_k(d)$ as a subset of the projective space $P(\wedge^k \mathbb{R}^d)$ obtained as the projection $\pi_k(\wedge_0^k \mathbb{R}^d)$ of the set of decomposable vectors $\wedge_0^k \mathbb{R}^d \subset \wedge^k \mathbb{R}^d$ (for details see Subsect. 3.2.3), where

$$\pi_k : \wedge^k \mathbb{R}^d \to P(\wedge^k \mathbb{R}^d)$$

is the canonical projection. Hence

$$G_k(d) = \pi_k(\wedge_0^k \mathbb{R}^d) \subset P(\wedge^k \mathbb{R}^d)$$

is considered as a $k(d-k)$-dimensional (proper if $1 < k < d-1$) compact connected submanifold of $P(\wedge^k \mathbb{R}^d)$.

A linear mapping $A \in Gl(d, \mathbb{R})$ induces naturally a smooth diffeomorphism $\hat{A}_k$ on $G_k(d)$ via

$$G_k(d) \ni x \mapsto \hat{A}_k(x) \in G_k(d),$$

where $\hat{A}_k(x)$ is the image of the subspace x under A. The mapping A also lifts naturally to $\wedge^k \mathbb{R}^d$ by linear extension of

$$\left(\wedge^k A \right) (u_1 \wedge u_2 \wedge \ldots \wedge u_k) = Au_1 \wedge Au_2 \wedge \ldots \wedge Au_k,$$

obviously leaving the set $\wedge_0^k \mathbb{R}^d$ of decomposable vectors invariant. $\wedge^k A$ in turn projects down to $P(\wedge^k \mathbb{R}^d)$ as $\pi_k(\wedge^k A)$ and the latter coincides with $\hat{A}_k$ from above on the invariant manifold $G_k(d)$.

A linear RDS Φ in $\mathbb{R}^d$ hence induces a smooth RDS φ_k on $G_k(d)$ by

$$\varphi_k = \pi_k(\wedge^k \Phi)|_{G_k(d)}.$$

For computational reasons we prefer to work on the oriented Grassmannian (in which every subspace appears twice, but with opposite orientation)

$$G_k^+(d) = \bar{\pi}_k(\wedge_0^k \mathbb{R}^d) \subset S(\wedge^k \mathbb{R}^d),$$

which is a submanifold of the unit sphere $S(\wedge^k \mathbb{R}^d)$ of $\wedge^k \mathbb{R}^d$ and a two-fold covering of $G_k(d) = G_k^+(d)/\mathbb{Z}_2$. As is well-known from linear algebra, two k-frames $(x_1, \ldots, x_k)$ and $(y_1, \ldots, y_k)$ spanning the same element of $G_k(d)$, hence satisfying $y_1 \wedge \ldots \wedge y_k = \lambda(x_1 \wedge \ldots \wedge x_k)$ for some $\lambda \neq 0$, have the same orientation if and only if $\lambda > 0$.

$G_k^+(d)$ and $G_k(d)$ are in exactly the same relation as $G_1^+(d) = S^{d-1}$ and $G_1(d) = P^{d-1}$ in Sect. 6.2, hence we will spare most details and abuse notation by using the same symbol for corresponding objects of $G_k^+(d)$ and $G_k(d)$. If we write $G_k^{(+)}(d)$, then the statement holds for both $G_k(d)$ and $G_k^+(d)$.

Invariant Measures for φ_k on $G_k(d)$

As in the particular case $k=1$ (see Subsect. 6.2.1), the invariant measures of φ_k can be classified by the MET – but now by the MET for the linear cocycle $\wedge^k \Phi$ on $\wedge^k \mathbb{R}^d$ (see Theorem 5.3.1). In fact, Theorem 6.2.3 holds verbatim for $\pi_k(\wedge^k \Phi)$, now based on the spectrum $\mathcal{S}(\theta, \wedge^k \Phi) = \{(\lambda_i^{(k)}, d_i^{(k)})_{i=1,\ldots,p(k)}\}$ and the splitting $\wedge^k \mathbb{R}^d = \oplus_{i=1}^{p(k)} E_i^{(k)}$.

But here we are not so much interested in the full system $\pi_k(\wedge^k \Phi)$, but in its restriction φ_k to $G_k(d)$.

Note first that since $E_i^{(k)}$ is the span of certain decomposable vectors,

$$G_i^{(k)}(\omega) := G_k(d) \cap \pi_k(E_i^{(k)}(\omega)), \quad i = 1, \ldots, p(k),$$

are nonvoid disjoint φ_k-invariant random compact submanifolds of $G_k(d)$, hence $\mathcal{I}_\mathbb{P}(\varphi_k|G_i^{(k)})$ is nonvoid. We can thus (but will not) repeat the arguments leading to Theorem 6.2.3 for $G_i^{(k)}$ and obtain the following theorem.

6.3.1 Theorem (Invariant Measures on $G_k(d)$). *Let Φ be a linear RDS on $\mathbb{R}^d$ over the ergodic metric DS θ with two-sided time satisfying the IC of the MET. Assume the situation described above. Then:*

(i) For any $i = 1, \ldots, p(k)$, the set $\mathcal{I}_\mathbb{P}(\varphi_k|G_i^{(k)})$ of φ_k-invariant measures supported by $G_i^{(k)}$ is nonvoid, compact and convex.

(ii) Each ergodic φ_k-invariant measure μ realizes one of the $\lambda_i^{(k)}$, i. e. for each μ there is a unique $i \in \{1, \ldots, p(k)\}$ such that

$$\lim_{t \to \pm\infty} \frac{1}{t} \log \frac{\| \wedge^k \Phi(t, \omega) u \|}{\|u\|} = \lambda_i^{(k)} \quad \mu\text{-a.s.},$$

equivalently $\mu \in \mathcal{I}_\mathbb{P}(\varphi_k|G_i^{(k)})$.

(iii) Conversely, for any $\lambda_i^{(k)}$ there exists a φ_k-invariant measure which realizes $\lambda_i^{(k)}$.

The spectrum of the induced RDS φ_k on $G_k(d)$ under μ can also be determined (see Crauel [111]).

6.3.2 Furstenberg-Khasminskii Formulas

We now generalize the procedure of Subsect. 6.2.3 and obtain representations of the sum $\lambda_i^{(k)} = \Lambda_{i_1} + \ldots + \Lambda_{i_k}$ of any of k exponents of Φ as a phase average of a function $Q^{(k)}$ on $G_k(d)$ with respect to a φ_k-invariant measure $\mu_i^{(k)}$ realizing $\lambda_i^{(k)}$. For $k > 1$, these formulas were first derived by Baxendale [57] (for the white noise and manifold case; see Sect. 6.5).

Let Φ be a linear cocycle in $\mathbb{R}^d$ satisfying the IC of the MET. Let $\varphi_k = \pi_k(\wedge^k \Phi)|_{G_k^+(d)}$ be the cocycle induced on the oriented Grassmannian $G_k^+(d)$. Then

$$q_k(t, \omega, u) := \log \| \wedge^k \Phi(t, \omega) u \| = \log \mathrm{vol}(\Phi(t,\omega)u_1, \ldots, \Phi(t,\omega)u_k),$$

$u = u_1 \wedge \ldots \wedge u_k \in G_k^+(d)$, is a helix over (θ, φ_k) and for each $\lambda_i^{(k)}$ there is a φ_k-invariant measure $\mu_i^{(k)}$ realizing $\lambda_i^{(k)}$, i.e.

$$\lim_{t \to \pm\infty} \frac{1}{t} q_k(t, \omega, u) = \lambda_i^{(k)}, \quad \mu_i^{(k)}\text{-a.s.}$$

We now assume that Φ has a generator. Then $\wedge^k \Phi$ has a generator (see Theorem 5.3.1(iv)), hence φ_k has a generator (see Subsect. 6.2.3) and the helix q_k has a generator whose explicit form will now be determined.

We will, however, restrict ourselves to the continuous time white noise case (from which the real noise case can be recovered by ignoring the sum $\sum_{j=1}^{m}$).

Let Φ be generated by the linear SDE (6.2.15). Then Φ, hence $\wedge^k \Phi$ satisfy the IC of the MET and $\wedge^k \Phi$ is generated by the linear SDE

$$du_t = \sum_{j=0}^{m} \hat{A}_j^k u_t \circ dW_t^j,$$

where

$$\hat{A}_j^k(u_1 \wedge \ldots \wedge u_k) = \sum_{i=1}^{k} u_1 \wedge \ldots \wedge u_{i-1} \wedge A_j u_i \wedge u_{i+1} \wedge \ldots \wedge u_k.$$

Hence φ_k on $G_k^+(d)$ is generated by

$$du_t = \sum_{j=0}^{m} h_j^{(k)}(u_t) \circ dW_t^j,$$

where, with

$$q_j^{(k)}(u) := \langle \hat{A}_j^k u, u \rangle = \mathrm{trace} A_j P_k(u),$$

$P_k(u)$ being the orthogonal projection onto $U := \mathrm{span}(u_1, \ldots, u_k)$,

$$h_j^{(k)}(u) = T\pi_k \hat{A}_j^k(u) = \hat{A}_j^k u - q_j^{(k)}(u)u = (A_j - \frac{1}{k}q_j^{(k)}(u)I)^{\wedge,k} u$$

(for convenience of notation, we sometimes write $B^{\wedge,k}$ instead of $\hat{B}^k$). Finally, the SDE for

$$r_t = \| \wedge^k \Phi(t,\omega)u \| = \text{vol}(\Phi(t,\omega)u_1, \ldots, \Phi(t,\omega)u_k)$$

is

$$dr_t = \sum_{j=0}^{m} q_j^{(k)}(u_t) r_t \circ dW_t^j.$$

Hence we obtain Liouville's formula

$$\log \det \Phi(t,\cdot)|_U = \sum_{j=0}^{m} \int_0^t q_j^{(k)}(\varphi_k(r,\cdot)u) \circ dW_r^j.$$

Converting Stratonovich to Itô integrals for $1 \le j \le m$ (and staying with the case $t > 0$) yields

$$\int_0^t q_j^{(k)}(\varphi_k(r,\cdot)u) \circ dW_r^j$$
$$= \int_0^t q_j^{(k)}(\varphi_k(r,\cdot)u) \, dW_r^j + \frac{1}{2} \int_0^t h_j^{(k)}(q_j^{(k)})(\varphi_k(r,\cdot)u) dr.$$

For the calculation of the correction terms we use the following lemma whose part (i) generalizes Lemma 3.2.6(vii) (note that $\hat{I}^k u = k\, u$).

6.3.2 Lemma. (i) For $A, B \in \mathbb{R}^{d \times d}$, $1 \le k \le d$ and $u = u_1 \wedge \ldots \wedge u_k \in \wedge^k \mathbb{R}^d$

$$\begin{aligned}\langle \hat{A}^k u, \hat{B}^k u \rangle &= (\text{trace} A^* B P_k(u) + (\text{trace} A P_k(u))(\text{trace} B P_k(u)) \\ &\quad - \text{trace} A^* P_k(u) B P_k(u)) \|u\|^2 \\ &= (\text{trace} A^* (I - P_k(u)) B P_k(u) \\ &\quad + (\text{trace} A P_k(u))(\text{trace} B P_k(u))) \|u\|^2,\end{aligned}$$

where $P_k(u)$ is the orthogonal projection onto $\text{span}(u_1, \ldots, u_k)$.

(ii) The correction terms $h_j^{(k)}(q_j^{(k)})$ have the form

$$h_j^{(k)}(q_j^{(k)})(u) = \text{trace}(A_j + A_j^*)(I - P_k(u))A_j P_k(u).$$

Proof. (i) (after notes of Schmalfuß (1994)): We first assume that $u_1, \ldots, u_k$ is an orthogonal basis of u. By the definition of $\hat{A}^k$

$$\langle \hat{A}^k u, \hat{B}^k u \rangle = \sum_{l,m} \langle u_1 \wedge \ldots \wedge A u_l \wedge \ldots \wedge u_k, u_1 \wedge \ldots \wedge B u_m \wedge \ldots \wedge u_k \rangle.$$

For each summand we use the definition of the scalar product in $\wedge^k \mathbb{R}^d$ as a determinant, which we expand along its nontrivial row/column.

If $u_1, \ldots, u_k$ is not orthogonal, note that both sides of the claimed equation remain unchanged if $u_1, \ldots, u_k$ are replaced with the orthogonal basis $v_1, \ldots, v_k$ obtained by the well-known Gram-Schmidt procedure (i. e. the v_j are orthogonal with $v_j \in u_j + \text{span}(u_1, \ldots, u_{j-1})$, $j = 1, \ldots, k$).

(ii) By (6.2.21)

$$h_j^{(k)}(q_j^{(k)})(u) = \langle (\hat{A}_j^k + (\hat{A}_j^k)^*)u, \hat{A}_j^k u \rangle - 2\langle \hat{A}_j^k u, u \rangle^2.$$

Since by Lemma 3.2.6(vii) $\hat{A}_j^k + (\hat{A}_j^k)^* = (A_j + A_j^*)^{\wedge,k}$,

$$h_j^{(k)}(q_j^{(k)})(u) = \langle (A_j + A_j^*)^{\wedge,k} u, \hat{A}_j^k u \rangle - 2\langle \hat{A}_j^k u, u \rangle^2.$$

We now use part (i) of this lemma for both terms on the right-hand side of the last equation and the fact that $\text{trace} A_j^* P_k(u) = \text{trace} A_j P_k(u)$. □

6.3.3 Theorem (Furstenberg-Khasminskii Formula for Sum of Lyapunov Exponents, White Noise Case). *Let Φ be generated by the linear SDE (6.2.15) and let φ_k be the RDS induced by Φ on $G_k^{(+)}(d)$. If $\mu_i^{(k)}$ is a φ_k-invariant measure on $G_k^{(+)}(d)$ realizing $\lambda_i^{(k)}$, then*

$$\lambda_i^{(k)} = \mathbb{E}_{\mu_i^{(k)}} Q^{(k)} + \sum_{j=1}^{m} \mathbb{E}_{\mu_i^{(k)}} \int_0^1 q_j^{(k)}(\varphi_k(t,\cdot)\cdot)\, dW_t^j, \tag{6.3.1}$$

where

$$Q^{(k)}(u) = \text{trace} A_0 P_k(u) + \frac{1}{2} \sum_{j=1}^{m} \text{trace}(A_j + A_j^*)(I - P_k(u))A_j P_k(u),$$

and $P_k(u)$ is the orthogonal projection onto $\text{span}(u_1, \ldots, u_k)$, $u = u_1 \wedge \ldots \wedge u_k$. If $\mu_i^{(k)}$ is a forward Markov measure with marginal $\rho_i^{(k)}$ on $G_k^{(+)}(d)$, then

$$\lambda_i^{(k)} = \int_{G_k^{(+)}(d)} Q^{(k)}(u)\, \rho_i^{(k)}(du).$$

In particular, there is always a forward Markov measure realizing the top exponent $\lambda_1^{(k)} = \Lambda_1 + \ldots + \Lambda_k$.

Proof. Just use Theorem 6.2.11 and Lemma 6.3.2. □

We could also derive a more explicit expression of the expectations of the second term in formula (6.3.1) using Malliavin calculus.

For $k = d$ we have $P_d(u) = I$, $q_j^{(d)} = \text{trace} A_j$, hence

$$\lambda_\Sigma := \frac{1}{d} \sum_{i=1}^{p} d_i \lambda_i = \frac{1}{d} \text{trace} A_0.$$

6.3.4 Exercise (Flag Manifolds). Let $\tau = (d_1, \ldots, d_r)$, $1 \leq d_1 < \ldots < d_r \leq d$, be a multi-index of dimensions and let

$$F_\tau(d) := \{f_\tau = (V_1 \subset \ldots \subset V_r) : V_i \subset \mathbb{R}^d, \dim V_i = d_i\}$$

be the flag manifold corresponding to τ. Determine the generator of the RDS induced on $F_\tau(d)$ by Φ. Develop Furstenberg-Khasminskii formulas for the r-vector whose i-th component consists of the sum of d_i Lyapunov exponents. In particular, for $\tau = (1, 2, \ldots, d)$ this gives a formula for the whole Lyapunov spectrum. ∎

6.4 Manifold Versions

For further use and ease of reference we will now briefly develop the manifold versions of Sects. 6.2 and 6.3: For a C^1 RDS φ with two-sided time on a Riemannian manifold M we study the RDS induced by the linearization $T\varphi$ on the unit sphere bundle SM, the projective bundle PM and the Grassmann bundles $G_k M$. The induced systems are denoted by $S\varphi$, $P\varphi$ and $G_k\varphi$, respectively.

Suppose μ is an invariant measure for φ for which $T\varphi$ satisfies the IC of the MET. Then we are able to describe all invariant measures $S\mu$ of $S\varphi$ on SM, etc., which are lifts of μ, i.e. which project down to μ under the canonical projection of $\pi : SM \to M$, $\pi(S\mu) = \mu$.

If φ has a generator, then all the above induced RDS have generators, which we will determine. By introducing polar coordinates in TM, respectively in $\wedge^k TM := \cup_{x \in M} \wedge^k T_x M$, we can derive Furstenberg-Khasminskii formulas for the Lyapunov exponents, respectively for the sum of Lyapunov exponents, of φ under μ.

6.4.1 Sphere Bundle and Projective Bundle

Let M be a Riemannian manifold of dimension $d \geq 1$ with tangent bundle $TM := \cup_{x \in M} T_x M$, unit sphere bundle $SM := \cup_{x \in M} S_x M$ and projective bundle $PM := \cup_{x \in M} P_x M$ ($S_x M$ the unit sphere and $P_x M$ the projective space of $T_x M$). To avoid redundancy we only treat the sphere bundle case here.

The following proposition is the nonlinear analogue of parts of Proposition 6.2.1 and Theorem 6.2.3, with analogous proofs (use $\Omega \times M$ and Θ in place of Ω and θ, and recall that SM has compact fibers).

6.4.1 Proposition. *Let φ be a C^1 RDS on M with two-sided time, let $T\varphi$ be the linearization of φ on TM and let*

$$S\varphi(t,\omega,x)s := \frac{T\varphi(t,\omega,x)s}{\|T\varphi(t,\omega,x)s\|_{\varphi(t,\omega)x}}, \quad s \in S_x M, \qquad (6.4.1)$$

be the RDS $S\varphi$ induced by $T\varphi$ on SM.

Suppose μ is an ergodic φ-invariant measure. Denote by $S\mu$ an invariant measure of $S\varphi$ which is a lift of μ, i. e. the image of which under the projection $\pi : SM \to M$ is μ, and let

$$\mathcal{I}_\mu(S\varphi) := \{S\mu : S\mu \text{ is } S\varphi\text{-invariant and a lift of } \mu\}.$$

Then the following hold:

(i) $\mathcal{I}_\mu(S\varphi)$ is a nonvoid convex compact set.

(ii) Let $T\varphi$ under μ satisfy the IC of the MET. Let $\Delta \subset \Omega \times M$ be the invariant set of full μ measure on which the MET holds and denote by $\mathcal{S}(\varphi, \mu) = \{(\lambda_i, d_i)_{i=1,\ldots,p}\}$ the Lyapunov spectrum and by $T_x M = \oplus_{i=1}^{p} E_i(\omega, x)$ the Oseledets splitting. Then

$$SE_i := \cup_{(\omega,x) \in \Delta} E_i(\omega, x) \cap S_x M, \quad i = 1, \ldots, p,$$

are disjoint strictly $S\varphi$-invariant subbundles of SM and

$$\mathcal{I}_\mu(S\varphi|SE_i) := \{S\mu \in \mathcal{I}_\mu(S\varphi) : S\mu_{(\omega,x)}(SE_i(\omega,x)) = 1, \mu\text{-a.s.}\}$$

is nonvoid.

The measures $S\mu \in \mathcal{I}_\mu(S\varphi|SE_i)$ realize λ_i in the sense that they are those for which

$$\lim_{t \to \pm\infty} \frac{1}{t} \log \|T\varphi(t, \omega, x)s\| = \lambda_i, \quad S\mu\text{-a.s.}$$

Further, for each ergodic $S\mu \in \mathcal{I}_\mu(S\varphi)$ there exists a unique $i \in \{1, \ldots, p\}$ such that $S\mu$ realizes λ_i.

Furstenberg-Khasminskii Formulas for RDS on Manifolds

We restrict ourselves to the white noise case.

Carverhill [90] was the first to carry over the classical Furstenberg-Khasminskii formula for the top Lyapunov exponent λ_1 of a linear SDE to the case of a nonlinear SDE on a manifold. Baxendale [57] and Arnold and San Martin [33] made further contributions. There is a wealth of deep and beautiful results about the "geometry of stochastic flows" based on the Lyapunov spectrum, of which we mention just Baxendale [55, 59, 62], Baxendale and Stroock [64], Carverhill [92], Carverhill and Elworthy [95], Elworthy [139, 140], and the references therein ([62], [92] and [140] are surveys).

Let

$$dx_t = \sum_{j=0}^{m} f_j(x_t) \circ dW_t^j \qquad (6.4.2)$$

be an SDE on a Riemannian manifold M of dimension $d \geq 1$, where $f_0 \in \mathcal{C}^{1,\delta}$ and $f_j \in \mathcal{C}^{2,\delta}$ for $j = 1, \ldots, m$, $\delta > 0$, so that (6.4.2) generates a local C^1 RDS φ (see Theorem 2.3.42). Further, $T\varphi$ is generated by

$$dv_t = \sum_{j=0}^{m} Tf_j(v_t) \circ dW_t^j, \qquad (6.4.3)$$

where Tf_j is the natural lift of f_j from M to TM. $Tf_j(v)$ has horizontal component $f_j(x)$ and vertical component $\nabla f_j(x)v$. Hence $T\varphi$ is generated by (6.4.2) and

$$Dv_t = \sum_{j=0}^{m} \nabla f_j(x_t)v_t \circ dW_t^j. \qquad (6.4.4)$$

We introduce polar coordinates in $T^0M = \cup_{x \in M} T_x^0 M$, $T_x^0 M := T_x M \setminus \{0\}$ via $T_x^0 M \ni v \leftrightarrow (r,s)$, $r = \|v\| \in (0,\infty)$, $s = \frac{v}{\|v\|} \in S_x M$. Then the local continuous RDS $S\varphi$ on SM induced by $T\varphi$ is the angular part of $T\varphi$ and is generated by the SDE

$$ds_t = \sum_{j=0}^{m} Sf_j(s_t) \circ dW_t^j, \qquad (6.4.5)$$

where Sf_j is the natural lift of f_j to SM (or: the projection of Tf_j from T^0M to SM). The decomposition of $Sf_j(s)$ into the horizontal component $f_j(x)$ and vertical component $h_j(x)(s)$ gives that (6.4.5) is equivalent to the two coupled SDE (6.4.2) and

$$Ds_t = \sum_{j=0}^{m} h_j(x_t)(s_t) \circ dW_t^j, \quad h_j(x)(s) = \nabla f_j(x)s - \langle \nabla f_j(x)s, s\rangle s. \qquad (6.4.6)$$

The radial part $r_t := \|v_t\|$ of $T\varphi$ is generated by

$$dr_t = \sum_{j=0}^{m} q_j(s_t) r_t \circ dW_t^j, \quad q_j(s) := \langle \nabla f_j(x)s, s\rangle, \qquad (6.4.7)$$

and (6.4.6) is decoupled from (6.4.7).

Therefore for any $v \in T^0 M$

$$\log \|v_t\| = \log \|v\| + \sum_{j=0}^{m} \int_0^t q_j(s_u) \circ dW_u^j, \qquad (6.4.8)$$

where s_t solves (6.4.5) with $s = \frac{v}{\|v\|}$.

Converting the Stratonovich integrals in (6.4.8) into Itô integrals yields

$$\log \|v_t\| = \log \|v\| + \int_0^t Q(s_u)du + \sum_{j=1}^{m} \int_0^t q_j(s_u)\, dW_u^j, \qquad (6.4.9)$$

where

272   Chapter 6. RDS on Homogeneous Spaces

$$Q(s) = q_0(s) + \frac{1}{2}\sum_{j=1}^{m} Sf_j(q_j)(s) \qquad (6.4.10)$$

and $Sf_j(q_j)$ is the function obtained by applying the vector field Sf_j to the function q_j. Carverhill [90] and Baxendale [55] have calculated the correction terms in (6.4.10) more explicitly, to which we refer for the purely geometric proof.

6.4.2 Lemma. *Let f be a C^2 vector field on M, Sf be the natural lift of f to the unit sphere bundle SM and let $q(s) = \langle \nabla f(x)s, s\rangle$. Then for $s \in S_x M$*

$$\begin{aligned} Sf(q)(x)(s) &= \langle \nabla(\nabla f(x)f(x))s, s\rangle + \|\nabla f(x)s\|^2 \qquad (6.4.11) \\ &\quad -2\langle \nabla f(x)s, s\rangle^2 + \langle R(f(x), s)f(x), s\rangle, \end{aligned}$$

where R is the Riemannian curvature tensor on M.

For the flat space $M = \mathbb{R}^d$, $TM \cong \mathbb{R}^d \times \mathbb{R}^d$, $SM \cong \mathbb{R}^d \times S^{d-1}$, $R \equiv 0$, $\nabla f(x)s = A(x)s$, $A(x) := Df(x)$ the Jacobian of f at x, $\nabla(\nabla f(x)f(x))s = B(x)s$, $B(x) = D(A(x)f(x)) = (D^2 f)(x)f(x) + A(x)^2$ the Jacobian of $g(x) := A(x)f(x)$ at x, and

$$Sf(q)(x)(s) = \langle B(x)s, s\rangle + \langle A(x)s, A(x)s\rangle - 2\langle A(x)s, s\rangle^2.$$

For the particular case $f(x) = Ax$ we have $A(x) \equiv A$ and $B(x) \equiv A^2$, reproducing the correction term of Theorem 6.2.11.

6.4.3 Theorem (Furstenberg-Khasminskii Formulas for SDE on Manifold). *Assume that the SDE (6.4.2) satisfies the conditions of Theorem 2.3.42 for $k=1$ so that it generates a local C^1 RDS φ on M. Let μ be an ergodic φ-invariant measure fulfilling the IC of the MET. Let $S\varphi$ denote the local continuous RDS on SM canonically induced by $T\varphi$ on TM and let $\mathcal{I}_\mu(S\varphi|SE_i)$ be the (nonvoid) set of $S\varphi$-invariant measures which realize λ_i and are lifts of μ.*

(i) If $S\mu \in \mathcal{I}_\mu(S\varphi|SE_i)$, then

$$\lambda_i = \int_{SM} Q(s)\, S\rho(ds) + \sum_{j=1}^{m} \int_{\Omega \times SM} \left(\int_{t=0}^{1} q_j(S\varphi(t,\cdot)\cdot)dW_t^j\right) dS\mu, \quad (6.4.12)$$

where $S\rho(ds) = \mathbb{E}\, S\mu.(ds)$ is the marginal of $S\mu$ on SM and Q is given by (6.4.10).

(ii) If $S\mu \in \mathcal{I}_\mu(S\varphi|SE_i)$ is $\mathcal{F}^-$-measurable, then

$$\lambda_i = \mathbb{E}_{S\mu} Q = \int_{SM} Q(s)\, S\rho(ds).$$

(iii) In particular, if φ has an invariant forward Markov measure μ (whose marginal ρ hence solves the Fokker-Planck equation $L^*\rho = 0$, where $L = f_0 + \frac{1}{2}\sum_{j=1}^{m} f_j^2$ is the generator of the diffusion process corresponding to the SDE (6.4.2)), then $S\varphi$ has a forward Markov measure $S\mu \in \mathcal{I}_\mu(S\varphi|SE_1)$. It hence realizes the top Lyapunov exponent λ_1 and the marginal $S\rho$ lifts ρ and solves the Fokker-Planck equation $(SL)^* S\rho = 0$, where

$$SL = Sf_0 + \frac{1}{2}\sum_{j=1}^{m}(Sf_j)^2$$

is the generator of the diffusion process corresponding to the SDE (6.4.5). Thus

$$\lambda_1 = \int_{SM} Q(s)\, S\rho(ds). \tag{6.4.13}$$

The statement under (iii) follows from Theorem 1.7.5 since SE_1 is an invariant $\mathcal{F}^-$-measurable random compact set. Formula (6.4.13) is the nonlinear analogue of the classical formula (6.2.24).

If $S\mu$ is not a Markov measure, the expectation of the second term in (6.4.12) can only be treated further by anticipative calculus, as in Theorem 6.2.14.

For the nonlinear analogue of Theorem 6.2.16 see Arnold and San Martin [33].

6.4.2 Grassmannian Bundles

We now very briefly develop the nonlinear version of Sect. 6.3 in the SDE case. For the geometry, we basically follow Baxendale [56].

Let M be a Riemannian manifold of dimension $d \geq 1$. We define the k-th exterior power of the tangent bundle by

$$\wedge^k TM := \cup_{x \in M} \wedge^k T_x M, \quad k = 1, \ldots, d,$$

and the k-th Grassmannian bundle by

$$G_k M := \cup_{x \in M} G_k(T_x M), \quad k = 1, \ldots, d.$$

We have $G_1 M = PM$, and $G_d M \cong M$.

As in Subsect. 6.3.1, we consider $G_k M$ as a submanifold of the projective bundle $P(\wedge^k TM)$,

$$G_k M \subset P(\wedge^k TM),$$

by representing the fiber $G_k(T_x M)$ as the submanifold of those elements of $P(\wedge^k T_x M)$ which are projections of decomposable elements in $\wedge^k T_x M$.

Assume that the SDE (6.4.2) on M satisfies the conditions of Theorem 2.3.42 for $k = 1$ which ensures the existence and uniqueness of a local C^1 RDS φ on M. Then the local continuous RDS $T\varphi$ generated by (6.4.3) acts naturally on $\wedge^k TM$ and induces a local continuous RDS $\wedge^k T\varphi$ generated by

$$dv_t^{(k)} = \sum_{j=0}^{m} \wedge^k Tf_j(v_t^{(k)}) \circ dW_t^j, \tag{6.4.14}$$

where the vector field $\wedge^k Tf_j$ on $\wedge^k TM$ is the lift of f_j.

Similarly, $T\varphi$ acts naturally on $G_k M$ and induces a local continous RDS $G_k\varphi$ generated by

$$ds_t^{(k)} = \sum_{j=0}^m G_k f_j(s_t^{(k)}) \circ dW_t^j, \qquad (6.4.15)$$

$G_k f_j$ being the vector field on $G_k M$ which is the lift of f_j.

The vertical components of the vector fields $\wedge^k T f_j$ and $G_k f_j$ can be explicitly written down by combining the expressions in (6.4.5) and (6.4.7) with what was done in Subsect. 6.3.2 for $\wedge^k \mathbb{R}^d$ and $G_k(d)$, which we leave as an exercise.

By introducing polar coordinates in $\wedge^k T^0 M$, a decomposable vector $u = u_1 \wedge \ldots \wedge u_k$ can be represented by the pair $(s,r) \in G_k M \times (0,\infty)$, where $s = \frac{u}{\|u\|}$ and

$$r = \|u\| = (\det\langle u_i, u_j\rangle_{k\times k})^{1/2} = \mathrm{vol}(u_1,\ldots,u_k)$$

is the volume of the k-dimensional parallelepiped spanned by $u_1,\ldots,u_k$ (see Subsect. 5.3.2). The SDE for

$$r_t^{(k)} = \|\wedge^k T\varphi(t,\omega,x)u\| = \mathrm{vol}(T\varphi(t,\omega,x)u_1,\ldots,T\varphi(t,\omega,x)u_k)$$

coupled to (6.4.15) is

$$dr_t^{(k)} = \sum_{j=0}^m q_j^{(k)}(s_t^{(k)}) r_t^{(k)} \circ dW_t^j, \qquad (6.4.16)$$

where

$$q_j^{(k)}(u) = \langle (\nabla f_j(x))^{\wedge,k} u, u\rangle = \mathrm{trace}\nabla f_j(x) P_k(u), \qquad (6.4.17)$$

$P_k(u)$ being the orthogonal projection onto $U := \mathrm{span}(u_1,\ldots,u_k)$, $u = u_1 \wedge \ldots \wedge u_k$.

Therefore for any decomposable $u \in \wedge^k T^0 M$ and $s = \frac{u}{\|u\|}$, we obtain Liouville's equation for $\det T\varphi(t,\omega,x)|_U$,

$$\begin{aligned}\log \det T\varphi(t,\cdot,x)|_U &= \log \frac{\mathrm{vol}(T\varphi(t,\cdot,x)u_1,\ldots,T\varphi(t,\cdot,x)u_k)}{\mathrm{vol}(u_1,\ldots,u_k)} \\ &= \sum_{j=0}^m \int_0^t q_j^{(k)}(s_\tau^{(k)}) \circ dW_\tau^j.\end{aligned} \qquad (6.4.18)$$

Converting the Stratonovich integrals in (6.4.18) into Itô integrals yields

$$\log\det T\varphi(t,\cdot,x)|_U = \int_0^t Q^{(k)}(s_\tau^{(k)})d\tau + \sum_{j=1}^m \int_0^t q_j^{(k)}(s_\tau^{(k)})\, dW_\tau^j, \qquad (6.4.19)$$

where

6.4 Manifold Versions

$$Q^{(k)}(s) = q_0^{(k)} + \frac{1}{2} \sum_{j=1}^{m} G_k f_j(q^{(k)}). \qquad (6.4.20)$$

The following lemma is a simultaneous generalization of Lemmas 6.3.2 and 6.4.2 and was proved by Baxendale [56], to whom we refer for the proof.

6.4.4 Lemma. *Let f be a C^2 vector field on M, $G_k f$ the natural lift of f to the Grassmann bundle $G_k M$ and $q^{(k)}(u) = \text{trace} \nabla f(x) P_k(u)$, where $P_k(u)$ denotes the orthogonal projection onto $U = \text{span}(u_1, \ldots, u_k)$ and $u = u_1 \wedge \ldots \wedge u_k$. Then*

$$\begin{aligned}
G_k f(q^{(k)})(x)(u) &= \text{trace}(\nabla(\nabla f(x) f(x)) P_k(u)) - (\text{trace}(\nabla f(x) P_k(u)))^2 \\
&\quad + \text{trace}((\nabla f(x))^*(I - P_k(u)) \nabla f(x) P_k(u)) \\
&\quad + \text{trace}(R(f(x), P_k(u)) f(x)), \qquad (6.4.21)
\end{aligned}$$

where R is the Riemannian curvature tensor on M.

Finally, assume that μ is an ergodic φ-invariant measure fulfilling the IC of the MET for $T\varphi$. Hence the IC of the MET for $\wedge^k T\varphi$ are also satisfied. Let the spectrum be $\mathcal{S}(\wedge^k T\varphi, \mu) = \{(\lambda_i^{(k)}, d_i^{(k)})_{i=1,\ldots,p(k)}\}$ and the splitting $\wedge^k T_x M = \oplus_{i=1}^{p(k)} E_i^{(k)}(\omega, x)$. Put $G_k E_i^{(k)}(\omega, x) := E_i^{(k)}(\omega, x) \cap G_k(T_x M)$.

We hence have

6.4.5 Theorem (Furstenberg-Khasminskii Formula for the Sum of Lyapunov Exponents, SDE on Manifold). *Assume that the SDE (6.4.2) satisfies the conditions of Theorem 2.3.42 for $k = 1$ which ensures the existence and uniqueness of a local C^1 RDS φ on M.*

Let μ be an ergodic φ-invariant measure fulfilling the IC of the MET. Let $G_k \varphi$ denote the local continuous RDS on $G_k M$ induced by $T\varphi$ on TM and let $\mathcal{I}_\mu(G_k \varphi | G_k E_i^{(k)})$ be the nonvoid set of $G_k \varphi$-invariant measures which realize

$$\lambda_i^{(k)} = \Lambda_{i_1} + \ldots + \Lambda_{i_k}$$

and are lifts of μ.

(i) If $G_k \mu \in \mathcal{I}_\mu(G_k \varphi | G_k E_i^{(k)})$, then

$$\begin{aligned}
\lambda_i^{(k)} &= \int_{G_k M} Q^{(k)}(s) \, G_k \rho(ds) \qquad (6.4.22) \\
&\quad + \sum_{j=1}^{m} \int_{\Omega \times G_k M} \left(\int_{t=0}^{1} q_j^{(k)}(G_k \varphi(t, \cdot) \cdot) dW_t^j \right) dG_k \mu,
\end{aligned}$$

where $G_k \rho(ds) = \mathbb{E} G_k \mu.(ds)$ is the marginal of $G_k \mu$ on $G_k M$ and $Q^{(k)}$ is given by (6.4.20).

(ii) If $G_k \mu \in \mathcal{I}_\mu(G_k \varphi | G_k E_i^{(k)})$ is $\mathcal{F}^-$-measurable, then

$$\lambda_i^{(k)} = \mathbb{E}_{G_k\mu} Q^{(k)} = \int_{G_kM} Q^{(k)}(s)\, G_k\rho(ds).$$

(iii) In particular, if φ has an invariant $\mathcal{F}^-$-measurable measure μ (whose marginal ρ hence solves the Fokker-Planck equation $L^*\rho = 0$, where $L = f_0 + \frac{1}{2}\sum_{j=1}^m f_j^2$ is the generator of the diffusion process corresponding to the SDE (6.4.2)), then $G_k\varphi$ has an $\mathcal{F}^-$-measurable $G_k\mu \in \mathcal{I}_\mu(G_k\varphi|G_kE_1^{(k)})$. It hence realizes the top Lyapunov exponent $\lambda_1^{(k)} = \Lambda_1 + \ldots + \Lambda_k$ and the marginal $G_k\rho$ lifts ρ and solves the Fokker-Planck equation $(G_kL)^*G_k\rho = 0$, where

$$G_kL = G_kf_0 + \frac{1}{2}\sum_{j=1}^m (G_kf_j)^2$$

is the generator of the diffusion process corresponding to the SDE (6.4.15). Thus

$$\lambda_1^{(k)} = \Lambda_1 + \ldots + \Lambda_k = \int_{G_kM} Q^{(k)}(s)\, G_k\rho(ds). \qquad (6.4.23)$$

For $k = 1$ we recover Theorem 6.4.3 (for PM instead of SM).

We finally have a closer look at the case $k = d$, where

$$q^{(d)}(x) = \text{trace}\,\nabla f(x) = \text{div}\, f(x),$$

$$Q^{(d)}(x) = \text{div}\, f_0(x) + \frac{1}{2}\sum_{j=1}^m f_j(\text{div}\, f_j)(x)$$

and (6.4.18) becomes Liouville's equation for $\det T\varphi(t,\omega,x)$,

$$\det T\varphi(t,\cdot,x) = \exp \sum_{j=0}^m \int_0^t \text{div}\, f_j(\varphi(s,\cdot)x) \circ dW_s^j$$

(see Theorem 2.3.42). This leads to the following formula for the average Lyapunov exponent.

6.4.6 Corollary (Average Lyapunov Exponent). *Under the assumptions of the above theorem*

$$\lambda_\Sigma := \frac{1}{d}\sum_{i=1}^p \lambda_i d_i \;=\; \frac{1}{d}\int_M \left(\text{div}\, f_0(x) + \frac{1}{2}\sum_{j=1}^m f_j(\text{div}\, f_j)(x)\right)\rho(dx)$$

$$+ \sum_{j=1}^m \int_{\Omega\times M} \left(\int_{t=0}^1 \text{div}\, f_j(\varphi(t,\cdot)\cdot)dW_t^j\right) d\mu.$$

If μ is $\mathcal{F}^-$-measurable with marginal ρ on M, then

$$\lambda_\Sigma = \frac{1}{d} \int_M (\mathrm{div} f_0(x) + \frac{1}{2} \sum_{j=1}^m f_j(\mathrm{div} f_j)(x)) \, \rho(dx). \qquad (6.4.24)$$

Further, if M is a compact manifold and if ρ has a density $p(x) > 0$, then

$$\lambda_\Sigma = -\frac{1}{2d} \int_M \sum_{j=1}^m \left(\frac{1}{p(x)} \mathrm{div}(p(x) f_j(x)) \right)^2 p(x) dx \leq 0, \qquad (6.4.25)$$

and $\lambda_\Sigma = 0$ if and only if $\varphi(t,\omega)(p(x)dx) = p(x)dx$ $\mathbb{P}$-a. s.

Formula (6.4.25) can be obtained from (6.4.24) using integration by parts and the divergence theorem, see Baxendale [57].

The formula for λ_Σ hence does not require the φ-invariant measure to be lifted, so quantitative information on λ_Σ is "easier" to obtain than for λ_1. See the literature quoted at the beginning of this section and Chappell [98].

6.5 Rotation Numbers

6.5.1 The Concept of Rotation Number of a Plane

Motivation

The MET furnishes through its Lyapunov exponents a random and dynamic analogue of "real part of an eigenvalue" of a deterministic matrix.

The aim of this section is to propose a random and dynamic analogue of "imaginary part of an eigenvalue", called rotation number.

The basic geometric idea, which is due to San Martin [300], is as follows: While the MET assigns to each tangent vector $v \in T_x M$ at a "good" base point $(\omega, x) \in \Omega \times M$ a number $\lambda(v)$ which measures the exponential growth rate of $T\varphi(t, \omega, x)v$ as $t \to \infty$, we will assign to each (two-dimensional) tangent plane $p \subset T_x M$ a number $\rho(p)$ which measures the rotation of $T\varphi(t, \omega, x)$ inside the moving plane $p(t, \omega, x) := T\varphi(t, \omega, x)p$ with respect to a (parallel-transported) reference direction. These numbers were shown to exist in the deterministic case for linear, conformal and Killing vector fields (San Martin [300]) and in the random case, provided the Lyapunov spectrum of φ under μ is simple (see Arnold and San Martin [34]). Here the analogy goes even further: As $\lambda(v)$ is a random variable which takes on only finitely many different basic values realized in canonical directions, similarly $\rho(p)$ takes on only finitely many basic values, the rotation numbers realized in canonical planes.

For $d = 2$, $\dot{x}_t = A(t) x_t$ written in polar coordinates $r = (x_1^2 + x_2^2)^{1/2}$, $\alpha = \arctan \frac{x_2}{x_1}$ reads

$$\dot{r}_t = \langle A(t) u_t, u_t \rangle r_t, \quad r_0 = r \in (0, \infty),$$

278   Chapter 6. RDS on Homogeneous Spaces

$$\dot{\alpha}_t = \langle A(t)u_t, v_t \rangle, \quad \alpha_0 = \alpha \in [0, 2\pi),$$

where

$$u_t = \begin{pmatrix} \cos \alpha_t \\ \sin \alpha_t \end{pmatrix} \in S^1, \quad v_t = \begin{pmatrix} -\sin \alpha_t \\ \cos \alpha_t \end{pmatrix} \in S^1, \quad \langle u_t, v_t \rangle = 0,$$

$\langle \cdot, \cdot \rangle$ the standard scalar product of $\mathbb{R}^2$.

The rotation number of the solution starting in (r, α) is defined to be the linear growth rate of the (unreduced, i.e. lifted) angular part, i.e. by

$$\rho := \lim_{T \to \infty} \frac{\alpha_T}{T} = \lim_{T \to \infty} \frac{1}{T} \int_0^T \langle A(t)u_t, v_t \rangle dt \qquad (6.5.1)$$

(if the limit exists), and its Lyapunov exponent is the exponential growth rate of the radial part, i.e.

$$\lambda := \lim_{T \to \infty} \frac{\log r_T}{T} = \lim_{T \to \infty} \frac{1}{T} \int_0^T \langle A(t)u_t, u_t \rangle dt \qquad (6.5.2)$$

(if the limit exists).

In case of a constant matrix A with eigenvalues $\mu_{1,2}(A)$, these limits exist for any initial values and yield

$$\rho = \pm \mathrm{Im}\mu_{1,2}(A) \quad \text{for any initial } (r, \alpha)$$

($\rho > 0$ if rotation is counterclockwise and if we use the canonical orientation of $\mathbb{R}^2$ by the standard frame (e_1, e_2)) and

$$\lambda = \lambda_{1,2} = \mathrm{Re}\mu_{1,2}(A),$$

the smaller one of these numbers being realized if and only if we start in the corresponding eigenspace.

The appearance of time averages in formulas (6.5.1) and (6.5.2) calls for using the ergodic theorem: If $\dot{x}_t = A(t)x_t$ and $A(t)$ is periodic, almost periodic or stationary stochastic with finite mean, then the limits in (6.5.1) and (6.5.2) exist. See e.g. Johnson and Moser [191] for the almost periodic case and Subsect. 6.2.2 for the stationary stochastic case.

The rotation number ρ plays a basic role in the theory of one-dimensional random Schrödinger operators as the "density of states" (see e.g. Bougerol and Lacroix [77], Carmona and Lacroix [86] and Pastur and Figotin [273]).

We will first develop the (deterministic) concept of rotation number for a vector field on a Riemannian manifold M of dimension $d \geq 2$. Whether the proposed concept is sensible or not will be tested against the following:

— for $M = \mathbb{R}^2$ and $\dot{x}_t = A(t)x_t$, it should reduce to the above elementary concept,

— for $M = \mathbb{R}^d$ and $\dot{x}_t = Ax_t$, the rotation number $\rho(p)$ of an eigenplane p corresponding to a pair of complex-conjugate eigenvalues of A should yield the imaginary part of the eigenvalues (modulo sign).

We stress that our concept is infinitesimal in the sense that we accumulate infinitesimal rotations in the time interval $[0, T]$, as already clearly visible in formula (6.5.1), thus

— it is a continuous time concept (i.e. does not work for discrete time, but see Remark 6.5.5(iii)),

— it needs a generator (vector field in the deterministic case, RDE or SDE in the random case), as the rotation numbers will be defined by Furstenberg-Khasminskii type formulas.

Linear Case

Let $M = \mathbb{R}^d$ with $d \geq 2$. Each $A \in \mathbb{R}^{d \times d}$ generates a flow $\Phi(t) = e^{tA} \in Gl(d, \mathbb{R})$ which by the left action of $Gl(d, \mathbb{R})$ naturally induces a flow on the Grassmann manifold $G_2(d)$ of two-dimensional subspaces (2-planes) of $\mathbb{R}^d$. Moreover, $Gl(d, \mathbb{R})$ also acts on the Grassmann manifold $G_2^+(d)$ of oriented 2-planes (see Subsect. 6.3.1). We have $\dim G_2(d) = \dim G_2^+(d) = 2d - 4$. $\Phi(t)$ induces a flow on $G_2^+(d)$ denoted by $p \mapsto p(t) := (G_2^+ \Phi)(t)p$.

We now want to measure the infinitesimal rotation by $\Phi(t)$ inside the plane $p(t)$. For this purpose, we pass to the principal bundle $\pi : St_2(d) \to G_2^+(d)$ with structure group $SO(2, \mathbb{R})$, where $St_2(d)$ is the Stiefel manifold of orthonormal 2-frames $n = (u, v)$ of $\mathbb{R}^d$, and $\pi(n) = p = \text{span}(u, v)$ is the canonical projection which maps n to the plane p oriented by the frame n. We have $\dim St_2(d) = 2d - 3$. By fixing the standard bases in $\mathbb{R}^2$ and $\mathbb{R}^d$, we obtain a one-to-one correspondence between $St_2(d)$ and the set of $d \times 2$ matrices n with $n^* n = I_{2 \times 2}$.

$Gl(d, \mathbb{R})$ also acts on the left on $St_2(d)$ by the action $n = (u, v) \mapsto g(n) := \perp (gu, gv)$, where

$$(u, v) \mapsto \perp (u, v) = \left(\frac{u}{\|u\|}, \frac{v - \frac{\langle v, u \rangle}{\langle u, u \rangle} u}{\|v - \frac{\langle v, u \rangle}{\langle u, u \rangle} u\|} \right) \in St_2(d) \qquad (6.5.3)$$

denotes orthonormalization of the linearly independent pair (u, v) in such a way that $\perp (u, v)$ and (u, v) have the same orientation. This action covers the above action of $Gl(d, \mathbb{R})$ on $G_2^+(d)$. Hence $\Phi(t)$ induces a flow $n \mapsto n(t) := (St_2 \Phi)(t)n$ on $St_2(d)$ which covers the flow $G_2^+ \Phi$ on $G_2^+(d)$.

In order to measure the rotation of the frame $n(t) = (St_2 \Phi)(t)n$ inside the plane $p(t) = (G_2^+ \Phi)(t)p = \pi(n(t))$, we need a reference frame $\bar{n}(t)$ in the fiber over $p(t)$. This reference frame $\bar{n}(t) = (\bar{u}(t), \bar{v}(t))$ should be chosen such that $\dot{\bar{u}}(t)$ and $\dot{\bar{v}}(t)$ are both orthogonal to the plane $p(t) := \text{span}(\bar{u}(t), \bar{v}(t))$, and hence, from the point of view of an observer inside $p(t)$, $\bar{n}(t)$ does not rotate. Such a reference frame can be obtained by "parallel transporting" n by means of a connection in $St_2(d) \to G_2^+(d)$. Suppose such a connection is chosen, and since $SO(2, \mathbb{R})$ acts transitively on the fibers of $St_2(d) \to G_2^+(d)$ from the right, we have

280   Chapter 6. RDS on Homogeneous Spaces

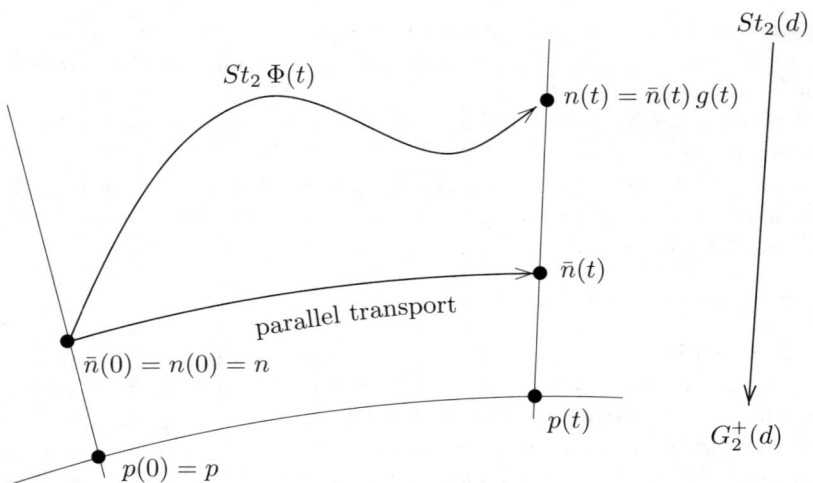

Fig. 6.2. Rotation of the frame $n(t)$ inside the plane $p(t)$ relative to $\bar{n}(t)$

$$n(t) = \bar{n}(t)g(t), \quad g(t) \in SO(2,\mathbb{R}), \tag{6.5.4}$$

see Fig. 6.2.

Now $g(t) \in SO(2,\mathbb{R})$ uniquely corresponds to an angle $\alpha(t)$ for which we need a differential equation analogous to $\dot{\alpha}_t = \langle Au_t, v_t \rangle$ in the case of $\mathbb{R}^2$. We then would define the rotation number of the plane p under the flow $\Phi(t)$ as in (6.5.1).

There is the following canonical choice of a connection on $St_2(d) \to G_2^+(d)$ (see Narasimhan and Ramanan [250]): If $n(t) \in St_2(d)$ is a differentiable path, then differentiating $n(t)^*n(t) = I$ yields

$$n(t)^*\dot{n}(t) + \dot{n}(t)^*n(t) = n(t)^*\dot{n}(t) + (n(t)^*\dot{n}(t))^* = 0.$$

Thus the tangent space $T_n St_2(d)$ at n can be identified with the set of $d \times 2$ matrices w such that n^*w is skew-symmetric. Therefore the expression $\omega_n = n^*dn : T_n St_2(d) \to so(2,\mathbb{R})$ defines an $so(2,\mathbb{R})$-valued 1-form in $St_2(d)$. This 1-form is the connection form of the canonical connection on $St_2(d) \to G_2^+(d)$ and its kernel is the horizontal subspace H_n of $T_n St_2(d) = H_n \oplus V_n$. The connection is invariant under the left action of $SO(d,\mathbb{R})$ on $St_2(d)$, i.e. $g_*H_n = H_{gn}$ for all $g \in SO(d,\mathbb{R})$.

The interpretation of $\omega_n(w) = n^*w$ is as follows: It measures the "infinitesimal deformation" of the 2-frame $n = (u,v) \in \pi^{-1}(p)$ without moving the plane p, in the "direction" of $w \in T_n St_2(d)$. A horizontal curve $\bar{n}(t) = (\bar{u}(t), \bar{v}(t))$ in $St_2(d)$ is thus one for which both $\dot{\bar{u}}(t)$ and $\dot{\bar{v}}(t)$ are orthogonal to the plane $\bar{p}(t) := \text{span}(\bar{u}(t), \bar{v}(t))$. Hence $\bar{n}(t)$ can be chosen as a reference frame.

Each $d \times d$ matrix A gives rise to a vector field $\tilde{A}$ on $St_2(d)$ by

$$\tilde{A}(n) = \frac{d}{dt}(St_2\Phi)(t)n|_{t=0},$$

so that
$$\dot{n}_t = (\dot{u}_t, \dot{v}_t) = \tilde{A}(n_t).$$

According to the above, $\omega_n(\tilde{A}(n))$ measures the infinitesimal rotation of $n \in \pi^{-1}(p)$ in the direction of $w = \tilde{A}(n)$, so that the following expression offers itself as the definition of the rotation number of the plane $p \in G_2^+(d)$ under the action of the flow $\Phi(t)$ generated by the matrix A:

$$\rho(p) := \lim_{T \to \infty} \frac{1}{T} \int_0^T \omega_{n(t)}(\tilde{A}(n(t))) dt, \qquad (6.5.5)$$

provided the limit exists and is independent of the element $n \in \pi^{-1}(p)$. Strictly speaking, $\rho(p) \in so(2, \mathbb{R})$ which is identified with $\mathbb{R}$ via $\begin{pmatrix} 0 & -\alpha \\ \alpha & 0 \end{pmatrix} \leftrightarrow \alpha$.

6.5.1 Lemma. *(i) The vector field on $St_2(d)$ induced by A is given for $n = (u, v) \in St_2(d)$ by the $d \times 2$ matrix*

$$\tilde{A}(n) = (Au - \langle Au, u\rangle u, Av - \langle Av, v\rangle v - (\langle Au, v\rangle + \langle Av, u\rangle)u). \qquad (6.5.6)$$

(ii) The connection form $\omega_n(\tilde{A}(n))$ is given by

$$\omega_n(\tilde{A}(n)) = \begin{pmatrix} 0 & -\langle \dot{u}, v\rangle \\ \langle \dot{u}, v\rangle & 0 \end{pmatrix} = \begin{pmatrix} 0 & -\langle Au, v\rangle \\ \langle Au, v\rangle & 0 \end{pmatrix}. \qquad (6.5.7)$$

(iii) If $g(t)$ in $n(t) = \bar{n}(t)g(t)$ (see (6.5.4)) is represented as $g(t) = \begin{pmatrix} \cos\alpha(t) & -\sin\alpha(t) \\ \sin\alpha(t) & \cos\alpha(t) \end{pmatrix}$, then

$$\dot{\alpha}(t) = \langle \dot{u}(t), v(t)\rangle = \langle Au(t), v(t)\rangle. \qquad (6.5.8)$$

Proof. (i) By definition, $\tilde{A}(n) = \frac{d}{dt} \perp (e^{tA}n)|_{t=0}$. To calculate this, replace (u, v) in (6.5.3) by $(e^{tA}u, e^{tA}v)$ and differentiate with respect to t.

(ii) This immediately follows from the formula $\omega_n(w) = n^*w$ and (6.5.6).

(iii) Differentiating (6.5.4) with respect to t and applying $\omega_{n(t)}(\cdot)$ to both sides gives the following differential equation in $SO(2, \mathbb{R})$:

$$g(t)^{-1}\dot{g}(t) = \omega_{n(t)}(\tilde{A}(n(t))), \quad g(0) = I_{2\times 2}.$$

But using the representation of $g(t)$, we obtain

$$g(t)^{-1}\dot{g}(t) = \begin{pmatrix} 0 & -\dot{\alpha}(t) \\ \dot{\alpha}(t) & 0 \end{pmatrix},$$

so that (6.5.8) follows from (6.5.7). □

Since only obvious alterations of this procedure are necessary to treat nonautonomous linear equations instead of autonomous ones, we finally arrive at the following definition.

6.5.2 Definition (Rotation Number). Consider the linear differential equation $\dot{x}_t = A(t)x_t$ in $\mathbb{R}^d$ for $d \geq 2$, with $t \mapsto A(t)$ locally integrable. The *rotation number* $\rho(p)$ of the oriented 2-plane $p \in G_2^+(d)$ under the flow generated by the above equation is defined by

$$\rho(p) := \lim_{T \to \infty} \frac{1}{T} \int_0^T \langle \dot{u}_t, v_t \rangle dt = \lim_{T \to \infty} \frac{1}{T} \int_0^T \langle A(t)u_t, v_t \rangle dt \quad (6.5.9)$$

(provided the limit exists and is independent of $n \in \pi^{-1}(p)$), where $\langle \cdot, \cdot \rangle$ is the standard scalar product in $\mathbb{R}^d$, and $n_t = (u_t, v_t)$ is the flow induced in $St_2(d)$ with arbitrary initial 2-frame $n_0 = n \in \pi^{-1}(p)$. ∎

Note that the flow n_t in $St_2(d)$ solves the equation $\dot{n}_t = (\dot{u}_t, \dot{v}_t) = \tilde{A}(t, n_t)$, more precisely

$$\dot{u}_t = A(t)u_t - \langle A(t)u_t, u_t \rangle u_t,$$
$$\dot{v}_t = A(t)v_t - \langle A(t)v_t, v_t \rangle v_t - (\langle A(t)u_t, v_t \rangle + \langle A(t)v_t, u_t \rangle)u_t.$$

It turns out that this definition passes our test.

6.5.3 Proposition (Rotation Number is Independent of Frame).
(i) The existence or nonexistence of the limit in (6.5.9) and its value in case of existence are independent of the particular choice of $n \in \pi^{-1}(p)$.
(ii) $\rho(p)$ changes sign with the orientation of p.

Proof. (i) Let $n, \bar{n} \in \pi^{-1}(p)$ with $n \neq \bar{n}$. Then

$$\bar{u}_t = \cos \beta_t u_t + \sin \beta_t v_t, \quad \bar{v}_t = -\sin \beta_t u_t + \cos \beta_t v_t,$$

where $t \to \beta_t$ is continuous with $0 < \beta_t < 2\pi$ since $n \mapsto n_t$ is injective.

Using $\langle \dot{u}_t, v_t \rangle = -\langle u_t, \dot{v}_t \rangle$ which follows from differentiating $\langle u_t, v_t \rangle \equiv 0$ we obtain

$$\langle \dot{\bar{u}}_t, \bar{v}_t \rangle = \dot{\beta}_t + \langle \dot{u}_t, v_t \rangle,$$

whence

$$\frac{1}{T} \int_0^T \langle \dot{\bar{u}}_t, \bar{v}_t \rangle dt = \frac{\beta_T - \beta_0}{T} + \frac{1}{T} \int_0^T \langle \dot{u}_t, v_t \rangle dt.$$

Since always $(\beta_T - \beta_0)/T \to 0$ as $T \to \infty$ the existence or non-existence of the limit (and its value in case of existence) is independent of the chosen frame.

(ii) If $\bar{n} = (\bar{u}, \bar{v})$ and $n = (u, v)$ have different orientation, repeat the calculation in (i) with $(\bar{u}, -\bar{v})$. ⌐

For $d = 2$, (6.5.9) obviously reduces to (6.5.1). Further, in the autonomous case $\dot{x}_t = Ax_t$, every 2-plane $p \subset \mathbb{R}^d$ has a rotation number; in particular, the imaginary parts of eigenvalues can be recovered as rotation numbers of their eigenplanes (for details see San Martin [300: Sect. 4]).

6.5.4 Example. Consider in $\mathbb{R}^4$

$$A = \begin{pmatrix} 0 & -1 & 0 & 0 \\ 1 & 0 & 0 & 0 \\ 0 & 0 & 0 & -2 \\ 0 & 0 & 2 & 0 \end{pmatrix}.$$

In this case the integrand in (6.5.9) is constant and equal to $\rho(p) = n^* A n$. If $\rho_{ij} := \rho(\text{span}(e_i, e_j))$, then $\rho_{12} = 1$, $\rho_{34} = 2$, $\rho_{ij} = 0$ for $i = 1, 2$ and $j = 3, 4$, but

$$\rho(\text{span}(\cos \beta e_1 + \sin \beta e_3, \cos \delta e_2 + \sin \delta e_4)) = \cos \beta \cos \delta + 2 \sin \beta \sin \delta,$$

always modulo the sign depending on the orientation of the corresponding plane. ∎

Nonlinear Case

Let f be a complete C^1 vector field on a Riemannian manifold M of dimension $\dim M = d \geq 2$. So $\dot{x}_t = f(x_t)$ generates a flow $(\varphi(t))_{t \in \mathbb{R}}$ of diffeomorphisms of M. The linearized flow $T\varphi$ on TM is generated by $\dot{v}_t = Tf(v_t)$, hence its vertical component by $v'_t = \nabla f(x_t) v_t$, with v'_t denoting the absolute derivative.

We now want to define the rotation number $\rho(p)$ of an oriented 2-plane $p \subset T_x M$ under $T\varphi$. $T\varphi$ induces a flow denoted by $G_2^+ \varphi$ on the Grassmann bundle

$$G_2^+ M := \cup_{x \in M} G_2^+(T_x M),$$

where $G_2^+(T_x M)$ is the Grassmann manifold of oriented 2-planes in $T_x M$.

We want to measure the infinitesimal rotation of $T\varphi(t)$ inside the plane $p(t)$. For this purpose, we pass to the principal bundle $\pi : St_2 M \to G_2^+ M$ with structure group $SO(2, \mathbb{R})$, where $St_2 M$ is the Stiefel bundle,

$$St_2 M := \cup_{x \in M} St_2(T_x M),$$

and $St_2(T_x M)$ is the Stiefel manifold of orthonormal 2-frames $n = (u, v)$ in $(T_x M, \langle \cdot, \cdot \rangle_x)$.

$T\varphi$ also induces a flow $St_2 \varphi$ on $St_2 M$ which covers $G_2^+ \varphi$ on $G_2^+ M$, defined for $n = (u, v) \in St_2 M$ by

$$n(t) = (u(t), v(t)) = St_2\varphi(t)n := \perp (T\varphi(t)u, T\varphi(t)v),$$

where $\perp$ is defined as in the linear case by (6.5.3).

Parallel transport of n is now accomplished by the following canonical choice of a connection on $St_2 M \to G_2^+ M$: We take the "composition" of the Levi-Civita connection on TM (for moving along the manifold) with the canonical connection from the linear case (for moving "inside" $St_2(T_xM) \to G_2^+(T_xM)$ for fixed x). It will be of fundamental importance that this connection is compatible with the Riemannian metric $\langle \cdot, \cdot \rangle$, i.e. we can differentiate the scalar product by the usual "product rule".

The comparison of $n(t) = St_2\varphi(t)n$ with the parallel-transported reference frame $\bar{n}(t)$ yields $n(t) = \bar{n}(t)g(t)$ with $g(t) \in SO(2,\mathbb{R})$ (see Figure 6.2 with $St_2(d)$, $St_2\Phi$ and $G_2^+(d)$ replaced with St_2M, $St_2\varphi$ and $G_2^+ M$) and the angle $\alpha(t)$ corresponding to $g(t)$ satisfies a differential equation which we will now determine.

The flow $n(t) = (u(t), v(t)) = St_2\varphi(t)n$ is generated by $\dot{n}_t = St_2 f(n_t)$, where $St_2 f$ is the lift of f to $St_2 M$, defined by

$$St_2 f(n) := \frac{d}{dt}(St_2\varphi(t)n)|_{t=0},$$

whose horizontal component is the horizontal lift of f to $St_2 M$ (which, by definition, does not contribute to rotation) and whose vertical component is

$$St_2 f(n)_v = (\nabla f)^\sim(n),$$

where $(\nabla f)^\sim$ is defined as in (6.5.6) with ∇f in place of A. Hence

$$n'_t = (\nabla f)^\sim(n_t)$$

with

$$(\nabla f)^\sim(n) = (\nabla fu - \langle \nabla fu, u\rangle u, \nabla fv - \langle \nabla fv, v\rangle v - (\langle \nabla fu, v\rangle + \langle \nabla fv, u\rangle)u). \tag{6.5.10}$$

From this, as above in the linear case,

$$\dot{\alpha}_t = \langle u'_t, v_t \rangle = \langle (\nabla f)u_t, v_t \rangle.$$

This suggests to define the rotation number of $p \in G_2^+ M$ by

$$\rho(p) := \lim_{T \to \infty} \frac{1}{T} \int_0^T \langle u'_t, v(t)\rangle dt = \lim_{T \to \infty} \frac{1}{T} \int_0^T \langle (\nabla f)u(t), v(t)\rangle dt \tag{6.5.11}$$

(provided the limit exists). Clearly we obtain the autonomous version of Definition 6.5.2 as a particular case: For $M = \mathbb{R}^d$ with the standard scalar product and $f(x) = Ax$ we have $(\nabla f)u = Au$.

Now Proposition 6.5.3 remains true since our connection is compatible with the Riemannian metric, meaning that $\frac{d}{dt}\langle u_t, v_t\rangle = \langle u'_t, v_t\rangle + \langle u_t, v'_t\rangle$, so that the proof for the linear case remains valid. Hence

— the existence or nonexistence of the limit in (6.5.11) and its value in case of existence are independent of the particular choice of $n \in \pi^{-1}(p)$,
— $\rho(p)$ changes sign with orientation of p.

We stress that even if M is compact, rotation numbers depend in general on the Riemannian metric (for examples see [300, 296]). They are therefore not as robust as the Lyapunov exponents.

6.5.5 Remark (Other Approaches). (i) There are other generalizations of the two-dimensional linear rotation number to some higher-dimensional systems, assigning, roughly speaking, a rotation number to a flow in the symplectic group, see Ruelle [294], Johnson [188, 190], Delyon and Foulon [125]. However, only one number is obtained even for higher dimensions.

(ii) To measure rotation inside a k-dimensional oriented subspace of the tangent space, we could consider the principal bundle $St_k M \to G_k^+ M$, $k = 2, \ldots, d$ with (for $k \geq 3$: non-Abelian) structure group $SO(k, \mathbb{R})$ which would lead to a skew-symmetric rotation matrix $\rho \in so(k, \mathbb{R})$. For the SDE case this is treated by Ruffino [296: Chap. 3].

(iii) Ruffino [296: Chap. 5] also built a bridge between Poincaré's original definition of rotation number for orientation-preserving homeomorphisms of S^1 and the continuous time linear concept for $\dot{x}_t = A(t)x_t$ in $\mathbb{R}^2$ by proving a "sampling theorem": If $\rho(h)$ is Poincaré's rotation number of the discretized system (with time increment h) on S^1 and if ρ is the rotation number of the linear system, then (modulo conditions) $\rho(h)/h \to \rho$ as $h \to 0$. ∎

6.5.2 Rotation Numbers for RDE

Let M be a Riemannian manifold of dimension $d \geq 2$ and assume that the random vector field $f(\omega, x)$ satisfies the conditions of Theorem 2.2.14 which ensures the existence and uniqueness of a local C^1 RDS φ on M generated by

$$\dot{x}_t = f(\theta_t \omega, x_t). \tag{6.5.12}$$

The linearized RDS $T\varphi$ solves

$$\dot{v}_t = (Tf(\Theta_t(\omega, x)))v_t, \tag{6.5.13}$$

equivalently, with v'_t denoting the absolute derivative,

$$v'_t = \nabla f(\Theta_t(\omega, x))v_t, \tag{6.5.14}$$

and is a linear bundle RDS over $\Theta = (\theta, \varphi)$ restricted to the Θ-invariant set graph$E(\cdot) \subset \Omega \times M$, where $E(\omega)$ is the set of never-exploding initial values $x \in M$ for equation (6.5.12) (see Subsect. 4.2.3).

We will define the rotation number of an oriented 2-plane $p \subset T_x M$ in complete analogy to the case of an autonomous vector field of Subsect. 6.5.1. The linear RDS $T\varphi$ induces an RDS $G_2^+ \varphi$ on the Grassmann bundle $G_2^+ M$ and an RDS $St_2 \varphi$ on the Stiefel bundle $St_2 M$ which covers $G_2^+ \varphi$.

286   Chapter 6. RDS on Homogeneous Spaces

The RDS $St_2\varphi$ is generated by an RDE $\dot{n}_t = St_2 f(\Theta_t(\omega,x), n_t)$, whose horizontal component is the horizontal lift of f and whose vertical component $(\nabla f)^\sim$ is explicitly given in (6.5.10) with

$$n'_t = (\nabla f)^\sim(\Theta_t(\omega,x), n_t).$$

In view of (6.5.11) it is evident to define the rotation number $\rho(p)$ of p by

$$\begin{aligned}\rho(p) &:= \lim_{T\to\infty} \frac{1}{T}\int_0^T \langle u'(t), v(t)\rangle dt \qquad (6.5.15)\\ &= \lim_{T\to\infty} \frac{1}{T}\int_0^T \langle \nabla f(\Theta_t(\cdot))u(t), v(t)\rangle dt\,,\end{aligned}$$

provided the limit exists. As made clear above, Proposition 6.5.3 remains valid.

Lifts of Invariant Measures to the Stiefel Bundle

Let μ be an ergodic invariant measure for φ (necessarily supported by graph$E$) which satisfies

$$\|\nabla f\| \in L^1(\mu). \qquad (6.5.16)$$

Then the IC of the MET are satisfied (see Theorem 4.2.10), providing us with the Lyapunov spectrum $\mathcal{S}(\varphi,\mu) = \{(\lambda_i, d_i)_{i=1,\dots,p}\}$, a Θ-invariant set $\Delta \subset \text{graph}E$ of full μ measure and a splitting

$$T_x M = E_1(\omega,x) \oplus \cdots \oplus E_p(\omega,x), \quad (\omega,x) \in \Delta.$$

As we want to use the ergodic theorem to prove the existence of the limit in (6.5.15), the first task is to determine all invariant measures $St_2\mu$ of $St_2\varphi$ which are lifts of μ, as every number

$$\rho(St_2\mu) := \int_{\Omega\times St_2 M} \langle (\nabla f)u, v\rangle d(St_2\mu)$$

qualifies as a possible rotation number.

As usual, we freely identify a measure $St_2\mu$ on $\Omega \times St_2 M$ with marginal μ on $\Omega \times M$ with its factorization $St_2\mu_{(\omega,x)}$ obtained from

$$St_2\mu(d\omega, dx, dn) = St_2\mu_{(\omega,x)}(dn)\mu(d\omega, dx).$$

Recall that $St_2\mu$ is invariant under $St_2\varphi$ if and only if for each $t \in \mathbb{R}$

$$St_2\varphi(t,\omega,x) St_2\mu_{(\omega,x)} = (St_2\mu)_{\Theta_t(\omega,x)} \quad \mu\text{-a.s.}$$

We also need to consider the RDS $F_{1,2}\varphi$ induced by $T\varphi$ on the flag bundle

$$F_{1,2}M := \cup_{x\in M} F_{1,2}(T_x M),$$

where $F_{1,2}(T_x M)$ is the manifold of flags $f = (V \subset p)$, in which V and p are subspaces of $T_x M$ of dimension 1 and 2, respectively.

6.5 Rotation Numbers

We now need for the remainder of this subsection the following

Assumption: Let the RDS φ under μ have simple Lyapunov spectrum.

See Remark 6.5.15 for a discussion.

Under this assumption all Oseledets spaces E_i are one-dimensional. For each pair (i,j), $1 \leq i,j \leq d$, $i < j$, the spaces $E_i(\omega, x)$ and $E_j(\omega, x)$ uniquely define a (measurable) random plane field $p_{ij}(\omega, x) := \text{span}(E_i(\omega, x), E_j(\omega, x))$ and a random flag field $f_{ij}(\omega, x) := (E_i(\omega, x) \subset p_{ij}(\omega, x))$. The invariance of the E_i's implies the invariance of p_{ij} and f_{ij} or, equivalently,

$$G_2\mu_{ij,(\omega,x)}(dp) := \delta_{p_{ij}(\omega,x)}(dp), \quad F_{1,2}\mu_{ij,(\omega,x)}(df) := \delta_{f_{ij}(\omega,x)}(df)$$

are lifts of μ. Moreover, $F_{1,2}\mu_{ij}$ lifts $G_2\mu_{ij}$ from $\Omega \times G_2 M$ to $\Omega \times F_{1,2}M$. We now lift $F_{1,2}\mu_{ij}$ further from $\Omega \times F_{1,2}M$ to $\Omega \times St_2 M$.

Given a flag $f = (V \subset p) \in F_{1,2}M$, we select $n = (u, v) \in St_2 M$ from f as follows: Choose $u \in V$, $\|u\| = 1$, and $v \in p$, $\|v\| = 1$, $\langle u, v \rangle = 0$. These two vectors give rise to four possible orthonormal frames of p selected from f, namely: $n_1 = n = (u, v)$, $n_2 = (-u, v)$, $n_3 = (-u, -v)$, $n_4 = (u, -v)$. Here n_1 and n_3 have the same orientation, while n_2 and n_4 also have the same, but opposite, orientation.

6.5.6 Lemma. *(i) Let $G := \{g_1 = \text{id}, g_2, g_3, g_4\}$ be the group of diffeomorphisms of $St_2 M$ defined by $g_i n = n_i$, $i = 1, \ldots, 4$. Then $G \cong \mathbb{Z}_2^2$ is an Abelian group which is isomorphic to the Klein group of four elements. G acts freely on the left on $St_2 M$.*

(ii) The group G commutes with the RDS $St_2\varphi$ and $F_{1,2}M \cong St_2 M / G$. Hence $St_2 M$ is a 4-fold covering of $F_{1,2}M$.

Proof. (i) can be easily checked.

(ii) follows from the definition of $St_2\varphi$ and the fact that G commutes with $\perp$. □

6.5.7 Lemma (Lift of Measure from Flag to Stiefel Bundle). *Let $f(\omega, x) = (V(\omega, x) \subset p(\omega, x)) \in F_{1,2}M$ be an invariant random flag, i.e. let*

$$F_{1,2}\varphi(t, \omega, x)f(\omega, x) = f(\Theta_t(\omega, x)).$$

Then the invariant measure $F_{1,2}\mu_{(\omega,x)} = \delta_{f(\omega,x)}$ can be lifted from $\Omega \times F_{1,2}M$ to $\Omega \times St_2 M$ to yield exactly $N(f)$ ergodic invariant measures, where $N(f) = 1, 2,$ or 4. More specifically:

(i) $N(f) = 1$: There is exactly one ergodic lift $St_2\mu$ given by

$$St_2\mu_{(\omega,x)} = \frac{1}{4}\sum_{g \in G} g\delta_{n(\omega,x)} = \frac{1}{4}\sum_{g \in G} \delta_{gn(\omega,x)},$$

where $n(\omega, x) = (u(\omega, x), v(\omega, x))$ is an arbitrary measurable selection from $f(\omega, x)$ with $u(\omega, x) \in V(\omega, x)$.

(ii) $N(f) = 2$: There are exactly two different ergodic lifts $St_2\mu^1$ and $St_2\mu^2$ given by

$$St_2\mu^1_{(\omega,x)} = \frac{1}{2}\sum_{g\in H}\delta_{gn_1(\omega,x)}, \quad St_2\mu^2_{(\omega,x)} = \frac{1}{2}\sum_{g\in G\setminus H}\delta_{gn_1(\omega,x)},$$

where $H = H_i = \{g_1, g_i\}$ for $i = 2, 3, 4$ is a two-element subgroup of G and $n_1(\omega, x) = (u_1(\omega, x), v_1(\omega, x))$ is some measurable selection from $f(\omega, x)$ with $u_1(\omega, x) \in V(\omega, x)$.

Moreover,

(a) for $H = \{g_1, g_2\}$ or $H = \{g_1, g_4\}$, $St_2\mu^k_{(\omega,x)}$, $k = 1, 2$, charges points with different orientation.

(b) for $H = \{g_1, g_3\}$, $St_2\mu^k_{(\omega,x)}$, $k = 1, 2$, charges points with the same orientation;

(iii) $N(f) = 4$: There are exactly four different ergodic lifts described by the Dirac measures

$$St_2\mu^k_{(\omega,x)} = \delta_{g_k n_0(\omega,x)}, \quad k = 1,\ldots,4,$$

where $n_0(\omega, x) = (u_0(\omega, x), v_0(\omega, x))$ is an invariant measurable selection from $f(\omega, x)$ with $u_0(\omega, x) \in V(\omega, x)$, i.e. $St_2\varphi(t, \omega, x)n_0(\omega, x) = n_0(\Theta_t(\omega, x))$.

Proof. The same reasoning as in Remark 6.2.2 gives the existence of an ergodic lift ν of $\delta_{f(\omega,x)}$ to $\Omega \times St_2 M$. By Lemma 6.5.6(ii) all elements of the set $\{g\nu\}_{g\in G}$ are also lifts. The set of $g \in G$ for which $g\nu = \nu$ is obviously a subgroup H of G.

Now the following possibilities arise:

(i) $N(f) = 1$: $H = G$: Take some measurable selection $n(\omega, x)$ from $f(\omega, x)$ and write

$$\nu_{(\omega,x)} = \sum_{i=1}^{4}\alpha_i(\omega, x)g_i\delta_{n(\omega,x)}, \quad 0 \leq \alpha_i(\omega, x) \leq 1, \quad \sum_{i=1}^{4}\alpha_i(\omega, x) = 1.$$

Equating weights in $g_2\nu = \nu$ yields $\alpha_1 = \alpha_2$ and $\alpha_3 = \alpha_4$ μ-a.s., while equating weights in $g_3\nu = \nu$ gives $\alpha_1 = \alpha_3$, hence $\alpha_i(\omega, x) = \frac{1}{4}$.

(ii) $N(f) = 2$: $H = H_i = \{g_1, g_i\}$, $i = 2, 3, 4$: To fix ideas, consider $H = H_2$ (the other cases are similar). Put $\nu^1 := \nu$ and $\nu^2 := g_3\nu = g_4\nu \neq \nu$. These two measures are ergodic, hence singular, so that there are subsets $\Gamma_k(\omega, x)$ in the 4-point fibre over $f(\omega, x)$ for which $\nu^i_{(\omega,x)}(\Gamma_k(\omega, x)) = \delta_{ik}$.

Now take a measurable selection $n_1(\omega, x) \in \Gamma_1(\omega, x)$ such that $u_1(\omega, x) \in V(\omega, x)$. Since $g_2\Gamma_1 = \Gamma_1$, $g_3\Gamma_1 = g_4\Gamma_1 = \Gamma_2$, ν^1 has the form

$$\nu^1_{(\omega,x)} = \alpha_1(\omega, x)\delta_{n_1(\omega,x)} + \alpha_2(\omega, x)g_2\delta_{n_1(\omega,x)}. \qquad (6.5.17)$$

Applying g_2 to both sides of (6.5.17) gives $\alpha_1 = \alpha_2 = \frac{1}{2}$. As a result,

$$\nu^1 = \frac{1}{2}(\delta_{n_1} + g_2\delta_{n_1}), \quad \nu^2 = \frac{1}{2}(g_3\delta_{n_1} + g_4\delta_{n_1}).$$

(iii) $N(f) = 4$: $H = \{\text{id}\}$: The four ergodic measures $\nu^i := g_i\nu$, $i = 1, \ldots, 4$, are all different, hence singular. Their support $\Gamma_i = \{n_i\}$ is thus one-point, and $\nu^i = \delta_{g_i n_1}$. In particular, $St_2\varphi(t, \omega, x)n_i(\omega, x) = n_i(\Theta_t(\omega, x))$ for $i = 1, \ldots, 4$. □

With this lemma, we have determined all possible lifts of the Dirac measure supported by the canonical flag $f = f_{ij} := (E_i \subset p_{ij})$, $p_{ij} := \text{span}(E_i, E_j)$, from the flag bundle $F_{1,2}M$ to the Stiefel bundle St_2M.

On the other hand, f_{ij} projects to the canonical invariant plane $p_{ij} \in G_2M$. Applying the usual argument, the possible lifts of the corresponding Dirac measure to G_2^+M are either

(1) one measure $\frac{1}{2}(\delta_{p_{ij}^+} + \delta_{p_{ij}^-})$, or

(2) two Dirac measures on p_{ij}^+ and p_{ij}^-, respectively.

Finally, on the principal $SO(2,\mathbb{R})$-bundle $St_2M \to G_2^+M$ (the projection identifies all frames of the same orientation),

— cases (i) and (ii)(a) of Lemma 6.5.7 on St_2M project to case (1) on G_2^+M, while

— cases (ii)(b) and (iii) project to case (2).

6.5.8 Proposition (Rotation Number of Invariant Measure). *Let φ be the local C^1 RDS on M generated by the RDE (6.5.12). Let μ be an ergodic φ-invariant measure satisfying the integrability condition (6.5.16). Assume that φ under μ has simple Lyapunov spectrum.*

Then $\langle \nabla f(\cdot), \cdot \rangle \in L^1(St_2\mu)$ for any lift $St_2\mu$ of μ from $\Omega \times M$ to $\Omega \times St_2M$., i.e. the rotation number of $St_2\mu$,

$$\rho(St_2\mu) := \int_{\Omega \times St_2 M} \langle (\nabla f)u, v \rangle d(St_2\mu),$$

is finite.

More specifically, consider the $N_{ij} := N(f_{ij})$ measures $St_2\mu_{ij}^k$ constructed in Lemma 6.5.7 from the canonical frame $f_{ij} = (E_i \subset p_{ij})$. Then we have

(i) for $N_{ij} = 1$, $\rho(St_2\mu_{ij}) = 0$,
(ii)(a) for $N_{ij} = 2$ and case (a), $\rho(St_2\mu_{ij}^k) = 0$ for $k = 1, 2$,
(ii)(b) for $N_{ij} = 2$ and case (b), $\rho(St_2\mu_{ij}^1) = -\rho(St_2\mu_{ij}^2)$,
(iii) for $N_{ij} = 4$, $\rho(St_2\mu_{ij}^1) = \rho(St_2\mu_{ij}^3) = -\rho(St_2\mu_{ij}^2) = -\rho(St_2\mu_{ij}^4)$.

Proof. The integrability assertion follows from $|\langle \nabla f u, v \rangle| \leq \|\nabla f\| \in L^1(\mu)$.

(i) We have

$$\rho(St_2\mu_{ij}) =$$
$$\frac{1}{4}\int_{\Omega \times M} (\langle \nabla f u, v \rangle + \langle \nabla f(-u), v \rangle + \langle \nabla f(-u), -v \rangle + \langle \nabla f u, -v \rangle) d\mu$$

which vanishes since $(u, v) \mapsto \langle \nabla f u, v \rangle$ is bilinear.

290   Chapter 6. RDS on Homogeneous Spaces

(ii) and (iii) are also simple consequences of bilinearity.   □

Rotation Numbers for Canonical Planes

As a first step towards an MET for rotation numbers we now prove that rotation numbers exist for canonical planes

$$p_{ij}(\omega, x) := \text{span}(E_i(\omega, x), E_j(\omega, x)), \quad i < j,\ 1 \leq i, j \leq d.$$

However, the MET does not supply these planes with a canonical frame (or orientation). We hence start by measurably selecting an arbitrary frame $n_{ij} = (u, v) \in \pi^{-1}(p_{ij})$ from p_{ij}. To determine to which invariant measure on St_2M the orbit $St_2\varphi(t)n_{ij}$ is attracted, we construct a canonical frame $\bar{n}_{ij} = (\bar{u}, \bar{v})$ in p_{ij} from $n_{ij} = (u, v)$ as follows:

If $\text{proj}_{ij} : p_{ij} \to E_i$ denotes projection onto E_i along E_j, take

$$\bar{n}_{ij} = (\bar{u}, \bar{v}) := \perp (\text{proj}_{ij}(u), v).$$

We can assume without loss of generality that $\text{proj}_{ij}(u) \neq 0$, otherwise we use $(-v, u)$ to start with. Note that since $\bar{u} \in E_i$, $\bar{n}_{ij}$ is a selection from the flag f_{ij}, i.e. it is one of the four (canonical) frames selected from f_{ij} in the way used in Lemma 6.5.6, and n_{ij} and $\bar{n}_{ij}$ give p_{ij} the same orientation.

6.5.9 Theorem (Rotation Numbers for Canonical Planes). *Let the RDE $\dot{x}_t = f(\theta_t\omega, x_t)$ on a Riemannian manifold M with $d = \dim M \geq 2$ satisfy the assumptions of Theorem 2.2.14 and denote by φ the local C^1 RDS generated by it. Let μ be an ergodic φ-invariant measure which satisfies $\|\nabla f\| \in L^1(\mu)$, implying the validity of the IC of the MET. Assume that the RDS φ under μ has simple Lyapunov spectrum.*

Then each canonical plane $p_{ij} = \text{span}(E_i, E_j)$, $1 \leq i < j \leq d$, oriented with an arbitrary frame n_{ij}, has a rotation number $\rho(p_{ij}) = \rho_{ij}$, the absolute value of which is a fixed non-random number which is uniquely determined by the unoriented p_{ij} and the sign of which is uniquely determined by the orientation.

More specifically, ρ_{ij} is determined as follows: Let $f_{ij} = (E_i \subset p_{ij})$ be the corresponding canonical flag and let $N_{ij} = 1, 2, 4$ be the number of ways it can be lifted to an invariant measure on St_2M.

(i) If $N_{ij} = 1$, or if $N_{ij} = 2$, but the measures charge frames with different orientation, then $\rho_{ij} = 0$ irrespective of any orientation.

(ii) If $N_{ij} = 2$, but the measures charge frames with the same orientation, or if $N_{ij} = 4$, then

$$\rho_{ij}(\omega, x) = \begin{cases} \rho(St_2\mu_{ij}^1), & (\omega, x) \in A_{ij}, \\ -\rho(St_2\mu_{ij}^1), & (\omega, x) \notin A_{ij}, \end{cases}$$

where

$$\rho(St_2\mu_{ij}^1) := \int_{\Omega \times St_2 M} \langle (\nabla f)u, v \rangle d(St_2\mu_{ij}^1),$$

$$A_{ij} = \begin{cases} \{(\omega, x) : \bar{n}_{ij}(\omega, x) \in \mathrm{supp} St_2\mu_{ij}^1\}, & N_{ij} = 2, \\ \{(\omega, x) : \bar{n}_{ij}(\omega, x) \in \mathrm{supp} St_2\mu_{ij}^1 \cup \mathrm{supp} St_2\mu_{ij}^3\}, & N_{ij} = 4, \end{cases}$$

and $\bar{n}_{ij}$ is the canonical frame associated with n_{ij}.

Proof. If $n_{ij}(\omega, x) = \bar{n}_{ij}(\omega, x)$, then $n_{ij}(\omega, x) \in \mathrm{supp} St_2\mu_{ij,(\omega,x)}^k$ for some measure, hence by the ergodic theorem

$$\rho_{ij} = \lim_{T \to \infty} \frac{1}{T} \int_0^T \langle \nabla f(\Theta_t(\omega, x))u(t), v(t) \rangle dt = \int_{\Omega \times St_2 M} \langle \nabla f u, v \rangle d(St_2\mu_{ij}^k)$$

exists.

If $n_{ij}(\omega, x) \neq \bar{n}_{ij}(\omega, x)$, we prove that, with $n_{ij}(t) := (u(t), v(t))$ and $\bar{n}_{ij}(t) := (\bar{u}(t), \bar{v}(t))$,

$$\lim_{T \to \infty} \frac{1}{T} \int_0^T \langle \nabla f(\Theta_t(\omega, x))u(t), v(t) \rangle dt \qquad (6.5.18)$$

$$= \lim_{T \to \infty} \frac{1}{T} \int_0^T \langle \nabla f(\Theta_t(\omega, x))\bar{u}(t), \bar{v}(t) \rangle dt.$$

We have

$$|\langle \nabla f u, v \rangle - \langle \nabla f \bar{u}, \bar{v} \rangle| \leq \|\nabla f\|(d(u, \bar{u}) + d(v, \bar{v}))$$
$$= 2\|\nabla f\| d(u, \bar{u}),$$

the latter being true because for two orthonormal frames (u, v) and $(\bar{u}, \bar{v})$ in the same plane, $d(v, \bar{v}) = d(u, \bar{u})$.

By Lemma 6.5.10(i) following this proof,

$$\lim_{t \to \infty} \frac{1}{t} \log d(u(t), \bar{u}(t)) = \begin{cases} \lambda_j - \lambda_i < 0, & u \neq \bar{u}, \\ -\infty, & u = \bar{u}, \end{cases}$$

so that part (ii) of Lemma 6.5.10 gives the equality (6.5.18).

The remaining assertions immediately follow from Lemma 6.5.7 and Proposition 6.5.8. □

To prepare the lemma needed in the above proof, we come back to the RDE (6.5.12) for φ and (6.5.13), respectively (6.5.14), for $T\varphi$.

By the real noise version of the statements preceding Theorem 6.4.3, using polar coordinates in $T^0 M = \cup_{x \in M} T_x^0 M$, (6.5.13) turns into the equivalent coupled RDE

$$\dot{s}_t = Sf(\Theta_t(\omega, x), s_t) \qquad (6.5.19)$$

for the angular part, where Sf is the vector field on SM induced by Tf on TM, equivalently

$$s'_t = \nabla f(\Theta_t(\omega, x))s_t - \langle \nabla f(\Theta_t(\omega, x))s_t, s_t \rangle s_t, \qquad (6.5.20)$$

and
$$\dot{r}_t = \langle \nabla f(\Theta_t(\omega, x))s_t, s_t \rangle r_t \qquad (6.5.21)$$

for the radial part. The RDE (6.5.19) generates an RDS $S\varphi$ on SM over Θ.

Inserting $S\varphi$ into (6.5.21) and solving it gives the Lyapunov exponent of $v \in T_x^0 M$ as a time average (writing $s = \frac{v}{\|v\|}$ and $s(t) = S\varphi(t)s$),

$$\lambda(\omega, x, v) = \lim_{T \to \infty} \frac{1}{T} \int_0^T \langle \nabla f(\Theta_t(\omega, x))s(t), s(t) \rangle dt. \qquad (6.5.22)$$

6.5.10 Lemma. *Let φ be generated by (6.5.12) and let μ be an ergodic φ-invariant measure for which (6.5.16), hence the MET, holds, with corresponding Θ-invariant set Δ of full μ measure. Let for $(\omega, x) \in \Delta$, $0 \neq v(\omega, x) = \oplus_{i=1}^p v_i(\omega, x) \in T_x M = \oplus_{i=1}^p E_i(\omega, x)$ and define*

$$i_0 := \min\{i \geq 1 : v_i \neq 0\}, \quad j_0 := \min\{i > i_0 : v_i \neq 0\}$$

($j_0 := p+1$ if the latter set is empty).

(i) If $d(s, \bar{s}) = \|s - \bar{s}\| = $ distance in $T_x M$ and if we write $s = \frac{v}{\|v\|}$, $s_{i_0} = \frac{v_{i_0}}{\|v_{i_0}\|}$, $s(t) = S\varphi(t)s$ and $s_{i_0}(t) = S\varphi(t)s_{i_0}$, then for $(\omega, x) \in \Delta$

$$\lim_{t \to \infty} \frac{1}{t} \log d(s(t), s_{i_0}(t)) = \lambda_{j_0} - \lambda_{i_0} < 0 \quad (\lambda_{p+1} := -\infty).$$

(ii) If $f : \Omega \times SM \to \mathbb{R}$ is such that

$$|f(\omega, x, s) - f(\omega, x, \bar{s})| \leq L(\omega, x) d(s, \bar{s}) \quad \text{with } L \in L^1(\mu),$$

and if for some lift $S\mu$ of μ

$$\lim_{T \to \infty} \frac{1}{T} \int_0^T f(\Theta_t(\omega, x), s_{i_0}(t)) dt = \int_{\Omega \times SM} f \, dS\mu \quad S\mu\text{-a.s.},$$

then also

$$\lim_{T \to \infty} \frac{1}{T} \int_0^T f(\Theta_t(\omega, x), s(t)) dt = \int_{\Omega \times SM} f \, dS\mu \quad S\mu\text{-a.s.}$$

Proof. (i) Since

$$d(s, s_{i_0}) = \|s - s_{i_0}\| = 2 \left| \sin \frac{\sphericalangle(s, s_{i_0})}{2} \right|,$$

where $\sphericalangle(s, s_{i_0})$ is the angle between s and s_{i_0}, and since

$$|\sin \sphericalangle(s, s_{i_0})| = \frac{\|v \wedge v_{i_0}\|}{\|v\| \|v_{i_0}\|},$$

the statement follows immediately from (the obvious manifold version of) Corollary 5.3.7.

(ii) By assumption,
$$\left| \frac{1}{T} \int_0^T (f(\Theta_t(\omega,x), s(t)) - f(\Theta_t(\omega,x), s_{i_0}(t))) dt \right|$$
$$\leq \frac{1}{T} \int_0^T L(\Theta_t(\omega,x)) d(s(t), s_{i_0}(t)) dt. \qquad (6.5.23)$$

By (i) there is, for any $\varepsilon > 0$, a $T_0(\omega, x) < \infty$ such that $d(s(t), s_{i_0}(t)) \leq \varepsilon/4c$ for all $t \geq T_0$, where $c = \int_{\Omega \times M} L(\omega, x) d\mu < \infty$. Thus for $T \geq T_0$

$$\frac{1}{T} \int_{T_0}^T L(\Theta_t(\omega,x)) d(s(t), s_{i_0}(t)) dt \leq \frac{\varepsilon}{4c} \frac{1}{T} \int_{T_0}^T L(\Theta_t(\omega,x)) dt.$$

By the ergodic theorem, $\frac{1}{T} \int_0^T L\, dt \to c$ μ-a.s., so there is a $T_1(\omega, x) < \infty$ for which
$$\frac{1}{T} \int_{T_0}^T L\, dt \leq 2c \quad \text{for all } T \geq T_1.$$

Finally, choose $T_2(\omega, x) < \infty$ such that for all $T \geq T_2$
$$\frac{1}{T} \int_0^{T_0} L(\Theta_t(\omega,x)) d(s(t), s_{i_0}(t)) dt < \frac{\varepsilon}{2}.$$

This makes the right-hand side of (6.5.23) less than ε for all $T \geq \max(T_1, T_2)$. □

Rotation Numbers for Arbitrary Planes

Our final steps towards an MET for rotation numbers are now (i) to prove the existence of the rotation number for an arbitrary plane, and (ii) to show that this rotation number is a random variable whose possible values are just the canonical rotation numbers ρ_{ij}.

Let now $p = p(x) \in G_2(T_x M)$ be an arbitrary plane. We will determine the "strongest" of the canonical planes $p_{ij}(\omega, x)$ in $T_x M$ such that p "has a component in direction of p_{ij}".

Let, as usual, $v = \oplus_{i=1}^d v_i \in T_x M = \oplus_{i=1}^d E_i(\omega, x)$ be the decomposition of a vector according to the Oseledets splitting. Introduce the subspaces

$$F_i(\omega, x) := \sum_{j=i}^d E_j(\omega, x), \quad F_{d+1} := \{0\}.$$

We define the following indices for $p \in G_2(T_x M)$:
$$i_0 = i_0(\omega, x, p) := \min\{i : \text{there exists } v \in p \text{ with } v_i \neq 0\},$$
$$i_1 = i_1(\omega, x, p) := \min\{i > i_0 : \text{there exists } v \in p \text{ with } v_i \neq 0\},$$

294   Chapter 6. RDS on Homogeneous Spaces

$$i_2 = i_2(\omega, x, p) := \begin{cases} \min\{i > i_0 : v_i \neq 0 \text{ for some } v \in p \text{ with } v_{i_0} \neq 0\}, \\ d+1, \quad \text{if above set is empty}, \end{cases}$$

$$j_0 = j_0(\omega, x, p) := \min\{i : \text{there exists } v \in p \cap F_{i_0+1} \text{ with } v_i \neq 0\},$$

$$j_1 = j_1(\omega, x, p) := \begin{cases} \min\{i > j_0 : \exists v \in p \cap F_{i_0+1} \text{ with } v_i \neq 0\}, \\ d+1, \quad \text{if above set is empty}. \end{cases}$$

We have

$$1 \leq i_0 < i_1 \leq j_0 < j_1 \leq d+1, \quad \text{and} \quad 2 \leq i_1 \leq i_2 \leq d+1.$$

The next three lemmas are analogues of Lemma 6.5.10 on G_2M, $F_{1,2}M$ and St_2M, respectively.

6.5.11 Lemma. *The orbits $p(t) := G_2\varphi(t)p$ and $p_{i_0 j_0}(t) := G_2\varphi(t)p_{i_0 j_0}$ satisfy*

$$\lim_{t \to \infty} \frac{1}{t} \log d(p(t), p_{i_0 j_0}(t)) = \begin{cases} \max(\lambda_{i_1} - \lambda_{i_0}, \lambda_{j_1} - \lambda_{j_0}) < 0, & i_1 < j_0, \\ \lambda_{j_1} - \lambda_{j_0} < 0, & i_1 = j_0, \end{cases} \quad (6.5.24)$$

where $d(p_1, p_2)$ is the distance of the two planes $p_1, p_2 \in T_xM$ considered as elements of the projective space $P(\wedge^2 T_xM)$.

Proof. We represent a plane $p = \text{span}(u,v)$ as the element in $P(\wedge^2 T_xM)$ corresponding to $u \wedge v \in \wedge^2 T_xM$. The Oseledets splitting in $\wedge^2 TM$ is given by $\oplus_{i,j} E_i \wedge E_j$ (see the manifold version of Theorem 5.3.1). The relative position of p to this splitting is described by our indices as

$$p = (u_{i_0} + u_{i_1} + \bar{u}) \wedge (v_{j_0} + v_{j_1} + \bar{v}),$$

where $u_{i_0} \in E_{i_0}$, $u_{i_1} \in E_{i_1}$, $\bar{u} \in F_{i_1+1}$, $v_{j_0} \in E_{j_0}$, $v_{j_1} \in E_{j_1}$ and $\bar{v} \in F_{j_1+1}$. Hence

$$p = u_{i_0} \wedge v_{j_0} + u_{i_0} \wedge v_{j_1} + u_{i_1} \wedge v_{j_0} + \ldots,$$

where ... represents an element whose exponential growth rate is strictly smaller than $\min(\lambda_{i_0} + \lambda_{j_1}, \lambda_{i_1} + \lambda_{j_0})$. The strongest component of p is $u_{i_0} \wedge v_{j_0}$. For the second strongest component, we have to distinguish the following cases:

(i) $i_1 < j_0$: Then either (a) $\lambda_{i_1} + \lambda_{j_0} \geq \lambda_{i_0} + \lambda_{j_1}$, or (b) $\lambda_{i_1} + \lambda_{j_0} < \lambda_{i_0} + \lambda_{j_1}$. In case (a), Lemma 6.5.10 (applied to $\wedge^2 TM$) yields

$$\lim_{t \to \infty} \frac{1}{t} \log d(p(t), p_{i_0 j_0}(t)) = (\lambda_{i_0} + \lambda_{j_0}) + (\lambda_{i_1} + \lambda_{j_0}) - 2(\lambda_{i_0} + \lambda_{j_0})$$

$$= \lambda_{i_1} - \lambda_{i_0} = \max(\lambda_{i_1} - \lambda_{i_0}, \lambda_{j_1} - \lambda_{j_0}) < 0.$$

In case (b), again by Lemma 6.5.10,

$$\lim_{t \to \infty} \frac{1}{t} \log d(p(t), p_{i_0 j_0}(t)) = \lambda_{j_1} - \lambda_{j_0} = \max(\lambda_{i_1} - \lambda_{i_0}, \lambda_{j_1} - \lambda_{j_0}) < 0.$$

(ii) $i_1 = j_0$: Now $u_{i_1} \wedge v_{j_0} = 0$, so that the second strongest component is $u_{i_0} \wedge v_{j_1}$; thus

$$\lim_{t\to\infty} \frac{1}{t} \log d(p(t), p_{i_0 j_0}(t)) = \lambda_{i_0} + \lambda_{j_1} - (\lambda_{i_0} + \lambda_{j_0}) = \lambda_{j_1} - \lambda_{j_0} < 0.$$

□

We now look at frames inside the planes and study their exponential convergence. Let $n = (u, v)$ be a frame in p. Assume without loss of generality that $\bar{u}_{i_0} := \text{proj}_{i_0} u \neq 0$. Let $\bar{v}$ be the only unit vector in $p \cap F_{j_0+1}$ such that (u, v) and $(\bar{u}_{i_0}, \bar{v})$ have the same orientation, put $\bar{v}_{j_0} := \text{proj}_{j_0} \bar{v}$ and

$$n_{i_0 j_0} = (u_{i_0}, v_{j_0}) := \perp (\bar{u}_{i_0}, \bar{v}_{j_0}).$$

Note that $n_{i_0 j_0}$ is one of our standard frames in $p_{i_0 j_0}$ since $u_{i_0} \in E_{i_0}$ and n and $n_{i_0 j_0}$ have the same orientation.

As an intermediate step, we consider the orbits of

$$f := (V \subset p) = (\text{span}(u) \subset p), \quad f_{i_0 j_0} := (E_{i_0} \subset p_{i_0 j_0}),$$

under $F_{1,2}\varphi$ in $F_{1,2}M$.

6.5.12 Lemma. *The orbits $f(t) := F_{1,2}\varphi(t)f$ and $f_{i_0 j_0}(t) := F_{1,2}\varphi(t) f_{i_0 j_0}$ satisfy*

$$\lim_{t\to\infty} \frac{1}{t} \log d(f(t), f_{i_0 j_0}(t)) = \max(\lambda_{i_2} - \lambda_{i_0}, \lambda_{j_1} - \lambda_{j_0}) < 0, \quad (6.5.25)$$

where

$$d(f_1, f_2) = d((V_1 \subset p_1), (V_2 \subset p_2)) := d_1(V_1, V_2) + d_2(p_1, p_2),$$

with d_1 the distance in $P(T_x M)$ and d_2 the distance in $P(\wedge^2 T_x M)$.

Proof. By definition, $i_1 \leq i_2$. We have to distinguish two cases:

(i) $i_2 \leq j_0$: Then $i_1 = i_2$. Indeed, in this case $V \subset E_{i_0} + F_{i_2}$. Since $p \cap F_{i_0+1} \subset F_{j_0}$ and p is spanned by V and $p \cap F_{i_0+1}$, it follows that $p \subset E_{i_0} + F_{i_2}$. Thus $i_1 \geq i_2$, so $i_1 = i_2$.

By Lemma 6.5.10,

$$\lim_{t\to\infty} \frac{1}{t} \log d_1(V(t), E_{i_0}(t)) = \lambda_{i_2} - \lambda_{i_0} = \lambda_{i_1} - \lambda_{i_0}.$$

This and Lemma 6.5.11 give the result.

(ii) $i_2 > j_0$: Then $i_1 = j_0$. In fact, $V \subset E_{i_0} + F_{i_2} \subset E_{i_0} + F_{j_0}$. Since $p \cap F_{i_0+1} \subset F_{j_0}$, it follows that $p \subset E_{i_0} + F_{j_0}$; hence $i_1 \geq j_0$ and finally $i_1 = j_0$. The result follows again by combining Lemma 6.5.10 for d_1 and Lemma 6.5.11 for d_2. □

The final step now consists of lifting this result from $F_{1,2}M$ to $St_2 M$.

6.5.13 Lemma. *The orbits $n(t) := St_2\varphi(t)n$ and $n_{i_0 j_0}(t) := St_2\varphi(t) n_{i_0 j_0}$ satisfy*

$$\lim_{t\to\infty} \frac{1}{t} \log d(n(t), n_{i_0 j_0}(t)) = \max(\lambda_{i_2} - \lambda_{i_0}, \lambda_{j_1} - \lambda_{j_0}) < 0, \qquad (6.5.26)$$

where $d(n_1, n_2) = d(f_1, f_2)$.

Proof. By Lemma 6.5.11, $d(f(t), f_{i_0 j_0}(t))$ is small for large t. For every such t, there exists an open subset U_t of $F_{1,2}M$ that trivializes the bundle $\pi : St_2 M \to F_{1,2}M$ and such that $f(t), f_{i_0 j_0}(t) \in U_t$. Let us check that $n(t)$ and $n_{i_0 j_0}(t)$ are in the same component of $\pi^{-1}(U_t)$.

For large t, $d_1(V(t), E_{i_0}(t))$ is small and, moreover, by our choice of u_{i_0}, $d(u(t), u_{i_0}(t))$ is small in $T_{\varphi(t,\omega,x)}M$. This implies that $n(t)$ is either in the component which contains $n_{i_0 j_0}(t) = (u_{i_0}(t), v_{j_0}(t))$ or in the component which contains $(u_{i_0}(t), -v_{j_0}(t))$. However, by our choice of $\bar{v}$, and since orientation is unaltered by projection and orthonormalization, the first possibility is true. Thus, for large t,

$$d(n(t), n_{i_0 j_0}(t)) = d(f(t), f_{i_0 j_0}(t)),$$

and the assertion follows from Lemma 6.5.11. □

We are finally in a position to derive the main result of this subsection.

6.5.14 Theorem (MET for Rotation Numbers, Real Noise Case). Let the RDE $\dot{x}_t = f(\theta_t \omega, x_t)$ on a Riemannian manifold M with $d = \dim M \geq 2$ satisfy the assumptions of Theorem 2.2.14 and denote by φ the local C^1 RDS generated by it. Let μ be an ergodic φ-invariant measure which satisfies $\|\nabla f\| \in L^1(\mu)$, implying the validity of the IC of the MET. Assume that the RDS φ under μ has simple Lyapunov spectrum.

Then $\langle \nabla f(\cdot)\cdot, \cdot\rangle \in L^1(St_2\mu)$ for any lift $St_2\mu$ of μ from $\Omega \times M$ to $\Omega \times St_2 M$ and there exists a Θ-invariant set $\Gamma \subset \Omega \times M$ of full μ measure such that on Γ for any plane $p \in G_2^+ M$ the rotation number of p,

$$\rho(p) = \lim_{T\to\infty} \frac{1}{T} \int_0^T \langle \nabla f(\Theta_t(\cdot))u(t), v(t)\rangle dt , \qquad (6.5.27)$$

exists. Here $n(t) = (u(t), v(t)) = St_2\varphi(t)n$ with an arbitrary $n \in \pi^{-1}(p)$.

The quantity $\rho(\cdot)$ is a random variable on $\Omega \times G_2^+ M$ which takes on only finitely many values, namely the $d(d-1)/2$ rotation numbers ρ_{ij} of the (properly oriented) canonical planes $p_{ij} = \text{span}(E_i, E_j)$ for $1 \leq i < j \leq d$ determined in Theorem 6.5.9.

More precisely,

$$\rho(p) = \rho_{i_0 j_0},$$

where $\rho_{i_0 j_0}$ is the rotation number for $p_{i_0 j_0}$, the "strongest" canonical plane "contained" in p and oriented by $n_{i_0 j_0}$.

Proof. Let Γ be the subset of the validity of the MET for which the ergodic theorem on $\Omega \times St_2 M$ for any lift $St_2\mu$ holds. The only statement which needs to be proved is that

$$\lim_{T \to \infty} \frac{1}{T} \int_0^T \langle \nabla f(\Theta_t(\omega, x)) u(t), v(t) \rangle dt$$
$$= \lim_{T \to \infty} \frac{1}{T} \int_0^T \langle \nabla f(\Theta_t(\omega, x)) u_{i_0}(t), v_{j_0}(t) \rangle dt.$$

But this follows from Lemma 6.5.10 since

$$|\langle (\nabla f) u, v \rangle - \langle (\nabla f) u_{i_0}, v_{j_0} \rangle| \leq \|\nabla f\| (d(u, u_{i_0}) + d(v, v_{j_0}))$$

and, for large t and a certain constant c, by Lemma 6.5.13,

$$d(u(t), u_{i_0}(t)) + d(v(t), v_{j_0}(t)) \leq c\, d(n(t), n_{i_0 j_0}(t)) \to 0 \text{ exponentially fast.}$$

$\square$

Theorem 6.5.14 tells us that the rotation numbers ρ_{ij} of the canonical planes $p_{ij} = \text{span}(E_i, E_j)$ are the basic ones and that any other plane p has a rotation number $\rho(p)$ that picks the value ρ_{ij} whenever p has p_{ij} as its "strongest" component. We thus have exactly the same situation as for the Lyapunov exponents of tangent vectors in the classical MET. Also note the striking similarity between (6.5.22) and (6.5.27).

Also, the rotation number $\rho(p)$ for an oriented plane $p \in G_2^+ M$ is uniquely determined, while for an unoriented plane it is only determined up to its sign.

Particular cases:

(i) Our results apply also to the deterministic situation $\dot{x}_t = f(x_t)$, where μ is a measure invariant under the local flow $\varphi(t)$ generated by f on M. There is, however, in general no simplification.

(ii) If $\mu(dx, d\omega) = \delta_{x_0}(dx) \mathbb{P}(d\omega)$ (i.e. x_0 is a nonrandom fixed point of $f(\omega, \cdot)$), we can restrict our considerations to the tangent space $T_{x_0} M \cong \mathbb{R}^d$, and thus to $G_2(\mathbb{R}^d)$, $F_{1,2}(\mathbb{R}^d)$ and $St_2(\mathbb{R}^d)$. In this case $v(t) = T\varphi(t, \omega, x_0) v$ satisfies

$$\dot{v}_t = A(\theta_t \omega) v_t, \quad A(\omega) := \left. \frac{\partial f(\omega, x)}{\partial x} \right|_{x=x_0}.$$

(iii) For $d = 2$, there exists a unique rotation number ρ_{12} for φ and μ. The particular case $\mu(dx, d\omega) = \delta_{x_0}(dx) \mathbb{P}(d\omega)$ leads back to the concept of rotation numbers for linear systems in $\mathbb{R}^2$ by which we started our discussion in Subsect. 6.5.1. Choosing the standard frame $n = (e_1, e_2)$ in $\mathbb{R}^2$ and an invariant measure μ_1 on S^1 realizing λ_1, $\rho(\mathbb{R}^2) = \rho_{12} = \mathbb{E}_{\mu_1} \langle A(\cdot) u, v \rangle$.

6.5.15 Remark (Lyapunov Spectrum not Simple). At this moment, we are not able to prove the existence of rotation numbers for a situation where the Lyapunov spectrum is not simple. This is due to the lack of knowledge about the dynamics of the cocycle $T\varphi$ inside a higher-dimensional Oseledets space E_i. As Example 6.5.4 shows, there can be continuously many different rotation numbers in one Oseledets space.

The assumption of simple spectrum is, however, "generically true" for "truly random" situations as noise tends to split multiplicities. Evidence for this claim has been found in many situations, see, e. g. Arnold and Nguyen Dinh Cong [30] for products of random matrices and Remark 6.2.15. ∎

6.5.3 Rotation Numbers for SDE

With the preparations of Subsect. 6.5.1 and with the real noise case of Subsect. 6.5.2 in mind, the concept of rotation number in the white noise case should now be rather evident. For more details, examples and discussions see Ruffino [296, 297] and Arnold and Imkeller [20].

We restrict ourselves to the linear case since it already reveals the necessary modifications of the real noise concept.

Let

$$dx_t = A_0 x_t dt + \sum_{j=1}^{m} A_j x_t \circ dW_t^j =: \sum_{j=0}^{m} A_j x_t \circ dW_t^j \qquad (6.5.28)$$

be a linear SDE in $\mathbb{R}^d$ with $d \geq 2$, generating the linear RDS Φ. Then the RDS $St_2\Phi$ induced on the Stiefel manifold $St_2(d)$ of orthonormal 2-frames $n = (u, v)$ is generated by the SDE

$$dn_t = \sum_{j=0}^{m} \tilde{A}_j(n_t) \circ dW_t^j, \qquad (6.5.29)$$

where the lift $\tilde{A}_j(n)$ to $St_2(d)$ of the linear vector field $A_j x$ in $\mathbb{R}^d$ is explicitly given by the $d \times 2$ matrix

$$\tilde{A}_j(u,v) = (A_j u - \langle A_j u, u\rangle u, A_j v - \langle A_j v, v\rangle v - (\langle A_j u, v\rangle + \langle A_j v, u\rangle)u). \qquad (6.5.30)$$

The SDE for u_t and v_t are hence, respectively,

$$du_t = \sum_{j=0}^{m} (A_j u_t - \langle A_j u_t, u_t\rangle u_t) \circ dW_t^j, \qquad (6.5.31)$$

$$dv_t = \sum_{j=0}^{m} (A_j v_t - \langle A_j v_t, v_t\rangle v_t - (\langle A_j u_t, v_t\rangle + \langle A_j v_t, u_t\rangle)u_t) \circ dW_t^j. \qquad (6.5.32)$$

If $n_t = \bar{n}_t g_t$, where $\bar{n}_t$ is the parallel transported initial frame n, and $g_t = \begin{pmatrix} \cos\alpha_t & -\sin\alpha_t \\ \sin\alpha_t & \cos\alpha_t \end{pmatrix} \in SO(2,\mathbb{R})$, then

$$d\alpha_t = \langle \circ du_t, v_t\rangle = v_t^* \circ du_t = \sum_{j=0}^{m} \langle A_j u_t, v_t\rangle \circ dW_t^j,$$

and we naturally arrive at the following definition of the rotation number of the oriented 2-plane $p \in G_2^+(d)$ under Φ:

$$\rho(p) := \lim_{T\to\infty} \frac{1}{T} \int_0^T \langle \circ du_t, v_t \rangle = \lim_{T\to\infty} \sum_{j=0}^m \frac{1}{T} \int_0^T \langle A_j u_t, v_t \rangle \circ dW_t^j, \quad (6.5.33)$$

provided the stochastic integrals in (6.5.33) make sense and the limit exists. We again have the statements of Proposition 6.5.3 since $\circ d\langle u_t, v_t \rangle = \langle \circ du_t, v_t \rangle + \langle u_t, \circ dv_t \rangle$.

It is more difficult to assure the existence of the rotation number in the SDE case because of the following technical problem which we already encountered in Subsect. 6.2.2: If we mimic the procedure in Subsect. 6.5.2 leading to the MET for rotation numbers, we have to solve the SDE (6.5.29) for random initial values, e. g. n_{ij} chosen in $p_{ij} = \mathrm{span}(E_i, E_j)$, which are in general not $\mathcal{F}^-$-measurable. This calls for the use of anticipative calculus. We quote the main result of work (see [20]) which is still in progress.

6.5.16 Theorem (MET for Rotation Numbers, Linear White Noise Case). *Consider the linear SDE (6.5.28) and assume that the smallest closed subgroup of $Gl(d, \mathbb{R})$ generated by the matrices $A_0, \ldots, A_m$ is equal to $Sl(d, \mathbb{R})$ or $Gl^+(d, \mathbb{R})$. Then:*

(i) The Lyapunov spectrum of Φ is simple. The distributions of the forward flag, the backward flag and of the Oseledets spaces $(E_1, \ldots, E_d)$ have C^∞ densities with respect to the canonical Riemannian volume of the respective state spaces.

(ii) The rotation numbers $\rho(p_{ij}) = \rho_{ij}$ of the canonical planes $p_{ij} := \mathrm{span}(E_i, E_j)$, $1 \le i < j \le d$, oriented by some n_{ij}, exist and are given by

$$\rho_{ij} = \int_{\Omega \times St_2(d)} (R(u,v) + D(u,v)) d(St_2 \mu_{ij}),$$

where $St_2 \mu_{ij}$ is a certain lift of the invariant Dirac measure in $f_{ij} := (E_i \subset p_{ij})$ from $F_{1,2}(d)$ to $St_2(d)$,

$$R(u,v) := \langle A_0 u, v \rangle + \frac{1}{2} \sum_{j=1}^m R_j(u,v),$$

$$\begin{aligned} R_j(u,v) &= \langle A_j u, A_j v \rangle - \langle A_j u, v \rangle \langle A_j v, v \rangle - \langle A_j u, u \rangle \langle A_j v, u \rangle + \langle A_j^2 u, v \rangle \\ &\quad - 2\langle A_j u, u \rangle \langle A_j u, v \rangle, \end{aligned} \quad (6.5.34)$$

and $D(u,v)$ is an additional Stratonovich-Skorokhod correction term.

(iii) The rotation number $\rho(p)$ for every plane $p \in G_2^+(d)$ exists, and $\rho(\cdot)$ is a random variable on $\Omega \times G_2^+(d)$ which takes on the values ρ_{ij}. More specifically,

$$\rho(p) = \rho_{i_0 j_0},$$

where i_0 and j_0 are determined as in Theorem 6.5.14.

(iv) In particular,
$$\rho_{12} = \int_{St_2(d)} R(u,v) d(St_2 \nu),$$
where $St_2\nu$ is the marginal of the $\mathcal{F}^-$-measurable invariant measure $St_2\mu_{12}$ from (ii).

Further, for each deterministic plane $p \in G_2^+(d)$
$$\rho(p) = \rho_{12} \quad \mathbb{P}\text{-a. s.}$$

(v) In dimension $d = 2$ the unique rotation number of $\mathbb{R}^2$ with canonical orientation is given by
$$\rho = \int_{S^1} \left(\langle A_0 u, v \rangle + \frac{1}{2} \sum_{j=1}^m \langle A_j u, v \rangle (\langle A_j v, v \rangle - \langle A_j u, u \rangle) \right) d\nu(u),$$
where ν is a solution of the Fokker-Planck equation on S^1 and $v = (-u_2, u_1)$.

Proof. We will only sketch the very technical proof and refer to [20] for details.

(i) For the simplicity of the spectrum see Remark 6.2.15. The existence of the densities was proved by Imkeller [179].

(ii) Hard work using Malliavin calculus is needed to prove the existence of $\rho(p_{ij})$ and an explicit form of the correction term $D(u,v)$.

If, however, n is $\mathcal{F}^-$-measurable (e. g. constant), then $\langle A_j u_t, v_t \rangle$ is a semi-martingale, and
$$\int_0^T \langle A_j u_t, v_t \rangle \circ dW_t^j = \int_0^T \langle A_j u_t, v_t \rangle dW_t^j + \frac{1}{2} \int_0^T R_j(u_t, v_t) dt,$$
where the Stratonovich-Itô correction term can be calculated by standard methods to be equal to (6.5.34).

Taking time averages, the Itô integral cancels due to the following lemma (for a proof see e. g. Liptser [243: Theorem 1]) applied to $M_t = \int_0^t \langle A_j u_s, v_s \rangle dW_s^j$.

6.5.17 Lemma. *Let M be a local square integrable martingale. Then*
$$\int_0^\infty \frac{d\langle M \rangle_t}{(1+t)^2} < \infty \quad \Rightarrow \quad \lim_{t \to \infty} \frac{M_t}{t} = 0 \quad \mathbb{P}\text{-a. s.}$$

For an $\mathcal{F}^-$-measurable $n = (u,v) \in \pi^{-1}(p)$ we arrive at
$$\rho(p) = \lim_{T \to \infty} \frac{1}{T} \int_0^T R(u_t, v_t) \, dt \qquad (6.5.35)$$

(existence provided).

(iii) The difficult part of the proof is to carry over Lemma 6.5.10 to time averages of anticipative Stratonovich integrals.

(iv) Since p_{12} is $\mathcal{F}^-$-measurable, the correction term $D(u,v)$ vanishes and we can use (6.5.35) and the ergodic theorem.

Further, due to the existence of densities (see (i)) for each fixed p, $i_0(p) = 1$ and $j_0(p) = 2$ $\mathbb{P}$-a.s., hence $\rho(p) = \rho_{12}$ $\mathbb{P}$-a. s.

(v) For $d = 2$ further terms in $R(u,v)$ vanish, as can be easily checked. Since by our assumptions the hypoellipticity condition 6.2.25 is valid, the Fokker-Planck equation on S^1 either has a unique solution ν, or has exactly two symmetric solutions which give the same average. □

Part III

Smooth Random Dynamical Systems

Chapter 7. Invariant Manifolds

Summary

This longest and technically most complicated chapter of the book is devoted to the theory of invariant manifolds of RDS. Our aim is to give an up-to-date picture which is as complete and as reliable as possible.

For the whole chapter we use a "dynamical" method which today is widely used in the deterministic case and was carried over to RDS by Wanner [340]. The idea is to work with a scale of Banach spaces of orbits with certain exponential growth rates by which invariant manifolds can be characterized. The advantage of this method is (besides yielding global invariant manifolds and their regularity) that it also gives the Hartman-Grobman theorem for RDS, thus technically unifying the whole chapter.

After explaining the problem (Sect. 7.1) and doing various preparations to simplify the main work (Sect. 7.2), the long Sect. 7.3 is devoted to the construction of global invariant manifolds (Theorem 7.3.1 for unstable manifolds, Theorem 7.3.10 for stable manifolds and Theorems 7.3.14 and 7.3.17 for center manifolds). The moral of the two key technical lemmas 7.3.3 and 7.3.6 is that all one has to do is solving affine difference equations by the variation of constants formula.

The outcome of these constructions are invariant manifolds which are just Lipschitz-continuous no matter how smooth the RDS under consideration is. We will improve the regularity by playing with the parameters in the scale of Banach spaces. To accomplish e. g. C^k stable manifolds for a C^k RDS one has to place two parameters $a < b$ into the gap between the two Lyapunov exponents at which the spectrum is split such that also $ka < b$ – which for $k > 1$ is not always possible and is the origin of the "gap conditions" (Theorem 7.3.19).

In Sect. 7.4 we complete the picture by showing that the splitting of the Lyapunov spectrum between any two exponents yields foliations of invariant manifolds (Theorem 7.4.1). These can be used as nonlinear coordinates to topologically decouple the nonlinear RDS into blocks according to the (linear) splitting of the MET (Theorem 7.4.4). There is just one more step to (topologically) linearize all blocks having non-zero Lyapunov exponents (Hartman-Grobman Theorem 7.4.12).

Due to very restrictive conditions, global invariant manifolds rarely exist. In contrast, local invariant manifolds usually do exist (Theorem 7.5.5), and some can be dynamically characterized and extended to global objects (Theorem 7.5.13).

We hope that after all this the reader will enjoy studying the examples in Sect. 7.6.

7.1 The Problem of Invariant Manifolds

Suppose φ is a C^k RDS, $k \geq 1$, with two-sided time on a d-dimensional Riemannian manifold M. Let μ be an ergodic invariant measure of φ such that $T\varphi$ satisfies the IC of the MET. Consider the Lyapunov spectrum $\mathcal{S}(\varphi, \mu) = \{(\lambda_i, d_i)_{i=1,...,p}\}$ of φ under μ and the invariant splitting $T_x M = \oplus_{i=1}^p E_i(\omega, x)$, where (ω, x) is from the invariant set $\Delta \subset \Omega \times M$ of "good" points, which has full μ measure.

Now choose $\emptyset \neq \Lambda \subset \{\lambda_1, \ldots, \lambda_p\}$ and consider
$$E_\Lambda(\omega, x) := \oplus_{i:\lambda_i \in \Lambda} E_i(\omega, x) \subset T_x M.$$
This is a subbundle of TM of dimension $d_\Lambda := \sum_{i:\lambda_i \in \Lambda} d_i$ which is invariant under $T\varphi$,
$$T\varphi(t, \omega, x) E_\Lambda(\omega, x) = E_\Lambda(\Theta(t)(\omega, x)),$$
where $\Theta(t)(\omega, x) = (\theta(t)\omega, \varphi(t, \omega)x)$ is the corresponding skew product flow.

The *problem of invariant manifolds* consists of "bending E_Λ down" from TM to the manifold M, that is, of "finding nonlinear analogues" of E_Λ. More precisely: We seek (local) C^k submanifolds $M_\Lambda(\omega, x)$ with $x \in M_\Lambda(\omega, x)$ with the following properties:

(i) Local closeness of M_Λ and E_Λ: $T_x M_\Lambda(\omega, x) = E_\Lambda(\omega, x)$.
(ii) Invariance under φ: $\varphi(t, \omega) M_\Lambda(\omega, x) = M_\Lambda(\Theta(t)(\omega, x))$ (locally).
(iii) Dynamical characterization: We aim at characterizing the points $y \in M_\Lambda(\omega, x)$ by the exponential growth rate of $d(\varphi(t, \omega)y, \varphi(t, \omega)x)$ as $t \to \pm\infty$, where this rate should be related to the Lyapunov exponents of of tangent vectors v under $T\varphi$ collected in Λ. See Fig. 7.1.

The invariant manifolds are of utmost importance for the study of the dynamical behavior of φ. For example, they allow us to reduce an RDS to a lower-dimensional manifold to study stability. Since we try to obtain the nonlinear object M_Λ from the linear object E_Λ, invariant manifold theory is part of what is called *local theory*.

Particular cases:

(i) If Λ consists of all negative, vanishing, positive, non-positive or non-negative Lyapunov exponents, we obtain the random versions of the following five "classical" invariant manifolds: the stable, center, unstable, center-stable, and center-unstable manifold, denoted respectively by M_s, M_c, M_u, M_{cs}, M_{cu}. We also call the random versions of these manifolds *classical stable manifold, classical center manifold*, etc.

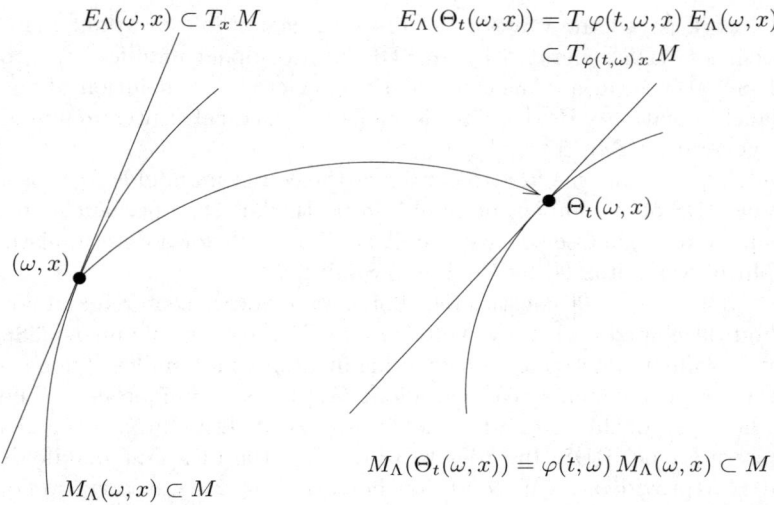

Fig. 7.1. Invariant manifold $M_\Lambda(\omega, x)$ tangent to $E_\Lambda(\omega, x)$

(ii) If Λ is any interval from the spectrum, i.e. if for some $1 \leq i \leq j \leq p$, $\Lambda_{ij} = \{\lambda_i > \ldots > \lambda_j\}$, then $E_{ij} = \oplus_{k:\lambda_i \geq \lambda_k \geq \lambda_j} E_k$ and M_{ij} tangent to E_{ij} is called the *center manifold*. For $i = 1$, we obtain the *unstable manifold* M_{1j}. For $j = p$, we obtain the *stable manifold* M_{ip} and for $i = j$, we obtain the *Oseledets manifold* M_i tangent to the Oseledets space E_i. The invariant manifold $M_{1p}(\omega, x)$ has dimension d and will be a random neighborhood of x, possibly equal to M. We sometimes add the attribute *generalized* to distinguish these manifolds from their classical versions.

(iii) We could select more general subsets Λ from the spectrum of φ under μ, e.g. $\Lambda = \{\lambda_1, \lambda_p\}$ if $p \geq 3$. However, it will in general not be possible to obtain smooth (or even Lipschitz continuous) invariant manifolds tangent to such an E_Λ, as deterministic counterexamples show (see Aulbach and Wanner [47: Sect. 7]). But see Remark 7.4.5.

Our program for this chapter is to construct the global and local center manifolds M_{ij} of a C^k RDS φ under an invariant measure μ. We stress that this construction is based on the information provided by the MET for $T\varphi$ and μ, hence depends in particular crucially on μ.

Invariant manifold theory for RDS based on the MET is an important part of *smooth ergodic theory*. It was started in 1976 with the pioneering work of Pesin [274, 275]. He constructed the classical stable and unstable manifolds of a (deterministic) diffeomorphism on a compact Riemannian manifold preserving a measure which is absolutely continuous with respect to the Riemannian volume. His technique to cope with the non-uniformity of the MET (random norms, ε-slowly varying functions) can be adapted to RDS and will also be used by us. Ruelle [292] removed the smoothness condition on the measure and generalized the construction to Hilbert manifolds [293].

Ruelle's work paved the way for the first classical stable invariant manifold theorem for RDS generated by an SDE on a compact manifold due to Carverhill [89] (the invariant measure is still restricted to a solution of the Fokker-Planck equation). Boxler [79] developed a classical center manifold theory for general RDS in $\mathbb{R}^d$.

Dahlke [120, 121] was the first to construct the center manifolds M_{ij} for a discrete time RDS on a compact manifold, in particular, the invariant manifolds tangent to a single Oseledets space. A similar result for a deterministic diffeomorphism was obtained by Pugh and Shub [282].

Wanner [339, 341, 340] constructed global (and local) center manifolds by a method developed earlier by Aulbach and Wanner (see Wanner [338], Aulbach and Wanner [46, 47]) for constructing invariant subbundles of nonautonomous dynamical systems. We will adopt Wanner's technique here. The drawback, however, of this method is that it only yields Lipschitz-continuous manifolds even for C^k RDS. In order to prove that the invariant manifolds are indeed C^k (provided a gap condition holds if $k \geq 2$), we carry over a method developed by Vanderbauwhede and Van Gils [333] and Hilger [171] (based on a contraction mapping theorem for a scale of Banach spaces) to the random case. The same method was also used by Siegmund [317] to prove the C^k property of integral manifolds for nonautonomous differential equations).

Classical stable and/or unstable manifolds for RDS under various conditions and for various purposes were also constructed by Brin and Kifer [82], Gundlach [164] and Arnold and Kloeden [28]. A new method using a random fixed point theorem is used by Schmalfuß [311]. For an account of Pesin theory see Katok and Strelcyn [201] (for the deterministic case) and Liu and Qian [244] (for the random case).

Let us mention some related work for nonautonomous dynamical systems: For skew product flows over a continuous DS on a compact space, a program analogous to the above can be formulated, but now based on exponential dichotomy and the Sacker-Sell spectral theory (see Remark 4.2.9) in place of the MET. Essential parts of such a program have been carried out, see e.g. Sell [316] for generalized center manifolds and Yi [348, 349] and Chow and Yi [99] for the classical manifolds. However, due to the built-in continuity and uniformity, constructions for skew-product flows are, though similar, technically simpler than those for RDS. See also Bronstein [83] and Colonius and Kliemann [105] for more nonautonomous constructions.

7.2 Reductions and Preparations

Our aim is to construct the invariant manifold $M_\Lambda(\omega, x)$ as a graph $E_\Lambda(\omega, x) \ni y \mapsto m_\Lambda(\omega, y) \in E_{\Lambda^c}(\omega, x)$ in the $E_\Lambda(\omega, x)$, $E_{\Lambda^c}(\omega, x)$ coordinate system, where Λ^c is the complement of Λ in the spectrum $\mathcal{S}(\varphi, \mu)$, i.e. $E_{\Lambda^c}(\omega, x) := \oplus_{i:\lambda_i \notin \Lambda} E_i(\omega, x)$ (on a manifold M, the graph describes $\exp_x^{-1} M_\Lambda(\omega, x)$).

7.2.1 Reductions

Reduction of the State Space from M to $\mathbb{R}^d$

By Nash's theorem, every smooth Riemannian manifold M of dimension d can be isometrically embedded into $\mathbb{R}^{2d+1}$ (into $\mathbb{R}^{2d}$ if M is compact) endowed with its standard Riemannian structure. We hence have a mapping

$$g : (M, \langle \cdot, \cdot \rangle) \to N := g(M) \subset (\mathbb{R}^{2d+1}, \langle \cdot, \cdot \rangle_S)$$

which is an isometric diffeomorphism between M and N. In particular, N is an embedded closed submanifold of $\mathbb{R}^{2d+1}$ with

$$\langle u, v \rangle = \langle T_x g(u), T_x g(v) \rangle_S \quad \text{for all } x \in M, \ u, v \in T_x M.$$

Now if φ is a C^k RDS on M with invariant measure μ satisfying the IC of the MET, then $\psi := g(\varphi)$ is a C^k RDS on $N = g(M)$ with invariant measure $\nu = g(\mu)$, also satisfying the IC. Moreover, the Lyapunov spectrum is preserved under g (since $\|Tg\| = 1$) and the Oseledets splitting of ψ is $Tg(E_i)$.

It is well-known that every closed smooth submanifold N of $\mathbb{R}^{2d+1}$ has a tubular neighborhood

$$U_r(N) := \cup_{x \in N} \{y \in \mathbb{R}^{2d+1} : \|y - x\| < r(x)\}, \quad r : N \to (0, \infty),$$

where r is a smooth function. If N is compact, then r can be taken to be constant. We want to construct an extension $\tilde{\psi}$ of ψ from N to $\mathbb{R}^{2d+1}$ which is trivial outside a tubular neighborhood $U_r(N)$, i.e. $\tilde{\psi}(t, \omega)x = x$, such that its Oseledets splitting at $T_{g(x)} \mathbb{R}^{2d+1}$ is $Tg(E_i(\omega, x))$, with the exception that the (possibly trivial) Oseledets space corresponding to the Lyapunov exponent 0 has to be augmented by the $(d+1)$-dimensional normal space $T^{\perp}_{g(x)} N$. As a result, $\mathcal{S}(\tilde{\psi}, \nu) = \mathcal{S}(\varphi, \mu) \cup \{0\}$, where the (possibly zero) multiplicity of the Lyapunov exponent 0 has to be increased by $d + 1$.

At present we only know how to do this extension for the case where N is compact, time is continuous and ψ has a generator, which we consider now.

For any $x \in \mathbb{R}^{2d+1}$ let $p(x)$ be the nearest point in N such that the line between x and $p(x)$ is orthogonal to $T_{p(x)} N$. Let $\chi : \mathbb{R} \to \mathbb{R}^+$ be a C^∞ bump function with $\chi(0) = 1$ and $\chi(\cdot) = 0$ outside $[-r, r]$. If an RDS ψ on N is generated by an RDE $\dot{x}_t = f(\theta_t \omega, x_t)$ or an SDE $dx_t = \sum_{j=0}^m f_j(x_t) \circ dW_t^j$, we extend the generator from N to $\mathbb{R}^{2d+1}$ as follows: In the RDE case define

$$\tilde{f}(\omega, x) := \chi(\|p(x) - x\|) f(\omega, p(x)), \quad x \in \mathbb{R}^{2d+1},$$

and in the SDE case define

$$\tilde{f}_j(x) := \chi(\|p(x) - x\|) f_j(p(x)), \quad x \in \mathbb{R}^{2d+1}, \ j = 0, \ldots, m.$$

The corresponding RDS $\tilde{\psi}$ has the same regularity as ψ, it leaves N and the measure ν invariant and has the Lyapunov spectrum described above. For details in the SDE case see Carverhill [89: pp. 281 and 285].

Reduction of General Invariant Measure to Dirac Measure at 0

For RDS in $\mathbb{R}^d$ there is the following improvement of Lemma 1.4.8.

7.2.1 Lemma (Every Invariant Measure is Dirac at 0). *Let φ be a C^k RDS in $\mathbb{R}^d$ over the DS θ with invariant measure μ. Consider the following extension of θ: $\bar{\Omega} := \Omega \times \mathbb{R}^d$, $\bar{\mathcal{F}} := \mathcal{F} \otimes \mathcal{B}$, $\bar{\theta}(t)\bar{\omega} := \Theta(t)(\omega, x)$, $\bar{\mathbb{P}} := \mu$. Then*

$$\bar{\varphi}(t, \bar{\omega})v := \varphi(t, \omega, x + v) - \varphi(t, \omega, x)$$

is a C^k RDS over $\bar{\theta}$ which has invariant measure δ_0.

Further, $T\varphi$ satisfies the IC of the MET with respect to μ if and only if $T\bar{\varphi}$ does for δ_0, and $\mathcal{S}(\varphi, \mu) = \mathcal{S}(\bar{\varphi}, \delta_0)$, while for the Oseledets splitting $E_i(\omega, x) = x + \bar{E}_i(\bar{\omega}, 0)$.

The simple proof is left as an exercise.

This lemma enables us to do the construction of the invariant manifolds at a fixed point of our cocycle. This is without loss of generality. For, if we have constructed the invariant manifold $\bar{M}_\Lambda(\bar{\omega}, 0)$ for $\bar{\varphi}$ at $x = 0$, we can recover the invariant manifold $M_\Lambda(\omega, x)$ for φ at any "good" point $(\omega, x) \in \Delta$ by putting

$$M_\Lambda(\omega, x) = x_0(\bar{\omega}) + \bar{M}_\Lambda(\bar{\omega}, 0) = x + \bar{M}_\Lambda(\omega, x, 0).$$

One easily checks that M_Λ is φ-invariant if and only if $\bar{M}_\Lambda$ is $\bar{\varphi}$-invariant.

The result is that we can assume that φ is a C^k RDS in $\mathbb{R}^d$ with $\varphi(t, \omega)0 = 0$, hence $D\varphi(t, \omega, 0) =: \Phi(t, \omega)$ is a linear cocycle over θ and we can split φ at 0 into a linear cocycle and a nonlinear part,

$$\varphi(t, \omega, x) = \Phi(t, \omega)x + \psi(t, \omega, x), \tag{7.2.1}$$

where $\psi(t, \omega, x) := \varphi(t, \omega, x) - \Phi(t, \omega)x$ consists of the terms of higher than linear order, satisfies $\psi(t, \omega, 0) = 0$, $D\psi(t, \omega, 0) = 0$ and $\psi(0, \omega, x) \equiv 0$.

Discretization of Time

As the main proofs will be done for discrete time only, we consider in Sects. 7.3 to 7.5 in case of continuous time the embedded discrete time cocycle

$$\varphi(n, \omega, x) = \Phi(n, \omega)x + \psi(n, \omega, x), \quad n \in \mathbb{Z}, \tag{7.2.2}$$

where $\psi(n, \omega, 0) = 0$ and $D\psi(n, \omega, 0) = 0$, and show that the discrete time objects (invariant manifolds and foliations, decoupling and linearizing homeomorphisms) are indeed also the continuous time ones.

7.2 Reductions and Preparations

The generators of the cocycles φ and Φ are given by their time-one mappings. With

$$\varphi(\omega) := \varphi(1,\omega), \quad A(\omega) := \Phi(1,\omega), \quad F(\omega,x) := \psi(1,\omega,x),$$

(7.2.2) is equivalent to the random difference equation

$$x_{n+1} = \varphi(\theta^n\omega, x_n) = A(\theta^n\omega)x_n + F(\theta^n\omega, x_n), \quad x_0 = x, \quad n \in \mathbb{Z}, \quad (7.2.3)$$

where $A : \Omega \to Gl(d, \mathbb{R})$, $F(\omega, 0) = 0$ and $DF(\omega, 0) = 0$.

7.2.2 Preparations

Random Norms, Block-Diagonalization of Linear Part

We assume throughout that the linear cocycle Φ in (7.2.1) satisfies the IC cf the MET and that (without loss of generality) the MET holds on all of Ω without exceptional set.

Our next aim is the improvement of contraction and expansion properties of Φ through the use of random norms (see Sect. 4.3) and its simplification through the choice of an adapted basis.

Let $\mathbb{R}^d = E_1(\omega) \oplus \cdots \oplus E_p(\omega)$ be the Oseledets splitting of Φ. Recall that for a given constant $\kappa > 0$, we can construct a new norm $\|x\|_{\kappa,\omega}$ which crucially improves the uniformity in the behavior of Φ because of

$$e^{\lambda_i t - \kappa |t|} \leq \|\Phi(t,\omega)|_{E_i(\omega)}\|_{\kappa,\omega,\theta(t)\omega} \leq e^{\lambda_i t + \kappa |t|} \quad \text{for all } t \in \mathbb{T}. \quad (7.2.4)$$

By choosing an adapted basis (see Corollary 4.3.12) we define a coordinate transformation $P(\omega)$ such that

$$\bar{\Phi}(t,\omega) = P(\theta(t)\omega)\Phi(t,\omega)P(\omega)^{-1} = \mathrm{diag}(\bar{\Phi}_1(t,\omega),\ldots,\bar{\Phi}_p(t,\omega))$$

is Lyapunov cohomologous to Φ and block-diagonal, where the blocks are cocycles on $\mathbb{R}^{d_i}$ with one-point spectrum $\{(\lambda_i, d_i)\}$. Note that the cocycle $\bar{\Phi}$ is for continuous time in general only measurable with respect to t. This is, however, irrelevant at this stage as the construction of random norms (which requires continuity in t) is done on the original Φ.

The main advantage of $\bar{\Phi}$ over Φ is that its Oseledets spaces are obviously deterministic, namely

$$\bar{E}_i(\omega) = \{0\} \times \cdots \times \{0\} \times \mathbb{R}^{d_i} \times \{0\} \times \cdots \times \{0\} \subset \mathbb{R}^d. \quad (7.2.5)$$

We will tacitly use the identification $\bar{E}_i \cong \mathbb{R}^{d_i}$ and $\bar{\Phi}|_{\bar{E}_i} \cong \bar{\Phi}_i|_{\mathbb{R}^{d_i}}$.

The nonlinear cocycle (7.2.1) in the new coordinates is

$$\bar{\varphi}(t,\omega,x) = \bar{\Phi}(t,\omega)x + \bar{\psi}(t,\omega,x).$$

Again, $\bar{\varphi}$ is in general only measurable with respect to t.

The random coordinate transformation $P(\omega)$ induces a random norm on $\bar{E}_i \cong \mathbb{R}^{d_i}$ via
$$\|\bar{x}_i\|'_{\kappa,\omega} = \|x_i\|'_{\kappa,\omega} := \|P(\omega)^{-1}\bar{x}_i\|_{\kappa,\omega}, \qquad (7.2.6)$$
where
$$\bar{E}_i \ni \bar{x}_i = (0,\ldots,0,x_i,0,\ldots,0) \cong x_i \in \mathbb{R}^{d_i},$$
and $P(\omega)^{-1}\bar{x}_i \in E_i(\omega)$. We extend this definition to $\mathbb{R}^d$ by taking the 1-norm (which allows simpler estimates than the 2-norm, but is related to it by deterministic constants)
$$\|\bar{x}\|'_{\kappa,\omega} := \sum_{i=1}^{p} \|x_i\|'_{\kappa,\omega}, \quad \bar{x} = \oplus_{i=1}^{p}\bar{x}_i, \quad x_i \in \mathbb{R}^{d_i}. \qquad (7.2.7)$$

By Lemma 4.3.13, there is for each $\varepsilon > 0$ an ε-slowly varying random variable $R_\varepsilon : \Omega \to [1,\infty)$ such that
$$\frac{1}{R_\varepsilon(\omega)} \leq \|P(\omega)^{\pm 1}\| \leq R_\varepsilon(\omega).$$

Hence the norm $\|\cdot\|'_{\kappa,\omega}$ defined in $\mathbb{R}^d$ by (7.2.6) and (7.2.7) is in the same relation to the standard norm $\|\cdot\|$ as the original random norm $\|\cdot\|_{\kappa,\omega}$: For each $\varepsilon > 0$ there exists an ε-slowly varying random variable $C_\varepsilon : \Omega \to [1,\infty)$ such that
$$\frac{1}{C_\varepsilon(\omega)}\|\bar{x}\| \leq \|\bar{x}\|'_{\kappa,\omega} \leq C_\varepsilon(\omega)\|\bar{x}\|.$$

By our definitions, the cocycle $\bar{\Phi}$ also satisfies the uniform estimate (7.2.4) under the norm (7.2.7).

As a result, for an interval
$$\Lambda_{ij} = \{\lambda_i > \ldots > \lambda_j\}, \quad 1 \leq i \leq j \leq p,$$
from the spectrum of $\mathcal{S}(\bar{\Phi}) = \mathcal{S}(\Phi)$ and with
$$\bar{E}_{\Lambda_{ij}} = \bar{E}_{ij} := \oplus_{k=i}^{j}\bar{E}_i \cong \mathbb{R}^{d_i} \times \cdots \times \mathbb{R}^{d_j},$$
we have the estimates
$$e^{(\lambda_j - \kappa)t} \leq \|\bar{\Phi}(t,\omega)|_{\bar{E}_{ij}}\|'_{\kappa,\omega,\theta(t)\omega} \leq e^{(\lambda_i + \kappa)t}, \quad t \geq 0, \qquad (7.2.8)$$
and
$$e^{(\lambda_i + \kappa)t} \leq \|\bar{\Phi}(t,\omega)|_{\bar{E}_{ij}}\|'_{\kappa,\omega,\theta(t)\omega} \leq e^{(\lambda_j - \kappa)t}, \quad t \leq 0. \qquad (7.2.9)$$

From now on *we drop* the bar over the transformed objects and the superscript $'$ of the norm. We hence have arrived at a nonlinear cocycle $\varphi(t,\omega,x) = \Phi(t,\omega)x + \psi(t,\omega,x)$ on the normed bundle $(\mathbb{R}^d, \|\cdot\|_{\kappa,\omega})$ which is measurable in (t,ω) and C^k in x. It satisfies $\varphi(t,\omega,0) = 0$ and $D\varphi(t,\omega,0) = \Phi(t,\omega)$, where the linear cocycle

$$\Phi(t,\omega) = \mathrm{diag}(\Phi_1(t,\omega),\ldots,\Phi_p(t,\omega))$$

is block-diagonal and satisfies the estimates (7.2.8) and (7.2.9) for any Λ_{ij}.

A Scale of Banach Spaces

We follow Wanner [340] (and many others) and define Banach spaces of functions with a geometrically weighted sup norm which allows for exponential growth of their elements.

7.2.2 Definition (Weighted Banach Spaces). Consider $(\mathbb{R}^d, \|\cdot\|_{\kappa,\omega})$, let $\mathbb{T}^\pm = \mathbb{T} \cap \mathbb{R}^\pm$ and $\alpha, \beta > 0$. Define for each $\omega \in \Omega$

$$X_{\beta^+,\omega} := \{g : \mathbb{T}^+ \to \mathbb{R}^d \text{ measurable}: \|g\|_{\beta^+,\omega} := \sup_{t \geq 0} \beta^{-t}\|g(t)\|_{\kappa,\theta(t)\omega} < \infty\},$$

$$X_{\alpha^-,\omega} := \{g : \mathbb{T}^- \to \mathbb{R}^d \text{ measurable}: \|g\|_{\alpha^-,\omega} := \sup_{t \leq 0} \alpha^{-t}\|g(t)\|_{\kappa,\theta(t)\omega} < \infty\},$$

$$X_{\alpha^-,\beta^+,\omega} :=$$
$$\{g : \mathbb{T} \to \mathbb{R}^d \text{ measurable}: \|g\|_{\alpha^-,\beta^+,\omega} := \sup(\|g\|_{\alpha^-,\omega},\|g\|_{\beta^+,\omega}) < \infty\},$$

and
$$X_{\alpha,\omega} := X_{\alpha^-,\alpha^+,\omega}.$$

We write $X_{\beta^+,\omega}(E_\Lambda)$ etc. if $\mathbb{R}^d$ is replaced by E_Λ etc. We write $X_{\beta^+,\omega,\mathbb{Z}^+}$ etc. if we want to emphasize that time is $\mathbb{Z}^+$, etc. We denote by

$$X_{\beta^+} := \{g : \mathbb{T}^+ \to \mathbb{R}^d \text{ measurable}: \|g\|_{\beta^+} := \sup_{t \geq 0} \beta^{-t}\|g(t)\| < \infty\},$$

etc. the corresponding Banach spaces based on the (nonrandom) Euclidean norm. ∎

An element $g \in X_{\beta^+,\omega}$ grows at most like β^t forward in time, i.e. there is a random variable $C^+(\omega) \geq 0$ such that

$$\|g(t)\|_{\kappa,\theta(t)\omega} \leq C^+(\omega)\beta^t, \quad t \geq 0,$$

and is said to be (β^+,ω)-*bounded*[1]. An element $g \in X_{\alpha^-,\omega}$ grows at most like α^t backward in time, i.e. there is a random variable $C^-(\omega) \geq 0$ such that

$$\|g(t)\|_{\kappa,\theta(t)\omega} \leq C^-(\omega)\alpha^t, \quad t \leq 0,$$

and is called (α^-,ω)-*bounded*. Both properties characterize an element $g \in X_{\alpha^-,\beta^+,\omega}$ as being $(\alpha^-,\beta^+,\omega)$-*bounded* (see Fig. 7.2). Note that the spaces also depend on $\kappa > 0$ which is suppressed in our notation.

---

[1] Aulbach and Wanner [47: Sect. 3] use the name β^+-*quasibounded*.

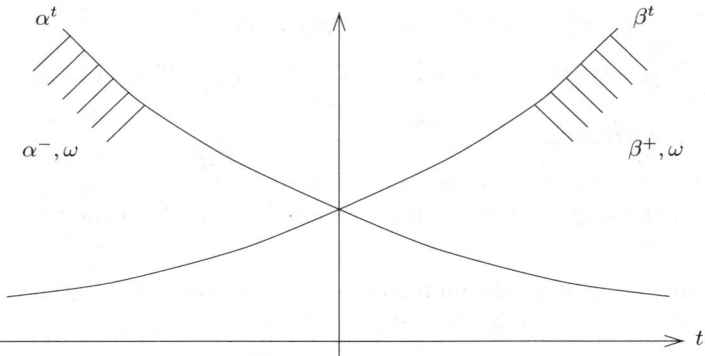

Fig. 7.2. The space $X_{\alpha^-,\beta^+,\omega}$ for $\alpha < 1 < \beta$

7.2.3 Lemma (Scale of Banach Spaces). *(i) For each α, $\beta > 0$ and each $\omega \in \Omega$, $X_{\alpha^-,\omega}$, $X_{\beta^+,\omega}$ and $X_{\alpha^-,\beta^+,\omega}$ are Banach spaces.*

(ii) For each $\omega \in \Omega$, they form a scale of Banach spaces in the following sense: If $\bar{\alpha} \leq \alpha$ and $\beta \leq \bar{\beta}$ we have the continuous embeddings

$$X_{\beta^+,\omega} \hookrightarrow X_{\bar{\beta}^+,\omega}, \quad \|\cdot\|_{\bar{\beta}^+,\omega} \leq \|\cdot\|_{\beta^+,\omega},$$

$$X_{\alpha^-,\omega} \hookrightarrow X_{\bar{\alpha}^-,\omega}, \quad \|\cdot\|_{\bar{\alpha}^-,\omega} \leq \|\cdot\|_{\alpha^-,\omega},$$

$$X_{\alpha^-,\beta^+,\omega} \hookrightarrow X_{\bar{\alpha}^-,\bar{\beta}^+,\omega}, \quad \|\cdot\|_{\bar{\alpha}^-,\bar{\beta}^+,\omega} \leq \|\cdot\|_{\alpha^-,\beta^+,\omega}.$$

(iii) If g is (β^+,ω)-bounded, then for each $\varepsilon > 0$ it is $(\beta+\varepsilon)^+$-bounded in the Euclidean norm of $\mathbb{R}^d$. Conversely, if g is β^+-bounded in the Euclidean norm, then for each $\varepsilon > 0$ it is $((\beta+\varepsilon)^+,\omega)$-bounded.

Similarly, if g is (α^-,ω)-bounded or $(\alpha^-,\beta^+,\omega)$-bounded.

Proof. (i) is clear.

(ii) The continuous embeddings follow from the inequalities between the respective norms, which immediately follow from their definitions.

(iii) We have $g \in X_{\beta^+,\omega}$ if and only if there is a random variable $C(\omega) \geq 0$ such that $\|g(t)\|_{\kappa,\theta(t)\omega} \leq C(\omega)\beta^t$ for all $t \geq 0$. Further, for each $\delta > 0$, the random and Euclidean norm are related by $B_\delta(\omega)^{-1}\|\cdot\|_{\kappa,\omega} \leq \|\cdot\| \leq B_\delta(\omega)\|\cdot\|_{\kappa,\omega}$, where B_δ is a δ-slowly varying function. Hence for any $t \geq 0$

$$\|g(t)\| \leq B_\delta(\theta(t)\omega)\|g(t)\|_{\kappa,\theta(t)\omega} \leq B_\delta(\omega)C(\omega)(\beta e^\delta)^t = K_\varepsilon(\omega)(\beta+\varepsilon)^t$$

if we choose $\delta = \log \frac{\beta+\varepsilon}{\beta}$ and $K_\varepsilon = B_\delta C$.

Similarly for all other cases. □

7.2.4 Remark (Weighted Banach Spaces and Lyapunov Index).
Let

$$\lambda(g) := \limsup_{n\to\infty} \frac{1}{n} \log \|g(n,\omega)\|, \quad \lambda(g)_\omega := \limsup_{n\to\infty} \frac{1}{n} \log \|g(n,\omega)\|_{\theta^n\omega}$$

be the Lyapunov index of $g(n,\omega)$ in the Euclidean and the random norm, respectively. Then we have the following elementary relations:

(i) If $g \in X_{\beta+,\omega}$ or $\in X_{\beta+}$, then the Lyapunov index of g satisfies $\lambda(g) \leq \log \beta$ and $\lambda(g)_\omega \leq \log \beta$.

(ii) $\lambda(g) \leq \log \beta \iff \lambda(g)_\omega \leq \log \beta \iff g \in X_{(\beta+\varepsilon)+,\omega}$ for all $\varepsilon > 0$ $\iff g \in X_{(\beta+\varepsilon)+}$ for all $\varepsilon > 0$. ∎

Selection of Constants

In this last preparatory step we choose (and then fix) two parameters related to the Lyapunov exponents $\lambda_1 > \ldots > \lambda_p$ of Φ.

Choice of κ. We first choose the constant $\kappa > 0$ which enters into the construction of the random norm $\|\cdot\|_{\kappa,\omega}$ small enough so that the intervals $[\lambda_i - \kappa, \lambda_i + \kappa]$, $i = 1, \ldots, p$, do not overlap, and in the hyperbolic case (all $\lambda_i \neq 0$) none of them contains 0. Henceforth we will *suppress* κ, writing just $\|\cdot\|_\omega$.

It turns out that it is more convenient to consider the exponentiated intervals

$$[\alpha_i, \beta_i] = [e^{\lambda_i - \kappa}, e^{\lambda_i + \kappa}], \quad i = 1, \ldots, p.$$

By our preparations, the block Φ_i of the linear cocycle Φ satisfies

$$\|\Phi_i(t,\omega)\|_{\omega,\theta(t)\omega} \leq (e^{\lambda_i + \kappa})^t = \beta_i^t, \quad t \geq 0,$$

and

$$\|\Phi_i(t,\omega)\|_{\omega,\theta(t)\omega} \leq (e^{\lambda_i - \kappa})^t = \alpha_i^t, \quad t \leq 0.$$

More generally, for any i, j with $1 \leq i \leq j \leq p$, $\Lambda = \Lambda_{ij} = \{\lambda_i > \ldots > \lambda_j\}$ and E_{ij} the corresponding deterministic invariant subspace,

$$\|\Phi(t,\omega)|_{E_{ij}}\|_{\omega,\theta(t)\omega} \leq \beta_i^t, \quad t \geq 0, \tag{7.2.10}$$

and

$$\|\Phi(t,\omega)|_{E_{ij}}\|_{\omega,\theta(t)\omega} \leq \alpha_j^t, \quad t \leq 0. \tag{7.2.11}$$

By the cocycle property for $t - t = 0$ (7.2.10) implies

$$\|\Phi(t,\omega)|_{E_{ij}}\|_{\omega,\theta(t)\omega} \geq \beta_i^t, \quad t \leq 0$$

and (7.2.11) implies

$$\|\Phi(t,\omega)|_{E_{ij}}\|_{\omega,\theta(t)\omega} \geq \alpha_j^t, \quad t \geq 0.$$

Choice of δ. In order to cope with the growth due to the nonlinearity F of the difference equation (7.2.3) we introduce a constant δ such that

$$0 < \delta < \min\left(\frac{\alpha_1 - \beta_2}{2}, \ldots, \frac{\alpha_{p-1} - \beta_p}{2}, \frac{\alpha_p - 0}{2}\right),$$

i. e. the closed intervals

$$[\alpha_i - \delta, \beta_i + \delta], \quad i = 1, \ldots, p,$$

do not overlap, and $\alpha_p - \delta > \delta$. In the hyperbolic case we assume in addition that none of these intervals contains 1.

The interval

$$\Gamma_i = [\beta_{i+1} + \delta, \alpha_i - \delta], \quad i = 1, \ldots, p-1,$$

is called the *spectral gap* between λ_{i+1} and λ_i. See Fig. 7.3. This choice of κ and δ remains unaltered during the proofs.

To describe this preliminary work with a single word we introduce the following notation.

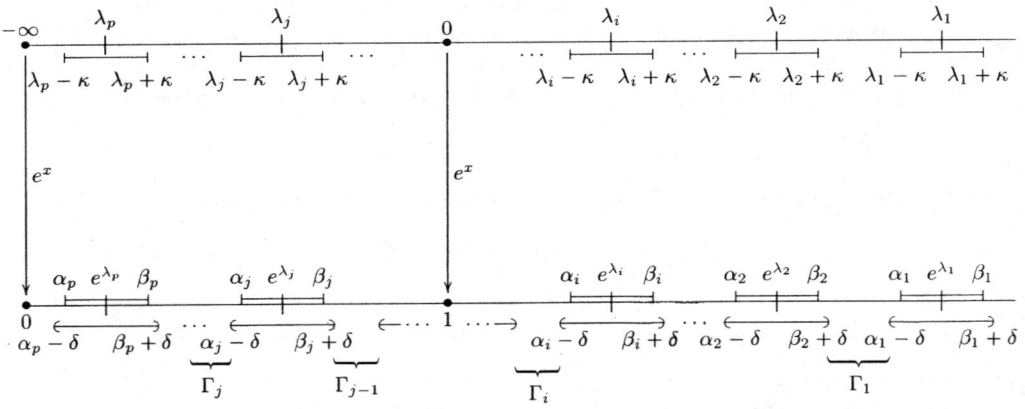

Fig. 7.3. Selection of constants for the construction of invariant manifolds

7.2.5 Definition (Prepared RDS). Let φ be a measurable RDS on the normed bundle $(\mathbb{R}^d, \|\cdot\|_\omega)$ with two-sided time and the property $\varphi(t, \omega, 0) = 0$. Let Φ denote a measurable linear RDS on $(\mathbb{R}^d, \|\cdot\|_\omega)$ (if φ is C^1 in x we always choose $\Phi(t, \omega) = D\varphi(t, \omega, 0)$) and define the "nonlinear part" ψ of φ by

$$\psi(t, \omega, x) := \varphi(t, \omega, x) - \Phi(t, \omega)x,$$

so that

$$\varphi(t, \omega, x) = \Phi(t, \omega)x + \psi(t, \omega, x), \quad \psi(t, \omega, 0) = 0.$$

Assume that $\Phi = \operatorname{diag}(\Phi_1, \ldots, \Phi_p)$ is block-diagonal and has a spectral theory such that (7.2.10) and (7.2.11) hold. Let the constants κ and δ be chosen as above. We call such an RDS (more precisely, the linear part Φ with respect to the random norm $\|\cdot\|_\omega$) *prepared*. ∎

We summarize our findings from above.

7.2.6 Proposition (Linear Equivalence to Prepared RDS).
Every C^1 RDS φ with two-sided time and fixed point 0 such that the linear RDS $\Phi(t,\omega) := D\varphi(t,\omega,0)$ satisfies the IC of the MET is linearly equivalent to a prepared RDS $\bar{\varphi}$, i. e. there is a measurable map $P: \Omega \to Gl(d, \mathbb{R})$ such that

$$P(\theta(t)\omega) \circ \varphi(t,\omega) = \bar{\varphi}(t,\omega) \circ P(\omega).$$

P defines a Lyapunov cohomology between the linear parts Φ and $\bar{\Phi}$.

More specifically, the random norm $\|\cdot\|_\omega$ and the random coordinate transformation $P(\omega)$ can be chosen in such a way that for each $\varepsilon > 0$ there is an ε-slowly varying random variable $B_\varepsilon: \Omega \to [1, \infty)$ such that

$$\frac{1}{B_\varepsilon(\omega)}\|x\| \le \|x\|_\omega \le B_\varepsilon(\omega)\|x\|$$

and

$$\frac{1}{B_\varepsilon(\omega)} \le \|P(\omega)^{\pm 1}\| \le B_\varepsilon(\omega).$$

In particular, if $\bar{\varphi}(\cdot,\omega,x) \in X_{\beta+,\omega}$ or $\in X_{\alpha-,\omega}$ then for each $\varepsilon > 0$ or $0 < \varepsilon < \alpha$, $\varphi(\cdot,\omega,P(\omega)^{-1}x) = P(\theta(\cdot)\omega)^{-1}\bar{\varphi}(\cdot,\omega,x) \in X_{(\beta+\varepsilon)+}$ or $\in X_{(\alpha-\varepsilon)-}$, respectively. Similarly in the reverse direction.

7.3 Global Invariant Manifolds

In this inevitably very technical section we construct unstable, stable and center manifolds for a C^k RDS with fixed point 0.

In a first step (Subsects. 7.3.1 to 7.3.4) we construct manifolds which are just Lipschitz-continuous even if the RDS is C^k. Later (Subsect. 7.3.5) we will show that these manifolds have in fact the same regularity as the RDS φ.

The constructions are first done for discrete time and then carried over to continuous time in Subsect. 7.3.4.

Throughout Subsects. 7.3.1 to 7.3.4, we do not need to assume that φ is C^1. Rather, we only need to assume now that φ is *prepared* in the sense of Definition 7.2.5. The corresponding generating random difference equation is

$$x_{n+1} = \varphi(\theta^n \omega, x_n) = A(\theta^n \omega)x_n + F(\theta^n \omega, x_n), \quad n \in \mathbb{Z}, \qquad (7.3.1)$$

where $F(\omega, 0) = 0$.

For given $\Lambda_{ij} \subset \mathcal{S}(\Phi)$, $1 \le i \le j \le p$, we seek an invariant manifold $M_{ij}(\omega)$ of φ which is a graph in the (deterministic) E_{ij}, $E_{ij}^\perp$ coordinate system, where

$$E_{ij} := \{0\} \times \ldots \times \{0\} \times \mathbb{R}^{d_i} \times \ldots \times \mathbb{R}^{d_j} \times \{0\} \times \ldots \times \{0\} \cong \mathbb{R}^{d_i} \times \ldots \times \mathbb{R}^{d_j},$$

with random norm

$$\|x_{ij}\|_\omega := \sum_{i \le k \le j} \|x_k\|_\omega, \quad x_{ij} = (x_i, \ldots, x_j), \quad x_k \in \mathbb{R}^{d_k},$$

and

$$E_{ij}^\perp := \mathbb{R}^{d_1} \times \ldots \times \mathbb{R}^{d_{i-1}} \times \{0\} \times \ldots \times \{0\} \times \mathbb{R}^{d_{j+1}} \times \ldots \times \mathbb{R}^{d_p},$$

with random norm

$$\|x_{ij}^\perp\|_\omega := \sum_{k \notin [i,j]} \|x_k\|_\omega,$$

and

$$\mathbb{R}^d = E_{ij} \oplus E_{ij}^\perp, \quad \|x\|_{\kappa,\omega} = \|x_{ij}\|_{\kappa,\omega} + \|x_{ij}^\perp\|_{\kappa,\omega}, \quad x = x_{ij} \oplus x_{ij}^\perp.$$

The basic idea of the construction of $M_{ij}(\omega)$ is to collect the initial values x of φ-orbits $\varphi(\cdot, \omega, x)$ of "similar" asymptotic behavior as the Φ-orbits $\Phi(\cdot, \omega)x_{ij}$ on E_{ij}.

7.3.1 Construction of Unstable Manifolds

Main Result. Structure of Proof

We will first construct the unstable manifolds M_{1j} for $1 \le j < p$ as Lipschitz continuous invariant graphs.

Here is our main result.

7.3.1 Theorem (Global Unstable Manifold Theorem). *Let the RDS φ with discrete time $\mathbb{T} = \mathbb{Z}$ be prepared and generated by the difference equation (7.3.1). Assume*

$$\|F(\omega, x) - F(\omega, y)\|_{\theta\omega} \le L\|x - y\|_\omega \quad \text{for all} \quad x, y \in \mathbb{R}^d, \tag{7.3.2}$$

where

$$0 \le L < \delta/2.$$

Choose any j with $1 \le j < p$, let $\Lambda^+ = \Lambda_{1j} = \{\lambda_1 > \ldots > \lambda_j\}$ be the corresponding spectral interval of Φ, put $E^+ := E_{1j}$, $x^+ := x_{1j} = (x_1, \ldots, x_j)$, $E^- := E_{1j}^\perp = E_{j+1,p}$, $x^- := x_{1j}^\perp = (x_{j+1}, \ldots, x_p)$, choose a constant a in the spectral gap to the left of Λ^+,

$$a := a_j \in \Gamma_{\text{left}} = [\beta_{j+1} + \delta, \alpha_j - \delta],$$

and define the set

$$M^+(\omega) = M_{1j}(\omega) := \{x \in \mathbb{R}^d : \varphi(\cdot, \omega, x) \in X_{a^-, \omega}\}. \tag{7.3.3}$$

Then

(i) $M^+(\omega)$ is a topological manifold of dimension $\dim E^+ = d_1 + \ldots + d_j$, called the unstable manifold *corresponding to the interval* Λ^+. *It is independent of the particular choice of* a.

(ii) $M^+(\omega)$ is a graph in $E^+ \oplus E^-$,

$$M^+(\omega) = \{x^+ \oplus m^+(\omega, x^+) : x^+ \in E^+\},$$

with the following properties:

1. $m^+ : \Omega \times E^+ \to E^-$ is measurable.
2. $m^+(\omega, 0) = 0$.
3. m^+ is globally Lipschitz continuous in the random norms of E^+ and E^-,

$$\|m^+(\omega, x^+) - m^+(\omega, y^+)\|_\omega \leq D(L)\|x^+ - y^+\|_\omega,$$

where the constant

$$D(L) = \frac{L(\delta - L)}{\delta(\delta - 2L)}$$

is independent of j.

(iii) $M^+(\omega)$ is invariant under φ, i. e.

$$\varphi(n, \omega) M^+(\omega) = M^+(\theta^n \omega), \quad n \in \mathbb{Z}.$$

See Fig. 7.4.

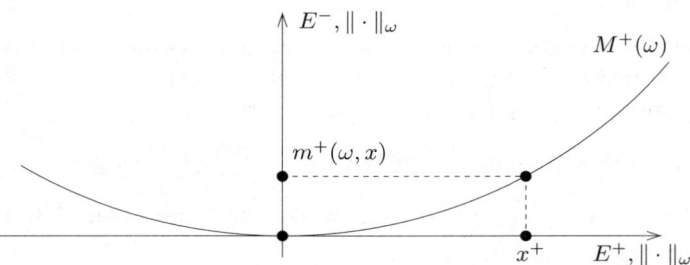

Fig. 7.4. The unstable manifold $M^+(\omega)$ and its graph $m^+(\omega, \cdot)$

7.3.2 Remark. (i) It is important to note that F is part of a cocycle, hence the norm $\|\cdot\|_{\theta\omega}$ on the left-hand side and $\|\cdot\|_\omega$ on the right-hand side of (7.3.2).

(ii) Since $F(\omega, 0) = 0$, the Lipschitz condition (7.3.2) implies sublinear growth of F,
$$\|F(\omega, x)\|_{\theta\omega} \leq L\|x\|_\omega, \quad x \in \mathbb{R}^d.$$

(iii) If $F(\omega, x) \equiv 0$, then $M^+(\omega) = E^+$ with trivial graph $m^+(\omega, x^+) \equiv 0$. In this case the result is true even for $\delta = 0$.

(iv) We will call a definition of the form (7.3.3), (7.3.47) and (7.3.50) a *dynamical characterization* of the corresponding set. Hence the unstable (stable, center) manifolds are dynamically characterized (by exponential growth conditions on their orbits). ∎

The proof of Theorem 7.3.1 is quite involved, so we will break it up into a number of simpler steps.

By our preparations, the linear part A is block-diagonal,
$$A = \begin{pmatrix} A^+ & 0 \\ 0 & A^- \end{pmatrix},$$
where
$$A^+ = \mathrm{diag}(A_1, \ldots, A_j), \quad A^- = \mathrm{diag}(A_{j+1}, \ldots, A_p),$$
and the nonlinear part can be written as
$$F = \begin{pmatrix} F^+ \\ F^- \end{pmatrix}, \quad F^+ = \begin{pmatrix} F_1 \\ \vdots \\ F_j \end{pmatrix}, \quad F^- = \begin{pmatrix} F_{j+1} \\ \vdots \\ F_p \end{pmatrix}.$$

Our random difference equation (7.3.1) takes the form
$$x_{n+1}^+ = A^+(\theta^n\omega)x_n^+ + F^+(\theta^n\omega, x_n^+, x_n^-), \quad x_0^+ = x^+, \quad (7.3.4)$$
$$x_{n+1}^- = A^-(\theta^n\omega)x_n^- + F^-(\theta^n\omega, x_n^+, x_n^-), \quad x_0^- = x^-. \quad (7.3.5)$$

The two equations are coupled through their nonlinear parts and from now on will be called, respectively, the *unstable* and *stable equation*.

Again by our preparations, putting $\alpha := \alpha_j$, $\beta := \beta_{j+1}$,
$$\|\Phi^+(n,\omega)\|_{\omega,\theta^n\omega} \leq \alpha^n, \quad n \leq 0, \quad \|\Phi^-(n,\omega)\|_{\omega,\theta^n\omega} \leq \beta^n, \quad n \geq 0,$$
where we have $\beta + \delta < \alpha - \delta$ for our choice of δ. By our assumption (7.3.2),
$$\|F^\pm(\omega, x) - F^\pm(\omega, y)\|_{\theta\omega} \leq L\|x - y\|_\omega.$$

We will now construct the unstable manifold $M^+(\omega)$ of (7.3.4) and (7.3.5) as a graph $x^+ \mapsto x^- = m^+(\omega, x^+)$ in the E^+, E^- coordinate system (the result of this construction is formulated in Proposition 7.3.7). The main idea is to assure, for each ω and $x^+ \in E^+$, the existence of a unique element $x^- = m^+(\omega, x^+) \in E^-$ which is characterized by the growth property $\varphi(\cdot, \omega, x^+, x^-) \in X_{a^-, \omega}$, where $a \in [\beta + \delta, \alpha - \delta]$ is arbitrary.

Here is a **6-step program** for constructing the unstable manifolds M^-:

1. Solve the unstable equation (7.3.4) for some initial value x^+, but for the generally "wrong" stable part $\xi^-(\omega) = (\xi^-(n,\omega))_{n \in \mathbb{Z}^-} \in X_{a^-,\omega}(E^-)$. The result is $\xi^+(\omega, x^+) \in X_{a^-,\omega}(E^+)$.
2. Insert this solution $\xi^+(\omega, x^+)$ into the stable equation (7.3.5). It turns out that there is exactly one solution $\eta^-(\omega, x^+) \in X_{a^-,\omega}(E^-)$ (which picks its initial value $x^- = \eta^-(0, \omega, x^+)$).
3. Prove that the operator $T_{\omega, x^+} : X_{a^-,\omega}(E^-) \to X_{a^-,\omega}(E^-)$ defined by $\xi^-(\omega) \mapsto \eta^-(\omega, x^+)$ is contracting, hence has a unique fixed point $\xi^-_*(\omega, x^+) \in X_{a^-,\omega}(E^-)$.
4. For each ω and $x^+ \in E^+$, the point $x^- = m^+(\omega, x^+) := \xi^-_*(0, \omega, x^+) \in E^-$ is the uniquely determined point for which $\varphi(\cdot, \omega, x^+, x^-) \in X_{a^-,\omega}$ (dynamical characterization).
5. The function $(\omega, x^+) \mapsto m^+(\omega, x^+)$ is measurable and independent of the particular value of $a \in [\beta + \delta, \alpha - \delta]$.
6. $M^+(\omega)$ has the other properties (Lipschitz-continuity, invariance) claimed in Theorem 7.3.1.

We will give an operator formulation of this program in Subsect. 7.3.5.

Two Key Technical Lemmas

We will first deal with the crucial step 2 of the above program.

7.3.3 Lemma (First Key Technical Lemma). *Consider the random difference equation*

$$x_{n+1} = A(\theta^n \omega) x_n + f(n, \omega, x_n) + f_0(n+1, \omega), \quad n \in \mathbb{Z}, \qquad (7.3.6)$$

in $(\mathbb{R}^d, \|\cdot\|_\omega)$ with measurable $Gl(d,\mathbb{R})$-valued A and measurable f and f_0.[2] Assume that there are constants $\beta > 0$ and $L \geq 0$ such that for each fixed ω

$$\|A(\omega)\|_{\omega, \theta\omega} \leq \beta, \qquad (7.3.7)$$

and

$$f(n, \omega, 0) = 0, \quad \|f(n, \omega, x) - f(n, \omega, y)\|_{\theta^{n+1}\omega} \leq L \|x - y\|_{\theta^n \omega}. \qquad (7.3.8)$$

Let

$$\gamma > \beta + L.$$

Then the following assertions hold:

---

[2] Note that this difference equation is in general only solvable forwards in time and does not necessarily generate an RDS.

(i) If $f_0(\cdot,\omega) \in X_{\gamma^-,\omega}$, then there exists exactly one solution of (7.3.6) which has the property $\xi(\cdot,\omega) \in X_{\gamma^-,\omega}$ and

$$\|\xi(\cdot,\omega)\|_{\gamma^-,\omega} \leq \frac{\gamma}{\gamma - (\beta + L)}\|f_0(\cdot,\omega)\|_{\gamma^-,\omega}. \tag{7.3.9}$$

Moreover, $\omega \mapsto \xi(n,\omega)$ is measurable for each $n \in \mathbb{Z}$.

(ii) If, moreover, $f_0(\cdot,\omega) \in X_{\gamma,\omega}$, then the solution of part (i) satisfies $\xi(\cdot,\omega) \in X_{\gamma,\omega}$ and

$$\|\xi(\cdot,\omega)\|_{\gamma,\omega} \leq \frac{\gamma}{\gamma - (\beta + L)}\|f_0(\cdot,\omega)\|_{\gamma,\omega}. \tag{7.3.10}$$

(iii) If $M(\omega) := \sup_{n \geq 1} \gamma^{-n}\|f_0(n,\omega)\|_{\theta^n\omega} < \infty$ (equivalently, $f_0(\cdot,\omega) \in X_{\gamma^+,\omega}$), then all solutions of (7.3.6) on $\mathbb{Z}^+$ satisfy $\varphi(\cdot,\omega,x) \in X_{\gamma^+,\omega}$ and

$$\|\varphi(\cdot,\omega,x)\|_{\gamma^+,\omega} \leq \|x\|_\omega + \frac{\gamma}{\gamma - (\beta + L)} M(\omega) \tag{7.3.11}$$

$$\leq \|x\|_\omega + \frac{\gamma}{\gamma - (\beta + L)}\|f_0(\cdot,\omega)\|_{\gamma^+,\omega}. \tag{7.3.12}$$

(iv) Two solutions ξ_1 and ξ_2 of (7.3.6) defined on $\mathbb{Z}^-$ satisfy

$$\|\xi_1(n,\omega) - \xi_2(n,\omega)\|_{\theta^n\omega} \geq \gamma^n \|\xi_1(0,\omega) - \xi_2(0,\omega)\|_\omega, \quad n \leq 0. \tag{7.3.13}$$

Proof. (i) We prove the existence and uniqueness of a (γ^-,ω)-bounded solution of (7.3.6) in three steps.

Step 1: Let $f_0 = 0$ and $L = 0$ (hence $f = 0$). Then (7.3.6) reduces to the linear random difference equation

$$x_{n+1} = A(\theta^n\omega)x_n, \quad x_0 = \xi(0,\omega),$$

whose solution is

$$\xi(n,\omega) = \Phi(n,\omega)\xi(0,\omega), \quad n \in \mathbb{Z},$$

where Φ is the linear cocycle generated by A. The trivial solution $\xi = 0$ is certainly (γ^-,ω)-bounded. We show that this is the only one.

Indeed, since by assumption $\|\Phi(n,\omega)\|_{\omega,\theta^n\omega} \leq \beta^n$ for all $n \geq 0$, any other (γ^-,ω)-bounded solution satisfies for each $n < 0$ by the cocycle property

$$\|\xi(0,\omega)\|_\omega \leq \|\Phi(n,\omega)^{-1}\|_{\theta^n\omega,\omega}\|\xi(n,\omega)\|_{\theta^n\omega} \leq \beta^{-n}\|\xi(n,\omega)\|_{\theta^n\omega}$$

$$\leq \left(\frac{\beta}{\gamma}\right)^{-n} \gamma^{-n}\|\xi(n,\omega)\|_{\theta^n\omega} \leq \left(\frac{\beta}{\gamma}\right)^{-n} \|\xi(\cdot,\omega)\|_{\gamma^-,\omega} < \infty.$$

Letting $n \to -\infty$ yields $\xi(0,\omega) = 0$. Hence the trivial solution is the only solution in $X_{\gamma^-,\omega}$. It is in fact in $X_{\gamma,\omega}$.

7.3 Global Invariant Manifolds

Step 2: Let now $f_0 \in X_{\gamma^-,\omega}$, but still $L = 0$ (hence $f = 0$). Now (7.3.6) is the affine equation

$$x_{n+1} = A(\theta^n \omega)x_n + f_0(n+1,\omega), \quad x_0 = \xi(0,\omega).$$

A (γ^-,ω)-bounded solution of this equation is certainly unique. For, if ξ_1, ξ_2 are solutions in $X_{\gamma^-,\omega}$, then $\xi := \xi_1 - \xi_2 \in X_{\gamma^-,\omega}$ satisfies $\xi(n+1,\omega) = A(\theta^n\omega)\xi(n,\omega)$, hence $\xi = 0$ by step 1.

As to the existence, the variation of constants formula gives the solution

$$\xi(n,\omega) := \sum_{i=-\infty}^{n} \Phi(n-i,\theta^i\omega)f_0(i,\omega), \quad n \in \mathbb{Z}. \tag{7.3.14}$$

In particular,

$$\xi(0,\omega) = \sum_{i=-\infty}^{0} \Phi(i,\omega)^{-1} f_0(i,\omega).$$

These expressions make sense. For example, for $n \leq 0$

$$\gamma^{-n}\|\xi(n,\omega)\|_{\theta^n\omega} \leq \sum_{i=-\infty}^{n}\left(\frac{\beta}{\gamma}\right)^{n-i} \gamma^{-i}\|f_0(i,\omega)\|_{\theta^i\omega}$$

$$\leq \sum_{n=0}^{\infty}\left(\frac{\beta}{\gamma}\right)^{n} \|f_0(\cdot,\omega)\|_{\gamma^-,\omega}$$

$$= \frac{\gamma}{\gamma-\beta}\|f_0(\cdot,\omega)\|_{\gamma^-,\omega} < \infty,$$

finally

$$\|\xi(\cdot,\omega)\|_{\gamma^-,\omega} \leq \frac{\gamma}{\gamma-\beta}\|f_0(\cdot,\omega)\|_{\gamma^-,\omega}. \tag{7.3.15}$$

Step 3: Consider now the full equation (7.3.6). We define the (nonlinear) operator

$$T_\omega : X_{\gamma^-,\omega} \to X_{\gamma^-,\omega}, \quad \xi(\omega) \mapsto T_\omega\xi(\omega) =: \eta(\omega), \tag{7.3.16}$$

where $\eta(\omega)$ is the unique solution in $X_{\gamma^-,\omega}$ of the affine equation

$$x_{n+1} = A(\theta^n\omega)x_n + \tilde{f}_0(n+1,\omega), \tag{7.3.17}$$

where

$$\tilde{f}_0(n+1,\omega) := f(n,\omega,\xi(n,\omega)) + f_0(n+1,\omega)$$

(i. e. with the possibly "wrong" sequence $(\xi(n,\omega))$ in the nonlinearity of the right-hand side).

324   Chapter 7. Invariant Manifolds

We have to make sure that $\tilde{f}_0(\cdot,\omega) \in X_{\gamma^-,\omega}$ provided $\xi(\cdot,\omega), f_0(\cdot,\omega) \in X_{\gamma^-,\omega}$. Indeed, for each $n \leq 0$, by (7.3.8),

$$\gamma^{-n}\|\tilde{f}_0(n,\omega)\|_{\theta^n\omega}$$
$$\leq \gamma^{-n}\|f_0(n,\omega)\|_{\theta^n\omega} + \gamma^{-n}\|f(n-1,\omega,\xi(n-1,\omega))\|_{\theta^n\omega}$$
$$\leq \gamma^{-n}\|f_0(n,\omega)\|_{\theta^n\omega} + \gamma^{-n}L\|\xi(n-1,\omega)\|_{\theta^{n-1}\omega}$$
$$\leq \gamma^{-n}\|f_0(n,\omega)\|_{\theta^n\omega} + \frac{L}{\gamma}\left(\gamma^{-(n-1)}\|\xi(n-1,\omega)\|_{\theta^{n-1}\omega} + \|\xi(0,\omega)\|_\omega\right).$$

Thus
$$\|\tilde{f}_0(\cdot,\omega)\|_{\gamma^-,\omega} \leq \|f_0(\cdot,\omega)\|_{\gamma^-,\omega} + \frac{L}{\gamma}\|\xi(\cdot,\omega)\|_{\gamma^-,\omega}. \qquad (7.3.18)$$

Hence by step 2, the "wrong" affine equation (7.3.17) has a unique (γ^-,ω)-bounded solution η, and thus the operator T_ω in (7.3.16) is well-defined. Step 2 also yields the estimate

$$\|T_\omega\xi(\cdot,\omega)\|_{\gamma^-,\omega} \leq \frac{\gamma}{\gamma-\beta}\|\tilde{f}_0(\cdot,\omega)\|_{\gamma^-,\omega},$$

hence together with (7.3.18)

$$\|T_\omega\xi(\cdot,\omega)\|_{\gamma^-,\omega} \leq \frac{\gamma}{\gamma-\beta}\left(\frac{L}{\gamma}\|\xi(\cdot,\omega)\|_{\gamma^-,\omega} + \|f_0(\cdot,\omega)\|_{\gamma^-,\omega}\right). \qquad (7.3.19)$$

The operator T_ω is for each ω a contraction in the Banach space $X_{\gamma^-,\omega}$. This can be seen as follows:

$$\zeta(\cdot,\omega) := T_\omega\xi_1(\cdot,\omega) - T_\omega\xi_2(\cdot,\omega) \in X_{\gamma^-,\omega}$$

solves the affine equation

$$\zeta(n+1,\omega) = A(\theta^n\omega)\zeta(n,\omega) + \bar{f}_0(n+1,\omega), \qquad (7.3.20)$$

where

$$\bar{f}_0(n+1,\omega) := f(n,\omega,\xi_1(n,\omega)) - f(n,\omega,\xi_2(n,\omega)).$$

An estimate completely analogous to the one leading to (7.3.18) yields

$$\|\bar{f}_0(\cdot,\omega)\|_{\gamma^-,\omega} \leq \frac{L}{\gamma}\|\xi_1(\cdot,\omega) - \xi_2(\cdot,\omega)\|_{\gamma^-,\omega}, \qquad (7.3.21)$$

hence $\bar{f}_0(\cdot,\omega) \in X_{\gamma^-,\omega}$.

Applying step 2 to the (γ^-,ω)-bounded solution ζ of (7.3.20) gives

$$\|\zeta(\cdot,\omega)\|_{\gamma^-,\omega} \leq \frac{\gamma}{\gamma-\beta}\|\bar{f}_0(\cdot,\omega)\|_{\gamma^-,\omega}. \tag{7.3.22}$$

Recalling the definition of ζ and combining (7.3.21) and (7.3.22) finally gives

$$\|T_\omega\xi_1(\cdot,\omega) - T_\omega\xi_2(\cdot,\omega)\|_{\gamma^-,\omega} \leq \frac{L}{\gamma-\beta}\|\xi_1(\cdot,\omega) - \xi_2(\cdot,\omega)\|_{\gamma^-,\omega}.$$

If $\gamma > \beta + L$, T_ω is a contraction, hence has a unique fixed point $\xi^*(\cdot,\omega) \in X_{\gamma^-,\omega}$. By definition of T_ω, $\xi^*(\cdot,\omega)$ solves equation (7.3.6) on $\mathbb{Z}^-$, hence on all of $\mathbb{Z}$.

It remains to estimate the norm of the fixed point $\xi^*(\cdot,\omega)$. By (7.3.19),

$$\|\xi^*(\cdot,\omega)\|_{\gamma^-,\omega} \leq \frac{\gamma}{\gamma-\beta}\left(\frac{L}{\gamma}\|\xi^*(\cdot,\omega)\|_{\gamma^-,\omega} + \|f_0(\cdot,\omega)\|_{\gamma^-,\omega}\right), \tag{7.3.23}$$

whence

$$\|\xi^*(\cdot,\omega)\|_{\gamma^-,\omega} \leq \frac{\gamma}{\gamma-(\beta+L)}\|f_0(\cdot,\omega)\|_{\gamma^-,\omega},$$

which is (7.3.9).

To complete the proof of part (i), we still have to show that $\omega \mapsto \xi^*(n,\omega)$ is measurable for each $n \in \mathbb{Z}$. We do this by appealing to the following useful lemma.

7.3.4 Lemma (Measurability of Fixed Point). *Let for all $\omega \in \Omega$, $(B_\omega, \|\cdot\|_\omega)$ be a Banach space of sequences $\xi = (\xi(n))$, $n \in \mathbb{Z}^{(\pm)}$, $\xi(n) \in \mathbb{R}^d$, such that the evaluation mapping $\xi \mapsto \xi(n)$ is continuous for all n. Let for each ω*

$$T_\omega : B_\omega \to B_\omega$$

be a contraction such that if $(\xi(n,\omega))$ is a sequence of measurable functions for which $\xi(\cdot,\omega) \in B_\omega$, then $\eta(n,\omega) := T_\omega\xi(\cdot,\omega)(n)$ is measurable for all n. Denote by $\xi^*(\cdot,\omega) \in B_\omega$ the unique fixed point of T_ω.

Then $\omega \mapsto \xi^*(n,\omega)$ is measurable for each n.

Proof. We start with $\xi_0(n,\omega) \equiv 0$ (which is obviously measurable) and recursively define

$$\xi_{k+1}(n,\omega) := T_\omega\xi_k(\cdot,\omega)(n),$$

whose right-hand side, hence left-hand side is measurable by assumption.

By the Banach fixed point theorem, for all ω

$$\xi^*(\cdot,\omega) = \lim_{k\to\infty} \xi_k(\cdot,\omega).$$

The continuity of the evaluation mapping $\xi \mapsto \xi(n)$ implies that also

$$\xi^*(n,\omega) = \lim_{k\to\infty} \xi_k(n,\omega),$$

and so as a pointwise limit of measurable functions, $\xi^*(n,\cdot)$ is measurable. $\square$

We apply this lemma as follows: Choose $B_\omega = X_{\gamma^-,\omega}$. The evaluation mapping is clearly continuous since

$$\|\xi(n)\|_{\theta^n\omega} \leq \gamma^n \|\xi\|_{\gamma^-,\omega}.$$

Further,

$$\eta(n,\omega) := T_\omega \xi(\cdot,\omega)(n) = \sum_{i=-\infty}^{n} \Phi(n-i,\theta^i\omega)\tilde{f}_0(i,\omega),$$

where

$$\tilde{f}_0(n+1,\omega) := f(n,\omega,\xi(n,\omega)) + f_0(n+1,\omega).$$

Since f and f_0 are assumed to be measurable, the measurability of $\xi(n,\cdot)$ implies that of $\eta(n,\cdot)$.

This completes the proof of part (i) of the lemma.

(ii) Let now $f_0(\cdot,\omega) \in X_{\gamma,\omega}$. In step 2, the (γ^-,ω)-bounded solution is defined on all of $\mathbb{Z}$ and the estimate leading to (7.3.15) also holds for all $n \in \mathbb{Z}$, implying

$$\|\xi(\cdot,\omega)\|_{\gamma,\omega} \leq \frac{\gamma}{\gamma-\beta} \|f_0(\cdot,\omega)\|_{\gamma,\omega}.$$

This can be used in step 3 to derive (7.3.23) with γ instead of γ^-, from which (7.3.10) follows.

(iii) We now solve equation (7.3.6) with initial value $x_0 = x = \xi(0,\omega)$ forwards in time and assume that $M(\omega) < \infty$.

By the variation of constants formula, for all $n \geq 0$,

$$\xi(n,\omega) = \Phi(n,\omega)\xi(0,\omega) + \sum_{i=1}^{n} \Phi(n-i,\theta^i\omega)(f(i-1,\omega,\xi(i-1,\omega)) + f_0(i,\omega)),$$

where the summation is not present if $n = 0$. By the assumptions (7.3.7) and (7.3.8),

$$\|\xi(n,\omega)\|_{\theta^n\omega}$$
$$\leq \beta^n \|\xi(0,\omega)\|_\omega + \sum_{i=1}^{n} \beta^{n-i} \left(L\|\xi(i-1,\omega)\|_{\theta^{i-1}\omega} + \|f_0(i,\omega)\|_{\theta^i\omega} \right),$$

hence

$$\beta^{-n}\|\xi(n,\omega)\|_{\theta^n\omega}$$
$$\leq \|\xi(0,\omega)\|_\omega + \frac{L}{\beta}\sum_{i=0}^{n-1}\beta^{-i}\|\xi(i,\omega)\|_{\theta^i\omega} + \frac{\gamma}{\beta}\sum_{i=0}^{n-1}\left(\frac{\gamma}{\beta}\right)^i M(\omega).$$

With the abbreviations
$$\begin{aligned} \Lambda(n) &:= \beta^{-n}\|\xi(n,\omega)\|_{\theta^n\omega}, \\ \Gamma(n) &:= \|\xi(0,\omega)\|_\omega + \frac{\gamma}{\beta}\sum_{i=0}^{n-1}\left(\frac{\gamma}{\beta}\right)^i M(\omega), \\ \Gamma(0) &:= \|\xi(0,\omega)\|_\omega, \end{aligned}$$

we arrive at
$$\Lambda(n) \leq \Gamma(n) + \frac{L}{\beta}\sum_{i=0}^{n-1}\Lambda(i), \quad n \geq 0. \tag{7.3.24}$$

We now require a discrete time version of the Gronwall-Bellman lemma.

7.3.5 Lemma (Discrete Time Gronwall-Bellman Lemma). *Let $b \geq 0$, $a(n) \geq 0$ and $c(n) \geq 0$ for $n \geq 0$, and suppose that*
$$a(n) \leq c(n) + b\sum_{i=0}^{n-1}a(i), \quad n \geq 0.$$

Then for all $n \geq 1$
$$a(n) \leq (1+b)^n c(0) + \sum_{i=1}^{n}(1+b)^{n-i}(c(i) - c(i-1)).$$

A proof can be found in Wanner [338: pp. 68–70].

Applying this to (7.3.24) yields for $n \geq 1$
$$\Lambda(n) \leq \left(1 + \frac{L}{\beta}\right)^n \Gamma(0) + \sum_{i=1}^{n}\left(1 + \frac{L}{\beta}\right)^{n-i}(\Gamma(i) - \Gamma(i-1)).$$

After some elementary manipulations we obtain for $n \geq 1$
$$\gamma^{-n}\|\xi(n,\omega)\|_{\theta^n\omega} \leq \|\xi(0,\omega)\|_\omega + \frac{\gamma}{\gamma-(\beta+L)}M(\omega).$$

This estimate holds trivially also for $n = 0$. Hence altogether, putting $\xi(0,\omega) = x$,
$$\|\xi(\cdot,\omega)\|_{\gamma^+,\omega} = \|\varphi(\cdot,\omega)x\|_{\gamma^+,\omega} \leq \|x\|_\omega + \frac{\gamma}{\gamma-(\beta+L)}M(\omega),$$

which is (7.3.11).

(iv) If ξ_1 and ξ_2 are solutions of (7.3.6) on $\mathbb{Z}^-$, hence on $\mathbb{Z}$, then $\xi := \xi_1 - \xi_2$ solves
$$x_{n+1} = A(\theta^n \omega) x_n + \tilde{f}(n, \omega, x_n), \quad n \in \mathbb{Z}, \qquad (7.3.25)$$
where
$$\tilde{f}(n, \omega, x) := f(n, \omega, x + \xi_2(n, \omega)) - f(n, \omega, \xi_2(n, \omega)).$$
Equation (7.3.25) is of the type of equation (7.3.6) with $f_0 = 0$, and $\tilde{f}$ clearly satisfies conditions (7.3.8). We thus have for all $n \in \mathbb{Z}$, in particular for $n < 0$,
$$\xi(n+1, \omega) = A(\theta^n \omega) \xi(n, \omega) + \tilde{f}(n, \omega, \xi(n, \omega)),$$
whence
$$\|\xi(n+1, \omega)\|_{\theta^{n+1}\omega} \leq (\beta + L) \|\xi(n, \omega)\|_{\theta^n \omega},$$
and after $-n$ steps
$$\|\xi(0, \omega)\|_\omega \leq (\beta + L)^{-n} \|\xi(n, \omega)\|_{\theta^n \omega},$$
equivalently
$$\|\xi(n, \omega)\|_{\theta^n \omega} \geq \left(\frac{1}{\beta + L}\right)^{-n} \|\xi(0, \omega)\|_\omega.$$
Now the assumption $\gamma > \beta + L$ implies $\left(\frac{1}{\gamma}\right)^{-n} < \left(\frac{1}{\beta+L}\right)^{-n}$ for all $n < 0$, consequently
$$\|\xi_1(n, \omega) - \xi_2(n, \omega)\|_{\theta^n \omega} \geq \gamma^n \|\xi_1(0, \omega) - \xi_2(0, \omega)\|_\omega, \quad \text{all } n \leq 0.$$
$\square$

For step 1 of the program for the construction of the unstable manifold we need the following lemma which is "dual" to Lemma 7.3.3.

7.3.6 Lemma (Second Key Technical Lemma). *Consider the random difference equation*
$$x_{n+1} = A(\theta^n \omega) x_n + f(n, \omega, x_n) + f_0(n+1, \omega), \quad n \in \mathbb{Z}, \qquad (7.3.26)$$
in $(\mathbb{R}^d, \|\cdot\|_\omega)$ with measurable $Gl(d, \mathbb{R})$-valued A and measurable f and f_0. Assume that there are constants $\alpha > 0$ and $L \in [0, \alpha)$ such that for each fixed ω
$$\|A(\omega)^{-1}\|_{\theta \omega, \omega} \leq \alpha^{-1}, \qquad (7.3.27)$$
and
$$f(n, \omega, 0) = 0, \quad \|f(n, \omega, x) - f(n, \omega, y)\|_{\theta^{n+1}\omega} \leq L\|x - y\|_{\theta^n \omega}. \qquad (7.3.28)$$
Then we have:
(o) For all $\omega \in \Omega$ and $n \in \mathbb{Z}$, $x \mapsto A(\theta^n \omega) x + f(n, \omega, x) + f_0(n+1, \omega)$ defines a bimeasurable bijection of $\mathbb{R}^d$, hence every initial value problem for equation (7.3.26) can be solved on all of $\mathbb{Z}$.

Let
$$0 < \gamma < \alpha - L.$$

Then the following assertions hold:

(i) If $M(\omega) := \sup_{n \geq 1} \gamma^{-n} \|f_0(n,\omega)\|_{\theta^n \omega} < \infty$ (equivalently, $f_0(\cdot,\omega) \in X_{\gamma^+,\omega}$), then there exists exactly one solution of (7.3.26) which has the property $\xi(\cdot,\omega) \in X_{\gamma^+,\omega}$ and

$$\|\xi(\cdot,\omega)\|_{\gamma^+,\omega} \leq \frac{\gamma}{\alpha - L - \gamma} M(\omega) \leq \frac{\gamma}{\alpha - L - \gamma} \|f_0(\cdot,\omega)\|_{\gamma^+,\omega}. \quad (7.3.29)$$

Moreover, $\omega \mapsto \xi(n,\omega)$ is measurable for each $n \in \mathbb{Z}$.

(ii) If, moreover, $f_0(\cdot,\omega) \in X_{\gamma,\omega}$, then the solution of part (i) satisfies $\xi(\cdot,\omega) \in X_{\gamma,\omega}$ and

$$\|\xi(\cdot,\omega)\|_{\gamma,\omega} \leq \frac{\gamma}{\alpha - L - \gamma} \|f_0(\cdot,\omega)\|_{\gamma,\omega}. \quad (7.3.30)$$

(iii) If $f_0(\cdot,\omega) \in X_{\gamma^-,\omega}$, then all solutions of (7.3.26) on $\mathbb{Z}^-$ satisfy $\varphi(\cdot,\omega,x) \in X_{\gamma^-,\omega}$ and

$$\|\varphi(\cdot,\omega,x)\|_{\gamma^-,\omega} \leq \|x\|_\omega + \frac{\gamma}{\alpha - L - \gamma} \|f_0(\cdot,\omega)\|_{\gamma^-,\omega}. \quad (7.3.31)$$

(iv) Two solutions ξ_1 and ξ_2 of (7.3.26) on $\mathbb{Z}^+$ satisfy

$$\|\xi_1(n,\omega) - \xi_2(n,\omega)\|_{\theta^n \omega} \geq \gamma^n \|\xi_1(0,\omega) - \xi_2(0,\omega)\|_\omega, \quad n \geq 0. \quad (7.3.32)$$

Proof. (o) We refer to Exercise 2.1.3 for the characterization of those time-one mappings which define an RDS with two-sided discrete time.

Let ω and $n \in \mathbb{Z}$ be fixed and for each $z \in \mathbb{R}^d$ define the operator

$$T_{n,\omega,z} : (\mathbb{R}^d, \|\cdot\|_{\theta^n \omega}) \to (\mathbb{R}^d, \|\cdot\|_{\theta^n \omega})$$

by

$$T_{n,\omega,z} x := A(\theta^n \omega)^{-1} z - A(\theta^n \omega)^{-1} f(n,\omega,x) - A(\theta^n \omega)^{-1} f_0(n+1,\omega).$$

We have

$$\|T_{n,\omega,z} x - T_{n,\omega,z} y\|_{\theta^n \omega}$$
$$\leq \|A(\theta^n \omega)^{-1}\|_{\theta^{n+1}\omega, \theta^n \omega} \|f(n,\omega,x) - f(n,\omega,y)\|_{\theta^{n+1}\omega}$$
$$\leq \frac{L}{\alpha} \|x - y\|_{\theta^n \omega}.$$

Since we have assumed that $L < \alpha$, the operator $T_{n,\omega,z}$ is contracting, hence has a unique fixed point $\xi(n,\omega,z)$. As a pointwise limit of measurable functions, this fixed point is measurable in (ω,z) for each n.

By definition, $\xi(n,\omega,z)$ fulfills

$$A(\theta^n\omega)\xi(n,\omega,z) = z - f(n,\omega,\xi(n,\omega,z)) - f_0(n+1,\omega),$$

hence for all n,ω,z there exists a unique measurable solution $\xi(n,\omega,z)$ of equation

$$z = A(\theta^n\omega)\xi(n,\omega,z) + f(n,\omega,\xi(n,\omega,z)) + f_0(n+1,\omega).$$

(i) The proof is analogous to the one for Lemma 7.3.3(i). Assumption (7.3.27) implies that the linear cocycle generated by A satisfies

$$\|\Phi(n,\omega)\|_{\omega,\theta^n\omega} \leq \alpha^n \quad \text{for all } n \leq 0.$$

Further, in the analogue of step 2 ($L = 0$), the unique solution $\xi(\omega) \in X_{\gamma^+,\omega}$ of the affine equation

$$x_{n+1} = A(\theta^n\omega)x_n + f_0(n+1,\omega)$$

is

$$\xi(n,\omega) = -\sum_{i=n+1}^{\infty} \Phi(n-i,\theta^i\omega)f_0(i,\omega), \quad n \in \mathbb{Z},$$

in particular

$$\xi(0,\omega) = -\sum_{i=1}^{\infty} \Phi(i,\omega)^{-1} f_0(i,\omega).$$

That this makes sense if $f_0 \in X_{\gamma^+,\omega}$ can be seen as follows: For $n \geq 0$

$$\gamma^{-n}\|\xi(n,\omega)\|_{\theta^n\omega} \leq \sum_{i=n+1}^{\infty} \alpha^{n-i}\gamma^{-n+i}\gamma^{-i}\|f_0(i,\omega)\|_{\theta^i\omega}$$

$$\leq \sum_{i=n+1}^{\infty} \left(\frac{\gamma}{\alpha}\right)^{-n+i} M(\omega)$$

$$= \frac{\gamma}{\alpha-\gamma} M(\omega) < \infty.$$

We thus obtain

$$\|\xi(\cdot,\omega)\|_{\gamma^+,\omega} \leq \frac{\gamma}{\alpha-\gamma} M(\omega), \tag{7.3.33}$$

which is the analogue of (7.3.15).

The proofs of (ii), (iii) and (iv) are analogous to their counterparts in Lemma 7.3.3, hence are omitted. □

Construction of the Unstable Manifold M^+

By means of the two Key Technical Lemmas, we are now able to construct a random graph $x^+ \mapsto x^- = m^+(\omega,x^+)$ in the E^+, E^- coordinate system, the elements of which are dynamically characterized in a way that qualifies them as members of the unstable manifold $M^+(\omega)$.

We collect the main part of the work in the following proposition.

7.3.7 Proposition (Graph of Unstable Manifold). *Let the conditions of Theorem 7.3.1 be satisfied. Fix $a \in \Gamma_{\text{left}} = [\beta_{j+1} + \delta, \alpha_j - \delta]$.*

Then for each ω and $x^+ \in E^+$ there exists a unique element $x^- = m^+(\omega, x^+) \in E^-$ which depends measurably on (ω, x^+) and is independent of the particular value of a such that the cocycle φ generated by the coupled pair of equations (7.3.4) and (7.3.5) has the property that $\varphi^-(\cdot, \omega, x^+, m^+(\omega, x^+)) \in X_{a^-,\omega}(E^-)$, equivalently $\varphi(\cdot, \omega, x^+, m^+(\omega, x^+)) \in X_{a^-,\omega}$.

Proof. We follow the program (steps 1 to 5) formulated at the beginning of this subsection.

Step 1: We choose $\xi^-(\cdot, \omega) \in X_{a^-,\omega}(E^-)$ and consider the unstable equation (7.3.4) with the "wrong" stable component in the right-hand side,

$$x_{n+1}^+ = A^+(\theta^n \omega)x_n^+ + F^+(\theta^n \omega, x_n^+, \xi^-(n, \omega)), \quad x_0^+ = x^+ \in E^+. \quad (7.3.34)$$

Equation (7.3.34) satisfies the assumptions of the Second Key Technical Lemma 7.3.6 with

$$f(n, \omega, x^+) := F^+(\theta^n \omega, x^+, \xi^-(n, \omega)) - F^+(\theta^n \omega, 0, \xi^-(n, \omega))$$

and

$$f_0(n+1, \omega) := F^+(\theta^n \omega, 0, \xi^-(n, \omega)).$$

Indeed, (7.3.27) is satisfied for A^+, furthermore $f(n, \omega, 0) = 0$ and

$$\|f(n, \omega, x^+) - f(n, \omega, y^+)\|_{\theta^{n+1}\omega} \leq L\|x^+ - y^+\|_{\theta^n \omega},$$

so that also (7.3.28) holds. Lemma 7.3.6(o) tells us that each initial value problem for (7.3.34) can be solved on all of $\mathbb{Z}$.

We now make sure that we can apply part (iii) of the Second Key Technical Lemma 7.3.6 by proving that our assumption $\xi^-(\cdot, \omega) \in X_{a^-,\omega}(E^-)$ implies that $f_0(\cdot, \omega) \in X_{a^-,\omega}(E^+)$. We have for all $n \in \mathbb{Z}$

$$a^{-(n+1)}\|f_0(n+1, \omega)\|_{\theta^{n+1}\omega} \leq \left(\frac{L}{a}\right) a^{-n}\|\xi^-(n, \omega)\|_{\theta^n \omega},$$

thus

$$\|f_0(\cdot, \omega)\|_{a^-,\omega} \leq \frac{L}{a}\|\xi^-(\cdot, \omega)\|_{a^-,\omega}.$$

Hence Lemma 7.3.6(iii) applies: If $0 < \gamma = a < \alpha - L$ (which holds true for all $a \in [\beta + \delta, \alpha - \delta]$ under the condition $L < \delta$), then any solution $\xi^+(\cdot, \omega, x^+)$ of (7.3.34) is in $X_{a^-,\omega}(E^+)$ and

$$\|\xi^+(\cdot, \omega, x^+)\|_{a^-,\omega} \leq \|x^+\|_\omega + \frac{a}{\alpha - L - a}\|f_0(\cdot, \omega)\|_{a^-,\omega}.$$

The last two estimates yield

$$\|\xi^+(\cdot,\omega,x^+)\|_{a^-,\omega} \le \|x^+\|_\omega + \frac{L}{\alpha - L - a}\|\xi^-(\cdot,\omega)\|_{a^-,\omega}. \qquad (7.3.35)$$

Step 2: The arbitrary $\xi^-(\cdot,\omega) \in X_{a^-,\omega}(E^-)$ and the corresponding solution $\xi^+(\cdot,\omega,x^+) \in X_{a^-,\omega}(E^+)$ of step 1 are now inserted into the nonlinear part of the stable equation (7.3.5). We obtain

$$x^-_{n+1} = A^-(\theta^n\omega)x^-_n + F^-(\theta^n\omega, \xi^+(n,\omega,x^+), \xi^-(n,\omega)), \quad n \in \mathbb{Z}. \qquad (7.3.36)$$

The assumptions of the First Key Technical Lemma 7.3.3 are satisfied for (7.3.36) with $f = 0$ (hence $L = 0$) and

$$f_0(n+1,\omega) := F^-(\theta^n\omega, \xi^+(n,\omega,x^+), \xi^-(n,\omega)),$$

provided $\gamma = a > \beta$ (which is automatically the case).

Thus Lemma 7.3.3(i) applies if $f_0(\cdot,\omega) \in X_{a^-,\omega}(E^-)$, which we are going to verify now.

We have for any $n \in \mathbb{Z}$

$$a^{-(n+1)}\|f_0(n+1,\omega)\|_{\theta^{n+1}\omega}$$
$$\le \frac{L}{a}\left(a^{-n}\|\xi^+(n,\omega,x^+)\|_{\theta^n\omega} + a^{-n}\|\xi^-(n,\omega)\|_{\theta^n\omega}\right)$$

implying

$$\|f_0(\cdot,\omega)\|_{a^-,\omega} \le \frac{L}{a}\left(\|\xi^+(\cdot,\omega,x^+)\|_{a^-,\omega} + \|\xi^-(\cdot,\omega)\|_{a^-,\omega}\right).$$

We have estimated the first term on the right-hand side of the last inequality in (7.3.35), resulting in

$$\|f_0(\cdot,\omega)\|_{a^-,\omega} \le \frac{L}{a}\|x^+\|_\omega + \frac{L(\alpha-a)}{a(\alpha-L-a)}\|\xi^-(\cdot,\omega)\|_{a^-,\omega}. \qquad (7.3.37)$$

Hence by Lemma 7.3.3(i) there exists a unique solution $\eta^-(\cdot,\omega,x^+) \in X_{a^-,\omega}(E^-)$ of (7.3.36) and its norm satisfies

$$\|\eta^-(\cdot,\omega,x^+)\|_{a^-,\omega} \le \frac{a}{a-\beta}\|f_0(\cdot,\omega)\|_{a^-,\omega}.$$

Together with the estimate (7.3.37) we obtain

$$\|\eta^-(\cdot,\omega,x^+)\|_{a^-,\omega} \le \frac{L}{a-\beta}\|x^+\|_\omega + \frac{L(\alpha-a)}{(a-\beta)(\alpha-L-a)}\|\xi^-(\cdot,\omega)\|_{a^-,\omega}. \qquad (7.3.38)$$

Step 3: Steps 1 and 2 allow us to define for each fixed ω and $x^+ \in E^+$ the operator

$$T_{\omega,x^+} : X_{a^-,\omega}(E^-) \to X_{a^-,\omega}(E^-)$$

by

7.3 Global Invariant Manifolds

$$T_{\omega,x^+}\xi^-(\cdot,\omega) = \eta^-(\cdot,\omega,x^+).$$

We will now prove that T_{ω,x^+} is contracting.

We choose initial values $x_1^+, x_2^+ \in E^+$ and $\xi_1^-, \xi_2^- \in X_{a^-,\omega}(E^-)$, and determine the solutions $\xi_1^+(\cdot,\omega,x_1^+)$ and $\xi_2^+(\cdot,\omega,x_2^+)$ of equation (7.3.34) according to step 1. The difference

$$\zeta^+ := \xi_2^+ - \xi_1^+$$

satisfies

$$y_{n+1}^+ = A^+(\theta^n\omega)y_n^+ + F^+(\theta^n\omega, y_n^+ + \xi_1^+, \xi_2^-) - F^+(\theta^n\omega, \xi_1^+, \xi_1^-), \quad (7.3.39)$$

with initial value $\zeta^+(0,\omega) = y_0^+ = x_2^+ - x_1^+$. Equation (7.3.39) satisfies the assumptions of the Second Key Technical Lemma 7.3.6 with

$$f(n,\omega,y^+) := F^+(\theta^n\omega, y^+ + \xi_1^+, \xi_2^-) - F^+(\theta^n\omega, \xi_1^+, \xi_2^-)$$

and

$$f_0(n+1,\omega) := -F^+(\theta^n\omega, \xi_1^+, \xi_1^-) + F^+(\theta^n\omega, \xi_1^+, \xi_2^-),$$

provided $L < \alpha - a$ (which is implied by $L < \delta$).

Hence Lemma 7.3.6(iii) applies to (7.3.39) and yields

$$\|\xi_2^+ - \xi_1^+\|_{a^-,\omega} \leq \|x_2^+ - x_1^+\|_\omega + \frac{L}{\alpha - L - a}\|\xi_2^- - \xi_1^-\|_{a^-,\omega}. \quad (7.3.40)$$

Now let $\eta_1^- = T_{\omega,x_1^+}\xi_1^-$ and $\eta_2^- = T_{\omega,x_2^+}\xi_2^-$. The difference $\eta_2^- - \eta_1^- \in X_{a^-,\omega}(E^-)$ solves

$$x_{n+1}^- = A^-(\theta^n\omega)x_n^- + F^-(\theta^n\omega, \xi_2^+, \xi_2^-) - F^-(\theta^n\omega, \xi_1^+, \xi_1^-). \quad (7.3.41)$$

Equation (7.3.41) satisfies the assumptions of the First Key Technical Lemma 7.3.3 with $f = 0$ and

$$f_0(n+1,\omega) := F^-(\theta^n\omega, \xi_2^+(n,\omega,x_2^+), \xi_2^-) - F^-(\theta^n\omega, \xi_1^+(n,\omega,x_1^+), \xi_1^-).$$

The same reasoning which lead to (7.3.37) yields

$$\|f_0(\cdot,\omega)\|_{a^-,\omega} \leq \frac{L}{a}\|x_2^+ - x_1^+\|_\omega + \frac{L(\alpha-a)}{a(\alpha - L - a)}\|\xi_2^- - \xi_1^-\|_{a^-,\omega}. \quad (7.3.42)$$

We hence can apply Lemma 7.3.3(i) which results in

$$\|T_{\omega,x_2^+}\xi_2^- - T_{\omega,x_1^+}\xi_1^-\|_{a^-,\omega} \leq \frac{a}{a-\beta}\|f_0(\cdot,\omega)\|_{a^-,\omega}.$$

The latter combined with (7.3.42) finally gives

$$\|T_{\omega,x_2^+}\xi_2^- - T_{\omega,x_1^+}\xi_1^-\|_{a^-,\omega} \leq \frac{L}{a-\beta}\|x_2^+ - x_1^+\|_\omega \qquad (7.3.43)$$
$$+\frac{L(\alpha-a)}{(a-\beta)(\alpha-L-a)}\|\xi_2^- - \xi_1^-\|_{a^-,\omega}.$$

We now look for an upper bound in (7.3.43) which holds for any value of $a \in [\beta+\delta, \alpha-\delta]$. An elementary calculation shows that for such an a

$$\frac{L}{a-\beta} \leq \frac{L}{\delta}, \quad \frac{L(\alpha-a)}{(a-\beta)(\alpha-L-a)} \leq \frac{L}{\delta-L}.$$

The final estimate is thus: For all $a \in [\beta+\delta, \alpha-\delta]$ and $L < \delta$

$$\|T_{\omega,x_2^+}\xi_2^- - T_{\omega,x_1^+}\xi_1^-\|_{a^-,\omega} \leq \frac{L}{\delta}\|x_2^+ - x_1^+\|_\omega + \frac{L}{\delta-L}\|\xi_2^- - \xi_1^-\|_{a^-,\omega}. \quad (7.3.44)$$

For the particular choice $x^+ := x_1^+ = x_2^+$, (7.3.43) reduces to

$$\|T_{\omega,x^+}\xi_2^- - T_{\omega,x^+}\xi_1^-\|_{a^-,\omega} \leq \frac{L}{\delta-L}\|\xi_2^- - \xi_1^-\|_{a^-,\omega}. \qquad (7.3.45)$$

Thus T_{ω,x^+} is contracting if $L < \delta/2$ which we have assumed.

Hence there is a unique fixed point $\xi_*^-(\cdot,\omega,x^+) \in X_{a^-,\omega}(E^-)$ of T_{ω,x^+}.

Step 4: Put $m^+(\omega,x^+) := \xi_*^-(0,\omega,x^+) \in E^-$.

Starting our procedure with ξ_*^-, determining the corresponding ξ_*^+ with initial value $\xi_*^+(0,\omega) = x^+$ and using the definition of T_{ω,x^+}, we obtain a solution

$$(\xi_*^+, \xi_*^-) = \varphi(\cdot, \omega, x^+, m^+(\omega, x^+))$$

of the pair of equations (7.3.4) and (7.3.5) such that the point $x^- := m^+(\omega, x^+) \in E^-$ is the uniquely determined point (see step 2) for which $\varphi^-(\cdot, \omega, x^+, x^-) \in X_{a^-,\omega}(E^-)$.

As $\varphi^+(\cdot, \omega, x^+, m^+(\omega, x^+)) \in X_{a^-,\omega}(E^+)$ is automatically true (see step 1) and as

$$\|(x^+, x^-)\|_\omega = \|x^+\|_\omega + \|x^-\|_\omega,$$

hence

$$\|\varphi^\pm(\cdot,\omega,x)\|_{a^-,\omega} \leq \|\varphi(\cdot,\omega,x)\|_{a^-,\omega} \leq \|\varphi^+(\cdot,\omega,x)\|_{a^-,\omega} + \|\varphi^-(\cdot,\omega,x)\|_{a^-,\omega},$$

it follows that even $\varphi(\cdot, \omega, x^+, m^+(\omega, x^+)) \in X_{a^-,\omega}$.

Conversely, $\varphi(\cdot, \omega, x^+, x^-) \in X_{a^-,\omega}$ implies that $\varphi^-(\cdot, \omega, x^+, x^-) \in X_{a^-,\omega}(E^-)$. The latter is the case if and only if $x^- = m^+(\omega, x^+)$. Consequently, for prescribed ω and $x^+ \in E^+$, $x = (x^+, m^+(\omega, x^+))$ is the uniquely determined initial value for which the orbit of the cocycle φ has the property $\varphi(\cdot, \omega, x) \in X_{a^-,\omega}$. We have thus proved the dynamical characterization of the graph,

$$M^+(\omega) := \{x^+ \oplus m^+(\omega, x^+) : x^+ \in E^+\} = \{x \in \mathbb{R}^d : \varphi(\cdot, \omega, x) \in X_{a^-, \omega}\}.$$

Step 5: The measurability of $(\omega, x^+) \mapsto m^+(\omega, x^+)$ follows again as in the proof of Lemma 7.3.3(i) by applying Lemma 7.3.4.

That $m^+(\omega, x^+)$ does not depend on $a \in [\beta + \delta, \alpha - \delta]$ can be seen as follows: Assume $a_1, a_2 \in [\beta+\delta, \alpha-\delta]$ with $a_1 > a_2$. By Lemma 7.2.3, $a_1 > a_2$ implies $X_{a_1^-, \omega} \hookrightarrow X_{a_2^-, \omega}$. Hence if ξ^- is (a_1^-, ω)-bounded, then it is (a_2^-, ω)-bounded. By the uniqueness of the (a_2^-, ω)-bounded solution, $m_{a_1}^+(\omega, x^+) = m_{a_2}^+(\omega, x^+)$. □

Step 6 of our program remains to be carried out.

Completion of the proof of Theorem 7.3.1:

(i) That $M^+(\omega)$ is a topological manifold with the corresponding dimension is a consequence of the fact that it is a Lipschitz-continuous graph in $E^+ \oplus E^-$.

(ii) The measurablity of m^+ was proved in Proposition 7.3.7.

The trivial solution $\varphi(\cdot, \omega, 0) = 0$ is (a^-, ω)-bounded, hence by uniqueness $m^+(\omega, 0) = 0$.

For the Lipschitz-continuity of $m^+(\omega, \cdot)$ we utilize the estimate (7.3.44) which for the fixed points takes the form

$$\|\xi_{2,*}^- - \xi_{1,*}^-\|_{a^-, \omega} \leq \frac{L}{\delta}\|x_2^+ - x_1^+\|_\omega + \frac{L}{\delta - L}\|\xi_{2,*}^- - \xi_{1,*}^-\|_{a^-, \omega}.$$

Our assumption $L < \delta/2$ implies

$$1 - \frac{L}{\delta - L} = \frac{\delta - 2L}{\delta - L} > 0,$$

hence

$$\|\xi_{2,*}^- - \xi_{1,*}^-\|_{a^-, \omega} \leq \frac{L(\delta - L)}{\delta(\delta - 2L)}\|x_2^+ - x_1^+\|_\omega.$$

This implies in particular that

$$\|m^+(\omega, x_2^+) - m^+(\omega, x_1^+)\|_\omega \leq \frac{L(\delta - L)}{\delta(\delta - 2L)}\|x_2^+ - x_1^+\|_\omega.$$

(iii) Invariance of $M^+(\omega)$: We have to prove that

$$\varphi(n, \omega)M^+(\omega) = M^+(\theta^n \omega) \quad \text{for all } n \in \mathbb{Z}, \, \omega \in \Omega. \tag{7.3.46}$$

This is equivalent to the two relations
(a) $\varphi(1, \omega)M^+(\omega) \subset M^+(\theta\omega)$,
(b) $\varphi(-1, \omega)M^+(\omega) \subset M^+(\theta^{-1}\omega)$.

Indeed, since $\varphi(n, \omega)$ is a cocycle of homeomorphisms of $\mathbb{R}^d$, (a) and (b) imply

$$M^+(\omega) \subset \varphi(1,\omega)^{-1}M^+(\theta\omega) = \varphi(-1,\theta\omega)M^+(\theta\omega) \subset M^+(\omega),$$

from which (7.3.46) follows for $n=1$ and $n=-1$, and thus for all $n \in \mathbb{Z}$.

Proof of (a): Let $x \in M^+(\omega)$. We want to show that $\varphi(1,\omega,x) \in M^+(\theta\omega)$. By the dynamical characterization, $x \in M^+(\omega)$ if and only if $\varphi(\cdot,\omega,x) \in X_{a^-,\omega}$. By the cocycle property,

$$\varphi(n+1,\omega,x) = \varphi(n,\theta\omega,\varphi(1,\omega,x)).$$

Since the left-hand side, hence the right-hand side is in $X_{a^-,\omega}$, $\varphi(1,\omega,x) \in M^+(\theta\omega)$.

Proof of (b): By the cocycle property

$$\varphi(n,\theta^{-1}\omega,\varphi(-1,\omega,x)) = \varphi(n-1,\omega,x).$$

If $x \in M^+(\omega)$ then the right-hand side, hence the left-hand side is in $X_{a^-,\omega}$, thus $\varphi(-1,\omega,x) \in M^+(\theta^{-1}\omega)$.

This completes the proof of Theorem 7.3.1. $\square$

As a by-product, we have the following dynamical properties of orbits of φ starting on $M^+(\omega)$.

7.3.8 Corollary. *Let the conditions of Theorem 7.3.1 be satisfied. Choose $x \in M^+(\omega)$. Then for any $0 < \gamma \leq \alpha_j - \delta$*

$$\|\varphi(n,\omega,x)\|_{\theta^n\omega} \leq \gamma^n \frac{\delta-L}{\delta-2L}\|x\|_\omega, \quad n \leq 0,$$

and hence

$$\|\varphi(n,\omega,x)\|_{\theta^n\omega} \geq \gamma^n \frac{\delta-2L}{\delta-L}\|x\|_\omega, \quad n \geq 0.$$

Proof. We again utilize the estimate (7.3.44) which for $n \leq 0$ and $x_1^+ = x^+$, $x_2^+ = 0$, $x = x^+ \oplus m^+(\omega,x^+)$ and γ instead of a yields

$$\|\varphi^-(n,\omega,x)\|_{\theta^n\omega} \leq \gamma^n \|\varphi^-(\cdot,\omega,x)\|_{\gamma^-,\omega} \leq \gamma^n D(L)\|x^+\|_\omega.$$

On the other hand, by (7.3.35),

$$\|\varphi^+(n,\omega,x)\|_{\theta^n\omega} \leq \gamma^n \left(\|x^+\|_\omega + \frac{L}{\delta-L}\|\varphi^-(\cdot,\omega,x)\|_{\gamma^-,\omega}\right)$$

$$\leq \gamma^n \left(1 + D(L)\frac{L}{\delta-L}\right)\|x^+\|_\omega.$$

Adding up, we obtain

$$\|\varphi(n,\omega,x)\|_{\theta^n\omega} \leq \gamma^n \frac{\delta-L}{\delta-2L}\|x^+\|_\omega \leq \gamma^n \frac{\delta-L}{\delta-2L}\|x\|_\omega,$$

since

7.3 Global Invariant Manifolds

$$1 + D(L) + D(L)\frac{L}{\delta - L} = \frac{\delta - L}{\delta - 2L}.$$

Now choose $n \geq 0$. By the cocycle property and the first estimate,

$$\begin{aligned}\|x\|_\omega = \|\varphi(0,\omega,x)\|_\omega &= \|\varphi(-n,\theta^n\omega)\varphi(n,\omega,x)\|_\omega \\ &\leq \gamma^{-n}\frac{\delta-L}{\delta-2L}\|\varphi(n,\omega,x)\|_{\theta^n\omega},\end{aligned}$$

which is equivalent to

$$\|\varphi(n,\omega,x)\|_{\theta^n\omega} \geq \gamma^n\frac{\delta-2L}{\delta-L}\|x\|_\omega.$$

□

7.3.9 Corollary (Flag of Unstable Manifolds). *Let the conditions of Theorem 7.3.1 be satisfied. Then we obtain a nested sequence of unstable manifolds*

$$M_{11} \subset M_{12} \subset \ldots \subset M_{1j} \subset \ldots \subset M_{1p} = \mathbb{R}^d$$

corresponding to the backward filtration of the MET for Φ,

$$E_{11} = V_p^- \subset \ldots \subset E_{1j} = V_{p+1-j}^- \subset \ldots \subset E_{1p} = V_1^- = \mathbb{R}^d.$$

The above inclusions are embeddings of submanifolds.

Proof. The inclusions follow from the dynamical characterization of the unstable manifolds and the embeddings of the corresponding Banach spaces (Lemma 7.2.3).

That the inclusions are embeddings follows from the representation of the manifolds as graphs. □

7.3.2 Construction of Stable Manifolds

The reader will certainly be able to imagine that the stable manifolds can be constructed in a completely analogous "dual" procedure with the help of the two Key Technical Lemmas (now we need the finer estimates using $M(\omega)$ in Lemma 7.3.3(iii) and 7.3.6(i)). We will hence omit the proof and immediately state the final result.

7.3.10 Theorem (Global Stable Manifold Theorem). *Let the RDS φ with discrete time $\mathbb{T} = \mathbb{Z}$ be prepared and generated by the difference equation (7.3.1). Assume*

$$\|F(\omega,x) - F(\omega,y)\|_{\theta\omega} \leq L\|x-y\|_\omega \quad \text{for all} \quad x,y \in \mathbb{R}^d,$$

where

$$0 \leq L < \delta/2.$$

Choose any i with $1 < i \leq p$, let $\Lambda^- = \Lambda_{ip} = \{\lambda_i > \ldots > \lambda_p\}$ be the corresponding spectral interval of Φ, put $E^- := E_{ip}$, $x^- = (x_i, \ldots, x_p)$, $E^+ := E_{ip}^\perp = E_{1,i-1}$, $x^+ = (x_1, \ldots, x_{i-1})$, choose a constant b in the spectral gap to the right of Λ^-,

$$b := b_i \in \Gamma_{\text{right}} = [\beta_i + \delta, \alpha_{i-1} - \delta],$$

and define the set

$$M^-(\omega) = M_{ip}(\omega) := \{x \in \mathbb{R}^d : \varphi(\cdot, \omega, x) \in X_{b+,\omega}\}. \tag{7.3.47}$$

Then

(i) $M^-(\omega)$ is a topological manifold of dimension $\dim E^- = d_i + \ldots + d_p$, called the stable manifold *corresponding to the interval Λ^-. It is independent of the particular choice of b.*

(ii) $M^-(\omega)$ is a graph in $E^- \oplus E^+$,

$$M^-(\omega) = \{x^- \oplus m^-(\omega, x^-) : x^- \in E^-\},$$

with the following properties:

1. $m^- : \Omega \times E^- \to E^+$ is measurable.
2. $m^-(\omega, 0) = 0$.
3. m^- is globally Lipschitz continuous in the random norms of E^- and E^+,

$$\|m^-(\omega, x^-) - m^-(\omega, y^-)\|_\omega \leq D(L)\|x^- - y^-\|_\omega,$$

where the constant

$$D(L) = \frac{L(\delta - L)}{\delta(\delta - 2L)}$$

is independent of i.

(iii) $M^-(\omega)$ is invariant under φ, i. e.

$$\varphi(n, \omega)M^-(\omega) = M^-(\theta^n \omega), \quad n \in \mathbb{Z}.$$

As another by-product, we have the following dynamical properties of orbits of φ starting on $M^-(\omega)$.

7.3.11 Corollary. *Let the conditions of Theorem 7.3.10 be satisfied. Choose $x \in M^-(\omega)$. Then for any $\gamma \geq \beta_i + \delta$*

$$\|\varphi(n, \omega, x)\|_{\theta^n \omega} \leq \gamma^n \frac{\delta - L}{\delta - 2L}\|x\|_\omega, \quad n \geq 0,$$

and hence

$$\|\varphi(n, \omega, x)\|_{\theta^n \omega} \geq \gamma^n \frac{\delta - 2L}{\delta - L}\|x\|_\omega, \quad n \leq 0.$$

7.3.12 Corollary (Flag of Stable Manifolds). *Let the conditions of Theorem 7.3.10 be satisfied. Then we obtain a nested sequence of stable manifolds*
$$M_{pp} \subset M_{p-1,p} \subset \ldots \subset M_{ip} \subset \ldots \subset M_{1p} = \mathbb{R}^d$$
corresponding to the forward filtration of the MET for Φ,
$$E_{pp} = V_p^+ \subset \ldots \subset E_{ip} = V_i^+ \subset \ldots \subset E_{1p} = V_1^+ = \mathbb{R}^d.$$
The above inclusions are embeddings of submanifolds.

We finally combine the Corollaries 7.3.8 and 7.3.11 in the following corollary, the result of which is shown in Fig. 7.5.

7.3.13 Corollary (Disjoint Decomposition of Spectrum).
Let the conditions of Theorems 7.3.1 and 7.3.10 be satisfied, assume that the Lyapunov spectrum of Φ is not one-point and choose $1 \leq j < p$. Denote by $M^+(\omega)$ the unstable manifold corresponding to the spectral interval $\Lambda^+ := \{\lambda_1 > \cdots > \lambda_j\}$ and by $M^-(\omega)$ the stable manifold corresponding to the complementary spectral interval $\Lambda^- := \{\lambda_{j+1} > \cdots > \lambda_p\}$.
Then
$$M^+(\omega) \cap M^-(\omega) = \{0\}.$$

Proof. Let $0 \neq x \in M^+(\omega) \cap M^-(\omega)$.

Since $x \in M^+(\omega)$, Corollary 7.3.8 applies: For $\gamma_1 = \alpha_j - \delta$,
$$\|\varphi(n,\omega,x)\|_{\theta^n \omega} \geq \frac{1}{C}\gamma_1^n \|x\|_\omega, \quad n \geq 0,$$
where $C = (\delta - L)/(\delta - 2L)$.

Since $x \in M^-(\omega)$, Corollary 7.3.11 applies: For $\gamma_2 = \beta_{j+1} + \delta < \gamma_1$,

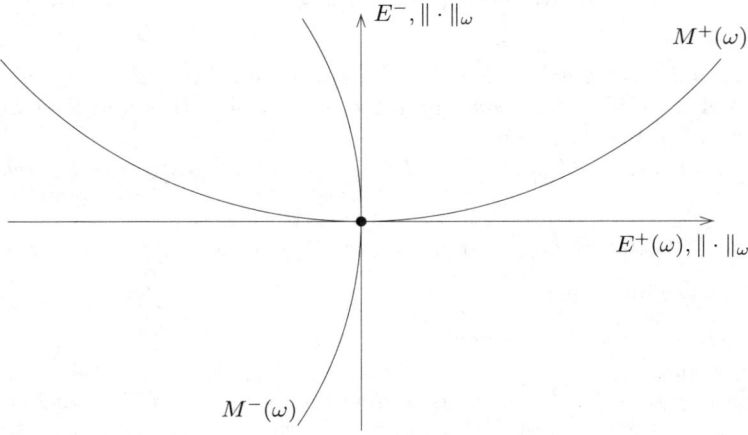

Fig. 7.5. Stable and unstable manifolds do not intersect except at 0

$$\|\varphi(n,\omega,x)\|_{\theta^n\omega} \leq C\gamma_2^n \|x\|_\omega, \quad n \geq 0.$$

Combining these estimates and using $\|x\|_\omega \neq 0$,

$$\left(\frac{\gamma_1}{\gamma_2}\right)^n \leq C^2 \quad \text{for all } n \geq 0,$$

which is a contradiction. □

7.3.3 Construction of Center Manifolds

We now complete our program of constructing global invariant manifolds by proving the following theorem.

7.3.14 Theorem (Global Center Manifold Theorem, Discrete Time). *Let the RDS φ with discrete time $\mathbb{T} = \mathbb{Z}$ be prepared and generated by the difference equation (7.3.1). Assume*

$$\|F(\omega,x) - F(\omega,y)\|_{\theta\omega} \leq L\|x-y\|_\omega \quad \text{for all} \quad x,y \in \mathbb{R}^d, \tag{7.3.48}$$

where

$$0 \leq L < \frac{3-\sqrt{5}}{2}\delta.^3 \tag{7.3.49}$$

Consider for any i, j with $1 \leq i \leq j \leq p$ the interval $\Lambda_{ij} = \{\lambda_i > \ldots > \lambda_j\}$ from the spectrum of Φ, choose constants

$$a_j \in \Gamma_{\text{left}} = [\beta_{j+1} + \delta, \alpha_j - \delta], \quad b_i \in \Gamma_{\text{right}} = [\beta_i + \delta, \alpha_{i-1} - \delta],$$

and define the set

$$M_{ij}(\omega) := \{x \in \mathbb{R}^d : \varphi(\cdot,\omega,x) \in X_{a_j^-,b_i^+,\omega}\} \tag{7.3.50}$$

(see Fig. 7.6) so that

$$M_{ij}(\omega) = M_{1j}(\omega) \cap M_{ip}(\omega). \tag{7.3.51}$$

Then:

(i) $M_{ij}(\omega)$ is a topological manifold of dimension $\dim E_{ij} = d_i + \ldots + d_j$, called the center manifold *corresponding to the interval Λ_{ij}. It is independent of the particular choice of a_j and b_i.*

(ii) For $(i,j) = (1,p)$, $M_{1p}(\omega) = \mathbb{R}^d$. For $(i,j) \neq (1,p)$, $M_{ij}(\omega)$ is a graph in $E_{ij} \oplus E_{ij}^\perp$,

$$M_{ij}(\omega) = \{x_{ij} \oplus m_{ij}(\omega,x_{ij}) : x_{ij} \in E_{ij}\},$$

with the following properties:

1. *$m_{ij} : \Omega \times E_{ij} \to E_{ij}^\perp$ is measurable.*
2. *$m_{ij}(\omega,0) = 0$.*
3. *m_{ij} is globally Lipschitz continuous in the random norms of E_{ij} and $E_{ij}^\perp$,*

---
[3] Note that $\frac{3-\sqrt{5}}{2} = 0.38197\ldots < \frac{1}{2}$.

$$\|m_{ij}(\omega, x_{ij}) - m_{ij}(\omega, y_{ij})\|_\omega \leq C(L)\|x_{ij} - y_{ij}\|_\omega,$$

where the constant

$$C(L) := \frac{2D(L)}{1 - D(L)}, \quad D(L) = \frac{L(\delta - L)}{\delta(\delta - 2L)} < 1,$$

is independent of (i, j).

(iii) $M_{ij}(\omega)$ is invariant under φ, i. e.

$$\varphi(n, \omega) M_{ij}(\omega) = M_{ij}(\theta^n \omega), \quad n \in \mathbb{Z}.$$

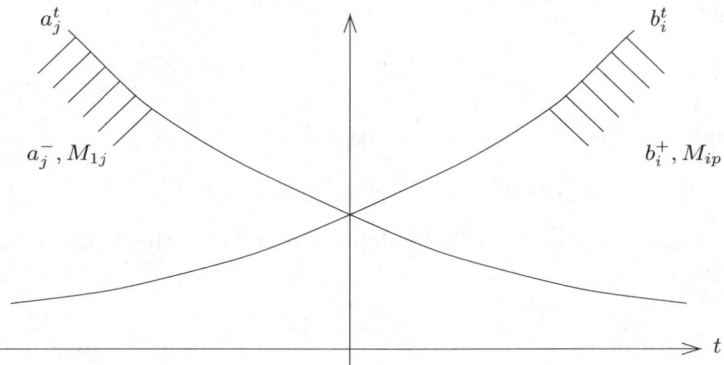

Fig. 7.6. Dynamical characterization of the center manifold M_{ij}

Proof. Assume the situation of Theorems 7.3.1 and 7.3.10.

To single-out nontrivial cases we assume that $1 < i \leq j < p$ and introduce the notations $E^+ := E_{1,i-1}$, $E^0 := E_{i,j}$ and $E^- := E_{j+1,p}$, so that $\mathbb{R}^d = E^+ \oplus E^0 \oplus E^-$. These are nontrivial subspaces.

Under the conditions $\mathrm{Lip}(F(\omega, \cdot))_{\omega, \theta\omega} \leq L$ and $L < \delta/2$ we have constructed the following invariant manifolds:

$$M^+ = M_{1j}: \quad m^+ : \Omega \times E^+ \oplus E^0 \to E^-, \quad \mathrm{Lip}(m^+(\omega, \cdot))_\omega \leq D(L),$$

and

$$M^- = M_{ip}: \quad m^- : \Omega \times E^0 \oplus E^- \to E^+, \quad \mathrm{Lip}(m^-(\omega, \cdot))_\omega \leq D(L).$$

We now define for each ω and $x^0 \in E^0$ the operator

$$T_{\omega, x^0} : (E^+ \oplus E^-, \|\cdot\|_\omega) \to (E^+ \oplus E^-, \|\cdot\|_\omega)$$

by

$$T_{\omega,x^0}(x^+, x^-) := (m^-(\omega, x^0, x^-), m^+(\omega, x^+, x^0)).$$

We check under which condition this is a contraction. Using the Lipschitz-continuity of m^- and m^+, we obtain

$$\begin{aligned}
&\|T_{\omega,x^0}(x^+, x^-) - T_{\omega,x^0}(y^+, y^-)\|_\omega \\
&\leq \|m^-(\omega, x^0, x^-) - m^-(\omega, x^0, y^-)\|_\omega \\
&\quad + \|m^+(\omega, x^+, x^0) - m^+(\omega, y^+, x^0)\|_\omega \\
&\leq D(L)\left(\|x^- - y^-\|_\omega + \|x^+ - y^+\|_\omega\right) \\
&= D(L)\|(x^+, x^-) - (y^+, y^-)\|_\omega.
\end{aligned}$$

Hence T_{ω,x^0} will be a contraction if $D(L) < 1$, equivalently $L^2 - 3\delta L + \delta^2 > 0$. Since we also need $L < \delta/2$, the latter is the case if and only if

$$L < \frac{3 - \sqrt{5}}{2}\delta.$$

Under this condition there exists a unique fixed point

$$(x_*^+, x_*^-) = m^0(\omega, x^0) = m^{0,+}(\omega, x^0) \oplus m^{0,-}(\omega, x^0) \in E^+ \oplus E^-$$

which depends measurably on (ω, x^0). By definition of T_{ω,x^0}, this fixed point has the properties

$$m^{0,+}(\omega, x^0) = m^-(\omega, x^0, m^{0,-}(\omega, x^0)) \tag{7.3.52}$$

and

$$m^{0,-}(\omega, x^0) = m^+(\omega, m^{0,+}(\omega, x^0), x^0). \tag{7.3.53}$$

(7.3.52) means that the point $x^0 \oplus m^0(\omega, x^0) \in \mathbb{R}^d$ lies on $M^-(\omega)$ and (7.3.53) means that it lies on $M^+(\omega)$. Thus

$$\begin{aligned}
M^0(\omega) &:= M_{ij}(\omega) = \{x^0 \oplus m^0(\omega, x^0) : x^0 \in E^0\} \\
&= M^+(\omega) \cap M^-(\omega) = \{x \in \mathbb{R}^d : \varphi(\cdot, \omega, x) \in X_{a_j^-,\omega} \cap X_{b_i^+,\omega}\}.
\end{aligned}$$

Clearly $m^0(\omega, 0) = 0$ by uniqueness.

We will now calculate the Lipschitz constant of $m^0(\omega, \cdot)$. Let $x^0, y^0 \in E^0$. Since the Lipschitz constants of $m^+(\omega, \cdot)$ and $m^-(\omega, \cdot)$ are known to be $D(L)$, we obtain

$$\begin{aligned}
&\|m^{0,+}(\omega, x^0) - m^{0,+}(\omega, y^0)\|_\omega \\
&\leq \|m^-(\omega, x^0, m^{0,-}(\omega, x^0)) - m^-(\omega, y^0, m^{0,-}(\omega, y^0))\|_\omega \\
&\leq D(L)\left(\|x^0 - y^0\|_\omega + \|m^{0,-}(\omega, x^0) - m^{0,-}(\omega, y^0)\|_\omega\right),
\end{aligned}$$

and

$$\|m^{0,-}(\omega,x^0) - m^{0,-}(\omega,y^0)\|_\omega$$
$$\leq \|m^+(\omega,m^{0,+}(\omega,x^0),x^0) - m^+(\omega,m^{0,+}(\omega,y^0),y^0)\|_\omega$$
$$\leq D(L)\left(\|x^0 - y^0\|_\omega + \|m^{0,+}(\omega,x^0) - m^{0,+}(\omega,y^0)\|_\omega\right).$$

Addition of both estimates yields
$$\|m^0(\omega,x^0) - m^0(\omega,y^0)\|_\omega$$
$$\leq D(L)\left(2\|x^0 - y^0\|_\omega + \|m^0(\omega,x^0) - m^0(\omega,y^0)\|_\omega\right).$$

Since we made sure that $D(L) < 1$, we finally obtain
$$\|m^0(\omega,x^0) - m^0(\omega,y^0)\|_\omega \leq \frac{2D(L)}{1-D(L)}\|x^0 - y^0\|_\omega.$$

We leave the proof of the invariance of $M^0(\omega)$ as an exercise to the reader.
□

7.3.15 Remark. (i) If $F(\omega,x) \equiv 0$, then $M_{ij}(\omega) = E_{ij}$ with trivial graph $m_{ij}(\omega,x_{ij}) \equiv 0$. In this case the result is true even for $\delta = 0$.

(ii) For $i = j$ we obtain an invariant manifold $M_i(\omega)$ "tangent" to the Oseledets space E_i, which we call *Oseledets manifold*.

(iii) Note that the restriction on the Lipschitz constant of F is less severe for the construction of the unstable and stable manifolds than it is for the center manifolds. ∎

We now combine Corollaries 7.3.8 and 7.3.11 and obtain a more precise dynamical description of the orbits on the center manifold (see Fig. 7.7).

7.3.16 Corollary. *Let the conditions of Theorem 7.3.14 be satisfied. Choose $x \in M_{ij}(\omega)$.*

Then for any choice of a_j and b_i such that $a_j \leq \alpha_j - \delta < \beta_i + \delta \leq b_i$

$$\frac{1}{C}a_j^n\|x\|_\omega \leq \|\varphi(n,\omega,x)\|_{\theta^n\omega} \leq Cb_i^n\|x\|_\omega, \quad n \geq 0,$$

and

$$\frac{1}{C}b_i^n\|x\|_\omega \leq \|\varphi(n,\omega,x)\|_{\theta^n\omega} \leq Ca_j^n\|x\|_\omega, \quad n \leq 0,$$

where $C = \frac{\delta-L}{\delta-2L} > 1$.

7.3.4 The Continuous Time Case

Our program is completed if the RDS has discrete time. However, the continuous time case can easily be reduced to the discrete time case.

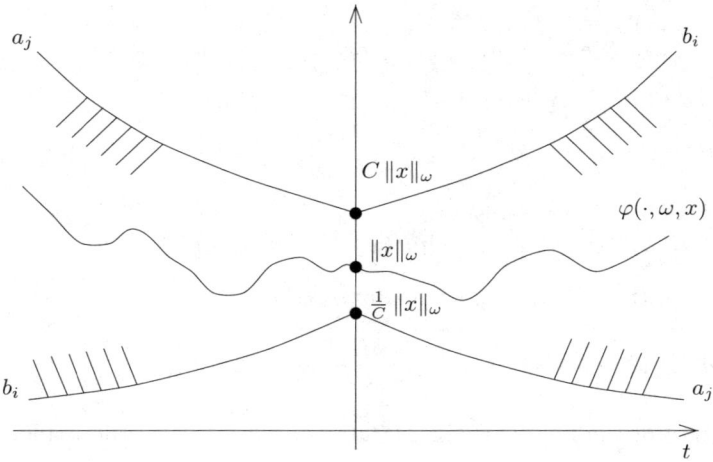

Fig. 7.7. Dynamical description of an orbit in M_{ij}

7.3.17 Theorem (Global Center Manifold Theorem, Continuous Time). Let

$$\varphi(t, \omega, x) = \Phi(t, \omega)x + \psi(t, \omega, x), \quad t \in \mathbb{R}, \tag{7.3.54}$$

be an RDS with $\psi(t, \omega, 0) = 0$ which is prepared according to Definition 7.2.5. Assume

$$\sup_{0 \leq t \leq 1} \|\psi(t, \omega, x) - \psi(t, \omega, y)\|_{\theta(t)\omega} \leq L\|x - y\|_{\omega}, \tag{7.3.55}$$

where

$$L < \frac{3 - \sqrt{5}}{2}\delta.$$

Then the invariant manifolds $M_{ij}(\omega)$ constructed for the embedded discrete time RDS

$$\varphi(n, \omega, x) = \Phi(n, \omega)x + \psi(n, \omega, x)$$

are also the invariant manifolds for the continuous time RDS (7.3.54). In particular,

$$\varphi(t, \omega)M_{ij}(\omega) = M_{ij}(\theta(t)\omega) \quad \text{for all } t \in \mathbb{R}$$

and

$$M_{ij}(\omega) = \{x \in \mathbb{R}^d : \varphi(\cdot, \omega, x) \in X_{b_i^+, \omega, \mathbb{R}^+} \cap X_{a_j^-, \omega, \mathbb{R}^-}\}.$$

Proof. (i) We first prove the invariance of $M_{ij}(\omega)$: First observe that

$$\sup_{0 \leq t \leq 1} \|\Phi(t, \omega)\|_{\omega, \theta(t)\omega} \leq e^{\lambda_1 + \kappa} =: c < \infty.$$

Consider $x \in M_{ij}(\omega)$. First take $n \in \mathbb{Z}^+$ and $0 \leq t \leq 1$. We write $\theta^n := \theta(n)$. By the cocycle property,

$$\begin{aligned}\varphi(n+t,\omega,x) &= \varphi(n,\theta(t)\omega,\varphi(t,\omega,x)) \\ &= \varphi(t,\theta^n\omega,\varphi(n,\omega,x)) \\ &= \Phi(t,\theta^n\omega)\varphi(n,\omega,x) + \psi(t,\theta^n\omega,\varphi(n,\omega,x)).\end{aligned}$$

Hence

$$\begin{aligned}\|\varphi(n,\theta(t)\omega,\varphi(t,\omega,x))\|_{\theta^n\theta(t)\omega} &\leq (c+L)\|\varphi(n,\omega,x)\|_{\theta^n\omega} \\ &\leq (c+L)b_i^n\|\varphi(\cdot,\omega,x)\|_{b_i^+,\omega,\mathbb{Z}^+}.\end{aligned}$$

Analogously, for $n \in \mathbb{Z}^-$,

$$\|\varphi(n,\theta(t)\omega,\varphi(t,\omega,x))\|_{\theta^n\theta(t)\omega} \leq (c+L)a_j^n\|\varphi(\cdot,\omega,x)\|_{a_j^-,\omega,\mathbb{Z}^-}.$$

By the discrete time dynamical characterization of $M_{ij}(\omega)$,

$$\varphi(t,\omega,x) \in M_{ij}(\theta(t)\omega) \quad \text{for all } t \in [0,1].$$

Represent a general $t \in \mathbb{R}$ as $t = [t] + s$, $0 \leq s < 1$. If $\varphi(s,\omega,x) \in M(\theta(s)\omega)$ then

$$\varphi(t,\omega,x) = \varphi([t],\theta(s)\omega,\varphi(s,\omega,x)) \in M_{ij}(\theta^{[t]}\theta(s)\omega) = M_{ij}(\theta(t)\omega),$$

by the discrete time invariance.

So far we have proved that

$$\varphi(t,\omega)M_{ij}(\omega) \subset M_{ij}(\theta(t)\omega) \quad \text{for all } t \in \mathbb{R}.$$

Since $\varphi(t,\omega)$ is a homeomorphism of $\mathbb{R}^d$, the invariance of $M_{ij}(\omega)$ follows as in the proof of Theorem 7.3.1(iii).

(ii) We finally prove the continuous time dynamical characterization. It suffices to show that

$$\sup_{n\in\mathbb{Z}^+} b_i^{-n}\|\varphi(n,\omega,x)\|_{\theta^n\omega} < \infty \iff \sup_{t\in\mathbb{R}^+} b_i^{-t}\|\varphi(t,\omega,x)\|_{\theta(t)\omega} < \infty \quad (7.3.56)$$

and the analogous statement on $\mathbb{Z}^-/\mathbb{R}^-$.

Now for $n \in \mathbb{Z}^+$ and $0 \leq t \leq 1$, as above,

$$b_i^{-(n+t)}\|\varphi(n+t,\omega,x)\|_{\theta^n\theta(t)\omega} \leq (c+L)B_i b_i^{-n}\|\varphi(n,\omega,x)\|_{\theta^n\omega},$$

where

$$B_i := \sup_{0\leq t\leq 1} b_i^{-t} = \max(1,\frac{1}{b_i}).$$

This means that (7.3.56) holds, which finishes the proof. $\square$

7.3.5 Higher Regularity

Main Result. Structure of Proof

Consider a prepared RDS with discrete time given by

$$\varphi(n,\omega,x) = \Phi(n,\omega)x + \psi(n,\omega,x), \quad n \in \mathbb{Z},$$

where $\psi(n,\omega,0) = 0$, equivalently, generated by the random difference equation

$$x_{n+1} = A(\theta^n \omega)x_n + F(\theta^n \omega, x_n), \quad n \in \mathbb{Z}, \quad x_0 = x, \quad (7.3.57)$$

where

$$A(\omega) = \Phi(1,\omega), \quad F(\omega,x) := \psi(1,\omega,x), \quad F(\omega,0) = 0.$$

In the previous subsections we have constructed the global center manifolds M_{ij} of φ "tangent" to E_{ij}, provided the nonlinear part F has a Lipschitz constant $\text{Lip}(F(\omega,\cdot))_{\omega,\theta\omega} \leq L$ with L "small enough". "Tangent" to E_{ij} meant so far that $M_{ij}(\omega) = \{x_{ij} \oplus m_{ij}(\omega, x_{ij}) : x_{ij} \in E_{ij}\}$ is a Lipschitz-continuous graph in the $E_{ij} \oplus E_{ij}^\perp$ coordinate system of $\mathbb{R}^d$ with a "small" Lipschitz constant.

Suppose our prepared RDS is such that $\varphi(1,\omega,\cdot) \in \text{Diff}^k(\mathbb{R}^d)$ for some $k \geq 1$, in particular, $F(\omega,\cdot)$ is C^k and $A(\omega) = D\varphi(1,\omega,0)$, hence $DF(\omega,0) = 0$. Our method does not allow for using this additional information to prove that $M_{ij}(\omega)$ is more regular than Lipschitz, and is in fact tangent to E_{ij} in the classical sense, i.e. $T_0 M_{ij}(\omega) = E_{ij}$, equivalently $Dm_{ij}(\omega,0) = 0$.

We will now show that the invariant manifolds M_{ij} are indeed C^k if the RDS is – for the case $k > 1$ under the provision that the spectral gaps to the left and to the right of the block $\Lambda_{ij} = \{\lambda_i > \ldots > \lambda_j\}$ are "wide enough" (gap condition). Siegmund [317] gives a detailed and clear exposition for the case of nonautonomous deterministic differential equations.

7.3.18 Definition. Let $F : \Omega \times \mathbb{R}^d \to \mathbb{R}^d$ be a measurable function.

(i) If $F(\omega,\cdot)$ is continuous, we say that $F(\omega,\cdot) \in \mathcal{C}_{b,\omega}^0$ if there is a finite random variable $L_0(\omega) \geq 0$ such that for all ω

$$\|F(\omega,\cdot)\|_\omega^0 := \sup_{x \in \mathbb{R}^d} \|F(\omega,x)\|_{\theta\omega} \leq L_0(\omega).$$

(ii) Let $F(\omega,\cdot)$ be C^k for some $k \geq 1$. We say that

$$F(\omega,\cdot) \in \mathcal{C}_{b,\omega}^k := \mathcal{C}_b^k((\mathbb{R}^d, \|\cdot\|_\omega), (\mathbb{R}^d, \|\cdot\|_{\theta\omega}))$$

if for each q with $1 \leq q \leq k$ there is a finite random variable $L_q(\omega) \geq 0$ such that for all ω

$$\|F(\omega,\cdot)\|_\omega^q := \sup_{x \in \mathbb{R}^d} \|D^q F(\omega,x)\|_{\omega,\theta\omega} \leq L_q(\omega).$$

Here $D^q F(\omega,x)$ is the q-th derivative of $F(\omega,\cdot)$ at x,

$$D^q F(\omega, x) \in \mathcal{L}^{(q)}((\mathbb{R}^d \times \cdots \times \mathbb{R}^d, \|\cdot\|_\omega), (\mathbb{R}^d, \|\cdot\|_{\theta\omega})) =: \mathcal{L}^{(q)},$$

where $\mathcal{L}^{(q)}$ is the space of continuous q-linear operators endowed with the operator norm, and $\mathbb{R}^d \times \cdots \times \mathbb{R}^d$ has norm $\|x\|_\omega := \sup_{1 \leq i \leq q} \|x_i\|_\omega$. We have $\mathcal{L}^{(1)} = \mathcal{L}((\mathbb{R}^d, \|\cdot\|_\omega), (\mathbb{R}^d, \|\cdot\|_{\theta\omega}))$, $\mathcal{L}^{(2)} := \mathcal{L}(\mathbb{R}^d \times \mathbb{R}^d, \mathbb{R}^d) \cong \mathcal{L}(\mathbb{R}^d, \mathcal{L}^{(1)})$ and $\mathcal{L}^{(q)} \cong \mathcal{L}(\mathbb{R}^d, \mathcal{L}^{(q-1)})$ in general.

(iii) The space
$$\mathcal{C}_b^k((E_{ij}, \|\cdot\|_\omega), (E_{ij}^\perp, \|\cdot\|_\omega))$$

is defined analogously as the space of random functions between the corresponding spaces for which the derivatives of order $1 \leq q \leq k$ are bounded by finite random variables. ∎

Suppose $F(\omega, \cdot) \in \mathcal{C}_{b,\omega}^1$ and there is a deterministic constant L_1 such that

$$\|F(\omega, \cdot)\|_\omega^1 = \sup_{x \in \mathbb{R}^d} \|DF(\omega, x)\|_{\omega, \theta\omega} \leq L_1.$$

By the mean value theorem, this implies

$$\|F(\omega, x) - F(\omega, y)\|_{\theta\omega} \leq L_1 \|x - y\|_\omega.$$

Hence the Unstable and Stable Manifold Theorems 7.3.1 and 7.3.10 hold provided $L_1 < \delta/2$, while the Center Manifold Theorem 7.3.14 holds provided $L_1 < \frac{3-\sqrt{5}}{2}\delta$.

7.3.19 Theorem (Global Smooth Invariant Manifold Theorem).

(A) Discrete time case: Let $k \geq 1$ and let φ be a prepared RDS with discrete time $\mathbb{T} = \mathbb{Z}$ such that $\varphi(1, \omega, \cdot) \in \text{Diff}^k(\mathbb{R}^d)$. Let the generating difference equation be

$$x_{n+1} = A(\theta^n \omega) x_n + F(\theta^n \omega, x_n), \quad n \in \mathbb{Z},$$

where $A(\omega) = D\varphi(1, \omega, 0)$, $F(\omega, 0) = 0$ and $DF(\omega, 0) = 0$. Assume that $F(\omega, \cdot) \in \mathcal{C}_{b,\omega}^k$ and that there is a deterministic constant L_1 such that

$$\|F(\omega, \cdot)\|_\omega^1 := \sup_{x \in \mathbb{R}^d} \|DF(\omega, x)\|_{\omega, \theta\omega} \leq L_1,$$

where the bound L_1 satisfies

$$L_1 < \delta_1 := \frac{3 - \sqrt{5}}{2} \delta. \tag{7.3.58}$$

Consider for any i, j with $1 \leq i \leq j \leq p$ the center manifold

$$M_{ij}(\omega) = \{x_{ij} \oplus m_{ij}(\omega, x_{ij}) : x_{ij} \in E_{ij}\}$$

corresponding to the interval $\Lambda_{ij} = \{\lambda_i > \ldots > \lambda_j\}$ from the Lyapunov spectrum of Φ.

Then the following assertions hold:
(i) $M_{ij}(\omega)$ is a C^1 manifold, more precisely

$$m_{ij}(\omega, \cdot) \in \mathcal{C}_b^1((E_{ij}, \|\cdot\|_\omega), (E_{ij}^\perp, \|\cdot\|_\omega))$$

with

$$\|m_{ij}(\omega,\cdot)\|_\omega^1 := \sup_{x_{ij} \in E_{ij}} \|Dm_{ij}(\omega, x_{ij})\|_{\omega,\omega} \leq 2\frac{\delta - L_1}{\delta - 2L_1}.$$

(ii) The manifold $M_{ij}(\omega)$ is tangent to E_{ij}, i. e. $T_0 M_{ij}(\omega) = E_{ij}$, equivalently $Dm_{ij}(\omega, 0) = 0$.

(iii) Higher regularity: Suppose $k \geq 2$ and the following gap conditions are satisfied:

1. *The spectral gap $\Gamma_{\text{right}} := [\beta_i + \delta, \alpha_{i-1} - \delta]$ to the right of Λ_{ij} is wide enough such that we can choose two numbers $b, \bar{b} \in \Gamma_{\text{right}}$ with $b < \bar{b}$ for which, moreover, also $b^q < \bar{b}$ for every $q = 2, \ldots, k$ (no condition for $i = 1$, i. e. for the unstable manifolds).*
2. *The spectral gap $\Gamma_{\text{left}} := [\beta_{j+1} + \delta, \alpha_j - \delta]$ to the left of Λ_{ij} is wide enough such that we can choose two numbers $a, \bar{a} \in \Gamma_{\text{left}}$ with $\bar{a} < a$ for which, moreover, also $\bar{a} < a^q$ for every $q = 2, \ldots, k$ (no condition for $j = p$, i. e. for the stable manifolds).*

Then there exists a constant $\delta_k \in (0, \delta_1]$ such that if $L_1 < \delta_k$, $M_{ij}(\omega)$ is a C^k manifold, more precisely,

$$m_{ij}(\omega, \cdot) \in \mathcal{C}_b^k((E_{ij}, \|\cdot\|_\omega), (E_{ij}^\perp, \|\cdot\|_\omega)).$$

In particular, if for $1 \leq q \leq k$, $D^q F(\omega, 0) = 0$, then also $D^q m_{ij}(\omega, 0) = 0$.

(B) Continuous time: The statements in part (A) are also valid for a prepared continuous time RDS

$$\varphi(t, \omega, x) = \Phi(t, \omega)x + \psi(t, \omega, x), \quad t \in \mathbb{R},$$

where $\varphi(t, \omega, \cdot) \in \text{Diff}^k(\mathbb{R}^d)$, $\Phi(t, \omega) = D\varphi(t, \omega, 0)$, $\psi(t, \omega, 0) = 0$ and $D\psi(t, \omega, 0) = 0$, provided $\psi(\cdot, \omega, \cdot) \in \mathcal{C}_{b,\omega}^k$, where now for $1 \leq q \leq k$

$$\|\psi(\cdot, \omega, \cdot)\|_\omega^q := \sup_{0 \leq t \leq 1} \sup_{x \in \mathbb{R}^d} \|D^q \psi(t, \omega, x)\|_{\omega, \theta(t)\omega},$$

and there is a deterministic constant L_1 such that

$$\|\psi(\cdot, \omega, \cdot)\|_\omega^1 := \sup_{0 \leq t \leq 1} \sup_{x \in \mathbb{R}^d} \|D\psi(t, \omega, x)\|_{\omega, \theta(t)\omega} \leq L_1$$

and L_1 satisfies the condition $L_1 < \delta_k$ from (A).

7.3.20 Remark (When are the Gap Conditions Satisfied?). (i) If $\text{int}\,\Gamma_{\text{right}} \cap (0, 1] \neq \emptyset$ then the first gap condition is automatically satisfied for all $k \in \mathbb{N}$. In particular, the stable manifolds M_{ip} with $\lambda_i < 0$ (in particular, the classical stable manifold M_s) are C^k (if the remaining conditions are met).

If $\text{int}\Gamma_{\text{left}} \cap [1,\infty) \neq \emptyset$ then the second gap condition is automatically satisfied for all $k \in \mathbb{N}$. In particular, the unstable manifolds M_{1j} with $\lambda_j > 0$ (in particular, the classical unstable manifold M_u) are C^k (if the remaining conditions are met).

(ii) Let some Lyapunov exponent vanish, $\lambda_i = 0$, and consider the classical center manifold $M_c = M_{i,i}$ tangent to E_i. Then $\alpha = \alpha_i = e^{-\kappa} < 1 < e^{\kappa} = \beta_i = \beta$, and $\Gamma_{\text{right}} = [\beta + \delta, \alpha_{i-1} - \delta]$, $\Gamma_{\text{left}} = [\beta_{i+1} + \delta, \alpha - \delta]$. It will hence not always be possible to satisfy the gap conditions for a given $k \geq 2$. ∎

Part (B) of Theorem 7.3.19 is a by-product of Theorem 7.3.17. We hence can restrict ourselves to the proof of the discrete time case, part (A).

We will prove Theorem 7.3.19(A) only for the unstable manifolds $M^+ = M_{1j}$ for $j < p$. There is a similar "dual" proof for the stable manifolds $M^- = M_{ip}$, hence the smoothness result for M_{ij} follows from $M_{ij} = M_{1j} \cap M_{ip}$.

The reader should recall the notations of Subsect. 7.3.1, in particular Λ^+, E^+, x^+, E^-, x^-, $\alpha := \alpha_j$, $\beta := \beta_{j+1}$ and
$$a := a_j \in \Gamma_{\text{left}} = [\beta + \delta, \alpha - \delta].$$

Our random difference equation (7.3.57) takes the form of the following coupled pair of the so-called unstable and stable equation (see (7.3.4) and (7.3.5)):

$$x_{n+1}^+ = A^+(\theta^n\omega)x_n^+ + F^+(\theta^n\omega, x_n^+, x_n^-), \quad x_0^+ = x^+, \tag{7.3.59}$$

$$x_{n+1}^- = A^-(\theta^n\omega)x_n^- + F^-(\theta^n\omega, x_n^+, x_n^-), \quad x_0^- = x^-. \tag{7.3.60}$$

The corresponding linear cocycles satisfy

$$\|\Phi^+(n,\omega)\|_{\omega,\theta^n\omega} \leq \alpha^n, \quad n \leq 0, \quad \|\Phi^-(n,\omega)\|_{\omega,\theta^n\omega} \leq \beta^n, \quad n \geq 0.$$

Our assumption (7.3.58) assures that Theorem 7.3.1 holds. We now give an operator formulation of its proof.

Operator formulation of the construction of the unstable manifold: In order to find the invariant manifold M^+ we have solved the following equation (which for convenience we write down for all $n \in \mathbb{Z}$) in $X_{a^-,\omega}$, given $\omega \in \Omega$ and $x^+ \in E^+$:

$$\xi^+(n,\omega) = \Phi^+(n,\omega)x^+ + \begin{cases} \sum_{i=1}^{n} \Phi^+(n-i, \theta^i\omega)F_{i,\omega}^+, & n > 0, \\ 0, & n = 0, \\ -\sum_{i=n+1}^{0} \Phi^+(n-i, \theta^i\omega)F_{i,\omega}^+, & n < 0, \end{cases}$$

$$\xi^-(n,\omega) = \sum_{i=-\infty}^{n} \Phi^-(n-i, \theta^i\omega)F^-(\theta^{i-1}\omega, \xi(i-1,\omega)),$$

where for reasons of display we have used the abbreviation

$$F^+_{i,\omega} := F^+(\theta^{i-1}\omega, \xi(i-1,\omega))$$

in the (+) equation.

This equation is derived from steps 1 and 2 of our 6-step program of Subsect. 7.3.1, where we have inserted the explicit solution of the initial value problem of the unstable equation (7.3.59) and the explicit form (7.3.14) of the unique solution in $X_{a^-,\omega}$ of the stable equation (7.3.60) (see First Key Technical Lemma 7.3.3). The graph of the invariant manifold is then obtained as $m^+(\omega, x^+) = \xi^-(0,\omega)$.

We write the above equation in the following succinct operator form:

$$\xi(\omega) = S(\omega)x^+ + K(\omega)N_F(\omega)\xi(\omega), \quad \xi(\omega) \in X_{a^-,\omega}. \tag{7.3.61}$$

The operators $S(\omega)$, $K(\omega)$ and $N_F(\omega)$ are defined as follows:

$$S(\omega) : E^+ \to X_{a^-,\omega}, \quad x^+ \mapsto \xi(\omega) := \begin{pmatrix} \Phi^+(\cdot,\omega)x^+ \\ 0 \end{pmatrix},$$

$$K(\omega) : X_{a^-,\omega} \to X_{a^-,\omega},$$

where (again for $n \in \mathbb{Z}$, but used only for $n \leq 0$)

$$K(\omega)\xi(\omega)(n) = \begin{pmatrix} \begin{cases} \sum_{i=1}^n \Phi^+(n-i,\theta^i\omega)\xi^+(i,\omega), & n > 0, \\ 0, & n = 0, \\ -\sum_{i=n+1}^0 \Phi^+(n-i,\theta^i\omega)\xi^+(i,\omega), & n < 0, \end{cases} \\ \sum_{i=-\infty}^n \Phi^-(n-i,\theta^i\omega)\xi^-(i,\omega) \end{pmatrix}.$$

Because of its importance for the study of regularity, the last operator $N_F(\omega)$ deserves a name of its own.

7.3.21 Definition (Nemytskii Operator). (i) Let $f(n,\cdot) : \mathbb{R}^d \to \mathbb{R}^d$ be a sequence of functions. The *Nemytskii operator* corresponding to this sequence is defined as

$$N_f : (\mathbb{R}^d)^{\mathbb{Z}} \to (\mathbb{R}^d)^{\mathbb{Z}}, \quad N_f(\xi)_{n+1} := f(n, \xi_n).$$

(ii) For the case $f(n,x) := F(\theta^n\omega, x)$ the Nemytskii operator $N_F(\omega)$ corresponding to the random function F is defined by $N_F(\omega)(\xi)_{n+1} := F(\theta^n\omega, \xi_n)$. Clearly $N_F(\omega)0 = 0$ if $F(\omega, 0) = 0$. ∎

Note that the Nemytskii operator $N_F(\omega)$ is solely determined by the nonlinear part F of the RDS.

In the language of these operators Proposition 7.3.7 states that for each ω and x^+ equation (7.3.61) has a unique solution $\xi(\omega, x^+) \in X_{a^-,\omega}$ which depends measurably on (ω, x^+). In other words, the operator $I - K(\omega)N_F(\omega)$ is invertible in $X_{a^-,\omega}$ and the unique solution can be represented as

$$\xi(\omega, x^+) = (I - K(\omega)N_F(\omega))^{-1}S(\omega)x^+. \tag{7.3.62}$$

We first prove that the definitions of $S(\omega)$ and $K(\omega)$ make sense.

7.3 Global Invariant Manifolds

7.3.22 Lemma. *(i) The operator $S(\omega) : E^+ \to X_{\gamma-,\omega}$ is a bounded linear operator for any $\gamma \leq \alpha$ (in particular, for $\gamma = a \in \Gamma_{\text{left}}$) and*

$$\|S(\omega)\|_{\gamma-,\omega} \leq 1.$$

(ii) The operator $K(\omega) : X_{\gamma-,\omega} \to X_{\gamma-,\omega}$ is a bounded linear operator for any $\gamma \in (\beta, \alpha)$ (in particular, for $\gamma = a \in \Gamma_{\text{left}}$) and

$$\|K(\omega)\|_{\gamma-,\omega} \leq \max\left(\frac{\gamma}{\alpha-\gamma}, \frac{\gamma}{\gamma-\beta}\right),$$

in particular

$$\|K(\omega)\|_{a-,\omega} \leq \frac{\alpha-\delta}{\delta} \quad \text{for all } a \in \Gamma_{\text{left}}.$$

Proof. (i) Since by Lemma 7.2.3(ii), $X_{\alpha-,\omega} \hookrightarrow X_{\gamma-,\omega}$ for $\gamma \leq \alpha$ and the inclusion mapping is a bounded linear operator with norm ≤ 1, it suffices to prove the claim for $\gamma = \alpha$.

$S(\omega)x^+(n) = \Phi^+(n,\omega)x^+$ is definitely linear. For $n \leq 0$

$$\|S(\omega)x^+(n)\|_{\theta^n\omega} \leq \|\Phi^+(n,\omega)\|_{\omega,\theta^n\omega}\|x^+\|_\omega \leq a^n\|x^+\|_\omega,$$

whence $\|S(\omega)\|_{\alpha-,\omega} \leq 1$.

(ii) First observe that

$$\|K(\omega)\xi\|_{\gamma-,\omega} \leq \|(K(\omega)\xi)^+\|_{\gamma-,\omega} + \|(K(\omega)\xi)^-\|_{\gamma-,\omega}.$$

Let $n < 0$.

(a) We first consider the (+) component: We have

$$\left\|\sum_{i=n+1}^{0} \Phi^+(n-i,\theta^i\omega)\xi^+(i,\omega)\right\|_{\theta^n\omega} \leq \sum_{i=n+1}^{0} a^{n-i}\|\xi^+(i,\omega)\|_{\theta^i\omega}.$$

Hence for $\gamma < \alpha$,

$$\gamma^{-n}\|\cdot\|_{\theta^n\omega} \leq \sum_{i=n+1}^{0} \left(\frac{\alpha}{\gamma}\right)^{n-i} \gamma^{-i}\|\xi^+(i,\omega)\|_{\theta^i\omega}.$$

It follows that

$$\|(K(\omega)\xi(\omega))^+\|_{\gamma-,\omega} \leq \frac{\gamma}{\alpha-\gamma}\|\xi^+(\omega)\|_{\gamma-,\omega}.$$

(b) We now consider the (−) component: We have for $\beta < \gamma$

$$\gamma^{-n}\left\|\sum_{i=-\infty}^{n} \Phi^-(n-i,\theta^i\omega)\xi^-(i,\omega)\right\|_{\theta^n\omega} \leq \sum_{i=-\infty}^{n} \left(\frac{\beta}{\gamma}\right)^{n-i} \gamma^{-i}\|\xi^-(i,\omega)\|_{\theta^i\omega},$$

whence

352   Chapter 7. Invariant Manifolds

$$\|(K(\omega)\xi(\omega))^-\|_{\gamma^-,\omega} \leq \frac{\gamma}{\gamma-\beta}\|\xi^-(\omega)\|_{\gamma^-,\omega}.$$

Altogether, for $\beta < \gamma < \alpha$,

$$\|K(\omega)\xi(\omega)\|_{\gamma^-,\omega} \leq \max\left(\frac{\gamma}{\alpha-\gamma},\frac{\gamma}{\gamma-\beta}\right)\|\xi(\omega)\|_{\gamma^-,\omega}.$$

For any $\gamma = a \in \Gamma_{\text{left}} = [\beta + \delta, \alpha - \delta]$, $(\alpha - \delta)/\delta$ is a uniform bound for the norm. □

The Nemytskii Operator

Since bounded linear operators are C^∞, equation (7.3.61) indicates that the smoothness of the Nemytskii operator $N_F(\omega)$ determines the smoothness of the unstable manifold $M^+(\omega)$.

Indeed, if $N_F(\omega) : X_{a^-,\omega} \to X_{a^-,\omega}$ were C^k, the implicit function theorem would imply that the right-hand side, hence the left-hand side of (7.3.61), hence $x^+ \mapsto \xi^-(\omega, x^+)(0) = m^+(\omega, x^+)$ is C^k.

However, even if F is C^k, $N_F(\omega)$ is in general not differentiable (for a deterministic counterexample see Siegmund [317: Sect. 5.2]). The way out is to play with the scale parameter $a \in \Gamma_{\text{left}}$ and consider $N_F(\omega)$ as an operator from $X_{a^-,\omega}$ into a bigger space $X_{b^-,\omega}$, where $b \in \Gamma_{\text{left}}$, but $b < a$. For $k > 1$ this is the origin of the gap condition.

7.3.23 Proposition (Regularity of Nemytskii Operator). *(i) Continuity: Let $F(\omega, \cdot) \in \mathcal{C}^0_{b,\omega}$ with deterministic bound L_0. Then for all $a > 0$ and $c \leq 1$, the operator $N_F(\omega) : X_{a^-,\omega} \to X_{c^-,\omega}$ is well-defined, and is continuous for $c < 1$.*

(ii) Lipschitz continuity: Let $\text{Lip}(F(\omega, \cdot))_{\omega,\theta\omega} \leq L_1$ for some deterministic constant L_1. Then for all $0 < b \leq a$, $N_F(\omega) : X_{a^-,\omega} \to X_{b^-,\omega}$ is Lipschitz continuous with

$$\|N_F(\omega)(\xi) - N_F(\omega)(\eta)\|_{b^-,\omega} \leq \frac{L_1}{a}\|\xi - \eta\|_{a^-,\omega}.$$

(iii) Differentiability: Let $F(\omega, \cdot) \in \mathcal{C}^1_{b,\omega}$ with deterministic bound L_1. Then for all $0 < b < a$, $N_F(\omega) : X_{a^-,\omega} \to X_{b^-,\omega}$ is C^1, more precisely $N_F(\omega) \in \mathcal{C}^1_b(X_{a^-,\omega}, X_{b^-,\omega})$, with

$$(DN_F(\omega)(\xi)\eta)_{n+1} = DF(\theta^n\omega, \xi_n)\eta_n, \quad \xi = (\xi_n),\, \eta = (\eta_n) \in X_{a^-,\omega},$$

and

$$\|DN_F(\omega)(\xi)\|_{L(X_{a^-,\omega}, X_{b^-,\omega})} \leq \frac{1}{a}\|DN_F(\omega)(\xi)\|_{\left(\frac{b}{a}\right)^-,\omega}$$

$$\leq \frac{1}{a}\|F(\omega,\cdot)\|^1_{b,\omega} \leq \frac{L_1}{a},$$

where on the right-hand side of the first inequality, $DN_F(\omega)(\xi)$ is interpreted as an $\mathcal{L}(\mathbb{R}^d)$-valued sequence in $X_{\left(\frac{b}{a}\right)^-,\omega}$.

7.3 Global Invariant Manifolds

(iv) C^k-smoothness: Let $F(\omega, \cdot) \in \mathcal{C}_{b,\omega}^k$ for some $k > 1$ with deterministic bound L_1 and random bounds $L_q(\omega)$ for $2 \leq q \leq k$. Choose a, b such that $0 < b < a^q$ for all $q = 1, \ldots, k$. Then $N_F(\omega) : X_{a^-,\omega} \to X_{b^-,\omega}$ is C^k, more precisely $N_F(\omega) \in \mathcal{C}_b^k(X_{a^-,\omega}, X_{b^-,\omega})$, with

$$D^q N_F(\omega)(\cdot) : X_{a^-,\omega} \to \mathcal{L}^{(q)}(X_{a^-,\omega} \times \ldots \times X_{a^-,\omega}, X_{b^-,\omega}) =: \mathcal{L}^{(q)}$$

given by

$$D^q N_F(\omega)(\xi)_{n+1} = D^q F(\theta^n \omega, \xi_n),$$

and we have

$$\|D^q N_F(\omega)(\xi)\|_{\mathcal{L}^{(q)}} \leq \frac{1}{a^q} \|D^q N_F(\omega)(\xi)\|_{(\frac{b}{a^q})^-,\omega} \leq \frac{1}{a^q} \|F(\omega, \cdot)\|_\omega^q \leq \frac{L_q(\omega)}{a^q}$$

for $1 \leq q \leq k$, where in the second term of the last inequalities, $D^q N_F(\omega)$ is interpreted as an $\mathcal{L}^{(q)}$-valued sequence in $X_{(\frac{b}{a^q})^-,\omega}$.

Proof. (i)(a) $N_F(\omega)$ well-defined: First note that since $\|F(\omega, \cdot)\|_\omega^0 \leq L_0$, for any sequence $\xi \in (\mathbb{R}^d)^\mathbb{Z}$

$$\|N_F(\omega)(\xi)\|_{1^-,\omega} = \sup_{n \leq 0} \|N_F(\omega)(\xi)_n\|_{\theta^n \omega} = \sup_{n \leq 0} \|F(\theta^{n-1}\omega, \xi_{n-1})\|_{\theta^n \omega}$$
$$\leq L_0 < \infty.$$

Thus

$$N_F(\omega) : (\mathbb{R}^d)^\mathbb{Z} \to X_{1^-,\omega} \hookrightarrow X_{c^-,\omega}, \quad 0 < c \leq 1,$$

and finally $N_F(\omega) : X_{a^-,\omega} \to X_{c^-,\omega}$ is well-defined for any $a > 0$.

(i)(b) Continuity: We want to prove continuity of $\xi \mapsto N_F(\omega)(\xi)$ at the arbitrary, but fixed reference point $\xi^0 \in X_{a^-,\omega}$.

For $\xi \in X_{a^-,\omega}$ and $N \in \mathbb{Z}^+$,

$$\|N_F(\omega)(\xi) - N_F(\omega)(\xi^0)\|_{c^-,\omega}$$
$$= \sup_{n \leq 0} c^{-n} \|F(\theta^{n-1}\omega, \xi_{n-1}) - F(\theta^{n-1}\omega, \xi_{n-1}^0)\|_{\theta^n \omega}$$
$$\leq \max\left(\max_{-N \leq n \leq 0} \|F(\theta^{n-1}\omega, \xi_{n-1}) - F(\theta^{n-1}\omega, \xi_{n-1}^0)\|_{\theta^n \omega}, 2c^N L_0\right),$$

where we have used the fact that $c \leq 1$.

Let $\varepsilon > 0$ be given. To control the infinitely long tail of the sequence $(F(\theta^n \omega, \xi_n))$, we now need to make the crucial assumption that $c < 1$, under which we can choose an $N \in \mathbb{Z}^+$ big enough such that $2c^N L_0 < \varepsilon$. The finitely many remaining terms are estimated by making use of the continuity of $F(\omega, \cdot)$.

Since $F(\omega, \cdot)$ is continuous, there exists a $\delta(\omega) > 0$ such that

$$\max_{-N \leq n \leq 0} \|F(\theta^{n-1}\omega, \xi_{n-1}) - F(\theta^{n-1}\omega, \xi_{n-1}^0)\|_{\theta^n \omega} < \varepsilon,$$

provided

354   Chapter 7. Invariant Manifolds

$$C(\omega) := \max_{-N \leq n \leq 0} \|\xi_{n-1} - \xi_{n-1}^0\|_{\theta^{n-1}\omega} < \delta(\omega).$$

Now
$$\begin{aligned}
\|\xi - \xi^0\|_{a-,\omega} &= \sup_{n \leq 0} a^{-n} \|\xi_n - \xi_n^0\|_{\theta^n\omega} \\
&\geq \sup_{-N-1 \leq n \leq -1} a^{-n} \|\xi_n - \xi_n^0\|_{\theta^n\omega} \\
&\geq C_1(a) C(\omega),
\end{aligned}$$

where $C_1(a) = \min(a^{N+1}, a) > 0$.

Summing up, for $\varepsilon > 0$ choose $\delta_1(\omega) := C_1(a)\delta(\omega)$. Then
$$\|\xi - \xi^0\|_{a-,\omega} < \delta_1(\omega) \Rightarrow \|N_F(\omega)(\xi) - N_F(\omega)(\xi^0)\|_{c-,\omega} < \varepsilon,$$

proving the continuity of $N_F(\omega) : X_{a-,\omega} \to X_{c-,\omega}$ for any $a > 0$ and $c < 1$, as claimed.

(ii) Lipschitz continuity: For $0 < b \leq a$
$$\begin{aligned}
\|N_F(\omega)(\xi) &- N_F(\omega)(\eta)\|_{b-,\omega} \\
&\leq \|N_F(\omega)(\xi) - N_F(\omega)(\eta)\|_{a-,\omega} \\
&= \sup_{n \leq 0} a^{-n} \|F(\theta^{n-1}\omega, \xi_{n-1}) - F(\theta^{n-1}\omega, \eta_{n-1})\|_{\theta^n\omega} \\
&\leq L_1 \sup_{n \leq 0} a^{-n} \|\xi_{n-1} - \eta_{n-1}\|_{\theta^{n-1}\omega} \\
&\leq \frac{L_1}{a} \|\xi - \eta\|_{a-,\omega}.
\end{aligned}$$

Since $N_F(\omega)(0) = 0$, it also follows that
$$\|N_F(\omega)(\xi)\|_{b-,\omega} \leq \frac{L_1}{a} \|\xi\|_{a-,\omega}.$$

(iii) Differentiability:

Step 1: Let again $0 < b \leq a$ and let $F(\omega, \cdot) \in \mathcal{C}_{b,\omega}^1$ with constant L_1. Hence $DF(\omega, \cdot) \in \mathcal{C}_{b,\omega}^0(\mathbb{R}^d, \mathcal{L}(\mathbb{R}^d))$. We can thus define a candidate for $DN_F(\omega)$, namely
$$N_F^{(1)}(\omega) : X_{a-,\omega}(\mathbb{R}^d) \to X_{\left(\frac{b}{a}\right)^-,\omega}(\mathcal{L}((\mathbb{R}^d, \|\cdot\|_\omega), (\mathbb{R}^d, \|\cdot\|_{\theta\omega}))),$$

where
$$\xi \mapsto N_F^{(1)}(\omega)(\xi) =: \eta, \quad \eta_{n+1} := DF(\theta^n \omega, \xi_n).$$

By part (i) of this theorem, $\xi \mapsto N_F^{(1)}(\omega)(\xi)$ is defined for all $0 < b \leq a$, and continuous for all $0 < b < a$. Further, again for all $0 < b \leq a$,
$$\|N_F^{(1)}(\omega)(\xi)\|_{\left(\frac{b}{a}\right)^-,\omega} \leq \|DF(\omega, \cdot)\|_{b,\omega}^0 \leq L_1. \tag{7.3.63}$$

Step 2: There is, however, a second interpretation of $N_F^{(1)}(\omega)(\xi)$ for fixed $\xi \in X_{a-,\omega}$, namely as a linear mapping between Banach spaces,

7.3 Global Invariant Manifolds

$$N_F^{(1)}(\omega)(\xi) : X_{a^-,\omega} \to X_{b^-,\omega},$$

defined by

$$\eta \mapsto N_F^{(1)}(\omega)(\xi)(\eta) =: \zeta, \quad \zeta_{n+1} := DF(\theta^n\omega, \xi_n)\eta_n.$$

We claim that for $b \leq a$, $N_F^{(1)}(\omega)(\xi) \in \mathcal{L}(X_{a^-,\omega}, X_{b^-,\omega})$ and, moreover,

$$\|N_F^{(1)}(\omega)(\xi)\|_{\mathcal{L}(X_{a^-,\omega}, X_{b^-,\omega})} \leq \frac{1}{a}\|N_F^{(1)}(\omega)(\xi)\|_{(\frac{b}{a})^-,\omega}. \tag{7.3.64}$$

Indeed,

$$\|N_F^{(1)}(\omega)(\xi)\eta\|_{b^-,\omega}$$
$$\leq \frac{1}{a}\sup_{n \leq 0}\left(\frac{b}{a}\right)^{-n}\|DF(\theta^{n-1}\omega, \xi_{n-1})\|_{\theta^{n-1}\omega,\theta^n\omega}a^{-n+1}\|\eta_{n-1}\|_{\theta^{n-1}\omega}$$
$$\leq \frac{\|\eta\|_{a^-,\omega}}{a}\|N_F^{(1)}(\omega)(\xi)\|_{(\frac{b}{a})^-,\omega}.$$

Step 3: Let now $0 < b < a$. We now prove that $N_F(\omega)(\cdot)$ is differentiable at any $\xi \in X_{a^-,\omega}$, and that $DN_F(\omega)(\xi) = N_F^{(1)}(\omega)(\xi)$.

Let $\xi, \eta \in X_{a^-,\omega}$, and $b \leq a$. By definition,

$$\|N_F(\omega)(\xi + \eta) - N_F(\omega)(\xi) - N_F^{(1)}(\omega)(\xi)\eta\|_{b^-,\omega}$$
$$= \sup_{n \leq 0} b^{-n}\|F(\theta^{n-1}\omega, \xi_{n-1} + \eta_{n-1}) - F(\theta^{n-1}\omega, \xi_{n-1})$$
$$\qquad - DF(\theta^{n-1}\omega, \xi_{n-1})\eta_{n-1}\|_{\theta^n\omega}$$
$$=: \sup_{n \leq 0} b^{-n}\|*\|_{\theta^n\omega}.$$

Since $F(\omega, \cdot) \in \mathcal{C}^1_{b,\omega}$ we can apply Taylor's formula in $\mathbb{R}^d$,

$$F(\omega, x + y) - F(\omega, x) = \left(\int_0^1 DF(\omega, x + ty)dt\right) y,$$

to obtain

$$\|F(\theta^{n-1}\omega, \xi_{n-1} + \eta_{n-1}) - F(\theta^{n-1}\omega, \xi_{n-1}) - DF(\theta^{n-1}\omega, \xi_{n-1})\eta_{n-1}\|_{\theta^n\omega}$$
$$= \left\|\int_0^1 \left(DF(\theta^{n-1}\omega, \xi_{n-1} + t\eta_{n-1}) - DF(\theta^{n-1}\omega, \xi_{n-1})\right) dt\, \eta_{n-1}\right\|_{\theta^n\omega}.$$

It follows that

$$\sup_{n\leq 0} b^{-n}\|*\|_{\theta^n\omega}$$

$$\leq \frac{\|\eta\|_{a^-,\omega}}{a} \int_0^1 \sup_{n\leq 0}\left(\frac{b}{a}\right)^{-n}$$

$$\|DF(\theta^{n-1}\omega, \xi_{n-1} + t\eta_{n-1}) - DF(\theta^{n-1}\omega, \xi_{n-1})\|_{\theta^{n-1}\omega, \theta^n\omega} dt$$

$$\leq \frac{\|\eta\|_{a^-,\omega}}{a} \sup_{0\leq t\leq 1} \|N_F^{(1)}(\omega)(\xi + t\eta) - N_F^{(1)}(\omega)(\xi)\|_{\left(\frac{b}{a}\right)^-,\omega}.$$

By step 1, $X_{a^-,\omega} \ni \xi \mapsto N_F^{(1)}(\omega)(\xi) \in X_{\left(\frac{b}{a}\right)^-,\omega}$ is continuous, provided $b < a$. That is, for $\varepsilon > 0$ there exists a $\delta > 0$ such that

$$\|N_F^{(1)}(\omega)(\xi + t\eta) - N_F^{(1)}(\omega)(\xi)\|_{\left(\frac{b}{a}\right)^-,\omega} < \varepsilon$$

whenever $\|\xi + t\eta - \xi\|_{a^-,\omega} \leq \|\eta\|_{a^-,\omega} < \delta$.

As a result, $N_F(\omega) : X_{a^-,\omega} \to X_{b^-,\omega}$ is differentiable for all $0 < b < a$ and $DN_F(\omega)(\xi) = N_F^{(1)}(\omega)(\xi)$. Further, since by step 1, $X_{a^-,\omega} \ni \xi \mapsto DN_F(\omega)(\xi) \in X_{\left(\frac{b}{a}\right)^-,\omega}$ is continuous, by (7.3.64), $X_{a^-,\omega} \ni \xi \mapsto DN_F(\omega)(\xi) \in \mathcal{L}(X_{a^-,\omega}, X_{b^-,\omega})$ is continuous. Finally, by (7.3.63) and (7.3.64), $N_F(\omega) \in \mathcal{C}_b^1(X_{a^-,\omega}, X_{b^-,\omega})$.

(iv) C^k-smoothness: By induction on $q = 1, \ldots, k$. □

Proof of the Smoothness of Invariant Manifolds

With the knowledge on the operators $S(\omega)$, $K(\omega)$ and $N_F(\omega)$ summarized in Lemma 7.3.22 and Proposition 7.3.23 we return to equation (7.3.61),

$$\xi(\omega) = S(\omega)x^+ + K(\omega)N_F(\omega)\xi(\omega).$$

7.3.24 Proposition. *Let* $\mathrm{Lip}(F(\omega,\cdot))_{\omega,\theta\omega} \leq L_1$ *and* $\gamma \in (\beta, \alpha)$.

(i) $K(\omega)N_F(\omega) : X_{\gamma^-,\omega} \to X_{\gamma^-,\omega}$ *is Lipschitz continuous with Lipschitz constant*

$$\mathrm{Lip}(K(\omega)\,N_F(\omega)) \leq l = \frac{L_1}{\min(\alpha - \gamma, \gamma - \beta)}.$$

In particular, for $\gamma = a \in \Gamma_{\mathrm{left}} = [\beta + \delta, \alpha - \delta]$, $l \leq \frac{L_1}{\delta}$.

(ii) *For* $l < 1$, $I - K(\omega)N_F(\omega) : X_{\gamma^-,\omega} \to X_{\gamma^-,\omega}$ *is a homeomorphism and its inverse*

$$\Psi(\omega) := (I - K(\omega)N_F(\omega))^{-1} : X_{\gamma^-,\omega} \to X_{\gamma^-,\omega}$$

has Lipschitz constant $\leq \frac{1}{1-l}$.

(iii) *The solution set* Σ^+ *of equation (7.3.61) in* $E^+ \times X_{\gamma^-,\omega}$ *is*

$$\{(x^+, \Psi(\omega)S(\omega)x^+) : x^+ \in E^+\}$$

and its projection onto $\mathbb{R}^d = E^+ \oplus E^-$,

$$M^+(\omega) = \{x^+ \oplus m^+(\omega, x^+) : x^+ \in E^+\},$$

is given by
$$m^+(\omega, x^+) = ((\Psi(\omega)S(\omega)x^+)(0))^-. \tag{7.3.65}$$

We also have
$$m^+(\omega, x^+) = \sum_{n=-\infty}^{0} \Phi^-(-n, \theta^n \omega) F^-(\theta^{n-1}\omega, (\Psi(\omega)S(\omega)x^+)_{n-1}).$$

Proof. (i) This immediately follows from Lemma 7.3.22(ii) and Proposition 7.3.23(ii) (putting $\gamma = b = a$).

(ii) If $\mathrm{Lip}(K(\omega)N_F(\omega)) \leq l < 1$, then $I - K(\omega)N_F(\omega)$ is invertible and

$$\Psi(\omega) = (I - K(\omega)N_F(\omega))^{-1} = \sum_{n=0}^{\infty} (K(\omega)N_F(\omega))^n : X_{\gamma^-,\omega} \to X_{\gamma^-,\omega}.$$

A global Lipschitz constant of $\Psi(\omega)$ is obtained from

$$\|(K(\omega)N_F(\omega))^n(\xi) - (K(\omega)N_F(\omega))^n(\eta)\|_{\gamma^-,\omega} \leq l^n \|\xi - \eta\|_{\gamma^-,\omega}$$

as $\sum_{n=0}^{\infty} l^n = \frac{1}{1-l}$. Hence $\Psi(\omega)$ is in particular continuous, thus a homeomorphism.

(iii) This is just the formal expression of our procedure leading to Theorem 7.3.1 to determine $M^+(\omega)$. Since $\Psi(\omega)$ is Lipschitz and all the other operations in (7.3.65) are linear and bounded (hence C^∞), the Lipschitz continuity of $m^+(\omega, \cdot)$ can be read-off. $\square$

For the C^k-smoothness of $m^+(\omega, \cdot)$ an inspection of (7.3.65) reveals that we need $\Psi(\omega)$ to be C^k. Our hope is that this follows from the fact that $N_F(\omega)$: $X_{a^-,\omega} \to X_{b^-,\omega}$ is C^k whenever $0 < b < a^q$ for $q = 1, \ldots, k$ (Proposition 7.3.23(iv)) – provided we can meet the gap condition, i.e. we can choose $b, a \in \Gamma_{\text{left}}$ with the properties of Theorem 7.3.19(iii)(2).

The smoothness of $\Psi(\omega)$ can be deduced from a contraction principle on a scale of Banach spaces which was developed for the deterministic case (which suffices since ω is fixed here), e.g. by Vanderbauwhede and Van Gils [333: Theorem 3], Hilger [171: Theorem 6.1] and Chow and Yi [99], see also Siegmund [317]. It allows to show smooth dependence of a fixed point on parameters.

We follow Vanderbauwhede and Van Gils [333] and quote their Theorem 3 without proof.

7.3.25 Theorem (Contraction Theorem on Scale of B-Spaces).
Let Y_0, Y, Y_1 and Λ be Banach spaces with norms denoted respectively by $\|\cdot\|_0$, $\|\cdot\|$, $\|\cdot\|_1$, $|\cdot|$, such that Y_0 is continuously embedded in Y and Y is continuously embedded in Y_1. We denote the embedding operators by $J_0 : Y_0 \to Y$ and $J : Y \to Y_1$. Consider a fixed point equation

358   Chapter 7. Invariant Manifolds

$$y = f(y, \lambda), \qquad (7.3.66)$$

where $f : Y \times \Lambda \to Y$ satisfies the following hypotheses:

(H1) $Jf : Y \times \Lambda \to Y_1$ has a continuous partial derivative $D_y(Jf) : Y \times \Lambda \to \mathcal{L}(Y, Y_1)$ with

$$D_y(Jf)(y, \lambda) = Jf^{(1)}(y, \lambda) = f_1^{(1)}(y, \lambda)J \quad \text{for all } (y, \lambda) \in Y \times \Lambda$$

for some $f^{(1)} : Y \times \Lambda \to \mathcal{L}(Y)$ and $f_1^{(1)} : Y \times \Lambda \to \mathcal{L}(Y_1)$.

(H2) $f_0 : Y_0 \times \Lambda \to Y$, $(y_0, \lambda) \mapsto f_0(y_0, \lambda) := f(J_0 y_0, \lambda)$ has a continuous partial derivative $D_\lambda f_0 : Y_0 \times \Lambda \to \mathcal{L}(\Lambda, Y)$.

(H3) There exists some $\kappa \in [0, 1)$ such that

$$\|f(y, \lambda) - f(\tilde{y}, \lambda)\| \leq \kappa \|y - \tilde{y}\| \quad \text{for all } y, \tilde{y} \in Y, \ \lambda \in \Lambda$$

and

$$\|f^{(1)}(y, \lambda)\| \leq \kappa, \quad \|f_1^{(1)}(y, \lambda)\|_1 \leq \kappa \quad \text{for all } (y, \lambda) \in Y \times \Lambda.$$

(H4) It follows from (H3) that (7.3.66) has for each $\lambda \in \Lambda$ a unique solution $y = \tilde{y}(\lambda) \in Y$. Suppose that $\tilde{y}(\lambda) = J_0 \tilde{y}_0(\lambda)$ for some continuous $\tilde{y}_0 : \Lambda \to Y_0$.

The hypotheses allow us to consider the following equation in $\mathcal{L}(\Lambda, Y)$,

$$A = f^{(1)}(\tilde{y}(\lambda), \lambda)A + D_\lambda f_0(\tilde{y}_0(\lambda), \lambda); \qquad (7.3.67)$$

because of (H3) this equation has for each $\lambda \in \Lambda$ a unique solution $\tilde{A}(\lambda) \in \mathcal{L}(\Lambda, Y)$.

Then under (H1) to (H4), the solution map $\tilde{y} : \Lambda \to Y$ of (7.3.66) is Lipschitz continuous and $\tilde{y}_1 := J\tilde{y} : \Lambda \to Y_1$ is C^1 with

$$D\tilde{y}_1(\lambda) = J\tilde{A}(\lambda), \quad \text{for all } \lambda \in \Lambda. \qquad (7.3.68)$$

As an application of this theorem, we can finally prove Theorem 7.3.19.

Proof of Theorem 7.3.19: We restrict ourselves to the C^1 case; the C^k case for $k \geq 2$ is in principle proved by induction on $q = 1, \ldots, k$ using Theorem 7.3.25 at its full strength (see Vanderbauwhede and Van Gils [333: Sect. 5] for the autonomous deterministic case and Siegmund [317] for the nonautonomous deterministic case).

We now fix $\omega \in \Omega$ and suppress it for ease of notation. Put $\Lambda = E^+$ with generic element $\lambda = x^+$, $Y_0 = Y = X_{a-}$ with generic element $y = \xi$, $Y_1 = X_{b-}$, where $a, b \in \Gamma_{\text{left}}$ with $b < a$. The embedding operators are $J_0 = \text{id} : X_{a-} \hookrightarrow X_{a-}$ and $J = J_{a,b} : X_{a-} \hookrightarrow X_{b-}$. Our equation (7.3.66) is (7.3.61),

$$\xi = f(\xi, x^+) = Sx^+ + KN_F(\xi).$$

We now verify the conditions (H1) to (H4) in our case.

(H1) The function $J_{a,b}f : X_{a^-} \times E^+ \to X_{b^-}$ defined by $J_{a,b}f(\xi, x^+) = KN_F(\xi) + Sx^+$ has a continuous partial derivative with respect to ξ which is even independent of x^+ and can be represented by

$$D_\xi(J_{a,b}f)(\xi, x^+) = J_{a,b}f^{(1)}(\xi, x^+) = f_1^{(1)}(\xi, x^+)J_{a,b},$$

where for each $\xi \in X_{a^-}$ and $x^+ \in E^+$

$$f^{(1)}(\xi, x^+) = K\, DN_F(\xi) \in \mathcal{L}(X_{a^-}, X_{a^-})$$

and

$$f_1^{(1)}(\xi, x^+) = K\, DN_F(\xi) \in \mathcal{L}(X_{b^-}, X_{b^-}).$$

This holds by Lemma 7.3.22(ii) and since by (7.3.64), $DN_F(\xi) \in \mathcal{L}(X_{\gamma^-}, X_{\gamma^-})$ for any $\gamma > 0$.

(H2) $f_0 = f : X_{a^-} \times E^+ \to X_{a^-}$ defined by $(\xi, x^+) \mapsto Sx^+ + KN_F(\xi)$ has a continuous partial derivative with respect to x^+ which is even independent of ξ and x^+ and is given by

$$D_{x^+}f : X_{a^-} \times E^+ \to \mathcal{L}(E^+, X_{a^-}), \quad (\xi, x^+) \mapsto S.$$

This follows from Lemma 7.3.22(i).

(H3) (a) We first prove that f is uniformly contracting. Indeed, by Proposition 7.3.24(i) for any $a \in \Gamma_{\text{left}}$

$$\begin{aligned} \|f(\xi, x^+) - f(\eta, x^+)\|_{a^-} &= \|KN_F(\xi) - KN_F(\eta)\|_{a^-} \\ &\leq \frac{L_1}{\delta}\|\xi - \eta\|_{a^-}, \end{aligned}$$

and $\kappa := L_1/\delta < 1$ if and only if $L_1 < \delta$, which follows from (7.3.58).

(b) The other two estimates follow by Lemma 7.3.22(ii) and from the estimates (7.3.63) and (7.3.64).

(H4) Since in our case $Y_0 = Y$, hypothesis (H4) follows from (H2).

Hence Theorem 7.3.25 applies. The result is as follows: Denote, as above, by $x^+ \mapsto \tilde{\xi}(x^+) = \Psi S(x^+) \in X_{a^-}$ the unique solution map of equation $\xi = Sx^+ + KN_F(\xi)$, i.e. $\Psi = (I - KN_F)^{-1}$. Then:

1. $\Psi S : E^+ \to X_{a^-}$ is Lipschitz continuous and $\tilde{\xi}_1 := J_{a,b}\Psi S : E^+ \to X_{b^-}$ is C^1.

2. The derivative of $x^+ \mapsto J_{a,b}\Psi S(x^+)$ is given by

$$D(J_{a,b}\Psi S(x^+)) = J_{a,b}\tilde{A}(x^+) \in \mathcal{L}(E^+, X_{b^-}),$$

where $\tilde{A}(x^+) \in \mathcal{L}(E^+, X_{a^-})$ is the unique solution of

$$(I - K\, DN_F(\Psi S(x^+)))A = S$$

i.e. $\tilde{A}(x^+) = (I - K\, DN_F(\Psi S(x^+)))^{-1}S$.

Denoting by $\pi_0 : X_{\gamma^-} \to \mathbb{R}^d$ the projection $\xi \mapsto \xi_0$ and by $\pi^- : \mathbb{R}^d \to E^-$ the projection $x = (x^+, x^-) \mapsto x^-$,

$$m^+(x^+) = \pi^- \pi_0(J_{a,b}\Psi S(x^+)),$$

the derivative at x^+ is

$$Dm^+(x^+) = \pi^- \pi_0 J_{a,b} \tilde{A}(x^+),$$

whence

$$\|Dm^+(x^+)\| \leq \|\tilde{A}(x^+)\| \leq \frac{1}{1-C},$$

where $C \leq \frac{L_1}{\delta - L_1}$, yielding part of the bound of Theorem 7.3.19(i). Moreover,

$$Dm^+(0) = \pi^- \pi_0 J_{a,b} \tilde{A}(0).$$

Since $S(0) = 0$, hence $\Psi S(0) = 0$ and thus $DN_F(\Psi S(0)) = (DF(\theta^n \omega, 0)) = 0$ by assumption, we have $\tilde{A}(0) = S$, and $Dm^+(0) = 0$, i.e. M^+ is tangent to E^+. □

7.3.26 Exercise (Higher Order Derivatives). We can derive higher order derivatives of m^+ at $x^+ = 0$ by differentiating the defining operator equation $\Psi(S(x^+)) = S(x^+) + K N_F(\Psi(S(x^+)))$ at $x^+ = 0$, cf. Siegmund [317: Korollar 8.4]. ∎

7.3.27 Remark (Hölder Continuity). Inspecting the proof of Hilger [171] for the deterministic case, one can convince oneself that the following result holds: If the RDS is $C^{k,\varepsilon}_{b,\omega}$ and the gap conditions are augmented by the requirements that also $b^{k+\varepsilon} < \bar{b}$ and $\bar{a} < a^{k+\varepsilon}$, then the invariant manifolds are also $C^{k,\varepsilon}$. The moral is that there is no loss of regularity in the construction of invariant manifolds. ∎

7.3.6 Final Global Invariant Manifold Theorem

In this subsection we "undo" our preparations: We carry over the main invariant manifold results from Subsects. 7.3.1 to 7.3.5 to our original, unprepared C^k RDS φ with fixed point 0, considered in $\mathbb{R}^d$ with Euclidean norm.

By Proposition 7.2.6, φ is linearly equivalent to a prepared RDS denoted by $\bar{\varphi}$,

$$P(\theta(t)\omega) \circ \varphi(t,\omega) = \bar{\varphi}(t,\omega) \circ P(\omega).$$

7.3.28 Theorem (Global Invariant Manifold Theorem). *Let φ be a C^k RDS ($k \geq 1$) with two-sided time and $\varphi(t, \omega, 0) = 0$, represented by*

$$\varphi(t, \omega, x) = \Phi(t, \omega)x + \psi(t, \omega, x),$$

where $\Phi(t, \omega) = D\varphi(t, \omega, 0)$, hence $\psi(t, \omega, 0) = 0$ and $D\psi(t, \omega, 0) = 0$.

Suppose that Φ satisfies the IC of the MET. Let $\mathcal{S}(\Phi) = \{(\lambda_i, d_i)_{i=1,\dots,p}\}$ be the Lyapunov spectrum, $\mathbb{R}^d = E_1(\omega) \oplus \cdots \oplus E_p(\omega)$ be the splitting of Φ and Λ_{ij} be any interval from $\mathcal{S}(\Phi)$

Consider the corresponding prepared RDS $\bar\varphi$ presented by

$$\bar\varphi(t,\omega,x) = \bar\Phi(t,\omega)x + \bar\psi(t,\omega,x),$$

and assume that $\bar\varphi$ satisfies the conditions of the Smooth Invariant Manifold Theorem 7.3.19. Let for any $1 \leq i \leq j \leq p$ $\bar M_{ij}(\omega)$ be the C^k center manifold of $\bar\varphi$ and put

$$M_{ij}(\omega) := P(\omega)^{-1}\bar M_{ij}(\omega).$$

Then

(i) $M_{ij}(\omega)$ is invariant under φ,

$$\varphi(t,\omega)M_{ij}(\omega) = M_{ij}(\theta(t)\omega).$$

(ii) $M_{ij}(\omega)$ is an embedded C^k submanifold of $\mathbb{R}^d$ tangent to $E_{ij}(\omega) = E_i(\omega) \oplus \cdots \oplus E_j(\omega)$ of dimension $d_i + \cdots + d_j$, given by a C^k graph in the $E_{ij}(\omega) \oplus F_{ij}(\omega)$ coordinate system, where $F_{ij}(\omega) := \oplus_{k\notin[i,j]} E_k(\omega)$.

(iii) $M_{ij}(\omega)$ is dynamically characterized as follows: For any $a_j \in \mathrm{int}\,\Gamma_{\mathrm{left}} = (\beta_{j+1} + \delta, \alpha_j - \delta)$, $b_i \in \mathrm{int}\,\Gamma_{\mathrm{right}} = (\beta_i + \delta, \alpha_{i-1} - \delta)$

$$M_{ij}(\omega) = \{x \in \mathbb{R}^d : \varphi(\cdot,\omega,x) \in X_{a_j^-,b_i^+}\}.$$

Proof. Everything carries over from the prepared RDS via the linear isomorphism $x = P(\omega)^{-1}\bar x$. The dynamical characterization follows from Proposition 7.2.6 and Lemma 7.2.3. □

7.4 Invariant Foliations, and the Hartman-Grobman Theorem

Our aims in this section are the following: (i) We first exhibit a whole foliation of invariant manifolds corresponding to a section of the Lyapunov spectrum of Φ into a "stable" and "unstable" part. (ii) We use this foliation as a new coordinate system to also decouple (block-diagonalize) the *nonlinear* part of the RDS. (iii) We finally linearize all blocks with a non-zero Lyapunov exponent (Hartman-Grobman theorem).

We basically follow Wanner [340: Sects. 3.3–3.4].

7.4.1 Invariant Foliations

We assume the situation of Sect. 7.3, in particular the conditions of the Global Center Manifold Theorems 7.3.14 ($\mathbb{T} = \mathbb{Z}$) and 7.3.17 ($\mathbb{T} = \mathbb{R}$).

Suppose the Lyapunov spectrum of Φ is not one-point and is split into the disjoint nonvoid intervals $\Lambda^+ = \{\lambda_1 > \cdots > \lambda_j\}$ and $\Lambda^- = \{\lambda_{j+1} > \cdots > \lambda_p\}$, with corresponding splitting $\mathbb{R}^d = E^+ \oplus E^-$. The unstable manifold $M^+(\omega)$ and stable manifold $M^-(\omega)$ are given by

$$M^+(\omega) = \{x \in \mathbb{R}^d : \varphi(\cdot, \omega, x) \in X_{\gamma^-,\omega}\},$$

$$M^-(\omega) = \{x \in \mathbb{R}^d : \varphi(\cdot, \omega, x) \in X_{\gamma^+,\omega}\},$$

where $\gamma \in \Gamma = [\beta_{j+1} + \delta, \alpha_j - \delta]$ and the manifolds do not depend on the particular value of γ. By Corollary 7.3.13,

$$M^+(\omega) \cap M^-(\omega) = \{0\}.$$

We now want to study the asymptotic behavior of orbits which start off those manifolds.

For example, consider the case where Φ is hyperbolic and the spectrum split at 0, equivalently in the gap Γ with $1 \in \text{int}\,\Gamma$. Then the orbits starting on $M^-(\omega)$ approach 0 exponentially fast forwards in time (choose $\gamma < 1$), while the orbits starting on $M^+(\omega)$ approach 0 exponentially fast backwards in time (choose $1 < \gamma$). However, if $x \notin M^-(\omega) \cup M^+(\omega)$ we expect that $\varphi(\cdot, \omega, x)$ approaches $M^+(\omega)$, in fact a certain reference orbit $\varphi(\cdot, \omega, P^+(\omega)x)$ on $M^+(\omega)$, exponentially fast forwards in time and a reference orbit $\varphi(\cdot, \omega, P^-(\omega)x)$ on $M^-(\omega)$ exponentially fast backwards in time. This is indeed the case and these reference points $P^\pm(\omega)x \in M^\pm(\omega)$ are uniquely determined (see Fig. 7.8).

We will prove that, given $x \in \mathbb{R}^d$, all initial values with the same reference point $y := P^+(\omega)x$ form an invariant manifold $M^-(\omega, y)$ and all initial values with the same reference point $z := P^-(\omega)x$ form an invariant manifold $M^+(\omega, z)$. These two manifolds have exactly the point x in common, so that $(P^+(\omega)x, P^-(\omega)x)$ are nonlinear coordinates of x. All these manifolds hence form a foliation (i.e. have local product structure). We will use the corresponding nonlinear coordinate system in Subsect. 7.4.2 to decouple the nonlinear RDS into a $(+)$ and a $(-)$ part.

7.4.1 Theorem (Foliation by Invariant Manifolds). *(A) Discrete time case:* Let φ be a prepared RDS with fixed point 0 generated by the random difference equation

$$x_{n+1} = A(\theta^n \omega)x_n + F(\theta^n \omega, x_n), \quad n \in \mathbb{Z}, \tag{7.4.1}$$

which satisfies the conditions of the Global Center Manifold Theorem 7.3.14. Assume that the Lyapunov spectrum of Φ is not one-point and split into nonvoid Λ^+ and Λ^- with the corresponding invariant manifolds M^+ and M^-.

Then there are uniquely defined measurable mappings $P^\pm : \Omega \times \mathbb{R}^d \to \mathbb{R}^d$ such that for each $\omega \in \Omega$:

(i) $P^\pm(\omega) := P^\pm(\omega, \cdot) : \mathbb{R}^d \to \mathbb{R}^d$ is continuous and satisfies $P^\pm(\omega)0 = 0$, $P^\pm(\omega)(\mathbb{R}^d) = M^\pm(\omega)$ and $P^\pm(\omega)^2 = P^\pm(\omega)$ (i.e. $P^\pm(\omega)$ are nonlinear projections). Further,

$$\|P^{\pm}(\omega,x)\|_\omega \le \frac{1}{1-D(L)}\|x\|_\omega\,, \quad D(L) = \frac{L(\delta-L)}{\delta(\delta-2L)}. \tag{7.4.2}$$

(ii) The mappings $P^{\pm}(\omega)$ are invariant, i. e.

$$P^{\pm}(\theta(t)\omega) \circ \varphi(t,\omega) = \varphi(t,\omega) \circ P^{\pm}(\omega) \quad \text{for all } t \in \mathbb{Z}.$$

(iii) The preimages of any point $x \in M^{\pm}(\omega)$, i. e. the sets

$$M^{-}(\omega,x) := P^{+}(\omega)^{-1}(\{x\}), \quad x \in M^{+}(\omega),$$

$$M^{+}(\omega,x) := P^{-}(\omega)^{-1}(\{x\}), \quad x \in M^{-}(\omega),$$

are graphs $E^{-} \ni z^{-} \mapsto m^{-}(\omega,x,z^{-}) \in E^{+}$ and $E^{+} \ni z^{+} \mapsto m^{+}(\omega,x,z^{+}) \in E^{-}$, respectively, which are measurable in $(\omega,x,z^{\pm})$, continuous in $(x,z^{\pm})$, and Lipschitz continuous in $z^{\pm}$ with constant $D(L)$, hence are embedded submanifolds of $\mathbb{R}^d$.

(iv) The manifolds $M^{\pm}(\omega,x)$, $x \in M^{\mp}(\omega)$, are dynamically characterized as follows: For arbitrary $\gamma \in \Gamma$

$$M^{+}(\omega,x) = \{z \in \mathbb{R}^d : \varphi(\cdot,\omega,z) - \varphi(\cdot,\omega,x) \in X_{\gamma^-,\omega}\}, \quad x \in M^{-}(\omega),$$

$$M^{-}(\omega,x) = \{z \in \mathbb{R}^d : \varphi(\cdot,\omega,z) - \varphi(\cdot,\omega,x) \in X_{\gamma^+,\omega}\}, \quad x \in M^{+}(\omega).$$

In particular, $M^{\pm}(\omega,0) = M^{\pm}(\omega)$.

(v) The manifolds $M^{\pm}$ are invariant, i. e.

$$\varphi(t,\omega)M^{\pm}(\omega,x) = M^{\pm}(\theta(t)\omega,\varphi(t,\omega,x)) \quad \text{for all } t \in \mathbb{Z}, \quad x \in M^{\mp}(\omega).$$

(vi) If the RDS φ is C^k, $k \ge 1$, and satisfies the hypotheses of Theorem 7.3.19 then the invariant manifolds $M^{\pm}(\omega,x)$ are embedded C^k submanifolds of $\mathbb{R}^d$ and the graphs $(x,z^{\pm}) \mapsto m^{\pm}(\omega,x,z^{\pm})$ are k times continuously differentiable with respect to $z^{\pm}$.

(B) Continuous time case: Let

$$\varphi(t,\omega,x) = \Phi(t,\omega)x + \psi(t,\omega,x), \quad t \in \mathbb{R},$$

be a prepared RDS satisfying the conditions of the Global Center Manifold Theorem 7.3.17. Assume the situation described in part (A).

Then all assertions of part (A) hold for $\mathbb{Z}$ replaced with $\mathbb{R}$.

Proof. We restrict ourselves to the proof of part (A). Part (B) can be reduced to part (A) by arguments similar to those used for the same purpose in Subsect. 7.3.4.

The fact that we want to study the RDS φ with respect to an arbitrary reference solution $\varphi(\cdot,\omega,x)$ (in general not the trivial solution) forces us to make an excursion to general nonautonomous random difference equations.

364   Chapter 7. Invariant Manifolds

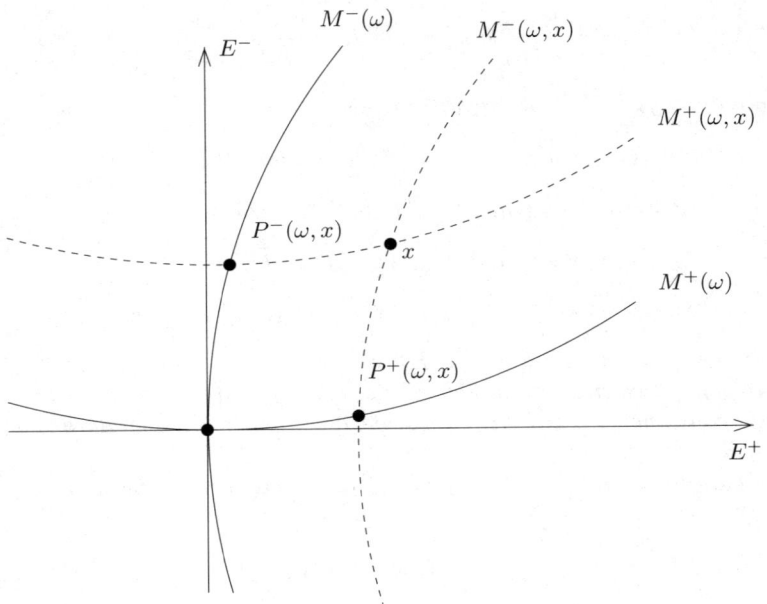

Fig. 7.8. Foliations by invariant manifolds

Step 1: Let $x \in \mathbb{R}^d$ be fixed. Our ultimate aim is to prove that there is a unique point $P^\pm(\omega, x) \in M^\pm(\omega)$ with the property that the difference of the orbits with initial values x and $P^\pm(\omega, x)$ satisfies

$$\varphi(\cdot, \omega, x) - \varphi(\cdot, \omega, P^\pm(\omega, x)) \in X_{\gamma^\pm, \omega}.$$

To see this, we have to study for the fixed parameter $(\omega, k, x) \in \Omega \times \mathbb{Z} \times \mathbb{R}^d$ the *difference equation of perturbed motion*, i.e.

$$z_{n+1} = A(\theta^n \omega) z_n + f(n, (\omega, k, x), z_n), \quad z_l = z, \quad n \in \mathbb{Z}, \quad (7.4.3)$$

where, denoting by $\varphi(\cdot, \omega, k, x)$ the solution of the original random difference equation (7.4.1) with initial value $x_k = x$,[4]

$$f(n, (\omega, k, x), z) := F(\theta^n \omega, z + \varphi(n, \omega, k, x)) - F(\theta^n \omega, \varphi(n, \omega, k, x)),$$

which is measurable in (ω, x, z), continuous in (x, z) and satisfies

$$f(n, (\omega, k, x), 0) = 0,$$

$$\|f(n, (\omega, k, x), z_1) - f(n, (\omega, k, x), z_2)\|_{\theta^{n+1} \omega} \leq L \|z_1 - z_2\|_{\theta^n \omega}.$$

Denote the solution of (7.4.3) with initial data (l, z) and parameters (ω, k, x) by $\tilde{\varphi}(\cdot, (\omega, k, x), l, z)$.

---

[4] We will make use of the fact that $\varphi(n, \omega, k, x) = \varphi(n - k, \theta^k \omega, x)$ later in step 4 of the proof.

7.4 Hartman-Grobman Theorem

Note that $\tilde{\varphi}$ is a solution of (7.4.3) with parameter values (ω, k, x) if and only if $\tilde{\varphi}+\varphi(\cdot, \omega, k, x)$ is a solution of (7.4.1). In particular, the solution $\tilde{\varphi} = 0$ of (7.4.3) corresponds to our reference solution $\varphi(\cdot, \omega, x)$ of (7.4.1).

With this well-known trick, the study of our original RDS in a neighborhood of the arbitrary reference solution $\varphi(\cdot, \omega, x)$ is transferred to the study of the neighborhood of the trivial reference solution $z = 0$ of the nonautonomous equation (7.4.3).

While this trick in general fails to simplify the situation, it is successful in our case: It can be easily checked that the two Key Technical Lemmas 7.3.3 and 7.3.6 and all constructions based on them remain valid if the "correct" difference equation (7.4.1) generating an RDS is replaced with (7.4.3), which has the "wrong" nonlinear part, hence does in general not generate an RDS (Wanner [340] has formulated all results for this more general case of a random difference equation in his Chap. 2). The price we have to pay is that the invariant manifolds for the fixed point $z = 0$ become time (and parameter) dependent, hence bundles. For instance, the unstable and stable manifolds are now the bundles

$$\tilde{M}^{\pm}(\omega, k, x) = \{(n, z^{\pm} \oplus \tilde{m}^{\pm}(n, (\omega, k, x), z^{\pm})) : n \in \mathbb{Z}, z^{\pm} \in E^{\pm}\},$$

of graphs and are dynamically characterized as

$$\tilde{M}^{\pm}(\omega, k, x) = \{(n, z) : \tilde{\varphi}(\cdot, (\omega, k, x), n, z) \in X_{\gamma^{\mp}, \omega}\}.$$

The graphs have the property $\tilde{m}^{\pm}(n, (\omega, k, x), 0) = 0$ and have the same Lipschitz constant, namely $D(L)$, for any (ω, k, x) with respect to the norm $\|\cdot\|_{\theta^k \omega}$.

In the remainder of the proof we need slight variations of arguments already presented in great depth in Sect. 7.3, so we take the liberty of omitting some of the details.

The measurability of $(\omega, x, z^{\pm}) \mapsto \tilde{m}^{\pm}(n, (\omega, k, x), z^{\pm})$ follows as before from Lemma 7.3.4, but now with ω replaced with (ω, x).

We now verify the continuity of $(x, z^{\pm}) \mapsto \tilde{m}^{\pm}(n, (\omega, k, x), z^{\pm})$. The continuity with respect to x is a consequence of Lemma 5.3 of Wanner [340] on the continuous dependence of a fixed point on a parameter. This fact and (Lipschitz) continuity with respect to $z^{\pm}$ uniformly in x yield joint continuity with respect to $(x, z^{\pm})$, as claimed.

The general flow property for the solution $\tilde{\varphi}$ of (7.4.3) and the dynamical characterization of $\tilde{M}^{\pm}(\omega, k, x)$ obviously imply that $\tilde{M}^{\pm}(\omega, k, x)$ is invariant in the following sense: If $(n, z) \in \tilde{M}^{\pm}(\omega, k, x)$, then $(p, \tilde{\varphi}(p, (\omega, k, x), n, z)) \in \tilde{M}^{\pm}(\omega, k, x)$ for any $p \in \mathbb{Z}$.

The dynamical characterization of $\tilde{M}^{\pm}(\omega, k, x)$ also has the consequence that for any (ω, k, x)

$$\tilde{M}^{+}(\omega, k, x) \cap \tilde{M}^{-}(\omega, k, x) = \{(n, 0) : n \in \mathbb{Z}\}.$$

Step 2: We now transfer the constructions from (7.4.3) back to the original equation (7.4.1), thereby dropping the tilde from $\tilde{M}$ and $\tilde{m}$.

We define manifolds $M^\pm(\omega, k, x)$ via the graphs

$$m^\pm(n, (\omega, k, x), y^\pm) := \tilde{m}^\pm(n, (\omega, k, x), y^\pm - \varphi^\pm(n, \omega, k, x)) + \varphi^\mp(n, \omega, k, x).$$

These graphs $m^\pm$ enjoy the same measurability, continuity and Lipschitz conditions as those proved above for $\tilde{m}^\pm$.

The manifolds defined by these graphs are dynamically characterized as

$$M^\pm(\omega, k, x) = \{(n, z) : \varphi(\cdot, \omega, n, z) - \varphi(\cdot, \omega, k, x) \in X_{\gamma^\mp, \omega}\}.$$

They are φ-invariant in the sense that if $(n, y) \in M^\pm(\omega, k, x)$, then $(p, \varphi(p, \omega, n, y)) \in M^\pm(\omega, k, x)$ for any $p \in \mathbb{Z}$.

In particular, since $\varphi(\cdot, \omega, k, 0) = 0$ for all $k \in \mathbb{Z}$,

$$\begin{aligned} M^\pm(\omega, k, 0) &= M^\pm(\omega, 0, 0) \\ &= \{(n, z) : \varphi(\cdot, \omega, n, z) \in X_{\gamma^\mp, \omega}\} \\ &= \{(n, z^\pm \oplus m^\pm(n, (\omega, 0, 0), z^\pm)) : n \in \mathbb{Z}, z^\pm \in E^\pm\}.\end{aligned}$$

Step 3: We now construct the "base points" $P^\pm(\omega, k, x)$ and for reasons of symmetry restrict ourselves to $P^+(\omega, k, x)$.

The idea is to just intersect the k-fiber of $M^-(\omega, k, x)$ with the k-fiber of $M^+(\omega, 0, 0)$. To detect points in the intersection, define an operator

$$T_{\omega, k, x} : E^+ \to E^+, \quad z^+ \mapsto m^-(k, (\omega, k, x), m^+(k, (\omega, 0, 0), z^+)).$$

It has contraction rate $D(L)^2 < 1$, hence has a unique fixed point $\xi^+(\omega, k, x) \in E^+$ which is measurable and depends continuously on x.

By the definition of the operator $T_{\omega, k, x}$, the fixed point ξ^+ fulfills

$$\xi^+(\omega, k, x) = m^-(k, (\omega, k, x), m^+(k, (\omega, 0, 0), \xi^+(\omega, k, x))),$$

thus
$$(k, \xi^+, m^+(k, (\omega, 0, 0), \xi^+)) \in M^-(\omega, k, x) \cap M^+(\omega, 0, 0).$$

Define now

$$P^+(\omega, k, x) := (\xi^+(\omega, k, x), m^+(k, (\omega, 0, 0), \xi^+(\omega, k, x))) \in \mathbb{R}^d.$$

Hence for every (ω, k, x), $P^+(\omega, k, x)$ is the uniquely defined point in $\mathbb{R}^d$ such that

- $(k, P^+(\omega, k, x)) \in M^+(\omega, 0, 0)$, and
- $(k, P^+(\omega, k, x)) \in M^-(\omega, k, x)$, i.e. for which

$$\varphi(\cdot, \omega, k, x) - \varphi(\cdot, \omega, k, P^+(\omega, k, x)) \in X_{\gamma^+, \omega}.$$

The base point $P^-(\omega, k, x)$ is similarly constructed.

For the proof of (7.4.2) note that

$$\|\xi^+(\omega,k,x)\|_{\theta^k\omega} \leq \frac{1}{1-D(L)^2}\|x\|_{\theta^k\omega},$$

so

$$\|P^+(\omega,k,x)\|_{\theta^k\omega} \leq (1+D(L))\|\xi^+(\omega,k,x)\|_{\theta^k\omega} \leq \frac{1}{1-D(L)}\|x\|_{\theta^k\omega}.$$

Analogously for $\|P^-(\omega,k,x)\|_{\theta^k\omega}$.

Step 4: We finally use the fact that (7.4.1) generates an RDS, i.e.

$$\varphi(n,\omega,k,x) = \varphi(n-k,\theta^k\omega,0,x) = \varphi(n-k,\theta^k\omega,x).$$

This implies for the graphs of the n-th fiber of $M^\pm(\omega,k,x)$ that

$$m^\pm(n,(\omega,k,x),\cdot) = m^\pm(n-k,(\theta^k\omega,0,x),\cdot). \tag{7.4.4}$$

We hence can put without loss of generality $k = 0$, thus consider just bundles $M^\pm$ depending on (ω,x) only. Their 0-fiber is denoted by

$$M_0^\pm(\omega,x) := \{z^\pm \oplus m_0^\pm(\omega,x,z^\pm) : z^\pm \in E^\pm\} \tag{7.4.5}$$

and is dynamically characterized by

$$M_0^\pm(\omega,x) = \{z \in \mathbb{R}^d : \varphi(\cdot,\omega,z) - \varphi(\cdot,\omega,x) \in X_{\gamma^\mp,\omega}\}. \tag{7.4.6}$$

In particular, we recover

$$M^\pm(\omega) = M_0^\pm(\omega,0) = \{z \in \mathbb{R}^d : \varphi(\cdot,\omega,x) \in X_{\gamma^\mp,\omega}\}.$$

Using (7.4.4), the construction of $P^\pm(\omega,k,x)$ in step 3 also yields

$$P^\pm(\omega,k,x) = P^\pm(\theta^k\omega,0,x).$$

We hence put

$$P^\pm(\omega,x) := P^\pm(\omega,0,x) \in M_0^\mp(\omega,x) \cap M^\pm(\omega). \tag{7.4.7}$$

Since obviously $P^\pm(\omega,x) = x$ for all $x \in M^\pm(\omega)$, $x \in M^\pm(\omega)$ is its own base point. We collect all $z \in \mathbb{R}^d$ with this prescribed base point $x \in M^\pm(\omega)$,

$$M^\mp(\omega,x) := \{z \in \mathbb{R}^d : P^\pm(\omega,z) = x\} = P^\pm(\omega)^{-1}(\{x\}). \tag{7.4.8}$$

By (7.4.6)

$$M^\mp(\omega,x) = \{z \in \mathbb{R}^d : \varphi(\cdot,\omega,z) - \varphi(\cdot,\omega,x) \in X_{\gamma^\pm,\omega}\},$$

and by (7.4.5), $M^\mp(\omega,x)$ is a graph,

$$z^{\mp} \mapsto m^{\mp}(\omega, x, z^{\mp}) = m_0^{\mp}(\omega, x, z^{\mp}).$$

With (7.4.7) and (7.4.8) we finally have obtained the objects figuring in Theorem 7.4.1.

The manifolds $M^{\pm}(\omega, x)$ have the regularity claimed in (iii). The dynamical characterization in (iv) also gives (v). Parts (i) and (ii) immediately follow from our construction.

Part (vi) basically follows, as the C^k property of $M^{\pm}(\omega)$ in Subsect. 7.3.5 does, from a contraction theorem in a scale of Banach spaces which gives smooth dependence of the fixed point on the parameter $z^{\pm}$. However, we refrain from giving more details. Recall that for $k > 1$, the spectral interval Γ has to be wide enough to satisfy the two gap conditions of Theorem 7.3.19. □

7.4.2 Remark. As deterministic counterexamples show, even for smooth RDS the continuous dependence of $M^{\pm}(\omega, x)$ on x can in general not be improved to C^1 dependence. ∎

7.4.2 Topological Decoupling

We continue to consider the situation of the last subsection. For each $x \in \mathbb{R}^d$ we have constructed the base points

$$P^+(\omega, x) = \xi^+(\omega, x) \oplus m^+(\omega, \xi^+(\omega, x)) \in M^+(\omega),$$

$$P^-(\omega, x) = \xi^-(\omega, x) \oplus m^-(\omega, \xi^-(\omega, x)) \in M^-(\omega),$$

where $\xi^{\pm}(\omega, x)$ are the fixed points of certain operators $T_{\omega,x}^{\pm} : E^{\pm} \to E^{\pm}$ (see step 3 of the last proof).

Since $P^+(\omega, x)$ lies on a graph in $E^+ \oplus E^-$, it is uniquely determined by its coordinate $\xi^+(\omega, x) \in E^+$. Similarly $P^-(\omega, x)$ is uniquely determined by $\xi^-(\omega, x)$.

We thus have a mapping

$$g(\omega) = g(\omega, \cdot) : \mathbb{R}^d \to \mathbb{R}^d,$$

$$x \mapsto g^+(\omega, x) \oplus g^-(\omega, x) = \xi^+(\omega, x) \oplus \xi^-(\omega, x) \in E^+ \oplus E^- = \mathbb{R}^d,$$

such that by Theorem 7.4.1 $(\omega, x) \mapsto g(\omega, x)$ is measurable and $x \mapsto g(\omega, x)$ is continuous.

Our original prepared RDS φ is for discrete time generated by the pair of random difference equations

$$x_{n+1}^+ = A^+(\theta^n \omega) x_n^+ + F^+(\theta^n \omega, x_n^+, x_n^-), \quad (7.4.9)$$

$$x_{n+1}^- = A^-(\theta^n \omega) x_n^- + F^-(\theta^n \omega, x_n^+, x_n^-), \quad (7.4.10)$$

coupled through their nonlinear parts F^+ and F^-.

7.4 Hartman-Grobman Theorem

We claim that if $\varphi(\cdot)$ is a solution of (7.4.9) and (7.4.10) with initial value $x_0 = x$, then
$$\bar\varphi(n) := g(\theta^n \omega, \varphi(n))$$
is a solution of
$$z_{n+1}^+ = A^+(\theta^n \omega) z_n^+ + F^+(\theta^n \omega, z_n^+, m^+(\theta^n \omega, z_n^+)), \quad (7.4.11)$$
$$z_{n+1}^- = A^-(\theta^n \omega) z_n^- + F^-(\theta^n \omega, m^-(\theta^n \omega, z_n^-), z_n^-), \quad (7.4.12)$$
with initial value $z_0 = g(\omega, x)$.

Indeed, let $x_n = \varphi(n, \omega, x)$ solve (7.4.9), (7.4.10). Then by the invariance of $P^\pm$ (Theorem 7.4.1(ii)), $P^\pm(\theta^n \omega, x_n) = z_n^\pm \oplus m^\pm(\theta^n \omega, z_n^\pm)$ are also solutions of (7.4.9), (7.4.10) on $M^\pm(\theta^n \omega)$. Hence z_n^+ solves (7.4.11) and z_n^- solves (7.4.12).

Now note that the pair (7.4.11), (7.4.12) is simpler than the original pair (7.4.9), (7.4.10) since their nonlinear parts, hence their right-hand sides, are decoupled from each other.

Summing up, we assign to each orbit $\varphi(\cdot, \omega, x)$ a pair of two orbits: the orbit $\varphi(\cdot, \omega, P^+(\omega, x))$ moving on M^+ and being identified with its E^+ coordinate given by (7.4.11) and the orbit $\varphi(\cdot, \omega, P^-(\omega, x))$ moving on M^- and being identified with its E^- coordinate given by (7.4.12).

We will now prove that this assignment is in fact one-to-one and onto; more precisely: $g(\omega)$ is continuously invertible, hence a random homeomorphism, so that it defines a topological equivalence between φ and $\bar\varphi$.

7.4.3 Proposition (Topological Decoupling into Two Blocks).

(A) Discrete time: Assume that the conditions of Theorem 7.4.1(A) are satisfied. Then:

(i) The mapping $g(\omega) : \mathbb{R}^d \to \mathbb{R}^d$ defined above is a random homeomorphism of $\mathbb{R}^d$ (i. e. $g(\omega) := g(\omega, \cdot) \in \mathrm{Homeo}(\mathbb{R}^d)$) with $g(\omega, 0) = 0$. For $z = z^+ \oplus z^-$, the inverse $g^{-1}(\omega, z)$ is the unique point in the intersection of the manifolds $M^-(\omega, z^+ \oplus m^+(\omega, z^+))$ and $M^+(\omega, z^- \oplus m^-(\omega, z^-))$ from the invariant foliation.

(ii) g defines a topological equivalence between the RDS φ generated by (7.4.9) and (7.4.10) and the RDS $\bar\varphi$ generated by (7.4.11) and (7.4.12), i. e. we have
$$g(\theta(t)\omega) \circ \varphi(t, \omega) = \bar\varphi(t, \omega) \circ g(\omega), \quad t \in \mathbb{Z}. \quad (7.4.13)$$

(iii) The random homeomorphism g satisfies the estimates
$$\frac{1}{B}\|x\|_\omega \leq \|g(\omega, x)\|_\omega \leq B\|x\|_\omega,$$
$$\frac{1}{B}\|x\|_\omega \leq \|g^{-1}(\omega, x)\|_\omega \leq B\|x\|_\omega,$$
where $B = B(L, \delta) \geq 1$. As a consequence, if $\gamma > 0$ is arbitrary, a solution $\varphi(\cdot, \omega, x)$ of (7.4.9) and (7.4.10) is $(\gamma^\pm, \omega)$-bounded if and only if the corresponding solution $\bar\varphi(\cdot, \omega, g(\omega, x))$ of (7.4.11) and (7.4.12) is $(\gamma^\pm, \omega)$-bounded.

(iv) *The invariant manifolds $M^{\pm}(\omega)$ of φ are mapped by $g(\omega)$ to the linear subspaces $E^{\pm}$,*

$$g(\omega)M^{\pm}(\omega) = E^{\pm},$$

hence $E^{\pm}$ are the corresponding (linear and deterministic) invariant manifolds of $\bar{\varphi}$.

(B) *Continuous time case: Assume that the conditions of Theorem 7.4.1(B) are satisfied.*

Then all assertions of part (A) hold for φ represented by the pair

$$\varphi^+(t,\omega,x^+,x^-) = \Phi^+(t,\omega)x^+ + \psi^+(t,\omega,x^+,x^-),$$
$$\varphi^-(t,\omega,x^+,x^-) = \Phi^-(t,\omega)x^- + \psi^-(t,\omega,x^+,x^-).$$

and $\bar{\varphi}$ represented by the pair

$$\bar{\varphi}^+(t,\omega,z^+) = \Phi^+(t,\omega)z^+ + \psi^+(t,\omega,z^+,m^+(\omega,z^+)),$$
$$\bar{\varphi}^-(t,\omega,z^-) = \Phi^-(t,\omega)z^- + \psi^-(t,\omega,m^-(\omega,z^-),z^-).$$

Proof. Since we again apply methods which have been elaborated in detail before we permit ourselves to be quite brief. In particular, we restrict ourselves to the discrete time case, leaving the reduction of the continuous time case to the discrete time case as an exercise to the reader.

(i) By definition, $g(\omega)$ is continuous. We now prove that for any given $z = z^+ \oplus z^-$,

$$M^-(\omega, z^+ \oplus m^+(\omega, z^+)) \cap M^+(\omega, z^- \oplus m^-(\omega, z^-))$$

is a one-point set. A point $y^+ \oplus y^-$ is in this intersection if and only if it is a fixed point of the operator

$$T_{\omega,z} : E^+ \oplus E^- \to E^+ \oplus E^-,$$

$$y^+ \oplus y^- \mapsto m^-(\omega, z^+ \oplus m^+(\omega, z^+), y^-) \oplus m^+(\omega, z^- \oplus m^-(\omega, z^-), y^+).$$

This operator is contracting with constant $D(L) < 1$, and, by the usual argument, its unique fixed point $\tilde{g}(\omega, z)$ is measurable with respect to (ω, z) and continuous with respect to z. From the way we constructed $P^{\pm}(\omega, x)$ it is clear that $g(\omega) \circ \tilde{g}(\omega) = \tilde{g}(\omega) \circ g(\omega) = \mathrm{id}_{\mathbb{R}^d}$. From this it follows that $g(\omega)$ and $\tilde{g}(\omega)$ are one-to-one and onto.

(ii) That (7.4.13) defines a topological equivalence follows from the fact that $g(\omega)$ is a random homeomorphism.

(iii) is derived in the usual manner from the fact that both $g(\omega, x)$ and $g^{-1}(\omega, z)$ are fixed points of contracting mappings.

(iv) is an immediate consequence of (iii) and the dynamical characterization of the unstable and stable manifold. □

By using Theorem 7.4.3 in an induction on $i = 2, \ldots, p$ we arrive at the best possible topological decoupling result based on the MET.

7.4.4 Theorem (Topological Decoupling).

(A) Discrete time case: Assume that the conditions of the Global Center Manifold Theorem 7.3.14 are satisfied. Let the prepared RDS φ be generated by the p coupled random difference equations

$$x_{i,n+1} = A_i(\theta^n \omega)x_{i,n} + F_i(\theta^n \omega, x_{1,n}, \ldots, x_{p,n}), \quad n \in \mathbb{Z}, \; i = 1, \ldots, p. \quad (7.4.14)$$

Then there exists a random homeomorphism $g(\omega) : \mathbb{R}^d \to \mathbb{R}^d$ with $g(\omega)0 = 0$ such that φ is topologically equivalent to the RDS $\bar{\varphi}$,

$$g(\theta(t)\omega) \circ \varphi(t, \omega) = \bar{\varphi}(t, \omega) \circ g(\omega), \quad t \in \mathbb{T},$$

where $\bar{\varphi}$ is generated by the following set of p decoupled random difference equations:

$$z_{i,n+1} = A_i(\theta^n \omega)z_{i,n} + F_i(\theta^n \omega, z_{i,n} \oplus m_i(\theta^n \omega, z_{i,n})), \quad n \in \mathbb{Z}, \; i = 1, \ldots, p. \quad (7.4.15)$$

Here

$$M_i(\omega) = \{z_i \oplus m_i(\omega, z_i) : z_i \in E_i\}$$

is the Oseledets manifold "tangent" to E_i.

(ii) The random homeomorphism g satisfies the estimates

$$\frac{1}{B}\|x\|_\omega \leq \|g(\omega, x)\|_\omega \leq B\|x\|_\omega,$$

$$\frac{1}{B}\|x\|_\omega \leq \|g^{-1}(\omega, x)\|_\omega \leq B\|x\|_\omega,$$

where $B = B(L, \delta) \geq 1$.

(iii) All invariant manifolds $M_{ij}(\omega)$, $1 \leq i \leq j \leq p$, of φ are mapped by $g(\omega)$ to the linear subspaces E_{ij},

$$g(\omega)M_{ij}(\omega) = E_{ij},$$

hence the E_{ij} are the corresponding (linear and deterministic) invariant manifolds of $\bar{\varphi}$.

(B) Continuous time case: Assume that the conditions of the Global Center Manifold Theorem 7.3.17 are satisfied. Let the prepared RDS φ be represented by the p coupled equations

$$\varphi_i(t, \omega, x_1, \ldots, x_p) = \Phi_i(t, \omega)x_i + \psi_i(t, \omega, x_1, \ldots, x_p), \quad i = 1, \ldots, p. \quad (7.4.16)$$

Then the statements of part (A) hold for (7.4.16), where $\bar{\varphi}$ is now represented by the following set of p decoupled equations:

$$\bar{\varphi}_i(t, \omega, z_i) = \Phi_i(t, \omega)z_i + \psi_i(t, \omega, z_i \oplus m_i(\omega, z_i)), \quad i = 1, \ldots, p. \quad (7.4.17)$$

7.4.5 Remark (More Invariant Manifolds, but not Lipschitz). In Theorem 7.3.14 we have constructed the $(p-1)(p+2)/2$ invariant manifolds $M_{ij}(\omega)$ with Lipschitz continuous graphs, corresponding to spectral intervals Λ_{ij}.

But the RDS $\bar{\varphi}$ generated by (7.4.15) has $2^p - 2$ nontrivial invariant subspaces $E_\sigma = \oplus_{i \in \sigma} E_i$, where $\sigma \subset \{1, \ldots, p\}$ is nontrivial. The homeomorphic images
$$g^{-1}(\omega) E_\sigma = M_\sigma(\omega)$$
are topological manifolds which are invariant for φ. However, these manifolds are in general not Lipschitz continuous, as deterministic counterexamples show (cf. Aulbach and Wanner [47: Sect. 7]). ■

We can now derive the random version of the reduction principle which is a key statement of center manifold theory: Suppose the Lyapunov spectrum of Φ is cut at a negative point ($\gamma < 1$). Then the stability behavior of equations (7.4.9) and (7.4.10) for $t \to \infty$ can be reduced to the one of the "unstable" equation (7.4.11).

7.4.6 Definition (Stability). Let φ be an RDS of continuous mappings in $\mathbb{R}^d$. A reference solution $\varphi(\cdot, \omega, x_0)$ is called *stable* (with respect to the random norm $\|\cdot\|_\omega$) if for any $\varepsilon : \Omega \to (0, \infty)$ there exists a $\delta : \Omega \to (0, \infty)$ such that
$$\sup_{0 \leq t < \infty} \|\varphi(t, \omega, x) - \varphi(t, \omega, x_0)\|_{\theta(t)\omega} < \varepsilon(\omega)$$
whenever $\|x - x_0\|_\omega < \delta(\omega)$. The reference solution is called *asymptotically stable* if it is stable and, in addition, if for any x with $\|x - x_0\|_\omega < \delta(\omega)$
$$\lim_{t \to \infty} \|\varphi(t, \omega, x) - \varphi(t, \omega, x_0)\|_{\theta(t)\omega} = 0.$$
If the reference solution is stable, and if there exists a $c < 0$ such that for any x with $\|x - x_0\|_\omega < \varepsilon(\omega)$
$$\|\varphi(t, \omega, x) - \varphi(t, \omega, x_0)\|_{\theta(t)\omega} \leq e^{ct} \quad \text{for } t \geq T(\omega, x),$$
it is called *exponentially stable*. ■

7.4.7 Corollary (Reduction Principle). *Assume that the top Lyapunov exponent of Φ satisfies*
$$\lambda_1 \leq 0$$
and take $\Lambda^+ = \{\lambda_1\}$ and $\Lambda^- = \{\lambda_2 > \ldots > \lambda_p\}$. Then the trivial solution of (7.4.9), (7.4.10) is stable (unstable, asymptotically stable, exponentially stable) if and only if the trivial solution of the reduced equation *(7.4.11) on E_1 is stable (unstable, asymptotically stable, exponentially stable).*

Proof. Consider discrete time. Proposition 7.4.3(iii) implies that 0 is stable, etc., for φ if and only if it is stable, etc., for $\bar{\varphi}$.

Equation (7.4.12) describes an RDS on the stable manifold $M^-(\omega)$. Since $\gamma < 1$, Corollary 7.3.11 implies that every solution $\bar{\varphi}^-$ of (7.4.12) tends to 0 exponentially fast, more precisely

$$\|\bar{\varphi}^-(n,\omega,x)\|_{\theta^n\omega} \leq C\gamma^n \|x\|_\omega, \quad n \geq 0, \ x \in M^-(\omega).$$

Because

$$\|\bar{\varphi}(n,\omega,x)\|_{\theta^n\omega} = \|\bar{\varphi}^+(n,\omega,x^+)\|_{\theta^n\omega} + \|\bar{\varphi}^-(n,\omega,x^-)\|_{\theta^n\omega},$$

the stability behavior of $\bar{\varphi}$ is exactly reflected by that of $\bar{\varphi}^+$. ⊐

7.4.3 Hartman-Grobman Theorem

We can now improve the result of the last subsection: Not only is the RDS φ topologically equivalent to an RDS $\bar{\varphi}$ which is block-diagonal, but even to an RDS whose blocks corresponding to non-vanishing Lyapunov exponents are linear. This is the key statement of a random version of the Hartman-Grobman theorem due to Wanner [340].

We prepare its proof by a lemma which is an application of the First Key Technical Lemma 7.3.3. Except for the final theorem we will restrict ourselves to discrete time.

7.4.8 Lemma. *Consider the random difference equations*

$$x_{n+1} = A(\theta^n\omega)x_n + f_1(n,\omega,x_n) \quad (7.4.18)$$

and

$$x_{n+1} = A(\theta^n\omega)x_n + f_2(n,\omega,x_n), \quad (7.4.19)$$

in $(\mathbb{R}^d, \|\cdot\|_\omega)$ with measurable $Gl(d,\mathbb{R})$-valued A and measurable f_1 and f_2. Assume that there are constants $\beta \in (0,1)$, $M \geq 0$ and $L \in [0, 1-\beta)$ such that for each fixed ω

$$\|A(\omega)\|_{\omega,\theta\omega} \leq \beta,$$

$$f_i(n,\omega,0) = 0, \quad \sup_{x \in \mathbb{R}^d} \|f_i(n,\omega,x)\|_{\theta^{n+1}\omega} \leq M, \quad i=1,2,$$

and

$$\|f_i(n,\omega,x) - f_i(n,\omega,y)\|_{\theta^{n+1}\omega} \leq L\|x-y\|_{\theta^n\omega}, \quad i=1,2.$$

Finally assume that all solutions of (7.4.18) and (7.4.19) exist on all of $\mathbb{Z}$ and that the solution with initial data (k,x), denoted by $\varphi^{(i)}(\cdot,\omega,k,x)$, is continuous with respect to x.

Then for every (ω,k,x) there exists a uniquely determined point $H(\omega,k,x) \in \mathbb{R}^d$ such that

$$\varphi^{(2)}(\cdot,\omega,k,H(\omega,k,x)) - \varphi^{(1)}(\cdot,\omega,k,x) \in X_{1,\omega},$$

where

$$X_{1,\omega} = \{\xi : \mathbb{Z} \to \mathbb{R}^d : \|\xi\|_{1,\omega} := \sup_{n \in \mathbb{Z}} \|\xi_n\|_{\theta^n \omega} < \infty\}.$$

The function $(\omega, k, x) \mapsto H(\omega, k, x)$ is measurable, continuous with respect to x and satisfies the estimate

$$\|H(\omega, k, x) - x\|_{\theta^k \omega} \leq \frac{2M}{1 - \beta - L}.$$

Further, if $\xi^{(1)}(\cdot)$ is a solution of (7.4.18), then $\xi^{(2)}(\cdot) = H(\omega, \cdot, \xi^{(1)}(\cdot))$ is a solution of (7.4.19).

Proof. Consider the random difference equation

$$x_{n+1} = A(\theta^n \omega) x_n + f(n, \omega, k, x_n) + f_0(n+1, \omega, k, x), \quad (7.4.20)$$

where

$$f(n, \omega, k, x) := f_2(n, \omega, x + \varphi^{(1)}(n, \omega, k, x)) - f_2(n, \omega, \varphi^{(1)}(n, \omega, k, x)),$$

and

$$f_0(n+1, \omega, k, x) = f_2(n, \omega, \varphi^{(1)}(n, \omega, k, x)) - f_1(n, \omega, \varphi^{(1)}(n, \omega, k, x)).$$

Equation (7.4.20) depends on the parameter (k, x), it is solvable on all of $\mathbb{Z}$ and satisfies all the assumptions of the First Key Technical Lemma 7.3.3 with $\gamma = 1$ and

$$\|f_0(\cdot, \omega, k, x)\|_{1,\omega} \leq 2M.$$

Hence by the Key Technical Lemma 7.3.3(ii) there is a uniquely determined initial data $(k, \eta(\omega, k, x))$ (depending measurably on (ω, x) and continuously on x) for which the solution $\varphi^{(3)}(\cdot, \omega, \eta(\omega, k, x), k, x)$ of (7.4.20) is in $X_{1,\omega}$ and

$$\|\varphi^{(3)}(\cdot, \omega, \eta(\omega, k, x), k, x)\|_{1,\omega} \leq \frac{2M}{1 - \beta - L},$$

hence

$$\|\eta(\omega, k, x)\|_{\theta^k \omega} \leq \frac{2M}{1 - \beta - L}. \quad (7.4.21)$$

Note that $\varphi^{(3)}(\cdot)$ is a solution of (7.4.20) for the parameter values (k, x) if and only if $\varphi^{(3)}(\cdot) + \varphi^{(1)}(\cdot, \omega, k, x)$ is a solution of (7.4.19). Hence

$$H(\omega, k, x) := x + \eta(\omega, k, x)$$

is the uniquely determined quantity for which

$$\varphi^{(2)}(\cdot, \omega, k, H(\omega, k, x)) - \varphi^{(1)}(\cdot, \omega, k, x) \in X_{1,\omega}.$$

Moreover, the mapping $(\omega, x) \mapsto H(\omega, k, x)$ is measurable and continuous with respect to x since $(\omega, x) \mapsto \eta(\omega, k, x)$ is so. By its definition and (7.4.21), H satisfies the required estimate.

7.4 Hartman-Grobman Theorem

As to the last assertion, let $\xi^{(1)}(\cdot)$ be an arbitrary solution of (7.4.18), set $x := \xi^{(1)}(k)$ and $\xi^*(n) := \varphi^{(2)}(n, \omega, k, H(\omega, k, x))$. Then $\xi^* - \xi^{(1)}$ is a solution of (7.4.20) which is $(1, \omega)$-bounded.

Now let $n \in \mathbb{Z}$ be arbitrary and put $x^* := H(\omega, n, \xi^{(1)}(n))$. We proved above that x^* is the unique point in $\mathbb{R}^d$ for which

$$\varphi^{(2)}(\cdot, \omega, n, x^*) - \varphi^{(1)}(\cdot, \omega, n, \xi^{(1)}(n)) \in X_{1,\omega}.$$

The identity $\xi^{(1)}(\cdot) = \varphi^{(1)}(\cdot, \omega, n, \xi^{(1)}(n))$ implies

$$\xi^*(\cdot) - \varphi^{(1)}(\cdot, \omega, n, \xi^{(1)}(n)) \in X_{1,\omega},$$

hence by uniqueness $\xi^*(\cdot) = \varphi^{(2)}(\cdot, \omega, n, x^*)$ and $x^* = H(\omega, n, \xi^{(1)}(n)) = \xi^*(n)$. Since $n \in \mathbb{Z}$ was arbitrary, we are done: $H(\omega, \cdot, \xi^{(1)}(\cdot))$ is indeed a solution of (7.4.19), namely $\xi^*(\cdot)$. □

As an intermediate step we linearize an RDS with a stable fixed point 0 by means of the last lemma.

7.4.9 Proposition (Topological Linearization of Stable RDS). *Let φ be a prepared RDS with discrete time and fixed point 0 generated by the random difference equation*

$$x_{n+1} = A(\theta^n \omega)x_n + F(\theta^n \omega, x_n), \quad n \in \mathbb{Z}, \quad F(\omega, 0) = 0. \quad (7.4.22)$$

Suppose that the linear RDS Φ generated by A has top Lyapunov exponent $\lambda_1 < 0$, so that by our preparations $\|A(\omega)\|_{\omega,\theta\omega} \leq \beta := \beta_1 = e^{\lambda_1 + \kappa} < 1$.

Assume further that the nonlinear part F of φ satisfies the following conditions: There are constants $L \in [0, 1-\beta)$ and $M \geq 0$ for which

$$\sup_{x \in \mathbb{R}^d} \|F(\omega, x)\|_{\theta\omega} \leq M$$

and

$$\|F(\omega, x) - F(\omega, y)\|_{\theta\omega} \leq L\|x - y\|_\omega, \quad x, y \in \mathbb{R}^d.$$

Then the RDS φ and Φ are topologically equivalent, i. e. there exists a random homeomorphism $h(\omega) : \mathbb{R}^d \to \mathbb{R}^d$ with $h(\omega)0 = 0$ such that

$$h(\theta(t)\omega) \circ \varphi(t, \omega) = \Phi(t, \omega) \circ h(\omega), \quad t \in \mathbb{Z}.$$

Moreover, h has the property of being "near-identity" in the following sense:

$$\|h(\omega, x) - x\|_\omega \leq \frac{2M}{1 - \beta - L}, \quad \|h^{-1}(\omega, x) - x\|_\omega \leq \frac{2M}{1 - \beta - L},$$

and $h(\omega, x) \in \mathbb{R}^d$ is the uniquely determined point for which

$$\Phi(\cdot, \omega)h(\omega, x) - \varphi(\cdot, \omega)x \in X_{1,\omega}.$$

Proof. We apply Lemma 7.4.8 twice: First with $f_1(n, \omega, x) = F(\theta^n \omega, x)$ and $f_2 = 0$ yielding $H(\omega, k, x) = H(\theta^k \omega, 0, x)$ and a second time with $f_1 = 0$ and $f_2(n, \omega, x) = F(\theta^n \omega, x)$ yielding $\tilde{H}(\omega, k, x) = \tilde{H}(\theta^k \omega, 0, x)$. Put

$$h(\omega, x) := H(\omega, 0, x), \quad \tilde{h}(\omega, x) := \tilde{H}(\omega, 0, x).$$

Both h and $\tilde{h}$ are measurable with respect to (ω, x) and continuous with respect to x.

We are done if we can prove that $\tilde{h}(\omega) = h^{-1}(\omega)$.

This is accomplished by applying Lemma 7.4.8 a third time, now with $f_1(n, \omega, x) = f_2(n, \omega, x) = F(\theta^n \omega, x)$. There is hence a unique mapping $h^*(\omega, x)$ such that

$$\varphi(\cdot, \omega, h^*(\omega, x)) - \varphi(\cdot, \omega, x) \in X_{1,\omega},$$

namely $h^*(\omega, x) = x$. On the other hand, the above definitions of h and $\tilde{h}$ say that also

$$\Phi(\cdot, \omega) h(\omega, x) - \varphi(\cdot, \omega, x) \in X_{1,\omega},$$

as well as

$$\varphi(\cdot, \omega, \tilde{h}(\omega, h(\omega, x))) - \Phi(\cdot, \omega) h(\omega, x) \in X_{1,\omega}.$$

But then

$$\varphi(\cdot, \omega, \tilde{h}(\omega, h(\omega, x))) - \varphi(\cdot, \omega, x) \in X_{1,\omega}.$$

By the uniqueness statement of Lemma 7.4.8,

$$\tilde{h}(\omega, h(\omega, x)) = h^*(\omega, x) = x, \quad \text{all } x \in \mathbb{R}^d.$$

Similarly

$$h(\omega, \tilde{h}(\omega, x)) = x, \quad \text{all } x \in \mathbb{R}^d.$$

Hence $\tilde{h}(\omega) = h^{-1}(\omega)$.

That h and h^{-1} map respective orbits to each other follows from the last statement of Lemma 7.4.8. □

7.4.10 Remark. There is an obvious dual version of Lemma 7.4.8 and Proposition 7.4.9, applying to an RDS φ generated by (7.4.22) and its linear part Φ, but now for the case that the smallest Lyapunov exponent of Φ satisfies $\lambda_p > 0$. Then $\alpha := \alpha_p = e^{\lambda_p - \kappa} > 1$, $\|A(\omega)^{-1}\|_{\theta\omega,\omega} \leq \alpha^{-1}$ and the constant L has to satisfy $L \in [0, \alpha - 1)$. Now the Second Key Technical Lemma 7.3.6 has to be used. We leave details to the reader. ∎

7.4.11 Proposition (Topological Linearization of Hyperbolic RDS).
Let φ be a prepared RDS with discrete time and fixed point 0 generated by the random difference equation

$$x_{n+1} = A(\theta^n \omega) x_n + F(\theta^n \omega, x_n), \quad n \in \mathbb{Z}, \quad F(\omega, 0) = 0.$$

Suppose that the linear RDS Φ generated by A is hyperbolic.

Assume further that there is a constant $M \geq 0$ such that

$$\sup_{x \in \mathbb{R}^d} \|F(\omega, x)\|_{\theta\omega} \leq M$$

and a constant $L \geq 0$ such that $L(1 + C(L)) < \frac{3-\sqrt{5}}{2}\delta$, where $C(L)$ is given in Theorem 7.3.14, and

$$\|F(\omega, x) - F(\omega, y)\|_{\theta\omega} \leq L\|x - y\|_\omega, \quad x, y \in \mathbb{R}^d.$$

Then the RDS φ and Φ are topologically equivalent, i.e. there exists a random homeomorphism $h(\omega) : \mathbb{R}^d \to \mathbb{R}^d$ with $h(\omega)0 = 0$ such that

$$h(\theta(t)\omega) \circ \varphi(t, \omega) = \Phi(t, \omega) \circ h(\omega), \quad t \in \mathbb{Z}.$$

Moreover, for any $1 \leq i \leq j \leq p$, the center manifolds $M_{ij}(\omega)$ of φ are homeomorphically mapped onto the linear subspaces E_{ij}, the center manifolds of Φ,

$$h(\omega)M_{ij}(\omega) = E_{ij}.$$

Further, the unique invariant measure of φ is δ_0, the Dirac measure at 0.

Proof. We first apply Theorem 7.4.4 according to which φ is topologically equivalent via a random homeomorphism $g(\omega)$ to the RDS $\bar{\varphi}$ generated by the system (7.4.15) of p decoupled equations.

Now each block satisfies the conditions of Proposition 7.4.9 or its dual (see Remark 7.4.10). Hence the RDS $\bar{\varphi}_i$ generated by the i-th equation is topologically equivalent to the linear RDS Φ_i generated by A_i. Denote the corresponding random homeomorphism by $\bar{h}_i(\omega)$. The random homeomorphism

$$h(\omega, x) := (\bar{h}_1(\omega, g_1(\omega, x)), \ldots, \bar{h}_p(\omega, g_p(\omega, x)))$$

serves our purpose.

A measure μ_ω is φ-invariant if and only if the measure $\nu_\omega := h(\omega)\mu_\omega$ is Φ-invariant. But the only invariant measure of a hyperbolic RDS is δ_0, cf. Corollary 5.6.3. □

We now treat the classical C^1 case and also "undo" the preparations.

7.4.12 Theorem (Hartman-Grobman Theorem). *Let φ be a C^1 RDS with discrete or continuous time and fixed point 0 so that $\varphi(t, \omega, x) = D\varphi(t, \omega, 0)x + \psi(t, \omega, x)$ with $\psi(t, \omega, 0) = 0$ and $D\psi(t, \omega, 0) = 0$. Suppose that the linear RDS $D\varphi$ satisfies the IC of the MET and is hyperbolic.*

Assume that the nonlinear part $\bar{\psi}(t, \omega, \cdot)$ (for discrete time $\bar{F}(\omega, \cdot) := \bar{\psi}(1, \omega, \cdot)$) of the corresponding prepared version of φ fulfills the following conditions:

(a) There exists a constant $M \geq 0$ such that

$$\text{for } \mathbb{T} = \mathbb{Z}: \quad \sup_{x \in \mathbb{R}^d} \|\bar{F}(\omega, x)\|_{\theta\omega} \leq M,$$

$$\text{for } \mathbb{T} = \mathbb{R}: \quad \sup_{0 \leq t \leq 1} \sup_{x \in \mathbb{R}^d} \|\bar{\psi}(t, \omega, x)\|_{\theta(t)\omega} \leq M.$$

(b) There exists a constant $L \geq 0$ such that $L(1 + C(L)) < \frac{3-\sqrt{5}}{2}\delta$ and

for $\mathbb{T} = \mathbb{Z}$: $\sup_{x \in \mathbb{R}^d} \|D\bar{F}(\omega, x)\|_{\omega, \theta\omega} \leq L$,

for $\mathbb{T} = \mathbb{R}$: $\sup_{0 \leq t \leq 1} \sup_{x \in \mathbb{R}^d} \|D\bar{\psi}(t, \omega, x)\|_{\omega, \theta(t)\omega} \leq L$.

Then φ is topologically equivalent to its linear part $D\varphi$ at $x = 0$, i. e. there exists a random homeomorphism $h(\omega)$ with $h(\omega)0 = 0$, such that

$$h(\theta(t)\omega) \circ \varphi(t, \omega) = D\varphi(t, \omega, 0) \circ h(\omega), \quad t \in \mathbb{T}.$$

Moreover, all invariant manifolds $M_{ij}(\omega)$ are homeomorphically mapped by $h(\omega)$ onto their linear counterparts $E_{ij}(\omega)$.

Further, the unique invariant measure of φ is δ_0, the Dirac measure at 0.

Proof. (i) We first assume that the RDS is prepared. The mean value theorem implies that the assumptions of Theorem 7.4.11 are satisfied. Hence there is a random homeomorphism $\bar{h}(\omega)$ such that

$$\bar{h}(\theta(t)\omega) \circ \bar{\varphi}(t, \omega) = D\bar{\varphi}(t, \omega, 0) \circ \bar{h}(\omega),$$

which also maps the invariant manifolds $\bar{M}_{ij}(\omega)$ homeomorphically onto E_{ij}.

(ii) If φ is not prepared, we use Theorem 7.2.6 and switch to the corresponding prepared RDS $\bar{\varphi}$. If $P(\omega)$ denotes the random linear isomorphism of Theorem 7.2.6 then

$$h(\omega) := P(\omega)^{-1} \circ \bar{h}(\omega) \circ P(\omega)$$

certainly defines a topological equivalence of φ and $D\varphi(0)$, as a simple calculation shows. Since $P(\omega)M_{ij}(\omega) = \bar{M}_{ij}(\omega)$, the proof is completed. □

7.4.13 Remark (Hartman-Grobman Theorem, Non-Hyperbolic Case). In the non-hyperbolic C^1 case, the equation of the decoupled system (7.4.15) and (7.4.17) which corresponds to the vanishing Lyapunov exponent cannot be simplified further. Under the assumptions of Theorem 7.4.12 with the only exception that now $\lambda_{i_0} = 0$ for some i_0, the RDS φ is topologically equivalent in general not to $D\varphi =: \Phi = (\Phi_s, \Phi_c, \Phi_u)$, but to $\tilde{\Phi} := (\Phi_s, \tilde{\Phi}_c, \Phi_u)$, where $\Phi_{s,c,u} := \Phi|_{E_{s,c,u}}$, $E_{s,c,u}$ the stable/center/unstable space of Φ and (e.g. for discrete time) $\tilde{\Phi}_c z_c := \Phi_c z_c + F_c(z_c + m_c(z_c))$, $E_c(\omega) \ni z_c \mapsto m_c(z_c) \in E_s(\omega) \oplus E_u(\omega)$ the C^1 graph of the classical center manifold $M_c(\omega)$ of φ.

A measure μ is φ-invariant if and only if the image measure $\tilde{\mu}_\omega$ is $\tilde{\Phi}$-invariant, which is the case if and only if $\operatorname{supp}\tilde{\mu}_\omega \subset E_c(\omega)$ and $\tilde{\mu}$ is $\tilde{\Phi}_c$-invariant. Hence nontrivial invariant measures of φ are supported by the classical center manifold $M_c(\omega)$ of φ. In particular, if φ has an invariant measure which is different from δ_0, then the point 0 cannot be hyperbolic. ∎

7.4.14 Remark (Topological Versus Smooth Linearization). We stress that even for smooth RDS, the linearizing homeomorphism is in general not a diffeomorphism. The problem of linearizing a C^∞ RDS by means of a C^∞ diffeomorphism is part of the normal form problem treated in Chap. 8. It turns out that even in the hyperbolic case, obstructions to linearization in the form of "resonances" appear. ∎

7.5 Local Invariant Manifolds

Although the outcome of our invariant manifold theory developed in Sect. 7.3 is very satisfactory, this is not at all the case for the assumptions we have to impose, as some of these drastically restrict its use in applications. Recall that for a C^k RDS φ with two-sided time and fixed point 0 we put $\Phi(t,\omega) := D\varphi(t,\omega,0)$ and write $\varphi(t,\omega,x) = \Phi(t,\omega)x + \psi(t,\omega,x)$, where ψ is C^k in x and has the properties $\psi(t,\omega,0) = 0$ and $D\psi(t,\omega,0) = 0$. We first need the not very restrictive and inevitable assumption that Φ satisfies the IC of the MET – without this assumption there would be no invariant manifold theory for RDS at all.

Our further assumptions and their drawbacks are:

– First we "prepare" the RDS φ: We block-diagonalize Φ by means of a linear random coordinate transformation $P(\omega)$ whose "construction" (i.e. measurable selection) needs the knowledge of the Oseledets spaces of Φ, which is hardly ever available. Then we introduce the technically very convenient random norm $\|\cdot\|_\omega$, the construction of which also needs the knowledge of Oseledets spaces as well as Lyapunov exponents – altogether an extremely unrealistic situation.

– Next we impose a global Lipschitz condition with a deterministic constant L on the nonlinearity $\bar\psi$ of the prepared RDS $\bar\varphi$, but formulated in terms of the random norm. This constant L has to be sufficiently small, where smallness is measured in terms of the gaps in the Lyapunov spectrum of Φ. For the Hartman-Grobman theorem we need in addition a uniform deterministic bound M on $\bar\psi$, also measured by means of the random norm.

However, there is a satisfactory way out: the construction of *local invariant manifolds* under conditions which are quite general and are met in most applications, to which this section is devoted. Our method will be to replace the nonlinearity ψ of φ by a new one, $\tilde\psi$, which is of the same regularity class and agrees with ψ in a neighborhood U of 0, but satisfies the assumptions for the existence of global invariant manifolds. We call the intersection of the global manifolds for $\tilde\varphi$ with the neighborhood U the local manifolds of φ. (The problem then is what those manifolds mean dynamically for the RDS φ.)

7.5.1 Local Manifolds for Discrete Time

The existence of local invariant manifolds can be proved in great generality for discrete time. We hence start again with a C^k RDS φ with fixed point 0 generated by
$$x_{n+1} = A(\theta^n \omega) x_n + F(\theta^n \omega, x_n), \quad n \in \mathbb{Z}, \tag{7.5.1}$$
where $A(\omega) = D\varphi(\omega, 0)$, $F(\omega, 0) = 0$ and $DF(\omega, 0) = 0$, and the linear cocycle Φ generated by A satisfies the IC of the MET. In view of Proposition 7.2.6 we can without loss of generality assume that φ is prepared in the sense of Definition 7.2.5.

Truncation of Nonlinear Part

7.5.1 Lemma. *Let $F(\omega, \cdot) : \mathbb{R}^d \to \mathbb{R}^d$ be a C^k function with $F(\omega, 0) = 0$ and $DF(\omega, 0) = 0$. Then for any fixed $M \in (0, \infty]$ and $L \in (0, \infty)$ there exists a random variable $\rho : \Omega \to (0, 1]$ and a function $\tilde{F}(\omega, \cdot) : \mathbb{R}^d \to \mathbb{R}^d$ such that*
(a) $F(\omega, x) = \tilde{F}(\omega, x)$ for all $x \in B_{\rho(\omega)}$, where
$$B_{\rho(\omega)} := \{x \in \mathbb{R}^d : \|x\|_\omega \leq \rho(\omega)\},$$
(b) $\tilde{F}(\omega, \cdot) \in \mathcal{C}^0_{b,\omega}$, more precisely[5]
$$\|\tilde{F}(\omega, \cdot)\|^0_\omega := \sup_{x \in \mathbb{R}^d} \|\tilde{F}(\omega, x)\|_{\theta\omega} \leq M,$$
(void if $M = \infty$) and $\tilde{F}(\omega, \cdot) \in \mathcal{C}^k_{b,\omega}$, more precisely
$$\|\tilde{F}(\omega, \cdot)\|^1_\omega := \sup_{x \in \mathbb{R}^d} \|D\tilde{F}(\omega, x)\|_{\omega, \theta\omega} \leq L.$$

Proof. (i) We fix a C^∞ cut-off function $\chi : \mathbb{R}^d \to \mathbb{R}^d$ with the following properties:
$$C_0 := \sup_{x \in \mathbb{R}^d} \|\chi(x)\| = 1, \quad \chi(x) = \begin{cases} x, & \|x\| \leq \frac{1}{2}, \\ \frac{x}{\|x\|}, & \|x\| \geq 1. \end{cases}$$
Note that all derivatives of χ are bounded, so
$$C_q := \sup_{x \in \mathbb{R}^d} \|D^q \chi(x)\| < \infty, \quad q \geq 1.$$
(ii) Now define for $\varepsilon > 0$
$$\chi_\varepsilon(x) : \mathbb{R}^d \to \mathbb{R}^d, \quad \chi_\varepsilon(x) := \varepsilon \chi(\frac{1}{\varepsilon} x) = \begin{cases} x, & \|x\| \leq \frac{\varepsilon}{2}, \\ \frac{\varepsilon}{\|x\|} x, & \|x\| \geq \varepsilon. \end{cases}$$
Clearly $\sup_{x \in \mathbb{R}^d} \|\chi_\varepsilon(x)\| = \varepsilon$ and the chain rule yields

---
[5] See Definition 7.3.18 for the spaces $\mathcal{C}^k_{b,\omega}$.

$$\sup_{x\in\mathbb{R}^d} \|D\chi_\varepsilon(x)\| \leq C_1.$$

(iii) We now construct a Euclidean ball $K_{R(\omega)} = \{x \in \mathbb{R}^d : \|x\| \leq R(\omega)\}$ in which $F(\omega, \cdot)$ and $DF(\omega, \cdot)$ have certain prescribed bounds.

Denote by $B_1 : \Omega \to [1, \infty)$ the 1-slowly varying function which connects the Euclidean and the random norm,

$$\frac{1}{B_1(\omega)}\|x\| \leq \|x\|_\omega \leq B_1(\omega)\|x\|, \quad \frac{1}{e}B_1(\omega) \leq B_1(\theta\omega) \leq eB_1(\omega).$$

First choose $R_1(\omega) > 0$ such that in case $M < \infty$

$$R_1(\omega) := \max\{r : \|F(\omega, x)\| \leq \frac{M}{eB_1(\omega)} \text{ for all } \|x\| \leq r\}.$$

This is possible since $F(\omega, 0) = 0$ and $F(\omega, \cdot)$ is continuous. Clearly R_1 is measurable. In case $M = \infty$ take $R_1 = \infty$.

Now choose $R_2(\omega) > 0$ such that

$$R_2(\omega) := \max\{r : \|DF(\omega, x)\| \leq \frac{L}{C_1 eB_1(\omega)^2} \text{ for all } \|x\| \leq r\}.$$

Again, this choice is possible since $DF(\omega, 0) = 0$ and $DF(\omega, \cdot)$ is continuous. In addition, R_2 is measurable.

(iv) We now define $R(\omega) := \min(R_1(\omega), R_2(\omega), 2B_1(\omega))$ and

$$\tilde{F}(\omega, x) := F(\omega, \chi_{R(\omega)}(x)),$$

and verify the assertions about $\tilde{F}$.

Since $\chi_{R(\omega)}$ is C^∞, $\tilde{F}$ is of the same regularity class as F, hence C^k. Clearly

$$\tilde{F}(\omega, x) = F(\omega, x) \quad \text{for all} \quad x \in K_{R(\omega)/2}.$$

By definition,

$$\|\tilde{F}(\omega, x)\|_{\theta\omega} \leq B_1(\theta\omega)\|\tilde{F}(\omega, x)\| \leq eB_1(\omega)\|\tilde{F}(\omega, x)\|$$

and

$$\sup_{x\in\mathbb{R}^d} \|\tilde{F}(\omega, x)\| = \sup_{\|x\|\leq R(\omega)} \|F(\omega, x)\| \leq \frac{M}{eB_1(\omega)},$$

hence

$$\|\tilde{F}(\omega, \cdot)\|_\omega^0 = \sup_{x\in\mathbb{R}^d} \|\tilde{F}(\omega, x)\|_{\theta\omega} \leq M.$$

Similarly,

$$\|\tilde F(\omega,\cdot)\|^1_\omega = \sup_{x\in\mathbb{R}^d} \|D\tilde F(\omega,x)\|_{\omega,\theta\omega} \le e B_1(\omega)^2 \sup_{x\in\mathbb{R}^d} \|D\tilde F(\omega,x)\|$$
$$= C_1\, e\, B_1(\omega)^2 \sup_{\|x\|\le R(\omega)} \|DF(\omega,x)\| \le L.$$

Putting $\rho(\omega) := \frac{R(\omega)}{2B_1(\omega)} \le 1$, $F(\omega,x) = \tilde F(\omega,x)$ also holds in the $\rho(\omega)$-ball $B_{\rho(\omega)}$ with respect to random norm since

$$B_{\rho(\omega)} = \{x \in \mathbb{R}^d : \|x\|_\omega \le \rho(\omega)\} \subset K_{R(\omega)/2}.$$

Also, for $2 \le q \le k$, $\|\tilde F(\omega,\cdot)\|^q_\omega = \sup_{x\in\mathbb{R}^d} \|D^q \tilde F(\omega,x)\|_{\theta\omega} \le L_q(\omega) < \infty$, hence $\tilde F(\omega,\cdot) \in C^k_{b,\omega}$. □

Construction of Local Manifolds

7.5.2 Proposition (Local Invariant Manifolds, Discrete Time).
Let φ be the prepared C^k RDS generated by (7.5.1) and let $L < \frac{3-\sqrt{5}}{2}\delta$. Choose $\rho = \rho(L) > 0$ according to Lemma 7.5.1 (case $M = \infty$) and put

$$U(\omega) := B_{\rho(\omega)} = \{x \in \mathbb{R}^d : \|x\|_\omega \le \rho(\omega)\}.$$

Then:
(i) $\tilde\varphi(\omega,x) := A(\omega)x + \tilde F(\omega,x)$ is the time-one mapping of a prepared C^k RDS which satisfies the assumptions of the Global Smooth Invariant Manifold Theorem 7.3.19. Denote the corresponding global manifolds by $\tilde M_{ij}(\omega)$.
(ii) We have

$$\varphi(n,\omega,x) = \tilde\varphi(n,\omega,x) \quad \text{for all } x \in U(\omega),\ n \in [T^-_U(\omega,x), T^+_U(\omega,x)],$$

where

$$T^+_U(\omega,x) := \sup\{n \in \mathbb{Z}^+ : \varphi(k,\omega,x) \in U(\theta^k\omega) \text{ for all } 0 \le k \le n\} \ge 0,$$

$$T^-_U(\omega,x) := \inf\{n \in \mathbb{Z}^- : \varphi(k,\omega,x) \in U(\theta^k\omega) \text{ for all } n \le k \le 0\} \le 0$$

denote the moment before the first exit forwards/backwards in time from the neighborhood U.
(iii) For each $N \in \mathbb{N}$ there is a radius $\rho_N(\omega) > 0$ such that, setting

$$U_N(\omega) := B_{\rho_N(\omega)} = \{x \in \mathbb{R}^d : \|x\|_\omega \le \rho_N(\omega)\},$$

$$\varphi(n,\omega,x) = \tilde\varphi(n,\omega,x) \quad \text{for all} \quad x \in U_N(\omega),\ |n| \le N.$$

Equivalently,

$$\lim_{x\to 0} T^+_U(\omega,x) = \infty, \quad \lim_{x\to 0} T^-_U(\omega,x) = -\infty.$$

(iv) We call

$$M_{ij}^{\mathrm{loc}}(\omega) := \tilde{M}_{ij}(\omega) \cap U(\omega)$$

the local invariant manifold *of φ corresponding to the spectral interval* $\Lambda_{ij} = \{\lambda_i > \ldots > \lambda_j\}$ *tangent to* E_{ij}. *It is locally invariant under φ in the following sense: If* $x \in M_{ij}^{\mathrm{loc}}(\omega)$ *then* $\varphi(n,\omega,x) \in M_{ij}^{\mathrm{loc}}(\theta^n\omega)$ *for all* $n \in [T_U^-(\omega,x), T_U^+(\omega,x)]$.

See Fig. 7.9.

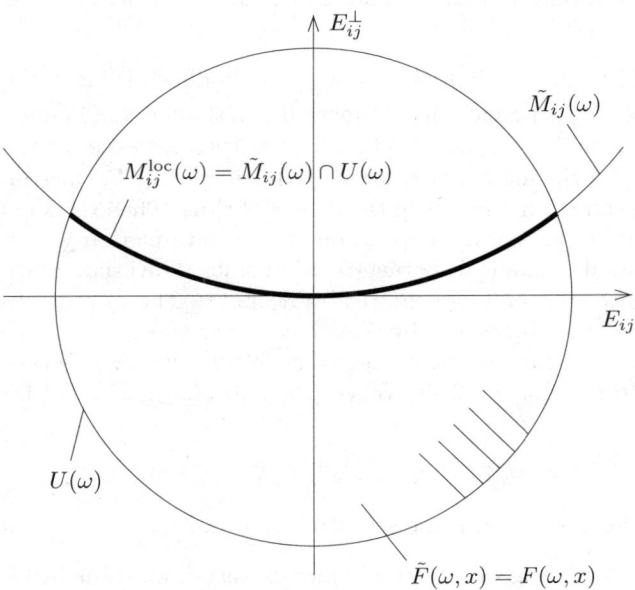

Fig. 7.9. Local invariant manifold

Proof. (i) We show that $\tilde{\varphi}(\omega,\cdot) := A(\omega) + \tilde{F}(\omega,\cdot) \in \mathrm{Diff}^k(\mathbb{R}^d)$.

As in the proof of part (o) of the Second Key Technical Lemma 7.3.6 we prove invertibility of $\tilde{\varphi}(\omega)$ by studying for each $(\omega, z) \in \Omega \times \mathbb{R}^d$ the operator

$$T_{\omega,z}: \mathbb{R}^d \to \mathbb{R}^d, \quad x \mapsto A(\omega)^{-1}z - A(\omega)^{-1}\tilde{F}(\omega,x).$$

Since $\|A(\omega)^{-1}\|_{\theta\omega,\omega} \leq \alpha_p^{-1}$ and δ was chosen in Subsect. 7.2.2 to satisfy $\delta < \frac{\alpha_p}{2}$, the operator is contracting with uniform rate $L/2\delta$.

By the well-known fact that the fixed point $x^*(z)$ of a uniform contraction $f: X \times Z \to X$ of class C^k depends in a C^k-way on the parameter z (see Chow and Hale [100: Theorem 2.2]), the result follows.

(ii) If $x_n \in U(\theta^n\omega)$ then $\tilde{F}(\theta^n\omega, x_n) = F(\theta^n\omega, x_n)$, implying $\tilde{x}_{n+1} = x_{n+1}$, similarly backwards in time. Hence the orbits of $\tilde{\varphi}$ and φ coincide at times $n \pm 1$ if they do so at time n, and if $x_n \in U(\theta^n\omega)$.

384   Chapter 7. Invariant Manifolds

(iii) Choose an arbitrary $N \in \mathbb{N}$ and consider for $|n| \leq N$ the mappings $x \mapsto \varphi(n, \omega, x)$. Since they are continuous (even C^k) and $\varphi(n, \omega, 0) = 0$, there exists a $\rho_N(\omega) > 0$ for which

$$\|\varphi(n, \omega, x)\|_{\theta^n \omega} \leq \rho(\theta^n \omega) \quad \text{for all} \quad |n| \leq N,$$

provided $x \in U_N(\omega) = B_{\rho_N(\omega)} = \{x \in \mathbb{R}^d : \|x\|_\omega \leq \rho_N(\omega)\}$. For those x clearly $T_U^+(\omega, x) \geq N$ and $T_U^-(\omega, x) \leq -N$.

(iv) The local invariance immediately follows from the $\tilde{\varphi}$-invariance of $\tilde{M}_{ij}(\omega)$ and the fact that $\tilde{\varphi}(n, \omega, x)$ and $\varphi(n, \omega, x)$ coincide for $n \in [T_U^-(\omega, x), T_U^+(\omega, x)]$. □

7.5.3 Remark. (i) Although the ball $U(\omega) = B_{\rho(\omega)}$ only depends on φ, the function $\tilde{F}$ depends on the particular cut-off function χ in the proof of Lemma 7.5.1, and so do its invariant manifolds $\tilde{M}_{ij}(\omega)$. This is the case even for the part of these manifolds in the intersection with $U(\omega)$, where $\tilde{F} = F$, since an orbit starting in $U(\omega)$ can go in and out in the course of time. Thus the local invariant manifolds $M_{ij}^{\text{loc}}(\omega)$ generally depend on the cut-off function χ.

(ii) There is no general dynamical characterization of local invariant manifolds. Hence so far they are purely geometrical objects tangent to the subspaces E_{ij} characterized (not uniquely) by their local invariance.

(iii) The local invariance can also be expressed as follows: For $N \in \mathbb{N}$ take $U_N(\omega)$ according to Proposition 7.5.2(iii). Since $\varphi(n, \omega)U_N(\omega) \subset U(\theta^n \omega)$ for $|n| \leq N$, we have

$$\varphi(n, \omega)M_{ij}^{\text{loc}}(\omega) = M_{ij}^{\text{loc}}(\theta^n \omega) \quad \text{for all } -N \leq n \leq N$$

on the intersection of both sides with the set $\varphi(n, \omega)U_N(\omega)$. ∎

There is, however, the following observation which is very important in bifurcation theory, namely that "small" solutions always lie on certain local invariant manifolds, irrespective of the cut-off function.

7.5.4 Proposition. *Let the conditions of Proposition 7.5.2 be satisfied. Let $x \in U(\omega)$.*

(i) If $T_U^-(\omega, x) = -\infty$,[6] then $x \in M_{1j}^{\text{loc}}(\omega)$ for any local unstable manifold for which $\lambda_j \leq 0$ and for any cut-off function.

(ii) If $T_U^+(\omega, x) = \infty$, then $x \in M_{ip}^{\text{loc}}(\omega)$ for any local stable manifold for which $\lambda_i \geq 0$ and for any cut-off function.

(iii) If $T_U^-(\omega, x) = -\infty$ and $T_U^+(\omega, x) = \infty$, then $x \in M_{ij}^{\text{loc}}(\omega)$ for any local center manifold for which $\lambda_j \leq 0 \leq \lambda_i$ and any cut-off function. In particular, if $\lambda_i = 0$ for some i and $T_U^{\pm}(\omega, x) = \pm\infty$, then $x \in M_c^{\text{loc}}(\omega) = M_{ii}^{\text{loc}}(\omega)$, a classical local center manifold, for any cut-off function.

Proof. (i) We have $T_U^-(\omega, x) = -\infty$ if and only if

---

[6] This, like the other conditions in parts (ii) and (iii) of this proposition, is a condition on φ only.

$$\|\varphi(n,\omega,x)\|_{\theta^n\omega} \le \rho(\theta^n\omega) \quad \text{for all } n \le 0,$$

hence $\tilde{\varphi}(n,\omega,x) = \varphi(n,\omega,x)$ for this $x \in U(\omega)$ and all $n \in \mathbb{Z}^-$. Now

$$\tilde{M}_{1j}(\omega) = \{x \in \mathbb{R}^d : \tilde{\varphi}(\cdot,\omega,x) \in X_{a^-,\omega}\},$$

where $a \in \Gamma_j = [\beta_{j+1}+\delta, \alpha_j-\delta] \subset (0,1)$, is the dynamical characterization of this manifold. We try our $x \in U(\omega)$ with $T_U^-(\omega,x) = -\infty$. Then

$$\begin{aligned}
\|\tilde{\varphi}(\cdot,\omega,x)\|_{a^-,\omega} &= \|\varphi(\cdot,\omega,x)\|_{a^-,\omega} = \sup_{n \le 0} a^{-n} \|\varphi(\cdot,\omega,x)\|_{\theta^n\omega} \\
&\le \sup_{n \le 0} a^{-n} \rho(\theta^n\omega) \le 1 < \infty,
\end{aligned}$$

where we have used the fact that $a \le 1$ and $\rho(\omega) \le 1$. As a result, $x \in M_{1j}^{\text{loc}}(\omega)$ no matter which cut-off function was used for $\tilde{\varphi}$.

The proofs of (ii) and (iii) are completely analogous. □

The moral of Proposition 7.5.4(iii) is that those initial values whose φ-orbits never escape the ball U forwards and backwards in time (this is the meaning of "small" here) are the common core of all local center manifolds.

We collect our findings in the following theorems.

7.5.5 Theorem (Local Invariant Manifold Theorem, Discrete Time). Let φ be the C^k RDS with discrete time and fixed point 0 generated by (7.5.1). Assume that $D\varphi$ satisfies the IC of the MET.

Then for each interval Λ_{ij} from the Lyapunov spectrum of $D\varphi$, φ has local invariant C^k manifolds M_{ij}^{loc} tangent to E_{ij}.

Proof. We switch to the prepared RDS $\tilde{\varphi}(n,\omega) = P(\theta^n\omega) \circ \varphi(n,\omega) \circ P(\omega)^{-1}$ and construct its local invariant manifolds $\tilde{M}_{ij}^{\text{loc}}(\omega)$ according to Proposition 7.5.2. Then $M_{ij}^{\text{loc}}(\omega) := P(\omega)^{-1}\tilde{M}_{ij}^{\text{loc}}(\omega)$ are the corresponding local invariant manifolds of φ. □

7.5.6 Theorem (Local Hartman-Grobman Theorem, Discrete Time). Let φ be a C^1 RDS with discrete time and fixed point 0 generated by (7.5.1). Assume that $\Phi := D\varphi$ satisfies the IC of the MET.

(i) Suppose Φ is hyperbolic. Then φ is locally topologically equivalent to Φ, i.e. there exists a random homeomorphism $h(\omega)$ with $h(\omega,0) = 0$ and a neighborhood $U(\omega)$ of 0 such that

$$h(\theta^n\omega, \varphi(n,\omega,x)) = \Phi(n,\omega)h(\omega,x) \quad \text{for all } x \in U(\omega)$$

and all $n \in [T_U^-(\omega,x), T_U^+(\omega,x)]$.

(ii) If Φ is not hyperbolic, then φ is locally topologically equivalent to the RDS $\tilde{\Phi} = (\Phi_s, \tilde{\Phi}_c, \Phi_u)$, where $\Phi_{s,c,u} = \Phi|_{E_{s,c,u}}$, $E_{s,c,u}$ denotes the stable/center/unstable space of Φ and $\tilde{\Phi}_c = \Phi_c + \tilde{F}_c(z_c + m_c(z_c))$. Here $E_c(\omega) \ni z_c \mapsto m_c(\omega, z_c) \in E_s(\omega) \oplus E_u(\omega)$ is the classical C^1 center manifold of a corresponding cut-off RDS $\tilde{\varphi}$.

Proof. A brief outline of the proof for the case (i) will suffice: We first move to a prepared RDS by a linear isomorphism $P(\omega)$, then use Lemma 7.5.1 and replace the nonlinearity $F = \varphi - \Phi$ with a cut-off $\tilde{F}$ satisfying the assumptions of the global Hartman-Grobman Theorem 7.4.12. We next linearize the RDS $\tilde{\varphi} = \Phi + \tilde{F}$ according to Theorem 7.4.12 by a random homeomorphism which maps all $\tilde{M}_{ij}(\omega)$ homeomorphically to E_{ij}. Let $B_{\rho(\omega)}$ be the closed ball inside of which $\tilde{\varphi} = \varphi$. Then $U(\omega) := P(\omega)^{-1} \text{int} B_{\rho(\omega)}$ serves our purpose. □

7.5.7 Remark. (i) The local topological equivalence in Theorem 7.5.6 can be brought into the following form: For each $N \in \mathbb{N}$ there exists a neighborhood $U_N(\omega)$ of 0 such that (in case (i))

$$h(\theta^n \omega, \varphi(n, \omega, x)) = \Phi(n, \omega) h(\omega, x) \quad \text{for all } x \in U_N(\omega),\ |n| \leq N.$$

(ii) *"Small" invariant measures:* We have the following extension of the global statements in Theorem 7.4.12 and Remark 7.4.13. If a φ-invariant measure μ is "small" in the sense that for some $N \in \mathbb{N}$, $\text{supp}\mu_\omega \subset U_N(\omega)$, then necessarily $\mu_\omega = \delta_0$ in the hyperbolic case, while $h(\omega)\mu_\omega$ is $\tilde{\Phi}_c$-invariant in the non-hyperbolic case. In particular, if φ has a "small" nontrivial invariant measure then Φ cannot be hyperbolic. This is relevant for bifurcation theory of RDS, see Theorem 9.2.3. ■

7.5.2 Dynamical Characterization and Globalization

In extension of Proposition 7.5.4, we would expect that small enough initial values x of φ-orbits which decay exponentially fast to 0 have $T_U^+(\omega, x) = \infty$, and are natural candidates for elements of any local classical stable manifold $M_s^{\text{loc}}(\omega)$ tangent to $E_s = \oplus_{i:\lambda_i < 0} E_i$. This is indeed the case, provided the size of $U(\theta^n \omega)$ can be controlled in the sense that it cannot shrink to zero at a geometric rate. This gives rise to an additional assumption on the nonlinear part F of φ which is not present in the deterministic case.

7.5.8 Definition (Tempered Ball). The ball $U(\omega) = B_{\rho(\omega)} = \{x \in \mathbb{R}^d : \|x\|_\omega \leq \rho(\omega)\}$ is called *tempered* (with respect to θ) if ρ is a tempered random variable in the sense of Definition 4.1.1, i.e. if

$$\lim_{n \to \pm\infty} \frac{1}{n} \log \rho(\theta^n \omega) = 0.$$

■

Since $\rho \leq 1$ our condition excludes the case $\liminf \log \rho(\theta^n \omega)/n = -\infty$. Also, the property of U being tempered or not is a property of the nonlinear part F of φ.

7.5.9 Proposition (Dynamical Characterization of Local Manifolds).
Consider the prepared C^k RDS φ generated by (7.5.1). Choose a constant $L < \frac{3-\sqrt{5}}{2}\delta$, construct the ball $U(\omega) = B_{\rho(\omega)}$ according to Lemma 7.5.1 and local invariant manifolds $M_{ij}^{\text{loc}}(\omega)$ according to Proposition 7.5.2.

Assume, in addition, that U is tempered.

(i) Consider the local stable manifold $M_{ip}^{\text{loc}}(\omega)$ for $\lambda_i < 0$ in $U(\omega)$ and select $b_i \in \Gamma_{\text{right}} \cap (0,1)$.

Then there exists a closed ball $V(\omega) := B_{r(\omega)} \subset U(\omega)$ such that

$$M_{ip}^{\text{loc,dyn}}(\omega) := M_{ip}^{\text{loc}}(\omega) \cap V(\omega)$$

has the following properties:

For any $x \in M_{ip}^{\text{loc,dyn}}(\omega)$, $T_U^+(\omega, x) = \infty$. Further, $M_{ip}^{\text{loc,dyn}}(\omega)$ is dynamically characterized by φ as follows: For $x \in V(\omega)$,

$$x \in M_{ip}^{\text{loc,dyn}}(\omega) \iff \varphi(\cdot, \omega, x) \in X_{b_i^+, \omega}.$$

Consequently, $M_{ip}^{\text{loc,dyn}}(\omega)$ is a piece of the local stable manifold which is dynamically characterized solely by φ, hence is independent of the choice of the cut-off function.

Finally, $M_{ip}^{\text{loc,dyn}}(\omega)$ has the following forward invariance property:

$$\varphi(n,\omega) M_{ip}^{\text{loc,dyn}}(\omega) \subset M_{ip}^{\text{loc}}(\theta^n \omega) \quad \text{for all } n \in \mathbb{Z}^+.$$

(ii) Consider the local unstable manifold $M_{1j}^{\text{loc}}(\omega)$ for $\lambda_j > 0$ in $U(\omega)$ and select $a_j \in \Gamma_{\text{left}} \cap (1, \infty)$.

Then there exists a closed ball $V(\omega) := B_{r(\omega)} \subset U(\omega)$ such that

$$M_{1j}^{\text{loc,dyn}}(\omega) := M_{1j}^{\text{loc}}(\omega) \cap V(\omega)$$

has the following properties:

For any $x \in M_{1j}^{\text{loc,dyn}}(\omega)$, $T_U^-(\omega, x) = -\infty$. Further, $M_{1j}^{\text{loc,dyn}}(\omega)$ is dynamically characterized by φ as follows: For $x \in V(\omega)$,

$$x \in M_{1j}^{\text{loc,dyn}}(\omega) \iff \varphi(\cdot, \omega, x) \in X_{a_j^-, \omega}.$$

Consequently, $M_{1j}^{\text{loc,dyn}}(\omega)$ is a piece of the local unstable manifold which is dynamically characterized solely by φ, hence is independent of the choice of the cut-off function.

Finally, $M_{1j}^{\text{loc,dyn}}(\omega)$ has the following backward invariance property:

$$\varphi(n,\omega) M_{1j}^{\text{loc,dyn}}(\omega) \subset M_{1j}^{\text{loc}}(\theta^n \omega) \quad \text{for all } n \in \mathbb{Z}^-.$$

Proof. (i) Fix i with $\lambda_i < 0$, choose $b_i < 1$ and a cut-off function and construct $\tilde{M}_{ip}(\omega)$. Choose $x \in \tilde{M}_{ip}(\omega)$. We have $x \in \tilde{M}_{ip}(\omega)$ if and only if $\tilde{\varphi}(\cdot, \omega, x) \in X_{b_i^+, \omega}$. By Corollary 7.3.16,

$$\|\tilde{\varphi}(n,\omega,x)\|_{\theta^n\omega} \leq Cb_i^n\|x\|_\omega, \quad n \geq 0,$$

where $C = \frac{\delta-L}{\delta-2L}$. Hence for $x \in \tilde{M}_{ip}(\omega) \cap B_{1/C}$

$$\|\tilde{\varphi}(n,\omega,x)\|_{\theta^n\omega} \leq b_i^n, \quad n \geq 0.$$

Clearly

$$b_i^n \leq \rho(\theta^n\omega) \iff \log b_i \leq \frac{1}{n}\log \rho(\theta^n\omega)$$

and $-\infty < \log b_i < 0$.

Now we use our assumption that U is tempered: We have $\frac{1}{n}\log \rho(\theta^n\omega) \to 0$ ($n \to \infty$), hence there exists an $n_i(\omega)$ such that $b_i^n \leq \rho(\theta^n\omega)$ for all $n > n_i(\omega)$, implying

$$\|\tilde{\varphi}(n,\omega,x)\|_{\theta^n\omega} \leq \rho(\theta^n\omega) \quad \text{for } x \in \tilde{M}_{ip}(\omega) \cap B_{1/C}, \quad n > n_i(\omega).$$

Now choose a ball $V(\omega) := B_{r(\omega)} \subset B_{1/C}$ such that for all $x \in V(\omega)$, the latter is true also for the remaining finitely many indices $0 \leq n \leq n_i(\omega)$. This is possible because $\tilde{\varphi}(n,\omega,0) = 0$ and $x \mapsto \tilde{\varphi}(n,\omega,x)$ is continuous.

We now prove that $M_{ip}^{\mathrm{loc,dyn}}(\omega) := V(\omega) \cap \tilde{M}_{ip}(\omega)$ serves our purpose: Take $x \in M_{ip}^{\mathrm{loc,dyn}}(\omega)$: Then by construction $\tilde{\varphi}(n,\omega,x) = \varphi(n,\omega,x) \in U(\theta^n\omega)$ for all $n \geq 0$, thus $T_U^+(\omega,x) = \infty$.

As to the dynamical characterization, note that $x \in M_{ip}^{\mathrm{loc,dyn}}(\omega) \subset \tilde{M}_{ip}(\omega)$ is equivalent to $\tilde{\varphi}(\cdot,\omega,x) \in X_{b_i^+,\omega}$, which because of $T_U^+(\omega,x) = \infty$ is equivalent to $\varphi(\cdot,\omega,x) \in X_{b_i^+,\omega}$.

The forward invariance property immediately follows from the forward invariance of $M_{ip}^{\mathrm{loc}}(\omega)$ and the fact that the orbit $\varphi(n,\omega,x)$ of $x \in M_{ip}^{\mathrm{loc,dyn}}(\omega)$ never leaves $U(\theta^n\omega)$.

The proof of (ii) is completely similar. □

7.5.10 Remark. (i) The proof shows that uniformly on $M_{ip}^{\mathrm{loc,dyn}}(\omega)$,

$$\|\varphi(n,\omega,x)\|_{\theta^n\omega} \leq b_i^n \quad \text{for all } n \geq 0. \tag{7.5.2}$$

(ii) We have "flags" of dynamically characterized local stable (for $\lambda_i < 0$) and local unstable manifolds (for $\lambda_j > 0$), similar to Corollaries 7.3.9 and 7.3.12.

(iii) We always have $M_{1p}^{\mathrm{loc}}(\omega) = U(\omega)$. For $\lambda_1 < 0$, $M_{1p}^{\mathrm{loc,dyn}}(\omega) = V(\omega) \subset U(\omega)$ is a ball around 0 such that (7.5.2) holds for some $b_1 < 1$ and uniformly for all $x \in V(\omega)$. This uniform exponential decay to 0 also holds in the Euclidean norm. ∎

It remains to be investigated when the nonlinearity F of φ admits a tempered ball $U(\omega) = B_{\rho(\omega)}$ such that

$$\sup_{\|x\|_\omega \leq \rho(\omega)} \|DF(\omega,x)\|_{\omega,\theta\omega} \leq L. \tag{7.5.3}$$

Put

$$M(\omega, r) := \sup_{\|x\|_\omega \leq r} \|DF(\omega, x)\|_{\omega, \theta\omega}.$$

The function $r \mapsto M(\omega, r)$ is continuous and increasing with $M(\omega, 0) = 0$, and the biggest possible ρ is

$$\rho(\omega) = \sup\{r \leq 1 : M(\omega, r) \leq L\}$$

It is hence the steep increase of $M(\omega, \cdot)$, thus of $DF(\omega, \cdot)$, near 0 which is dangerous as it forces ρ to be small.

7.5.11 Lemma (Sufficient Condition for Temperedness). *Let*

$$M(\omega, r) := \sup_{\|x\|_\omega \leq r} \|DF(\omega, x)\|_{\omega, \theta\omega} \leq \alpha(\omega) r^q$$

for some $q > 0$ and $0 \leq r \leq r_0$, where $\log^+ \alpha \in L^1(\mathbb{P})$. Then $\log \rho \in L^1(\mathbb{P})$, hence $U = B_\rho$ is tempered. We can in particular choose $q = 1$ and

$$\alpha(\omega) := \sup_{\|x\|_\omega \leq r_0} \|D^2 F(\omega, x)\|_{\omega, \theta\omega}.$$

Proof. We have $\log \rho \in L^1(\mathbb{P})$ if and only if for any $\varepsilon > 0$

$$\sum_{n=1}^\infty \mathbb{P}\{\log \frac{1}{\rho} \geq n\varepsilon\} = \sum_{n=1}^\infty \mathbb{P}\{M(\cdot, e^{-n\varepsilon}) \geq L\} < \infty.$$

But for each $\varepsilon > 0$ there is some $\varepsilon_1 > 0$ such that

$$\begin{aligned}
\mathbb{P}\{M(\cdot, e^{-n\varepsilon}) \geq L\} &\leq \mathbb{P}\{\alpha \geq L e^{n\varepsilon q}\} \\
&= \mathbb{P}\{\log \alpha \geq \log L + n\varepsilon q\} \\
&\leq \mathbb{P}\{\log^+ \alpha \geq n\varepsilon_1\},
\end{aligned}$$

the latter from a certain index n_0 on.

The second statement follows from the first one and the mean value theorem. □

Dahlke [120, 121] introduced an integrability condition on the second derivative of a random diffeomorphism φ on a compact manifold for exactly the same reason as we do: to obtain a dynamical characterization of local stable and unstable invariant manifolds (see his Lemma 3.2).

In Proposition 7.5.9 we have obtained a dynamical characterization of certain local invariant manifolds for RDS which do not satisfy the global assumptions. This dynamical characterization is exactly the same as the one by which we defined invariant manifolds at the outset in Sect. 7.3. We hence could try the definition

$$M'_{ip}(\omega) := \{x \in \mathbb{R}^d : \varphi(\cdot, \omega, x) \in X_{b_i^+, \omega}\}, \ \lambda_i < 0, \ b_i \in \Gamma_{\text{right}} \cap (0, 1), \quad (7.5.4)$$

and

390   Chapter 7. Invariant Manifolds

$$M'_{1j}(\omega) := \{x \in \mathbb{R}^d : \varphi(\cdot, \omega, x) \in X_{a_j^-, \omega}\}, \; \lambda_j > 0, \; a_j \in \Gamma_{\text{left}} \cap (1, \infty),$$
(7.5.5)

for stable or unstable invariant sets[7] for φ. By the cocycle property, these sets are indeed invariant, i.e. $\varphi(n, \omega) M'_{ip}(\omega) = M'_{ip}(\theta^n \omega)$ and $\varphi(n, \omega) M'_{1j}(\omega) = M'_{1j}(\theta^n \omega)$ for all $n \in \mathbb{Z}$.

In particular, if $\lambda_i < 0$ is the biggest negative Lyapunov exponent or if $\lambda_j > 0$ is the smallest positive exponent, we can define the classical stable and classical unstable set $M'_s(\omega)$ and $M'_u(\omega)$ as particular cases of (7.5.4) and (7.5.5).

The above definition also makes formally sense for $\lambda_i \geq 0$ and $\lambda_j \leq 0$, but then the corresponding orbits are permitted to grow exponentially fast, hence no connection with our local constructions in a neighborhood of 0 can be expected.

There is, however, a second concept to arrive at stable and unstable invariant sets: namely to *globalize* a local manifold, i.e. to start from a local manifold as a nucleus and collect all initial values whose orbits eventually enter this local manifold.

7.5.12 Proposition (Globalization of Local Manifolds). *Assume that the conditions of Proposition 7.5.9 are satisfied.*
(i) The set

$$M''_{ip}(\omega) := \bigcup_{n \geq 0} \varphi(n, \omega)^{-1} M^{\text{loc,dyn}}_{ip}(\theta^n \omega)$$

of initial values whose orbits eventually enter $M^{\text{loc,dyn}}_{ip}$ forwards in time is an immersed submanifold of $\mathbb{R}^d$ of the same regularity and dimension as $M^{\text{loc,dyn}}_{ip}(\omega)$.
(ii) The set

$$M''_{1j}(\omega) := \bigcup_{n \leq 0} \varphi(n, \omega)^{-1} M^{\text{loc,dyn}}_{1j}(\theta^n \omega)$$

of initial values whose orbits eventually enter $M^{\text{loc,dyn}}_{1j}$ backwards in time is an immersed submanifold of $\mathbb{R}^d$ of the same regularity and dimension as $M^{\text{loc,dyn}}_{1j}(\omega)$.

Proof. We only consider (i). We need to prove that $(W_n(\omega))_{n \geq 0}$ contains an increasing subsequence, where the closed sets

$$\begin{aligned} W_n(\omega) &:= \varphi(n, \omega)^{-1} M^{\text{loc,dyn}}_{ip}(\theta^n \omega) \\ &= \varphi(-n, \theta^n \omega) M^{\text{loc,dyn}}_{ip}(\theta^n \omega), \; n \geq 0, \end{aligned}$$

consist of those initial values whose orbits are in $M^{\text{loc,dyn}}_{ip}(\theta^n \omega)$ at the fixed time $n \geq 0$ (and then never leave M^{loc}_{ip} afterwards forwards in time).

---
[7] We do not know yet whether these sets are manifolds.

7.5 Local Invariant Manifolds   391

It suffices to show that for $W_0(\omega) = M_{ip}^{\mathrm{loc,dyn}}(\omega)$ there exists an $n_0 = n_0(\omega)$ such that

$$M_{ip}^{\mathrm{loc,dyn}}(\omega) \subset \varphi(n_0,\omega)^{-1} M_{ip}^{\mathrm{loc,dyn}}(\theta^{n_0}\omega)$$

and then to repeat the argument.

By the construction of $V(\omega) = B_{r(\omega)}$ in Proposition 7.5.9

$$\|\varphi(n,\omega,x)\|_{\theta^n\omega} \leq b_i^n \quad \text{for all } x \in M_{ip}^{\mathrm{loc,dyn}}(\omega).$$

We now choose $n_0(\omega)$ such that $b_i^{n_0} < r(\theta^{n_0}\omega)$ which is (infinitely often) possible since $\limsup_{n\to\infty} \log r(\theta^n\omega)/n = 0$ (we possibly have $\liminf_{n\to\infty} \log r(\theta^n\omega)/n = -\infty$, as V need not be tempered). For this n_0, the above is true.

Consequently, for each ω, $M_{ip}''(\omega)$ is the monotone union of the "discs" $\varphi(n_k,\omega)^{-1} M_{ip}^{\mathrm{loc,dyn}}(\theta^{n_k}\omega)$ for some increasing subsequence $\{n_k\}$ which satisfies $\lim_{k\to\infty} \log r(\theta^{n_k}\omega)/n_k = 0$. Hence $M_{ip}''(\omega)$ is an injective immersion of the Euclidean space E_{ip} such that $T_0 M_{ip}''(\omega) = E_{ip}$, see Pugh and Shub [282: p. 3]. □

The next theorem is the "climax" of invariant manifold theory for RDS: It says that the above two concepts of defining stable and unstable invariant sets are the same.

7.5.13 Theorem (Stable and Unstable Invariant Manifold Theorem). *Let φ be a C^k RDS ($k \geq 1$) with discrete time and fixed point 0 generated by (7.5.1). Suppose that $D\varphi$ satisfies the IC of the MET. Let $\mathcal{S}(D\varphi) = \{\lambda_1 > \cdots > \lambda_p\}$ be the Lyapunov spectrum and let $\mathbb{R}^d = E_1(\omega) \oplus \cdots \oplus E_p(\omega)$ be the splitting of $D\varphi$.*

Consider the corresponding prepared RDS $\bar\varphi(n,\omega) = P(\theta^n\omega) \circ \varphi(n,\omega) \circ P(\omega)^{-1}$, choose a constant $L < \frac{3-\sqrt{5}}{2}\delta$ and construct the ball $U(\omega) = B_{\rho(\omega)}$ according to Lemma 7.5.1, assume that U is tempered, construct the dynamically defined local stable and unstable invariant manifolds $\bar M_{ip}^{\mathrm{loc,dyn}}(\omega)$ and $\bar M_{1j}^{\mathrm{loc,dyn}}(\omega)$ according to Proposition 7.5.9 and consider $M_{ip}^{\mathrm{loc,dyn}}(\omega) := P(\omega)^{-1} \bar M_{ip}^{\mathrm{loc,dyn}}(\omega)$ and $M_{1j}^{\mathrm{loc,dyn}}(\omega) := P(\omega)^{-1} \bar M_{1j}^{\mathrm{loc,dyn}}(\omega)$.

(i) For each i such that $\lambda_i < 0$,

$$M_{ip}'(\omega) = M_{ip}''(\omega) =: M_{ip}(\omega),$$

and $M_{ip}(\omega)$ is called the stable manifold *of the RDS φ corresponding to the spectral interval $\Lambda_{ip} = \{\lambda_i > \ldots > \lambda_p\}$. It is of the same regularity class as $M_{ip}^{\mathrm{loc}}(\omega)$ and tangent to E_{ip} at 0. It is dynamically characterized by*

$$M_{ip}(\omega) = \{x \in \mathbb{R}^d : \varphi(\cdot,\omega,x) \in X_{b_i^+,\omega}\},$$

where b_i is any number less than 1 in the spectral gap $\Gamma_{\mathrm{right}} = [\beta_i+\delta, \alpha_{i-1}-\delta]$ to the right of λ_i. $M_{ip}(\omega)$ is invariant under φ.

(ii) For each j such that $\lambda_j > 0$,
$$M'_{1j}(\omega) = M''_{1j}(\omega) =: M_{1j}(\omega),$$
and $M_{1j}(\omega)$ is called the unstable manifold of the RDS φ corresponding to the spectral interval $\Lambda_{1j} = \{\lambda_1 > \ldots > \lambda_j\}$. It is of the same regularity class as $M_{1j}^{\mathrm{loc}}(\omega)$ and tangent to E_{1j} at 0. It is dynamically characterized by
$$M_{1j}(\omega) = \{x \in \mathbb{R}^d : \varphi(\cdot, \omega, x) \in X_{a_j^-, \omega}\},$$
where a_j is any number bigger than 1 in the spectral gap $\Gamma_{\mathrm{left}} = [\beta_{j+1} + \delta, \alpha_j - \delta]$ to the left of λ_j. $M_{1j}(\omega)$ is invariant under φ.

(iii) We have "flags" of stable (for $\lambda_i < 0$) and unstable (for $\lambda_j > 0$) manifolds similar to Corollaries 7.3.9 and 7.3.12, but now immersed in each other.

Proof. We only deal with part (i) and can assume without loss of generality that φ is prepared.

We first prove that $M'_{ip}(\omega) \subset M''_{ip}(\omega)$. Let $x \in M'_{ip}(\omega)$. Then by definition of $X_{b_i^+, \omega}$ there exists a number $C(\omega, x)$ such that
$$\|\varphi(n, \omega, x)\|_{\theta^n \omega} \le b_i^n C(\omega, x) \quad \text{for all } n \ge 0.$$
It again follows from the fact that $\limsup_{n \to \infty} \log r(\theta^n \omega)/n = 0$ that there exists an $n_0 = n_0(\omega, x)$ for which $b_i^{n_0} C(\omega, x) \le r(\theta^{n_0} \omega)$, hence
$$\|\varphi(n_0, \omega, x)\|_{\theta^{n_0} \omega} \le r(\theta^{n_0} \omega),$$
i.e. $y := \varphi(n_0, \omega, x) \in V(\theta^{n_0} \omega)$.

Now by the cocycle property, $\varphi(\cdot, \theta^{n_0}\omega, y) \in X_{b_i^+, \theta^{n_0}\omega}$. But then by Proposition 7.5.9(i), $y \in M_{ip}^{\mathrm{loc,dyn}}(\theta^{n_0}\omega)$, implying $x \in M''_{ip}(\omega)$.

Conversely, let $x \in M''_{ip}(\omega)$. Then there exists an index $n_0 = n_0(\omega, x)$ for which $y := \varphi(n_0, \omega, x) \in M_{ip}^{\mathrm{loc,dyn}}(\theta^{n_0}\omega)$. Hence, again by Proposition 7.5.9(i), $\varphi(\cdot, \theta^{n_0}\omega, y) \in X_{b_i^+, \theta^{n_0}\omega}$ and by the cocycle property $\varphi(\cdot, \omega, x) \in X_{b_i^+, \omega}$, which proves $x \in M'_{ip}(\omega)$. □

7.5.14 Remark (Alternative Dynamical Characterizations). (i) We also have
$$M_{ip}(\omega) = \{x \in \mathbb{R}^d : \varphi(\cdot, \omega, x) \in X_{b_i^+}\}, \quad b_i \in \mathrm{int}\,\Gamma_{\mathrm{right}} \cap (0, 1),$$
and
$$M_{1j}(\omega) = \{x \in \mathbb{R}^d : \varphi(\cdot, \omega, x) \in X_{a_j^-}\}, \quad a_j \in \mathrm{int}\,\Gamma_{\mathrm{left}} \cap (1, \infty).$$

(ii) The classical stable and unstable manifolds $M_s(\omega)$ and $M_u(\omega)$ are particular cases of the above. If $D\varphi(n, \omega, 0)$ is hyperbolic, the following dynamical characterization is equivalent to the ones given in Theorem 7.5.13 and part (i) of this remark:

$$M_s(\omega) = \{x \in \mathbb{R}^d : \varphi(n,\omega,x) \to 0 \text{ as } n \to \infty \text{ exponentially fast}\}$$

(i.e. $x \in M_s(\omega)$ if and only if $\limsup_{n\to\infty} \frac{1}{n} \log \|\varphi(n,\omega,x)\| < 0$) and

$$M_u(\omega) = \{x \in \mathbb{R}^d : \varphi(n,\omega,x) \to 0 \text{ as } n \to -\infty \text{ exponentially fast}\}$$

(either in the Euclidean or the random norm). ∎

7.5.15 Corollary (Asymptotic Stability in Probability). *Assume the conditions of Theorem 7.5.13 and $\lambda_1 < 0$. Then the classical stable manifold $M_s(\omega)$ is a neighborhood of 0 (its domain of attraction). Moreover, 0 is asymptotically stable in probability.*

Proof. We only need to prove that 0 is asymptotically stable in probability (for stochastic stability theory for SDE see Khasminskii [206] and Arnold [8]).

Using Remark 7.5.10(ii) for the Euclidean norm, we have for some $b_1 < 1$ uniformly for $x \in V(\omega)$

$$\|\varphi(n,\omega,x)\| \leq C(\omega) b_1^n \quad \text{for all } n \geq 0.$$

For each given $\varepsilon > 0$ choose first n_0 such that $C(\omega) b_1^{n_0} < \varepsilon$ with high probability, then choose x small enough to make the finitely many remaining terms $< \varepsilon$ with high probability (using continuity in x and $\varphi(n,\omega,0) = 0$). This implies

$$\lim_{x \to 0} \mathbb{P}\{\omega : \sup_{n \geq 0} \|\varphi(n,\omega,x)\| > \varepsilon\} = 0 \quad \text{for all } \varepsilon > 0,$$

i.e. stability in probability.

Further, since $V(\omega) = B_{r(\omega)}$ with $r(\omega) > 0$ $\mathbb{P}$-a.s. ,

$$\lim_{x \to 0} \mathbb{P}\{\omega : \lim_{n \to \infty} \varphi(n,\omega,x) = 0\} \geq \lim_{x \to 0} \mathbb{P}\{\omega : x \in V(\omega)\} = 1,$$

i.e. we have asymptotic stability in probability. ◻

7.5.3 Local Manifolds for Continuous Time

As local invariant manifolds can be constructed for local RDS, we start for $\mathbb{T} = \mathbb{R}$ with a local C^k RDS

$$\varphi : D \to \mathbb{R}^d, \quad (t,\omega,x) \mapsto \varphi(t,\omega,x), \quad D \subset \mathbb{R} \times \Omega \times \mathbb{R}^d$$

(see Definition 1.2.1), hence

$$\varphi(t,\omega) : D(t,\omega) \to R(t,\omega) = D(-t, \theta(t)\omega)$$

is a C^k diffeomorphism. Define the strictly invariant set

$$E(\omega) := \cap_{t \in \mathbb{R}} D(t,\omega) \subset \mathbb{R}^d$$

of never exploding initial values (see Sect. 1.8).

394   Chapter 7. Invariant Manifolds

To go on, we assume that φ possesses an invariant measure μ (for a sufficient condition see Corollary 1.6.13). Then necessarily $\mu_\omega(E(\omega)) = 1$ $\mathbb{P}$-a.s., in particular, $E(\omega) \neq \emptyset$.

We can reduce μ_ω to the Dirac measure δ_0 by Proposition 7.2.1, but now the extension is not to $\Omega \times \mathbb{R}^d$, but to $\bar{\Omega} = \text{graph} E$. We have $v = 0 \in D(t, \bar{\omega}) = D(t, \omega) - x$ for all $t \in \mathbb{R}$, hence $0 \in E(\bar{\omega})$. We now write again ω instead of $\bar{\omega}$, and x instead of v.

We hence can assume that $\varphi(t, \omega) : D(t, \omega) \to R(t, \omega)$ is a local C^k RDS with $\varphi(t, \omega, 0) = 0$, hence both $D(t, \omega)$ and $R(t, \omega)$ are open neighborhoods of 0. We put $\Phi(t, \omega) := D\varphi(t, \omega, 0)$ which is a global linear RDS for which we assume that the IC of the MET are satisfied and write

$$\varphi(t, \omega, x) = \Phi(t, \omega) + \psi(t, \omega, x), \quad \psi(t, \omega, 0) = 0, \ D\psi(t, \omega, 0) = 0. \quad (7.5.6)$$

In view of Proposition 7.2.6 we can always pass to a prepared φ if necessary.

7.5.16 Theorem (Local Invariant Manifold Theorem, Local Hartman-Grobman Theorem, Continuous Time). *Let the continuous time RDS φ given by (7.5.6) satisfy the following conditions:*
(i) There exists a global C^k RDS $\tilde{\varphi}$,

$$\tilde{\varphi}(t, \omega, x) = \Phi(t, \omega)x + \tilde{\psi}(t, \omega, x), \quad \tilde{\psi}(t, \omega, 0) = 0, \quad D\tilde{\psi}(t, \omega, 0) = 0,$$

which satisfies the sufficient conditions of the Global Invariant Manifold Theorem 7.3.28.
(ii) On a closed ball $U(\omega) = B_{\rho(\omega)} = \{x \in \mathbb{R}^d : \|x\|_\omega \leq \rho(\omega)\}$ we have

$$\varphi(t, \omega, x) = \tilde{\varphi}(t, \omega, x) \quad \text{for all } x \in U(\omega), \ t \in (T_U^-(\omega, x), T_U^+(\omega, x)),$$

and

$$\lim_{x \to 0} T_U^+(\omega, x) = \infty, \quad \lim_{x \to 0} T_U^-(\omega, x) = -\infty. \quad (7.5.7)$$

Here

$$T_U^+(\omega, x) := \sup\{t \in \mathbb{R}^+ : \varphi(r, \omega, x) \in U(\theta_r \omega) \text{ for all } 0 \leq r \leq t\},$$

and

$$T_U^-(\omega, x) := \inf\{t \in \mathbb{R}^- : \varphi(r, \omega, x) \in U(\theta_r \omega) \text{ for all } t \leq r \leq 0\}.$$

Then the continuous time analogue of the Local Invariant Manifold Theorem 7.5.5 holds for φ, and its local invariant manifolds $M_{ij}^{\text{loc}}(\omega)$ enjoy all the properties stated in Subsects. 7.5.1 and 7.5.2 for the discrete time case.

Further, if $\tilde{\varphi}$ satisfies the conditions of the Global Hartman-Grobman Theorem 7.4.12, then the continuous time analogue of the Local Hartman-Grobman Theorem 7.5.6 holds for φ.

The discrete time proofs hold verbatim for the continuous time case, with only one exception: The fact that φ is now a local RDS causes the following modification of Invariant Manifold Theorem 7.5.13:

(a) The stable manifold M_{ip} satisfies
$$M_{ip}(\omega) \subset E^+(\omega) := \cap_{t \geq 0} D(t, \omega)$$
and fulfills (instead of the full invariance property)
$$\varphi(t, \omega) M_{ip}(\omega) = M_{ip}(\theta(t)\omega) \cap R(t, \omega), \quad t \geq 0.$$

(b) The unstable manifold M_{1j} satisfies
$$M_{1j}(\omega) \subset E^-(\omega) := \cap_{t \leq 0} D(t, \omega)$$
and
$$\varphi(t, \omega) M_{1j}(\omega) = M_{1j}(\theta(t)\omega) \cap R(t, \omega), \quad t \leq 0.$$

Note that it follows from assumption (ii) of the last theorem that for each $T > 0$ there exists a radius $\rho_T(\omega)$ such that, putting $U_T(\omega) = B_{\rho_T(\omega)}$,
$$\varphi(t, \omega, x) = \tilde{\varphi}(t, \omega, x) \text{ for all } x \in U_T(\omega) \text{ and } |t| \leq T.$$

The question remains as to which RDS meet the requirements of Theorem 7.5.16. In the discrete time case, every RDS has local invariant manifolds of the same regularity and satisfies the local Hartman-Grobman theorem. This is due to the availability of truncation techniques for the nonlinear part in the neighborhood of a fixed point. Those truncation techniques do not seem to be available for a general continuous time RDS.[8]

However, what we really truncate in the discrete time case is the nonlinear part of the generator (time-one mapping) of the RDS. This procedure has of course its continuous time analogue. We will hence investigate whether RDE and SDE can be cut-off on the level of their vector fields so that the RDS they generate meet the requirements of Theorem 7.5.16.

For the RDE case, a quite satisfactory answer is given by Wanner [340: Sect. 4.2] for the Lipschitz case which can be carried over to the C^k case (we use the less clumsy θ_t instead of $\theta(t)$).

7.5.17 Theorem (Local Invariant Manifolds for RDE). *Let*

$$\dot{x}_t = A(\theta_t \omega) x_t + f(\theta_t \omega, x_t), \quad f(\omega, 0) = 0, \; Df(\omega, 0) = 0, \quad (7.5.8)$$

where $f \in L^1_{\mathbb{P}}(\Omega, \mathcal{C}^k)$ and $A \in L^1(\mathbb{P})$, entailing that (7.5.8) generates a local C^k RDS of type (7.5.6) and the RDE $\dot{x}_t = A(\theta_t \omega) x_t$ generates a linear RDS Φ which satisfies the IC of the MET.

Assume further that $\Phi = \text{diag}(\Phi_1, \ldots, \Phi_p)$ is prepared with

---

[8] It is not known to the author whether such techniques exist for deterministic flows.

$$\dot{\Phi}_i(t,\omega) = A_i(\theta_t\omega)\Phi_i(t,\omega), \quad i = 1,\ldots,p. \tag{7.5.9}$$

Then all assertions of Theorem 7.5.16 hold.

We refrain from reproducing the very technical and long proof. Its key idea is as follows: Replace f with a specially manufactured $\tilde{f}$ in such a way that $\dot{x}_t = A(\theta_t\omega)x_t + \tilde{f}(\theta_t\omega, x_t)$ generates a global C^k RDS $\tilde{\varphi}$. By the variation of constants formula $\tilde{\varphi}$ can be represented as

$$\tilde{\varphi}(t,\omega)x = \Phi(t,\omega)\left(x + \int_0^t \Phi(s,\omega)^{-1}\tilde{f}(\theta_s\omega, \tilde{\varphi}(s,\omega)x)ds\right).$$

The Gronwall-Bellman lemma applied to the last expression and to the corresponding expression for $D\tilde{\varphi}$ shows that the assumptions (i) and (ii) of Theorem 7.5.16 are satisfied.

We have discussed in Subsect. 6.2.2 (see Corollary 6.2.9) when assumption (7.5.9) can be satisfied.

For the case of an SDE Carverhill [89] (following Ruelle [292]) established the existence and regularity of classical local stable manifolds for SDE on compact Riemannian manifolds and for invariant Markov measures.

Carverhill's result was used by Baxendale [60: Theorem 5.1] to prove the existence of classical local stable manifolds for

$$dx_t = \sum_{j=1}^m f_j(x_t) \circ dW_t^j, \quad f_j(0) = 0, \; j = 1,\ldots,m.$$

in $\mathbb{R}^d$ for the Dirac measure δ_0 by replacing the f_j with compactly supported $\tilde{f}_j$ which agree with the f_j inside a ball.

However, general criteria are still missing, but work is in progress (e. g. by Mohammed and Scheutzow [258]).

7.6 Examples

In this section we collect several mainly pedagogical examples in which invariant manifolds can be constructed explicitly without reference to general existence theorems. In cases where we write generators, we restrict ourselves to the white noise case.

7.6.1 Example (Explicitly Solvable Scalar SDE). The explicitly solvable scalar SDE

$$dx_t = (\alpha x_t - x_t^N)dt + \sigma x_t \circ dW_t, \quad \alpha, \sigma \in \mathbb{R}, \; N \geq 2,$$

is treated in great detail in Subsect. 2.3.7, Example 4.2.15 and Subsect. 9.3.1, to which we refer. All invariant measures, their Lyapunov exponents and their invariant manifolds can be explicitly determined. ∎

7.6.2 Example (Affine RDS with Hyperbolic Linear Part).

We refer to Subsect. 5.6.2 for a detailed presentation of the following: Let $\varphi(t,\omega)x = \Phi(t,\omega)x + \psi(t,\omega)$ be an affine RDS in $\mathbb{R}^d$ whose linear part Φ satisfies the IC of the MET and is hyperbolic, and whose additive part ψ satisfies the assumptions of Theorem 5.6.1(i). Recall that all these assumptions are automatically satisfied for the affine RDS generated by

$$dx_t = \sum_{j=0}^{m} (A_j x_t + b_j) \circ dW_t^j$$

with $dx_t = \sum_{j=0}^{m} A_j x_t \circ dW_t^j$ hyperbolic. Then there exists a unique φ-invariant measure $\mu_\omega = \delta_{\kappa(\omega)}$ whose spectrum, splitting, stable and unstable invariant manifolds and invariant foliation were explicitly determined in Corollary 5.6.4. ∎

7.6.3 Example (Induced RDS on S^{d-1}, P^{d-1}, and $G_k(d)$).

Let Φ be a linear RDS with two-sided time and let $\bar{\Phi}$ be the nonlinear C^∞ RDS induced on S^{d-1} by Φ via

$$\bar{\Phi}(t,\omega)s := \frac{\Phi(t,\omega)s}{\|\Phi(t,\omega)s\|}, \quad s \in S^{d-1}.$$

This RDS is extensively treated in Sect. 6.2. We determine all $\bar{\Phi}$-invariant measures in Theorem 6.2.3 and for each such measure we explicitly determine the spectrum and the splitting in Theorem 6.2.20. On this basis one can easily construct the corresponding invariant manifolds.

Suppose for simplicity that the ergodic $\bar{\Phi}$-invariant measure $\bar{\mu}$ realizes the simple Lyapunov exponent λ_{i_0}, and $\bar{\mu}_\omega = \delta_{s(\omega)}$, $s(\omega) \in S^{d-1} \cap E_{i_0}(\omega)$ (the latter is always possible over an extension of θ, see Theorem 6.2.6). Hence $\bar{\Phi}$ is hyperbolic under $\bar{\mu}$ with spectrum $\mathcal{S}(\theta, \bar{\Phi}, \bar{\mu}) = \{(\lambda_j - \lambda_{i_0}, d_j)_{j=1,\ldots,p; j \neq i_0}\}$ and Oseledets spaces

$$\bar{E}_j(\omega, s(\omega)) = \text{span}(v - \langle v, s(\omega) \rangle s(\omega) : v \in E_j(\omega)) \subset T_{s(\omega)} S^{d-1} = s(\omega)^\perp,$$

$j \neq i_0$.

The Oseledets manifold tangent to $\bar{E}_j(\omega, s(\omega))$ clearly is

$$\bar{M}_j(\omega, s(\omega)) = S^{d-1} \cap \text{span}^+(E_j(\omega), s(\omega)), \quad j \neq i_0,$$

where $\text{span}^+(E, s) := \text{span}(\alpha v + \beta s : v \in E, \alpha \in \mathbb{R}, \beta > 0)$. The stable manifold is

$$\bar{M}_s(\omega, s(\omega)) = S^{d-1} \cap \text{span}^+(\oplus_{j>i_0} E_j(\omega), s(\omega))$$

and the unstable manifold is

$$\bar M_u(\omega, s(\omega)) = S^{d-1} \cap \mathrm{span}^+(\oplus_{j<i_0} E_j(\omega), s(\omega)).$$

We leave the analogous case of the RDS induced by a linear one on projective space P^{d-1} to the reader. The RDS induced on the Grassmannian manifold $G_k(d)$ can be treated similarly on the basis of Subsect. 6.3.1. ∎

7.6.4 Example (Fractional Linear Transformations). The mapping τ which assigns to each one-dimensional subspace spanned by $\binom{x}{y} \in \mathbb{R}^2 \backslash \{0\}$ the value $\frac{x}{y} \in \bar{\mathbb{R}} := \mathbb{R} \cup \{\infty\}$ is a homeomorphism between P^1 and $\bar{\mathbb{R}}$. The action $\hat{A}$ induced by a matrix $A = \begin{pmatrix} a & b \\ c & d \end{pmatrix} \in Gl(2, \mathbb{R})$ on P^1 and the corresponding fractional linear transformation $z \mapsto f_A(z) := \frac{az+b}{cz+d}$ on $\bar{\mathbb{R}}$ commute, $\tau \circ \hat{A} = f_A \circ \tau$. Assume that the linear cocycle $\Phi(t, \omega)$ on $\mathbb{R}^2$ with two-sided time satisfies the IC of the MET and is hyperbolic with spectrum $\lambda_1 > \lambda_2$ and splitting $\mathbb{R}^2 = E_1(\omega) \oplus E_2(\omega)$. Then $\hat{\Phi}$ has exactly two invariant measures, namely the random Dirac measures at $\hat{E}_1(\omega)$ (with exponent $\lambda_2 - \lambda_1 < 0$ and stable manifold $P^1 \setminus \{\hat{E}_2(\omega)\}$) and at $\hat{E}_2(\omega)$ (with exponent $\lambda_1 - \lambda_2 > 0$ and unstable manifold $P^1 \setminus \{\hat{E}_1(\omega)\}$). Hence the cocycle $\varphi(t, \omega) = f_{\Phi(t,\omega)}$ on $\bar{\mathbb{R}}$ (with values in the three-dimensional Lie group of fractional linear transformations) has the two "random fixed points" $z_i(\omega) = \tau(\hat{E}_i(\omega))$, which are the only φ-invariant measures, with corresponding stable and unstable manifolds. ∎

7.6.5 Example (Three RDS on the Unit Circle). (A) Additive noise: Let $S^1 = \mathbb{R}/2\pi\mathbb{Z} \cong [0, 2\pi]$ (boundary points identified) and consider the SDE

$$dx_t = f_0(x_t)dt + f_1(x_t) \circ dW_t = b(x_t)dt + \sigma dW_t \quad (7.6.1)$$

on S^1, where $f_0(x) = b(x)\frac{d}{dx}$, $f_1(x) = \sigma\frac{d}{dx}$ and where b is smooth with $b(2\pi) = b(0)$ and $\sigma \neq 0$. Equation (7.6.1) generates a global C^∞ RDS φ.

(i) The generator of the corresponding diffusion process for time $\mathbb{R}^+$,

$$L^+ = b(x)\frac{d}{dx} + \frac{\sigma^2}{2}\frac{d^2}{dx^2},$$

is elliptic, hence there is a unique stationary measure with a smooth positive density $\rho_1(x)$ solving $(L^+)^* \rho_1 = 0$ given by

$$\rho_1(x) = C \exp\left(\frac{2}{\sigma^2}\int_0^x b(y)dy\right)\left(1 + \frac{1-\beta}{\beta\gamma}\int_0^x \exp\left(-\frac{2}{\sigma^2}\int_0^y b(u)du\right)dy\right), \quad (7.6.2)$$

where

$$\beta = \exp\left(\frac{2}{\sigma^2}\int_0^{2\pi} b(x)dx\right), \quad \gamma = \int_0^{2\pi} \exp\left(-\frac{2}{\sigma^2}\int_0^x b(y)dy\right)dx.$$

The Lyapunov exponent of $T\varphi$ (which solves $dv_t = b'(x_t)v_t dt$) under ρ_1 is

$$\lambda(\rho_1) = \int_0^{2\pi} b'(x)\rho_1(x)dx = -\frac{\sigma^2}{2}\int_0^{2\pi}\frac{\rho_1'(x)^2}{\rho_1(x)}dx \leq 0, \qquad (7.6.3)$$

where the second equality follows by integration by parts (or from formula (6.4.25)).

We have $\lambda(\rho_1) < 0$ if and only if b is not constant, in which case the $\mathcal{F}^-$-measurable φ-invariant measure $\mu_\omega^1(dx)$ corresponding to ρ_1 consists of finitely many atoms of equal mass (Le Jan [227: Proposition 2], and Crauel [110: Proposition 2.6]). The stable manifold is an open neighborhood of these points, which cannot contain the whole segment between two adjacent points (if μ_ω^1 is a Dirac measure, the stable manifold is a proper subset of S^1).

We have $\lambda(\rho_1) = 0$ if and only if $b(x) \equiv b \in \mathbb{R}$ if and only if $\rho_1(x) = \frac{1}{2\pi}$, in which case the RDS consists of the random rotations

$$\varphi(t,\omega)x = x + bt + W_t(\omega) \mod 2\pi. \qquad (7.6.4)$$

The φ-invariant Markov measure corresponding to $\rho_1(x)dx$ is hence $\mu_\omega^1(dx) \equiv \rho_1(x)dx$. It is the unique φ-invariant measure, and all of S^1 is the center manifold for any x. For the particular case $b = 0$ and $\sigma = 1$, the generator of the diffusion process is $L^+ = \frac{1}{2}\Delta = \frac{1}{2}\frac{d^2}{dx^2}$, hence $dx_t = dW_t$ defines Brownian motion on S^1.

(ii) Similarly, the diffusion process defined by (7.6.1) for time $\mathbb{R}^-$ has generator

$$L^- = -b(x)\frac{d}{dx} + \frac{\sigma^2}{2}\frac{d^2}{dx^2}$$

and its unique stationary measure has density $\rho_2(x)$ given by (7.6.2), with b replaced by $-b$. The formula for the corresponding Lyapunov exponent for $\varphi(-\cdot)$ is given by (7.6.3) with $-b$ instead of b. Finally, the Lyapunov exponent of φ is (see Sect. 5.5)

$$\lambda(\rho_2) = \int_0^{2\pi} b'(x)\rho_2(x)dx = \frac{\sigma^2}{2}\int_0^{2\pi}\frac{\rho_2'(x)^2}{\rho_2(x)}dx \geq 0. \qquad (7.6.5)$$

We have $\lambda(\rho_2) > 0$ if and only if b is not constant, in which case the $\mathcal{F}^+$-measurable μ_ω^2 corresponding to ρ_2 is concentrated on finitely many atoms of equal weight, and $\lambda(\rho_2) = 0$ if and only if $b(x) \equiv b$ if and only if $\rho_1(x) = \rho_2(x) = \frac{1}{2\pi}$, leading back to (7.6.4).

(B) Gradient Brownian flow on S^1: The following example is a particular case of Example 7.6.6, but studying it together with (A) is quite instructive. Consider the SDE

$$dx_t = \sin x_t \circ dW_t^1 + \cos x_t \circ dW_t^2 \qquad (7.6.6)$$

on S^1, which generates the gradient Brownian flow on S^1. The (forward as well as backward) generator is, with $f_1(x) = \sin x \frac{d}{dx}$, $f_2(x) = \cos x \frac{d}{dx}$,

400   Chapter 7. Invariant Manifolds

$$L^{\pm} = \frac{1}{2}(f_1^2(x) + f_2^2(x)) = \frac{1}{2}\frac{d^2}{dx^2} = \frac{1}{2}\Delta,$$

so that the one-point motions are Brownian motions on S^1, and thus are statistically identical with the one-point motions generated by (7.6.1) for $b = 0$ and $\sigma = 1$.

Both Fokker-Planck equations have the unique solution $\rho(x) = \frac{1}{2\pi}$, but the Lyapunov exponents of the corresponding φ-invariant measures $\mu_\omega^1 = \delta_{x_1(\omega)}$ and $\mu_\omega^2 = \delta_{x_2(\omega)}$ are, directly by formula (6.4.25), or by Theorem 7.6.7 below,

$$\lambda(\mu^1) = -\frac{1}{2}\int_0^{2\pi}((\text{div}\sin x\frac{d}{dx})^2 + (\text{div}\cos x\frac{d}{dx})^2)\frac{1}{2\pi}dx = -\frac{1}{2},$$

and $\lambda(\mu^2) = +\frac{1}{2}$.

The stable manifold of μ_ω^1 is $M_s(\omega, x_1(\omega)) = S^1 \setminus \{x_2(\omega)\}$, and the unstable manifold of μ_ω^2 is $M_u(\omega, x_2(\omega)) = S^1 \setminus \{x_1(\omega)\}$.

Comparing cases (A) ($b = 0$ and $\sigma = 1$) and (B) we learn that Brownian motion on S^1 can be the one-point motion of completely different RDS and that invariant measures and their Lyapunov exponents cannot be "seen" by the generator of the corresponding diffusion process. Case (B) also shows that the same RDS can be stable under one invariant measure, but unstable under another one.

(C) Carverhill's noisy north-south flow: An example related to (A) and (B) was treated by Carverhill [89]: Consider on $S^1 \cong \mathbb{R}/2\pi\mathbb{Z}$ the SDE

$$dx_t = -\sin x_t\, dt + \sigma dW_t.$$

The two stationary measures ρ_1 and ρ_2 are given by (7.6.2) with $\beta = 1$ and $b(x) = -\sin x$ for ρ_1, respectively $b(x) = \sin x$ for ρ_2. By (7.6.3) and (7.6.5), $\lambda(\rho_1) < 0$ and $\lambda(\rho_2) > 0$.

But for this example one can prove, as in (B), that the corresponding φ-invariant measures $\mu_\cdot^1$ and $\mu_\cdot^2$ are Dirac measures at $x_1(\cdot) \in \mathcal{F}^-$ and $x_2(\cdot) \in \mathcal{F}^+$ and the stable and unstable manifolds are as in (B). ∎

7.6.6 Example (Gradient Brownian Flow on Compact Manifold Embedded in $\mathbb{R}^d$). Let M be a k-dimensional compact Riemannian manifold isometrically embedded in $\mathbb{R}^d$, $d \geq 2$. A standard way of obtaining Brownian motion on M is to consider the vector fields $f_j(x)$ which are orthogonal projections to T_xM of the unit vector fields $e_j(x)$ restricted to M, equivalently $f_j(x) = \text{grad} e_j(x)$. The SDE

$$dx_t = \sum_{j=1}^d f_j(x_t) \circ dW_t^j \qquad (7.6.7)$$

on M has generator $L^+ = \frac{1}{2}\Delta$ (see Elworthy [138]), hence the one-point motions are Brownian motions on M, and the unique stationary measure is normalized Riemannian volume $\rho(dx) = \frac{1}{\text{vol}M}dx$.

The global flow of C^∞ diffeomeorphisms of M (global C^∞ RDS φ) generated by (7.6.7) is called gradient Brownian flow, and its ergodic theory and "stochastic geometry" has been the subject of numerous studies up-to date. See, for example, Baxendale [56, 55], Carverhill, Chappell and Elworthy [93], Chappell [98], Arnold and San Martin [33], Bougerol [75], Elworthy [139, 140], Elworthy and Yor [145], Elworthy and Rosenberg [144]. As the material would easily fill a small monograph, we cannot be self-contained here, but will only quote some pertinent results.

We first deal with the Lyapunov spectrum of φ under the forward Markov measure μ^1 corresponding to $\rho(dx)$. Corresponding formulas are given in Sect. 6.4. The calculations and estimates are facilitated by the fact that for the gradient Brownian flow

$$\sum_{j=1}^{d} \nabla f_j(x) f_j(x) \equiv 0.$$

Formula (6.4.25) for the sum of the Lyapunov exponents gives

$$\lambda_\Sigma(\mu^1) = -\frac{1}{2\mathrm{vol}M} \int_M \sum_{j=1}^{d} (\mathrm{div} f_j(x))^2 dx = -\frac{1}{2\mathrm{vol}M} \int_M \sum_{j=1}^{d} (\Delta e_j(x))^2 dx,$$

from which Chappell's result [98] follows:

$$\lambda_\Sigma(\mu^1) = -\frac{1}{2\mathrm{vol}M} \int_M (\mathrm{trace}\alpha_x)^2 dx \leq \frac{l}{2} \dim M, \qquad (7.6.8)$$

where $\alpha_x : T_xM \times T_xM \to (T_xM)^\perp$ is the second fundamental form of M, l is the largest non-zero eigenvalue of Δ (which is negative) and we have equality in (7.6.8) if and only if $\frac{1}{2}\Delta e_j = l e_j$ for all $j = 1, \ldots, d$.

For the top Lyapunov exponent we obtain the formula

$$\lambda_1(\mu^1) = \int_{PM} \left(\frac{1}{2} |\alpha_x(s, \cdot)|^2 - |\alpha_x(s, s)|^2 - \frac{1}{2}\mathrm{Ric}(s, s) \right) d(P\rho), \qquad (7.6.9)$$

where $P\rho$ is an ergodic lift of ρ to PM (which exists by compactness; if M has dimension $d-1$, then $P\rho$ is unique with $\mathrm{supp} P\rho = PM$, see Arnold and San Martin [33: p. 61]) and $\mathrm{Ric}(u, v)$ denotes the Ricci curvature.

Now assume $M = S^{d-1}(r) = \{x \in \mathbb{R}^d : \|x\| = r\}$. Then (7.6.7) reads

$$dx_t = \sum_{j=1}^{d} \left(e_j - \frac{1}{r^2}\langle x_t, e_j\rangle x_t \right) \circ dW_t^j, \qquad (7.6.10)$$

$\alpha_x(u, v) = -\frac{1}{r}\langle u, v\rangle \frac{x}{r}$, $\mathrm{Ric}(u, v) = \frac{d-2}{r^2}\langle u, v\rangle$, $l = -(d-1)/r^2$, and

402   Chapter 7. Invariant Manifolds

$$\lambda_1(\mu^1) = -\frac{1}{2}\frac{d-1}{r^2} = \frac{1}{d-1}\lambda_\Sigma(\mu^1) < 0, \qquad (7.6.11)$$

i. e. φ has a negative one-point spectrum under μ^1. Bougerol [75] has shown that among all hypersurfaces it is only the spheres which possess a one-point spectrum and that the spectrum is simple in all other cases.

However, all this holds for the $\mathcal{F}^-$-measurable invariant measure $\mu^1_\omega := \lim_{t\to\infty} \varphi(-t,\omega)^{-1}\rho$ (which for the unit sphere will be determined below). Taking into account that the diffusion on $\mathbb{R}^-$ of the SDE (7.6.7) also has generator $L^- = \frac{1}{2}\Delta$ (hence $(\varphi(-t,\cdot)x)_{t\in\mathbb{R}^+}$ are Brownian motions), there is a second φ-invariant measure μ^2 which is $\mathcal{F}^+$-measurable and obtained by

$$\mu^2_\omega = \lim_{t\to\infty} \varphi(t,\omega)^{-1}\rho.$$

Since $L^+ = L^-$ we also have $PL^+ = PL^-$, hence

$$\lambda_1(\mu^2) = -\lambda_1(\mu^1) \quad \text{and} \quad \lambda_\Sigma(\mu^2) = -\lambda_\Sigma(\mu^1) \geq 0.$$

For $M = S^{d-1} \subset \mathbb{R}^d$,

$$\lambda_1(\mu^2) = \frac{1}{d-1}\lambda_\Sigma(\mu^2) = \frac{1}{2}\frac{d-1}{r^2} > 0.$$

Interpreting and augmenting the results of Baxendale [55] within this framework gives the following.

7.6.7 Theorem (Ergodic Theory of Gradient Brownian Flow on S^{d-1}). *Let $M = S^{d-1} = \{x \in \mathbb{R}^d : \|x\| = 1\} \subset \mathbb{R}^d$ and let the SDE*

$$dx_t = \sum_{j=1}^{d}(e_j - \langle x_t, e_j\rangle x_t) \circ dW_t^j$$

generate the gradient Brownian flow φ on S^{d-1}. Then

(i) $\varphi(t,\omega) \in G$, where G is the Lie group of orientation preserving conformal diffeomorphisms of S^{d-1}.

(ii) φ has exactly two invariant measures, namely

$$\mu^1_\omega(dx) = \delta_{x_1(\omega)}(dx), \quad x_1 \in \mathcal{F}^-, \quad \mathbb{P}\{x_1 \in \cdot\} = \rho(\cdot),$$

and

$$\mu^2_\omega(dx) = \delta_{x_2(\omega)}(dx), \quad x_2 \in \mathcal{F}^+, \quad \mathbb{P}\{x_2 \in \cdot\} = \rho(\cdot),$$

where $\rho(dx)$ is the normalized Riemannian volume of S^{d-1}. In particular, x_1 and x_2 are independent.

Further, μ^1 is stable with one-point spectrum $\lambda_1(\mu^1) = -\frac{d-1}{2}$, while μ^2 is unstable with one-point spectrum $\lambda_1(\mu^2) = \frac{d-1}{2}$.

(iii) The stable manifold of $x_1(\omega)$ is $M_s(\omega, x_1(\omega)) = S^{d-1} \setminus \{x_2(\omega)\}$, while the unstable manifold of $x_2(\omega)$ is $M_u(\omega, x_2(\omega)) = S^{d-1} \setminus \{x_1(\omega)\}$.

Proof. Parts (i) and (iii) are taken from Baxendale [55]. For part (ii) (and μ^2, say) all we have to prove is that for any $f \in \mathcal{C}(S^{d-1}, \mathbb{R})$

$$\lim_{t\to\infty} \int_{S^{d-1}} f(\varphi(t,\omega)^{-1}x)\rho(dx) = f(x_2(\omega)), \quad \mathbb{P}\text{-a.s.} \qquad (7.6.12)$$

The limit on the left-hand side exists by the usual martingale argument and defines a φ-invariant measure (Crauel [110: Proposition 3.1]).

For $d \geq 4$, (7.6.12) follows from the fact that for each fixed $x \in S^{d-1}$,

$$\lim_{t\to\infty} \varphi(t, \cdot)^{-1}x = x_2(\cdot), \quad \mathbb{P}\text{-a.s.}$$

For $d = 2$ and 3, however, the latter is not true [55: Theorem 6.4]. Property (7.6.12) nevertheless holds: It follows from the considerations prior to equation (6.2) in [55] that for t large enough, $\varphi(t,\omega)^{-1}$ compresses "most" of S^{d-1} (in the sense of Riemannian volume) into any given small neighborhood of $x_2(\omega)$, which implies (7.6.12).

Suppose there were another ergodic invariant measure μ besides $\mu^{1,2}$. Then its support $C(\omega) := \mathrm{supp}\mu_\omega$ is, on the one hand, an invariant random closed set, but on the other hand μ_ω is supported by the stable/unstable manifold of $x_{1,2}(\omega)$, which is a contradiction. □

∎

Chapter 8. Normal Forms

Summary

Normal form theory is at the heart of the theory of nonlinear deterministic and random DS. Its aim is to simplify (ultimately linearize) a DS by means of a smooth coordinate transformation. This is fundamental e. g. in bifurcation theory. However, obstructions against transforming away certain terms called "resonances" appear, so that the "simplest possible" form is in general nonlinear.

Here we develop normal form theory for RDS which was initiated by engineers and physicists almost 20 years ago. It turns out that the cohomological equations that need to be solved to eliminate terms are now random affine difference or differential equations, so that we can use our results from Sect. 5.6. Resonances now take the form of integer relations between Lyapunov exponents.

As a preparation we give a brief introduction into deterministic normal form theory in Sect. 8.1. Then normal form theory for discrete time RDS is presented in Sect. 8.2, the main statement being Theorem 8.2.11. The RDE case is dealt with in Sects. 8.3 (Theorem 8.3.7 for the nonresonant case and Theorem 8.3.10 for the resonant case). In Sect. 8.4 we present the random analogue of a very successful procedure for simultaneously obtaining the normal form, eliminating the stable variables from the center equations and determining the center manifold (Theorems 8.4.1 and 8.4.3). We apply this procedure to the noisy Duffing-van der Pol oscillator in Subsect. 8.4.3. The SDE case is treated in Sect. 8.5, where we have to cope with anticipative data (Theorem 8.5.1).

8.1 Deterministic Prerequisites

In 1892 Poincaré initiated a technique for simplifying a nonlinear system in the neighborhood of a reference solution by a smooth change of coordinates, today called theory of normal forms. In the deterministic case, the reference solution is almost exclusively assumed to be a fixed point (sometimes a periodic solution), while in the random case the reference "solution" can be an arbitrary invariant measure.

406   Chapter 8. Normal Forms

For later use we present a brief introduction into the deterministic theory (for recent presentations see e.g. Anosov and V. I. Arnold [2], Vanderbauwhede [332], Wiggins [346] or Katok and Hasselblatt [200]).

Suppose $\varphi : \mathbb{R}^d \to \mathbb{R}^d$ is a C^∞ diffeomorphism with $\varphi(0) = 0$.[1] The normal form problem for φ is to find a C^∞ diffeomorphism $h : \mathbb{R}^d \to \mathbb{R}^d$ with $h(0) = 0$ such that
$$\psi = h^{-1} \circ \varphi \circ h$$
is "as simple as possible" (to be specified later) in a neighborhood of 0. Since $D\psi(0) = (Dh(0))^{-1} \circ D\varphi(0) \circ Dh(0)$, and $D\varphi(0)$ is considered simplest possible if the matrix representation is in Jordan canonical form, it is reasonable to assume without loss of generality that $D\varphi(0)$ is in Jordan canonical form, and h is near-identity, i.e. $Dh(0) = \mathrm{id}$. Then $D\psi(0) = D\varphi(0)$ for any such h.

Put in terms of the C^∞ classification of diffeomorphisms, we look for that element ψ in the C^∞ equivalence class of φ which is "simplest possible" in a neighborhood of 0. The result, called normal form of φ, is the natural generalization of the Jordan canonical form to a diffeomorphism at a fixed point.

The ultimate aim of normal form theory is the linearization of φ, i.e. to find an h such that $\psi = D\varphi(0)$. The Hartman-Grobman theorem states that near a hyperbolic fixed point a local C^1 (hence in particular: C^∞) diffeomorphism φ is topologically equivalent to its linear part by means of a local homeomorphism h. The question whether h can be chosen to be smooth has a negative answer in general (Sternberg's theorem, see below).

The general deterministic procedure along which we develop the random case is as follows: Since φ and h, hence ψ, are C^∞, they have Taylor expansions at 0 of any order, and also formal Taylor series expansions $\tilde\varphi$, $\tilde h$ and $\tilde\psi$ (which might have radius of convergence equal to 0 since C^∞ functions need not be analytic – the reason these power series are called "formal").

Our *first step* is an inductive procedure to determine for each $N \geq 2$ a near-identity N-jet H_N such that if we use it as a coordinate transformation then in the Taylor expansion of ψ, $\psi(x) = \Psi_N(x) + O(|x|^{N+1})$, the N-jet $\Psi_N(x) = Ax + \sum_{n=2}^{N} \psi_n(x)$ is "as simple as possible". In applications, this N-jet Ψ_N is the main object of interest.

The *second step* is to determine full formal power series $\tilde h$ and $\tilde\psi$ which fit into the formal power series relation (also called "formal equivalence") $\tilde\psi = \tilde h^{-1} \circ \tilde\varphi \circ \tilde h$, such that $\tilde\psi$ is "as simple as possible". $\tilde\psi$ is then called the formal normal form of φ. It will turn out that all invariants of formal equivalence are already determined by the linear part $D\varphi(0)$ of the diffeomorphism φ.

---

[1] Deterministic as well as stochastic normal form theory makes statements about germs of C^∞ diffeomorphisms or vector fields, i.e. equivalence classes of C^∞ diffeomorphisms or vector fields which coincide in a neighborhood of 0. However, for ease of presentation we ignore this point.

The *third step* is to assure that there are indeed C^∞ diffeomorphisms h and ψ having the formal power series expansions determined in step 2 as their formal Taylor series. The answer is always affirmative (Borel's lemma).

The *fourth* and final *step* is to prove that if two C^∞ diffeomorphisms have the same formal Taylor series expansion at 0 they are indeed C^∞ equivalent by a near-identity h. The main classical result is Chen's theorem: If $D\varphi(0)$ is hyperbolic and $\varphi \sim \psi$ formally, then $\varphi \sim \psi$ smoothly. In particular, we obtain Sternberg's smooth linearization theorem provided the nonresonance conditions (8.1.5) hold.

Let for $n \geq 2$

$$\mathbb{N}_n^d := \{\tau = (\tau_1, \ldots, \tau_d) \in (\mathbb{Z}^+)^d : |\tau| := \sum_{i=1}^d \tau_i = n\}.$$

Denote by $x^\tau := x_1^{\tau_1} x_2^{\tau_2} \cdots x_d^{\tau_d}$ the scalar monomial in d variables of degree $|\tau| = n$. Then

$$H_{n,d} := H_{n,d}(\mathbb{R}^d) = \{f = \sum_{\tau \in \mathbb{N}_n^d} x^\tau f_\tau : f_\tau \in \mathbb{R}^d\}$$

is the vector space of homogeneous polynomials of degree n in d variables with values in $\mathbb{R}^d$. We also write $|\tau| = n$ for $\tau \in \mathbb{N}_n^d$. Observe that

$$\Delta := \Delta(n,d) = \#\mathbb{N}_n^d = \binom{n+d-1}{n},$$

so

$$\Delta = \dim H_{n,d}(\mathbb{R}^1), \quad D = \dim H_{n,d}(\mathbb{R}^d) = \Delta d.$$

A basis $F = (u_1, \ldots, u_d)$ in $\mathbb{R}^d$ and the basis $(x^\tau)_{|\tau|=n}$ of $H_{n,d}(\mathbb{R}^1) \cong \mathbb{R}^\Delta$ give a basis $(x^\tau F) := (x^\tau u_i)_{i=1,\ldots,d;|\tau|=n}$ in $H_{n,d}$, while

$$H_{n,d} \ni f = \sum_{|\tau|=n} \sum_{i=1}^d f_{i,\tau} x^\tau u_i \cong k_F(f) = (k_F(f_\tau))_{|\tau|=n} = (f_{i,\tau}) \in \mathbb{R}^D$$

(column vectors, ordered lexicographically) identifies $H_{n,d}$ with $\mathbb{R}^D$, $D = \Delta d$, where k_F is the mapping which assigns F coordinates (respectively $(x^\tau F)$ coordinates) to an element of $\mathbb{R}^d$ (respectively $H_{n,d}$).

There is the following "canonical" choice of a scalar product in $H_{n,d}$ (see Vanderbauwhede [332]): If F is a basis in $\mathbb{R}^d$, put

$$\langle f, g \rangle_F := \sum_{|\tau|=n} \tau! \langle f_\tau, g_\tau \rangle_F = \sum_{|\tau|=n} \tau! (k_F(f_\tau), k_F(g_\tau))_S,$$

where $(x,y)_S := \sum_{i=1}^d x_i y_i$ is the standard scalar product in $\mathbb{R}^d$ and $\tau! := \tau_1! \cdots \tau_d!$. This makes $(H_{n,d}, \langle \cdot, \cdot \rangle_F)$ a Euclidean space in which the basis $(x^\tau F)$ is orthogonal.

408   Chapter 8. Normal Forms

It will turn out to be very convenient for our purpose to identify $H_{n,d} = H_{n,d}(\mathbb{R}^d)$ with the tensor product of $H_{n,d}(\mathbb{R}^1)$ and $\mathbb{R}^d$ (for basic facts see Sect. 5.4),

$$H_{n,d}(\mathbb{R}^d) = H_{n,d}(\mathbb{R}^1) \otimes \mathbb{R}^d \cong \mathbb{R}^\Delta \otimes \mathbb{R}^d,$$

where the above choice of bases induces the basis with elements $x^\tau \otimes u_i$ in $H_{n,d}(\mathbb{R}^d)$ and the isomorphism induced by the coordinate mappings maps this basis to the standard basis $f_j \otimes e_i \in \mathbb{R}^\Delta \otimes \mathbb{R}^d$. The above scalar product in $H_{n,d}$ can also be obtained by lifting the scalar products $\langle x, y \rangle := \sum_{|\tau|=n} \tau! x_\tau y_\tau$ in $H_{n,d}(\mathbb{R}^1) \cong \mathbb{R}^\Delta$ and $\langle x, y \rangle_F := (k_F(x), k_F(y))_S$ on $\mathbb{R}^d$ to the tensor product (see Lemma 5.4.1(vii)).

Let a linear mapping in $\mathbb{R}^d$ be represented in the basis F by a $d \times d$ matrix A. We now define the basic cohomological operator on $H_{n,d}$ by $C(A)_n f(x) := Af(x) - f(Ax)$ (the definition is motivated later by the relation (8.1.4)). It is important to note (but often not clearly said) that this operator depends on the choice of the basis F in $\mathbb{R}^d$. This fact is often disguised by the assumption that A is in Jordan canonical form in the standard basis of $\mathbb{R}^d$. The point is that the definition of $f(Ax)$, $f \in H_{n,d}$, depends on the particular matrix representation of the linear operator.

If a basis F in $\mathbb{R}^d$ is chosen and the above identification $H_{n,d} \cong \mathbb{R}^\Delta \otimes \mathbb{R}^d$ is made, the *cohomological operator* $C(A)_n$ of a $d \times d$ matrix A on the space $H_{n,d}$ is given as the following Kronecker product (see Sect. 5.4)

$$C(A)_n = I_1 \otimes A - N(A)_n \otimes I_2, \tag{8.1.1}$$

where I_1 and I_2 are $\Delta \times \Delta$ and $d \times d$ unit matrices. The $\Delta \times \Delta$ matrix $N(A)_n$ is defined by

$$N(A)_n = (N^{(n)}_{\tau\sigma}(A)), \quad (Ax)^\tau = \sum_{|\sigma|=n} N^{(n)}_{\sigma\tau}(A) x^\sigma. \tag{8.1.2}$$

In particular, the entries of $N(A)_n$ depend nonlinearly on the entries of A.

8.1.1 Example. For $d = 2$ and $n = 2$ we have $\Delta = 3$, hence $D = 6$. If $F = (u_1, u_2)$ is a basis of $\mathbb{R}^2$ the corresponding six basis vectors of $H_{2,2}(\mathbb{R}^2) = H_{2,2}(\mathbb{R}^1) \otimes \mathbb{R}^2 \cong \mathbb{R}^6$ are $x_1^2 \otimes e_i$ for $\tau = (2,0)$, $x_1 x_2 \otimes e_i$ for $\tau = (1,1)$ and $x_2^2 \otimes e_i$ for $\tau = (0,2)$, where $i = 1, 2$. Further,

$$I_1 \otimes A = \begin{pmatrix} A & 0 & 0 \\ 0 & A & 0 \\ 0 & 0 & A \end{pmatrix},$$

$$N(A)_2 = \begin{pmatrix} a_{11}^2 & a_{11}a_{21} & a_{21}^2 \\ 2a_{11}a_{12} & a_{11}a_{22} + a_{12}a_{21} & 2a_{21}a_{22} \\ a_{12}^2 & a_{12}a_{22} & a_{22}^2 \end{pmatrix},$$

hence

$$N(A)_2 \otimes I_2 =$$

$$\begin{pmatrix} a_{11}^2 & 0 & a_{11}a_{21} & 0 & a_{21}^2 & 0 \\ 0 & a_{11}^2 & 0 & a_{11}a_{21} & 0 & a_{21}^2 \\ 2a_{11}a_{12} & 0 & a_{11}a_{22}+a_{12}a_{21} & 0 & 2a_{21}a_{22} & 0 \\ 0 & 2a_{11}a_{12} & 0 & a_{11}a_{22}+a_{12}a_{21} & 0 & 2a_{21}a_{22} \\ a_{12}^2 & 0 & a_{12}a_{22} & 0 & a_{22}^2 & 0 \\ 0 & a_{12}^2 & 0 & a_{12}a_{22} & 0 & a_{22}^2 \end{pmatrix}$$

∎

8.1.2 Lemma. *Let A, B be $d \times d$ matrices and let $n \geq 2$. Then:*
(i)
$$\|N(A)_n\| \leq c\|A\|^n$$
for the operator and Hilbert-Schmidt norm, where the constant c is independent of A.
(ii)
$$N(I_2)_n = I_1, \quad N(AB)_n = N(B)_n N(A)_n,$$
hence
$$N(A^{-1})_n = N(A)_n^{-1} \quad \text{for } A \in Gl(d, \mathbb{R}).$$
(iii) *With the scalar product $\langle x, y \rangle := \sum_{|\tau|=n} \tau! x_\tau y_\tau$ in $\mathbb{R}^\Delta$,*
$$N(A)_n^* = N(A^*)_n,$$
where $A^* = A'$ is the transpose of A. In particular, if U is orthogonal then so is $N(U)_n$.
(iv)
$$N(AB)_n \otimes I_2 = (N(B)_n \otimes I_2)(N(A)_n \otimes I_2)$$
and
$$N(A^{-1})_n \otimes I_2 = (N(A)_n \otimes I_2)^{-1} \quad \text{for } A \in Gl(d, \mathbb{R}).$$

Proof. (i) By the definition of $N(A)_n$,

$$N_{\tau\sigma}^{(n)} = \sum_{|\mu|=n} c_\mu \prod_{i,j=1}^d a_{ij}^{\mu_{ij}},$$

where $\mu = (\mu_{ij})$, $|\mu| = \sum_{i,j=1}^d \mu_{ij}$, $\mu_{ij} \in \mathbb{N}$, and the matrices μ and constants c_μ depend only on τ and σ, but not on A.

Consequently,

$$(N_{\tau\sigma}^{(n)})^2 \leq c_1 (\sum_{|\mu|=n} \prod_{i,j=1}^d a_{ij}^{\mu_{ij}})^2 \leq c_2 \sum_{|\mu|=n} (\prod_{i,j=1}^d (a_{ij}^2)^{\mu_{ij}/n})^n$$

$$\leq c_3 \sum_{|\mu|=n} (\sum_{i,j=1}^d \frac{\mu_{ij}}{n} a_{ij}^2)^n$$

since the geometric mean is less than or equal to the arithmetic mean. Thus

$$(N_{\tau\sigma}^{(n)})^2 \leq c_4 (\sum_{i,j=1}^{d} a_{ij}^2)^n = c_4 \|A\|_{HS}^n$$

and
$$\|N(A)_n\|_{HS} \leq c\|A\|_{HS}^n.$$

Similarly for the operator norm.

(ii) By the definition of $N(AB)_n$,
$$((AB)x)^\tau = \sum_{|\sigma|=n} N(AB)_{\sigma\tau} x^\sigma = \sum_{|\rho|=n} \sum_{|\mu|=n} N(A)_{\rho\tau} N(B)_{\mu\rho} x^\mu.$$

Equating coefficients gives
$$N(AB)_{\sigma\tau} = \sum_{|\rho|=n} N(B)_{\sigma\rho} N(A)_{\rho\tau}.$$

(iii) follows by an elementary calculation, see Vanderbauwhede [332].
(iv) follows from (ii) and Lemma 5.4.1(iii). □

8.1.3 Remark. (i) In case $A = \text{diag}(a_1, \ldots, a_d)$,
$$I_1 \otimes A = \text{diag}(A, \ldots, A) = \text{diag}(a_1, \ldots, a_d, \ldots, a_1, \ldots, a_d),$$
$$N(A)_n = \text{diag}(a^\tau)_{|\tau|=n}, \quad N(A)_n \otimes I_2 = \text{diag}(a^\tau I_2)_{|\tau|=n},$$
where $a^\tau = a_1^{\tau_1} \cdots a_d^{\tau_d}$.

(ii) If A is block-diagonal, then so is $N(A)_n \otimes I_2$.

(iii) If A is upper (or lower) triangular then $N(A)_n \otimes I_2$ is lower (or upper) triangular. ■

For example, for $d = n = 2$ and $A = \text{diag}(a_1, a_2)$
$$C(A)_2 = \text{diag}(a_1 - a_1^2, a_2 - a_1^2, a_1 - a_1 a_2, a_2 - a_1 a_2, a_1 - a_2^2, a_2 - a_2^2).$$

We come back to the equation $\psi = h^{-1} \circ \varphi \circ h$, or
$$h(\psi(x)) = \varphi(h(x)), \tag{8.1.3}$$

where φ, ψ and h are C^∞ diffeomorphisms with $\varphi(0) = \psi(0) = h(0) = 0$ and $Dh(0) = \text{id}$, hence $A := D\varphi(0) = D\psi(0)$. In this relation φ is given and h has to be chosen to make the resulting ψ as simple as possible.

Expansion into formal Taylor series at 0 yields
$$\varphi(x) \sim \sum_{n=1}^{\infty} \varphi_n(x) = Ax + \sum_{n=2}^{\infty} \varphi_n(x),$$

$$\psi(x) \sim \sum_{n=1}^{\infty} \psi_n(x) = Ax + \sum_{n=2}^{\infty} \psi_n(x),$$

and

$$h(x) \sim \sum_{n=1}^{\infty} h_n(x) = x + \sum_{n=2}^{\infty} h_n(x),$$

where $\varphi_n, \psi_n, h_n \in H_{n,d}$. Inserting these expansions into equation (8.1.3) and equating coefficients yields the *cohomological equations*

$$C(A)_n h_n = \psi_n - k_n, \quad n = 2, 3, \ldots, \tag{8.1.4}$$

where $k_2 = \varphi_2$, and for $n \geq 3$, $k_n \in H_{n,d}$ only depends on the terms $\varphi_2, \ldots, \varphi_n$, $h_2, \ldots, h_{n-1}$, and $\psi_2, \ldots, \psi_{n-1}$, and $C(A)_n$ is the cohomological operator introduced above. Hence a recursive algorithm is possible for determining $h_2, \psi_2, \ldots, h_n, \psi_n, \ldots$ from successively solving equation (8.1.4) for $n = 2, 3, \ldots, n, \ldots$ on the basis of data either given at step n or determined in previous steps.

The cohomological equation (8.1.4) is a linear equation in $H_{n,d}$. The spectrum of $C(A)_n$ is

$$\Sigma(C(A)_n) = \{\lambda_i - \lambda^{\tau} : i = 1, \ldots, d; |\tau| = n\},$$

where $\lambda_1, \ldots, \lambda_d$ are the eigenvalues of A, and $\lambda^{\tau} := \lambda_1^{\tau_1} \cdots \lambda_d^{\tau_d}$. If $0 \notin \Sigma(C(A)_n)$, equivalently if the "nonresonance conditions" of order n

$$\lambda_i \neq \lambda_1^{\tau_1} \cdots \lambda_d^{\tau_d}, \quad 1 \leq i \leq d, \quad \tau_1, \ldots, \tau_d \in \mathbb{Z}^+, \quad |\tau| := \sum_{j=1}^{d} \tau_j = n \geq 2, \tag{8.1.5}$$

are satisfied, then the cohomological equation can be uniquely solved for any right-hand side $\psi_n - k_n$, in particular for our favorite choice $\psi_n = 0$ for any k_n.

Otherwise, if $0 \in \Sigma(C(A)_n)$, we have "resonance" of order n. In this case we split the space

$$H_{n,d} = \mathcal{R}(C(A)_n) \oplus \mathcal{N},$$

where $\mathcal{R}(C(A)_n)$ denotes the range of $C(A)_n$ and $\mathcal{N}$ is any complement of $\mathcal{R}(C(A)_n)$. This arbitrariness in the choice of $\mathcal{N}$ makes the normal form nonunique. Now the choice $\psi_n = 0$ is in general not possible anymore. Write $k_n = k_n^{\mathcal{R}} \oplus k_n^{\mathcal{N}}$, determine h_n by solving $C(A)_n h_n = -k_n^{\mathcal{R}}$ and put $\psi_n = k_n^{\mathcal{N}}$. The latter term cannot be eliminated if $k_n \notin \mathcal{R}(C(A)_n)$, irrespective of the choice of $\mathcal{N}$, and is part of the normal form ψ of φ.

There is the following "canonical" choice of $\mathcal{N}$ (see Vanderbauwhede [332]): Use the scalar product in $H_{n,d}$ introduced above. Then $C(A)_n^* = C(A^*)_n$, where $A^* = A'$ is the transpose of A. We hence have the orthogonal decomposition

$$H_{n,d} = \mathcal{R}(C(A)_n) \oplus \mathcal{N}(C(A^*)_n),$$

where $\mathcal{N}(C(A^*)_n)$ denotes the kernel of $C(A^*)_n$.

8.2 Normal Forms for Random Diffeomorphisms

8.2.1 The Random Cohomological Equation

We follow Arnold and Xu [40]. Let φ be a C^∞ RDS in $\mathbb{R}^d$ with time $\mathbb{T} = \mathbb{Z}$, time one mapping $\varphi(\omega) := \varphi(1,\omega)$ and with fixed point 0, $\varphi(\omega)0 = 0$.

In contrast to the deterministic case, the assumption of a fixed point at 0 is without much loss of generality for an RDS with an invariant measure, see Lemma 7.2.1.

Assume throughout that $D\varphi(\cdot, 0)$ satisfies the IC of the MET.

8.2.1 Definition (Random Coordinate Transformation). A measurable mapping $h : \Omega \times \mathbb{R}^d \to \mathbb{R}^d$ is called a *(near-identity) random coordinate transformation* if $h(\omega, \cdot) \in \mathrm{Diff}^\infty(\mathbb{R}^d)$, $h(\omega, 0) = 0$ and $Dh(\omega, 0) = \mathrm{id}$. ∎

The normal form problem for the RDS φ is now to determine a random coordinate transformation h such that the smooth random equivalence $\psi(\omega) = h^{-1}(\theta\omega) \circ \varphi(\omega) \circ h(\omega)$, equivalently

$$h(\theta\omega, \psi(\omega, x)) = \varphi(\omega, h(\omega, x)), \qquad (8.2.1)$$

makes ψ "as simple as possible", with the ultimate aim of smooth linearization, $\psi(\omega, x) = D\varphi(\omega, 0)x$.

We stress that φ, $D\varphi(\cdot, 0)$ and ψ are generators of cocycles (in particular, $\varphi(\omega) : \{\omega\} \times \mathbb{R}^d \to \{\theta\omega\} \times \mathbb{R}^d$), while $h(\omega)$ is a "static" coordinate transformation in the fiber $\{\omega\} \times \mathbb{R}^d$. This will be important for the interpretation of the cohomological equation.

8.2.2 Remark. Equation (8.2.1) implies that also

$$\psi(n, \omega) = h(\theta^n \omega)^{-1} \circ \varphi(n, \omega) \circ h(\omega), \quad n \in \mathbb{Z}.$$

It is, however, in general not true that normal forms perpetuate from time one to arbitrary times $n \in \mathbb{Z}$. So if $\psi(1,\omega) = \psi(\omega)$ is "as simple as possible", this might not be the case for $\psi(2,\omega) = \psi(\theta\omega) \circ \psi(\omega)$. For a discussion and a deterministic counterexample see Remark 5.9 of Arnold and Xu [40]. ∎

Let

$$\mathcal{H}_{n,d} := \{f : \Omega \to H_{n,d} \text{ measurable}\}$$

be the vector space of random homogeneous polynomials of degree $n \geq 2$ in d variables with values in $\mathbb{R}^d$.

8.2 Normal Forms for Random Diffeomorphisms

We next select (and then fix) a random basis $F(\omega) = (u_1(\omega), \ldots, u_d(\omega))$ of $\mathbb{R}^d$ adapted to the Oseledets splitting of $D\varphi(\cdot, 0)$ to "simplify" $D\varphi(\omega, 0)$ so that in the $F(\omega)$, $F(\theta\omega)$ bases the matrix representation $A(\omega)$ of $D\varphi(\omega, 0)$ is block-diagonal and the change from the standard basis to the random basis $F(\omega)$ is a Lyapunov cohomology (see Corollary 4.3.12).

The random basis $F(\omega)$ of $\mathbb{R}^d$ gives rise to the random basis $(x^\tau F(\omega))_{|\tau|=n}$ on $H_{n,d}$, so a random homogeneous polynomial $f \in \mathcal{H}_{n,d}$ has the coordinate representation

$$f(\omega) = \sum_{|\tau|=n} x^\tau f_\tau(\omega) = \sum_{|\tau|=n} \sum_{i=1}^{d} f_{i,\tau}(\omega) x^\tau u_i(\omega),$$

hence

$$k_{F(\omega)}(f(\omega)) = (k_{F(\omega)}(f_\tau(\omega)))_{|\tau|=n} \text{ with } k_{F(\omega)}(f_\tau(\omega)) = \begin{pmatrix} f_{1,\tau}(\omega) \\ \vdots \\ f_{d,\tau}(\omega) \end{pmatrix}.$$

The function $k_F : \mathcal{H}_{n,d} \to \mathcal{R}^D =$ random variables with values in $\mathbb{R}^D$, $D = \Delta d$, given by $f(\cdot) \mapsto k_{F(\cdot)}(f(\cdot))$, is a linear isomorphism and allows us to identify $\mathcal{H}_{n,d}$ and $\mathcal{R}^D$. Starting with $H_{n,d} \cong \mathbb{R}^\Delta \otimes \mathbb{R}^d$, we also have $\mathcal{H}_{n,d} \cong \mathcal{R}^\Delta \otimes \mathcal{R}^d$.

Since the matrix representation of $D\varphi(\omega, 0)$ in the bases $F(\omega)$ and $F(\theta\omega)$ is denoted by $A(\omega)$

$$k_{F(\theta\omega)}(D\varphi(\omega,0)x) = A(\omega) k_{F(\omega)}(x).$$

8.2.3 Definition (Random Cohomological Operator and Equation).
Let $F(\omega)$ be a random basis of $\mathbb{R}^d$. The linear operator

$$C(A)_n : \mathcal{H}_{n,d} \to \mathcal{H}_{n,d}$$

defined by

$$f(\cdot, x) \mapsto g(\theta\cdot, x) := A(\cdot) f(\cdot, x) - f(\theta\cdot, A(\cdot)x) = C(A)_n f(\cdot, x)(\theta\cdot)$$

is called the *(random) cohomological operator*. It depends on the random basis $F(\omega)$ and is given in $\mathcal{H}_{n,d} \cong \mathcal{R}^\Delta \otimes \mathcal{R}^d$ by

$$k_{F(\theta\omega)}(C(A)_n f) = $$
$$(I_1 \otimes A(\omega)) k_{F(\omega)}(f(\omega)) - (N(A(\omega))_n \otimes I_2) k_{F(\theta\omega)}(f(\theta\omega)).$$

Here the $\Delta \times \Delta$ matrix $N(A(\omega))_n = (N(A(\omega))_{\tau\sigma})$ is defined via $(A(\omega)x)^\tau = \sum_{|\sigma|=n} N(A(\omega))_{\sigma\tau} x^\sigma$.

Any equation of the type $C(A)_n f = g$ is called a *(random) cohomological equation*. ∎

We stress that the appearance of θ makes $C(A)_n$ a true bundle mapping, i. e. we cannot define $C(A)_n$ "fiberwise" as a mapping $f(\omega) \mapsto C(A(\omega))_n f(\omega)$.

Introducing the linear operator

$$U : \mathcal{H}_{n,d} \to \mathcal{H}_{n,d}, \quad f(\omega) \mapsto (Uf)(\omega) := f(\theta\omega),$$

a succinct form of the cohomological operator (suppressing ω and n) is

$$C(A) = I_1 \otimes A - (N(A) \otimes I_2) \circ U. \tag{8.2.2}$$

Note that, due to their origin, the two terms in (8.2.2) are of different character:
- $I_1 \otimes A(\omega)$ is the generator of a cocycle and maps $\{\omega\} \times \mathcal{H}_{n,d}$ to $\{\theta\omega\} \times \mathcal{H}_{n,d}$, while
- $N(A(\omega))_n \times I_2$ has its origin in a coordinate transformation and maps $\{\theta\omega\} \times \mathcal{H}_{n,d}$ to itself.

We now follow the deterministic procedure and expand the three mappings φ, h and ψ in equation (8.2.1) as formal Taylor series at 0. For this we write the coordinate transformation $h(\omega)$ in the basis $F(\omega)$, $F(\omega)$, but the cocycles $\varphi(\omega)$ and $\psi(\omega)$ in the bases $F(\omega)$ and $F(\theta\omega)$, so $x = (x_1, \ldots, x_d)'$ will be the coordinates in $F(\omega)$. This yields

$$\varphi(\omega, x) \sim \sum_{n=1}^{\infty} \varphi_n(\omega, x) = A(\omega)x + \sum_{n=2}^{\infty} \varphi_n(\omega, x),$$

$$\psi(\omega, x) \sim \sum_{n=1}^{\infty} \psi_n(\omega, x) = A(\omega)x + \sum_{n=2}^{\infty} \psi_n(\omega, x)$$

and

$$h(\omega, x) \sim \sum_{n=1}^{\infty} h_n(\omega, x) = x + \sum_{n=2}^{\infty} h_n(\omega, x),$$

where $\varphi_n, \psi_n, h_n \in \mathcal{H}_{n,d}$.

Inserting these expansions into equation (8.2.1), equating coefficients and keeping track of the fibers yields, analogously to the deterministic case, the following result.

8.2.4 Lemma. *The terms of order n of the formal power series expansion of ψ are successively given by the terms of order n of φ and h and lower order terms. More specifically, $\psi_1(\omega) = \varphi_1(\omega) = A(\omega)$ and*

$$C(A)_n h_n = \psi_n - k_n, \quad n \geq 2.$$

Here

$$k_n(\omega) = T_n\left((\varphi(\omega) - A(\omega)) \circ S_{n-1}h(\omega) + S_{n-1}h(\theta\omega) \circ S_{n-1}\psi(\omega)\right),$$

where $T_n(f)$ denotes the terms of order n in the formal Taylor expansion of f and $S_N(f) := \sum_{n=1}^{N} T_n(f)$ is the N-jet of f.

8.2.5 Example. For $d = 1$ we have $\Delta = D = 1$, hence $\mathcal{H}_{n,1} \cong \mathbb{R}^1$ for all $n \geq 2$. For some random basis $F(\omega) = (u(\omega))$,

$$\varphi(\omega, x) \sim A(\omega)x + \varphi_2(\omega)x^2 + \varphi_3(\omega)x^3 + \ldots, \quad A(\omega) \neq 0,$$

$$h(\omega, x) \sim x + h_2(\omega)x^2 + h_3(\omega)x^3 + \ldots,$$

$$\psi(\omega, x) \sim A(\omega)x + \psi_2(\omega)x^2 + \psi_3(\omega)x^3 + \ldots.$$

Hence

$$k_2(\omega) = \varphi_2(\omega), \quad k_3(\omega) = 2\varphi_2(\omega)h_2(\omega) + 2A(\omega)h_2(\theta\omega)\psi_2(\omega),$$

and

$$(C(A)_n h_n)(\theta\omega) = A(\omega)^n (A(\omega)^{1-n} - U)h_n(\omega).$$

∎

As a final step, we remedy the fact that the cohomological operator is not the generator of a cocycle.

8.2.6 Proposition. *Let $C(A)_n$ be the cohomological operator in $\mathcal{H}_{n,d}$ corresponding to A.*

(i) $C(A)_n$ can be written as

$$C(A)_n = (N(A)_n \otimes I_2)(M(A)_n - U),$$

where

$$M(A(\omega))_n := N(A(\omega)^{-1})_n \otimes A(\omega)$$

maps ω to $\theta\omega$, hence is the generator of a cocycle, while $N(A(\omega))_n \otimes I_2$ maps $\theta\omega$ to $\theta\omega$.

(ii) If A generates the cocycle Φ, then $N(A^{-1})_n$ generates the cocycle $N(\Phi^{-1})_n$ and $M(A)_n := N(A^{-1})_n \otimes A$ generates the cocycle

$$M(\Phi)_n := N(\Phi^{-1})_n \otimes \Phi.$$

(iii) Let the Lyapunov spectrum of the cocycle Φ generated by A be $\mathcal{S}(A) = (\Lambda_i)$ (the Lyapunov exponents listed with their multiplicities) and its splitting be (E_i). Then the MET holds for $N(A^{-1})_n$ and its spectrum is $\mathcal{S}(N(A^{-1})_n) = \{-(\Lambda, \tau) : |\tau| = n\}$, where $(\Lambda, \tau) := \sum_{i=1}^d \Lambda_i \tau_i$. Further, the MET also holds for $M(A)_n$ which has spectrum

$$\mathcal{S}(M(A)_n) = \{\Lambda_i - (\Lambda, \tau) : \Lambda_i \in \mathcal{S}(A), |\tau| = n\}$$

and splitting

$$H_\Lambda = \bigoplus_{(\mu,i):\Lambda_i+\mu=\Lambda, \Lambda_i \in \mathcal{S}(A), \mu \in \mathcal{S}(N(A^{-1})_n)} G_\mu \otimes E_i, \quad \Lambda \in \mathcal{S}(M(A)_n),$$

where G_μ denotes the splitting of $N(A^{-1})_n$.

Proof. (i) By Lemma 8.1.2 and the rules for the Kronecker product given in Lemma 5.4.1 (omitting the subscript n, but indicating the fibers of the mappings)

$$\begin{aligned} C(A) &= (N(A) \otimes I_2)_{\theta\omega,\theta\omega} \left((N(A) \otimes I_2)^{-1}_{\theta\omega,\theta\omega}(I_1 \otimes A)_{\omega,\theta\omega} - U_{\omega,\theta\omega} \right) \\ &= (N(A) \otimes I_2)_{\theta\omega,\theta\omega} \left((N(A)^{-1} \otimes I_2)_{\theta\omega,\theta\omega}(I_1 \otimes A)_{\omega,\theta\omega} - U_{\omega,\theta\omega} \right) \\ &= (N(A) \otimes I_2)_{\theta\omega,\theta\omega} \left((N(A^{-1}) \otimes A)_{\omega,\theta\omega} - U_{\omega,\theta\omega} \right). \end{aligned}$$

Note that $N(A^{-1})$ in the last line maps ω to $\theta\omega$ since it is the product of I_1 (which is part of a cocycle) and $N(A)^{-1}_{\theta\omega,\theta\omega}$.

(ii) By Theorem 5.4.2(iii), $A_1 \otimes A_2$ generates $\Phi_1 \otimes \Phi_2$ if A_i generate Φ_i, $i = 1, 2$. Thus it remains to prove that $N(A^{-1})$ generates $N(\Phi(n,\omega)^{-1}) =: B(n, \omega)$. But by Lemma 8.1.2

$$\begin{aligned} B(n+k,\omega) &= N(\Phi(n+k,\omega)^{-1}) \\ &= N(\Phi(k,\omega)^{-1}\Phi(n,\theta^k\omega)^{-1}) \\ &= N(\Phi(n,\theta^k\omega)^{-1}) N(\Phi(k,\omega)^{-1}) \\ &= B(n,\theta^k\omega)B(k,\omega). \end{aligned}$$

(iii) In view of Theorem 5.4.2(ii) and our preparations it remains to prove the statement on $N(A(\omega)^{-1}) : \mathbb{R}^\Delta \to \mathbb{R}^\Delta$. If

$$A(\omega) = P(\theta\omega)^{-1} B(\omega) P(\omega)$$

is a Lyapunov cohomology, then by Lemma 8.1.2(i) so is

$$N(A(\omega)^{-1}) = N(P(\theta\omega)) N(B(\omega)^{-1}) N(P(\omega))^{-1}.$$

Hence the MET holds for $N(A^{-1})$ in the basis F if it holds in the standard basis, which is again assured by Lemma 8.1.2(i).

To calculate the spectrum of $N(A^{-1})$ we use singular value decomposition in $\mathbb{R}^d$ and $\mathbb{R}^\Delta$, both endowed with the scalar product introduced above. Hence there are orthogonal $d \times d$ matrices U_k and V_k for which

$$\Phi(k,\omega) = U_k(\omega)\mathrm{diag}(\delta_1(k,\omega),\ldots,\delta_d(k,\omega))V_k(\omega),$$

where $\delta_1(k,\omega) \geq \cdots \geq \delta_d(k,\omega) > 0$, and

$$\lim_{k \to \infty} \frac{1}{k} \log \delta_i(k,\omega) = \Lambda_i, \quad i = 1,\ldots,d.$$

Thus

$$\Phi(k,\omega)^{-1} = V_k(\omega)^* \mathrm{diag}(\frac{1}{\delta_i(k,\omega)}) U_k(\omega)^*$$

and, applying the operation $N(\cdot)$ to the last line and using Remark 8.1.3(i),

$$N(\Phi(k,\omega)^{-1}) = N(U_k(\omega)^*)\mathrm{diag}(\delta^{-\tau}(k,\omega))_{|\tau|=n} N(V_k(\omega)^*).$$

By Lemma 8.1.2(iii),

$$N(\Phi(k,\omega)^{-1}) = N(U_k(\omega))^* \mathrm{diag}(\delta^{-\tau}(k,\omega))_{|\tau|=n} N(V_k(\omega))^*,$$

i.e. we have a singular value decomposition of $N(\Phi(k,\omega)^{-1})$. By the Furstenberg-Kesten theorem, the Lyapunov exponents of this cocycle are

$$\lim_{k\to\infty} \frac{1}{k} \log \delta^{-\tau}(k,\omega) = \lim_{k\to\infty} \frac{1}{k}\left(-\sum_{i=1}^{d} \tau_i \log \delta_i(k,\omega)\right)$$

$$= -\sum_{i=1}^{d} \tau_i \Lambda_i = -(\Lambda, \tau).$$

□

8.2.7 Example. (i) For $d = 2$ and $n \geq 2$ we have $\Delta = n+1$, $\mathbb{N}_n^d = \{\tau = (k, n-k) : k = 0, \ldots, n\}$ and $D = 2(n+1)$. If A has simple spectrum $\mathcal{S}(A) = \{\lambda_1 > \lambda_2\}$, then the spectrum

$$\mathcal{S}(N(A^{-1})_n) = \{-k\lambda_1 - (n-k)\lambda_2 : k = 0, \ldots, n\}$$

is simple, but in the spectrum

$$\mathcal{S}(M(A)_n) = \{(1-k)\lambda_1 - (n-k)\lambda_2, (1-l)\lambda_2 - (n-l)\lambda_1 : k, l = 0, \ldots, n\}$$

only the exponents for $k = 0$ and $l = n$ are simple, while the remaining $2n$ exponents have multiplicity 2 (put $l = k-1$).

Also, hyperbolicity does in general not carry over from A to $M(A)_n$: Even if $\lambda_1 > 0 > \lambda_2$, $0 \in \mathcal{S}(M(A)_n)$ if and only if $-\lambda_2/\lambda_1 = (k-1)/(n-k)$ for some $k = 2, \ldots, n-1$. Further, $0 \in \mathcal{S}(M(A)_n)$ for some $n \geq 2$ if and only if $-\lambda_2/\lambda_1 \in \mathbb{Q}^+$.

(ii) If, however, $M(A)_n$ is hyperbolic for some $n \geq 2$, then necessarily A is hyperbolic, since by Proposition 8.2.6(iii), $(1-n)\Lambda_i \in \mathcal{S}(M(A)_n)$ for any $i = 1, \ldots, d$. ∎

8.2.8 Definition (Resonance, Nonresonance). The linear cocycle Φ generated by A in $\mathbb{R}^d$ is called *resonant of order* $n \geq 2$ if $0 \in \mathcal{S}(M(A)_n)$. Otherwise it is called *nonresonant of order* n. ∎

8.2.2 Nonresonant Case

We now formally linearize φ by determining a formal power series of a coordinate transformation h for which we can solve

$$\psi_n = C(A)_n h_n + k_n$$

successively for $n = 2, 3, \ldots$ in $\mathcal{H}_{n,d}$ for the choice $\psi_n = 0$. This leads to the successive cohomological equations

$$C(A)_n h_n = -k_n, \quad n = 2, 3, \ldots.$$

With the representation of $C(A)_n$ from Proposition 8.2.6(i), this is equivalent to

$$(U - M(A)_n) h_n = (N(A^{-1})_n \otimes I_2) k_n =: g_n, \quad n = 2, 3, \ldots,$$

where $L(A)_n := U - M(A)_n$ is also called *cohomological operator*, and finally

$$U h_n = M(A)_n h_n + g_n, \quad n = 2, 3, \ldots \quad (8.2.3)$$

or, fiberwise,

$$h_n(\theta \omega) = M(A(\omega))_n h_n(\omega) + g_n(\omega), \quad n = 2, 3, \ldots.$$

Solving for it amounts to looking for a stationary solution of the random affine difference equation in $\mathbb{R}^D$

$$x_{k+1} = M(A(\theta^k \omega))_n x_k + g_n(\theta^k \omega).$$

For the nonresonant case we can readily use our results from Theorem 5.6.5(i).

Let us recall Definition 4.1.1 according to which a random variable f is called *tempered from above* with respect to θ if

$$\lim_{t \to \pm \infty} \frac{1}{|t|} \log^+ \|f(\theta(t)\omega)\| = 0$$

holds $\mathbb{P}$-a.s. We will also make use of Lemma 4.1.2.

8.2.9 Proposition. *If A is nonresonant for some $n \geq 2$ and if $g_n \in \mathcal{H}_{n,d}$ is tempered[2] (equivalently, if k_n is tempered), then there exists a unique $h_n \in \mathcal{H}_{n,d}$ (explicitly given in Theorem 5.6.5(i)) which solves equation (8.2.3). Moreover, h_n is also tempered.*

Proof. The existence and uniqueness statement is from Theorem 5.6.5(i).

That the statements about g_n and k_n being tempered are equivalent follows by Lemma 8.1.2(i), which gives

$$\log \|g_n\| \leq \log c + n \log \|A^{-1}\| + \log \|k_n\|,$$

$$\log \|k_n\| \leq \log c + n \log \|A\| + \log \|g_n\|,$$

and the fact that A and $D\varphi(\cdot, 0)$ are Lyapunov cohomologous. $\square$

---

[2] Henceforth we omit the words "from above" throughout Chap. 8.

8.2.10 Corollary. *Let $N \geq 2$. Let A be nonresonant of order n for $2 \leq n \leq N$ and suppose that in the Taylor expansion $\varphi(\omega, x) = A(\omega)x + \sum_{n=2}^{N} \varphi_n(\omega, x) + O(|x|^{N+1})$ at 0 of order N, all φ_n are tempered. Then there is a unique near-identity N-jet $H_N(\omega, x) = x + \sum_{n=2}^{N} h_n(\omega, x)$ with tempered coefficients h_n (hence is a local analytic diffeomorphism) such that φ is transformed by H_N to ψ, whose N-jet in the Taylor expansion of order N is linear, $\psi(\omega, x) = A(\omega)x + O(|x|^{N+1})$.*

Proof. All we have to prove is that all h_n are tempered provided all φ_n are. But this follows inductively from the formula for k_n in Lemma 8.2.4, the fact that sums and products of tempered random variables are tempered (Lemma 4.1.2) and from Proposition 8.2.9. □

8.2.11 Theorem (Formal Linearization of Random Diffeomorphism). *Suppose that the linear cocycle generated by $D\varphi(\cdot, 0)$ is nonresonant of any order $n \geq 2$, i.e.*

$$\Lambda_i \neq (\Lambda, \tau), \quad \Lambda_i \in \mathcal{S}(D\varphi(\cdot, 0)), \quad |\tau| \geq 2,$$

and that in the formal random Taylor series of the random diffeomorphism $\varphi(\omega) \sim A(\omega)x + \sum_{n \geq 2} \varphi_n(\omega, x)$ at 0, all φ_n are tempered. Then there is a random coordinate transformation $h(\omega)$ with a uniquely determined formal random Taylor series $h(\omega, x) \sim x + \sum_{n \geq 2} h_n(\omega, x)$ where all h_n are tempered and h formally linearizes φ, i.e. the formal random power series of $\psi(\omega) = h(\theta\omega)^{-1} \circ \varphi(\omega) \circ h(\omega)$ is $A(\omega)x$.

This is clear provided we can solve the third step of the normal form procedure, namely find a random diffeomorphism with the prescribed formal random power series. This is assured by the following random version of a result of Borel.

8.2.12 Lemma. *Given $h_0 = 0$, $h_1(\omega, x) = x$, $h_n \in \mathcal{H}_{n,d}$, $n \geq 2$. Then there is a random coordinate transformation h whose formal random Taylor series expansion has the coefficients $T_n h = h_n$.*

The proof is just an ω-wise version of the deterministic proof given by Vanderbauwhede [332: p. 142] and is thus omitted.

With the above procedure, infinitely flat terms of φ at 0 cannot be detected as they do not appear in formal power series. A random version of Sternberg's linearization theorem would assert that in the above situation $\varphi(\omega, x)$ and $A(\omega)x$ are indeed smoothly equivalent by a random coordinate transformation $h(\omega, x)$. However, such a theorem is still lacking.

8.2.3 Resonant Case

We now treat the case where we have resonance of order n, i.e. $0 \in \mathcal{S}(M(A)_n)$. In our adapted random basis

$$M(A)_n = \begin{pmatrix} M(A)_n^+ & 0 & 0 \\ 0 & M(A)_n^- & 0 \\ 0 & 0 & M(A)_n^0 \end{pmatrix},$$

where $M(A)_n^+$, $M(A)_n^-$ and $M(A)_n^0$ are the unstable, stable and neutral block, respectively. In this basis, the terms of order n of the formal Taylor series fulfill

$$\begin{pmatrix} \psi_n^+ \\ \psi_n^- \\ \psi_n^0 \end{pmatrix} = \left(\begin{pmatrix} M(A)_n^+ & 0 & 0 \\ 0 & M(A)_n^- & 0 \\ 0 & 0 & M(A)_n^0 \end{pmatrix} - U \right) \begin{pmatrix} h_n^+ \\ h_n^- \\ h_n^0 \end{pmatrix} + \begin{pmatrix} g_n^+ \\ g_n^- \\ g_n^0 \end{pmatrix}.$$

These are three decoupled equations, of which we can solve the first two by the techniques of the preceding subsection: Put $\psi_n^+ = 0$ and $\psi_n^- = 0$, then there is a unique h_n^+ and h_n^- provided g_n^+ and g_n^- are tempered, which we assume.

There remains to be considered the neutral part

$$\psi_n^0 = (M(A)_n^0 - U)h_n^0 + g_n^0, \quad \mathcal{S}(M(A)_n^0) = \{0\}. \tag{8.2.4}$$

The corresponding cohomological equation is

$$Uh_n^0 = M(A)_n^0 h_n^0 + g_n^0 - \psi_n^0 \tag{8.2.5}$$

and the cohomological operator $L_n := U - M(A)_n^0$ maps tempered random variables into tempered random variables, hence

$$L_n(\mathcal{T}_n) \subset \mathcal{T}_n, \quad \mathcal{T}_n := \{f \in \mathcal{H}_{n,d} : f \text{ tempered}\}.$$

The "only" problem hence is to describe $L_n(\mathcal{T}_n)$. This is an extremely complicated question. For the case $A = I$ we obtain $M(A) = I$, hence the classical cohomological equation $(U - I)h = g$ which is important in various contexts and has a rich literature (see Katok and Hasselblatt [200: Sects. 2.9 and 19.2]).

Arnold and Xu [40] have worked out a Hilbert space set-up for equations (8.2.4) and (8.2.5). However, we prefer to present the cleaner continuous time version of it in Sect. 8.3.

In practice, one need not worry so much about solving equation (8.2.5): We can always satisfy it by choosing $h_n^0 = 0$, hence $\psi_n^0 = g_n^0$.

8.3 Normal Forms for Random Differential Equations

Let f and g be vector fields in $\mathbb{R}^d$ with $f(0) = g(0) = 0$. They are called smoothly equivalent, $f \sim g$, if the local flows φ and ψ generated by $\dot{x}_t = f(x_t)$ and $\dot{x}_t = g(x_t)$ are smoothly equivalent, i.e. if there is a coordinate transformation h for which locally

$$\varphi(t) \circ h = h \circ \psi(t).$$

Differentiating this gives the equivalent infinitesimal form

$$f \circ h = Dh \circ g. \qquad (8.3.1)$$

Deterministic normal form theory for vector fields now seeks an h for which g is "simplest possible".

Normal forms for vector fields which are periodic in time have been worked out using Floquet theory by V. I. Arnold [44] and Iooss [180]. For the quasiperiodic case see Chow, Lu and Shen [101] and the references therein.

Analogously, if our RDS φ is generated by an RDE $\dot{x}_t = f(\theta_t \omega, x_t)$, there are obvious practical reasons for transforming the RDE (and not the resulting cocycle) into normal form. This problem will be addressed here, based on Arnold and Xu [43]. For the SDE case see Sect. 8.5.

8.3.1 The Random Cohomological Equation

We refer to Sect. 8.1 for basic facts about deterministic normal form theory, which we will freely use. We add that in the vector field case the formal Taylor series have to be inserted into equation (8.3.1) and the deterministic cohomological operator is

$$\mathrm{ad}_n A : H_{n,d} \to H_{n,d}, \quad h_n \mapsto (\mathrm{ad}_n A) h_n(x) := A h_n(x) - D h_n(x) A x. \qquad (8.3.2)$$

This operator depends linearly on the entries of A. Introducing a basis F in $\mathbb{R}^d$ and using the basis $(x^\tau)_{|\tau|=n}$ in $H_{n,d}(\mathbb{R}^1)$ gives a basis $(x^\tau F)$ in $H_{n,d}(\mathbb{R}^d) = H_{n,d}(\mathbb{R}^1) \otimes \mathbb{R}^d \cong \mathbb{R}^\Delta \otimes \mathbb{R}^d \cong \mathbb{R}^D$ and the $D \times D$ matrix representation of $\mathrm{ad}_n(A)$ is

$$\mathrm{ad}_n A = I_1 \otimes A - T(A)_n \otimes I_2, \qquad (8.3.3)$$

where I_1 and I_2 are unit matrices in $\mathbb{R}^\Delta$ and $\mathbb{R}^d$, respectively, and $T(A)_n$ is the $\Delta \times \Delta$ matrix describing the linear mapping on $H_{n,d}(\mathbb{R}^1) \cong \mathbb{R}^\Delta$ given by

$$h = \sum_{|\tau|=n} h_\tau x^\tau \mapsto \sum_{|\tau|=n} \sum_{j,k=1}^d h_\tau \frac{\partial (x^\tau)}{\partial x_j} a_{jk} x_k =: T(A)_n(h).$$

8.3.1 Example. For $d = n = 2$, $\Delta = 3$ and $D = 6$

$$I_1 \otimes A = \begin{pmatrix} A & 0 & 0 \\ 0 & A & 0 \\ 0 & 0 & A \end{pmatrix}, \quad T(A)_2 = \begin{pmatrix} 2a_{11} & a_{21} & 0 \\ 2a_{12} & a_{11} + a_{22} & 2a_{21} \\ 0 & a_{12} & 2a_{22} \end{pmatrix},$$

hence

$$T(A)_2 \otimes I_2 = \begin{pmatrix} 2a_{11} & 0 & a_{21} & 0 & 0 & 0 \\ 0 & 2a_{11} & 0 & a_{21} & 0 & 0 \\ 2a_{12} & 0 & a_{11}+a_{22} & 0 & 2a_{21} & 0 \\ 0 & 2a_{12} & 0 & a_{11}+a_{22} & 0 & 2a_{21} \\ 0 & 0 & a_{12} & 0 & 2a_{22} & 0 \\ 0 & 0 & 0 & a_{12} & 0 & 2a_{22} \end{pmatrix}.$$

∎

Let now
$$\dot{x}_t = f(\theta_t \omega, x_t) \tag{8.3.4}$$
be an RDE in $\mathbb{R}^d$ for which $f(\omega, 0) = 0$ and $f_\omega(\cdot, \cdot) := f(\theta.\omega, \cdot) \in L_{\text{loc}}(\mathbb{R}, \mathcal{C}^\infty)$. By Theorem 2.2.1 (8.3.4) generates a local C^∞ RDS φ over θ, more precisely, $\varphi(t, \omega) : D(t, \omega) \to R(t, \omega)$ is a local C^∞ diffeomorphism with fixed point 0.

We want to classify RDE by smooth equivalence. Let
$$\dot{x}_t = g(\theta_t \omega, x_t)$$
be a second RDE generating a local RDS ψ. We say f and g are smoothly equivalent, $f \sim g$, if the corresponding local RDS φ and ψ are smoothly equivalent by a random coordinate transformation h,
$$\varphi(t, \omega) \circ h(\omega) = h(\theta_t \omega) \circ \psi(t, \omega). \tag{8.3.5}$$

It is crucial to realize that at time t we have to apply the coordinate transformation $h(\theta_t \omega, \cdot)$.

Assuming for the moment that $\dot{h} = (d/dt)h(\theta_t \omega, x)$ exists, we obtain by differentiating (8.3.5) with respect to t the equivalent infinitesimal form
$$f(\theta_t \omega, h(\theta_t \omega, x)) = \dot{h}(\theta_t \omega, x) + Dh(\theta_t \omega, x)g(\theta_t \omega, x)$$
or
$$g(\theta_t \omega, x) = Dh(\theta_t \omega, x)^{-1}(f(\theta_t \omega, h(\theta_t \omega, x)) - \dot{h}(\theta_t \omega, x)). \tag{8.3.6}$$

As in the mapping case, normal form theory for RDE now consists of choosing h so as to make g "as simple as possible", the ultimate aim again being linearization, $g(\omega, x) = A(\omega)x$.

The linear cocycle $\Phi(t, \omega) := D\varphi(t, \omega, 0)$ on $T_0 \mathbb{R}^d \cong \mathbb{R}^d$ is generated by the linearized RDE
$$\dot{v}_t = A(\theta_t \omega)v_t, \quad A(\omega) := Df(\omega, 0). \tag{8.3.7}$$

We assume throughout that $A \in L^1(\mathbb{P})$, so that the MET holds for Φ.

Now the first obstacle appears on the linear level. As a preparatory step we try to make (8.3.7) "as simple as possible", in particular to choose a random basis adapted to the splitting of Φ which block-diagonalizes not only Φ, but also A, so that on each of the Oseledets spaces we would have an RDE which is decoupled from all others.

This might, however, not always be possible. But

(i) if $A(\omega) \equiv A$ is deterministic, A as well as $\exp(tA)$ are in Jordan canonical form in the same basis;

(ii) if $A(\theta_t \omega)$ is periodic with respect to t, then thanks to Floquet theory there is a random basis $F(\omega)$ in which the generator of the cocycle is constant and in Jordan canonical form (see Example 4.1.13 and Iooss [180: Lemma 4]);

(iii) if Φ has simple spectrum, then (over a possibly extended DS) we can find a basis of random eigenvectors which diagonalizes (8.3.7) (see Corollary 6.2.9).

We assume from now on that the best possible choice $F(\omega)$ for simplifying the linear RDE (8.3.7) has been made. The nonlinear random transformation $y = h(\omega, x)$ by which we want to simplify (8.3.4) can hence be near-identity, i.e. a random coordinate transformation in the sense of Definition 8.2.1.

The functions $f(\omega, \cdot)$, $g(\omega, \cdot)$ and $h(\omega, \cdot)$ have formal random Taylor series expansions at 0 given by

$$f(\omega, x) \sim \sum_{n=1}^{\infty} f_n(\omega, x) = A(\omega)x + \sum_{n=2}^{\infty} f_n(\omega, x),$$

$$g(\omega, x) \sim \sum_{n=1}^{\infty} g_n(\omega, x) = A(\omega)x + \sum_{n=2}^{\infty} g_n(\omega, x)$$

and

$$h(\omega, x) \sim \sum_{n=1}^{\infty} h_n(\omega, x) = x + \sum_{n=2}^{\infty} h_n(\omega, x),$$

where $f_n, g_n, h_n \in \mathcal{H}_{n,d}$.

Inserting these expansions into equation (8.3.6) and equating coefficients yields (the calculations are as in the deterministic case) the following result.

8.3.2 Lemma. *The term of order $n \geq 2$ of the formal power series expansion of g is given by*

$$g_n(\theta_t \omega, x) = \mathrm{ad}_n A(\theta_t \omega) h_n(\theta_t \omega, x) - \frac{d}{dt} h_n(\theta_t \omega, x) + k_n(\theta_t \omega, x), \quad (8.3.8)$$

where

$$k_n = f_n + p_n(f_k, g_k, h_k, 2 \leq k \leq n - 1)$$

and

$$p_n = T_n\{S_{n-1}(f - A) \circ S_{n-1}(h - \mathrm{id}) - DS_{n-1}(h - \mathrm{id})S_{n-1}(g - A)\}$$

is a polynomial of the lower order terms f_k, g_k, h_k, $2 \leq k \leq n - 1$, of f, g and h, respectively. Here $T_n(f)$ denotes the term of order n of f, $S_n(f)$ is the n-jet of f and $\mathrm{ad}_n A(\omega)$ is the generator of a linear cocycle in $\mathcal{H}_{n,d}$ defined by (8.3.2) and (8.3.3).

For example, $k_2 = f_2$, $k_3 = f_3 - Dh_2 g_2$, $k_4 = f_4 + f_2 \circ g_2 - Dh_2 g_3 - Dh_3 g_2$, etc.

8.3.3 Definition (Random Cohomological Operator, RDE Case).
The linear operator
$$L_n := \frac{d}{dt} - \operatorname{ad}_n A(\theta_t \omega)$$
acting on $\mathcal{H}_{n,d}$-valued stationary stochastic processes of the form $h_n(\theta_t \omega)$, which are absolutely continuous with respect to time t via
$$h_n(\theta_t \omega) \mapsto \frac{d}{dt} h_n(\theta_t \omega) - \operatorname{ad}_n A(\theta_t \omega) h_n(\theta_t \omega),$$
is called the *random cohomological operator* in $\mathcal{H}_{n,d}$.

With this definition, the system of equations (8.3.8) reads
$$L_n h_n = k_n - g_n, \quad n = 2, 3, \ldots, \tag{8.3.9}$$
called the system of *random cohomological equations*. ∎

We have to solve (8.3.9) successively for known k_n, and h_n chosen to make g_n simplest possible. The most desirable choice of g_n is $g_n = 0$, resulting in the task to solve the random cohomological equation
$$L_n h_n = \left(\frac{d}{dt} - \operatorname{ad}_n A(\theta_t \omega) \right) h_n(\theta_t \omega) = k_n(\theta_t \omega)$$
in $\mathcal{H}_{n,d}$, equivalently, to search for stationary solutions of the hierarchical system of affine RDE
$$\dot{h}_n = \operatorname{ad}_n A(\theta_t \omega) h_n + k_n(\theta_t \omega), \quad n = 2, 3, \ldots$$

In case $g_n = 0$ for all n, $k_n = f_n + \sum_{2 \leq p, q : pq = n} f_p \circ h_q$.

In order to apply our basic Theorem 5.6.5(ii) we need to know the spectrum $\mathcal{S}(\operatorname{ad}_n(A))$ of the cocycle generated by $\operatorname{ad}_n A$ in terms of the spectrum $\mathcal{S}(A)$ of A.

8.3.4 Proposition. *Let Φ_A denote the linear cocycle generated by the RDE $\dot{v}_t = A(\theta_t \omega) v_t$ with Lyapunov spectrum $\mathcal{S}(A) = (\Lambda_i)$ (the Lyapunov exponents listed with their multiplicities). Then:*

(i) The linear cocycle $\Phi_{-T(A)_n}$ in $\mathbb{R}^\Delta$ generated by $-T(A)_n$ is
$$\Phi_{-T(A)_n} = N(\Phi_{-A})_n,$$
where $N(A)_n$ is defined in (8.1.2). Moreover, the MET holds for $\Phi_{-T(A)_n}$ and
$$\mathcal{S}(-T(A)_n) = \{-(\Lambda, \tau) : |\tau| = n\}, \quad (\Lambda, \tau) := \sum_{i=1}^d \Lambda_i \tau_i.$$

(ii) The linear cocycle $\Phi_{\operatorname{ad}_n A}$ in $\mathbb{R}^D \cong \mathbb{R}^\Delta \otimes \mathbb{R}^d$ generated by $\operatorname{ad}_n A = I_1 \otimes A - T(A)_n \otimes I_2$ is

$$\Phi_{\mathrm{ad}_n A} = N(\Phi_{-A})_n \otimes \Phi_A = \Phi_{-T(A)_n} \otimes \Phi_A.$$

Moreover, the MET holds for $\Phi_{\mathrm{ad}_n A}$ and

$$\mathcal{S}(\mathrm{ad}_n A) = \{\Lambda_i - (\Lambda, \tau) : \Lambda_i \in \mathcal{S}(A), |\tau| = n\}.$$

Proof. (i) We have

$$\begin{aligned}
\frac{d}{dt} h(\Phi_{-A}(t,\omega)x) &= Dh(\Phi_{-A}(t,\omega)x) \frac{d}{dt}(\Phi_{-A}(t,\omega)x) \\
&= Dh(\Phi_{-A}(t,\omega)x)(-A(\theta_t\omega)\Phi_{-A}(t,\omega)x) \\
&= -Dh(y)A(\theta_t\omega)y|_{y=\Phi_{-A}(t,\omega)x}.
\end{aligned}$$

The matrix version of this states that $u = N(\Phi_{-A})_n$ solves $\dot{u}_t = -T(A(\theta_t\omega))_n u_t$ and thus by uniqueness $N(\Phi_{-A})_n = \Phi_{-T(A)_n}$.

The statements about the MET and the spectrum follow from Proposition 8.2.6(iii).

(ii) The generator of the cocycle $\Phi_{-T(A)_n} \otimes \Phi_A$ is $\mathrm{ad}_n A$ by Theorem 5.4.2(iii) (RDE case). The statements about the validity of the MET and the spectrum follow from Proposition 8.2.6(iii). □

Let $n \geq 2$. We call the matrix A *nonresonant of order n* if $0 \notin \mathcal{S}(\mathrm{ad}_n A)$, *resonant of order n* otherwise.

8.3.2 Nonresonant Case

We can immediately apply Theorem 5.6.5(ii) and obtain

8.3.5 Proposition. *Let A be nonresonant of some order $n \geq 2$ and assume that $k_n \in \mathcal{H}_{n,d}$ is tempered. Then the cohomological equation*

$$\dot{h}_n(\theta_t\omega) = \mathrm{ad}_n A(\theta_t\omega) h_n(\theta_t\omega) + k_n(\theta_t\omega)$$

has a unique solution $h_n(\theta_t\omega)$ which is absolutely continuous with respect to t (explicitly given in Theorem 5.6.5(ii)). In addition, h_n is also tempered.

8.3.6 Corollary. *Suppose $N \geq 2$, and A is nonresonant of order n for all $2 \leq n \leq N$. Assume further that in the Taylor expansion $f(\omega,x) = A(\omega)x + \sum_{n=2}^{N} f_n(\omega,x) + O(|x|^{N+1})$ of order N, all f_n are tempered. Then there is a unique near-identity N-jet $H_N(\omega,x) = x + \sum_{n=2}^{N} h_n(\omega,x)$ (hence is a local analytic diffeomorphism) with tempered coefficients h_n, such that f is transformed by H_N into g given by equation (8.3.6) with a linear N-jet of the Taylor expansion of order N, $g(\omega,x) = A(\omega)x + O(|x|^{N+1})$.*

The proof just repeats the arguments used for Corollary 8.2.10.

8.3.7 Theorem (Formal Linearization of RDE). *Suppose A is nonresonant of any order $n \geq 2$ and all f_n in the formal Taylor series expansion of f are tempered. Then there is a random coordinate transformation with a uniquely defined formal random Taylor series, $h(\omega, x) \sim x + \sum_{n=2}^{\infty} h_n(\omega, x)$, where all h_n are tempered, and h formally linearizes f, i. e. the formal power series expansion of g in equation (8.3.6) is $A(\omega)x$.*

The proof follows from the last corollary and Borel's lemma 8.2.12.

The full Sternberg linearization which would also transform away infinitely flat terms of f at 0 is not yet available.

8.3.3 Resonant Case

In the case of resonance we propose a functional analytic setting for the random cohomological operator which also allows for duality theory similar to the deterministic case.

For any (deterministic or random) basis $F(\omega)$ in $\mathbb{R}^d$ (and thus $(x^\tau F(\omega))_{|\tau|=n}$ in $H_{n,d}$) we endow the fiber $H_{n,d}$ over ω with the scalar product

$$\langle f(\omega), g(\omega) \rangle_{F(\omega)} := \sum_{|\tau|=n} \tau! (k_{F(\omega)}(f_\tau(\omega)), k_{F(\omega)}(g_\tau(\omega)))_S,$$

where $f(\omega) = \sum_{|\tau|=n} x^\tau f_\tau$, $g(\omega) = \sum_{|\tau|=n} x^\tau g_\tau$, $(x,y)_S := \sum_{i=1}^d x_i y_i$ is the standard scalar product in $\mathbb{R}^d$ and $k_F(x)$ are the coordinates of $x \in \mathbb{R}^d$ in the basis F. We can as usual identify $f(\omega) \in (H_{n,d}, \langle \cdot, \cdot \rangle_{F(\omega)})$ with $(k_{F(\omega)}(f_\tau(\omega)))_{|\tau|=n} \in (\mathbb{R}^D, \sum \tau!(\cdot, \cdot)_S)$, where this identification is an isometry. The above random scalar product makes $\Omega \times H_{n,d}$ a non-trivial Euclidean bundle.

We now endow the space $\mathcal{H}_{n,d}$ of $H_{n,d}$-valued random variables with the topology of convergence in probability with distance $\delta(f,g)$ given by

$$\delta(f,g) = |||f - g||| := \mathbb{E} \frac{|f(\omega) - g(\omega)|^2_{F(\omega)}}{1 + |f(\omega) - g(\omega)|^2_{F(\omega)}}.$$

This makes $(\mathcal{H}_{n,d}, \delta)$ a Fréchet space. The subset

$$\mathcal{L}^2_{n,d} := \{f \in \mathcal{H}_{n,d} : \|f\|^2 = \langle\langle f, f \rangle\rangle := \mathbb{E}\langle f(\cdot), f(\cdot)\rangle_{F(\cdot)} < \infty\}$$

is a real Hilbert space with scalar product $\langle\langle f, g \rangle\rangle := \mathbb{E}\langle f(\cdot), g(\cdot)\rangle_{F(\cdot)}$ which is densely and continuously embedded in $(\mathcal{H}_{n,d}, \delta)$ with $|||f||| \leq \|f\|$.

8.3.8 Proposition. *Assume $A \in L^2(\mathbb{P})$. Then:*

(i) The random cohomological operator

$$L_n := \frac{d}{dt} - \mathrm{ad}_n A(\theta_t \omega),$$

$$h(\theta_t\omega) \mapsto L_n h(\theta_t\omega) = \frac{d}{dt} h(\theta_t\omega) - \mathrm{ad}_n A(\theta_t\omega) h(\theta_t\omega),$$

is a densely defined linear operator in $\mathcal{L}^2_{n,d}$.

(ii) The adjoint L_n^* of L_n in $\mathcal{L}^2_{n,d}$ is

$$L_n^* = -\frac{d}{dt} - \mathrm{ad}_n A(\theta_t\omega)^*,$$

i. e. $\langle\langle L_n f, g \rangle\rangle = \langle\langle f, L_n^* g \rangle\rangle$, $f, g \in \mathcal{D}(L_n) = \mathcal{D}(L_n^*)$.

(iii) We have an orthogonal decomposition

$$\mathcal{L}^2_{n,d} = \overline{\mathcal{R}(L_n)} \oplus \mathcal{N}(L_n^*).$$

Proof. (i) $A \in L^2(\mathbb{P})$ implies $\mathrm{ad}_n A \in L^2(\mathbb{P})$. The multiplication of h by $\mathrm{ad}_n A$ is thus in $\mathcal{L}^2_{n,d}$ for any essentially[3] bounded h. The bounded polynomials are obviously dense in $\mathcal{L}^2_{n,d}$. It thus suffices to prove that there is a dense set of bounded elements $h \in \mathcal{L}^2_{n,d}$ for which $t \mapsto h(\theta_t\omega)$ is C^∞ with all derivatives bounded $\mathbb{P}$-a. s.

Take an arbitrary bounded $\mathbb{R}^D$-valued random variable h. Choose for each $k \in \mathbb{N}$ a C^∞ bump function $\varphi_k : \mathbb{R} \to \mathbb{R}^+$ with $\mathrm{supp}\,\varphi_k \subset [-1/k, 1/k]$ and $\int_{-1/k}^{1/k} \varphi_k(t)dt = 1$ and put

$$h_k(\omega) := \int_{-1/k}^{1/k} \varphi_k(t) h(\theta_t\omega) dt.$$

The integral exists $\mathbb{P}$-a. s. since $h \in L^1(\mathbb{P})$, $\|h_k\|_\infty \leq \|h\|_\infty$, and $h_k \to h$ in L^2. For,

$$\begin{aligned}
\mathbb{E}\|h_k - h\|^2 &= \mathbb{E}\left\|\int_{-1/k}^{1/k} \varphi_k(t)(h(\theta_t\omega) - h(\omega))dt\right\|^2 \\
&\leq \int_{-1/k}^{1/k} \varphi_k(t) \mathbb{E}\|h(\theta_t\omega) - h(\omega)\|^2 dt \\
&\to 0 \quad \text{for } k \to \infty.
\end{aligned}$$

The latter holds since due to the measurablity of $(t, \omega) \mapsto \theta_t\omega$, the group of isometries $U_t h := h \circ \theta_t$ is strongly continuous with respect to $t \in \mathbb{R}$ in L^2 (i. e. continuous in mean square) (see Appendix A.1).

We now prove that $t \mapsto h_k(\theta_t\omega)$ is C^∞ with bounded derivatives of any order. Differentiating $h_k(\theta_t\omega)$ m times with respect to t for fixed k and ω gives, using standard theorems,

$$\frac{d^m}{dt^m} h_k(\theta_t\omega) = g_m(\theta_t\omega), \quad g_m(\omega) := \int_{-1/k}^{1/k} (-1)^m \frac{d^m \varphi_k}{dt^m}(s) h(\theta_s\omega) ds,$$

and g_m is bounded.

---

[3] The word "esentially" will henceforth be supressed in the proof.

(ii) We check the formula for L_n^* for the dense set of test functions f, g of the type constructed in step (i).

The scalar product $\langle \cdot, \cdot \rangle_{F(\omega)}$ on $H_{n,d}$ is tailor-made to yield $\langle \mathrm{ad}_n A(\omega) f(\omega), g(\omega) \rangle_{F(\omega)} = \langle f(\omega), \mathrm{ad}_n A(\omega)^* g(\omega) \rangle_{F(\omega)}$ (see Vanderbauwhede [332]). Taking $\mathbb{E}(\cdot)$ on both sides gives

$$\langle\langle \mathrm{ad}_n A f, g \rangle\rangle = \langle\langle f, \mathrm{ad}_n A^* g \rangle\rangle.$$

Further, by definition

$$\begin{aligned}
\langle\langle \frac{d}{dt} f, g \rangle\rangle &= \sum_{|\tau|=n} \tau! \mathbb{E} (\frac{d}{dt} f(\theta_t \omega)|_{t=0}, g(\omega))_S \\
&= \sum_{|\tau|=n} \tau! \lim_{t \to 0} \frac{1}{t} \mathbb{E} (f(\theta_t \omega) - f(\omega), g(\omega))_S \\
&\quad (\mathbb{E} \lim_{t \to 0} = \lim_{t \to 0} \mathbb{E} \text{ by } L^1 \text{ continuity, see step (i)}) \\
&= \sum_{|\tau|=n} \tau! \lim_{t \to 0} \frac{1}{t} \mathbb{E} (f(\omega), g(\theta(-t)\omega) - g(\omega))_S \\
&= \sum_{|\tau|=n} \tau! \mathbb{E} (f(\omega), -\frac{d}{dt} g(\theta_t \omega)|_{t=0})_S \\
&= \langle\langle f, -\frac{d}{dt} g \rangle\rangle.
\end{aligned}$$

The fact that $\mathcal{D}(L_n) = \mathcal{D}(L_n^*)$ is proved by standard methods.

(iii) This is a well-known fact from Hilbert space theory. □

8.3.9 Corollary. *If $A \in L^2(\mathbb{P})$ is nonresonant of order n, then $\mathcal{N}(L_n^*) = \{0\}$.*

Proof. Clearly $f \in \mathcal{N}(L_n^*)$ if and only if f is a stationary solution of $\dot{f}_t = -\mathrm{ad}_n A(\theta_t \omega)^* f_t$. Suppose that $0 \notin \mathcal{S}(-\mathrm{ad}_n A^*)$. Then $f \equiv 0$ is the only stationary solution of this equation (see Corollary 5.6.3). By Theorem 5.1.1(i), $\mathcal{S}(-\mathrm{ad}_n A^*) = -\mathcal{S}(\mathrm{ad}_n A)$, so A is nonresonant of order n if and only if $-A^*$ is. □

The converse statement "$\mathcal{N}(L_n^*) = \{0\}$ implies $0 \notin \mathcal{S}(\mathrm{ad}_n A)$" is false in general, see the examples in Subsect. 8.3.4. Also remember that nonresonance of some order implies hyperbolicity of A.

The orthogonal decomposition

$$\mathcal{L}_{n,d}^2 = \overline{\mathcal{R}(L_n)} \oplus \mathcal{N}(L_n^*)$$

is marred by the fact that $\mathcal{R}(L_n)$ is in general not closed. Furthermore, the term $k_n \in \mathcal{H}_{n,d}$ in the equation $g_n + L_n h_n = k_n$ to be solved is in general not in $\mathcal{L}_{n,d}^2$. This makes the concept of ε-normal form (defined in the next theorem) necessary. To simplify the situation, we restrict ourselves to the case $\mathcal{S}(A) = \{0\}$ of total resonance, hence $\mathcal{S}(\mathrm{ad}_n A) = \{0\}$ for all $n \geq 2$.

8.3.10 Theorem (Approximate Normal Form, Resonant Case).
Consider the RDE
$$\dot{x}_t = f(\theta_t \omega, x_t),$$
where in some basis $A \in L^2(\mathbb{P})$ *and* $\mathcal{S}(A) = \{0\}$. *Then for each* $\varepsilon > 0$ *there is a random coordinate transformation* $x \mapsto h(\omega, x)$ *with formal Taylor coefficients* $T_n h \in \mathcal{L}_{n,d}^2$ *for all* $n \geq 2$ *such that the formal Taylor series of the right-hand side of*
$$\dot{x}_t = g(\theta_t \omega, x_t) = Dh(\theta_t \omega, x_t)^{-1}(f(\theta_t \omega, h(\theta_t \omega, x_t)) - \dot{h}(\theta_t \omega, x_t))$$
is in ε-*normal form, i. e. for each* $n \geq 2$ *there is a* $g_n = g_n^\mathcal{R} \oplus g_n^\mathcal{N} \in \mathcal{L}_{n,d}^2$ *with* $g_n^\mathcal{R} \in \overline{\mathcal{R}(L_n)}$, $\|g_n^\mathcal{R}\| \leq \varepsilon$, $g_n^\mathcal{N} \in \mathcal{N}(L_n^*)$ *and* $\||g_n - T_n g\|| \leq \varepsilon$.

Proof. By Lemma 8.3.2, $T_n g + L_n T_n h = k_n$, where k_n was determined by preceding steps, and $T_n h$ can be chosen to make $T_n g$ simplest possible. First move the equation from $\mathcal{H}_{n,d}$ into $\mathcal{L}_{n,d}^2$ by replacing k_n with $\bar{k}_n = \bar{k}_n^\mathcal{R} \oplus \bar{k}_n^\mathcal{N} \in \overline{\mathcal{R}(L_n)} \oplus \mathcal{N}(L_n^*) = \mathcal{L}_{n,d}^2$ which satisfies $\||k_n - \bar{k}_n\|| \leq \varepsilon$. Approximate $\bar{k}_n^\mathcal{R}$ by $\tilde{k}_n^\mathcal{R} \in \mathcal{R}(L_n)$ with $\|\bar{k}_n^\mathcal{R} - \tilde{k}_n^\mathcal{R}\| \leq \varepsilon$ and solve $L_n T_n h = \tilde{k}_n^\mathcal{R}$ in $\mathcal{L}_{n,d}^2$. Finally put $T_n g = (\bar{k}_n^\mathcal{R} - \tilde{k}_n^\mathcal{R}) \oplus \bar{k}_n^\mathcal{N}$.

Having determined all $T_n h \in \mathcal{L}_{n,d}^2$ we again invoke Borel's lemma 8.2.12 to find a diffeomorphism h. □

Random Averaging

The idea is to define an averaging operator (projection) on $\mathcal{L}_{n,d}^2$ which, applied to the right-hand side of a cohomological equation $L_n h = f$, singles out the part of f that is not in $\overline{\mathcal{R}(L_n)}$. The ideal situation would be an orthogonal projection onto $\mathcal{N}(L_n^*)$ with null space $\overline{\mathcal{R}(L_n)}$.

The following two possibilities offer themselves as generalizations of the deterministic case (see Vanderbauwhede [332]):

$$(\pi^* f)(\omega, x) = \lim_{T \to \infty} \frac{1}{T} \int_0^T \Phi_{-A^*}(t, \omega)^{-1} f(\theta_t \omega, \Phi_{-A^*}(t, \omega)x) dt, \quad (8.3.10)$$

$$(\pi f)(\omega, x) = \lim_{T \to \infty} \frac{1}{T} \int_0^T \Phi_A(t, \omega)^{-1} f(\theta_t \omega, \Phi_A(t, \omega)x) dt. \quad (8.3.11)$$

The interpretation of (8.3.10) and (8.3.11) is as follows: We average in the ω-fiber over all values that we obtain by

(i) moving (ω, x) in the ω-fiber with the skew-product flow to $(\theta_t \omega, \Phi_{-A^*}(t, \omega)x)$ and $(\theta_t \omega, \Phi_A(t, \omega)x)$, respectively, both in the $\theta_t \omega$-fiber,

(ii) and then moving back to the ω-fiber by applying the inverse of the cocycle Φ_{-A^*} and Φ_A, respectively, to the result of (i).

430   Chapter 8. Normal Forms

In the deterministic case, averaging works only under special conditions on A (e. g. semi-simplicity and spectrum on the imaginary axis). Similarly in the random case, see the examples in Subsect. 8.3.4 for illustration. Observe that even in the case where A is deterministic and $A = -A^*$ (for which (8.3.10) and (8.3.11) coincide)

$$(\pi f)(\omega, x) = \lim_{T \to \infty} \frac{1}{T} \int_0^T e^{-tA} f(\theta_t \omega, e^{tA} x) dt$$

is a *random* transformation.

8.3.4 Examples

8.3.11 Example (One-Dimensional Case). For $d = 1$ we have $\Delta = D = 1$, hence $H_{n,1}(\mathbb{R}^1) \cong \mathbb{R}^1$ (basis $x^n e_1$) for all $n \geq 2$, and

$$\operatorname{ad}_n A = \operatorname{ad}_n A^* = -(n-1)A,$$

$$L_n = \frac{d}{dt} + (n-1)A(\theta_t \omega), \quad L_n^* = -\frac{d}{dt} + (n-1)A(\theta_t \omega)$$

and

$$\Phi_{\operatorname{ad}_n A}(t, \omega) = \exp\left(-(n-1) \int_0^t A(\theta_s \omega) ds\right).$$

Assuming $A \in L^1(\mathbb{P})$, the MET yields $\mathcal{S}(A) = \{\lambda\}$, $\lambda = \mathbb{E} A$ and $\mathcal{S}(\operatorname{ad}_n A) = \{(n-1)\lambda\}$. We have resonance of all orders if and only if $\lambda = 0$.

We first consider the nonresonant case $\lambda \neq 0$. By Proposition 8.3.5 and Theorem 8.3.7, we can formally linearize the RDE by a random coordinate transformation h with tempered coefficients h_n, provided the coefficients f_n of f are tempered. To be more specific, the cohomological equation of order n

$$\dot{h}_n = -(n-1)A(\theta_t \omega) h_n + k_n(\theta_t \omega)$$

has a unique (tempered) stationary solution given by

$$h_n(\omega) = \begin{cases} \int_{-\infty}^0 \exp\left((n-1) \int_0^t A(\theta_s \omega) ds\right) k_n(\theta_t \omega) dt, & \lambda > 0, \\ -\int_0^{\infty} \exp\left((n-1) \int_0^t A(\theta_s \omega) ds\right) k_n(\theta_t \omega) dt, & \lambda < 0. \end{cases}$$

We next consider the resonant case $\lambda = 0$ and assume $A \in L^2(\mathbb{P})$. We determine

$$\mathcal{N}(L_n^*) = \{\text{stationary solutions of } \dot{f}_t = (n-1)A(\theta_t \omega) f_t\},$$

which amounts to finding initial random variables f (> 0 without loss of generality by ergodicity) for which

$$f(\theta_t\omega) = f(\omega)\exp\left((n-1)\int_0^t A(\theta_s\omega)ds\right), \qquad (8.3.12)$$

or, putting $g = \log f$, for which $\dot{g}_t = (n-1)A(\theta_t\omega)$, or

$$g(\theta_t\omega) - g(\omega) = (n-1)\int_0^t A(\theta_s\omega)ds. \qquad (8.3.13)$$

We have hence arrived at the following basic problem of algebraic ergodic theory: When is the helix $H(t,\omega) = \int_0^t A(\theta_s\omega)ds$ a coboundary (see Remark 5.6.7)? It is clear that two solutions of (8.3.13) differ only by a constant, hence either $\dim \mathcal{N}(L_n^*) = 0$ for all $n \geq 2$, or $= 1$ for all $n \geq 2$.
Consequently,

$$\mathcal{N}(L_n^*) = \begin{cases} \{0\}, & \text{if } H \text{ is not a coboundary,} \\ \mathbb{R}\, f^{n-1}, & \text{if } H \text{ is a coboundary.} \end{cases}$$

Here f is the solution of (8.3.12) for $n=2$ (unique up to a constant factor).

For necessary and sufficient conditions on A for the solvability of (8.3.13) see Orey [267] and Arnold and Wihstutz [38]. A typical case for which (8.3.12) has a solution is when $t \mapsto A(\theta_t\omega)$ is the derivative of an absolutely continuous stationary process, $A(\theta_t\omega) = (d/dt)B(\theta_t\omega)$, in which case $f(\omega) = e^{B(\omega)}$, hence $\mathcal{N}(L_n^*) = \mathbb{R}\, e^{(n-1)B}$. We can use the approach developed in Subsect. 8.3.3 if we assume that $B \in L^\infty$.

We now try random averaging for the case $A = \dot{B}$ with $B \in L^\infty$ and put $k = f^{n-1}$. One easily checks that the "Ansatz" (8.3.10) gives a projection π^* with range $\mathcal{R}(\pi^*) = \mathcal{N}(L_n^*) = \mathbb{R}\, k$, but with the wrong null space $\mathcal{N}(\pi^*) = \overline{\mathcal{R}(L_n^*)} = (k^{-1})^\perp$. However, (8.3.11) yields

$$\pi f(\omega) = \lim_{T\to\infty} \frac{1}{T} \int_0^T f(\theta_t\omega)\exp((n-1)(B(\theta_t\omega) - B(\omega)))dt = k^{-1}(\omega)\mathbb{E}\, fk$$

with $\mathcal{R}(\pi) = \mathcal{N}(L_n) = \mathbb{R}\, k^{-1}$ and $\mathcal{N}(\pi) = k^\perp = \overline{\mathcal{R}(L_n)}$. Thus, although π is in general not orthogonal, $f - \pi f \in \overline{\mathcal{R}(L_n)}$, $\pi f \in \mathcal{N}(L_n)$, and we have the (in general non-orthogonal) splitting $\mathcal{L}_{n,1}^2 \cong L^2(\mathbb{P}) = \overline{\mathcal{R}(L_n)} \oplus \mathcal{N}(L_n)$. Further, π^* is the adjoint of π and the projection is orthogonal ($\pi^* = \pi$) if and only if $B = $ const if and only if $A = 0$.

For a nonrandom A we have $\lambda = A$. In the nonresonant case $A \neq 0$ we can linearize the RDE, so the normal form of $\dot{x}_t = f(\theta_t\omega, x_t) \sim Ax_t + \sum_{n=2}^\infty f_n(\theta_t\omega)x_t^n$ is $\dot{x}_t = Ax_t$. In the resonant case $A = 0$, $\mathcal{N}(L_n^*) = \mathbb{R}\, 1$, and the ε-normal form is deterministic, $\dot{x}_t = g(x_t) \sim \sum_{n=2}^\infty (\mathbb{E}\, k_n)x_t^n$.

Note that this example also covers the quite frequently observed case where A has simple spectrum. We can then diagonalize A and thus all matrices appearing in our theory by choosing a random basis of eigenvectors. The problem hence splits into D scalar problems of the type just considered. ∎

8.3.12 Example (A Two-Dimensional Resonant Case). Assume that $d = 2$ (hence $\Delta = n + 1$ and $D = 2(n+1)$) and

$$A = \begin{pmatrix} 0 & \alpha \\ -\alpha & 0 \end{pmatrix} = -A^*, \quad \alpha \neq 0,$$

is deterministic. We identify $\mathbb{R}^2$ with $\mathbb{C}$ via $x = (x_1, x_2)' \mapsto x_1 + ix_2 = T(x) = z$, so that $T(Ax) = -i\alpha T(x)$. We identify $\mathcal{H}_{n,2}$ with the space of $\mathbb{C}^{n+1}$-valued random variables via

$$f_n(\omega, z) = \sum_{k=1}^{n} f_n^{(k)} z^{n-k} \bar{z}^k.$$

As $\mathcal{S}(A) = \{0\}$ we have $\mathcal{S}(\operatorname{ad}_n A) = \{0\}$ (with multiplicity D) for all $n \geq 2$, hence we are in the resonant case.

One easily checks that $L_n^* f = 0$ translates into $n+1$ scalar complex RDE

$$\dot{f}_n^{(k)} = (n - 2k - 1)i\alpha f_n^{(k)}, \quad k = 0, 1, \ldots, n.$$

Hence a stationary solution of such an equation satisfies

$$f_n^{(k)}(\theta_t \omega) = e^{i\alpha(n-2k-1)t} f_n^{(k)}(\omega), \quad k = 0, \ldots, n, \qquad (8.3.14)$$

i.e. $f_n^{(k)}$ is an eigenfunction of the group $U_t f(\omega) := f(\theta_t \omega)$ of unitary operators in $L^2(\mathbb{P}, \mathbb{C})$ corresponding to the eigenvalue $\alpha(n - 2k - 1)$. Recall that the set of eigenvalues of an ergodic U_t is a subgroup of the additive group $\mathbb{R}$ (countable if L^2 is separable) and that every eigenvalue has multiplicity 1, the eigenfunctions satisfy $|f| = \text{const}$ and are mutually orthogonal (see Cornfeld, Fomin and Sinai [106: p. 327]).

Thus

$$\mathcal{N}(L_n^*) = \{\sum_{k=0}^{n} f_n^{(k)} z^{n-k} \bar{z}^k : f_n^{(k)} \text{ eigenfunction of } U_t$$

corresponding to eigenvalue $\alpha(n - 2k - 1), k = 0, \ldots, n\}$.

We have nontrivial solutions certainly for $n = 2k+1$, namely $\mathbb{C}\,1$, but possibly also for other cases, as e.g. $\theta_t = $ rotation by the angle βt in $\Omega = S^1$ shows (exercise!). Such terms do not exist in the deterministic case. This clearly illustrates that random normal form theory depends on the interplay between the dynamical system θ and the linear part A of the RDE.

Averaging according to (8.3.10) and (8.3.11) is identical in this case and yields for the components of $f_n \in L^2(\mathbb{P}, \mathbb{C}^{n+1})$

$$\lim_{T\to\infty} \frac{1}{T} \int_0^T f_n^{(k)}(\theta_t\omega) e^{-i\alpha(n-2k-1)t} dt = \pi f_n^{(k)}(\omega)$$

$$= \begin{cases} \mathbb{E} f_n^{(k)}, & n = 2k+1, \\ \bar{f}_n^{(k)}, & n \neq 2k+1. \end{cases}$$

This limit exists in L^2 and $\mathbb{P}$-a.s., and $\bar{f}_n^{(k)}$ has mean zero and satisfies (8.3.14) (see Doob [128: p. 517]). In this case π is indeed the orthogonal projection onto $\mathcal{N}(L_n^*)$ with null space $\overline{\mathcal{R}(L_n)}$. ∎

8.4 Simultaneous Normal Form and Center Manifold Reduction for Random Differential Equations

The desire to simplify engineering and physics DS which are perturbed by noise has prompted numerous publications (see, e.g. Coullet, Elphick and Tirapegui [107], Nicolis and Nicolis [262], Schöner and Haken [314][315], Sri Namachchivaya and Lin [323], and the references therein). All authors have worked exclusively in a center/stable situation and with a smallness parameter multiplying the noise terms, thus providing a stochastic normal form as a small perturbation of the deterministic one. See Sri Namachchivaya [322] for a survey, which also comments on the older literature, and Subsect. 8.4.3.

We will now follow Arnold and Xu [41] and connect our general approach developed in Sects. 8.2 and 8.3 (which is independent of any smallness assumptions) with the existing physics and engineering literature by presenting a random analogue of a very successful procedure proposed by Elphick et al. [135] for simultaneously obtaining the normal form, eliminating the stable variables from the center equations and determining the center manifold (Subsect. 8.4.1). This connects also with the invariant manifold theory for RDS (see Chap. 7).

In Subsect. 8.4.2 we will deal with the case of a smallness parameter where the undisturbed system is allowed to be random. In Subsect. 8.4.3 we treat the particular case of a deterministic differential equation perturbed by small noise, and calculate the normal form of the Duffing-van der Pol oscillator with small noise, both for the pitchfork and the Hopf regime.

8.4.1 The Reduction Procedure

As in Sect. 8.3 we consider the RDE in $\mathbb{R}^d$

$$\dot{x}_t = f(\theta_t\omega, x_t) \tag{8.4.1}$$

for which $f(\omega, 0) = 0$ and $f_\omega \in L_{\text{loc}}(\mathbb{R}, \mathcal{C}^\infty)$ (if we only go up to a fixed order $N \geq 2$, then $f_\omega \in L_{\text{loc}}(\mathbb{R}, \mathcal{C}^{N+1})$ suffices). Then (8.4.1) generates a local C^∞ (or C^{N+1}) RDS with fixed point $x = 0$ and the linearized RDE

$$\dot{v}_t = A(\theta_t \omega) v_t, \quad A(\omega) := Df(\omega, 0), \qquad (8.4.2)$$

generates the linear cocycle Φ. We assume throughout that $A \in L^1(\mathbb{P})$, so that the MET holds for Φ.

We follow the deterministic philosophy and assume that we are in a center/stable situation, i.e. the Lyapunov spectrum of Φ is

$$\lambda_1 = 0 > \lambda_2 > \ldots > \lambda_r.$$

The general case requires obvious modifications of our procedure.

We define the center and stable invariant spaces by $E_c(\omega) := E_1(\omega)$, $E_s(\omega) := E_2(\omega) \oplus \cdots \oplus E_r(\omega)$, so that we have the invariant splitting

$$\mathbb{R}^d = E_c(\omega) \oplus E_s(\omega), \quad \Phi(t, \omega) E_{c,s}(\omega) = E_{c,s}(\theta_t \omega),$$

and $\dim E_{c,s}(\omega) = d_{c,s}$, $d_c + d_s = d$.

We next assume that a basis $F(\omega) = (F_c(\omega), F_s(\omega))$ exists which block-diagonalizes the linear part $A(\theta_t \omega)$ of $f(\theta_t \omega, x)$ into a center block and a stable block (for conditions see Subsect. 8.3.1) and the basis transfer matrix is a Lyapunov cohomology. In this basis

$$A(\theta_t \omega) = \begin{pmatrix} A_c(\theta_t \omega) & 0 \\ 0 & A_s(\theta_t \omega) \end{pmatrix}.$$

The coordinates in this basis are $x = \begin{pmatrix} x_c \\ x_s \end{pmatrix}$ and by the coordinate mapping $E_c(\omega)$, $E_s(\omega)$ and $\mathbb{R}^d = E_c(\omega) \oplus E_s(\omega)$ are identified with $\mathbb{R}^{d_c}$, $\mathbb{R}^{d_s}$ and $\mathbb{R}^d = \mathbb{R}^{d_c} \times \mathbb{R}^{d_s}$ with the standard bases, respectively.

The homogeneous polynomials in $E_c(\omega)$, $E_s(\omega)$ and $\mathbb{R}^d$ are written in the bases $x_c^\tau F_c(\omega)$, $x_s^\tau F_s(\omega)$ and $x^\tau F(\omega)$, where τ is a multi-index, by which we identify those polynomials with the corresponding ones in $\mathbb{R}^{d_c} \cong \mathbb{R}^{d_c} \times \{0\}$, $\mathbb{R}^{d_s} \cong \{0\} \times \mathbb{R}^{d_s}$ and $\mathbb{R}^d = \mathbb{R}^{d_c} \times \mathbb{R}^{d_s}$, respectively.

We now formulate the elimination procedure of order $N \geq 2$ for which we need the following Taylor expansions of order N (suppressing the argument ω):

$$\begin{aligned} f(x) &= Ax + \sum_{n=2}^{N} f_n(x) + O(|x|^{N+1}) \\ &= \begin{pmatrix} A_c x_c \\ A_s x_s \end{pmatrix} + \sum_{(p,q):\, p+q=2}^{N} \begin{pmatrix} f_{pq}^c(x_c, x_s) \\ f_{pq}^s(x_c, x_s) \end{pmatrix} + O(|x|^{N+1}), \end{aligned}$$

where $f_n(x)$ is a random homogeneous polynomial of degree n in x with values in $\mathbb{R}^d$ and the $f_{pq}^{c,s}(x_c, x_s)$ are random homogeneous polynomials of degree $n = p + q$ which have degree $p \in \{0, 1, \ldots, N\}$ in x_c, degree $q \in \{0, 1, \ldots, N\}$ in x_s and with values in $\mathbb{R}^{d_{c,s}}$.

Similarly, for the random near-identity polynomial $h(x) = x + H(x)$ of degree N we have

$$h(x) = x + H(x) = \begin{pmatrix} x_c \\ x_s \end{pmatrix} + \sum_{(p,q):\, p+q=2}^{N} \begin{pmatrix} H_{pq}^c(x_c, x_s) \\ H_{pq}^s(x_c, x_s) \end{pmatrix}.$$

We replace x in (8.4.1) by $h(x) = x + H(x)$ and obtain the new RDE

$$\dot{x}_t = g(\theta_t \omega, x_t),$$

where

$$g(\theta_t \omega, x) = Dh(\theta_t \omega, x)^{-1}(f(\theta_t \omega, h(\theta_t \omega, x)) - \frac{d}{dt} h(\theta_t \omega, x)). \quad (8.4.3)$$

If f is C^{N+1} then so is g, with corresponding Taylor expansion. However, we try to choose h such that g has Taylor expansion

$$g(x) = \begin{pmatrix} A_c x_c \\ A_s x_s \end{pmatrix} + \begin{pmatrix} \sum_{n=2}^{N} g_n^c(x_c) + O((|x_c| + |x_s|)^{N+1}) \\ \sum_{(p,q):\, p+q=2, q\geq 1}^{N} g_{pq}^s(x_c, x_s) + O((|x_c| + |x_s|)^{N+1}) \end{pmatrix}. \quad (8.4.4)$$

8.4.1 Theorem (Normal Form and Center Manifold for RDE).
Let $N \geq 2$ and assume that the coefficients of the Taylor expansion of f of order N are tempered. Then there are random polynomials of degree N with tempered coefficients, namely

- *$H(\omega, x)$ with values in $\mathbb{R}^d$ and $H(\omega, x_c, x_s) = O((|x_c| + |x_s|)^2)$,*
- *$g^c(\omega, x_c)$ with values in $E_c(\omega) \cong \mathbb{R}^{d_c} \times \{0\}$ and $g^c(\omega, x_c) = O(|x_c|^2)$,*
- *$g^s(\omega, x_c, x_s)$ with values in $E_s(\omega) \cong \{0\} \times \mathbb{R}^{d_s}$ and $g^s(\omega, x_c, x_s) = O(|x_s|(|x_c| + |x_s|))$,*

such that:
(i) The near-identity transformation

$$x \mapsto h(\omega, x) = x + H(\omega, x) = \begin{pmatrix} x_c \\ x_s \end{pmatrix} + \begin{pmatrix} H^c(\omega, x_c, x_s) \\ H^s(\omega, x_c, x_s) \end{pmatrix}$$

transforms $\dot{x}_t = f(\theta_t \omega, x_t)$ into

$$\dot{x}_c = A_c(\theta_t \omega) x_c + g^c(\theta_t \omega, x_c) + O((|x_c| + |x_s|)^{N+1}),$$

$$\dot{x}_s = A_s(\theta_t \omega) x_s + g^s(\theta_t \omega, x_c, x_s) + O((|x_c| + |x_s|)^{N+1}),$$

where g^c can be made "as simple as possible".
(ii) The manifold in $\mathbb{R}^d$ given by

$$M_c(\omega) := \{ \begin{pmatrix} x_c \\ 0 \end{pmatrix} + H(\omega, x_c, 0) : x_c \in \mathbb{R}^{d_c} \}$$

- *is locally invariant for the truncated normal form equations*

Chapter 8. Normal Forms

$$\dot{x}_c = A_c(\theta_t\omega)x_c + g^c(\theta_t\omega, x_c), \tag{8.4.5}$$

$$\dot{x}_s = A_s(\theta_t\omega)x_s + g^s(\theta_t\omega, x_c, x_s), \tag{8.4.6}$$

— has the property $T_0 M_c(\omega) = \mathbb{R}^{d_c}$, in particular $\dim M_c(\omega) = d_c$, and $M_c(\omega)$ approximates a local center manifold at $x = 0$ of the original RDS up to terms of order $|x_c|^N$ and is locally attracting.

Proof. (i) As in Sect. 8.3 (and suppressing ω), we replace x in (8.4.3) by $h(x) = x + H(x)$ and obtain

$$f(x + H(x)) = \dot{x} + (D_{x_c}H)\dot{x}_c + (D_{x_s}H)\dot{x}_s + \frac{d}{dt}H(x_c, x_s). \tag{8.4.7}$$

Now we replace $\dot{x}$ by $g(x)$, plug our Taylor expansions for f, h and g into equation (8.4.7), equate coefficients and separate the center and the stable components. For $(p, 0) = (1, 0)$ and $(0, 1)$ we recover $A_c x_c = A_c x_c$ and $A_s x_s = A_s x_s$. For $(p, 0)$ with $2 \le p \le N$ we obtain for the center component

$$\frac{d}{dt}H^c_{p0}(x_c) - A_c H^c_{p0}(x_c) + D_{x_c}H^c_{p0}(x_c)A_c x_c = -g^c_p(x_c) + R^c_{p0}(x_c) \tag{8.4.8}$$

and for the stable component

$$\frac{d}{dt}H^s_{p0}(x_c) - A_s H^s_{p0}(x_c) + D_{x_c}H^s_{p0}(x_c)A_c x_c = R^s_{p0}(x_c). \tag{8.4.9}$$

For (p, q) with $2 \le p + q \le N$ and $q \ge 1$ we get for the center component

$$\frac{d}{dt}H^c_{pq}(x_c, x_s) - A_c H^c_{pq}(x_c, x_s) + D_{x_c}H^c_{pq}(x_c, x_s)A_c x_c$$
$$+ D_{x_s}H^c_{pq}(x_c, x_s)A_s x_s = R^c_{pq}(x_c, x_s) \tag{8.4.10}$$

and for the stable component

$$\frac{d}{dt}H^s_{pq}(x_c, x_s) - A_s H^s_{pq}(x_c, x_s) + D_{x_c}H^s_{pq}(x_c, x_s)A_c x_c$$
$$+ D_{x_s}H^s_{pq}(x_c, x_s)A_s x_s = -g^s_{pq}(x_c, x_s) + R^s_{pq}(x_c, x_s). \tag{8.4.11}$$

Here the $R^{c,s}_{pq}$ depend only on f_{pq} and on $H_{p',q'}$, $g^c_{p'}$, $g^s_{p',q'}$ for $p' + q' \le p + q - 1$. The strategy is hence to solve equations (8.4.8) to (8.4.11) step by step, starting with $p + q = 2$ and then increasing $p + q$ by 1 at each step.

We now look at the four equations (8.4.8) to (8.4.11) separately.

1. Equation (8.4.8): This is our well-known cohomological equation for A_c,

$$\left(\frac{d}{dt} - \mathrm{ad}_p A_c(\theta_t\omega)\right) H^c_{p0} = -g^c_p + R^c_{p0}.$$

Since we are in the resonant case, we could use our theory of ε-normal forms from Subsect. 8.3.3 and choose H^c_{p0} such that $g^c_p \in \mathcal{N}\left(\frac{d}{dt} + \mathrm{ad}_p A^*_c\right)$. But we are also free to choose

$H_{p0}^c = 0$, hence $g_p^c = R_{p0}^c$, $2 \le p \le N$.

2. Equation (8.4.9): This is also a cohomological equation corresponding to the linear cocycle Φ_{p0}^s generated by $I_1 \otimes A_s - T(A_c)_p \otimes I_2$ having Lyapunov spectrum $\{\lambda_2, \ldots, \lambda_r\}$, λ_i with multiplicity $d_i \binom{p+d_c-1}{p}$. It is thus stable and equation (8.4.9) has the unique stationary solution

$$H_{p0}^s(\omega) = \int_{-\infty}^0 \Phi_{p0}^s(t,\omega)^{-1} R_{p0}^s(\theta_t\omega)dt, \quad 2 \le p \le N.$$

3. Equation (8.4.10): This is again a cohomological equation, corresponding to the linear cocycle Φ_{pq}^c generated by $I_1 \otimes I_2 \otimes A_c - T(A_c)_p \otimes I_2 \otimes I_3 - I_1 \otimes T(A_s)_q \otimes I_3$ in the polynomial space $H_{p,d_c}(\mathbb{R}^1) \otimes H_{q,d_s}(\mathbb{R}^1) \otimes \mathbb{R}^{d_c}$ and having Lyapunov spectrum $\{-\sum_{i=2}^r \tau_i \lambda_i : |\tau| = q\}$. So all exponents are positive, and the unique stationary solution of (8.4.10) is thus

$$H_{pq}^c(\omega) = -\int_0^\infty \Phi_{pq}^c(t,\omega)^{-1} R_{pq}^c(\theta_t\omega)dt, \quad 2 \le p+q \le N, q \ge 1.$$

4. Equation (8.4.11): Since we do not care about an optimal choice of g_{pq}^s we simply choose

$$H_{pq}^s = 0, \quad \text{hence } g_{pq}^s = R_{pq}^s, \quad 2 \le p+q \le N, q \ge 1.$$

(However, we could also solve the corresponding cohomological equation to obtain $g_{pq}^s = 0$ in the non-resonant case and an element in the corresponding kernel in the resonant case, hence a nontrivial H_{pq}^s. We will do this in Example 8.4.2).

(ii) We have to show that if $\varphi(t,\omega)$ is the local C^∞ RDS generated by the truncated equations (8.4.5) and (8.4.6) then locally

$$\varphi(t,\omega) M_c(\omega) = M_c(\theta_t\omega) .$$

More precisely (assuming for the moment that we made the trivial choice $H_{p0}^c = 0$ in equation (8.4.8)) we have to show that

$$\varphi(t,\omega) \begin{pmatrix} x_c \\ H^s(\omega, x_c, 0) \end{pmatrix} = \begin{pmatrix} \varphi_c(t,\omega) x_c \\ H^s(\theta_t\omega, \varphi_c(t,\omega) x_c, 0) \end{pmatrix},$$

where φ_c is the local RDS generated by the (decoupled) center equation (8.4.5). While this is clear for the center component, differentiating the stable component yields

$$A_s H^s(x_c, 0) + g^s(x_c, H^s(x_c, 0)) = \frac{dH^s}{dt}(x_c, 0) + D_{x_c} H^s(A_c x_c + g^c(x_c)).$$

Inserting the expansions for H^s, g^s and g^c and equating coefficients gives the cohomological equations (8.4.9) for $2 \le p \le N$, which are satisfied by our construction.

If H^c is nontrivial we also recover the cohomological equations (8.4.8).

The statement about the tangent space follows from $H(\omega, x_c, 0) = O(|x_c|^2)$. For the claim that M_c approximates a local center manifold of (8.4.1) we use the criterion of Theorem 5.1 of Boxler [80: Theorem 5.1]. It is locally attracting due to the negative spectrum of A_s. □

8.4.2 Example (Two-Dimensional Case). Assume $d = 2$, $\lambda_1 = 0 > \lambda_2$. Hence we can diagonalize the linear part A,

$$A(\omega) = \begin{pmatrix} A_c(\omega) & 0 \\ 0 & A_s(\omega) \end{pmatrix},$$

with $\lambda_1 = \mathbb{E} A_c = 0 > \lambda_2 = \mathbb{E} A_s$. All equations are scalar.

For (8.4.8) we can make the trivial choice $H^c_{p0} = 0$ or proceed as in Example 8.3.11.

For (8.4.9) we always have a unique solution given by

$$H^s_{p0}(\omega) = \int_{-\infty}^{0} \left(\exp \int_0^t (pA_c(\theta_u \omega) - A_s(\theta_u \omega)) du \right) R^s_{p0}(\theta_t \omega) dt.$$

For (8.4.10) we have the unique solution

$$H^c_{pq}(\omega) =$$
$$- \int_0^{\infty} \left(\exp \int_0^t ((p-1)A_c(\theta_u \omega) + qA_s(\theta_u \omega)) du \right) R^c_{pq}(\theta_t \omega) dt.$$

For (8.4.11) we made the trivial choice $H^s_{pq} = 0$ which made the terms $g^s_{pq} = R^s_{pq}$ survive. However, we could do better as follows: The corresponding cocycle has spectrum

$$(1-q)\lambda_2 \begin{cases} = 0, & q = 1, \\ > 0, & q \geq 2. \end{cases}$$

We hence keep the choice $H^s_{p1} = 0$ (or solve the resonant cohomological equation), but choose for $q \geq 2$

$$H^s_{pq}(\omega) =$$
$$- \int_0^{\infty} \left(\exp \int_0^t ((q-1)A_s(\theta_u \omega) + pA_c(\theta_u \omega)) du \right) R^s_{pq}(\theta_t \omega) dt,$$

which implies $g^s_{pq} = 0$ for $2 \leq q \leq N$.

The final truncated equations are thus

$$\dot{x}_c = A_c x_c + \sum_{n=2}^{N} g_n^c x_c^n, \quad \dot{x}_s = A_s x_s + \sum_{n=2}^{N} g_{n-1,1}^s x_c^{n-1} x_s,$$

and the approximate center manifold (for the choice $H_{n0}^c = 0$) is the graph $x_c \mapsto \sum_{n=2}^{N} H_{n0}^s x_c^n$. ∎

8.4.2 Parametrized RDE

For applications in stochastic bifurcation theory it is of considerable importance to treat the case

$$\dot{x}_t = f(\theta_t \omega, x_t, \alpha) \tag{8.4.12}$$

of an RDE in $\mathbb{R}^d$ which smoothly depends on a parameter $\alpha \in \mathbb{R}^m$. We assume that α is in the neighborhood of a critical value $\alpha_c = 0$ (say). We will obtain the approximate normal form and center manifold of (8.4.12) (augmented by the trivial equation $\dot{\alpha} = 0$) as a polynomial in the variables $(x, \alpha) \in \mathbb{R}^d \times \mathbb{R}^m$.

For simplicity we assume that

$$f(\omega, 0, \alpha) = 0 \quad \text{for all } \alpha \in \mathbb{R}^m \tag{8.4.13}$$

and that at $\alpha = 0$ we are in a center/stable situation, i.e. the cocycle $\Phi(t, \omega)$ generated by

$$\dot{v}_t = A(\theta_t \omega) v_t, \quad A(\omega) := D_x f(\omega, 0, 0), \tag{8.4.14}$$

has Lyapunov spectrum $\lambda_1 = 0 > \lambda_2 > \cdots > \lambda_r$. These simplifying assumptions are basically without loss of generality. The general case requires straightforward modifications of our procedure.

Assume finally that the decoupling of A as described in the last subsection has been successfully carried out. We then have the following random version of a result of Elphick et al. [135: Chap. III]:

8.4.3 Theorem Normal Form and Center Manifold for RDE with Parameters). *Let $N \geq 2$ and $M \geq 0$. Assume that the coefficients of the Taylor expansion of f in the variable (x, α) of order N in x and order M in α are tempered. Then there are random polynomials of degree N in x and M in α with tempered coefficients, namely*

- *$H(\omega, x, \alpha)$ with values in $\mathbb{R}^d$ and $H(\omega, x_c, x_s, \alpha) = O(|\alpha|(|x_c| + |x_s|) + (|x_c| + |x_s|)^2)$,*
- *$g^c(\omega, x_c, \alpha)$ with values in $E_c(\omega) \cong \mathbb{R}^{d_c} \times \{0\}$ and $g^c(\omega, x_c, \alpha) = O(|\alpha| |x_c| + |x_c|^2)$,*
- *$g^s(\omega, x_c, x_s, \alpha)$ with values in $E_s(\omega) \cong \{0\} \times \mathbb{R}^{d_s}$ and $g^s(\omega, x_c, x_s, \alpha) = O(|x_s|(|\alpha| + |x_c| + |x_s|))$,*

such that:

440   Chapter 8. Normal Forms

(i) *The near-identity transformation*

$$x \mapsto h(\omega, x, \alpha) = x + H(\omega, x, \alpha) = \begin{pmatrix} x_c \\ x_s \end{pmatrix} + \begin{pmatrix} H^c(\omega, x_c, x_s, \alpha) \\ H^s(\omega, x_c, x_s, \alpha) \end{pmatrix}$$

transforms $\dot{x}_t = f(\theta_t\omega, x_t, \alpha)$ *into*

$$\dot{x}_c = A_c(\theta_t\omega)x_c + g^c(\theta_t\omega, x_c, \alpha) + O((|x_c| + |x_s|)^{N+1} + |\alpha|^{M+1}),$$
$$\dot{x}_s = A_s(\theta_t\omega)x_s + g^s(\theta_t\omega, x_c, x_s, \alpha) + O((|x_c| + |x_s|)^{N+1} + |\alpha|^{M+1}),$$

where g^c can be made "as simple as possible".

(ii) *The manifold in $\mathbb{R}^d \times \mathbb{R}^m$ given by*

$$M_c(\omega) := \left\{ \begin{pmatrix} (x_c, \alpha) \\ 0 \end{pmatrix} + H(\omega, x_c, 0, \alpha) : (x_c, \alpha) \in \mathbb{R}^{d_c} \times \mathbb{R}^m \right\}$$

– *is locally invariant for the* truncated normal form equations

$$\dot{x}_c = A_c(\theta_t\omega)x_c + g^c(\theta_t\omega, x_c, \alpha), \quad \dot{\alpha} = 0, \quad (8.4.15)$$

$$\dot{x}_s = A_s(\theta_t\omega)x_s + g^s(\theta_t\omega, x_c, x_s, \alpha), \quad (8.4.16)$$

– *has the property $T_0 M_c(\omega) = \mathbb{R}^{d_c} \times \mathbb{R}^m$, in particular $\dim M_c(\omega) = d_c + m$, and $M_c(\omega)$ approximates a local center manifold at $(x, \alpha) = (0, 0)$ of the original RDS augmented by $\dot{\alpha} = 0$, up to terms of order $|x_c|^N + |\alpha|^M$, and is locally attracting.*

Proof. (i) Let the corresponding Taylor expansions of f, h and $g^{c,s}$ (with ω suppressed) be

$$f(x, \alpha) = Ax + \sum_{\substack{1 \leq n \leq N \\ 0 \leq r \leq M}} f_{nr}(x, \alpha) + O(|x|^{N+1} + |\alpha|^{M+1}),$$

$$H(x_c, x_s, \alpha) = \sum_{\substack{1 \leq p+q \leq N, \, 0 \leq r \leq M \\ (p+q, r) \neq (1, 0)}} H_{pqr}(x_c, x_s, \alpha) + O(|x|^{N+1} + |\alpha|^{M+1}),$$

$$g^c(x_c, \alpha) = \sum_{\substack{1 \leq n \leq N, \, 0 \leq r \leq M \\ (n, r) \neq (1, 0)}} g^c_{nr}(x_c, \alpha) + O(|x_c|^{N+1} + |\alpha|^{M+1}),$$

$$g^s(x_c, x_s, \alpha) = \sum_{\substack{1 \leq p+q \leq N, \, 0 \leq r \leq M \\ q \geq 1, \, (p, r) \neq (0, 0)}} g^s_{pqr}(x_c, x_s, \alpha) + O(|x|^{N+1} + |\alpha|^{M+1}).$$

Here f_{nr} is a random homogeneous polynomial of degree $n + r$ which is homogeneous of degree n in x and of degree r in α with values in $\mathbb{R}^d$; H_{pqr} is a random homogeneous polynomial of degree $p + q + r$ which is of degree p, q, r in x_c, x_s, α, respectively, with values in $\mathbb{R}^d = \mathbb{R}^{d_c} \times \mathbb{R}^{d_s}$; g^c_{nr} is of degree $n + r$, but of degree n and r in x_c and α, respectively, with values in $\mathbb{R}^{d_c} \times \{0\}$; and finally g^s_{pqr} is of degree $p + q + r$, but of degree p, q, r in x_c, x_s, α, respectively, with values in $\{0\} \times \mathbb{R}^{d_s}$.

As in the proof of Theorem 8.4.7, we insert our expansions into equation (8.4.7), equate coefficients and separate center and stable components. We obtain structurally the same equations, the difference being that the appearance of α gives rise to a factor $\binom{r+n-1}{r}$ for the dimensions of the corresponding polynomial spaces. The result is as follows:

For $(p, 0, r)$ with $2 \leq p \leq N$ and $0 \leq r \leq M$ we obtain

$$\frac{d}{dt} H^c_{p0r}(x_c, \alpha) - A_c H^c_{p0r}(x_c, \alpha)$$
$$+ D_{x_c} H^c_{p0r}(x_c, \alpha) A_c x_c = -g^c_{pr}(x_c, \alpha) + R^c_{p0r}(x_c, \alpha) \qquad (8.4.17)$$

and

$$\frac{d}{dt} H^s_{p0r}(x_c, \alpha) - A_s H^s_{p0r}(x_c, \alpha) + D_{x_c} H^s_{p0r}(x_c, \alpha) A_c x_c = R^s_{p0r}(x_c, \alpha). \qquad (8.4.18)$$

For (p, q, r) with $1 \leq p + q \leq N$, $q \geq 1$, $0 \leq r \leq M$, $(p+q, r) \neq (1, 0)$, we get

$$\frac{d}{dt} H^c_{pqr}(x_c, x_s, \alpha) - A_c H^c_{pqr}(x_c, x_s, \alpha)$$
$$+ D_{x_c} H^c_{pqr}(x_c, x_s, \alpha) A_c x_c + D_{x_s} H^c_{pqr}(x_c, x_s, \alpha) A_s x_s$$
$$= R^c_{pqr}(x_c, x_s, \alpha) \qquad (8.4.19)$$

and

$$\frac{d}{dt} H^s_{pqr}(x_c, x_s, \alpha) - A_s H^s_{pqr}(x_c, x_s, \alpha)$$
$$+ D_{x_c} H^s_{pqr}(x_c, x_s, \alpha) A_c x_c + D_{x_s} H^s_{pqr}(x_c, x_s, \alpha) A_s x_s$$
$$= -g^s_{pqr}(x_c, x_s, \alpha) + R^s_{pqr}(x_c, x_s, \alpha). \qquad (8.4.20)$$

Here the $R^{c,s}_{pqr}$ depend only on f_{pqr} and on $H_{p+1,q,r'-1}$, $H_{p'q'r'}$, $g^c_{p'r'}$, $g^s_{p'q'r'}$ for $p' \leq p$, $q' \leq q$, $r' \leq r$, $p' + q' + r' \leq p + q + r - 1$.

The step by step strategy to solve these equations is as follows: Start with $r = 0$. Then equations (8.4.17) to (8.4.20) reduce to equations (8.4.8) to (8.4.11) which have been solved above. This determines H_{pq0}, g^c_{p0} and g^s_{pq0}. Note that for $(p, q, 0) = (1, 0, 0)$ and $(0, 1, 0)$ we recover the linear parts. In the next step put $r = 1$ and solve all equations for $1 \leq p + q \leq N$, and so forth to $r = M$.

For $r \geq 1$, the above equations have the same structure as those for $r = 0$, so h_{pqr}, g^c_{pr} and g^s_{pqr} will have the same structure as h_{pq0}, g^c_{p0} and g^s_{pq0}. The presence of α amounts to having $\binom{r+n-1}{r}$ equations with the same cohomological operator instead of one for $r = 0$.

Note finally that due to our assumption (8.4.13), $R_{00r} = 0$ for all $r \geq 0$, which results in the fact that we do not have terms of the type $(0, 0, r)$ in H and g^s and of type $(0, r)$ in g^c. Also, as g^s only appears in equation (8.4.20) for which $q \geq 1$, there are no terms of the form g^s_{p0r}.

(ii) See the proof of the corresponding fact of Theorem 8.4.1. □

8.4.4 Remark (New Terms in Parameter Case). For use in Subsect. 8.4.3 we look at those equations among (8.4.17) to (8.4.20) that do not appear for $\alpha = 0$. For $r = 1$, the lowest possible triples are $(1, 0, 1)$ and $(0, 1, 1)$. These will lead to corrections of order 1 in α of the linear parts of the transformation h (so it will be near-identity only for $\alpha = 0$) as well as of the linear parts of the resulting equations. More precisely, the linear part of h has the form

$$\begin{pmatrix} x_c + \sum_{1 \leq i \leq m} \alpha_i H_i^{cs} x_s \\ x_s + \sum_{1 \leq i \leq m} \alpha_i H_i^{sc} x_c \end{pmatrix},$$

while the RDE start with

$$\dot{x}_c = \left(A_c(\theta_t \omega) + \sum_{1 \leq i \leq m} \alpha_i R_i^c(\theta_t \omega) \right) x_c + \cdots,$$

$$\dot{x}_s = \left(A_s(\theta_t \omega) + \sum_{1 \leq i \leq m} \alpha_i R_i^s(\theta_t \omega) \right) x_s + \cdots,$$

and the center manifold starts with $(x_c, \alpha) \mapsto (\sum_{1 \leq i \leq m} \alpha_i H_i^{sc}) x_c + \cdots$ ∎

8.4.3 Small Noise: A Case Study

By "small noise" we mean that the parametrized random vector field $f(\omega, x, \alpha)$ in the RDE (8.4.12) is deterministic at the critical value $\alpha = 0$. This is of course a particular case and we will not reiterate our procedure.

Let us stress that in the small noise case the linearized equation (8.4.14) is also deterministic. Consequently, all cohomological operators appearing in our procedure are of the form $\frac{d}{dt} - B$, where B is a deterministic linear operator. But note that the corresponding cohomological equations are of the form $(\frac{d}{dt} - B)h(\theta_t \omega) = f(\theta_t \omega)$, hence have stationary processes as solutions.

The prototypical *Duffing-van der Pol oscillator*

$$\ddot{y} = \alpha y + \beta \dot{y} - y^3 - y^2 \dot{y} \tag{8.4.21}$$

under the influence of parametric and additive noise has been the subject of numerous investigations (see Arnold, Sri Namachchivaya and Schenk-Hoppé [37], Schenk-Hoppé [302, 303, 304] and the references therein). For $\alpha < 0$ fixed and β the bifurcation parameter, the system (8.4.21) exhibits a Hopf bifurcation for $\beta = 0$. For $\beta < 0$ fixed and α the bifurcation parameter, it undergoes a pitchfork bifurcation at $\alpha = 0$.

To reduce complexity, we will treat a particular case of parametric noise: Let the parameter α be replaced by $\alpha + \sigma \xi(t)$, where $\xi(t, \omega) = \xi(\theta_t \omega)$ is a zero mean stationary stochastic process and σ is an intensity parameter. With $x = \binom{y}{\dot{y}}$, the perturbed version of (8.4.21) is

8.4 Normal Form and Center Manifold

$$\dot{x}_t = \begin{pmatrix} 0 & 1 \\ 0 & \beta \end{pmatrix} x_t + (\alpha + \sigma\xi(t)) \begin{pmatrix} 0 & 0 \\ 1 & 0 \end{pmatrix} x_t + \begin{pmatrix} 0 \\ -x_1^3 - x_1^2 x_2 \end{pmatrix}. \quad (8.4.22)$$

The linearization of (8.4.22) at $x = 0$ is

$$\dot{v}_t = \begin{pmatrix} 0 & 1 \\ \alpha & \beta \end{pmatrix} v_t + \sigma\xi(t) \begin{pmatrix} 0 & 0 \\ 1 & 0 \end{pmatrix} v_t, \quad (8.4.23)$$

in particular for $\sigma = 0$,

$$\dot{v}_t = \begin{pmatrix} 0 & 1 \\ \alpha & \beta \end{pmatrix} v_t. \quad (8.4.24)$$

The Pitchfork Scenario Under Small Random Perturbations

Put for simplicity $\beta = -1$. Then the eigenvalues of (8.4.24) are $(-1 \pm \sqrt{1+4\alpha})/2$. We treat (α, σ) as a small two-dimensional parameter and will determine the simultaneous stochastic normal form and center manifold reduction of (8.4.22) for small x, α and σ.

As a first step we diagonalize the linear part of (8.4.22) at $\alpha = \sigma = 0$ yielding (writing again x for the new coordinates)

$$\begin{aligned}
\dot{x}_t &= \begin{pmatrix} 0 & 0 \\ 0 & -1 \end{pmatrix} x + f_{1010} x_1 \alpha + f_{1001} x_1 \sigma + f_{0110} x_2 \alpha + f_{0101} x_2 \sigma \\
&\quad + f_{3000} x_1^3 + f_{2100} x_1^2 x_2 + f_{1200} x_1 x_2^2 \\
&=: f(\theta_t \omega, x_t, \alpha, \sigma),
\end{aligned} \quad (8.4.25)$$

where, putting $b := \begin{pmatrix} 1 \\ -\sqrt{2} \end{pmatrix}$,

$$f_{1010} = b, \quad f_{1001} = \xi(t) b, \quad f_{0110} = \frac{1}{\sqrt{2}} b, \quad f_{0101} = \frac{1}{\sqrt{2}} \xi(t) b,$$

$$f_{3000} = -b, \quad f_{2100} = -\sqrt{2} b, \quad f_{1200} = -\frac{1}{2} b,$$

f_{pqkl} denoting the coefficient of $x_1^p x_2^q \alpha^k \sigma^l$. Note that here $d = 2$, $d_c = d_s = 1$, $x_c = x_1$, $x_s = x_2$, $A_c \equiv 0$, $A_s \equiv -1$. We seek the transformation (depending on time and chance)

$$x \mapsto x + H(\theta_t \omega, x, \alpha, \sigma) = \begin{pmatrix} x_c + H^c(\theta_t \omega, x_c, x_s, \alpha, \sigma) \\ x_s + H^s(\theta_t \omega, x_c, x_s, \alpha, \sigma) \end{pmatrix}$$

which tranforms (8.4.25) into

$$\dot{x}_c = g^c(\theta_t \omega, x_c, \alpha, \sigma) + O((|x_c| + |x_s|)^{N+1} + (|\alpha| + |\sigma|)^{M+1}),$$

$$\dot{x}_s = -x_s + g^s(\theta_t \omega, x_c, x_s, \alpha, \sigma) + O((|x_c| + |x_s|)^{N+1} + (|\alpha| + |\sigma|)^{M+1}).$$

The result for $N = 3$ and $M = 2$ is: The transformation $H = \begin{pmatrix} H^c \\ H^s \end{pmatrix}$ has the form

444   Chapter 8. Normal Forms

$$H^c = \left(-\frac{\alpha}{\sqrt{2}} + H^c_{0101}\sigma + H^c_{0102}\sigma^2 + H^c_{0111}\alpha\sigma + \frac{2}{\sqrt{2}}\alpha^2\right)x_s$$
$$+ \left(H^c_{0301}\sigma + H^c_{0302}\sigma^2 - \frac{1}{6\sqrt{2}}\alpha + H^c_{0311}\alpha\sigma + \frac{29\sqrt{2}}{72}\alpha^2\right)x_s^3$$
$$+ \left(\frac{1}{4} + H^c_{1201}\sigma + H^c_{1202}\sigma^2 - \alpha + H^c_{1211}\alpha\sigma + \frac{13}{4}\alpha^2\right)x_c x_s^2$$
$$+ \left(\sqrt{2} + H^c_{2101}\sigma - \frac{7\sqrt{2}}{2}\alpha + H^c_{2111}\alpha\sigma + 14\sqrt{2}\alpha^2 + H^c_{2102}\sigma^2\right)x_c^2 x_s,$$

$$H^s = (H^s_{1001}\sigma + H^s_{1002}\sigma^2 - \sqrt{2}\alpha + H^s_{1011}\alpha\sigma + 2\sqrt{2}\alpha^2)x_c$$
$$+ \left(-\frac{\sqrt{2}}{2} + H^s_{1201}\sigma + H^s_{1202}\sigma^2 + \frac{9\sqrt{2}}{4}\alpha + H^s_{1211}\alpha\sigma - 8\sqrt{2}\alpha^2\right)x_c x_s^2$$
$$+ \left(\sqrt{2} + H^s_{3001}\sigma + H^s_{3002}\sigma^2 - 7\sqrt{2}\alpha + H^s_{3011}\alpha\sigma - 47\sqrt{2}\alpha^2\right)x_c^3$$
$$+ \left(H^s_{0301}\sigma + H^s_{0302}\sigma^2 + \frac{1}{4}\alpha + H^s_{0311}\alpha\sigma - \frac{4}{3}\alpha^2\right)x_s^3.$$

The truncated stochastic normal forms are

$$\dot{x}_c = g^c = (\xi\sigma + g^c_{1002}\sigma^2 + \alpha + g^c_{1011}\alpha\sigma - \alpha^2)x_c \qquad (8.4.26)$$
$$+ (-1 + g^c_{3001}\sigma + g^c_{3002}\sigma^2 + 3\alpha + g^c_{3011}\alpha\sigma - 18\alpha^2)x_c^3,$$

$$\dot{x}_s = -x_s + g^s = (-1 + g^s_{0102}\sigma^2 - \xi\sigma - \alpha + g^s_{0111}\alpha\sigma + \alpha^2)x_s$$
$$+ (2 + g^s_{2101}\sigma + g^s_{2102}\sigma^2 - 7\alpha + g^s_{2111}\alpha\sigma + 26\alpha^2)x_c^2 x_s.$$

In equation (8.4.26) the coefficients are as follows:

$$g^c_{1002}(\theta_t\omega) = \frac{\xi(t,\omega)}{\sqrt{2}}H^s_{1001}(\theta_t\omega),$$

$$g^c_{1011}(\theta_t\omega) = \frac{1}{\sqrt{2}}H^s_{1001}(\theta_t\omega) - \xi(t,\omega),$$

$$g^c_{3001}(\theta_t\omega) = \xi(t,\omega) - \sqrt{2}H^s_{1001}(\theta_t\omega),$$

$$g^c_{3011}(\theta_t\omega) = -10\xi(t,\omega) + \frac{1}{\sqrt{2}}H^s_{3001}(\theta_t\omega) + 4\sqrt{2}H^c_{0101}(\theta_t\omega)$$
$$+ \frac{7}{\sqrt{2}}H^s_{1001}(\theta_t\omega) - \sqrt{2}H^s_{1011}(\theta_t\omega),$$

$$g^c_{3002}(\theta_t\omega) = \frac{1}{\sqrt{2}}\xi(t,\omega)H^s_{3001}(\theta_t\omega) - 3H^c_{0101}(\theta_t\omega)H^s_{1001}(\theta_t\omega)$$
$$-\sqrt{2}H^s_{1002}(\theta_t\omega) + \sqrt{2}\xi(t,\omega)H^c_{0101}(\theta_t\omega)$$
$$+\sqrt{2}H^s_{1001}(\theta_t\omega) - \frac{1}{2}(H^s_{1001}(\theta_t\omega))^2,$$

where

$$H^s_{1001}(\omega) = \int_{-\infty}^0 e^s(-\sqrt{2}\xi(s,\omega))ds,$$

$$H^c_{0101}(\omega) = -\int_0^\infty e^{-s}\frac{\xi(s,\omega)}{\sqrt{2}}ds,$$

$$H^s_{3001}(\omega) = \int_{-\infty}^0 e^s(-4\sqrt{2}\xi(s,\omega) + 3H^s_{1001}(\theta_s\omega))ds,$$

$$H^s_{1011}(\omega) = \int_{-\infty}^0 e^s(2\sqrt{2}\xi(s,\omega) - 2H^s_{1001}(\theta_s\omega))ds,$$

$$H^s_{1002}(\omega) = \int_{-\infty}^0 e^s(-2\xi(s,\omega)H^s_{1001}(\theta_s\omega))ds.$$

The approximate stochastic center manifold is the graph

$$\mathbb{R} \times \mathbb{R}^2 \ni (x_c, \alpha, \sigma) \mapsto x_s = m_c(\omega, x_c, \alpha, \sigma) \in \mathbb{R}$$

given by

$$m_c(\omega, x_c, \alpha, \sigma) =$$
$$(H^s_{1001}(\omega)\sigma + H^s_{1002}(\omega)\sigma^2 - \sqrt{2}\alpha + H^s_{1011}(\omega)\alpha\sigma + 2\sqrt{2}\alpha^2)x_c +$$
$$(\sqrt{2} + H^s_{3001}(\omega)\sigma + H^s_{3002}(\omega)\sigma^2 - 7\sqrt{2}\alpha + H^s_{3011}(\omega)\alpha\sigma - 47\sqrt{2}\alpha^2)x_c^3.$$

The computational effort for these results is enormous and could only be accomplished by using the computer algebra program MAPLE. There are 106 cohomological equations to be solved to determine the coefficients. These cohomological equations fill 18 pages and can be found in an appendix of a reprint version of [41]. The very complicated explicit expressions of the coefficients H^s_{3002} and H^s_{3011} as functions of lower order terms can also be found there.

The scalar center equation (8.4.26) can now be utilized for stochastic bifurcation theory of the original two-dimensional system (8.4.22), similarly as in Xu [347]. See Subsect. 9.4.3.

The Hopf Scenario Under Small Random Perturbations

We now assume that $\alpha < 0$ is fixed and β is the bifurcation parameter which varies in the interval $|\beta| < \sqrt{-4\alpha}$ so that the "damped eigenfrequency"

446   Chapter 8. Normal Forms

$$\omega_d := \sqrt{-\alpha - \beta^2/4}$$

is well defined. We follow [37] and obtain the normal form for small x, β and σ.

The eigenvalues of (8.4.24) are $\frac{\beta}{2} \pm i\omega_d$ which are purely imaginary for $\beta = 0$. We have $d = d_c = 2$, while the stable component is not present. All cohomological equations are thus resonant (in the random sense, i.e. all Lyapunov exponents vanish). Our recipe for their solution is as follows: If in a cohomological equation $(\frac{d}{dt} - B)H = -G + R$, R is a random process, we choose $H = 0$ and $G = R$, while if R is nonrandom we search for a nonrandom H (we solve $-BH = -G + R$) for which G is "as simple as possible" by means of deterministic normal form theory (see Sect. 8.1).

The truncated stochastic normal form for $N = 3$ and $M = 1$ in polar coordinates (r, γ), $x_1 = r\cos\gamma$, $x_2 = r\sin\gamma$, is (after formidable computational efforts which we omit) as follows:

$$\dot{r}_t = \frac{\beta}{2}r_t - \frac{1}{2}r_t^3 + \frac{r_t}{8\omega_d^2}\left(-2r_t^2 + \left(\frac{5r_t^2}{\omega_d} - 4\omega_d\right)\sin 2\gamma_t\right.$$
$$\left. - \frac{2r_t^2}{\omega_d}\sin 4\gamma_t + r_t^2\cos 2\gamma_t\right)\sigma\xi(t),$$

$$\dot{\gamma}_t = \omega_d + \frac{6 + 3\beta}{4\omega_d}r_t^2 + \frac{1}{2\omega_d}\left(-1 + \frac{3r_t^2}{\omega_d^2} + \left(-1 + \frac{5r_t^2}{2\omega_d^2}\right)\cos 2\gamma_t\right.$$
$$\left. - \frac{r_t^2}{2\omega_d^2}\cos 4\gamma_t - \frac{3r_t^2}{2\omega_d}\sin 2\gamma_t\right)\sigma\xi(t).$$

For $\sigma = 0$ we recover the deterministic truncated normal form, from which the Hopf bifurcation at $\beta = 0$ can be read-off.

8.5 Normal Forms for Stochastic Differential Equations

Normal form theory for SDE is confronted with the technical problem of the appearance of terms which at time t are not measurable with respect to the information available at that moment, but rather anticipate it. These terms originate as stationary solutions of affine SDE with a hyperbolic linear part and a possibly anticipative additive part (see Theorem 5.6.5 for the simplest of such cases), but such SDE as well as their solutions are ruled out in classical stochastic analysis.

This fact has been clearly seen (but not rigorously handled) by the pioneers of stochastic normal form theory (see Coullet, Elphick and Tirapegui [107], Nicolis and Nicolis [262], Schöner and Haken [314, 315], Sri Namachchivaya and Lin [323]). We follow Arnold and Imkeller [21] and present a rigorous treatment.

8.5.1 The Random Cohomological Equation

We consider the SDE

$$dx_t = \sum_{j=0}^{m} f_j(x_t) \circ dW_t^j, \quad f_j(0) = 0, \tag{8.5.1}$$

in $\mathbb{R}^d$, where as usual, dW_t^0 stands for dt and where $f_j \in C^\infty$ for $0 \leq j \leq m$. Then (8.5.1) uniquely generates a local C^∞ RDS φ (Theorem 2.3.36) and the domain $D(t, \omega)$ and range $R(t, \omega)$ of $\varphi(t, \omega) : D(t, \omega) \to R(t, \omega)$ are neighborhoods of 0. How many different such local RDS exist modulo a smooth conjugacy?

The linear cocycle $\Phi_1(t, \omega) := \Phi(t, \omega) = D\varphi(t, \omega, 0)$ on $T_0 \mathbb{R}^d \cong \mathbb{R}^d$ is generated by the linearized SDE

$$dv_t = \sum_{j=0}^{m} A_j v_t \circ dW_t^j, \quad A_j := Df_j(0). \tag{8.5.2}$$

The MET always holds for (8.5.2) giving the Lyapunov spectrum

$$\mathcal{S}(\Phi_1) = \{\Lambda_1 \geq \ldots \geq \Lambda_d\}.$$

We now conjugate the local RDS φ generated by (8.5.1) with another local RDS ψ by means of a (near-identity) random coordinate transformation (see Definition 8.2.1),

$$\varphi(t, \omega) \circ h(\omega) = h(\theta_t \omega) \circ \psi(t, \omega) \quad \text{(locally)}, \tag{8.5.3}$$

where ψ is generated by an SDE

$$dx_t = \sum_{j=0}^{m} g_j(\theta_t \cdot, x_t) \circ dW_t^j \tag{8.5.4}$$

and h is chosen such that the SDE (8.5.4) makes sense and its coefficients g_j are "as simple as possible", the ultimate aim being linearization, i. e.

$$dx_t = \sum_{j=0}^{m} A_j x_t \circ dW_t^j.$$

Although it will turn out that the transformation $\bar{h}(t, \cdot, x) = h(\theta_t \cdot, x)$ to be applied at time t is in general not adapted to the filtration of W at t, we proceed formally (and justify later): Applying the Stratonovich lemma to (8.5.3) gives

$$d\varphi_t = dh_t + Dh(\theta_t \cdot, \psi_t) \circ d\psi_t, \tag{8.5.5}$$

where dh_t denotes the t differential of $h(\theta_t \cdot, x)$. Inserting the differentials of φ_t and ψ_t into (8.5.5) yields

$$\sum_{j=0}^{m} f_j(h(\theta_t\cdot, x_t)) \circ dW_t^j = dh_t + \sum_{j=0}^{m} Dh(\theta_t\cdot, x_t) g_j(\theta_t\cdot, x_t) \circ dW_t^j, \quad (8.5.6)$$

equivalently

$$\begin{aligned} dx_t &= \sum_{j=0}^{m} g_j(\theta_t\cdot, x_t) \circ dW_t^j \quad (8.5.7)\\ &= -Dh(\theta_t\cdot, x_t)^{-1} dh_t + \sum_{j=0}^{m} Dh(\theta_t\cdot, x_t)^{-1} f_j(h(\theta_t\cdot, x_t)) \circ dW_t^j. \end{aligned}$$

This is an equation for h and the g_j, where the choice of h is made such that the g_j are the simplest possible.

We now make the simplifying assumption that we choose the canonical basis in $\mathbb{R}^d$. In other words, we leave (8.5.2) untouched and refrain from making it "as simple as possible" by choosing an appropriate (non-adapted) random basis (cf. Corollary 6.2.17 for the diagonalization of a linear SDE in case of a simple spectrum).

We again make the formal Taylor series Ansatz

$$f_j(x) \sim A_j x + \sum_{n=2}^{\infty} f_{j,n}(x), \quad j = 0, \ldots, m,$$

$$g_j(\omega, x) \sim A_j x + \sum_{n=2}^{\infty} g_{j,n}(\omega, x), \quad j = 0, \ldots, m,$$

$$h(\omega, x) \sim x + \sum_{n=2}^{\infty} h_n(\omega, x),$$

where $f_{j,n} \in H_{n,d}$, while $g_{j,n}(\cdot), h_n(\cdot) \in \mathcal{H}_{n,d}$.

Plugging this into equation (8.5.6) and equating coefficients yields an identity for the linear part and for $n \geq 2$

$$\sum_{j=0}^{m} g_{j,n}(\theta_t\cdot) \circ dW_t^j = \sum_{j=0}^{m} (\text{ad}_n A_j) h_n(\theta_t\cdot) \circ dW_t^j - dh_n(\theta_t\cdot)$$

$$+ \sum_{j=0}^{m} k_{j,n}(\theta_t\cdot) \circ dW_t^j, \quad (8.5.8)$$

where $\text{ad}_n A_j : H_{n,d} \to H_{n,d}$ is the linear operator defined by $h_n \mapsto (\text{ad}_n A_j) h_n(x) := A_j h_n(x) - Dh_n(x) A_j x$, in matrix form $\text{ad}_n A_j = I_1 \otimes A_j - T(A_j)_n \otimes I_2$ on $H_{n,d}(\mathbb{R}^d) \cong H_{n,d}(\mathbb{R}^1) \otimes \mathbb{R}^d$ (see equations (8.3.4) and (8.3.7)) and

$$k_{j,n} = f_{j,n} + P_n(f_{j,k}, g_{j,k}, h_k, 2 \leq k \leq n-1), \quad j = 0, \ldots, m. \quad (8.5.9)$$

Here P_n is a deterministic polynomial of the lower order terms of f_j, g_j and h which is independent of j and given explicitly in Lemma 8.3.2.

8.5 Normal Forms for SDE

We now define the *random cohomological operator (SDE case)* by

$$dL_n(h_n) := dh_n - \sum_{j=0}^{m}(\mathrm{ad}_n A_j)h_n \circ dW_t^j, \qquad (8.5.10)$$

acting on those $\mathcal{H}_{n,d}$-valued stationary stochastic processes $h_n(\theta_t\omega)$ for which the expression in (8.5.10) makes sense. With this definition, (8.5.8) turns into the system of *random cohomological equations* (suppressing the $(\theta_t\omega)$ argument)

$$dL_n(h_n) = \sum_{j=0}^{m}(k_{j,n} - g_{j,n}) \circ dW_t^j =: dK_n - dG_n, \qquad (8.5.11)$$

which have to be solved successively for $n = 2, \ldots,$ for dK_n known from previous steps, and h_n choosen to make dG_n the simplest possible. The most desirable choice is again $dG_n \equiv 0$, resulting in the task of solving the cohomological equations

$$dL_n(h_n) = dK_n, \quad n \geq 2,$$

in $\mathcal{H}_{n,d}$ or, equivalently, looking for stationary solutions of the affine SDE

$$dh_n = \sum_{j=0}^{m}((\mathrm{ad}_n A_j)h_n + k_{j,n}(\theta_t \cdot)) \circ dW_t^j, \quad n \geq 2, \qquad (8.5.12)$$

where $k_{j,n}(\theta_t\omega)$ depends on the solutions of (8.5.12) of lower order $2 \leq k \leq n - 1$. We have $k_{j,2} = f_{j,2}$ deterministic, but for $n \geq 3$, $k_{j,n}$ is typically random and not adapted to the filtration of W.

Let Φ_n be the linear cocycle on $\mathcal{H}_{n,d}$ generated by the SDE

$$dv_t = \sum_{j=0}^{m}(\mathrm{ad}_n A_j)v_t \circ dW_t^j, \quad n \geq 2. \qquad (8.5.13)$$

The MET holds for Φ_n (without additional assumptions) and gives the spectrum

$$\mathcal{S}(\Phi_n) = \{\Lambda_i - (\Lambda, \tau) : \Lambda_i \in \mathcal{S}(\Phi_1), |\tau| = n\}.$$

This follows from Theorem 5.4.2(iii) (SDE case) and (ii), and from Proposition 8.3.4.

We know from looking at Theorem 5.6.5 that we have a chance of finding a (unique) stationary solution of (8.5.12) provided the linear SDE (8.5.13) is hyperbolic. We call the linear cocycle $\Phi = \Phi_1$ generated by (8.5.2) *non-resonant of order n* if this is the case, i.e. if $0 \notin \mathcal{S}(\Phi_n)$, *resonant of order n* otherwise.

8.5.2 Nonresonant Case

8.5.1 Theorem (Formal Linearization of SDE). *Given the SDE (8.5.1), assume that the linear cocycle generated by (8.5.2) is nonresonant of any order $n \geq 2$. Then there is a random coordinate transformation h whose formal Taylor series $h(\omega, x) \sim x + \sum_{n \geq 2} h_n(\omega, x)$ is uniquely determined and has tempered coefficients h_n, such that h formally linearizes equation (8.5.1).*

8.5.2 Remark. It is quite remarkable that at the beginning and at the end of the above procedure we have two bona fide classical SDE, while the transformation converting the solutions of the first into the solutions of the second is anticipative. ∎

The proof of Theorem 8.5.17 is quite complicated and is divided into three steps, the first two of which are of technical nature, while the third one (on invariant measures of affine SDE) is of independent interest and also finishes the proof of Theorem 5.6.5 for the white noise case.

Step 1: Boundedness of Moments of Solutions of a Hierarchical System of Affine SDE

Our main task here will consist in proving that all processes in our hierarchical system of affine SDE obtained by solving the cohomological equations step by step for fixed initial conditions satisfy the conditions of the following lemma.

8.5.3 Lemma. *Let $u = (u(t,x))_{t \in [0,1], x \in \mathbb{R}^d}$ be an $\mathbb{R}^k$-valued stochastic process which for fixed x is $\mathbb{P}$-a. s. continuous with respect to t, is adapted and satisfies the following two conditions:*

Conditions (C): *For any $p \geq 2$ and any compact set $K \subset \mathbb{R}^d$ there exist constants c_p and $c_{p,K} \in \mathbb{R}^+$ such that*

$$\mathbb{E}\left(\sup_{0 \leq t \leq 1} |u(t,0)|^p\right) \leq c_p,$$

$$\mathbb{E}\left(\sup_{0 \leq t \leq 1} |u(t,x) - u(t,y)|^p\right) \leq c_{p,K}|x - y|^{p/2} \quad \text{for all } x, y \in K.$$

Then u is $\mathbb{P}$-a. s. jointly continuous with respect to (t,x), and for $p > d$ and any compact set $K \subset \mathbb{R}^d$ there exist constants $C_p \in \mathbb{R}^+$ and $q \geq 1$ such that

$$\mathbb{E}\left(\sup_{x \in K} \sup_{0 \leq t \leq 1} |u(t,x)|^p\right) \leq C_p(\text{diam}K)^q. \tag{8.5.14}$$

Proof. The joint continuity of u follows from (C) by Kolmogorov's continuity criterion applied to the $C[0,1]$-valued process $x \mapsto u(\cdot, x)$ with $p/2 > d$.

(8.5.14) is a well-known implication of the fundamental continuity lemma of Garsia, Rodemich and Rumsey, see for example Barlow and Yor [48], formula (3.b), or Arnold and Imkeller [19]. □

We now prove that conditions (C) are passed on from u to processes obtained from u by reasonable operations.

8.5.4 Lemma. *Let u satisfy conditions (C), and let*

$$v(t,x) = \int_0^t u(s,x)\, dW_s, \quad w(t,x) = \int_0^t u(s,x)\, ds,$$

where W is a scalar Wiener process. Then v and w satisfy conditions (C).

Proof. This is an immediate consequence of the Burkholder and Hölder inequalities. □

The following considerations will be crucial for the algorithm by which we solve our hierarchical system of affine SDE. Let $d_1, d_2 \in \mathbb{N}$, and for $0 \leq i \leq m$ suppose that $(u_i(t, x_1))_{t \in [0,1], x_1 \in \mathbb{R}^{d_1}}$ is a parametrized semimartingale with decomposition

$$u_i(t, x_1) = \int_0^t w_i(s, x_1)\, ds + \sum_{j=1}^m \int_0^t v_i^j(s, x_1)\, dW_s^j \qquad (8.5.15)$$

with values in $\mathbb{R}^{d_2}$. Let, moreover, $B_0, B_1, \ldots, B_m$ be $d_2 \times d_2$ matrices and denote by $(\Phi(t))_{t \in \mathbb{R}}$ the linear flow in $\mathbb{R}^{d_2}$ generated by the linear SDE

$$dy_t = \sum_{j=0}^m B_j y_t \circ dW_t^j. \qquad (8.5.16)$$

We now consider the SDE

$$dx_t = \sum_{j=0}^m (B_j x_t + u_j(t, x_1)) \circ dW_t^j, \quad x_0 = x_2 \in \mathbb{R}^{d_2}. \qquad (8.5.17)$$

Let $(\varphi(t, x_1))_{t \in \mathbb{R}}$ be the flow generated by (8.5.17), the value of which at $x_2 \in \mathbb{R}^{d_2}$ will be written as $\varphi(t, x_1) x_2$.

8.5.5 Lemma. *Let $u_i = (u_i(t, x_1))_{t \in [0,1], x_1 \in \mathbb{R}^{d_1}}$ be given by (8.5.15). Assume that w_i and v_i^j (hence u_i), $0 \leq i \leq m$, $1 \leq j \leq m$, satisfy conditions (C). Let*

$$\varphi(t, x_1) x_2 = \int_0^t \rho(s, x_1, x_2)\, ds + \sum_{j=1}^m \int_0^t \psi^j(s, x_1, x_2)\, dW_s^j$$

be the semimartingale decomposition of the solution flow of (8.5.17). Then φ, ρ and ψ^j, $1 \leq j \leq m$, satisfy conditions (C).

Proof. (i) We prove only the second of the conditions (C) for φ.

Let us start by writing the Itô form of (8.5.17). We have

$$dy_t = \sum_{j=1}^{m}(B_j y_t + u_j(t, x_1))dW_t^j \tag{8.5.18}$$

$$+ \left((B_0 + \frac{1}{2}\sum_{j=1}^{m} B_j^2)y_t + \frac{1}{2}\sum_{j=1}^{m} v_j^j(t, x_1)\right) dt.$$

Now fix $p \geq 2$ and a compact set $K \subset \mathbb{R}^{d_1} \times \mathbb{R}^{d_2}$. Then for any (x_1, x_2), $(y_1, y_2) \in \mathbb{R}^{d_1} \times \mathbb{R}^{d_2}$, Burkholder's and Jensen's inequalities as well as the hypothesis yield, with suitable constants c_K and with the abbreviation

$$f(t) := \mathbb{E}\left(\sup_{s \in [0,t]} |\varphi(s, x_1)x_2 - \varphi(s, y_1)y_2|^p\right)^{1/p},$$

$$f(1) \leq c_K \left(|(x_1, x_2) - (y_1, y_2)|^{1/2} + \int_0^1 f(t)dt\right). \tag{8.5.19}$$

To (8.5.19) we have to apply Gronwall's lemma to obtain with a suitable constant $c_{p,K} \in \mathbb{R}^+$

$$\mathbb{E}\left(\sup_{t \in [0,1]} |\varphi(t, x_1)x_2 - \varphi(t, y_1)y_2|^p\right) \leq c_{p,K}|(x_1, x_2) - (y_1, y_2)|^{p/2}. \tag{8.5.20}$$

This is the second of the conditions (C) for φ.

(ii) Next observe that according to (8.5.18) for any $t \in [0, 1]$ and $(x_1, x_2) \in \mathbb{R}^{d_1} \times \mathbb{R}^{d_2}$

$$\psi^j(t, x_1, x_2) = B_j \varphi(t, x_1) x_2 + u_j(t, x_1),$$

$$\rho(t, x_1, x_2) = \left(B_0 + \frac{1}{2}\sum_{j=1}^{m} B_j^2\right) \varphi(t, x_1) x_2 + \frac{1}{2}\sum_{j=1}^{m} v_j^j(t, x_1).$$

Hence by our hypotheses and (8.5.20), ψ^j and ρ satisfy the conditions (C) as well. □

Next we show that conditions (C) are inherited from u and its characteristics to a polynomial of u and its characteristics.

8.5.6 Lemma. *Let*

$$u(t, x) = \sum_{j=1}^{m} \int_0^t v^j(s, x)dW_s^j + \int_0^t w(s, x)ds, \quad t \in [0, 1], x \in \mathbb{R}^{d_1},$$

take values in $\mathbb{R}^{d_2}$ and assume that u, v^j and w satisfy conditions (C). Let p be a polynomial in the variable $y \in \mathbb{R}^{d_2}$. If

$$(p \circ u)(t,x) = \sum_{j=1}^{m} \int_0^t q^j(s,x) dW_s^j + \int_0^t r(s,x) ds,$$

then $p \circ u$, q^j and r also satisfy conditions (C).

Proof. First note that, by Itô's formula, q^j and r are of the same structure as $p \circ u$. Hence it suffices to prove the conditions for $p \circ u$.

Due to Hölder's inequality, it is evidently enough to consider the case of a real-valued u and a polynomial p of the form $p(y) = y^l$ for some $l \in \mathbb{N}$. Then for $y_1, y_2 \in \mathbb{R}$

$$p(y_1) - p(y_2) = (y_1 - y_2) \sum_{k=0}^{l-1} y_1^k y_2^{l-1-k}. \tag{8.5.21}$$

Now observe that by conditions (C) for any $x \in \mathbb{R}^{d_1}$

$$\mathbb{E}\left(\sup_{t \in [0,1]} |u(t,x)|^p\right) \leq c_p \tag{8.5.22}$$

for a suitable constant $c_p \in \mathbb{R}^+$. Then (8.5.21) and an application of Hölder's inequality obviously allow us to deduce conditions (C) for $p \circ u$ from (8.5.22) and the conditions (C) for u. □

8.5.7 Lemma. Let $u = (u(t,x))_{t \in [0,1], x \in \mathbb{R}^d}$ with values in $\mathbb{R}^d$ satisfy conditions (C), let $(\Phi(t))_{t \in \mathbb{R}}$ be the linear flow generated by (8.5.16) for $d_2 = d$, and let W be a scalar Wiener process. Then the processes

$$v(t,x) = \int_0^t \Phi(s)^{-1} u(s,x) \, dW_s, \quad w(t,x) = \int_0^t \Phi(s)^{-1} u(s,x) ds$$

satisfy conditions (C).

Proof. $\Phi(t)^{-1}$ satisfies the SDE

$$d\Phi(t)^{-1} = -\sum_{j=0}^{m} \Phi(t)^{-1} B_j \circ dW_t^j, \quad \Phi(0) = I.$$

Hence for any $p \geq 1$ we have (see Remark 6.2.12)

$$\mathbb{E}\left(\sup_{t \in [0,1]} \|\Phi(t)^{-1}\|^p\right) < \infty. \tag{8.5.23}$$

Then it is clear that Burkholder's inequality, Hölder's inequality and (8.5.23) imply that v and w satisfy conditions (C). □

We finally come to the announced hierarchical system of affine SDE. Let $(d_n)_{n \in \mathbb{N}}$ be a sequence of integers. For $0 \leq j \leq m$ and each $n \in \mathbb{N}$, let $A_{j,n}$ be a $d_n \times d_n$ matrix and let $p_{j,n}$ be a polynomial in the variables $(x_1, \ldots, x_{n-1}) \in \mathbb{R}^{d_1} \times \cdots \times \mathbb{R}^{d_{n-1}}$ with $p_{j,1} = b_j \in \mathbb{R}^{d_1}$ a fixed vector. Then our *hierarchical system of affine SDE* is as follows:

$$dx_t^1 = \sum_{j=0}^{m}(A_{j,1}x_t^1 + p_{j,1}) \circ dW_t^j, \quad x_0^1 = x_1 \in \mathbb{R}^{d_1}, \quad (8.5.24)$$

$$dx_t^2 = \sum_{j=0}^{m}(A_{j,2}x_t^2 + p_{j,2}(x_t^1)) \circ dW_t^j, \quad x_0^2 = x_2 \in \mathbb{R}^{d_2}, \quad (8.5.25)$$

$$\ldots \quad \ldots$$

$$dx_t^n = \sum_{j=0}^{m}(A_{j,n}x_t^n + p_{j,n}(x_t^1, \ldots, x_t^{n-1})) \circ dW_t^j, \quad (8.5.26)$$

$$x_0^n = x_n \in \mathbb{R}^{d_n},$$

$$\ldots \quad \ldots$$

The algorithm for successively solving this system is as follows: Let $(\Phi_n(t))_{t \in \mathbb{R}}$ be the linear cocycle in $\mathbb{R}^{d_n}$ generated by the linear SDE

$$dy_t^n = \sum_{j=0}^{m} A_{j,n} y_t^n \circ dW_t^j, \quad y_0^n = y_n \in \mathbb{R}^{d_n}. \quad (8.5.27)$$

We first solve the first affine SDE (8.5.24). Denote the resulting affine cocycle by $(\varphi_1(t))_{t \in \mathbb{R}}$ which by the variation of constants formula is given by

$$\varphi_1(t)x = \Phi_1(t)\left(x + \sum_{j=0}^{m} \int_0^t \Phi_1(s)^{-1} p_{j,1} \circ dW_s^j\right).$$

Then we insert the cocycle $\varphi_1(t)x_1$ into the second equation (8.5.25) in place of x_t^1 which gives a non-autonomous affine SDE whose solution flow is denoted by $\varphi_2(t, x_1)$, etc. At step n insert $\varphi_1(t)x_1, \ldots, \varphi_{n-1}(t, x_1, \ldots, x_{n-2})x_{n-1}$ into the nth equation (8.5.26) which gives a non-autonomous affine SDE

$$dx_t^n = \quad (8.5.28)$$
$$\sum_{j=0}^{m}(A_{j,n}x_t^n + p_{j,n}(\varphi_1(t)x_1, \ldots, \varphi_{n-1}(t, x_1, \ldots, x_{n-2})x_{n-1})) \circ dW_t^j,$$

with initial conditions $x_0^n = x_n \in \mathbb{R}^{d_n}$. Denote the solution flow of this equation by $\varphi_n(t, x_1, \ldots, x_{n-1})$.

Note that the first n affine SDE considered as one equation generate the C^∞ RDS

$$\varphi(t)(x_1, \ldots, x_n) = (\varphi_1(t)x_1, \varphi_2(t, x_1)x_2, \ldots, \varphi_n(t, x_1, \ldots, x_{n-1})x_n). \quad (8.5.29)$$

8.5.8 Lemma. *For $n \in \mathbb{N}$ represent the solution of the nth equation (8.5.28) as*

$$\varphi_n(t, x_1, \ldots, x_{n-1})x_n = \int_0^t v^n(s, x_1, \ldots, x_n)\,ds$$
$$+ \sum_{j=1}^m \int_0^t u_j^n(s, x_1, \ldots, x_n)\,dW_s^j.$$

Then for any $n \in \mathbb{N}$, φ_n as well as v^n and u_j^n, $1 \leq j \leq m$, satisfy conditions (C).

Proof. We use induction on n.

The assertion holds for $n = 1$. This is an immediate consequence of Lemma 8.5.5, choosing $u_j = b_j$.

Let us now assume that the assertion holds for $\varphi_1, \ldots, \varphi_{n-1}$ and their characteristics. Consequently, Lemma 8.5.6 yields that all the components of $p_{j,n}(\varphi_1(t)x_1, \ldots, \varphi_{n-1}(t, x_1, \ldots, x_{n-2})x_{n-1})$ and their semimartingale characteristics satisfy conditions (C). Hence Lemma 8.5.5 applies and gives conditions (C) for φ_n as well as for u_j^n and v^n. □

8.5.9 Lemma. *Let for $t \in \mathbb{R}$*

$$X_1(t) = \sum_{j=0}^m \int_0^t \Phi_1(s)^{-1} b_j \circ dW_s^j$$

and for $n \geq 2$ and $(x_1, \ldots, x_{n-1}) \in \mathbb{R}^{d_1} \times \cdots \times \mathbb{R}^{d_{n-1}}$ let

$$X_n(t, x_1, \ldots, x_{n-1}) = \qquad (8.5.30)$$
$$\sum_{j=0}^m \int_0^t \Phi_n(s)^{-1} p_{j,n}(\varphi_1(t)x_1, \ldots, \varphi_{n-1}(t, x_1, \ldots, x_{n-2})x_{n-1}) \circ dW_s^j.$$

Then X_n satisfies conditions (C).

Proof. This is an immediate consequence of Lemmas 8.5.8, 8.5.6 and 8.5.7. □

Here is our final and main result of step 1.

8.5.10 Proposition. *Consider the hierarchical system of affine SDE introduced in (8.5.24) to (8.5.26). Then for any $n \geq 2$ and $p > d_1 + \cdots + d_{n-1}$ and any compact set $K \subset \mathbb{R}^{d_1} \times \cdots \times \mathbb{R}^{d_{n-1}}$ there exist constants $c_{n,p} \in \mathbb{R}^+$ and $q \geq 1$ such that*

$$\mathbb{E}\left(\sup_{(x_1, \ldots, x_{n-1}) \in K} \sup_{t \in [0,1]} |X_n(t, x_1, \ldots, x_{n-1})|^p \right) \leq c_{n,p}(\mathrm{diam} K)^q, \quad (8.5.31)$$

where X_n is defined by (8.5.30).

Proof. Combine Lemma 8.5.9 with Lemma 8.5.3. □

Step 2: Inheritance of Temperedness

We first study the inheritance of temperedness (from above) (Definition 4.1.1) in case tempered vectors are inserted into random fields. Property (8.5.14) will play a crucial role hereby, as indicated by the following lemma.

8.5.11 Lemma. *Let $(X(y))_{y \in \mathbb{R}^{d_1}}$ be a $\mathbb{P}$-a. s. continuous random field with values in $\mathbb{R}^{d_2}$ for which the following condition holds: For $p > d_1$ and any compact set $K \subset \mathbb{R}^{d_1}$ there exist $c_p \in \mathbb{R}^+$ and $q \geq 1$ such that*

$$\mathbb{E}\left(\sup_{y \in K} |X(y)|^p\right) \leq c_p (\operatorname{diam} K)^q. \tag{8.5.32}$$

Let Y be a random vector with values in $\mathbb{R}^{d_1}$ and for $\varepsilon > 0$, $m \in \mathbb{N}$ let

$$A_{\varepsilon,m} := \{\omega \in \Omega : |Y(\theta_t \omega)| \leq m \exp(t\varepsilon), \, t \in \mathbb{R}^+\}.$$

Then there exists a constant $c_{\varepsilon,m} \in \mathbb{R}^+$ such that for any $n \in \mathbb{Z}^+$

$$\mathbb{E}\left(1_{A_{\varepsilon,m}} \sup_{t \in [n,n+1]} |(X(Y))(\theta_t \cdot)|^p\right) \leq c_{\varepsilon,m} \exp(n \varepsilon q),$$

where p and q are related by (8.5.32).

Proof. For $\omega \in A_{\varepsilon,m}$ and $t \in [n, n+1]$ we have

$$|Y(\theta_t \omega)| \leq m \exp(t\varepsilon) \leq m \exp(\varepsilon) \exp(n\varepsilon).$$

Hence due to (8.5.32)

$$\mathbb{E}\left(1_{A_{\varepsilon,m}} \sup_{t \in [n,n+1]} |X(Y)(\theta_t \cdot)|^p\right) \leq \mathbb{E}\left(1_{A_{\varepsilon,m}} \sup_{y \in B_{\varepsilon,m,n}} |X(y)|^p\right)$$
$$\leq c_p (\operatorname{diam} B_{\varepsilon,m,n})^q, \tag{8.5.33}$$

where

$$B_{\varepsilon,m,n} := \{y \in \mathbb{R}^{d_1} : |y| \leq m \exp(\varepsilon) \exp(n\varepsilon)\}.$$

But $\operatorname{diam} B_{\varepsilon,m,n} \leq c_{d_1,\varepsilon,m} \exp(n\varepsilon)$, with a constant depending just on d_1, ε and m. Hence the desired inequality follows readily from (8.5.33). □

8.5.12 Proposition. *Let $(X(y))_{y \in \mathbb{R}^{d_1}}$ be a $\mathbb{P}$-a. s. continuous random field with values in $\mathbb{R}^{d_2}$ for which the following condition holds: For $p > d_1$ and any compact set $K \subset \mathbb{R}^{d_1}$ there exist $c_p \in \mathbb{R}^+$ and $q \geq 1$ such that*

$$\mathbb{E}\left(\sup_{y \in K} |X(y)|^p\right) \leq c_p (\operatorname{diam} K)^q. \tag{8.5.34}$$

Let Y be a tempered random vector with values in $\mathbb{R}^{d_1}$. Then the $\mathbb{R}^{d_2}$-valued random vector $X(Y)$ is tempered.

Proof. Define $A_{\varepsilon,m}$ as in Lemma 8.5.11. We shall prove that

$$\lim_{t\to\infty} \frac{1}{t} \log^+ |X(Y)(\theta_t \omega)| = 0,$$

remarking that the behavior for $t \to -\infty$ can be treated similarly. Since Y is tempered, $A_{\varepsilon,m} \uparrow \Omega$ $(m \uparrow \infty)$ $\mathbb{P}$-a.s. for any $\varepsilon > 0$.

Let $\varepsilon > 0$ be given. We have to prove that there exists a $\delta(\varepsilon) > 0$ such that for any $m \in \mathbb{N}$ we have

$$1_{A_{\delta(\varepsilon),m}} \limsup_{t\to\infty} \frac{1}{t} \log^+ |X(Y)(\theta_t \omega)| \leq \varepsilon, \quad \mathbb{P}\text{-a.s.} \qquad (8.5.35)$$

(8.5.35) will indeed imply temperedness since $A_{\delta(\varepsilon),m} \uparrow \Omega$, $\mathbb{P}$-a.s., for $m \uparrow \infty$ and any $\varepsilon > 0$.

To prove (8.5.35), let $\varepsilon, \delta > 0$, $m, n \in \mathbb{N}$ be given. Then by (8.5.34) and Lemma 8.5.11

$$\begin{aligned}
&\mathbb{P}(A_{\delta,m} \cap \{\sup_{t\in[n,n+1]} \frac{1}{t} \log^+ |X(Y)(\theta_t \cdot)| > \varepsilon\}) \\
&\leq \mathbb{P}(A_{\delta,m} \cap \{\sup_{t\in[n,n+1]} |X(Y)(\theta_t \cdot)| > \exp(n\varepsilon)\}) \\
&\leq \mathbb{E}\left(1_{A_{\delta,m}} \exp(-np\varepsilon) \sup_{t\in[n,n+1]} |X(Y)(\theta_t \cdot)|^p\right) \\
&\leq c_{m,\delta} \exp(nq\delta - np\varepsilon) = c_{m,\delta} \exp(n(q\delta - p\varepsilon)).
\end{aligned}$$

Now choose $\delta < \frac{p\varepsilon}{q}$. Then for all $m \in \mathbb{N}$ the Borel Cantelli lemma yields

$$1_{A_{\delta,m}} \limsup_{n\to\infty} \sup_{t\in[n,n+1]} \frac{1}{t} \log^+ |X(Y)(\theta_t \cdot)| \leq \varepsilon.$$

This clearly implies (8.5.35) and completes the proof. □

As a second issue we study the inheritance of temperedness by geometric series.

8.5.13 Lemma. *Suppose X is an $\mathbb{R}^d$-valued tempered random vector. Let $(T_n)_{n\in\mathbb{N}}$ be a sequence of random linear operators in $\mathbb{R}^d$ with negative Lyapunov index, i.e. for some deterministic $\beta > 0$*

$$\limsup_{n\to\infty} \frac{1}{n} \log \|T_n\| \leq -\beta.$$

Then

$$Y = \sum_{n=0}^{\infty} T_n(X \circ \theta(n))$$

is absolutely and geometrically convergent and tempered.

Proof. Choose $\varepsilon > 0$ such that $2\varepsilon < \beta$. Then by the assumptions, from a certain index n on, $\|T_n\|(X \circ \theta(n)) < \exp((2\varepsilon - \beta)n)$ from which the convergence statements follow. As to the temperedness of Y, there exists a random variable R_ε such that for $t \in \mathbb{R}$

$$|Y \circ \theta_t| \leq \sum_{n=0}^{\infty} \|T_n\| \, |X \circ \theta(n+t)|$$

$$\leq R_\varepsilon \sum_{n=0}^{\infty} \exp(\varepsilon n - \beta n) \exp(\varepsilon(n+|t|))$$

$$= R_\varepsilon \frac{1}{1-\exp(2\varepsilon-\beta)} \exp(\varepsilon|t|),$$

which clearly implies that Y is tempered. $\square$

Step 3: Invariant Measures of a Hierarchical System of Affine SDE

We return to the setting of a hierarchical system of affine SDE introduced above and determine its invariant measures.

So let $(\Phi_n(t))_{t \in \mathbb{R}}$ be the linear cocycle in $\mathbb{R}^{d_n}$ generated by (8.5.27). The MET holds for Φ_n, giving its Lyapunov spectrum $\mathcal{S}(\Phi_n)$. Having in mind our cohomological equations in the nonresonant case, we assume that all these cocycles are hyperbolic, i.e. $0 \notin \mathcal{S}(\Phi_n)$ for all $n \in \mathbb{N}$. Denote by π_n^u (π_n^s) the projection whose range is the unstable (stable) space and whose kernel is the stable (unstable) space of Φ_n. For convenience we recall the following facts (see Sect. 4.3).

8.5.14 Lemma. *Assume that the cocycle Φ_n is hyperbolic. Then there exists a constant $\beta_n > 0$ such that*

$$\limsup_{k \to \infty} \frac{1}{k} \log \|\pi_n^u \Phi_n(k)^{-1}\| \leq -\beta_n$$

and

$$\limsup_{k \to \infty} \frac{1}{k} \log \|\pi_n^s \Phi_n(-k)^{-1}\| \leq -\beta_n.$$

We first consider the affine SDE of order one, thereby proving part (iii) of Theorem 5.6.5.

8.5.15 Theorem. *Consider the affine SDE in $\mathbb{R}^{d_1}$*

$$dx_t^1 = \sum_{j=0}^{m} (A_{j,1} x_t^1 + b_j) \circ dW_t^j \tag{8.5.36}$$

and assume that the corresponding linear cocycle Φ_1 generated by $dy_t^1 = \sum_{j=0}^{m} A_{j,1} y_t^1 \circ dW_t^j$ is hyperbolic. Then the affine cocycle φ_1 generated by (8.5.36) has a unique invariant measure, namely the random Dirac measure $\mu_\omega = \delta_{\xi_1(\omega)}$, i.e. we have $\varphi_1(t,\cdot)\xi_1 = \xi_1 \circ \theta_t$. Here $\xi_1 = \xi_1^s \oplus \xi_1^u$ with

$$\xi_1^s = \sum_{k=0}^{\infty} \pi_1^s \circ \Phi_1(-k)^{-1}(X_1^- \circ \theta_{-k}), \tag{8.5.37}$$

$$\xi_1^u = -\sum_{k=0}^{\infty} \pi_1^u \circ \Phi_1(k)^{-1}(X_1^+ \circ \theta_k), \qquad (8.5.38)$$

where

$$X_1^- := \sum_{j=0}^{m} \int_{-1}^{0} \Phi_1(t)^{-1} b_j \circ dW_t^j, \quad X_1^+ := \sum_{j=0}^{m} \int_{0}^{1} \Phi_1(t)^{-1} b_j \circ dW_t^j.$$

The sequences in (8.5.37) and (8.5.38) converge $\mathbb{P}$-a. s. absolutely and geometrically, and the limits ξ_1^s and ξ_1^u as well as $\xi_1 = \xi_1^s \oplus \xi_1^u$ are tempered random vectors.

Proof. (i) We first prove the statements on temperedness. X_1^- and X_1^+ are tempered by Proposition 8.5.12 since a constant vector is tempered. Hence Lemmas 8.5.13 and 8.5.14 yield that ξ_1^s and ξ_1^u exist and are tempered. Thus ξ_1 is tempered.

(ii) We now check that δ_{ξ_1} is invariant, i. e. that $\varphi_1(t,\cdot)\xi_1 = \xi_1 \circ \theta_t$. This is equivalent with $\pi_1^{s,u}(\theta_t \cdot)\varphi_1(t,\cdot)\xi_1 = \xi_1^{s,u}(\theta_t \cdot)$. Using the variation of constants representation of φ_1 and the fact that for $k \in \mathbb{Z}^+$

$$\sum_{j=0}^{m} \int_{k}^{k+1} \Phi_1(s)^{-1} b_j \circ dW_s^j = \Phi_1(k)^{-1}(X_1^+ \circ \theta_k),$$

we obtain, after some rather lengthy but elementary manipulations,

$$\varphi_1(t,\cdot)\xi_1^u = \lim_{N\to\infty} \varphi_1(t,\cdot)\left(-\sum_{k=0}^{N-1} \pi_1^u \circ \Phi_1(k)^{-1}(X_1^+ \circ \theta_k)\right)$$

$$= \xi_1^u \circ \theta_t + \pi_1^s \sum_{j=0}^{m} \int_{0}^{t} \Phi_1(s)^{-1} b_j \circ dW_s^j.$$

A similar argument holds for the stable component.

(iii) The uniqueness of the invariant measure is a consequence of hyperbolicity and was proved in Theorem 5.6.1. □

Now suppose that $\xi_1, \ldots, \xi_{n-1}$ are tempered random vectors with values in $\mathbb{R}^{d_1}, \ldots, \mathbb{R}^{d_{n-1}}$, respectively, such that $\mu^i = \delta_{\xi_i}$, $1 \leq i \leq n-1$ is the unique invariant measure of the ith equation of our hierarchical system of affine SDE. We now study the nth SDE

$$dx_t^n = \sum_{j=0}^{m} (A_{j,n} x_t^n + p_{j,n}(\varphi_1(t)\xi_1, \ldots, \varphi_{n-1}(t, \xi_1, \ldots, \xi_{n-2})\xi_{n-1})) \circ dW_t^j$$

(8.5.39)

in $\mathbb{R}^{d_n}$. Note the fundamental fact that the validity of Lemma 8.5.3 implies the substitution rule for Stratonovich integrals (see Arnold and Imkeller [19: Corollary 1]): If η is any random variable, then $u(t, \eta)$ (though nonadapted) is Stratonovich integrable and

$$\int_0^t u(s,\eta) \circ dW_s = \left.\int_0^t u(s,x) \circ dW_s\right|_{x=\eta}. \qquad (8.5.40)$$

Hence the anticipative affine SDE (8.5.39) makes sense, and it generates an affine cocycle $\varphi_n(t, \xi_1, \ldots, \xi_{n-1})$. We now determine its invariant measure.

8.5.16 Theorem. *Suppose all equations of the hierarchical system of affine SDE have a hyperbolic linear part. Then each of the equations has a unique invariant measure. This measure is a random Dirac measure (stationary solution) whose supporting random variable can be obtained step by step by determining ξ_1 according to Theorem 8.5.15, inserting $\varphi_1(t)\xi_1$ into the second SDE, etc., where the stable and unstable component of ξ_n of the nth equation (8.5.39) are given by*

$$\xi_n^s = \sum_{k=0}^{\infty} \pi_n^s \circ \Phi_n(-k)^{-1}(X_n^- \circ \theta_{-k}), \qquad (8.5.41)$$

$$\xi_n^u = -\sum_{k=0}^{\infty} \pi_n^u \circ \Phi_n(k)^{-1}(X_n^+ \circ \theta_k), \qquad (8.5.42)$$

where

$$X_n^- := \sum_{j=0}^m \int_{-1}^0 \Phi_n(t)^{-1} p_{j,n}(\varphi_1(t)\xi_1, \ldots, \varphi_{n-1}(t,\xi_1,\ldots,\xi_{n-2})\xi_{n-1}) \circ dW_t^j,$$

$$X_n^+ := \sum_{j=0}^m \int_0^1 \Phi_n(t)^{-1} p_{j,n}(\varphi_1(t)\xi_1, \ldots, \varphi_{n-1}(t,\xi_1,\ldots,\xi_{n-2})\xi_{n-1}) \circ dW_t^j.$$

The sequences in (8.5.41) and (8.5.42) converge $\mathbb{P}$-a.s. absolutely and geometrically, and the limits ξ_n^s and ξ_n^u as well as $\xi_n = \xi_n^s \oplus \xi_n^u$ are tempered random vectors.

Proof. Since Proposition 8.5.10 holds, and since $(\xi_1, \ldots, \xi_{n-1})$ is a tempered vector, Proposition 8.5.12 applies, thus X_n^- and X_n^+ are tempered. Hence by Lemmas 8.5.14 and 8.5.13, the series in equations (8.5.41) and (8.5.42) have the convergence properties claimed, and the limits $\xi_n^{s,u}$ are tempered. Hence finally ξ_n is tempered.

Invariance of ξ_n is proved as in Theorem 8.5.15, and uniqueness as in Theorem 5.6.1. □

8.5.17 Remark. (i) By means of the substitution rule (8.5.40), ξ_1 can also be written as

$$\xi_1^s = \sum_{j=0}^m \int_{-\infty}^0 \Phi_1(t)^{-1}(\pi_1^s(\theta_t \cdot)b_j) \circ dW_t^j,$$

$$\xi_1^u = -\sum_{j=0}^{m} \int_0^\infty \Phi_1(t)^{-1}(\pi_1^u(\theta_t \cdot)b_j) \circ dW_t^j,$$

where the integrals exist as the ℙ-a.s. limits of the non-adapted Stratonovich integrals $\int_{-T}^0$ and $\int_0^T$ for $T \to \infty$. Similarly for ξ_n.

(ii) Again by the substitution rule (8.5.40), the cocycle (8.5.29) evaluated at $\xi := (\xi_1, \ldots, \xi_n)$ is equal to the solution of the first n equations with non-adapted initial value ξ, and is the unique stationary solution of these equations, $\varphi(t)\xi = \xi \circ \theta_t$. ∎

This completes the proof of Theorem 8.5.1. We close with an example.

8.5.18 Example (One-Dimensional Case). This is the white noise version of Example 8.3.11. For $d = 1$, $H_{n,1} \cong \mathbb{R}^1$ (choose the basis $x^n e_1$) for all $n \geq 2$. The linearization of the SDE $dx = \sum_{j=0}^m f_j(x) \circ dW^j$ is $dv = \sum_{j=0}^m A_j v \circ dW^j$ whose explicit solution is

$$\Phi(t, \omega) = \exp(A_0 t + \sum_{j=1}^m A_j W_t^j(\omega)).$$

Hence $\lambda = A_0$ is the Lyapunov exponent. Further, $\mathrm{ad}_n A_j = -(n-1)A_j$, and $\mathcal{S}(\Phi_n) = \{-(n-1)\lambda\}$. We have nonresonance for all n if and only if $\lambda \neq 0$. The unique stationary solution of the nth cohomological SDE

$$dh_n = \sum_{j=0}^m ((\mathrm{ad}_n A_j)h_n + k_{j,n}(\theta_t \cdot)) \circ dW_t^j$$

is

$$h_n = \begin{cases} \sum_{j=0}^m \int_{-\infty}^0 \exp((n-1)\sum_{l=0}^m A_l W_t^l) k_{j,n}(\theta_t \cdot) \circ dW_t^j, & \lambda > 0, \\ -\sum_{j=0}^m \int_0^\infty \exp((n-1)\sum_{l=0}^m A_l W_t^l) k_{j,n}(\theta_t \cdot) \circ dW_t^j, & \lambda < 0. \end{cases}$$

∎

8.5.3 Small Noise Case

Simultaneous Normal Form and Center Manifold Reduction

We are now in a position to carry over the procedure of Sect. 8.4 from the RDE to the SDE case. We will, however, not repeat details and just emphasize that all nonresonant (anticipative Stratonovich) cohomological equations in the SDE analogues of Theorems 8.4.1 and 8.4.3 are solved as in the nonresonant case of Subsect. 8.5.2. We make the convention that for all resonant equations we make the trivial choice $H_{pq} = 0$ and $H_{pqrs} = 0$. This mathematically rigorous procedure finally justifies earlier important work on normal forms for SDE mentioned at the beginning of this section.

Small Noise on Parametrized SDE

In stochastic bifurcation theory one needs to study an SDE

$$dx_t = f_0(x_t, \alpha)dt + \sigma \sum_{j=1}^{m} f_j(x_t, \alpha) \circ dW_t^j \qquad (8.5.43)$$

in $\mathbb{R}^d$, where the smooth vector fields $f_j(x, \alpha)$ smoothly depend on a parameter $\alpha \in \mathbb{R}^m$ and $\sigma \in \mathbb{R}$ is a (small) intensity parameter. We assume that $f_j(0, \alpha) = 0$ for all $\alpha \in \mathbb{R}^m$ so that the linearization of (8.5.43) is

$$dv_t = A_0(\alpha)v_t dt + \sigma \sum_{j=1}^{m} A_j(\alpha)v_t \circ dW_t^j, \ A_j(\alpha) = D_x f_j(x, \alpha)|_{x=0}, \ 0 \le j \le m.$$
$$(8.5.44)$$

Assume further that for $(\alpha, \sigma) = (0, 0)$ the deterministic equation

$$\dot{v}_t = A_0 v_t, \quad A_0 = D_x f_0(x, 0)|_{x=0},$$

is in a center/stable situation and has been brought into Jordan canonical form by a (deterministic) coordinate transformation, so that $\mathbb{R}^d = \mathbb{R}^{d_c} \times \mathbb{R}^{d_s}$ is the invariant splitting for $A_0 = \begin{pmatrix} A_0^c & 0 \\ 0 & A_0^s \end{pmatrix}$.

We treat $(\alpha, \sigma) \in \mathbb{R}^{m+1}$ as a small parameter and seek the normal form and center manifold reduction for small (x, α, σ). All cohomological operators (hence all linear cocycles) are now deterministic and of the form

$$dL_n(h_n) = dh_n - (\operatorname{ad}_n A_0)h_n dt.$$

We add the trivial equations $d\alpha = 0$ and $d\sigma = 0$. The result is given by the (we hope rather obvious) white noise analogue of Theorem 8.4.1: The truncated center equation is

$$dx_c = A_0^c x_c dt + \sum_{j=1}^{m} g_j^c(\theta_t \cdot, x_c, \alpha, \sigma) \circ dW_t^j, \quad d\alpha = 0, \quad d\sigma = 0,$$

and the approximate center manifold is the graph

$$\mathbb{R}^{d_c} \times \mathbb{R}^{m+1} \ni (x_c, \alpha, \sigma) \mapsto m_c(\cdot, x_c, \alpha, \sigma) = H^s(\cdot, x_c, 0, \alpha, \sigma) \in \mathbb{R}^{d_s},$$

where g_j^c and m_c are random polynomials of order N in x_c and M in (α, σ).

8.5.19 Example (Duffing-van der Pol Oscillator with White Noise).

We now consider the white noise version of the example in Subsect. 8.4.3, to which we refer for notations.

Replacing α by $\alpha + \sigma \dot{W}_t$ ($\dot{W}$ symbolically stands, as usual, for white noise) and putting $\beta = -1$ in $\ddot{y} = \alpha y + \beta \dot{y} - y^3 - y^2 \dot{y}$ gives the SDE in $\mathbb{R}^2$

8.5 Normal Forms for SDE

$$dx_t = \left(\begin{pmatrix} 0 & 1 \\ \alpha & -1 \end{pmatrix} x_t + \begin{pmatrix} 0 \\ -x_1^3 - x_1^2 x_2 \end{pmatrix}\right) dt + \sigma \begin{pmatrix} 0 & 0 \\ 1 & 0 \end{pmatrix} x_t \circ dW_t,$$
(8.5.45)

with linearization at $x = 0$ given by

$$dv_t = \begin{pmatrix} 0 & 1 \\ \alpha & -1 \end{pmatrix} v_t\, dt + \sigma \begin{pmatrix} 0 & 0 \\ 1 & 0 \end{pmatrix} v_t \circ dW_t.$$
(8.5.46)

The calculations are formally the same as in the RDE case. We introduce the auxiliary processes U, V, X and Y which are the stationary solutions of the scalar SDE

$$dU_t = -U_t dt + dW_t, \quad dV_t = V_t dt + dW_t,$$

and

$$dX_t = -X_t dt + dU_t, \quad dY_t = -Y_t dt + U_t \circ dW_t.$$

Note that V anticipates the future of W. The analogues of the RDE expressions given above are now

$$\xi = \dot{W}, \quad H^s_{1001} = -\sqrt{2} U, \quad H^c_{0101} = -\frac{1}{\sqrt{2}} V,$$

$$H^s_{3001} = -4\sqrt{2} U - 3\sqrt{2} X, \quad H^s_{1002} = 2\sqrt{2} Y,$$

$$g^c_{1002} = -U\dot{W}, \quad g^c_{1011} = -U - \dot{W}, \quad g^c_{3001} = 2U + \dot{W},$$
$$g^c_{3002} = (-3UV - 4Y - 2U - U^2) + (-4U - 3X - V)\dot{W},$$

and

$$g^c_{3011} = (-15U - 7X - 4V) - 10\dot{W}.$$

Inserting these expressions into the center equation (up to terms of order 3 in x_c and 2 in (α, σ)) yields

$$\begin{aligned}
dx_c = {} & (\alpha - \alpha\sigma U - \alpha^2) x_c dt + (\sigma - \sigma^2 U - \alpha\sigma) x_c \circ dW_t \\
& + (-1 + 2\sigma U - \sigma^2(3UV + 4Y + 2U + U^2) \\
& + 3\alpha - \alpha\sigma(15U + 7X + 4V) - 18\alpha^2) x_c^3 dt \\
& + (\sigma - \sigma^2(4U + 3X + V) - 10\alpha\sigma) x_c^3 \circ dW_t.
\end{aligned}$$
(8.5.47)

Note that this is an anticipative SDE since V is anticipative.

The approximate center manifold is the graph $(x_c, \alpha, \sigma) \mapsto m_c(\omega, x_c, \alpha, \sigma) = x_s$ given by

$$\begin{aligned}
m_c(\cdot, x_c, \alpha, \sigma) = {} & (-\sqrt{2}\sigma U_0 + 2\sqrt{2}\sigma^2 Y_0 - \sqrt{2}\alpha \\
& + 2\sqrt{2}\alpha\sigma(U_0 + X_0) + 2\sqrt{2}\alpha^2) x_c \\
& + (\sqrt{2} - \sqrt{2}\sigma(4U_0 + 3X_0) \\
& + \sigma^2 H^s_{3002} - 7\sqrt{2}\alpha + \alpha\sigma H^s_{3011} - 47\sqrt{2}\alpha^2) x_c^3,
\end{aligned}$$

where H^s_{3002} and H^s_{3011} are certain anticipative random variables.

The scalar center SDE (8.5.47) can now be utilized for stochastic bifurcation theory of the original two-dimensional SDE (8.5.45) in the neighborhood of $\alpha = 0$, see Subsect. 9.4.3. ∎

Chapter 9. Bifurcation Theory

Summary

In this chapter we investigate "qualitative changes" in parametrized families of RDS and call these studies "stochastic bifurcation theory".

The first problem is to develop a mathematical formalization (called D-bifurcation) of "qualitative change" which is connected to the stability of an RDS under an invariant measure (i.e. to the Lyapunov exponents) and reduces to the deterministic definition of bifurcation in the absence of noise (Subsect. 9.2.1). We propose as a first task to study the branching of new invariant measures from a family of reference measures at a parameter value at which the reference measure has a vanishing Lyapunov exponent. We also discuss an older concept on the level of densities in state space (called P-bifurcation) and discuss its relation with D-bifurcation (Subsects. 9.2.2 and 9.5.1).

As the theory of stochastic bifurcation is still in its infancy, we proceed mainly by way of instructive examples.

Sect. 9.3 treats explicitly solvable one-dimensional examples of stochastic transcritical and pitchfork bifurcation. In Subsect. 9.3.4 we prove a general criterion for a pitchfork bifurcation in dimension one using the theory of random attractors, to which we give a brief introduction.

Sect. 9.4 is devoted to the study of the prototypical noisy Duffing-van der Pol oscillator. We report on the state of the art as far as rigorous results are concerned and also present numerical findings supporting conjectures about the "correct" scenario of stochastic Hopf bifurcation.

Sect. 9.5 is devoted to the only available general condition (due to Baxendale) for a D-bifurcation out of the fixed point $x = 0$ (and an associated P-bifurcation of the new branch of measures) in a family of SDE in $\mathbb{R}^d$.

9.1 Introduction

Despite of its rapid development in the last decade and the fact that it is probably the most relevant part of the whole theory of RDS, bifurcation theory of RDS is still in its infancy. There are few rigorous general theorems and criteria, and many phenomena have been "proved" only by computer simulations, or for particular models. Nevertheless, this chapter was included although it is less self-contained and less mathematically rigorous than the rest of the book.

The interest in understanding the changes encountered in a deterministic bifurcation scenario "in the presence of noise" is historically at the beginning of the theory of RDS, and has been one of its strong driving forces.

This interest arose in the 1960's and prompted numerous investigations by engineers (S. T. Ariaratnam, F. Kozin, Y. K. Lin, N. Sri Namachchivaya, C. W. S. To, W. Wedig, and many others), physicists and chemists (W. Ebeling, P. Coullet, R. Graham, H. Haken, K. H. Hoffmann, W. Horsthemke, R. Lefever, M. Lücke, F. Moss, G. Nicolis, E. Tirapegui, K. Wiesenfeld, and many others), and mathematicians (L. Arnold, P. Baxendale, R. Khasminskii, W. Kliemann, P. Kloeden, G. Papanicolaou, M. Scheutzow, among others).

In particular, bifurcations "delayed" or "advanced" by noise, "bifurcation intervals", and the "disappearance" of branches from the deterministic bifurcation diagram were observed numerically, experimentally, and by approximations based on averaging and scaling methods.

The literature is so vast that the reader is referred to the book by Horsthemke and Lefever [175] (for the phenomenological approach, see Subsect. 9.2.2), and to the review papers by Arnold and Boxler [10] (for the literature prior to 1986), Sri Namachchivaya [322], Wedig [342] (for the literature prior to 1990) and To et al. [330, 331], to Moss and McClintock [259] and the references therein.

What is stochastic bifurcation theory (or more precisely: bifurcation theory of RDS)? As with other notions for RDS, it should be an extension and generalization of deterministic bifurcation theory, to which it should reduce in the absence of noise.

We hence ask what bifurcation theory of deterministic dynamical systems is about. Of the numerous books on this subject, several provide excellent background, including the now classical book by Guckenheimer and Holmes [162], and those of V. I. Arnold [45], Chow and Hale [100], Hale and Koçak [167], Ruelle [295] and Wiggins [346].

The general term "bifurcation" is used to describe "qualitative changes" in the phase portrait (orbit structure) of parametrized families (φ_α) of dynamical systems, e.g. those generated by a family of differential equations

$$\dot{x}_t = f_\alpha(x_t), \quad \alpha \in \mathbb{R}^k, \tag{9.1.1}$$

which occur when the parameter α is (slowly) varied. The vague notion of "qualitative change" has been successfully formalized by using the concepts of topological equivalence and structural stability: A parameter value α_0 is called an *abstract bifurcation point* of the family φ_α if the family is not structurally stable at α_0, i.e. if in any neighborhood of α_0 there are parameter values α such that φ_α is not topologically equivalent to φ_{α_0}.

9.1 Introduction

If we replace topological equivalence by *local* topological equivalence (for example, in the neighborhood of a particular point), we arrive at the definition of a *local abstract bifurcation point*.

This abstract definition of "qualitative change" via structural instability does not indicate which "qualitative change" actually happens at a bifurcation point of a family. For example, when there is "bifurcation" in the narrow sense of the word, i. e. when some invariant object splits into several such objects (after all, "bifurcation" means "splitting into two"), then we drop the word "abstract" and just speak of a *bifurcation point*.

We now turn to the most elementary and classical local bifurcations, namely those of equilibria (fixed points), in particular the saddle-node, transcritical, pitchfork, and Hopf bifurcation. The modest aim of this chapter is to discover their stochastic analogues.

To begin with, assume from now on that $f_\alpha(0) = 0$ for all $\alpha \in \mathbb{R}^k$, and that f_α depends smoothly on α. How do we detect the bifurcation of the family of reference fixed points $x = 0$ at α_0, meaning the coexistence of 0 with at least one other small solution $x_1 \neq 0$ (e. g. another fixed point, limit cycle, etc.), in a neighborhood of α_0?

It is futil to search for this phenomenon at parameter values α_0 for which $x = 0$ is a hyperbolic fixed point of φ_{α_0}, i. e. for which the linearized system $\Phi_\alpha(t) := D\varphi_\alpha(t, 0)$ generated by

$$\dot{v}_t = A_\alpha v_t, \quad A_\alpha := Df_\alpha(0), \quad \alpha \in \mathbb{R}^k, \tag{9.1.2}$$

is hyperbolic for $\alpha = \alpha_0$. The reason is that hyperbolic fixed points are structurally stable for any family (φ_α). Since for all α close to α_0 by the local Hartman-Grobman theorem φ_α and Φ_α are locally topologically equivalent in a neighborhood $U_\alpha \subset \mathbb{R}^d$ of 0, the family (φ_α) is locally structurally stable at α_0.

Moreover, $x = 0$ cannot "bifurcate" at α_0 in the narrow sense either: Since, in the case of hyperbolicity, the only bounded solution of (9.1.2) is the trivial solution, the local Hartman-Grobman theorem implies that the only solution of (9.1.1) which stays in U_α for all time is also the trivial solution $\varphi_\alpha(t, 0) \equiv 0$.

Consequently, a necessary (but not sufficient) condition for the bifurcation of small solutions out of the fixed point $x = 0$ at $\alpha = \alpha_0$ is that the matrix A_{α_0} is not hyperbolic, i. e. has eigenvalues on the imaginary axis. If, in addition, A_α is hyperbolic for some α in each neighborhood of α_0, then α_0 is a local abstract bifurcation point which may also be a local bifurcation point.

The simplest possible cases (for a scalar parameter $\alpha \in \mathbb{R}$, and at $\alpha_0 = 0$ here) are the following: As α increases from negative to positive values,

(a) either a simple real eigenvalue $\lambda(\alpha)$ of A_α increases from negative to positive values, with $\lambda(0) = 0$ and $d\lambda(\alpha)/d\alpha|_{\alpha=0} > 0$, while the rest of the spectrum of A_α remains in the left-hand side of the complex plane $\mathbb{C}$ (allowing for the transcritical and the pitchfork bifurcations),

(b) or a pair of complex-conjugate eigenvalues $\lambda_{1,2}(\alpha) = \beta(\alpha) \pm i\omega(\alpha)$ of A_α moves from the left-hand side to the right-hand side of $\mathbb{C}$, crossing the imaginary axis at $\alpha = 0$ with positive speed $d\beta(\alpha)/d\alpha|_{\alpha=0} > 0$ (allowing for the Hopf bifurcation).

There are very efficient means of further reducing the search for bifurcations of the trivial solution at $\alpha = \alpha_0$:

(i) Center manifold theory tells us that it suffices to investigate the system on the center manifold, which is decribed by a differential equation in the center space of A_{α_0} with dimension 1 in case (a) and dimension 2 in case (b).

(ii) This low-dimensional system can be simplified further by normal form theory (in fact, steps (i) and (ii) can be carried out simultaneously, see Elphick et al. [135]).

(iii) A persistence theorem assures that without loss of generality the normal form can be truncated beyond low order nonlinear x terms, from which the bifurcation behavior can be often easily read off.

The simplest possible truncated normal forms for the above mentioned elementary bifurcation scenarios of the fixed point $x = 0$ are (for a scalar parameter $\alpha \in \mathbb{R}$ and at $\alpha_0 = 0$):
- saddle-node bifurcation[1]: $\dot{x} = \alpha - x^2$ in $\mathbb{R}^1$,
- transcritical bifurcation: $\dot{x} = \alpha x - x^2$ in $\mathbb{R}^1$,
- pitchfork bifurcation: $\dot{x} = \alpha x - x^3$ in $\mathbb{R}^1$,
- Hopf bifurcation: $\dot{r} = \alpha r - r^3$, $\dot{\theta} = 1 + ar^2$ in polar coordinates of $\mathbb{R}^2$.

In all scenarios with a trivial reference solution, bifurcation is accompanied by an "exchange of stability". The reference solution loses its stability at the bifurcation point, while the bifurcating solutions are stable.

An analogous situation holds for families of mappings, where the period-doubling (or flip) bifurcation has to be added to the catalogue of elementary bifurcations of a fixed point, whose truncated normal form is $x \mapsto \varphi_\alpha(x) = (-1-\alpha)x + x^3$ in $\mathbb{R}^1$, the second iterate of which has two stable fixed points $\pm\sqrt{\alpha}$ for $\alpha > 0$.

9.2 What is Stochastic Bifurcation?

9.2.1 Definition of a Stochastic Bifurcation Point

In complete analogy to the deterministic case, stochastic bifurcation theory studies "qualitative changes" in parametrized families φ_α of RDS, e.g. those generated by a family of SDE

$$dx_t = f_0^\alpha(x_t)dt + \sum_{j=1}^m f_j^\alpha(x_t) \circ dW_t^j, \quad \alpha \in \mathbb{R}^k. \tag{9.2.1}$$

How can we formalize "qualitative change" in this situation?

---
[1] Here $x = 0$ is a fixed point of φ_α only for the parameter value $\alpha = 0$.

9.2 What is Stochastic Bifurcation?

In order to mimic deterministic reasoning we need the definition of (local) topological equivalence of RDS introduced in connection with the (Local) Hartman-Grobman Theorems in Chap. 7. For a systematic study of topological dynamics of RDS see Nguyen Dinh Cong [261].

As in the deterministic case, we add the word "abstract" if the bifurcation point of a family of RDS is a point of structural instability and omit it if we have a bifurcation point in the narrow sense, see below.

9.2.1 Definition (Abstract Bifurcation Point). (i) Let $(\varphi_\alpha)_{\alpha \in \mathbb{R}^k}$ be a family of continuous RDS on $\mathbb{R}^d$. A parameter value α_0 is called an *abstract bifurcation point* of the family if the family is not structurally stable at α_0, i.e. if in any neighborhood of α_0 there are parameter values α such that φ_α and φ_{α_0} are not topologically equivalent.

(ii) Let $(\varphi_\alpha)_{\alpha \in \mathbb{R}^k}$ be a family of local continuous RDS on $\mathbb{R}^d$ with fixed point 0. A parameter value α_0 is called a *local abstract bifurcation point* of the family, if the family is not locally structurally stable at α_0 in a neighborhood of 0, i.e. if in any neighborhood of α_0 there are parameter values α such that φ_α and φ_{α_0} are not locally topologically equivalent in a neighborhood of 0. ∎

We now turn to stochastic bifurcation in the narrow sense. As invariant measures are the random analogues of deterministic fixed points, we should start our investigations by finding necessary and sufficient conditions for the bifurcation of a family of invariant measures μ_α. What objects do we expect to bifurcate out of μ_α at a (local) bifurcation point α_0? Again analogous to the deterministic case, the simplest possible choice is a new family $\nu_\alpha \neq \mu_\alpha$ of (small) invariant measures.

9.2.2 Definition (D-Bifurcation Point). Let $(\varphi_\alpha)_{\alpha \in \mathbb{R}^k}$ be a family of local C^1 RDS in $\mathbb{R}^d$ with a respective family of ergodic invariant measures (μ_α). A parameter value α_D is called a *local dynamical*[2], or *D-bifurcation point* of $(\varphi_\alpha, \mu_\alpha)_{\alpha \in \mathbb{R}^k}$, if in each neighborhood of α_D there is an α for which there is an invariant measure $\nu_\alpha \neq \mu_\alpha$ with $\nu_\alpha \to \mu_{\alpha_D}$ (in the topology of weak convergence) as $\alpha \to \alpha_D$. One could also require that ν_α converges to μ_{α_D} in the Hausdorff sense, i.e. that in addition to the above, $\lim_{\alpha \to \alpha_D} d(\mathrm{supp}\nu_\alpha | \mathrm{supp}\mu_{\alpha_D}) = 0$, where $d(A|B) := \sup_{x \in A} d(x, B)$ is the Hausdorff semi-metric. ∎

See Fig. 9.1.

We further simplify the situation by assuming that $(\varphi_\alpha)_{\alpha \in \mathbb{R}^k}$ is a family of local C^1 RDS all having the fixed point $x = 0$. The family $\mu_\alpha = \delta_0$, $\alpha \in \mathbb{R}^k$, is thus a family of ergodic reference measures.

We linearize the local RDS φ_α at $x = 0$ and obtain the family of linear cocycles $\Phi_\alpha(t, \omega) := D\varphi_\alpha(t, \omega, 0)$. We assume that the IC of the MET are satisfied by Φ_α for all α, providing us with the Lyapunov spectrum

---

[2] See the next subsection for the reason for using of the word "dynamical".

Chapter 9. Bifurcation Theory

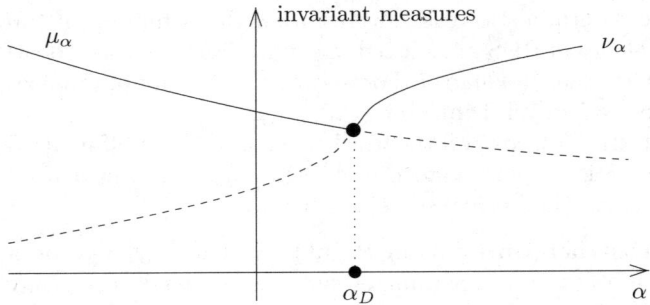

Fig. 9.1. D-bifurcation point of a family of measures

$$\mathcal{S}(\varphi_\alpha, \delta_0) = \{(\lambda_i(\alpha), d_i(\alpha))_{i=1,\ldots,p(\alpha)}\}$$

and the splitting

$$\mathbb{R}^d = E_1(\omega, \alpha) \oplus \cdots \oplus E_{p(\alpha)}(\omega, \alpha).$$

One basic difficulty is that even if we assume that φ_α depends smoothly on α, the Lyapunov exponents, their multiplicities and the Oseledets spaces will in general depend only measurably on α. Perhaps the most extreme expression for this is a result by Arnold and Nguyen Dinh Cong [30].

On the other hand, in many situations in which small perturbations of a given cocycle contain "enough randomness", the Lyapunov exponents, their multiplicities and even (in a certain sense) the Oseledets spaces depend continuously on the size of the perturbation. The most general result supporting this is one by Ochs [265] (generalizing an earlier result of Ledrappier and Young [231]), and such continuity is found in many applications (see the examples in this chapter).

We now prove that a parameter value α_0 for which the linearization Φ_{α_0} is hyperbolic can never be a local D-bifurcation point.

To be more specific, consider a family $(\varphi_\alpha)_{\alpha \in \mathbb{R}^k}$ of C^1 RDS with discrete or continuous two-sided time $\mathbb{T}$ and fixed point 0 such that $(x, \alpha) \mapsto \varphi_\alpha(t, \omega, x)$ is C^1 and the derivatives are continuous with respect to (t, x, α).

We mimic the deterministic procedure and study the augmented RDS on $\mathbb{R}^d \times \mathbb{R}^k$ defined by

$$\tilde{\varphi}(t, \omega, (x, \alpha)) := (\varphi_\alpha(t, \omega, x), \alpha), \quad (x, \alpha) \in \mathbb{R}^d \times \mathbb{R}^k, \; t \in \mathbb{T}. \qquad (9.2.2)$$

The RDS $\tilde{\varphi}$ is C^1, it has fixed points $(0, \alpha)$ and its linearization at $(0, 0)$ is given by

$$\tilde{\Phi}(t,\omega) := D_{(x,\alpha)}\tilde{\varphi}(t,\omega,(0,0)) = \begin{pmatrix} \Phi_0(t,\omega) & 0 \\ 0 & \mathrm{id}_{\mathbb{R}^k} \end{pmatrix}, \qquad (9.2.3)$$

where $\Phi_0(t,\omega) := D_x\varphi_0(t,\omega,0)$. $\tilde{\Phi}$ satisfies the IC of the MET if and only if Φ_0 does, which we assumed. The spectrum and splitting of $\tilde{\Phi}$ can be obtained from that of Φ_0 in an obvious manner by increasing the (possibly zero) multiplicity of a vanishing Lyapunov exponent by k and augmenting the (possibly trivial) center space $E_c(\omega)$ of Φ_0 to $E_c(\omega) \times \mathbb{R}^k$.

If U is a set, we call a measure ρ U-small if supp$\rho \subset U$.

9.2.3 Theorem (Necessary Condition for D-Bifurcation Point).
Assume the situation just described. Assume further that $\tilde{\varphi}$ satisfies the Local Hartman-Grobman Theorem 7.5.6 (discrete time) or 7.5.16 (continuous time) at $(0,0)$, respectively. Suppose there exists a family $\nu_\omega^\alpha(dx) \neq \delta_0(dx)$ of φ_α-invariant measures such that the $\tilde{\varphi}$-invariant measure $\tilde{\nu}_\omega(dx,d\beta) = \nu_\omega^{\alpha_1}(dx) \times \delta_{\alpha_1}(d\beta)$ is $U_T(\omega)$-small[3] for some α_1 and some $T > 0$.

Then necessarily Φ_0 is non-hyperbolic, i.e. $\lambda = 0$ is among its Lyapunov exponents.

Proof. All invariant measures of $\tilde{\varphi}$ clearly have the form $\tilde{\nu}_\omega(dx,d\beta) = \nu_\omega^\alpha(dx) \times \delta_\alpha(d\beta)$, where $\nu_\omega^\alpha(dx)$ is φ_α-invariant. If $\nu_\omega^{\alpha_1}(dx) \times \delta_{\alpha_1}(d\beta) \neq \delta_0(dx) \times \delta_{\alpha_1}(d\beta)$ is $U_T(\omega)$-small for some α_1 then Φ_0 cannot be hyperbolic for the following reason: In the hyperbolic case the only (ergodic) invariant measures of $\tilde{\Phi}$ are of the form $\delta_0(dx) \times \delta_\alpha(d\beta)$. By the local Hartman-Grobman theorem, this carries over to $U_T(\omega)$-small measures of $\tilde{\varphi}$. □

This theorem convinces us that our approach towards stochastic bifurcation is "correct" in the sense that
- it reduces to the deterministic concept in the absence of noise, and
- it ties the phenomenon of local branching of a family of invariant reference measures at a parameter value α_D to the stability of the reference measures in a neighborhood of α_D.

Hence we will have to search for parameter values for which one (typically the largest, for the stochastic Hopf bifurcation also the second largest) Lyapunov exponent changes its sign.

9.2.2 The Phenomenological Approach

We now present an alternative and older approach towards stochastic bifurcation.

Scientists have been observing in numerous experimental, numerical and analytic studies "qualitative changes" of probability densities p_α which are stationary for the forward Markov semigroup corresponding to the SDE (9.2.1), i.e. which are solutions of the Fokker-Planck equation

---
[3] $U_T(\omega)$ is the neighborhood of $(0,0)$ defined in the text following the Hartman-Grobman Theorems.

472   Chapter 9. Bifurcation Theory

$$L_\alpha^* p_\alpha = 0, \quad L_\alpha = f_0^\alpha + \frac{1}{2}\sum_{j=1}^m (f_j^\alpha)^2, \qquad (9.2.4)$$

where L_α^* is the formal adjoint of the generator L_α; here we have written L_α in Hörmander form.

For example, the stationary density can exhibit transition from a unimodal (one peak) to a bimodal (two peaks) or crater-like form, see Fig. 9.2. Also, the number and location of extrema of p_α as functions of α and their "bifurcations" have been carefully studied. For a systematic treatment and survey see Horsthemke and Lefever [175] (who call such phenomena "noise-induced transitions"), Arnold and Boxler [10], and Sri Namachchivaya [322]. Many case studies can also be found in the proceedings volume [174] edited by Horsthemke and Kondepudi.

This concept can be formalized based on the ideas of Zeeman [352, 351] according to which two probability densities are called equivalent if there are two diffeomorphisms β, γ such that $p = \beta \circ q \circ \gamma$. This gives rise to a notion of structural stability based on this equivalence and thus of a point of structural instability of families p_α of densities which captures the above-mentioned observations. For example, the transition point α_P from a unimodal density p_α to a bimodal density is such a point of structural instability.

We call such a phenomenon *phenomenological*, or *P-bifurcation*, and the corresponding parameter value α_P a phenomenological, or P-bifurcation point – in distinction to the concept introduced in Definition 9.2.2, which we called dynamical, or D-bifurcation[4], and denote the corresponding parameter value by α_D.

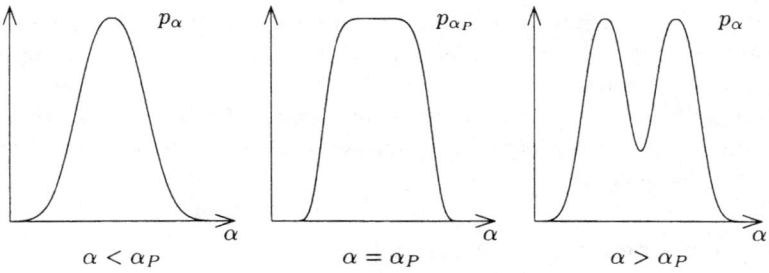

Fig. 9.2. P-bifurcation point of a family of densities

Besides being basically restricted to the white or Markovian noise case, the concept of P-bifurcation has the following severe drawbacks.

---

[4] The term "dynamical bifurcation" has also been used in a different context, namely for phenomena obtained by letting the bifurcation parameter $\alpha = \alpha(t)$ deterministically depend on time, for example, by augmenting $\dot{x}_t = f_\alpha(x_t)$ by $\dot\alpha_t = \varepsilon$ for small ε. See Benoît [66] for a survey.

(i) First, the concept is static since the stationary density $p_\alpha(x)dx$ measures the asymptotic proportion of time spent by a typical solution $\varphi_\alpha(t,\omega)x$ of (9.2.1) in the volume element dx. Hence, p_α has figuratively "forgotten" any dynamics of φ_α.[5]

(ii) Second, p_α is an object determined by the one-point motions $t \mapsto \varphi_\alpha(t,\omega)x$ of φ_α and can thus in principle not be related to the stability of φ_α which is a property of the two-point motions $t \mapsto (\varphi_\alpha(t,\omega)x, \varphi_\alpha(t,\omega)y)$ for $x \neq y$, in infinitesimal form of $D\varphi_\alpha(t,\omega,x)v$, hence is a property of the Lyapunov exponents.

(iii) The underlying concept of invariant measure is too narrow. Stationary measures for the forward Markov semigroup are in a one-to-one correspondence to invariant measures which only depend on the past (forward Markov measures, see Sect. 1.7), but there are typically many more invariant measures. Restricting ourselves to stationary measures leads to "missing branches" in bifurcation diagrams (representing non-Markovian invariant measures), and prevents us from understanding stochastic bifurcation.

Is there a connection beween the two types of bifurcation? In general, there is not. We will next present an example of a family of RDS undergoing a P-bifurcation, but not a D-bifurcation, and then an example where the converse occurs.

In Sect. 9.5, however, we will reveal a surprising connection between the two concepts.

Additive Noise: P-Bifurcation, but not D-Bifurcation

If $x_0(t)$ is a solution of a differential equation, and if this equation is "disturbed by noise", we call the noise *multiplicative* with respect to $x_0(t)$ if $x_0(t)$ also solves the disturbed equation. Otherwise the noise is called *additive* with respect to this solution. It is just called additive if it is additive with respect to any solution of the undisturbed equation. For example, the noise ξ_t in $\dot{x}_t = (\alpha + \sigma\xi_t)x_t - x_t^3$ is multiplicative with respect to $x = 0$, but additive with respect to $x = \pm\sqrt{\alpha}$ for $\alpha > 0$.

Consider the scalar SDE

$$dx_t = f(x_t)dt + g(x_t) \circ dW_t = (f(x_t) + \frac{1}{2}g(x_t)g'(x_t))dt + g(x_t)\,dW_t \quad (9.2.5)$$

with smooth coefficients f and g. Assume further that the conditions of Theorem 2.3.38 are satisfied which ensure that $\pm\infty$ are non-exit boundaries and hence that (9.2.5) generates a local smooth RDS φ which is forward, hence strictly forward complete. The generator of the forward Markov semigroup of (9.2.5) is

$$L = (f(x) + \frac{1}{2}g(x)g'(x))\frac{d}{dx} + \frac{1}{2}g(x)^2\frac{d^2}{dx^2},$$

so the Fokker-Planck equation is

---
[5] Zeeman [352] himself says at the end of his paper: "It seems a pity to have to represent a dynamical system by a static picture".

$$L^*p = -((f + \frac{1}{2}gg')p)' + \frac{1}{2}(g^2p)'' = 0. \tag{9.2.6}$$

We assume for simplicity that $g(x) > 0$ for all $x \in \mathbb{R}$ (ellipticity, implying that there is at most one stationary probability) and also assume positive recurrence implying that there is exactly one stationary probability with smooth density given by

$$p(x) = \frac{N}{g(x)} \exp\left(\int_0^x \frac{2f(y)}{g(y)^2} dy\right), \tag{9.2.7}$$

where N is a normalization constant.

We can explicitly calculate the Lyapunov exponent of the RDS φ generated by (9.2.5) under the stationary measure with density p (more precisely, under the invariant measure μ_ω which corresponds to p – but μ_ω is not needed for the calculation). For this we solve the linearized SDE

$$dv_t = f'(x_t)v_t dt + g'(x_t)v_t \circ dW_t$$

coupled to (9.2.5) to obtain

$$v_t = v_0 \exp\left(\int_0^t (f' + \frac{1}{2}gg'')(x_s)ds + \int_0^t g'(x_s)\, dW_s\right)$$

and finally (for example, under the assumption that g' is bounded, and $f' + \frac{1}{2}gg'' \in L^1(p(x)dx)$)

$$\lambda = \int_\mathbb{R} (f'(x) + \frac{1}{2}g(x)g''(x))p(x)dx. \tag{9.2.8}$$

Integrating the last expression by parts and using the explicit expression (9.2.7) for p we finally arrive at formula (9.2.9) of the following theorem.

9.2.4 Theorem (Scalar RDS with Additive White Noise is Always Stable). *Assume that the Fokker-Planck equation (9.2.6) of the SDE (9.2.5) with $g(x) > 0$ (all $x \in \mathbb{R}$) has a (necessarily unique) solution $\rho(dx) = p(x)dx$ with density (9.2.7). Then:*

(i) The Lyapunov exponent of ρ is given by

$$\lambda = -2 \int_\mathbb{R} \left(\frac{f(x)}{g(x)}\right)^2 p(x)dx < 0. \tag{9.2.9}$$

Hence the RDS φ generated by (9.2.5) is always exponentially stable under ρ.

(ii) The φ-invariant forward Markov measure $\mu_\omega = \lim_{t\to\infty} \varphi(-t,\omega)^{-1}\rho$ corresponding to ρ is a Dirac measure, $\mu_\omega = \delta_{a(\omega)}$, and $A(\omega) := \{a(\omega)\}$ is the random attractor of φ in $\mathbb{R}$ in a universe of sets which contains all bounded deterministic sets (see Definition 9.3.1).

(iii) The RDS φ has no other invariant measure besides μ_ω.

9.2 What is Stochastic Bifurcation?

(iv) For any initial measure $\pi \in \mathcal{P}(\mathbb{R})$ and $\pi_t(\cdot) := \int_\mathbb{R} P(t, x, \cdot)\pi(dx)$,

$$\lim_{t \to \infty} \pi_t = \rho = \mathbb{E}\,\delta_a = \mathbb{P}\{a \in \cdot\}. \qquad (9.2.10)$$

Proof. (ii) The measure μ_ω is ergodic since ρ is (Theorem 1.7.2), and is a Dirac measure by Theorem 1.8.4. The remaining statement is proved by Crauel and Flandoli [118: Corollaries 3.3 and 3.5]. See also Subsect. 9.3.3.

(iii) By a result of Crauel [114: Corollary 4.4] all invariant measures of a continuous RDS having a random attractor in the universe of bounded deterministic sets are supported by this attractor. Since the attractor is a one-point set, it can carry only a Dirac measure.

(iv) By dominated convergence and the fact that by the definition of the attractor, $\varphi(t, \theta_{-t}\omega, x) \to a(\omega)$ for all $x \in \mathbb{R}$, for any $f \in \mathcal{C}_b(\mathbb{R})$

$$\lim_{t \to \infty} \pi_t(f) = \int_\mathbb{R} \mathbb{E} \lim_{t \to \infty} f(\varphi(t, \theta_{-t}\cdot, x))\pi(dx) = \mathbb{E}\,f(a(\cdot)).$$

□

The fact that ρ is globally stable in the sense that it attracts arbitrary initial measures has been observed much earlier (see Horsthemke and Lefever [175: Sect. 6.1]) by using the relative entropy

$$V(t) := -\int \log \frac{d\rho}{d\pi_t}(x)\pi_t(dx) \geq 0$$

as a Lyapunov function and proving that $\frac{dV(t)}{dt} \leq 0$ always, and < 0 if and only if $\pi_t \neq \rho$.

9.2.5 Example (Effect of Additive Noise on a Pitchfork Bifurcation). For

$$dx_t = (\alpha x_t - x_t^3)dt + \sigma dW_t, \quad \alpha \in \mathbb{R},\ \sigma \neq 0,$$

the unique stationary measure has density

$$p_{\alpha,\sigma}(x) = N_{\alpha,\sigma} \exp\left(\frac{1}{\sigma^2}\left(\alpha x^2 - \frac{x^4}{2}\right)\right).$$

For fixed $\sigma \neq 0$, the density is unimodal for $\alpha < 0$ and bimodal for $\alpha > 0$. Hence the family $(p_{\alpha,\sigma})_{\alpha \in \mathbb{R}}$ undergoes a P-bifurcation at $\alpha_P = 0$ for each $\sigma \neq 0$.

The IC of the MET is satisfied, and the Lyapunov exponent of the RDS $\varphi_{\alpha,\sigma}$ under $p_{\alpha,\sigma}$ is

$$\lambda_{\alpha,\sigma} = \alpha - 3 \int_\mathbb{R} x^2 p_{\alpha,\sigma}(x) dx,$$

which, by formula (9.2.8), is strictly negative for all values of $\alpha \in \mathbb{R}$ and $\sigma \neq 0$.

476   Chapter 9. Bifurcation Theory

The P-bifurcation of the densities $p_{\alpha,\sigma}$ at $\alpha_P = 0$ is what remains of the pitchfork bifurcation at $\alpha = 0$ present for $\sigma = 0$. However, there is no D-bifurcation, since the measure $\mu_\omega^{\alpha,\sigma} = \delta_{a_{\alpha,\sigma}(\omega)}$ corresponding to $p_{\alpha,\sigma}$ is the unique invariant measure of $\varphi_{\alpha,\sigma}$ and always has negative Lyapunov exponent. One can thus say that "additive noise destroys a pitchfork bifurcation" – which is the title of the paper by Crauel and Flandoli [118]. ∎

D-Bifurcation, but not P-Bifurcation

Baxendale [56] considers the following family of SDE on the two-dimensional torus $M = T^2 = S^1 \times S^1$:

$$dx_t = \sum_{j=1}^{4} f_j^\alpha(x_t) \circ dW_t^j, \qquad (9.2.11)$$

where, with $x = (x_1, x_2) \in M$,

$$f_1^\alpha(x) := \sin x_1 (\cos\alpha \frac{\partial}{\partial x_1} + \sin\alpha \frac{\partial}{\partial x_2}),$$

$$f_2^\alpha(x) := \cos x_1 (\cos\alpha \frac{\partial}{\partial x_1} + \sin\alpha \frac{\partial}{\partial x_2}),$$

$$f_3^\alpha(x) := \sin x_2 (-\sin\alpha \frac{\partial}{\partial x_1} + \cos\alpha \frac{\partial}{\partial x_2}),$$

$$f_4^\alpha(x) := \cos x_2 (-\sin\alpha \frac{\partial}{\partial x_1} + \cos\alpha \frac{\partial}{\partial x_2}),$$

and $\alpha \in [0, \frac{\pi}{2}]$. For all α we have $L = \frac{1}{2}\Delta$, hence all of the one-point motions are Brownian motions, and the Fokker-Planck equation does not depend on α. Its unique solution ρ is the normalized Lebesgue measure on M.

However, the smooth global RDS φ_α generated by (9.2.11) does depend on α. For example, the sum of the two Lyapunov exponents of φ_α under ρ is

$$\lambda_1(\alpha) + \lambda_2(\alpha) = -(\cos\alpha)^2,$$

and the top exponent satisfies $\lambda_1(\alpha) <$ or $=$ or > 0 if and only if $\alpha <$ or $=$ or > 0.845. Hence $\alpha_D = 0.845$ is an abstract bifurcation point. Computer simulations indicate that it is a D-bifurcation point. Thus through this example it is shown that a system can exhibit a D-bifurcation without any phenomenological change in the invariant measure for all values of α.

9.3 Dimension One

We continue the study of the example treated in Subsect. 2.3.7 and Example 4.2.15 from the point of view of stochastic bifurcation. It suffices to treat the prototypical cases $N = 2$ and $N = 3$ (see Arnold and Boxler [12]), which amounts to investigating the effect of white noise added to the bifurcation parameter α in the truncated normal forms $\dot{x}_t = \alpha x_t - x_t^2$ and $\dot{x}_t = \alpha x_t - x_t^3$ of the transcritical and pitchfork bifurcations, respectively. Then we will consider the saddle-node case, $\dot{x}_t = \alpha - x_t^2$, and the effect of real noise.

9.3.1 Transcritical Bifurcation

For $N = 2$ we have
$$dx_t = (\alpha x_t - x_t^2)dt + \sigma x_t \circ dW_t,$$
which is solved by
$$\varphi_\alpha(t,\omega)x = \frac{xe^{\alpha t + \sigma W_t(\omega)}}{1 + x\int_0^t e^{\alpha s + \sigma W_s(\omega)}ds}.$$

The **ergodic invariant measures** are:
 (i) $\mu_\omega^\alpha = \delta_0$ for all α,
 (ii) $\nu_\omega^\alpha = \delta_{\kappa_\alpha(\omega)}$, where
$$\kappa_\alpha(\omega) := \begin{cases} -\left(\int_0^\infty e^{\alpha t + \sigma W_t(\omega)} dt\right)^{-1}, & \alpha < 0, \\ \left(\int_{-\infty}^0 e^{\alpha t + \sigma W_t(\omega)} dt\right)^{-1}, & \alpha > 0. \end{cases}$$

We have $\mathbb{E}\kappa_\alpha = \alpha$. Solving the forward Fokker-Planck equation
$$L^* p_\alpha(x) = -\left(\left(\alpha x + \frac{\sigma^2}{2}x - x^2\right)p_\alpha(x)\right)' + \frac{\sigma^2}{2}(x^2 p_\alpha(x))'' = 0$$
yields
 (i) $p_\alpha = \delta_0$ for all α,
 (ii) for $\alpha > 0$
$$q_\alpha(x) = \begin{cases} N_\alpha x^{\frac{2\alpha}{\sigma^2} - 1} \exp(-\frac{2x}{\sigma^2}), & x > 0, \\ 0, & x \leq 0, \end{cases}$$
where
$$N_\alpha^{-1} = \Gamma\left(\frac{2\alpha}{\sigma^2}\right)\left(\frac{\sigma^2}{2}\right)^{2\alpha/\sigma^2}.$$

We do not find a solution for $\alpha < 0$ corresponding to $\mathbb{E}\nu^\alpha(dx) = q_\alpha(-x)dx$. This is due to the fact that for $\alpha < 0$, κ_α is $\mathcal{F}^+$-measurable (we could, however, find ν^α by solving the backward Fokker-Planck equation).

The family of marginal densities $(q_\alpha)_{\alpha>0}$ of $(\nu^\alpha)_{\alpha>0}$ clearly undergoes a P-bifurcation, as shown in Fig. 9.3, at the parameter value $\alpha_P = \frac{\sigma^2}{2}$: While q_α is decreasing in x with pole at $x = 0$ for $0 < \alpha < \alpha_P$, it is a unimodal function with $q_\alpha(0) = 0$ for $\alpha > \alpha_P$.

A similar phenomenon holds for the marginal densities for $\alpha < 0$ at $-\alpha_P$.

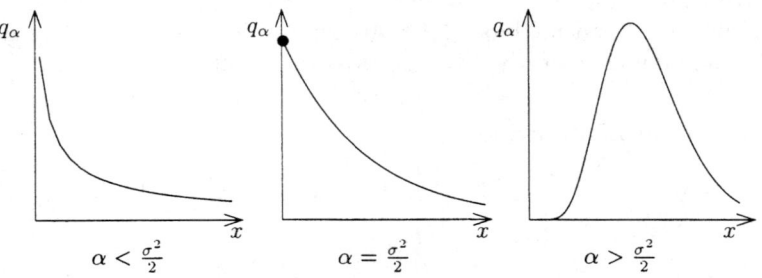

Fig. 9.3. P-bifurcation of q_α at $\alpha_P = \frac{\sigma^2}{2}$

The Lyapunov exponent of the linearized SDE

$$dv_t = (\alpha - 2x_t)v_t dt + \sigma v_t \circ dW_t$$

with respect to the two invariant measures is (see Example 4.2.15):

(i) For $\mu^\alpha = \delta_0$, $\lambda(\mu^\alpha) = \alpha$,
(ii) for $\nu^\alpha = \delta_{\kappa_\alpha}$, $\lambda(\nu^\alpha) = \alpha - 2\mathbb{E}\,\kappa_\alpha = -\alpha$.

We hence have a D-bifurcation of the trivial reference measure δ_0 of the family of local RDS $(\varphi_\alpha)_{\alpha \in \mathbb{R}}$ at $\alpha_D = 0$. The final bifurcation diagram of the stochastic transcritical bifurcation is given in Fig. 9.4.

For each of the invariant measures we can also easily determine the **invariant manifolds**:

(i) For $\mu^\alpha = \delta_0$, the stable manifold for $\alpha < 0$ is the random interval $M_s^\alpha(\omega) = (\kappa_\alpha(\omega), \infty)$ (which is $\mathcal{F}^+$-measurable), the center manifold for $\alpha = 0$ is $M_c^0(\omega) = \mathbb{R}$, and the unstable manifold for $\alpha > 0$ is $M_u^\alpha(\omega) = (-\infty, \kappa_\alpha(\omega))$ (which is $\mathcal{F}^-$-measurable). For $\alpha = 0$, the orbit behavior is very interesting: While for any $x \neq 0$, $\varphi_0(t, \theta_{-t}\omega, x) \to 0$ as $t \to \infty$, which can be obtained e.g. for $x > 0$ from

$$\varphi_0(t, \theta_{-t}\omega, x) \leq \frac{1}{\int_{-t}^0 e^{\sigma W_s(\omega)} ds}$$

and Lemma 2.3.41, we have, again for $x > 0$,

$$\liminf_{t \to \infty} \varphi_0(t, \omega, x) = 0, \quad \limsup_{t \to \infty} \varphi_0(t, \omega, x) = \infty,$$

since the corresponding diffusion process on $(0, \infty)$ is (null) recurrent. For details see Remark 9.3.7.

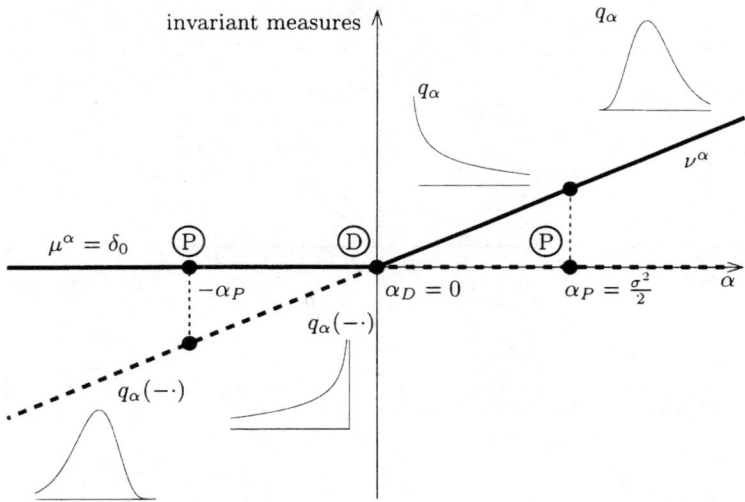

Fig. 9.4. Bifurcation diagram of the transcritical bifurcation

(ii) For the second measure ν^α, the unstable manifold for $\alpha < 0$ is $N_u^\alpha(\omega) = (-\infty, 0)$, and the stable manifold for $\alpha > 0$ is $N_s^\alpha(\omega) = (0, \infty)$.

Why do we have a P-bifurcation of $(q_\alpha)_{\alpha>0}$ at $\alpha_P = \frac{\sigma^2}{2}$? This turns out to be a large deviations phenomenon which is explained in Subsect. 9.5.1: Define the p-th moment Lyapunov exponent of a linear RDS Φ by

$$g(p) := \lim_{t\to\infty} \frac{1}{t} \log \mathbb{E} \, \|\Phi(t,\omega)v\|^p, \quad p \in \mathbb{R}.$$

In our case, using the explicit solution of the linearized SDE at $x = 0$,

$$g_\alpha(p) = \alpha p + \frac{\sigma^2}{2} p^2.$$

It has been shown by Baxendale [60, 63] that the marginal distributions $(q_\alpha)_{\alpha>0}$ of the new branch $(\nu^\alpha)_{\alpha>0}$ of invariant measures undergo a P-bifurcation in a neighborhood of $x = 0$ at that parameter value α_P for which the second zero of $p \mapsto g_\alpha(p)$ is at $p = -\dim \mathbb{R}^1 = -1$, hence at $\alpha_P = \frac{\sigma^2}{2}$ (see Fig. 9.5). The intuitive reason is as follows: Although the trivial solution $x = 0$ becomes unstable at the D-bifurcation point $\alpha_D = 0$, the repulsion from 0 or the attraction by κ_α are too weak for small $\alpha > 0$ to move the "most probable value" of a typical trajectory of the nonlinear system φ_α away from 0. The new branch ν^α is finally "freed" from the old reference branch δ_0 for $\alpha > \alpha_P$.

480   Chapter 9. Bifurcation Theory

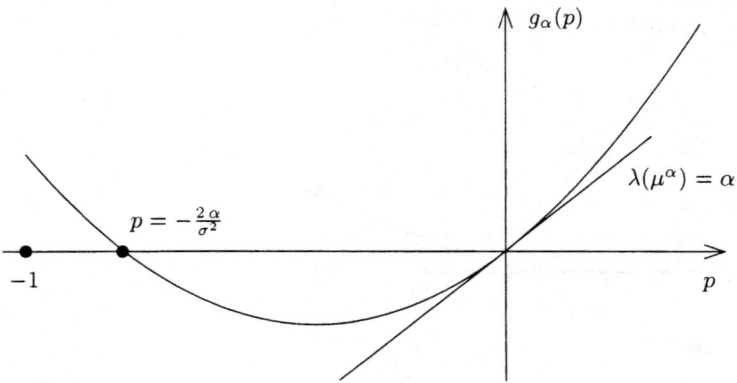

Fig. 9.5. The p-th moment Lyapunov exponent

9.3.2 Pitchfork Bifurcation

For $N = 3$ we have
$$dx_t = (\alpha x_t - x_t^3)dt + \sigma x_t \circ dW_t,$$
which is solved by
$$\varphi_\alpha(t,\omega)x = \frac{xe^{\alpha t+\sigma W_t(\omega)}}{\left(1 + 2x^2 \int_0^t e^{2(\alpha s+\sigma W_s(\omega))}ds\right)^{1/2}}. \qquad (9.3.1)$$

The **ergodic invariant measures** are:
 (i) For $\alpha \leq 0$, the only invariant measure is $\mu_\omega^\alpha = \delta_0$.
 (ii) For $\alpha > 0$, we have the three invariant forward Markov measures $\mu_\omega^\alpha = \delta_0$ and $\nu_{\pm,\omega}^\alpha = \delta_{\pm\kappa_\alpha(\omega)}$, where
$$\kappa_\alpha(\omega) := \left(2 \int_{-\infty}^0 e^{2\alpha t + 2\sigma W_t(\omega)}dt\right)^{-\frac{1}{2}}.$$

We have $\mathbb{E}\kappa_\alpha^2 = \alpha$. Solving the forward Fokker-Planck equation
$$L^* p_\alpha(x) = -\left((\alpha x + \frac{\sigma^2}{2}x - x^3)p_\alpha(x)\right)' + \frac{\sigma^2}{2}(x^2 p_\alpha(x))'' = 0$$
yields
 (i) $p_\alpha = \delta_0$ for all α,
 (ii) for $\alpha > 0$

$$q_\alpha^+(x) = \begin{cases} N_\alpha x^{\frac{2\alpha}{\sigma^2}-1} \exp(-\frac{x^2}{\sigma^2}), & x > 0, \\ 0, & x \leq 0, \end{cases} \quad \text{and} \quad q_\alpha^-(x) = q_\alpha^+(-x),$$

where $N_\alpha^{-1} = \Gamma\left(\frac{\alpha}{\sigma^2}\right) \sigma^{-2\alpha/\sigma^2}$.

Naturally the invariant measures $\nu_{\pm,\omega}^\alpha = \delta_{\pm\kappa_\alpha(\omega)}$ are those corresponding to the stationary measures $q_\alpha^\pm$. Hence all invariant measures are Markov measures.

The two families of densities $(q_\alpha^\pm)_{\alpha>0}$ clearly undergo a P-bifurcation at the parameter value $\alpha_P = \frac{\sigma^2}{2}$ – which is the same value as for the transcritical case, since the SDE linearized at $x = 0$ is the same in both cases.

The Lyapunov exponent of the linearized SDE

$$dv_t = (\alpha - 3x_t^2)v_t dt + \sigma v_t \circ dW_t$$

with respect to the three measures is (see Example 4.2.15):

(i) For $\mu^\alpha = \delta_0$, $\lambda(\mu^\alpha) = \alpha$,
(ii) for $\nu_{\pm,\omega}^\alpha = \delta_{\pm\kappa_\alpha(\omega)}$, $\lambda(\nu^\alpha) = \alpha - 3\mathbb{E}\,\kappa_\alpha^2 = -2\alpha$.

Hence, we have a D-bifurcation of the trivial reference measure δ_0 at $\alpha_D = 0$ and a P-bifurcation of $\nu_\pm^\alpha$ at $\alpha_P = \frac{\sigma^2}{2}$. The final bifurcation diagram of the stochastic pitchfork bifurcation is given in Fig. 9.6.

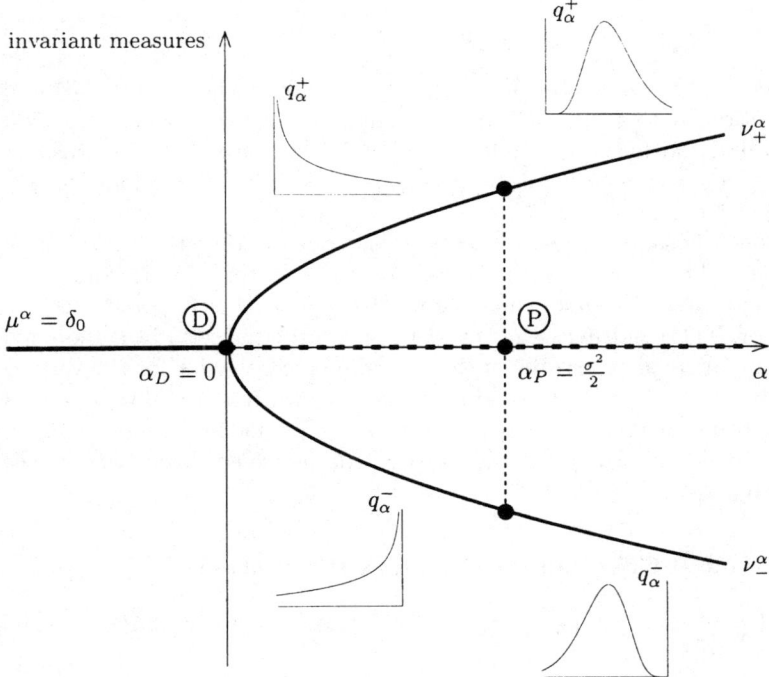

Fig. 9.6. Bifurcation diagram of the pitchfork bifurcation

The **invariant manifolds** are as follows:

(i) For $\mu^\alpha = \delta_0$, the stable manifold for $\alpha < 0$ is $M_s^\alpha(\omega) = \mathbb{R}$, the center manifold for $\alpha = 0$ is $M_c^0(\omega) = \mathbb{R}$ (for the orbit behavior see the transcritical case and Remark 9.3.7) and the unstable manifold for $\alpha > 0$ is the random interval $M_u^\alpha(\omega) = (-\kappa_\alpha(\omega), \kappa_\alpha(\omega))$, which is $\mathcal{F}^-$-measurable.

(ii) For the two other measures $\nu_\pm^\alpha$ which exist for $\alpha > 0$, the stable manifolds are $N_s^{\alpha,+}(\omega) = (0, \infty)$ and $N_s^{\alpha,-}(\omega) = (-\infty, 0)$, respectively.

9.3.3 Saddle-Node Case

The perturbed version of $\dot{x}_t = \alpha - x_t^2$ is

$$dx_t = (\alpha - x_t^2)dt + \sigma \circ dW_t, \quad \alpha \in \mathbb{R},\ \sigma > 0. \tag{9.3.2}$$

We put formally $x = \frac{\dot{u}}{u}$ and use $\dot{x} = (\alpha + \sigma \dot{W}_t) - x^2$ to arive at the undamped stochastic linear oscillator $\ddot{u} - (\alpha + \sigma \dot{W}_t)u = 0$. For $z = \binom{z_1}{z_2} = \binom{u}{\dot{u}}$, we obtain the SDE

$$dz_t = \begin{pmatrix} 0 & 1 \\ \alpha & 0 \end{pmatrix} z_t dt + \begin{pmatrix} 0 & 0 \\ \sigma & 0 \end{pmatrix} z_t \circ dW_t. \tag{9.3.3}$$

Explosion of (9.3.2) is equivalent to $u = z_1$ becoming 0, or the projection of (9.3.3) onto S^1,

$$d\gamma_t = (-\sin^2 \gamma_t + \alpha \cos^2 \gamma_t)dt + \sigma \cos^2 \gamma_t \circ dW_t, \tag{9.3.4}$$

crossing the angles $\gamma = \arctan \frac{\dot{u}}{u} = \pm \frac{\pi}{2}$. Linear SDE of the type (9.3.3) and their projections onto S^1 have been extensively investigated, see e. g. Arnold, Oeljeklaus and Pardoux [32]. It follows that in our case (9.3.4) is positive recurrent on S^1, equivalently all solutions $\varphi_\alpha(t, \omega)x$ of (9.3.2) explode $\mathbb{P}$-a. s. , i. e. $E_\alpha(\omega) = \emptyset$ for all $\alpha \in \mathbb{R}$.

Why does the saddle-node bifurcation "disappear" after perturbation by white noise? The general intuitive reason is that even if σ is small, the perturbed parameter $\alpha + \sigma \dot{W}_t$ can become "big", hence has a global effect on the behavior. The resulting behavior of the perturbed system at some fixed value α is a nontrivial "mixture" of the deterministic behavior for the various frozen parameter values. In the saddle-node case, white noise always pushes the system to negative parameter values where it explodes.

We can avoid this by using real noise properly scaled down with α, see Subsect. 9.3.5.

9.3.4 A General Criterion for Pitchfork Bifurcation

As we will now use the theory of random attractors, we first give a brief introduction to this subject.

The investigation of random sets which attract many other (deterministic or random) sets is an important new line of research and part of the topological dynamics of RDS. As in the deterministic case, random attractors encapsulate information about the long term behavior of the system and the study of their "qualitative change" is part of bifurcation theory.

Random attractors have been introduced and studied simultaneously by Crauel and Flandoli [117, 119] and Schmalfuß [307, 309, 310]. For further results see also Arnold and Schmalfuß [36], Crauel [115], Crauel, Debussche and Flandoli [116], Debussche [124], and Flandoli and Schmalfuß [151]. A useful synopsis and applications to the bifurcation theory of the noisy Duffing-van der Pol oscillator (see Sect. 9.4) are given by Schenk-Hoppé ([304: Chap. 2] and [306]).

We follow the approach of Schmalfuß and define a random attractor only relative to a "universe" of sets which are attracted. This more flexible notion allows us to capture local phenomena by separating disjoint invariant sets.

The following definitions are not the most general, but are adapted to our purposes.

9.3.1 Definition (Universe, Absorbing Set, Attractor, Domain of Attraction). Let φ be a continuous two-sided local RDS on a Polish space X.

1. A *universe* $\mathcal{D}$ is a collection of families $(D(\omega))_{\omega \in \Omega}$ of non-empty subsets of X which is closed with respect to set inclusion (i.e. if $D_1 \in \mathcal{D}$ and $D_2(\omega) \subset D_1(\omega)$ for all ω then $D_2 \in \mathcal{D}$).
2. A set $B \in \mathcal{D}$ is called *absorbing* in $\mathcal{D}$ for the RDS φ, if B absorbs all sets in $\mathcal{D}$, i.e. for any $D \in \mathcal{D}$ there exists a time $t_0 = t_0(\omega, D)$ such that
$$\varphi(t, \theta_{-t}\omega)D(\theta_{-t}\omega) \subset B(\omega) \quad \text{for all } t \geq t_0.$$
3. A (global) *(random) attractor* of φ in $\mathcal{D}$ is a random compact set $A \in \mathcal{D}$ with the following properties:
    a) A is strictly invariant, $\varphi(t,\omega)A(\omega) = A(\theta_t\omega)$ for $t \in \mathbb{T}$;
    b) A attracts all sets in $\mathcal{D}$, i.e. for all $D \in \mathcal{D}$,
$$\lim_{t\to\infty} d(\varphi(t, \theta_{-t}\omega)D(\theta_{-t}\omega)|A(\omega)) = 0,$$

where $d(A|B) := \sup_{x \in A} d(x, B)$ is the Hausdorff semi-metric. The universe $\mathcal{D}$ is called a *domain of attraction* of A.

∎

We stress that the definitions of absorbing and attracting are based on moving points from time $-t$ to time 0 (and not from 0 to t). This enables us to study the asymptotic behavior as $t \to \infty$ in the fixed fiber at time 0. Note that 3(b) implies

$$\lim_{t\to\infty} d(\varphi(t,\cdot)D(\cdot)|A(\theta_t\cdot)) = 0 \quad \text{in probability.}$$

In general this cannot be improved to ℙ-a. s. convergence, see Remark 9.3.7.

Here are some basic results of Schmalfuß [308] and Schenk-Hoppé [304, 306].

9.3.2 Proposition (Existence and Uniqueness of Random Attractor). *Let φ be a continuous local two-sided RDS on the Polish space X, and let $\mathcal{D}$ be a universe. If there exists a random compact absorbing set $B \in \mathcal{D}$ which is strictly forward invariant, i. e. which satisfies $\varphi(t,\omega)B(\omega) \subset B(\theta_t\omega)$ for $t > 0$, then there exists a random attractor A with domain of attraction $\mathcal{D}$, given by*

$$A(\omega) := \cap_{t\geq 0}\varphi(t, \theta_{-t}\omega)B(\theta_{-t}\omega) = \cap_{n\geq 0}\varphi(n, \theta_{-n}\omega)B(\theta_{-n}\omega).$$

Furthermore:

(i) $A(\omega) \subset E(\omega)$, E being the set of never exploding initial values of φ, and $D(\omega) \subset E^+(\omega) := \cap_{t\geq 0} D(t,\omega)$ for any $D \in \mathcal{D}$, where $D(t,\omega)$ denotes the domain of $\varphi(t,\omega)$.

(ii) If $C \in \mathcal{D}$ is strictly invariant, then $C(\omega) \subset A(\omega)$. In particular, the attractor is unique in $\mathcal{D}$ and if $E \in \mathcal{D}$, then $A(\omega) = E(\omega)$ for all $\omega \in \Omega$.

(iii) If $B(\omega)$ is connected for any ω, then so is $A(\omega)$.

(iv) If μ is φ-invariant and if for any $\varepsilon > 0$ there exists a $D_\varepsilon \in \mathcal{D}$ for which $\mu(D_\varepsilon) \geq 1 - \varepsilon$, then $\mu(A) = 1$.

(v) If $\varphi(t,\theta_{-t}\cdot)$ is $\mathcal{F}^-$-measurable for all $t \geq 0$ and B is $\mathcal{F}^-$-measurable, then A is $\mathcal{F}^-$-measurable too.

9.3.3 Theorem (Sufficient Condition for Stochastic Pitchfork Bifurcation). *Consider the scalar SDE*

$$dx_t = (\alpha x_t - x_t^3 + g(x_t))dt + \sigma x_t \circ dW_t, \quad \sigma \neq 0, \tag{9.3.5}$$

with parameter $\alpha \in \mathbb{R}$, where g is smooth and satisfies

$$C_1 x^4 \leq xg(x) \leq C_2 x^4 \quad \text{for some } C_1 \leq C_2 < 1 \text{ and all } x \in \mathbb{R}. \tag{9.3.6}$$

Then (9.3.5) generates for any α a smooth local RDS φ_α which is strictly forward complete. Moreover, the family $(\varphi_\alpha)_{\alpha\in\mathbb{R}}$ undergoes a stochastic pitchfork bifurcation at $\alpha_D = 0$. More precisely:

(i) The family $\mu^\alpha = \delta_0$ of invariant measures has Lyapunov exponent $\lambda = \alpha$, hence loses its stability at $\alpha_D = 0$. For $\alpha \leq 0$, δ_0 is the unique invariant measure, and $A_\alpha(\omega) = \{0\}$ is the random attractor of φ_α on $\mathbb{R}$ in the universe of all families $(C(\omega))_{\omega \in \Omega}$ of subsets of $\mathbb{R}$.

(ii) For $\alpha > 0$, there are exactly two additional invariant measures for φ_α: $\nu^\alpha_{+,\omega} = \delta_{\kappa^+_\alpha(\omega)}$ with $\kappa^+_\alpha > 0$, and $\nu^\alpha_{-,\omega} = \delta_{\kappa^-_\alpha(\omega)}$ with $\kappa^-_\alpha < 0$. Both measures are $\mathcal{F}^-$-measurable and have negative Lyapunov exponents. Further, $A^+_\alpha(\omega) := \{\kappa^+_\alpha(\omega)\}$ is the random attractor of φ_α on the state space $(0,\infty)$ in the universe of all families $(C(\omega))_{\omega \in \Omega}$ of subsets of $(0,\infty)$ which are tempered from below (in particular, containing all deterministic sets bounded away from 0).

Similarly, $A_\alpha^-(\omega) := \{\kappa_\alpha^-(\omega)\}$ is the random attractor of φ_α on the state space $(-\infty, 0)$ in the universe of all families $(C(\omega))_{\omega \in \Omega}$ of subsets of $(-\infty, 0)$ which are tempered from below.

We call a family $(C(\omega))_{\omega \in \Omega}$ of subsets of $(0, \infty)$ *tempered from below* if there exists a random variable $r > 0$ which is tempered from below and $C(\omega) \subset [r(\omega), \infty)$ for all ω, similarly for $(-\infty, 0)$. As before, a family $(C(\omega))_{\omega \in \Omega}$ of subsets of $\mathbb{R}^d$ is called tempered from above if it is a subset of a random ball $B_{r(\omega)} = \{x \in \mathbb{R}^d : \|x\| \leq r(\omega)\}$ with a radius r tempered from above (see Subsect. 4.1.1).

9.3.4 Remark. (i) Condition (9.3.6) means that g represents the higher order terms. In particular, it implies that $g(0) = g'(0) = g''(0) = 0$.

(ii) In contrast to the deterministic case, where one only needs local conditions on g in a neighborhood of 0, we have to impose the global condition (9.3.6). ∎

We prove this theorem in two steps:
(i) We first prove the existence of the random attractors.
(ii) Then we show that they are one-point sets.

9.3.5 Lemma. *Consider the scalar SDE*

$$dx_t = (ax_t + f(x_t))dt + \sigma x_t \circ dW_t, \qquad (9.3.7)$$

where f is smooth and satisfies

$$xf(x) \leq bx^2 + c \quad \text{for all } x \in \mathbb{R}, \quad \text{where } a + b < 0, \quad c \geq 0. \qquad (9.3.8)$$

Then:

(i) (9.3.7) generates a smooth local RDS φ which is strictly forward complete.

(ii) φ possesses an attractor $A = [a^-, a^+]$ in the universe of random compact subsets of $\mathbb{R}$ tempered from above.

Proof. (i) Uniqueness of the solution follows from the local Lipschitz property of the coefficients. Forward completeness, hence (in $\mathbb{R}$) strict forward completeness follows from

$$x(ax + \frac{\sigma^2}{2}x + f(x)) + \sigma^2 x^2 \leq K^2(1 + |x|^2),$$

see e. g. Gihman and Skorohod [159: Chap. 2, § 6, Remark 3].

(ii) We construct an absorbing set as follows: Observe that x_t^2 satisfies

$$dx_t^2 = 2(ax_t^2 + x_t f(x_t))dt + 2\sigma x_t^2 \circ dW_t.$$

Since $ax^2 + xf(x) \leq (a+b)x^2 + c$, and by the comparison theorem (Ikeda and Watanabe [178: Chap. 6]) we have

$$|\varphi(t,\omega,x)|^2 \leq \psi(t,\omega,|x|^2),$$

where ψ is the RDS generated by

$$dz_t = 2((a+b)z_t + c)dt + 2\sigma z_t \circ dW_t.$$

The last SDE is affine with stable linear part, hence its unique invariant measure is the Dirac measure supported by

$$\kappa = 2c \int_{-\infty}^{0} e^{-2(a+b)t - 2\sigma W_t} dt,$$

and κ attracts all points with exponential speed (see Subsect. 5.6.3).

It follows that in case $c > 0$ (which is without loss of generality)

$$B := [-2\kappa, 2\kappa]$$

is an absorbing set in the universe of random compact sets which are tempered from above. □

Proof of Theorem 9.3.3. We basically follow [36].

(i) We use Lemma 9.3.5. For $\alpha < 0$, put $a = \alpha$ and $f(x) = -x^3 + g(x)$. Then $xf(x) = -x^4 + xg(x) \leq (-1 + C_2)x^4 \leq 0$, so that condition (9.3.8) holds with $b = c = 0$, but $a + b < 0$. For $\alpha = 0$, choose any $a < 0$ and put $f(x) = -ax - x^3 + g(x)$. Since $xf(x) \leq -ax^2 + (-1 + C_2)x^4$, there exists a constant $c = c(a, C_2)$ such that $xf(x) \leq c$. For $\alpha > 0$ choose $a = -\alpha$ and $f(x) = 2\alpha x - x^3 + g(x) \leq c$, hence $b = 0$, but $a + b < 0$.

The result is that for all $\alpha \in \mathbb{R}$, (9.3.5) is strictly forward complete and has an attractor in the universe of random compact sets which are tempered from above.

We will now show that A_α actually attracts any family $C(\omega)$. Assume $x > 0$ and use the estimate $-x^3 + g(x) \leq -(1 - C_2)x^3$ to compare (9.3.5) with the explicitly solvable SDE

$$dy_t = (\alpha y_t - (1 - C_2)y_t^3)dt + \sigma y_t \circ dW_t$$

to obtain (see (9.3.1))

$$|\varphi_\alpha(t,\omega,x)|^2 \leq \frac{x^2 e^{2\alpha t + 2\sigma W_t(\omega)}}{1 + 2(1 - C_2)x^2 \int_0^t e^{2\alpha s + 2\sigma W_s(\omega)} ds}$$

$$\leq \frac{e^{2\alpha t + 2\sigma W_t(\omega)}}{2(1 - C_2) \int_0^t e^{2\alpha s + 2\sigma W_s(\omega)} ds}.$$

Hence

$$|\varphi_\alpha(t, \theta_{-t}\omega, x)|^2 \leq \frac{1}{2(1-C_2) \int_{-t}^{0} e^{2\alpha s + 2\sigma W_s(\omega)} ds} =: r_\alpha(t, \omega). \qquad (9.3.9)$$

It follows from the fact that the right-hand side of (9.3.9) is independent of x that φ_α maps any set after positive time $t > 0$ into a subset of the compact interval $[-r_\alpha(t, \omega), r_\alpha(t, \omega)]$.

For $\alpha < 0$, $r_\alpha(t, \omega) \to 0$ exponentially fast as $t \to \infty$, thus any set is shrunk to $A_\alpha = \{0\}$ exponentially fast.

For $\alpha = 0$, $r_0(t, \omega) \to 0$ as $t \to \infty$ by Lemma 2.3.41.

(ii) For $\alpha > 0$,

$$r_\alpha(t, \omega) \downarrow b_2(\omega)^2 := \frac{1}{2(1-C_2) \int_{-\infty}^{0} e^{2\alpha t + 2\sigma W_t(\omega)} dt},$$

so that, by the comparison theorem,

$$A_\alpha(\omega) = [a_\alpha^-(\omega), a_\alpha^+(\omega)] \subset [-b_2(\omega), b_2(\omega)].$$

Furthermore, $[-b_2, b_2]$ is strictly forward invariant under φ_α since $\varphi_\alpha(t, \omega)b_2(\omega) \leq \psi_\alpha(t, \omega)b_2(\omega) = b_2(\theta_t\omega)$, where ψ_α is the dominating RDS.

Let now $\alpha > 0$ and $x > 0$. We have just proved that $(0, b_2]$ is strictly forward invariant.

On the other hand, applying the transformation $y = 1/x^2$, (9.3.5) is transformed into

$$dy_t = (-2\alpha y_t + 2 + h(y_t))dt - 2\sigma y_t \circ dW_t, \qquad (9.3.10)$$

where $h(y) = -2g(x)/x^3 = -2y^{3/2}g(y^{-1/2}) \leq -2C_1$, where we have used the left-hand side of condition (9.3.6). Also, $2(1 - C_1) \geq 2(1 - C_2) > 0$. Comparing (9.3.10) with the corresponding stable affine SDE

$$dz_t = (-2\alpha z_t + 2(1 - C_1))dt - 2\sigma z_t \circ dW_t,$$

we conclude that the interval

$$(0, \frac{1}{b_1(\omega)^2}] := \left(0, 2(1 - C_1) \int_{-\infty}^{0} e^{2\alpha t + 2\sigma W_t(\omega)} dt\right]$$

is strictly forward invariant for (9.3.10), hence

$$[b_1(\omega), \infty) := \left[\frac{1}{(2(1 - C_1) \int_{-\infty}^{0} e^{2\alpha t + 2\sigma W_t(\omega)} dt)^{1/2}}, \infty\right)$$

is strictly forward invariant for φ_α. Finally, the intersection $[b_1, b_2]$ of the two intervals is non-void since $C_1 \leq C_2$, strictly forward invariant and obviously $\mathcal{F}^-$-measurable.

Again by the comparison theorem, $B = [\frac{1}{2}b_1, 2b_2]$ is an absorbing set for all families tempered from below. Hence there exists an attractor $A_\alpha^+ = [a_1, a_2]$ in this universe which is $\mathcal{F}^-$-measurable.

Similarly for $x < 0$.

The following result of Crauel and Flandoli [118] implies that the above attractors A_0 and $A_\alpha^\pm$ ($\alpha > 0$) are one-point sets. Since by Proposition 9.3.2 the attractor supports all invariant measures, the Dirac measure on the attractor is the unique invariant measure. It is a Markov measure. For $\alpha > 0$, the Lyapunov exponent of the two non-trivial measures is negative by Theorem 9.2.4. This concludes the proof of Theorem 9.3.3. □

9.3.6 Proposition. *Suppose φ is a continuous local RDS generated by an SDE on an interval $I \subset \mathbb{R}$ such that the associated Markov semigroup has a unique stationary probability measure. Suppose further that C is a strictly invariant random compact set which is $\mathcal{F}^-$-measurable. Then C is a one-point set.*

Proof. Since C is a random compact set which is $\mathcal{F}^-$-measurable, $x^+(\omega) := \max C(\omega)$ and $x^-(\omega) := \min C(\omega) \le x^+(\omega)$ are $\mathcal{F}^-$-measurable (choose a countable dense set of $\mathcal{F}^-$-measurable selections, see Proposition 1.6.3). Continuity of $t \mapsto \varphi(t,\omega)$ and injectivity of $\varphi(t,\omega)$ for all t and ω imply that $\varphi(t,\omega)$ is order preserving. Thus, by the strict invariance of C,

$$\varphi(t,\omega)x^+(\omega) = x^+(\theta_t\omega) \quad \text{and} \quad \varphi(t,\omega)x^-(\omega) = x^-(\theta_t\omega)$$

for all $t \in \mathbb{R}$. Hence the two invariant measures $\mu_\omega^\pm := \delta_{x^\pm(\omega)}$ are Markov measures. Their expectations give rise to two stationary measures which, by assumption, have to be identical. This would be contradicted by $\mathbb{P}\{\omega : x^-(\omega) < x^+(\omega)\} > 0$. Hence $x := x^- = x^+$, and $C(\omega) = \{x(\omega)\}$. □

9.3.7 Remark. Caution: For $\alpha = 0$ (and $g = 0$, say), the attractor $A_0 = \{0\}$ is not attracting forward in time. In fact, 0 and ∞ are both natural (repelling) boundaries for the diffusion process on $(0,\infty)$ since $\int_0^1 R(x)dx = \infty$ and $\int_1^\infty R(x)dx = \infty$, where

$$R(x) = \exp(-\int_1^x \frac{a(y)}{b(y)}dy) = \frac{C}{x}\exp(x^2/\sigma^2)$$

and $a(x) = \frac{\sigma^2}{2}x - x^3$ and $b(x) = \frac{\sigma^2}{2}x^2$ are derived from the Itô form of (9.3.5), see Gihman and Skorohod [159: Chap. 5, § 21, Theorem 1]. Hence the diffusion process is (null) recurrent in $(0,\infty)$, and for any $x > 0$, $\mathbb{P}$-a.s.

$$\liminf_{t\to\infty} \varphi_0(t,\omega,x) = 0, \quad \limsup_{t\to\infty} \varphi_0(t,\omega,x) = \infty.$$

However, $\lim_{t\to\infty} \varphi(t,\theta_{-t}\omega,x) = 0$, which implies that $\varphi(t,\omega,x) \to 0$ in probability. This supports the sensibleness of our definition of an attractor. ∎

9.3.8 Exercise (Criterion for Stochastic Transcritical Bifurcation). By similar arguments, we can establish a transcritical bifurcation at $\alpha = 0$ for the family

$$dx_t = (\alpha x_t - x_t^2 + g(x_t))dt + \sigma x_t \circ dW_t,$$

where g is smooth and satisfies

$$C_1 x^2 \leq g(x) \leq C_2 x^2, \quad x \in \mathbb{R}, \quad C_1 \leq C_2 < 1.$$

For $\alpha > 0$, we find a one-point attractor in $(0, \infty)$, and for $\alpha < 0$ we find a one-point repellor (an attractor under inverse time) in $(-\infty, 0)$. The details are left to the reader as an exercise. ∎

9.3.5 Real Noise Case

Transcritical and Pitchfork Bifurcation

If the parameters α and β in $\dot{x}_t = \alpha x_t - \beta x_t^N$, $N \geq 2$, are perturbed by real noise, we obtain the RDE

$$\dot{x}_t = (\alpha + \xi(\theta_t \omega))x_t - \eta(\theta_t \omega) x_t^N. \tag{9.3.11}$$

We assume that $\xi, \eta \in L^1$, $\mathbb{E}\xi = 0$, and $\mathbb{P}(\eta \neq 0) > 0$.

The RDE (9.3.11) can be explicitly solved by the same technique as its white noise version (see Subsect. 2.3.7), and the local C^∞ RDS φ_α it generates is

$$x \mapsto \varphi_\alpha(t,\omega)x = \frac{xe^{\alpha t + S_t(\omega)}}{\left(1 + (N-1)x^{N-1}\int_0^t \eta(\theta_s\omega)e^{(N-1)(\alpha s + S_s(\omega))}ds\right)^{1/(N-1)}}, \tag{9.3.12}$$

where we use the abbreviation

$$S_t(\omega) := \int_0^t \xi(\theta_s \omega)ds.$$

We can read off from the expression (9.3.12) the explosion time, the domain and the range of $\varphi_\alpha(t,\omega)$, the set E_α, and the invariant measures. We can also calculate their Lyapunov exponents and finally determine the bifurcation behavior of the family $(\varphi_\alpha)_{\alpha \in \mathbb{R}}$.

For a thorough study of the transcritical case $N = 2$ and pitchfork case $N = 3$ see Xu [347].

If we confine ourselves here to the case $\eta \equiv 1$, hence to the RDE

$$\dot{x}_t = (\alpha + \xi(\theta_t\omega))x_t - x_t^N, \quad \mathbb{E}\xi = 0,$$

we obtain exactly the same results as for the SDE (2.3.13), with W_t replaced with S_t.

The corresponding result for $\alpha = 0$ is not so easy to obtain, but follows from the following lemma which also is of independent interest.

9.3.9 Lemma. *Let $\xi \in L^1(\mathbb{P})$ with $\mathbb{E}\xi = 0$. Then*

$$\int_0^\infty e^{S_t(\omega)}dt = \infty \quad \mathbb{P}\text{-a.s.,} \quad \text{where } S_t(\omega) := \int_0^t \xi(\theta_s\omega)ds.$$

Stochastic Saddle-Node Bifurcation

The saddle-node scenario of $\dot{x}_t = \alpha - x_t^2$ disappears if α is perturbed by white noise, see Subsect. 9.3.2.

If, however, we consider

$$\dot{x}_t = \alpha(1 + \xi(\theta_t\omega)) - x_t^2, \qquad (9.3.13)$$

where $\xi(\omega) \in [a, b]$, $a > -1$, the situation is different: Putting again $x = \frac{\dot{u}}{u}$, we obtain the undamped random linear oscillator $\ddot{u} - \alpha(1 + \xi(\theta_t\omega))u = 0$, whose first order form is

$$\dot{z}_t = \begin{pmatrix} 0 & 1 \\ \alpha(1 + \xi(\theta_t\omega)) & 0 \end{pmatrix} z_t, \quad z = \begin{pmatrix} u \\ \dot{u} \end{pmatrix}, \qquad (9.3.14)$$

with angular part

$$\dot{\gamma}_t = -\sin^2 \gamma_t + \alpha(1 + \xi(\theta_t\omega))\cos^2 \gamma_t. \qquad (9.3.15)$$

Explosion of (9.3.13) is equivalent to γ crossing the angles $\pm\frac{\pi}{2}$. It follows that $E_\alpha = \emptyset$ for $\alpha < 0$ and that $E_0 = \{0\}$.

If we assume that $\xi = f(\eta)$, where η is an elliptic diffusion process on a compact manifold and f is smooth and not constant, we can use the results of Arnold, Kliemann and Oeljeklaus [27: Example 6.1] saying that (9.3.14) has a simple spectrum $\lambda_1 > 0 > \lambda_2 = -\lambda_1$. Then

$$E_\alpha = [d_\alpha^-, d_\alpha^+] \quad \text{for } \alpha > 0,$$

where

$$d_\alpha^\pm = \tan \gamma_\alpha^\pm, \quad -\frac{\pi}{2} < \gamma_\alpha^-(\omega) < 0 < \gamma_\alpha^+(\omega) < \frac{\pi}{2},$$

and $\gamma_\alpha^\pm$ are the angles of the two different Oseledets spaces of (9.3.14). The two ergodic invariant measures of (9.3.13) for $\alpha > 0$ are

$$\mu_{\alpha,\omega}^\pm = \delta_{d_\alpha^\pm(\omega)} \quad \text{with } \lambda(\mu_\alpha^\pm) = -2\mathbb{E}\tan\gamma_\alpha^\pm.$$

Since $d_\alpha^\pm \to 0$ as $\alpha \downarrow 0$ we can say that (9.3.13) exhibits a stochastic saddle-node bifurcation with the familiar bifurcation diagram.

9.3.6 Discrete Time

In discrete time $\mathbb{T} = \mathbb{Z}^{(+)}$, the first task should be to study the effect of noise on the bifurcation scenarios of the families of difference equations $x_{n+1} = \varphi_\alpha(x_n)$, where $\varphi_\alpha(x) = x + \alpha x - x^N$ (in particular, $N = 2$ (transcritical), $N = 3$ (pitchfork)), $\varphi_\alpha(x) = -x - \alpha x + x^3$ (period-doubling), and $\varphi_\alpha(x) = x + \alpha - x^2$ (saddle-node). Obvious choices would be additive noise or parametric noise in the sense that the deterministic bifurcation parameter α is replaced by $\alpha + \sigma\xi_n$, (ξ_n) a zero mean stationary sequence. Unfortunately, there is no explicit expression for the nth iterate of such a mapping here.

There is an abundance of case studies on the effect of parametric noise on the logistic map, i. e. of the random difference equation

$$x_{n+1} = r_n x_n (1 - x_n),$$

where (r_n) is a stationary sequence, of which we only quote the following: In a survey by Lücke [245: Chap. 6], the case $r_n = r(1 + \sigma \xi_n)$ with small σ is studied around $r = 1$ (deterministic transcritical bifurcation of a new fixed point $x_0 = \frac{r-1}{r}$) and $r = 3$ (period-doubling bifurcation of a limit cycle from the fixed point x_0). Lücke gives (physicist's style) expansions of all interesting quantities in powers of the noise intensity σ. Additive noise on the logistic map is also discussed.

Hess, Kiwi, Markus and Rössler [170] study the case where r_n is a two-state Markov chain or a periodic function with states A and B, and explore the incredibly complex behavior in the A-B plane. See also Lasota and Mackey [226: Chap. 10] and the references therein.

9.4 Dimension Two: The Noisy Duffing-van der Pol Oscillator

9.4.1 Introduction. Completeness. Linearization

The deterministic nonlinear *Duffing-van der Pol equation*

$$\ddot{y} = \alpha y + \beta \dot{y} + \gamma y^3 + \delta y^2 \dot{y} + \varepsilon \dot{y}^3, \quad \alpha, \beta, \gamma, \delta, \varepsilon \in \mathbb{R}, \tag{9.4.1}$$

has become a paradigm for mathematicians, physicists and engineers. There are numerous physical and engineering problems whose dynamics are described by (9.4.1) for some parameter values (some of these problems have been listed in [37]).

In particular, for $\gamma = \varepsilon = 0$ we obtain the *van der Pol equation* and for $\delta = \varepsilon = 0$ we obtain the *Duffing equation*.

The bifurcation behavior of this equation with parameter $(\alpha, \beta) \in \mathbb{R}^2$, which exhibits pitchfork, Hopf and global bifurcations, was investigated by Holmes and Rand [173].

By putting for simplicity $\gamma = \delta = -1$ and $\varepsilon = 0$, i. e. by considering

$$\ddot{y} = \alpha y + \beta \dot{y} - y^3 - y^2 \dot{y}, \quad \alpha, \beta \in \mathbb{R}, \tag{9.4.2}$$

and perturbing the remaining parameters and the right-hand side of (9.4.2) by real or white noise, we arrive at the noisy, more specifically, the *random* (for real noise), or *stochastic* (for white noise) *Duffing-van der Pol equation*, respectively,

$$\ddot{y} = (\alpha + \sigma_1 \xi_1(t))y + (\beta + \sigma_2 \xi_2(t))\dot{y} - y^3 - y^2\dot{y} + \sigma_3 \xi_3(t), \tag{9.4.3}$$

where σ_1, σ_2, and σ_3 are intensity parameters, $\xi(t) := (\xi_1(t), \xi_2(t), \xi_3(t))$ is a stationary stochastic process or white noise, and $\alpha, \beta \in \mathbb{R}$ are bifurcation parameters.

As a first order system for $x = \binom{x_1}{x_2} = \binom{y}{\dot{y}}$, (9.4.3) takes the form

$$\begin{aligned}\dot{x}_1 &= x_2, \\ \dot{x}_2 &= (\alpha + \sigma_1 \xi_1(t))x_1 + (\beta + \sigma_2 \xi_2(t))x_2 - x_1^3 - x_1^2 x_2 + \sigma_3 \xi_3(t).\end{aligned} \tag{9.4.4}$$

The bifurcation behavior of this and related equations has received considerable attention since the early 1980's. Of the earlier work on P-bifurcation we mention just Ebeling, Herzel, Richert and Schimansky-Geier [134] (for the case $\sigma_1 = \sigma_2 = 0$ and white noise) and the references therein, and for more recent reviews Arnold, Sri Namachchivaya and Schenk-Hoppé [37], and Schenk-Hoppé [304, 306].

(Strict) Completeness of the Noisy Duffing-van der Pol Equation

We first will study the possible explosive behavior of the local RDS φ generated by (9.4.4) and recall the notion of (strict) (forward/backward) completeness from Definition 2.3.37. We take the following results from Schenk-Hoppé [303, 304].

9.4.1 Proposition (Random Duffing-van der Pol Equation).
Suppose the stationary stochastic process $\xi(t) = (\xi_1(t), \xi_2(t), \xi_3(t))$ has locally integrable trajectories. Then the random Duffing-van der Pol equation (9.4.4) generates a local C^∞ RDS φ which has the following properties:

(i) φ is strictly forward complete for any $(\alpha, \beta, \sigma_1, \sigma_2, \sigma_3)$, i.e. $D(t, \omega) = \mathbb{R}^2$ for all $t \geq 0$;

(ii) Assume there exist constants $c_1, c_2 > 0$ such that

$$\mathbb{P}(A) = \mathbb{P}\{\omega : |\sigma_1 \xi_1(t)| + |\sigma_2 \xi_2(t)| + |\sigma_3 \xi_3(t)| \leq c_1, \forall t \in [0, c_2]\} > 0.$$

Then for arbitrary α and β, φ is not backward complete, i.e. $R(t, \omega) \subsetneq \mathbb{R}^2$ for all $t > 0$.

Proof. (i) The idea of the (pathwise) proof is as follows: For a given RDE of the type $\dot{x}_t = f(\xi(t, \omega), x_t)$ with locally integrable $\xi(\cdot, \omega)$, we try to find a Lyapunov function $V : \mathbb{R}^d \to \mathbb{R}^+$ such that $V(x) \to \infty$ as $\|x\| \to \infty$, and for which

$$\frac{dV(x_t)}{dt} := \langle DV(x_t), f(\xi(t, \omega), x_t) \rangle \leq c_1(c_2 + \|\xi(t, \omega)\|)(c_3 + V(x_t)),$$

with constants $c_1, c_2, c_3 \geq 0$. Then the Gronwall-Bellman Lemma B.3.4 applied to the integrated form of the last inequality in the time interval $[0, T]$ yields that x_t exists up to the arbitrary time T, thus establishing strict forward completeness.

In our case, the Lyapunov function $V(x) = x_1^4 + 2x_2^2$ can be used. The details are left to the reader.

(ii) The reasoning is based on the following deterministic argument (Proposition 5.1 in [303]): Given the differential equation $\dot{x}_t = f(t, x_t)$ (in the sense of Carathéodory) possessing a unique local continuous flow, assume that there exists an unbounded domain $G \subset \mathbb{R}^d$ and a function $V \in \mathcal{C}^1(G, \mathbb{R}^+)$ such that (a) G is forward invariant under the local flow and (b) $LV(x,t) := \langle DV(x), f(t,x) \rangle \leq -1$ for all $x \in G$ and $t \geq 0$. Then $\tau(x) \leq V(x) < \infty$ for all $x \in G$, where $\tau(x)$ is the explosion time of the solution with initial value x at $t = 0$.

Indeed, define $\tau_n := \inf\{t \geq 0 : \|x_t\| \geq n\}$. Then $\tau_n \uparrow \tau$, and by (b)

$$V(x_{t \wedge \tau_n}) - V(x) = \int_0^{t \wedge \tau_n} LV(x_s, s) ds \leq -(t \wedge \tau_n).$$

Letting $n \to \infty$, the positivity of V gives $-V(x) \leq -(t \wedge \tau)$ for all $t > 0$, hence $\tau \leq V(x)$.

We apply this result to the random Duffing-van der Pol equation with reversed time (t replaced by $-t$) as follows: Choose $G = \{x \in \mathbb{R}^2 : x_1 \geq c, x_2 \geq x_1^{3-\varepsilon}\}$ where $\varepsilon \in (0,1)$ is arbitrary, but fixed, $c = c(c_1, c_2, \varepsilon, \alpha, \beta)$ is sufficiently large and $V(x) = \frac{1}{x_1}$. The result is that

$$\mathbb{P}\{\tau^-(\omega, x) \geq -c_2\} \geq \mathbb{P}(A) > 0 \quad \text{for all } x \in G.$$

The details are once again left to the reader. □

9.4.2 Proposition (Stochastic Duffing-van der Pol Equation). *The stochastic Duffing-van der Pol equation*

$$\begin{aligned} dx_1 &= x_2 dt \\ dx_2 &= (\alpha x_1 + \beta x_2 - x_1^3 - x_1^2 x_2) dt \\ &\quad + \sigma_1 x_1 \circ dW_t^1 + \sigma_2 x_2 \circ dW_t^2 + \sigma_3 \circ dW_t^3 \end{aligned} \tag{9.4.5}$$

generates a local C^∞ RDS φ which has the following properties:

(i) φ is forward complete for any $(\alpha, \beta, \sigma_1, \sigma_2, \sigma_3)$;

(ii) If $\sigma_2 = 0$, or if $\sigma_1 = \sigma_3 = 0$ then for any value of the remaining parameters, φ is strictly forward complete, but φ is not backward complete.

Proof. (i) We apply the following generator condition of Schenk-Hoppé [304: Theorem 1.2.3] for the completeness of a dissipative second order SDE: Consider a d-dimensional second order stochastic Itô differential equation formally written as

$$\ddot{y} = f(y, \dot{y}) + g(y, \dot{y}) \dot{W}, \tag{9.4.6}$$

where $f : \mathbb{R}^d \times \mathbb{R}^d \to \mathbb{R}^d$ and $g : \mathbb{R}^d \times \mathbb{R}^d \to \mathbb{R}^{d \times m}$ are locally Lipschitz continuous and W is an m-dimensional Brownian motion. We rewrite (9.4.6) as a $2d$-dimensional Itô SDE for $x = (x_1, x_2) = (y, \dot{y})$ as

$$dx_1 = x_2 dt, \qquad (9.4.7)$$
$$dx_2 = f(x)dt + g(x)dW_t,$$

which has the infinitesimal generator

$$L = \frac{\partial}{\partial t} + \langle x_2, \frac{\partial}{\partial x_1}\rangle + \langle f(x), \frac{\partial}{\partial x_2}\rangle + \frac{1}{2}\sum_{i,j=1}^{d}(g(x)g(x)^*)_{ij}\frac{\partial^2}{\partial x_{2i}\partial x_{2j}}.$$

Let $V : \mathbb{R}^d \times \mathbb{R}^d \to \mathbb{R}^+$, $V(x_1, x_2) := E(x_1, x_2) + h_1(\|x_2\|)$, be a C^2 function, where $E(x_1, x_2) \geq 0$ for all $(x_1, x_2) \in \mathbb{R}^d \times \mathbb{R}^d$ and h_1 is a non-negative increasing function with $h_1(t) \uparrow \infty$ as $t \uparrow \infty$. Assume that there exist constants $c_1, c_2, c_3 \geq 0$ and a non-negative increasing function $h_2 : \mathbb{R}^+ \to \mathbb{R}^+$ such that for all $(x_1, x_2) \in \mathbb{R}^d \times \mathbb{R}^d$

$$LV(x_1, x_2) \leq c_1 + c_2 V(x_1, x_2) + c_3 h_2(\|x_1\|)$$

and that for all sufficiently large fixed $c > 0$

$$\lim_{t \to \infty} \frac{h_2(ct)}{h_1(t)} = 0.$$

Then the maximal solution of (9.4.7) is complete.

The proof of this theorem consists in applying Dynkin's formula and the estimate for LV to $V(x_1(t \wedge \tau_n), x_2(t \wedge \tau_n))$, where $\tau_n := \inf\{t \geq 0 : \|x_2(t)\| \geq n\}$, and then the Gronwall-Bellman Lemma to arrive at $\lim_{n \to \infty} \mathbb{P}\{\tau_n < T\} = 0$ for all $T > 0$.

We apply this theorem to our SDE (9.4.5), whose Itô form is obtained by replacing β by $\beta + \frac{1}{2}\sigma_2^2$. We choose $E(x_1, x_2) = \frac{x_1^4}{2}$, $h_1(\|x_2\|) = \|x_2\|^2$, $h_2 \equiv 0$. Then the estimate for LV is satisfied for $c_1 = |\alpha| + \sigma_1^2 + \sigma_3^2$, $c_2 = 2(|\alpha| + |\beta| + \sigma_1^2 + \sigma_2^2)$, and $c_3 = 0$. The details are left to the reader.

(ii) The proof of strict forward completeness is accomplished by converting the SDE into an RDE and then by proving the strict forward completeness for the RDE as follows:

Consider the case $\sigma_2 = 0$. Then the Stratonovich SDE (9.4.5) can be written as a *Liénard equation*

$$dx_1 = (x_2 + \beta x_1 - \frac{1}{3}x_1^3)dt, \qquad (9.4.8)$$
$$dx_2 = (\alpha x_1 - x_1^3)dt + \sigma_1 x_1 \circ dW_t^1 + \sigma_3 \circ dW_t^3,$$

where $y = x_1$ and $\dot{y} = x_2 + \beta x_1 - \frac{1}{3}x_1^3$, whose Itô-Stratonovich correction term is zero. Taking $\sigma_1 = \sigma_3 = 1$ for simplicity and applying the transformation

$$z(t) := x_2(t) - x_1(t)W_t^1 - W_t^3,$$

we convert (9.4.8) into the (pathwise) nonautonomous ordinary differential equation

$$\dot{x}_1 = z + x_1 W_t^1 + W_t^3 + \beta x_1 - \frac{1}{3}x_1^3, \qquad (9.4.9)$$

$$\dot{z} = \alpha x_1 - x_1^3 - (z + x_1 W_t^1 + W_t^3 + \beta x_1 - \frac{1}{3}x_1^3)W_t^1.$$

We apply the idea of the proof of Proposition 9.4.1(i) to (9.4.9), using the Lyapunov function $V(x_1, z) = x_1^4 + 2z^2$. Details are again left to the reader.

Similarly, if $\sigma_1 = \sigma_3 = 0$, by using the transformation $z(t) = \exp(-\sigma_2 W_t^2)x_2(t)$, the SDE (9.4.5) can be converted into an RDE whose strict forward completeness can be checked.

We finally have to prove that (9.4.5) is not backward complete. This can be seen for the case $\sigma_2 = 0$ by reversing time in (9.4.9) and applying the proof of Proposition 9.4.1(ii). Here

$$\mathbb{P}(A) = \mathbb{P}\{|\sigma_1 W_t^1| + |\sigma_3 W_t^3| \leq c_1 \text{ for all } t \in [0, c_2]\} > 0$$

holds for any $c_1, c_2 > 0$,

$$G = \{(x_1, z) \in \mathbb{R}^2 : x_1 \geq c, z \geq x_1^{3-\varepsilon}\},$$

where $\varepsilon \in (0, 1)$ and c is big enough, and $V(x_1, z) = \frac{1}{x_1}$. The result is that

$$\mathbb{P}\{\omega : -c_2 \leq \tau^-(\omega, x)\} \geq \mathbb{P}(A) > 0$$

for all $x = (x_1, z) \in G$. □

9.4.3 Remark. (i) It is an open problem as to whether or not the general stochastic Duffing-van der Pol equation is strictly forward complete.

(ii) *Noisy Duffing equation:* By omitting the term $-y^2\dot{y}$ in equation (9.4.3) we obtain the random/stochastic Duffing equation

$$\ddot{y} = (\alpha + \sigma_1 \xi_1(t))y + (\beta + \sigma_2 \xi_2(t))\dot{y} - y^3 + \sigma_3 \xi_3(t). \qquad (9.4.10)$$

In the random case, if ξ_1, ξ_2, ξ_3 are locally integrable, then equation (9.4.10) is strictly forward complete and strictly backward complete for all parameter values $\alpha, \beta, \sigma_1, \sigma_2, \sigma_3$.

In the stochastic case, (9.4.10) is forward and backward complete for arbitrary parameter values. Moreover, it is strictly forward complete and strictly backward complete if $\sigma_2 = 0$ or if $\sigma_1 = \sigma_3 = 0$.

(iii) *Noisy van der Pol equation:* By omitting the term $-y^3$ in equation (9.4.3) we obtain the random/stochastic van der Pol equation

$$\ddot{y} = (\alpha + \sigma_1 \xi_1(t))y + (\beta + \sigma_2 \xi_2(t))\dot{y} - y^2\dot{y} + \sigma_3 \xi_3(t). \qquad (9.4.11)$$

In the random case, if ξ_1, ξ_2, ξ_3 are locally integrable, then equation (9.4.11) is strictly forward complete, but not backward complete, for all parameter values $\alpha, \beta, \sigma_1, \sigma_2, \sigma_3$.

In the stochastic case, (9.4.11) is forward complete for arbitrary parameter values. Moreover, it is strictly forward complete, but not backward complete if $\sigma_2 = 0$, or if $\sigma_1 = \sigma_3 = 0$. ∎

Assumption: To reduce complexity, we will henceforth treat only the case $\sigma_2 = \sigma_3 = 0$ and write $\sigma_1 = \sigma$, $\xi_1 = \xi$ (assumed locally integrable) and $W^1 = W$. Thus, for the remainder of this section, our equations become

$$\dot{x} = \begin{pmatrix} 0 \\ -x_1^3 - x_1^2 x_2 \end{pmatrix} + \begin{pmatrix} 0 & 1 \\ \alpha & \beta \end{pmatrix} x + \sigma \xi(t) \begin{pmatrix} 0 & 0 \\ 1 & 0 \end{pmatrix} x \quad (9.4.12)$$

for the real noise case and

$$dx = \begin{pmatrix} 0 \\ -x_1^3 - x_1^2 x_2 \end{pmatrix} dt + \begin{pmatrix} 0 & 1 \\ \alpha & \beta \end{pmatrix} x\, dt + \sigma \begin{pmatrix} 0 & 0 \\ 1 & 0 \end{pmatrix} x \circ dW_t \quad (9.4.13)$$

for the white noise case.

Both systems generate a local C^∞ RDS φ which is strictly forward complete, but not backward complete.

Linear Analysis

The linearized RDS $D\varphi$ is generated by the linearization of the above equations, namely in the real noise case by

$$\dot{v} = \begin{pmatrix} 0 & 0 \\ -3x_1^2 - 2x_1 x_2 & -x_1^2 \end{pmatrix} v + \begin{pmatrix} 0 & 1 \\ \alpha & \beta \end{pmatrix} v + \sigma \xi(t) \begin{pmatrix} 0 & 0 \\ 1 & 0 \end{pmatrix} v \quad (9.4.14)$$

and in the white noise case by

$$dv = \begin{pmatrix} 0 & 0 \\ -3x_1^2 - 2x_1 x_2 & -x_1^2 \end{pmatrix} v\, dt + \begin{pmatrix} 0 & 1 \\ \alpha & \beta \end{pmatrix} v\, dt + \sigma \begin{pmatrix} 0 & 0 \\ 1 & 0 \end{pmatrix} v \circ dW_t. \quad (9.4.15)$$

For any invariant measure μ, the trace formula gives $\lambda_1(\mu) + \lambda_2(\mu) = \beta - \mathbb{E}_\mu x_1^2$. In particular, for $\mu = \delta_0$ we obtain $\lambda_1(\delta_0) + \lambda_2(\delta_0) = \beta$.

We also need the eigenvalues of the linearization of the deterministic system ($\sigma = 0$) at $x = 0$,

$$\dot{v}_t = \begin{pmatrix} 0 & 1 \\ \alpha & \beta \end{pmatrix} v_t,$$

which are

$$\lambda_{1,2} = \frac{\beta}{2} \pm \sqrt{\frac{\beta^2}{4} + \alpha}.$$

For $\frac{\beta^2}{4} + \alpha > 0$, we have two real eigenvalues, while for $\frac{\beta^2}{4} + \alpha < 0$ we have a pair of complex-conjugate eigenvalues

$$\lambda_{1,2} = \frac{\beta}{2} \pm i \omega_d, \quad \omega_d := \sqrt{-\frac{\beta^2}{4} - \alpha},$$

where ω_d represents the "damped eigenfrequency" of the linear system $\ddot{y} - \beta \dot{y} - \alpha y = 0$.

We recall the following well-known facts for the deterministic Duffing-van der Pol equation (9.4.2):
- For the frozen parameter $\alpha < 0$ and the bifurcation parameter β increasing across $\beta = 0$, the trivial solution $x = 0$ undergoes a Hopf bifurcation at $\beta = 0$ with a stable limit cycle for $\beta > 0$ bifurcating out of $x = 0$.
- For the frozen parameter $\beta < 0$ and the bifurcation parameter α increasing across $\alpha = 0$, it undergoes a pitchfork bifurcation at $\alpha = 0$ with two stable steady states $(\pm\sqrt{\alpha}, 0)$ for $\alpha > 0$ bifurcating out of $x = 0$.

In the remaining part of this section we want to study the changes occurring in these scenarios when the noise is "switched on".

9.4.2 Hopf Bifurcation

What is Stochastic Hopf Bifurcation?

Consider, for example, equation (9.4.13) for $\alpha < 0$ frozen and with β increasing across $\beta = 0$.

On the phenomenological level of the Fokker-Planck equation, computer simulations (Schenk-Hoppé [302]) show that for β sufficiently negative, the Dirac measure δ_0 is the only stationary measure. Increasing β, we witness the premature birth of a nontrivial stationary density p_β at $\beta = \beta_{D1} < 0$. This density has a pole at $x = 0$. If we increase β even further, this density undergoes a P-bifurcation at $\beta = \beta_P$, where the pole disappears and from which on it becomes crater-like (the parameter interval $[\beta_{D1}, \beta_P]$ was called "bifurcation interval" by Ebeling et al. [134]) (see Fig. 9.7 for which $\beta_{D1} = -0.06\ldots$ and $\beta_P = 0$).

As will be explained in Subsect. 9.5.1, β_P is that parameter value for which the moment Lyapunov exponent

$$g(p) := \lim_{t\to\infty} \frac{1}{t} \log \mathbb{E} \,\|\Phi(t,\omega)v\|^p$$

has its second zero at $p = -2 = -\dim \mathbb{R}^2$. By Proposition 9.5.6 applied to our case (9.4.15), $\beta_P = 0$ (see also Example 9.5.11).

What happens on the dynamical level? Extensive simulations of the RDS φ_β, its random attractors and invariant measures [302, 37] lead us to believe in the following D-Hopf bifurcation scenario:

1. It is a general observation that noise splits deterministic multiplicities of eigenvalues. For $\beta^2 < -4\alpha$ (which we assume) and $\sigma = 0$, the deterministic linear system has two complex-conjugate eigenvalues $\frac{\beta}{2} \pm i\omega_d$, which amounts to just one Lyapunov exponent $\lambda_1(0,\beta) = \beta/2$ with multiplicity 2. For $\sigma \neq 0$, however, the linearized SDE

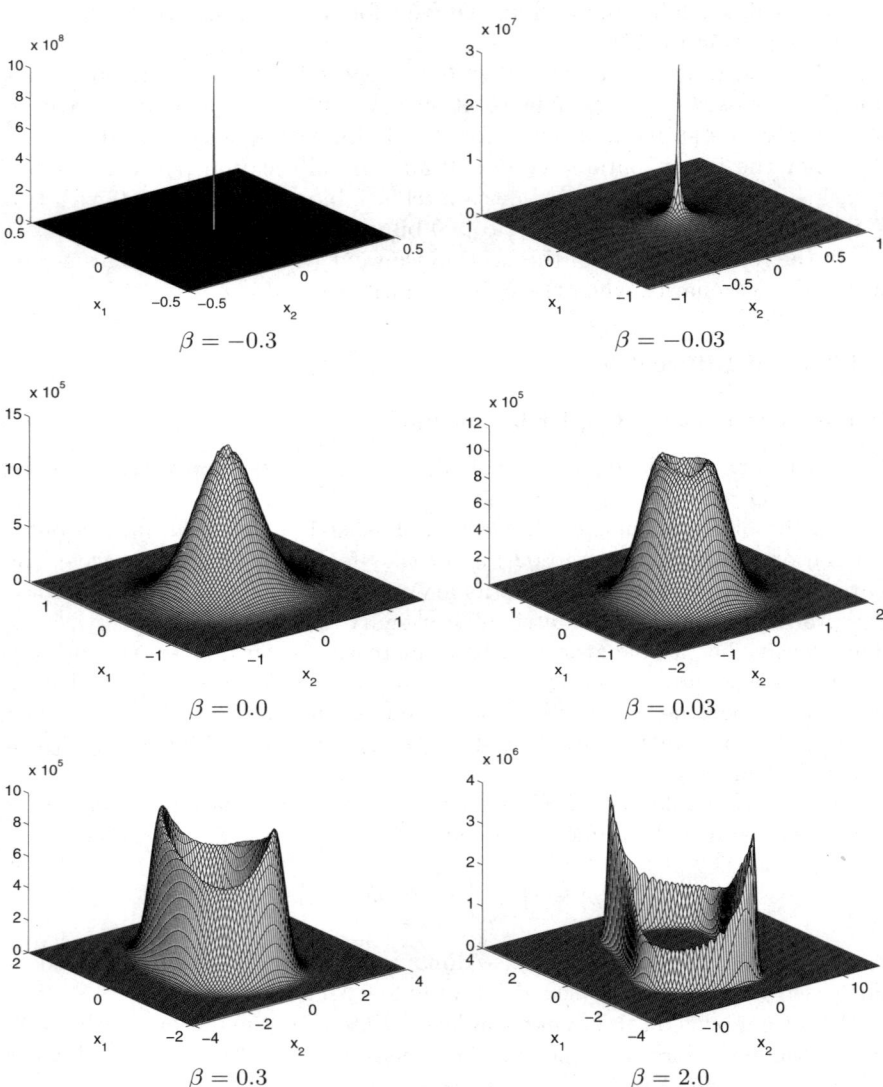

Fig. 9.7. Stationary density in the Hopf regime for $\alpha = -1$ and $\sigma = 0.5$

$$dv_t = \begin{pmatrix} 0 & 1 \\ \alpha & \beta \end{pmatrix} v_t\, dt + \sigma \begin{pmatrix} 0 & 0 \\ 1 & 0 \end{pmatrix} v_t \circ dW_t \qquad (9.4.16)$$

has two different simple Lyapunov exponents $\lambda_i(\sigma, \beta)$ (see Arnold, Oeljeklaus and Pardoux [32]), which satisfy $\lambda_1(\sigma, \beta) + \lambda_2(\sigma, \beta) = \beta$. For small σ we can use the asymptotic expansions given by Pardoux and Wihstutz [270: Theorem 4.1] and others, namely

9.4.4 Theorem (Small Noise Expansion of Lyapunov Exponent, Case of Complex Eigenvalues). *Let*

$$dx_t = \begin{pmatrix} a & -b \\ b & a \end{pmatrix} x_t dt + \sigma \sum_{j=1}^{m} A_j x_t \circ dW_t^j,$$

where $a, b \in \mathbb{R}$, $b \neq 0$ *and* $A_j = (a_{kl}^j)_{2\times 2}$, $j = 1, \ldots, m$. *Then as* $\sigma \to 0$ *the top Lyapunov exponent* $\lambda_1(\sigma)$ *satisfies*

$$\lambda_1(\sigma) = a + \frac{\sigma^2}{8} \sum_{j=1}^{m} \left((a_{11}^j - a_{22}^j)^2 + (a_{12}^j + a_{21}^j)^2 \right) + O(\sigma^4) \quad (9.4.17)$$

and the rotation number satisfies

$$\rho(\sigma) = b + c\sigma^4 + O(\sigma^6),$$

where c is some constant such that $bc \leq 0$.

For our case this yields

$$\lambda_{1,2}(\sigma, \beta) = \frac{\beta}{2} \pm \frac{\sigma^2}{8\omega_d^2} + O(\sigma^4) \quad (9.4.18)$$

and for the rotation number

$$\rho(\sigma, \beta) = \omega_d - \frac{c\sigma^4}{\omega_d} + O(\sigma^6)$$

for some $c > 0$.

For the real noise case a similar expansion holds: For example, if $\xi(t) = f(\eta(t))$, where $\eta(t)$ is an elliptic diffusion on a compact manifold and f is smooth and not constant, then the spectrum of the linear RDE is simple (Arnold, Kliemann and Oeljeklaus [27]), and for $\mathbb{E}\xi = 0$

$$\lambda_{1,2}(\sigma, \beta) = \frac{\beta}{2} \pm \frac{\sigma^2}{8\omega_d^2} S(2\omega_d) + O(\sigma^4), \quad (9.4.19)$$

where $S(\cdot)$ is the spectral density of $\xi(t)$ (Ariaratnam and Sri Namachchivaya [5]). Expression (9.4.19) matches the result (9.4.18) for the white noise case where $S(u) \equiv 1$. Furthermore, the rotation number expands as

$$\rho(\sigma, \beta) = \omega_d - \frac{c\sigma^2}{8\omega_d^2} T(2\omega_d) + O(\sigma^4),$$

where $T(\cdot)$ is the sine spectral density of $\xi(t)$ (see Example 9.5.12).

Consequently, at the deterministic bifurcation point $\beta = 0$ we have for the white noise case, neglecting $O(\sigma^4)$ terms,

$$\lambda_2(\sigma,0) = -\frac{\sigma^2}{8\omega_d^2} < 0 < \lambda_1(\sigma,0) = \frac{\sigma^2}{8\omega_d^2},$$

i.e. at $\beta = 0$ the top exponent has already crossed 0 and is positive.

It follows from (9.4.18) that the top Lyapunov exponent changes sign at $\beta_{D1} = \frac{\sigma^2}{4\alpha} + O(\sigma^4) < 0$ and the second Lyapunov exponent changes sign at $\beta_{D2} = -\frac{\sigma^2}{4\alpha} + O(\sigma^4) > 0$.

2. For $\beta < \beta_{D1}$, δ_0 is stable, it is the unique invariant measure, and $A_\beta = \{0\}$ is the attractor of the RDS φ_β in $\mathbb{R}^2$.

3. We have a first bifurcation from δ_0 at $\beta = \beta_{D1} < 0$ of a stable ergodic measure $\nu^1(\sigma,\beta)_\omega = \frac{1}{2}(\delta_{x^{(1)}(\omega)} + \delta_{-x^{(1)}(\omega)})$, which is a convex combination of two Dirac measures, sitting on the two "boundary points" of the one-dimensional unstable manifold of $x = 0$, which is a saddle point. This situation persists for $\beta \in (\beta_{D1}, \beta_{D2})$. As shown above, at $\beta_P = 0 \in (\beta_{D1}, \beta_{D2})$, ν^1 undergoes a P-bifurcation. δ_0 and ν^1 are both Markov measures. Hence $\rho = \mathbb{E}\nu^1$ solves the Fokker-Planck equation. The attractor A_β in (the universe of tempered sets of) $\mathbb{R}^2$ is the closure of the unstable manifold of $x = 0$, the two "boundary points" supporting the measure ν^1. In the punctured plane $\mathbb{R}^2 \setminus \{0\}$, the attractor A_β^0 (in the universe of simply connected tempered sets) consists of the two-point set $\mathrm{supp}\,\nu^1$.

4. At $\beta = \beta_{D2} > 0$, we have a second bifurcation from δ_0 of a measure $\nu^2(\sigma,\beta)_\omega = \frac{1}{2}(\delta_{x^{(2)}(\omega)} + \delta_{-x^{(2)}(\omega)})$, which is again a convex combination of two Dirac measures. ν^2 is a saddle point, i.e. has a positive and a negative Lyapunov exponent, while δ_0 has two positive exponents.

For $\beta > \beta_{D2}$, the stable measure ν^1 is supported by the "boundary" of the unstable manifold of ν^2. The closure of this unstable manifold is an invariant "circle" around $x = 0$ ("random limit cycle") and supports both measures ν^1 and ν^2. On this "circle", we have hyperbolic dynamics (ν^1 is attracting, and ν^2 is repelling), in contrast to the deterministic case, where the dynamics on the limit cycle is just rotation, and the invariant measure is unique. The interior of the "circle" is the two-dimensional unstable manifold of 0. Its closure is the attractor A_β in $\mathbb{R}^2$. It carries all three invariant measures. In the punctured plane $\mathbb{R}^2 \setminus \{0\}$, however, the attractor A_β^0 is the invariant "circle", on which we have two invariant measures, in particular, the unique Markov measure ν^1. See Fig. 9.8 (where $\alpha = -1$ and $\sigma = 1.5$).

The next four Figs. 9.9 to 9.12 show the deformation of a grid of points ("discrete Lebesgue measure") by the mapping $x \mapsto \varphi(-t,\omega)^{-1}x = \varphi(t,\theta_{-t}\omega)x$ as $t \to \infty$ (with $\alpha = -1$, $\sigma = 0.5$ and time increment 0.001).

The whole scenario (see Fig. 9.13) has been very well established numerically for real and white noise (Schenk-Hoppé [302, 305, 304], Arnold, Sri Namachchivaya and Schenk-Hoppé [37]). The task is now to prove it mathematically, but so far this has been only partly accomplished.

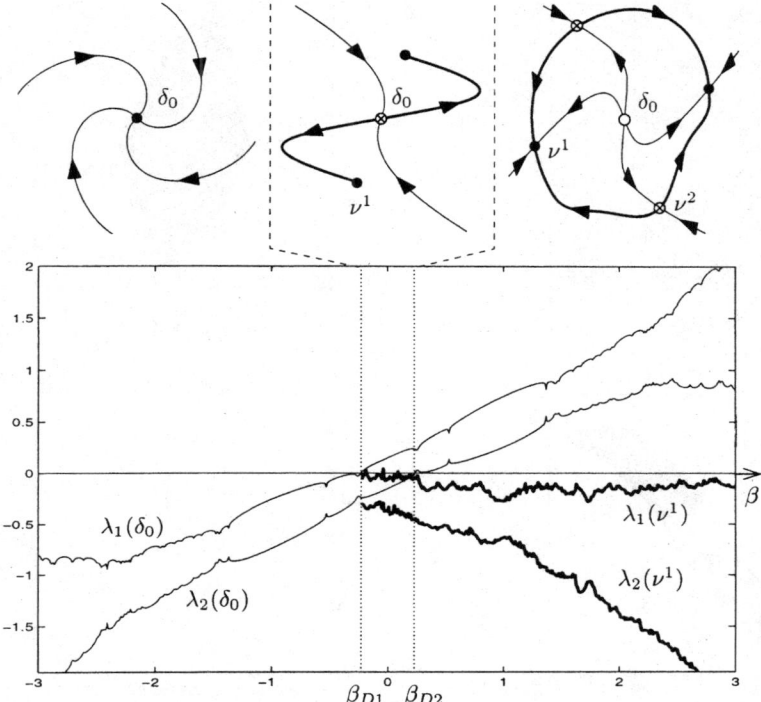

Fig. 9.8. Lyapunov exponents, invariant measures and attractors in the Hopf regime

Attractors and Invariant Measures

We first present a general theorem on the existence of a random attractor in the real noise case obtained by Schenk-Hoppé [304, 306].

9.4.5 Theorem (Existence of Attractor, Real Noise Case).
Consider the local C^∞ RDS φ generated by the random Duffing-van der Pol equation (9.4.12), where $m := \mathbb{E}|\xi(t)| < \infty$. Then for arbitrary parameter values $(\alpha, \beta) \in \mathbb{R}^2$ and $|\sigma| < \frac{2}{7m}$, φ possesses a random attractor $A(\omega)$ in $\mathbb{R}^2$ in the universe of all families $(C(\omega))_{\omega \in \Omega}$ of subsets of $\mathbb{R}^2$ which are tempered from above. Moreover, the attractor is measurable with respect to the past $\mathcal{F}^-$ and supports all φ-invariant measures.

Proof. By Proposition 9.3.2, all we have to construct is an absorbing set which is tempered from above. We follow the procedure described and carried out in [304, ?]:

(i) We first find a "Lyapunov function" $V : \mathbb{R}^2 \to \mathbb{R}^+$ which is continuous, surjective, has the property that pre-images of bounded sets are bounded, and satisfies a differential inequality of the form

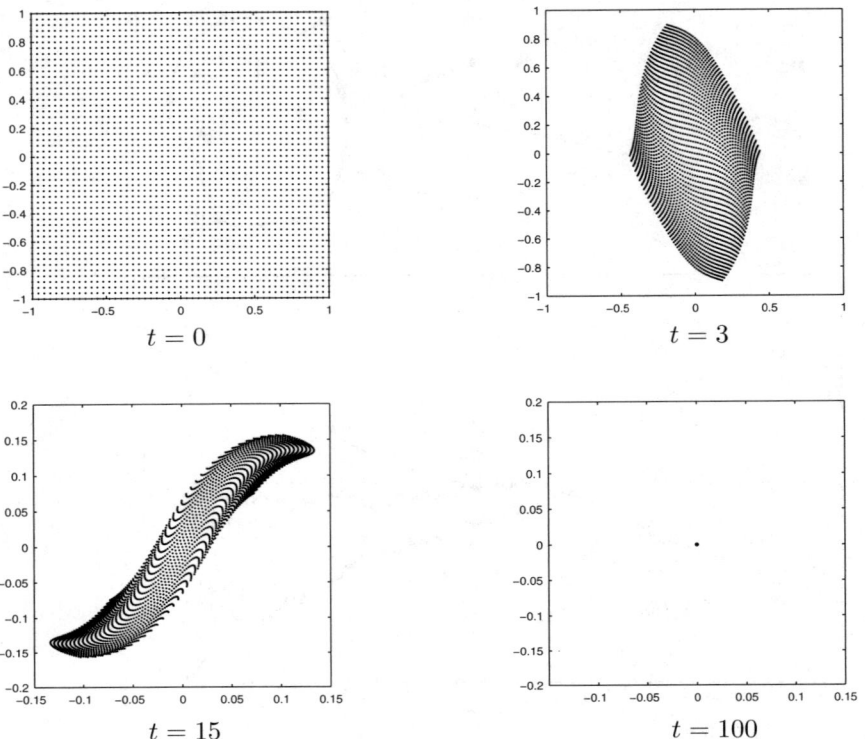

Fig. 9.9. Supports of invariant measures and attractors in the Hopf regime for $\beta = -0.3 < \beta_{D1} = -0.06\ldots$

$$\frac{dV(\varphi(t,\omega,x))}{dt} \leq \eta_1(t)V(\varphi(t,\omega,x)) + \eta_2(t),$$

where $\mathbb{E}\,\eta_1(t) < 0$ and $\mathbb{E}\,|\eta_2(t)| < \infty$.

In our case, we can use

$$V(x) = \frac{7}{24}x_1^4 + x_1^2 - x_1 x_2 + \frac{3}{4}x_2^2.$$

After some elementary estimates (exercise) we obtain that

$$\frac{dV(\varphi(t,\omega,x))}{dt} \leq (-1 + \frac{7|\sigma|}{2}|\xi(t)|)V(\varphi(t,\omega,x)) + c, \tag{9.4.20}$$

where $c = c(\alpha,\beta)$ is a positive constant.

(ii) The affine RDS $\psi(t,\omega)$ generated by the RDE

$$\dot{z}_t = (-1 + \frac{7|\sigma|}{2}|\xi(t)|)z_t + c$$

associated with (9.4.20) satisfies

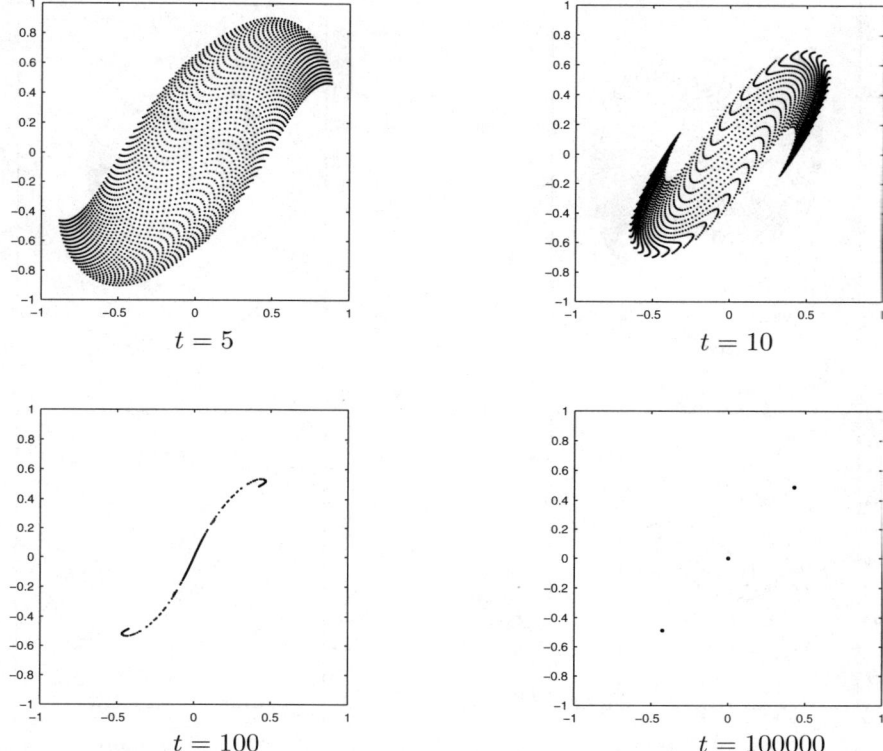

Fig. 9.10. Supports of invariant measures and attractors in the Hopf regime for $\beta_{D1} = -0.06\ldots < \beta = -0.03 < \beta_P = 0$

$$0 \leq V(\varphi(t,\omega,x)) \leq \psi(t,\omega,V(x)),$$

and has a unique stationary solution with initial value

$$0 < r(\omega) := c \int_{-\infty}^{0} e^{t - \int_{t}^{0} \frac{7|\sigma|}{2} |\xi(\theta_s \omega)| ds} dt < \infty,$$

which is exponentially stable, hence is the attractor of ψ in the universe of sets in $\mathbb{R}^+$ (or $\mathbb{R}$) which are tempered from above. Further, for any $\varepsilon > 0$, the set $[0,(1+\varepsilon)r(\omega)]$ is absorbing (exercise) for ψ in this universe.

(iii) We now "lift" these findings to the original RDS φ: By our assumptions on V (fulfilled for our particular choice),

$$B_\varepsilon(\omega) := V^{-1}([0,(1+\varepsilon)r(\omega)])$$

is a compact random set which is absorbing in the universe of pre-images $D = V^{-1}(D_1)$ of sets D_1 in $\mathbb{R}^+$ tempered from above. This universe contains in particular all sets in $\mathbb{R}^2$ tempered from above, since their image under the polynomial V is tempered from above.

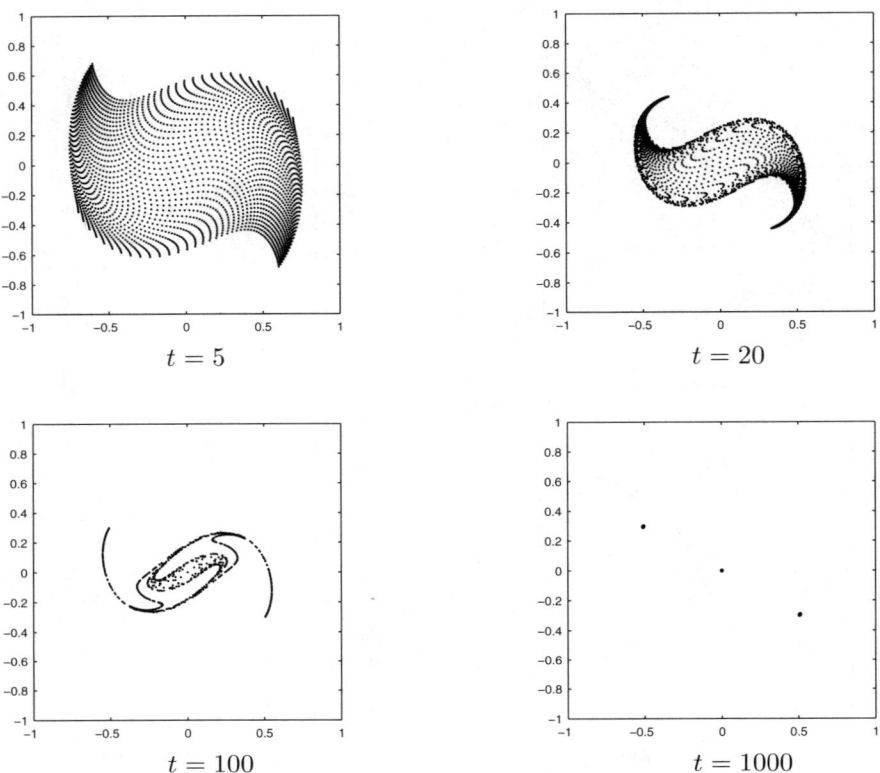

Fig. 9.11. Supports of invariant measures and attractors in the Hopf regime for $\beta = \beta_P = 0$

Hence, we can apply Proposition 9.3.2 and the proof is complete. □

9.4.6 Remark. (i) The proof furnishes the estimate

$$A(\omega) \subset V^{-1}([0, r(\omega)]).$$

(ii) There is so far no analogue of the above theorem for the white noise case. The existence of an attractor was established for the purely additive white noise case ($\sigma_1 = \sigma_2 = 0$) of (9.4.5) by Schenk-Hoppé [304, ?]. ∎

9.4.7 Remark (Invariant Measures). Under the conditions of Theorem 9.4.5, all invariant measures are supported by A. There exists at least one invariant measure supported by A (Theorem 1.6.13), and since A is $\mathcal{F}^-$-measurable there is at least one invariant forward Markov measure supported by A (Theorem 1.7.5).

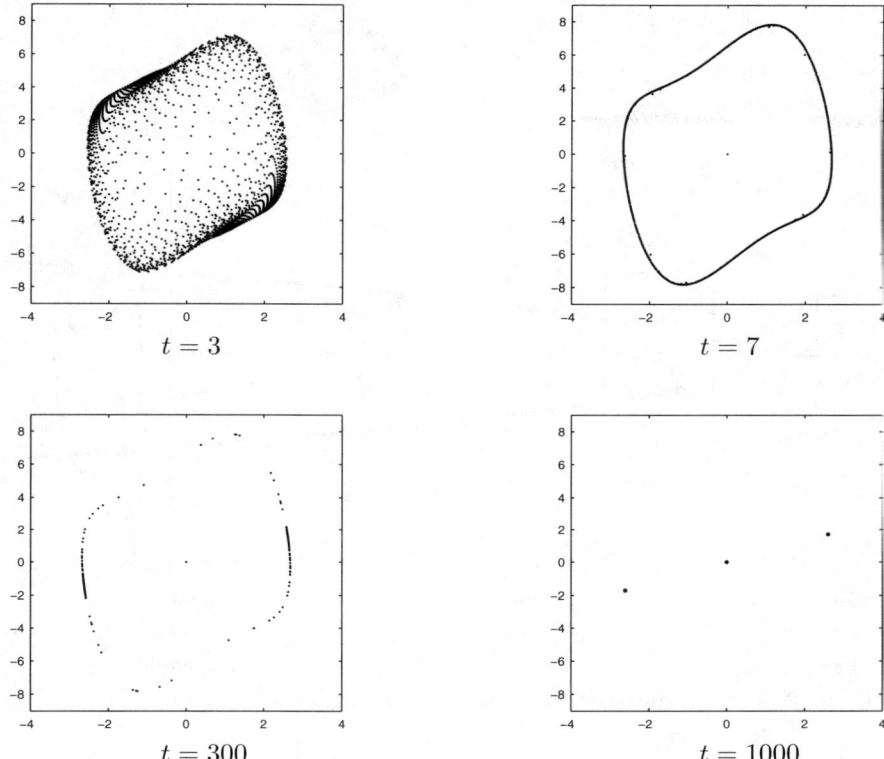

Fig. 9.12. Supports of invariant measures and attractors in the Hopf regime for $\beta = 2 > \beta_{D2} = 0.06\ldots$

Further, since φ is a local homeomorphism, the boundary $\partial A(\omega)$ of $A(\omega)$ is a random (see Schenk-Hoppé [304: Lemma 2.7.4]) compact invariant set which also supports an invariant measure. Since ∂A is $\mathcal{F}^-$-measurable, it also supports an invariant forward Markov measure. ∎

We now give partial answers to the question of the structure of A in the first regime $\beta < \beta_{D1}$, and in the third regime $\beta > \beta_{D2}$ (again taken from [306]).

Unfortunately, as yet there are no rigorous results for the intermediate regime $\beta_{D1} < \beta < \beta_{D2}$.

9.4.8 Theorem (Attractor for $\beta < \beta_{D1}$). *Consider the local C^∞ RDS φ_β generated by the random Duffing-van der Pol equation (9.4.12), where $m := \mathbb{E}|\xi(t)| < \infty$. Take for simplicity $\alpha = -1$. Then for arbitrary parameter values $\beta \in (-2, 0)$ and σ which satisfy*

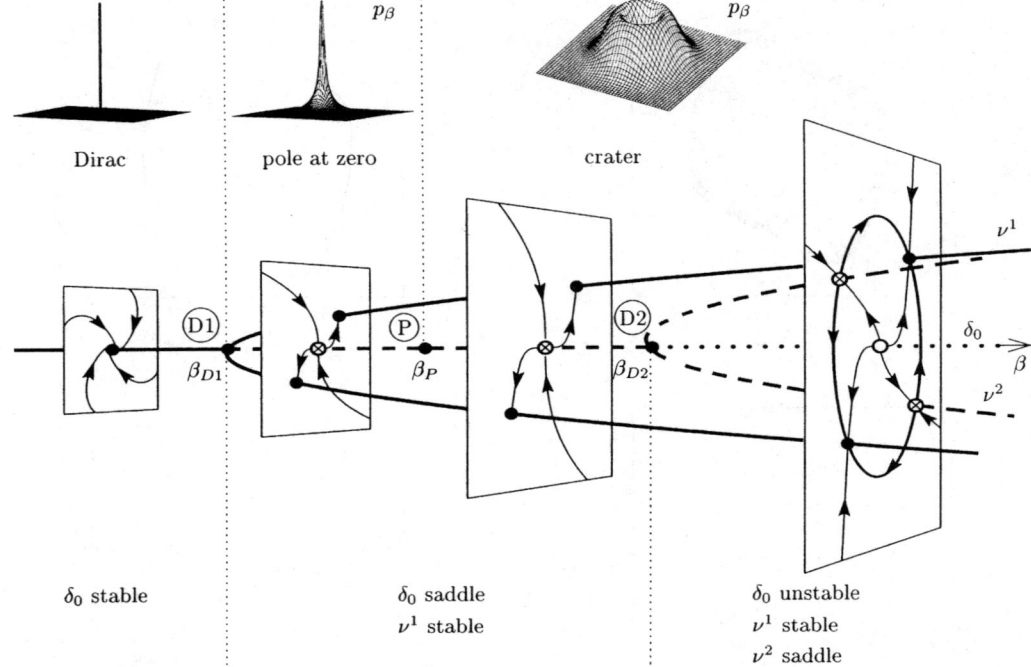

Fig. 9.13. The full stochastic Hopf bifurcation diagram

$$|\sigma| < \frac{|\beta|}{2m} \frac{2 - |\beta|}{1 + |\beta|}, \tag{9.4.21}$$

the attractor A_β of φ_β in the universe of all subsets of $\mathbb{R}^2$ tempered from above is $A_\beta = \{0\}$, and δ_0 is the unique invariant measure of φ_β.

Moreover, all orbits $\varphi_\beta(t, \theta_{-t}\omega, x)$ and $\varphi_\beta(t, \omega, x)$ tend to zero exponentially fast as $t \to \infty$.

Proof. (i) We need the extra condition on β and σ for bounds on the Lyapunov exponents $\lambda_{1,2}$ of the linearized RDE

$$\dot{v}_t = \begin{pmatrix} 0 & 1 \\ -1 + \sigma\xi(t) & \beta \end{pmatrix} v_t,$$

which is the random harmonic oscillator. This becomes clear through the estimates

$$\frac{\beta}{2} - \gamma \leq \lambda_2 \leq \lambda_1 \leq \frac{\beta}{2} + \gamma, \tag{9.4.22}$$

where

$$\gamma := m|\sigma|\frac{1+|\beta|}{2-|\beta|},$$

which hold for any $\beta \in (-2, 2)$ and $\sigma \neq 0$.[6]

To prove (9.4.22), we use the Lyapunov function $V(v) = v_1^2 - \beta v_1 v_2 + v_2^2$, for which $V(v) \geq 0$ since $\beta \in (-2, 2)$, and $V(v) = 0$ if and only if $v = 0$. This yields

$$\frac{dV(v(t))}{dt} = \beta V(v(t)) + \sigma(2v_1(t)v_2(t) - \beta v_1(t)^2)\xi(t).$$

Since

$$\frac{2-|\beta|}{2}\|v\|^2 \leq V(v) \leq \frac{2+|\beta|}{2}\|v\|^2 \qquad (9.4.23)$$

and $|2v_1 v_2 - \beta v_1^2| \leq (1+|\beta|)\|v\|^2$ we obtain

$$(\beta - \frac{2\gamma}{m}|\xi(t)|)V(v(t)) \leq \frac{dV(v(t))}{dt} \leq (\beta + \frac{2\gamma}{m}|\xi(t)|)V(v(t)).$$

Comparing $V(v(t))$ with the solutions of the exponentially stable linear RDE dominating from above and below, and using (9.4.23) gives

$$\frac{\beta}{2} - \gamma \leq \liminf_{t\to\infty} \frac{1}{t}\log\|v(t)\| \leq \limsup_{t\to\infty} \frac{1}{t}\log\|v(t)\| \leq \frac{\beta}{2} + \gamma,$$

for any (random, possibly non-adapted) initial value v_0. This proves (9.4.22).

(ii) We now prove that the attractor (which exists by Theorem 9.4.5) is equal to $A_\beta = \{0\}$. By step (i) and conditions (9.4.21), both Lyapunov exponents of the invariant measure δ_0 are negative.

Now we use the non-negative Lyapunov function

$$V(x) = \frac{6+\beta}{24}x_1^4 + \frac{1}{2}x_1^2 + \frac{\beta}{2}x_1 x_2 + \frac{1}{2}x_2^2$$

for the nonlinear equation which yields

$$\frac{dV(x_t)}{dt} \leq \left(\beta + \frac{2\gamma}{m}|\xi(t)|\right)V(x_t),$$

since $V(x) \geq \frac{1}{4}(2-|\beta|)\|x\|^2$ for $-2 < \beta < 0$.

Comparing with the associated linear RDE we see that $x = 0$ is exponentially stable with rate $\frac{\beta}{2} + \gamma < 0$. □

It is much harder to show that in the third regime, $\beta > \beta_{D2}$, there exists an attractor which does not contain $x = 0$.

9.4.9 Theorem (Attractor for $\beta > \beta_{D2}$). *Consider the local C^∞ RDS φ_β generated by the random Duffing-van der Pol equation (9.4.12), where $m := \mathbb{E}|\xi(t)| < \infty$. Take for simplicity $\alpha = -1$. Then for arbitrary parameter values $\beta \in (0, 2)$ and σ which satisfy*

---

[6] We gave conditions under which $\lambda_2 < \lambda_1$ at the beginning of this subsection, see (9.4.19).

$$|\sigma| < \frac{\beta}{2m} \frac{2-\beta}{1+\beta}, \qquad (9.4.24)$$

there exists an attractor A_β^0 for φ_β restricted to the punctured plane $\mathbb{R}^2 \setminus \{0\}$, in the universe of all tempered subsets of $\mathbb{R}^2 \setminus \{0\}$.

Furthermore, A_β^0 supports all invariant measures of φ_β in $\mathbb{R}^2 \setminus \{0\}$.

A family $(C(\omega))_{\omega \in \Omega}$ of subsets of $\mathbb{R}^d \setminus \{0\}$ is called tempered if it is contained in an annulus $\{x \in \mathbb{R}^d \setminus \{0\} : r(\omega) \leq \|x\| \leq R(\omega)\}$ such that r is tempered from below and R is tempered from above.

Proof. In order to avoid unnecessary repetition, we will just sketch the proof and refer the reader to [306] for details.

(i) By using an appropriate Lyapunov function V_1 and comparing with a dominating exponentially stable affine RDE we obtain an absorbing set $B_1 = V_1^{-1}([0, (1+\varepsilon_1)R_1])$ tempered from above. This ensures the existence of an attractor in the sets of $\mathbb{R}^2$ tempered from above.

(ii) We now use another Lyapunov function V_2, and apply the procedure of step (i) to $W(x_t)$, where

$$W := \frac{1}{V_2} : \mathbb{R}^2 \setminus \{0\} \to \mathbb{R}^+ \setminus \{0\}.$$

We arrive at an exponentially stable affine RDE dominating $W(x_t)$, hence at a (non-compact) forward invariant set for the original $V_2(x_t)$ given by $[(1-\varepsilon_2)R_2, \infty)$ which absorbs all sets tempered from below. Hence $B_2 = V_2^{-1}([(1-\varepsilon_2)R_2, \infty))$ is an absorbing set for all sets tempered from below.

(iii) Finally, the intersection $B := B_1 \cap B_2$ is a non-empty absorbing set for all sets in $\mathbb{R}^2 \setminus \{0\}$ tempered from below and from above. That B is non-void follows from the fact that both B_1 and B_2 absorb fixed points.

By comparing the Lyapunov functions V_1 and V_2 it can be shown that there are values of $\varepsilon_1, \varepsilon_2 \in (0,1)$ for which $B(\omega)$ is C^∞-diffeomorphic to a random annulus around 0. $\square$

Normal Form, Real Noise Case

Applying the linear transformation

$$T := \frac{1}{2} \begin{pmatrix} 1 & 0 \\ \frac{\beta}{\omega_d} & -\frac{1}{\omega_d} \end{pmatrix}, \quad \omega_d := \sqrt{-\frac{\beta^2}{4} - \alpha},$$

to (9.4.12), linearizing at $x = 0$, and writing the linearized RDE in polar coordinates $v_1 = r\cos\gamma$, $v_2 = r\sin\gamma$ gives

$$\dot{r}_t = \left(\frac{\beta}{2} - \frac{\sigma}{2\omega_d}(\sin 2\gamma_t)\xi(t)\right) r_t, \quad \dot{\gamma}_t = \omega_d - \frac{\sigma}{2\omega_d}(1 + \cos 2\gamma_t)\xi(t). \quad (9.4.25)$$

We immediately obtain the estimates

$$\frac{\beta}{2} - \frac{m|\sigma|}{2\omega_d} \leq \lambda_2 \leq \lambda_1 \leq \frac{\beta}{2} + \frac{m|\sigma|}{2\omega_d}, \quad m := \mathbb{E}|\xi(t)|, \qquad (9.4.26)$$

which (for $\alpha = -1$) are sharper than the estimate (9.4.22).

In Subsect. 8.4.3 we calculated the (truncated) normal form of the random Duffing-van der Pol equation (9.4.12) for fixed $\alpha < 0$ and small (x, β, σ) in polar coordinates.

In order to obtain a more tractable, simpler RDE we omit further higher order terms from the normal form of Subsect. 8.4.3 and arrive at the simplified normal form equations

$$\dot{r}_t = \left(\frac{\beta}{2} - \frac{\sigma}{2\omega_d}(\sin 2\gamma_t)\xi(t)\right) r_t - \frac{1}{2} r_t^3, \qquad (9.4.27)$$

$$\dot{\gamma}_t = \omega_d - \frac{\sigma}{2\omega_d}(1 + \cos 2\gamma_t)\xi(t) + \frac{3\beta + 6}{4\omega_d} r_t^2. \qquad (9.4.28)$$

These equations were derived by a scaling argument in [37], where extensive numerical simulations were also made. It was shown that the bifurcation behavior of the simplified normal form equations (9.4.27) and (9.4.28) is qualitatively the same as that of the full equation (9.4.12).

To demonstrate that equations (9.4.27) and (9.4.28) are indeed more tractable than the original equations (after all, they differ from the linearized equations (9.4.25) by only one term), we study their attractor.

We write

$$\eta_\beta^\pm(t) := \frac{\beta}{2} \pm \frac{\sigma}{2\omega_d} |\xi(t)|,$$

and note that by (9.4.26) $\lambda_1 < 0$ if $\mathbb{E}\eta_\beta^+ < 0$, and $\lambda_2 > 0$ if $\mathbb{E}\eta_\beta^- > 0$.

The radial part obviously satisfies the estimates

$$\eta_\beta^-(t) r_t - \frac{1}{2} r_t^3 \leq \dot{r}_t \leq \eta_\beta^+(t) r_t - \frac{1}{2} r_t^3.$$

Using the right-hand side it follows from Subsect. 9.3.5 that $\{0\}$ is the attractor, provided $\mathbb{E}\eta_\beta^+ \leq 0$, which by (9.4.26) assures that both Lyapunov exponents of δ_0 are less than or equal to zero.

Further, the stochastic process $y = 1/r^2$ satisfies the RDE $\dot{y} = -2\dot{r}/r^3$ which yields

$$-2\eta_\beta^+(t)y_t + 1 \leq \dot{y}_t \leq -2\eta_\beta^-(t)y_t + 1.$$

Solving the corresponding dominating affine RDE and transforming back to r we see that the attractor A_β^0 of the normal form (9.4.27) and (9.4.28) is contained in the compact forward invariant set

$$\frac{1-\varepsilon}{\sqrt{\int_{-\infty}^0 e^{\int_0^t \eta_\beta^-(s)ds} dt}} \leq \|x\| \leq \frac{1+\varepsilon}{\sqrt{\int_{-\infty}^0 e^{\int_0^t \eta_\beta^+(s)ds} dt}}, \qquad (9.4.29)$$

where $\varepsilon \in (0,1)$, and the right-hand side, respectively the left-hand side of this estimate is omitted if $\mathbb{E}\eta_\beta^+ \leq 0$, respectively if $\mathbb{E}\eta_\beta^- \leq 0$.

This leads to the following theorem (see Schenk-Hoppé [304: Sect. 3.4]).

9.4.10 Theorem (Attractor for Normal Form). *Suppose $\xi \in L^1$. Then the following holds:*

(i) The truncated normal form equations (9.4.27) and (9.4.28) generate a local C^∞ RDS φ_β which is strictly forward complete.

(ii) φ_β has a random attractor A_β which attracts all families of subsets of $\mathbb{R}^2$. Further, $A_\beta = E_\beta$, E_β the set of never exploding initial values of φ_β, and A_β supports all invariant measures. We have

$$A_\beta \begin{cases} = \{0\}, & \mathbb{E}\eta_\beta^+ \leq 0, \\ \subset B_{\beta,\varepsilon}, & \mathbb{E}\eta_\beta^+ > 0, \end{cases} \quad (9.4.30)$$

where $B_{\beta,\varepsilon}$ is the random ball with radius defined by the right-hand side of (9.4.29). Moreover, for $\mathbb{E}\eta_\beta^+ \downarrow 0$, the upper bound on the attractor tends to zero.

(iii) If $\mathbb{E}\eta_\beta^- > 0$, the RDS has an attractor A_β^0 in the subsets of the punctured plane $\mathbb{R}^2 \setminus \{0\}$ which are tempered from below. The attractor is contained in the annulus defined by (9.4.29), and $A_\beta^0 \subsetneq A_\beta$.

Proof. (i) Strict forward completeness can be proved as above using the Lyapunov function $V(r) = r^2$.

(ii) The ball whose radius is the right-hand side of (9.4.29) is tempered from above, and attracts all families tempered from above. Hence there is a unique attractor which is tempered from above.

To prove that $A_\beta = E_\beta$ it suffices to show that E_β belongs to the corresponding universe, i.e. is tempered from above. Due to strict forward completeness, $E_\beta(\omega) = E_\beta^+(\omega) := \cap_{t<0} D(t,\omega)$. Using $\dot{r} \leq (\eta_{|\beta|}^+(t) + 1)r - r^3/2$ we see that E_β is contained in the set of never exploding initial values of the dominating RDE. The latter set, hence the former, is tempered from above.

It follows that $A_\beta = E_\beta$ attracts $\mathbb{R}^2$, hence all other families of subsets of $\mathbb{R}^2$.

(iii) The annulus (9.4.29) is an absorbing set which is tempered in the punctured plane $\mathbb{R}^2 \setminus \{0\}$. □

Computer simulations show that the RDS φ_β behaves exactly as described at the beginning of Subsect. 9.4.2: The attractor is $A_\beta = \{0\}$ for $\lambda_1(\beta) \leq 0$, it is a one-dimensional set if $\lambda_2(\beta) < 0 < \lambda_1(\beta)$, and finally for $\lambda_2(\beta) > 0$, A_β is a (topological) random ball around $x = 0$, while A_β^0 is a (topological) circle, with $\partial A_\beta = A_\beta^0$. We are, however, not able to prove this yet.

9.4.3 Pitchfork Bifurcation

What is Stochastic Pitchfork Bifurcation?

Consider now, for example, the SDE (9.4.13) for $\beta < 0$ frozen, while α moves up across $\alpha = 0$. Denote the corresponding RDS by φ_α.

On the phenomenological level of the Fokker-Planck equation, computer simulations [302] show that for α up to (and to some extent beyond) $\alpha = 0$, the Dirac measure δ_0 is the only stationary measure. Increasing α well beyond 0, we witness the delayed birth of a non-trivial stationary density p_α at $\alpha = \alpha_D > 0$. This density has a pole at $x = 0$ (which persists for all $\alpha > \alpha_D$) and two pronounced peaks shifted away from 0 as α increases (see Fig. 9.14, for which $\alpha_D = 0.125\ldots$).

On the dynamical level, the trivial measure δ_0 is the only invariant measure for $\alpha \leq \alpha_D$, and $A_\alpha = \{0\}$ is the attractor of the RDS φ_α in $\mathbb{R}^2$. δ_0 loses its stability at $\alpha = \alpha_D$, where the top Lyapunov exponent $\lambda_1(\sigma, \alpha)$ becomes positive, and we witness the bifurcation of one (!) new ergodic invariant measure $\nu^1 = \frac{1}{2}(\delta_{x(\omega)} + \delta_{-x(\omega)})$ which is stable. It is supported by the two "boundary points" of the one-dimensional unstable manifold of $x = 0$, the closure of which forms the attractor A_β in $\mathbb{R}^2$. The attractor in the punctured plane $\mathbb{R}^2 \setminus \{0\}$ (in the universe of simply connected tempered sets) is $A_\alpha^0(\omega) = \{\pm x(\omega)\}$ (see Fig. 9.15, where $\beta = -1$ and $\sigma = 1.5$).

The deformation of a grid of points by $x \mapsto \varphi(-t, \omega)^{-1}x$ as $t \to \infty$ is shown in Figs. 9.16 and 9.17 (with $\beta = -1$, $\sigma = 0.5$ and time increment 0.001).

Why does only one measure bifurcate, in contrast to the bifurcation of two new steady states $(\pm\sqrt{\alpha}, 0)$ in the deterministic case? The intuitive reason is as follows: Assume $\alpha > -\frac{\beta^2}{4}$, so that the eigenvalues of the deterministic system are real and distinct. It might, however, happen that $\alpha + \sigma\xi(t) < -\frac{\beta^2}{4}$ with positive probability (which is always the case for white noise, $\xi(t) = \dot{W}_t$, irrespective of the size of σ). This means that the noisy system is pushed into parameter regions where there are complex-conjugate eigenvalues, hence it picks up rotation which makes it impossible to distinguish between $x(\omega)$ and $-x(\omega)$. This will not occur if we assume that the real noise $\xi(t)$ satisfies

$\sigma\xi(t) > -\frac{\beta^2}{4} - \alpha$, in which case we have the bifurcation of two new stable Dirac measures $\delta_{\pm x(\omega)}$. This phenomenon demonstrates again that stochastic bifurcation is typically not local with respect to parameters.

What can we say about the location of α_D? The two different Lyapunov exponents of the linearized SDE (9.4.16) satisfy $\lambda_1(\sigma, \alpha) + \lambda_2(\sigma, \alpha) = \beta < 0$. Their second order asymptotic expansion for $\alpha > -\frac{\beta^2}{4}$ (which we assume) for small σ was obtained by Pardoux and Wihstutz [270: Theorem 5.3], and others.

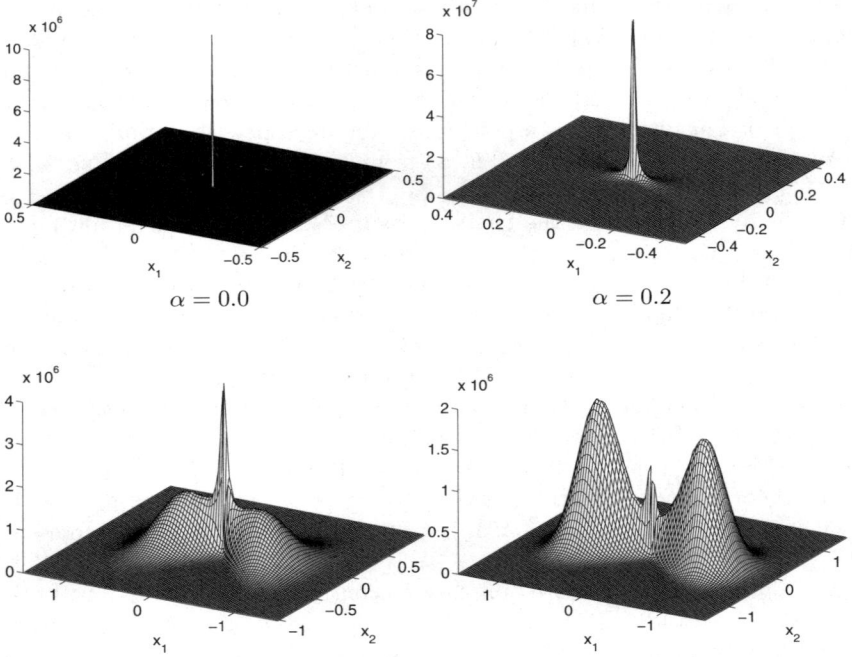

Fig. 9.14. Stationary density in the pitchfork regime for $\beta = -1$ and $\sigma = 0.5$

9.4.11 Theorem (Small Noise Expansion of Lyapunov Exponent, Case of Real Eigenvalues). *Let*

$$dx_t = \begin{pmatrix} a_1 & 0 \\ 0 & a_2 \end{pmatrix} x_t dt + \sigma \sum_{j=1}^{m} A_j x_t \circ dW_t^j,$$

where $a_1 > a_2$ and $A_j = (a_{kl}^j)_{2 \times 2}$, $j = 1, \ldots, m$. Then as $\sigma \to 0$ the top Lyapunov exponent $\lambda_1(\sigma)$ satisfies

$$\lambda_1(\sigma) = a_1 + \frac{\sigma^2}{2} \sum_{j=1}^{m} a_{12}^j a_{21}^j + O(\sigma^4). \tag{9.4.31}$$

In our case,

$$\lambda_{1,2}(\sigma, \alpha) = \frac{\beta}{2} \pm \sqrt{\frac{\beta^2}{4} + \alpha} \mp \frac{\sigma^2}{8(\frac{\beta^2}{4} + \alpha)} + O(\sigma^4). \tag{9.4.32}$$

For the real noise case, a similar expansion holds: For example, if $\xi(t) = f(\eta(t))$, where $\eta(t)$ is an elliptic diffusion on a compact manifold and f is smooth and not constant, then the spectrum is simple, and for $\mathbb{E}\,\xi = 0$

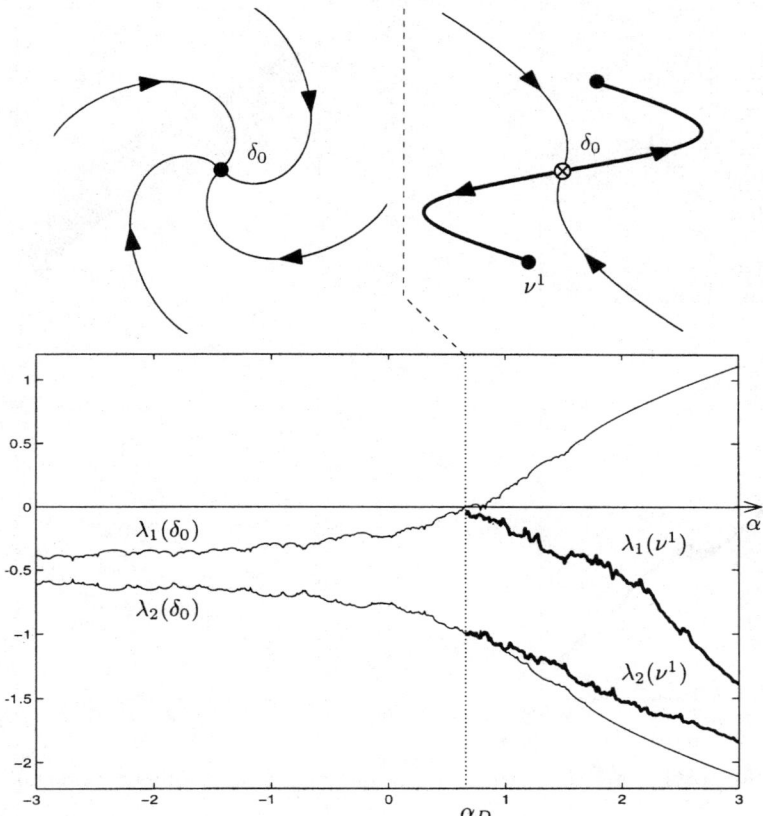

Fig. 9.15. Lyapunov exponents, invariant measures and attractors in the pitchfork regime

$$\lambda_{1,2}(\sigma,\alpha) = \frac{\beta}{2} \pm \sqrt{\frac{\beta^2}{4}+\alpha} \mp \frac{\sigma^2}{8(\frac{\beta^2}{4}+\alpha)} S_{\text{hyp}}(\sqrt{\beta^2+4\alpha}) + O(\sigma^4), \quad (9.4.33)$$

where $S_{\text{hyp}}(a) := 2\int_0^\infty C(t)e^{-at}dt$, and $C(t) = \mathbb{E}\,\xi(t)\xi(0)$ is the correlation function of $\xi(t)$ (see Lücke and Schank [246] and others [16]). Note that, with F denoting the spectral distribution of $\xi(t)$,

$$\int_0^\infty C(t)e^{-at}dt = \frac{a}{\pi}\int_0^\infty \frac{1}{a^2+\lambda^2}dF(\lambda) > 0.$$

(9.4.33) matches the white noise result (9.4.32) where $S_{\text{hyp}}(a) \equiv 1$.

Consequently, at the deterministic bifurcation point $\alpha = 0$ we still have (for the white noise case) $\lambda_1(\sigma,0) = -\frac{\sigma^2}{2\beta^2} + O(\sigma^4) < 0$, and the top Lyapunov exponent changes sign at

514   Chapter 9. Bifurcation Theory

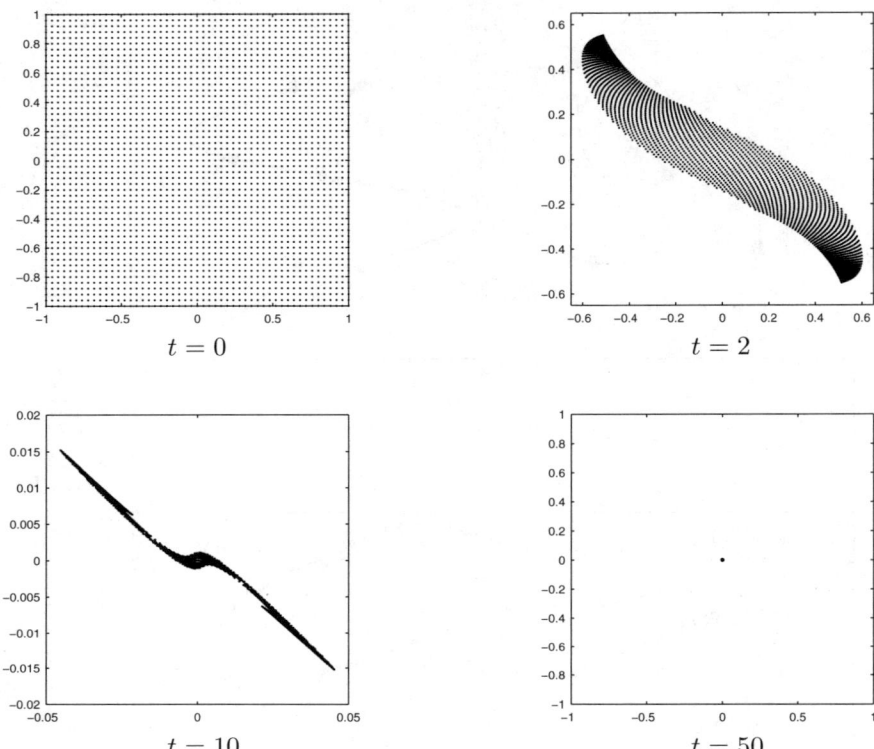

Fig. 9.16. Supports of invariant measures and attractors in the pitchfork regime for $\alpha = 0 < \alpha_D = 0.125\ldots$

$$\alpha_D = -\frac{1}{2\beta}\sigma^2 + O(\sigma^4) > 0,$$

while the second exponent remains always less than $\beta < 0$.

We now search for a P-bifurcation point α_P of p_α, i.e. for a parameter value α at which the moment Lyapunov exponent satisfies $g_{\sigma,\alpha}(-2) = 0$. Using Proposition 9.5.6 (see Example 9.5.11) such a value does not exist since $g_{\sigma,\alpha}(-2) = -\beta > 0$.

Unfortunately, not much of the above numerically observed bifurcation scenario has been rigorously proved.

Normal Forms, Real Noise

In Subsect. 8.4.3 we have derived the truncated scalar center RDE for the pitchfork scenario under small real noise perturbations for the parameter value $\beta = -1$. In a new coordinate system (again denoted by (x_1, x_2)) given

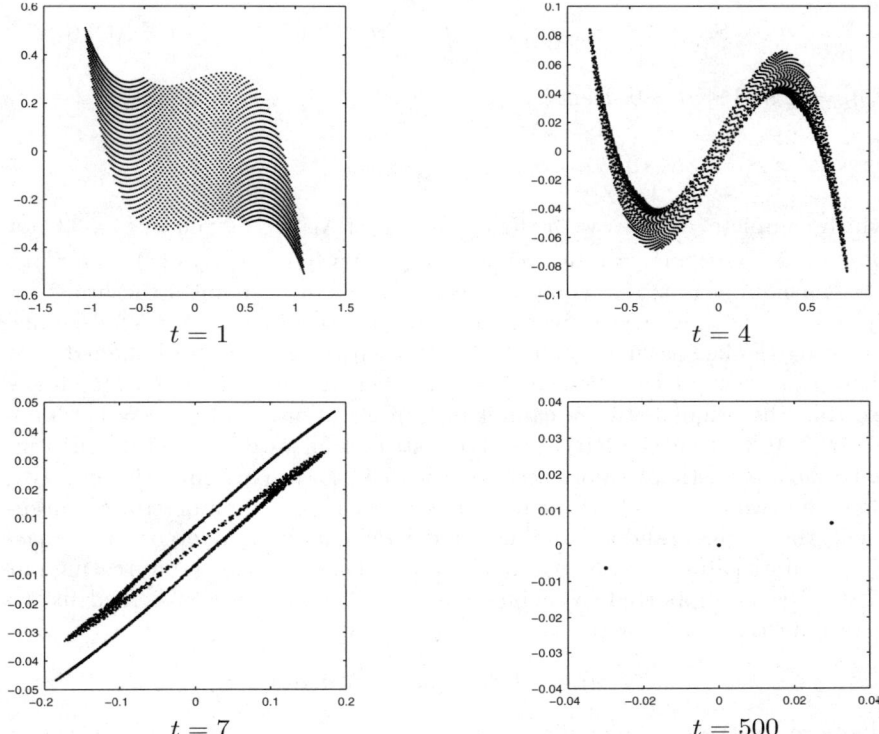

Fig. 9.17. Supports of invariant measures and attractors in the pitchfork regime for $\alpha = 0.2 > \alpha_D = 0.125\ldots$

by the eigenvectors $v_1 = (1,0)'$ and $v_2 = \frac{1}{\sqrt{2}}(1,-1)'$ corresponding to the eigenvalues 0 and -1 of the deterministic linearized equation at the bifurcation point $\alpha = 0$, the truncated RDE for the center variable $z = x_1$ is

$$\dot{z}_t = (\alpha + \sigma\xi(t) - \alpha^2 + \sigma^2\eta_1(t) + \alpha\sigma\eta_2(t))z_t + (-1 + 3\alpha + \sigma\eta_3(t))z_t^3 \quad (9.4.34)$$

(we have skipped for simplicity the σ^2, $\alpha\sigma$ and α^2 terms in the coefficient of z^3) where $\xi \in L^2$ (we want to use spectral theory of stationary processes) and $\mathbb{E}\,\xi = 0$, $\eta_1(t) = \frac{1}{\sqrt{2}}\xi(t)\eta(t)$, $\eta_2(t) = \frac{1}{\sqrt{2}}\eta(t) - \xi(t)$, $\eta_3(t) = \xi(t) - \sqrt{2}\eta(t)$, and

$$\eta(t) = -\sqrt{2}e^{-t}\int_{-\infty}^{t} e^s \xi(s)\,ds.$$

The scalar RDE (9.4.34) can now be utilized to study the bifurcation behavior of the original two-dimensional random Duffing-van der Pol equation (9.4.12). We first calculate the Lyapunov exponent $\lambda_c(\sigma, \alpha)$ of (9.4.34) for δ_0. We have

$$\mathbb{E}\,\eta_1 = -\frac{1}{2}S_{\text{hyp}}(1), \quad S_{\text{hyp}}(a) := 2\int_0^\infty C(t)e^{-at}dt, \quad C(t) = \mathbb{E}\,\xi(t)\xi(0),$$

$\mathbb{E}\,\eta_2 = 0$, and $\mathbb{E}\,\eta_3 = 0$. Hence

$$\lambda_c(\sigma,\alpha) = \alpha - \alpha^2 - \frac{\sigma^2}{2}S_{\text{hyp}}(1) + \text{h.o.t.},$$

which completely agrees with the expansion of $\lambda_1(\sigma,\alpha)$ given in (9.4.33) for $\beta = -1$. In particular, we recover $\lambda_c(\sigma,\alpha) = 0$ if $\alpha_D = \frac{\sigma^2}{2}S_{\text{hyp}}(1) + O(\sigma^4)$.

The nonlinear scalar equation (9.4.34) has the general structure $\dot{z}_t = \xi_1(t)z_t - \xi_2(t)z_t^3$, hence can be explicitly solved, see Subsect. 9.3.5 for details.

In particular, assume $|\xi(t)| \leq M$. Then all the η_i are also bounded, and there is a constant $C > 0$ such that whenever $|\alpha|, |\sigma| \leq C$, $\alpha + \sigma\xi(t) > -\frac{1}{4}$ (so that the original system cannot pick up rotation), and $\xi_2(t) = 1 - 3\alpha - \sigma\eta_3(t) > 0$ (so that the RDE (9.4.34) is strictly forward complete). Then we have the stochastic pitchfork scenario for (9.4.34). In particular, for $\alpha > \alpha_D$, there are two new stable Dirac measures $\nu^\pm = \delta_{\pm z(\omega)}$. By a persistence argument, the original random Duffing-van der Pol equation undergoes a stochastic pitchfork bifurcation at α_D, and the new Dirac measures bifurcating out of $x = 0$ are supported by points $\pm x(\omega)$ which can be represented in the deterministic $(x_1, x_2) = (x_c, x_s)$ coordinate system as

$$x(\omega) = (z(\omega), m_c(\omega, z(\omega), \sigma, \alpha)),$$

where m_c is the center manifold determined in Subsect. 8.4.3. A treatment of this situation including computer simulations was done by Xu [347].

Normal Forms, White Noise

In Example 8.5.19 we have derived the truncated scalar center SDE (8.5.47) for the pitchfork scenario of the stochastic Duffing-van der Pol equation for $\beta = -1$ and small α and σ. In the coordinate system described in the real noise case, after dropping the α^2, $\alpha\sigma$, and σ^2 terms in the z^3 coefficient we obtain

$$\begin{aligned}dz_t &= (\alpha - \alpha^2 - \alpha\sigma U_t)z_t dt + (\sigma - \alpha\sigma - \sigma^2 U_t)z_t \circ dW_t \quad (9.4.35)\\ &\quad + (-1 + 3\alpha + 2\sigma U_t)z_t^3 dt + \sigma z_t^3 \circ dW_t.\end{aligned}$$

Here U_t is the stationary Ornstein-Uhlenbeck process solving $dU_t = -U_t\,dt + dW_t$, which is adapted.

Equation (9.4.35) can now be utilized for stochastic bifurcation theory of the original two-dimensional SDE (9.4.13).

Linearizing the SDE (9.4.35) at $z = 0$ gives

$$dv_t = (\alpha - \alpha^2 - \alpha\sigma U_t)v_t\,dt + (\sigma - \alpha\sigma - \sigma^2 U_t)v_t \circ dW_t,$$

which yields the Lyapunov exponent

$$\lambda_c(\sigma,\alpha) = \alpha - \alpha^2 - \sigma^2 \lim_{t\to\infty} \frac{1}{t} \int_0^t U_s \circ dW_s = \alpha - \alpha^2 - \frac{\sigma^2}{2},$$

where we have used

$$\int_0^t U_s \circ dW_s = \int_0^t U_s\, dW_s + \frac{t}{2}, \quad \text{and} \quad \lim_{t\to\infty} \frac{1}{t} \int_0^t U_s\, dW_s = 0.$$

This is in full agreement with the asymptotic formula (9.4.32) for the top Lyapunov exponent $\lambda_1(\alpha,\sigma)$. In particular $\lambda_c(\sigma,\alpha) = 0$ if $\alpha_D = \frac{\sigma^2}{2} + O(\sigma^4)$.

The SDE (9.4.35) can be explicitly solved by converting it by means of $y = -\frac{1}{2z^2}$ into an affine SDE, see Subsect. 2.3.7. However, the presence of the term $\sigma z^3 \circ dW$ entails that all forward orbits explode, and $E_{\sigma,\alpha} = \{0\}$. We believe that this reflects the (inevitable) presence of rotations.

The Noisy Duffing Equation

Beginning already at an early stage of the subject, great efforts were devoted to the study of the bifurcation behavior of the parameter-driven damped anharmonic oscillator (also known as the Duffing equation)

$$\ddot{y} = (\alpha + \sigma\xi(t))y + \beta\dot{y} - y^3, \qquad (9.4.36)$$

which is (9.4.10) for the particular case $\sigma_2 = \sigma_3 = 0$, $\sigma_1 = \sigma$ and $\xi_1(t) = \xi(t)$ (locally integrable) real or white noise.

By Remark 9.4.3 (ii), the RDS generated by (9.4.36) is global in both cases, for all parameter values (α,β,σ).

For $\beta < 0$ frozen, $\sigma = 0$ and α the bifurcation parameter, (9.4.36) undergoes a pitchfork bifurcation at $\alpha = 0$, where the two new steady states $(\pm\sqrt{\alpha}, 0)$ bifurcate out of $x = 0$. For $\sigma \neq 0$, the linear analysis is the same as for the noisy Duffing-van der Pol equation. Physicist's style small noise expansions for the "stability threshold" $\alpha_D > 0$, the bifurcating "solutions" and their moments are given by Lücke and Schank [246], and Lücke [245]. Ariaratnam and Xie [6, 7] give, for the white noise case, an exact formula (in terms of the stationary measure on S^1) and a small noise expansion of the top Lyapunov exponent, and investigate for $\alpha > \alpha_D$ the bifurcated solution of the Fokker-Planck equation by the method of stochastic averaging.

The top Lyapunov exponent as a function of β and σ for $\alpha = -1$ and the bifurcation diagram with respect to the parameter σ was numerically calculated for the white noise case by Wedig [344, 343] and investigated analytically by Baxendale [63: Example 8.6].

9.5 General Dimension. Further Studies

9.5.1 Baxendale's Sufficient Conditions for D-Bifurcation and Associated P-Bifurcation

We will now present a general criterion for a D-bifurcation of a branch of forward Markov measures out of the fixed point 0 for a family of SDE in $\mathbb{R}^d$, and for a subsequent P-bifurcation of the new branch in a neighborhood of 0. These results were obtained by Baxendale in two profound papers [60, 63], on which this subsection is based. As the matter is very technical, and since most proofs in the above papers are based on earlier results from [64] and [61], we have decided not to reproduce the proofs here, but rather to give a complete and self-contained formulation of the results with some intuitive explanations.

Let

$$dx_t = \sum_{j=0}^{m} f_j^\alpha(x_t) \circ dW_t^j$$

$$= (f_0^\alpha(x_t) + \frac{1}{2}\sum_{j=1}^{m} D_{f_j^\alpha} f_j^\alpha(x_t))dt + \sum_{j=1}^{m} f_j^\alpha(x_t)\, dW_t^j \quad (9.5.1)$$

$(dW_t^0 = dt)$ be a family of SDE in $\mathbb{R}^d$ with $\alpha \in \mathbb{R}$ (which is only chosen to simplify presentation – all results are true for α in an open set of some $\mathbb{R}^k$) for which $(\alpha, x) \mapsto f_j^\alpha(x)$ is C^∞, and where $D_f g(x) := \sum_{i=1}^{d} f^i(x) \frac{\partial g}{\partial x_i}(x)$. Hence (9.5.1) generates for each α a local C^∞ RDS φ_α.

The associated (possibly explosive) Markov diffusion process $x_t^\alpha(x) = \varphi_\alpha(t, \cdot, x)$, $t \in \mathbb{R}^+$, in $\mathbb{R}^d$ has generator

$$L^\alpha = f_0^\alpha + \frac{1}{2}\sum_{j=1}^{m}(f_j^\alpha)^2 = \sum_{i=1}^{d} b_\alpha^i \frac{\partial}{\partial x_i} + \frac{1}{2}\sum_{k,l=1}^{d} a_\alpha^{kl} \frac{\partial^2}{\partial x_k \partial x_l}, \quad (9.5.2)$$

where

$$b_\alpha^i(x) = (f_0^\alpha(x))^i + \frac{1}{2}\sum_{j=1}^{m}\sum_{k=1}^{d}(f_j^\alpha(x))^k \frac{\partial (f_j^\alpha)^i}{\partial x_k}(x)$$

and

$$a_\alpha^{kl}(x) = \sum_{j=1}^{m}(f_j^\alpha(x))^k (f_j^\alpha(x))^l.$$

We now assume that

$$f_0^\alpha(0) = f_1^\alpha(0) = \ldots = f_m^\alpha(0) = 0 \quad \text{for all } \alpha \in \mathbb{R}, \quad (9.5.3)$$

so that $\mu_\alpha = \delta_0$ is an invariant measure for φ_α for all $\alpha \in \mathbb{R}$. This branch of trivial measures will serve as a reference branch from which we will study

9.5 General Dimension. Further Studies

the bifurcation of a new branch of measures ν_α supported by $\mathbb{R}^d \setminus \{0\}$ at the parameter value α_D at which the reference measure becomes unstable.

To this end we linearize the SDE (9.5.1) at $x = 0$ and obtain the linear SDE

$$\begin{aligned} dv_t &= A_0^\alpha v_t dt + \sum_{j=1}^m A_j^\alpha v_t \circ dW_t^j \\ &= (A_0^\alpha + \frac{1}{2}\sum_{j=1}^m (A_j^\alpha)^2) v_t dt + \sum_{j=1}^m A_j^\alpha v_t\, dW_t^j, \end{aligned} \quad (9.5.4)$$

where $A_j^\alpha = Df_j^\alpha(0)$ is the Jacobian of f_j^α at $x = 0$. The SDE generates the linear RDS $\Phi_\alpha(t,\omega) = D\varphi_\alpha(t,\omega,0)$.

The diffusion process $v_t^\alpha(v) = \Phi_\alpha(t,\cdot)v$, $t \in \mathbb{R}^+$, has generator DL^α, say, given by

$$DL^\alpha = \sum_{i=1}^d \left(\sum_{k=1}^d \frac{\partial b_\alpha^i}{\partial x_k}(0) v_k \right) \frac{\partial}{\partial v_i} + \frac{1}{2} \sum_{k,l=1}^d \left(\sum_{p,q=1}^d \frac{\partial^2 a_\alpha^{kl}}{\partial x_p \partial x_q}(0) v_p v_q \right) \frac{\partial^2}{\partial v_k \partial v_l}. \quad (9.5.5)$$

Note that DL^α only depends on the values of the coefficients a_α and b_α of L^α (hence on the values of the vector fields f_j^α) in an arbitrarily small neighborhood of $x = 0$.

We refer to Sect. 6.2 for a detailed treatment of Φ_α. In particular, the MET holds for Φ_α without further integrability conditions, and under the Lie algebra (or hypoellipticity) condition

$$\dim \mathcal{LA}(h_0^\alpha,\ldots,h_m^\alpha)(s) = d-1 \quad \text{for all } s \in P^{d-1},$$

where $h_j^\alpha(s) := A_j^\alpha s - \langle A_j^\alpha s, s\rangle s$ on P^{d-1} are the projected vector fields, the top Lyapunov exponent $\lambda^\alpha = \lambda_1^\alpha$ satisfies

$$\lambda^\alpha = \lim_{t\to\infty} \frac{1}{t} \log \|\Phi_\alpha(t,\omega)v\| \quad \text{for all } v \neq 0,\ \mathbb{P}\text{-a.s.}, \quad (9.5.6)$$

and the moment Lyapunov exponent

$$g^\alpha(p) := \lim_{t\to\infty} \frac{1}{t} \log E\|\Phi_\alpha(t,\cdot)v\|^p, \quad p \in \mathbb{R}, \quad (9.5.7)$$

is well-defined (i.e. the limit exists and is independent of $v \neq 0$). It is well-known that g^α is a convex analytic function of p satisfying $g^\alpha(0) = 0$ and $(dg^\alpha/dp)(0) = \lambda^\alpha$, see Fig. 9.18. Clearly λ^α and $g^\alpha(p)$ control the $\mathbb{P}$-a.s. and p-th moment stability of the linearized RDS Φ_α. In addition, g^α is the Legendre transform of the rate function for large deviations of $\frac{1}{t}\log\|\Phi_\alpha(t,\omega)v\|$ away from λ^α (see Arnold, Oeljeklaus and Pardoux [32], Arnold and Kliemann [26], Stroock [325], Baxendale [58], and Baxendale and Stroock [64]).

It turns out that λ^α and g^α also control the behavior near 0 of the nonlinear process $\varphi_\alpha(t, \cdot, x)$. For the following theorem see Baxendale [63: Theorem 2.8].

9.5.1 Theorem. *Assume that for some fixed $\alpha_0 \in \mathbb{R}$ the following conditions on the SDE (9.5.1) are satisfied:*

(H1) *Condition "at infinity": There exist functions $f, g \in \mathcal{C}(\mathbb{R}^d)$ with $g \geq 1$, positive constants c and R and a neighborhood $N = N(\alpha_0)$ of α_0 such that for each $\alpha \in N$ the Markov process $\varphi_\alpha(t, \cdot, x)$, $t \geq 0$, is complete and there exists $f^\alpha \in \mathcal{C}^2(\mathbb{R}^d)$ satisfying $0 \leq f^\alpha \leq f$, $L^\alpha f^\alpha + g \leq c$, and $L^\alpha f^\alpha(x) + g(x) \leq 0$ for $\|x\| \geq R$.*

(H2) *Condition "between infinity and zero": For all $r > 0$ and $x \neq 0$ there exists $T > 0$ such that $P^{\alpha_0}(T, x, B_r) > 0$, where P^α denotes the Markov transition probability with generator L^α, and $B_r := \{x \in \mathbb{R}^d : \|x\| \leq r\}$.*

(H3) *Condition "at zero": Let $\mathcal{LA}(A_1^{\alpha_0}, \ldots, A_m^{\alpha_0})(v) = \mathbb{R}^d$ for all $v \neq 0$.*[7]
Then the following hold:

(o) There exists an open interval $U = (\beta_1, \beta_2) \ni \alpha_0$ such that for all $\alpha \in U$, λ^α and g^α are well-defined through (9.5.6) and (9.5.7), $\alpha \mapsto \lambda^\alpha$ is continuous in U, and $\alpha \mapsto g^\alpha(p)$ is continuous in U for each $p \in \mathbb{R}$.

(i) If $\lambda^\alpha < 0$ for some $\alpha \in U$, then the local RDS φ_α satisfies

$$\mathbb{P}\left\{\omega : \limsup_{t \to \infty} \frac{1}{t} \log \|\varphi_\alpha(t, \omega, x)\| \leq \lambda^\alpha\right\} = 1 \quad \text{for all } x \neq 0. \quad (9.5.8)$$

In particular, there is no other stationary measure in $\mathbb{R}^d \setminus \{0\}$ for the Markov process generated by L^α and the stable manifold of $x = 0$ is dense in $\mathbb{R}^d$.

(ii) If $\lambda^\alpha > 0$ for some $\alpha \in U$, then there exists a unique probability measure ν_α on $\mathbb{R}^d \setminus \{0\}$ such that

$$\mathbb{P}\left\{\omega : \lim_{T \to \infty} \frac{1}{T} \int_0^T u(\varphi_\alpha(t, \omega, x)) dt = \int_{\mathbb{R}^d \setminus \{0\}} u(y) \nu_\alpha(dy)\right\} = 1 \quad (9.5.9)$$

for all bounded measurable functions $u : \mathbb{R}^d \setminus \{0\} \to \mathbb{R}$ and all $x \neq 0$. In particular, ν_α is the unique stationary measure on $\mathbb{R}^d \setminus \{0\}$ for the Markov process generated by L^α.

Finally, there exists a unique $\gamma_\alpha > 0$ such that $g^\alpha(-\gamma_\alpha) = 0$, and constants $\delta_\alpha > 0$ and $K_\alpha \in (0, \infty)$ for which

$$\frac{1}{K_\alpha} r^{\gamma_\alpha} \leq \nu_\alpha(B_r \setminus \{0\}) \leq K_\alpha r^{\gamma_\alpha} \quad \text{for } 0 < r < \delta_\alpha. \quad (9.5.10)$$

---

[7] For a weaker version of condition (H3) see Baxendale [63: Sect. 2].

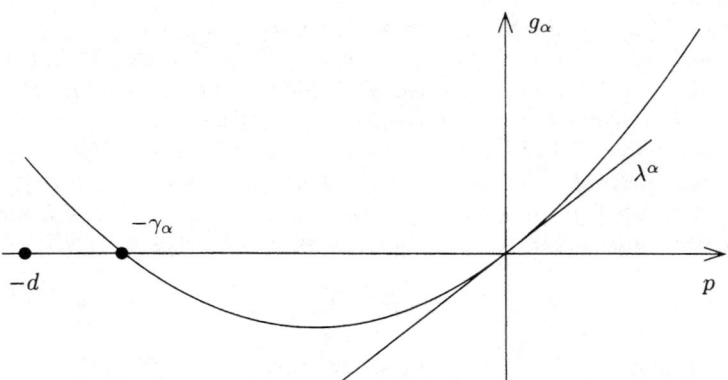

Fig. 9.18. The p-th moment Lyapunov exponent

(iii) *If $\lambda^\alpha = 0$ for some $\alpha \in U$, then there exists a σ-finite (but not finite) measure ν_α on $\mathbb{R}^d \setminus \{0\}$, unique up to a multiplicative constant, such that*

$$\mathbb{P}\left\{\omega : \lim_{T \to \infty} \frac{\int_0^T u(\varphi_\alpha(t,\omega,x))dt}{\int_0^T v(\varphi_\alpha(t,\omega,x))dt} = \frac{\int_{\mathbb{R}^d \setminus \{0\}} u(y)\,\nu_\alpha(dy)}{\int_{\mathbb{R}^d \setminus \{0\}} v(y)\,\nu_\alpha(dy)}\right\} = 1 \quad (9.5.11)$$

for all bounded measurable ν_α-integrable functions $u, v : \mathbb{R}^d \setminus \{0\} \to \mathbb{R}$ with $\int v\,d\nu_\alpha \neq 0$ and for all $x \neq 0$. In particular, ν_α is the unique, up to a multiplicative constant, stationary measure on $\mathbb{R}^d \setminus \{0\}$ for the Markov process generated by L^α. Moreover, $\nu_\alpha(\mathbb{R}^d \setminus B_r) < \infty$ for all $r > 0$, and there exists a $C_\alpha \in (0, \infty)$ such that

$$\lim_{r \to 0} \frac{\nu_\alpha(\mathbb{R}^d \setminus B_r)}{|\log r|} = C_\alpha. \quad (9.5.12)$$

Suppose that in addition to the assumptions (H1), (H2) and (H3) the SDE (9.5.1) is non-degenerate in $\mathbb{R}^d \setminus \{0\}$, for example in the sense that

$$\dim \mathcal{LA}(f_1^\alpha, \ldots, f_m^\alpha)(x) = d \quad \text{for all } x \neq 0, \; \alpha \in \mathbb{R}.$$

Then Theorem 9.5.1 gives a complete classification of the diffusion process $\varphi_\alpha(t, \cdot, x)$ on $\mathbb{R}^d \setminus \{0\}$ as positive recurrent, null recurrent, and transient according to λ^α being positive, zero, or negative.

For $d = 1$ the statements of the theorem apply to φ_α restricted to one of the invariant sets $(-\infty, 0)$ and $(0, \infty)$.

D-Bifurcation

We will now specify a certain scenario in which ν_α actually bifurcates out of the trivial reference measure. The following is a reformulation of part of Baxendale's Theorem 2.13 in [63].

9.5.2 Theorem (Sufficient Condition for D-Bifurcation).

Suppose that there exists an $\alpha_D \in \mathbb{R}$ for which the conditions (H1), (H2), and (H3) of Theorem 9.5.1 are satisfied, and that the function g in (H1) satisfies $g(x) \to \infty$ as $\|x\| \to \infty$. Then the following holds:

Assume that $\lambda^{\alpha_D} = 0$, and that there is an open interval $(\beta_1, \beta_2) \ni \alpha_D$ (contained in the interval U of Theorem 9.5.1(o)) such that $\lambda^\alpha < 0$ for $\alpha \in (\beta_1, \alpha_D)$, and $\lambda^\alpha > 0$ for $\alpha \in (\alpha_D, \beta_2)$. Then $\lim_{\alpha \to \alpha_D} \lambda^\alpha = 0$, and if for $\alpha > \alpha_D$, ν_α denotes the unique stationary measure of Theorem 9.5.1(ii)

$$\lim_{\alpha \downarrow \alpha_D} \nu_\alpha = \mu_{\alpha_D} = \delta_0 \qquad (9.5.13)$$

in the topology of weak convergence of probability measures in $\mathbb{R}^d$. Hence the family φ_α undergoes a D-bifurcation of the new branch of forward Markov measures ν_α from the branch of trivial measures $\mu_\alpha = \delta_0$ at the parameter value α_D where the trivial measure loses its stability.

Further, for $\alpha \in (\alpha_D, \beta_2)$ the nontrivial zero $-\gamma_\alpha$ of g^α figuring in Theorem 9.5.1(iii) fulfills

$$\lim_{\alpha \downarrow \alpha_D} \gamma_\alpha = 0. \qquad (9.5.14)$$

9.5.3 Remark.
Assume the situation of the last theorem. Why can the generator L^α of the one-point motions $\varphi_\alpha(t, \cdot, x)$ "feel" that $v = 0$ becomes unstable for the linearized RDS Φ_α at $\alpha = \alpha_D$, and give birth to a non-trivial solution of the Fokker-Planck equation $(L^\alpha)^* \nu_\alpha = 0$ for $\alpha > \alpha_D$? The principal reason is that the quantities λ^α and g^α are determined by the law of the one-point motions $\Phi_\alpha(t, \cdot)v$ which is determined by DL^α given by (9.5.5) and hence by L^α (in fact, by the coefficients of L^α in an arbitrarily small neighborhood of 0). The smaller Lyapunov exponents are, however, not determined by the one-point motions of Φ_α. See Arnold [8] for a discussion of one-point versus flow quantities. ∎

9.5.4 Remark (Rate of Convergence to Dirac Measure).
Assume the situation of Theorem 9.5.2. Baxendale [63: Theorem 2.13] also determined the rate at which ν_α approaches the Dirac measure in (9.5.13): Let ν_{α_D} denote the unique σ-finite stationary measure of φ_{α_D} on $\mathbb{R}^d \setminus \{0\}$ for which the constant in (9.5.12) is

$$C_{\alpha_D} = \frac{2}{V_{\alpha_D}}, \quad \text{where } V_{\alpha_D} := \frac{d^2 g^{\alpha_D}}{dp^2}(0) > 0.$$

Then

$$\lim_{\alpha \downarrow \alpha_D} \frac{1}{\lambda^\alpha} \nu_\alpha = \nu_{\alpha_D} \qquad (9.5.15)$$

in the sense that

$$\lim_{\alpha \downarrow \alpha_D} \frac{1}{\lambda^\alpha} \int u(x)\nu_\alpha(dx) = \int u(x)\nu_{\alpha_D}(dx)$$

for all continuous functions $u : \mathbb{R}^d \setminus \{0\} \to \mathbb{R}$ satisfying $u(x)/g(x) \to 0$ as $\|x\| \to \infty$ and $u(x)/\|x\|^p \to 0$ as $x \to 0$ for some $p > 0$.

This result incorporates the σ-finite stationary measure ν_{α_D} into the bifurcation scenario at $\alpha = \alpha_D$. ∎

It is still unknown whether the measures ν_α, $\alpha \in (\alpha_D, \beta_2)$, are stable, i. e. have negative top Lyapunov exponent.

P-Bifurcation

Assume again the situation of Theorem 9.5.2. Then (9.5.10) holds for all $\alpha \in (\alpha_D, \beta_2)$. We now compare the mass assigned by ν_α to $B_r \setminus \{0\}$ with that assigned by the Lebesgue measure,

$$\mathrm{Leb}(B_r \setminus \{0\}) = \mathrm{Leb}(B_1) \, r^d,$$

where $\mathrm{Leb}(B_1)$ is the volume of the unit ball B_1 in $\mathbb{R}^d$. Theorem 9.5.1(ii) yields (possibly with a different constant K_α)

$$\frac{1}{K_\alpha} r^{\gamma_\alpha - d} \leq \frac{\nu_\alpha(B_r \setminus \{0\})}{\mathrm{Leb}(B_r \setminus \{0\})} \leq K_\alpha r^{\gamma_\alpha - d} \quad \text{for } 0 < r < \delta_\alpha, \tag{9.5.16}$$

implying that

$$\lim_{r \to 0} \frac{\nu_\alpha(B_r \setminus \{0\})}{\mathrm{Leb}(B_r \setminus \{0\})} = \begin{cases} \infty, & \text{if } \gamma_\alpha < d, \\ 0, & \text{if } \gamma_\alpha > d, \end{cases} \tag{9.5.17}$$

while the ratio remains bounded away from 0 and ∞ for $\gamma_\alpha = d$. Hence, Baxendale's result allows the following interpretation, where we use the term "P-bifurcation" in the sense of the qualitative change of ν_α in a neighborhood of 0 just described (for a discussion of the concept see Subsect. 9.2.2).

9.5.5 Theorem (Sufficient Condition for P-Bifurcation). *Assume the situation of Theorem 9.5.2. Assume further that there exists an $\alpha_P \in (\alpha_D, \beta_1)$ for which $\gamma_{\alpha_P} = d$, but $\gamma_\alpha < d$ for $\alpha < \alpha_P$, and $\gamma_\alpha > d$ for $\alpha > \alpha_P$, and let ν_α be the unique stationary measure of φ_α in $\mathbb{R}^d \setminus \{0\}$.*

Then the family of ν_α undergoes a P-bifurcation at $\alpha = \alpha_P$ in the sense expressed by (9.5.17).

Baxendale's findings are collected in Fig. 9.19.

There is a simple criterion for $g^\alpha(-d) = 0$, for which we recall the following facts:

Consider the linear SDE

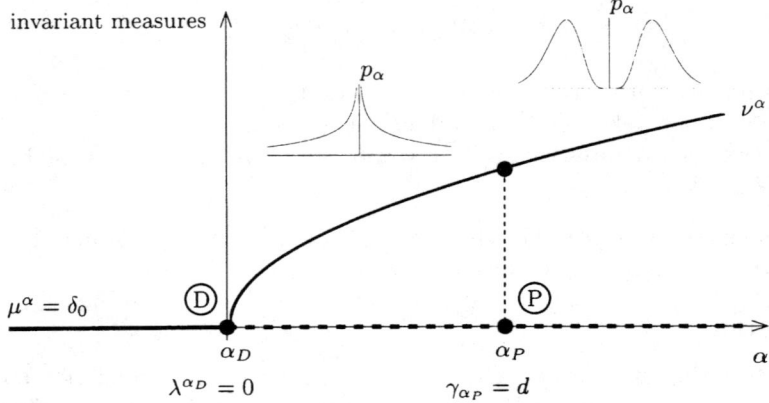

Fig. 9.19. D-bifurcation and P-bifurcation

$$dx_t = A_0 x_t dt + \sum_{j=1}^{m} A_j x_t \circ dW_t^j$$

in $\mathbb{R}^d$ and assume that $\dim \mathcal{LA}(h_0, \ldots, h_m)(s) = d - 1$ for all $s \in P^{d-1}$, where $h_j(s) := A_j s - \langle A_j s, s \rangle s$ is the projection of the linear vector field $x \mapsto A_j x$ onto P^{d-1}. Then the p-th moment Lyapunov exponent $g(p) = \lim_{t \to \infty} \frac{1}{t} \log \mathbb{E} \|\Phi(t, \cdot) v\|^p$, $p \in \mathbb{R}$, is well-defined and independent of $v \neq 0$. Further, the determinant $\Delta(t, \omega) := \det \Phi(t, \omega)$ satisfies the scalar SDE

$$d\Delta_t = (\mathrm{trace}\, A_0) \Delta_t dt + \sum_{j=1}^{m} (\mathrm{trace}\, A_j) \Delta_t \circ dW_t^j$$

which has the solution (Liouville's equation)

$$\det \Phi(t, \omega) = \exp\left((\mathrm{trace}\, A_0) t + \sum_{j=1}^{m} (\mathrm{trace}\, A_j) W_t^j(\omega) \right)$$

and the p-th moment Lyapunov exponent

$$h(p) = \lim_{t \to \infty} \frac{1}{t} \log \mathbb{E} |\det \Phi(t, \cdot)|^p = (\mathrm{trace}\, A_0) p + \frac{p^2}{2} \sum_{j=1}^{m} (\mathrm{trace}\, A_j)^2.$$

9.5.6 Proposition (Criterion for $g(-d) = 0$). *Let the above Lie algebra condition be satisfied. Then*

$$g(-d) = h(-1) = -\mathrm{trace}\, A_0 + \frac{1}{2} \sum_{j=1}^{m} (\mathrm{trace}\, A_j)^2, \qquad (9.5.18)$$

hence

$$g(-d) = 0 \iff \operatorname{trace} A_0 = \frac{1}{2} \sum_{j=1}^{m} (\operatorname{trace} A_j)^2. \tag{9.5.19}$$

Proof. For equation (9.5.18) see Baxendale and Stroock [64: Corollary 2.14)(i)]. □

9.5.7 Remark (P-Bifurcation on the Level of Densities). Let ν_α have a smooth density p_α in $\mathbb{R}^d \setminus \{0\}$ [8]. Then under certain circumstances it follows from Theorem 9.5.5 that the family of densities p_α undergoes a P-bifurcation at $\alpha = \alpha_P$ in a neighborhood of 0 in the sense of Subsect. 9.2.2: For $\alpha_D < \alpha < \alpha_P$, p_α has an (integrable) pole at $x = 0$, p_{α_P} approaches a finite positive limit as $x \to 0$, and for $\alpha_P < \alpha < \beta_1$, $\lim_{x \to 0} p_\alpha(x) = 0$ (see Fig. 9.2). ∎

9.5.8 Remark (Local Dimension of ν_α at $x = 0$). The qualitative change of ν_α at α_P can also be predicted by looking at the local dimension of ν_α at $x = 0$, defined by

$$\lim_{r \to 0} \frac{\log \nu_\alpha(B_r \setminus \{0\})}{\log r} = \gamma_\alpha,$$

and comparing it to the local dimension d of the Lebesgue measure. ∎

9.5.9 Remark (Why is There a P-Bifurcation at α_P?). Consider the scenario assumed in Theorem 9.5.5. When α passes through α_D from left to right, the top Lyapunov exponent λ^α of the reference measure $\mu_\alpha = \delta_0$ becomes positive, hence $x = 0$ becomes "repelling" for $\varphi_\alpha(t, \cdot, x)$. Since by condition (H1), the Markov process cannot explode, it is bound to build up a stationary measure ν_α in $\mathbb{R}^d \setminus \{0\}$. For $\alpha - \alpha_D > 0$ small (hence $\lambda^\alpha > 0$ small), however, the estimates show that $\varphi_\alpha(t, \cdot, x)$ still spends a large amount of time in small punctured neighborhoods of 0, hence the mass of the occupation measure is large around 0 compared to the mass of the Lebesgue measure.

Only if $\alpha > \alpha_P$, hence $\gamma_\alpha > d$, delicate estimates show that the percentage of time spent in a small punctured neighborhood of 0 becomes small compared to its volume, and the new stationary measure of the nonlinear RDS φ_α is finally "freed" from the control by the linear RDS Φ_α.

We find it quite remarkable that the condition for the P-bifurcation of ν_α is a condition on the one-point motions of the linear RDS Φ_α, obtained by linearizing φ_α at $x = 0$.

---

[8] A sufficient condition under which ν_α has a smooth density in $\mathbb{R}^d \setminus \{0\}$ is that L^α is hypoelliptic, which is implied by the Lie algebra condition $\dim \mathcal{LA}(f_0^\alpha, \ldots, f_m^\alpha)(x) = d$ for all $x \neq 0$.

The sequence of two bifurcations at $\alpha = \alpha_D$ and at some point $\alpha = \alpha_P$ was observed by physicists, and the interval $[\alpha_D, \alpha_P]$ was called *bifurcation interval* (see e.g. Ebeling et al. [134]). ∎

Examples

9.5.10 Example (Stochastic Pitchfork and Transcritical Bifurcation in Dimension One). We can verify the assumptions and statements of this subsection for the SDE

$$dx_t = (\alpha x_t - x_t^3)dt + \sigma x_t \circ dW_t,$$

which is treated in great detail in Subsect. 9.3.2. Let us only add that (9.5.19) gives $\alpha_P = \frac{\sigma^2}{2}$, and the σ-finite measure at $\alpha_D = 0$ to which $\frac{1}{\alpha}q_\alpha^+(x)dx$ converges in the sense of Remark 9.5.4 as $\alpha \downarrow 0$ has density

$$q_0^+(x) = \frac{2}{\sigma^2}\frac{1}{x}\exp(-\frac{x^2}{\sigma^2})$$

on $(0, \infty)$, and $\frac{1}{\alpha}q_\alpha^+(x) \to q_0^+(x)$ as $\alpha \downarrow 0$ for all $x > 0$, with the analogous situation on $(-\infty, 0)$. Remember that for $\alpha = 0$, the diffusion is null recurrent on $(0, \infty)$, but nevertheless $\lim_{\alpha \downarrow 0} q_\alpha^+ = \delta_0$, and $\{0\}$ is the attractor in $\mathbb{R}^2$.

For the analogous results for the transcritical case (see Subsect. 9.3.1)

$$dx_t = (\alpha x_t - x_t^2)dt + \sigma x_t \circ dW_t$$

we restrict our considerations to the state space $[0, \infty)$. ∎

9.5.11 Example (Hopf and Pitchfork Bifurcation for the Stochastic Duffing-van der Pol Equation). We briefly return to the bifurcation scenarios of the stochastic Duffing-van der Pol equation dealt with at length in Subsects. 9.4.2 (Hopf bifurcation) and 9.4.3 (pitchfork bifurcation). Due to the lack of rigorous results we have to rely on a small noise analysis of λ^α to determine the (first and second) D-bifurcation point. The point of P-bifurcation can, however, be calculated explicitly by means of (9.5.19): Since here $m = 1$, trace$A_0 = \beta$ and trace$A_1 = 0$, we obtain $h(p) = \beta p$, and $h(-1) = 0 \iff \beta = 0$. We thus have a P-bifurcation at $\beta_P = 0$ in the Hopf case and no P-bifurcation in the neighborhood of $x = 0$ in the pitchfork case since $g(-2) = h(-1) = -\beta > 0$, hence $d = 2 > \gamma_\alpha > 0$ for all values of $\alpha > \alpha_D$. ∎

9.5.12 Example (Averaged Duffing-van der Pol Equation in the Hopf Scenario). Starting with the random Duffing-van der Pol equation (9.4.12) with $\alpha < 0$ fixed and β the bifurcation parameter, assuming that ξ_t satisfies $\mathbb{E}\,\xi_t = 0$ and certain strong mixing conditions (see Khasminskii [204], or Freidlin and Wentzell [152: Chap. 7, §9]), scaling the state, the noise intensity and the dissipation as $x \to \varepsilon x$, $\sigma \to \varepsilon \sigma$, and $\beta \to \varepsilon^2 \beta$, respectively, the *method of stochastic averaging* states that the solution of the so transformed original RDE (9.4.12) can be approximated in distribution on a time interval of length $O(\frac{1}{\varepsilon^2})$ by the solution of an averaged SDE which in polar coordinates is

$$dr_t = (\lambda_0 - Rr_t^3)dt + \sigma_1 r_t \circ dW_t^1, \quad (9.5.20)$$
$$d\phi_t = (-a + Sr_t^2)dt + \sigma_2 \circ dW_t^2 \quad (9.5.21)$$

(for details see Arnold, Sri Namachchivaya and Schenk-Hoppé [37]). The parameters here are

$$\lambda_0 = \frac{\beta}{2} + \sigma_1^2, \quad R = \frac{1}{2}, \quad S = \frac{6+3\beta}{4\omega_d}, \quad a = \kappa_2(2\omega_d), \quad \sigma_1^2 = \kappa_1(2\omega_d),$$

$$\sigma_2^2 = \kappa_1(2\omega_d) + 2\kappa_2(0), \quad \kappa_1(\cdot) = \frac{1}{2}\left(\frac{\sigma}{2\omega_d}\right)^2 S(\cdot), \quad \kappa_2(\cdot) = \frac{1}{2}\left(\frac{\sigma}{2\omega_d}\right)^2 T(\cdot),$$

where the sine and cosine spectral densities are defined, respectively, as

$$T(z) = 2\int_0^\infty C(t)\sin zt\, dt, \quad S(z) = 2\int_0^\infty C(t)\cos zt\, dt,$$

$C(t) = \mathbb{E}\,\xi_0\xi_t$ being the covariance function of ξ_t. The case $a = -1$, $S = 0$ and $\sigma_2 = 0$ was treated by Baxendale [63: Example 8.4].

The system (9.5.20), (9.5.21) can be explicitly solved by first solving (9.5.20) via the transformation $u = \frac{1}{r^2}$ and then inserting this solution into (9.5.21). This permits a complete explicit analysis of the averaged system.

The only invariant measures are the following two Markov measures:

(i) For all $\beta \in \mathbb{R}$, $\mu_\beta = \delta_0$, with Lyapunov exponents $\lambda_1(\delta_0) = \lambda_2(\delta_0) = \lambda_0 = \frac{\beta}{2} + \sigma_1^2$, and $g(p) = p\lambda_0 + \frac{p^2}{2}\sigma_1^2$. Hence

$$\beta_D = -2\sigma_1^2 < 0, \quad \beta_P = 0.$$

The rotation number is $\rho(\delta_0) = -a$.

(ii) For $\beta > \beta_D$,

$$\nu_{\beta,\omega}(dr, d\phi) = \delta_{\kappa_\beta(\omega)}(dr)\frac{d\phi}{2\pi\kappa_\beta(\omega)},$$

where

$$\kappa_\beta(\omega) := \frac{1}{\sqrt{2R\int_{-\infty}^0 \exp(2(\lambda_0 t + \sigma_1 W_t^1))dt}},$$

which is the disintegration of the solution of the Fokker-Planck equation with density in Euclidean coordinates given by

$$p_\beta(x_1, x_2) = \frac{2R}{\sigma_1^2\Gamma(\gamma)}(x_1^2 + x_2^2)^{\gamma-1}\exp\left(-\frac{R}{\sigma_1^2}(x_1^2 + x_2^2)\right), \quad \gamma = 1 + \frac{\beta}{2\sigma_1^2}.$$

The density has an integrable pole at $x = 0$ for $\beta_D < \beta < \beta_P$, the finite maximum $\frac{2R}{\sigma_1^2}$ at the origin for β_P, vanishes at zero and has a maximum value away from zero for $\beta > \beta_P$, as predicted. The σ-finite stationary measure at β_D has density

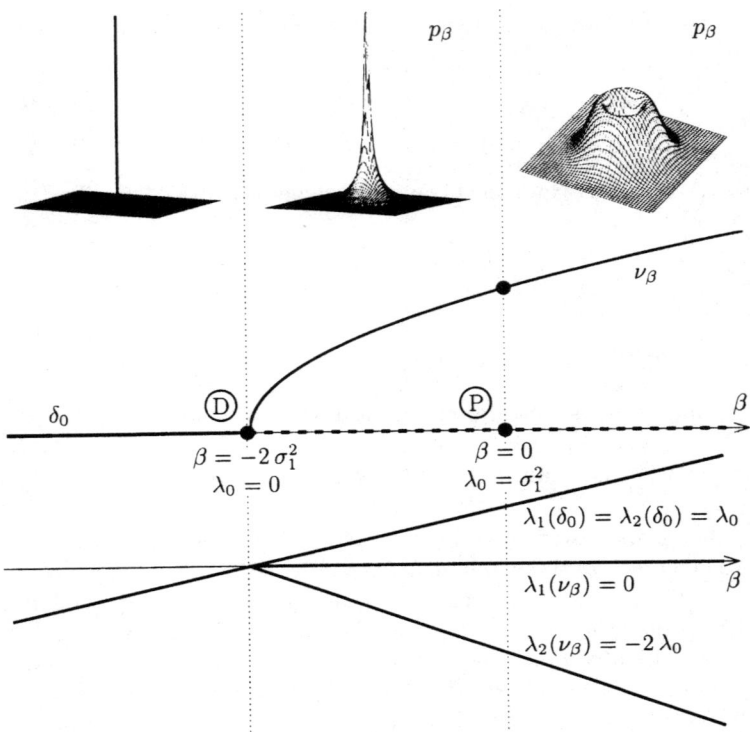

Fig. 9.20. Bifurcation diagram of the averaged system

$$p_{\beta_D}(x_1, x_2) = \frac{2}{\sigma_1^2}(x_1^2 + x_2^2)^{-1} \exp\left(-\frac{R}{\sigma_1^2}(x_1^2 + x_2^2)\right).$$

The linearization of the averaged equations in $(\mathbb{R}^2 \setminus \{0\}) \times \mathbb{R}^2$ can also be explicitly solved (the details are left as an exercise, see [37]). This yields the Lyapunov exponents $\lambda_1(\nu_\beta) = 0$ (since angular distances are preserved) and $\lambda_2(\nu_\beta) = -2\lambda_0 < 0$.

The overall picture is thus the following:

(i) For $\beta < \beta_D$, all trajectories converge to 0 with exponential speed λ_0 and rotation number $-a$. The attractor in $\mathbb{R}^2$ is $\{0\}$. For $\beta = \beta_D$, the attractor is still $\{0\}$.

(ii) For $\beta > \beta_D$, the attractor in the punctured plane $\mathbb{R}^2 \setminus \{0\}$ (for sets tempered from below) is the random circle with radius $\kappa_\beta(\omega)$, and the rotation number of the nonlinear system is

$$\lim_{t \to \infty} \frac{\phi(t)}{t} = -a + \frac{S}{R}\lambda_0$$

(this rate vanishes for $\lambda_0 = \frac{R}{S}a$). Further, $\mathbb{E}\kappa_\beta^2 = \frac{\lambda_0}{R}$. The final bifurcation diagram for the averaged system is shown in Fig. 9.20.

While the original random Duffing-van der Pol equation in the Hopf regime undergoes a sequence of two D-bifurcations out of the trivial solution at points where the two different Lyapunov exponents of δ_0 change their sign, the averaged equation only has one D-bifurcation since, due to averaging, δ_C has a one-point Lyapunov spectrum. ∎

9.5.2 Further Studies

The Stochastic Lorenz Equation

The deterministic Lorenz equation

$$\begin{aligned}\dot{x}_1 &= -sx_1 + sx_2 \\ \dot{x}_2 &= rx_1 - x_2 - x_1 x_3 \\ \dot{x}_3 &= -bx_3 + x_1 x_2\end{aligned} \quad (9.5.22)$$

in $\mathbb{R}^3$ with parameters $s > 0$ (the Prandtl number), $r > 0$ (the Rayleigh number), and $b > 0$ (a geometric factor), is extremely well investigated (see e. g. Guckenheimer and Holmes [162], Sparrow [321], or almost any other book on dynamical systems, and the references therein). The equation is dissipative, hence forward complete, and has a global attractor in $\mathbb{R}^3$ which for certain parameter values (e. g. for $s = 10$, $r = 28$ and $b = 8/3$) is exceedingly complicated (often called "strange attractor").

For s and b fixed and r the bifurcation parameter, we have the following elementary local bifurcations in (9.5.22):
– For $0 < r < 1$, the origin is the global attractor.
– At $r = 1$, the origin undergoes a pitchfork bifurcation, and two nontrivial fixed points

$$x^+ = (\sqrt{b(r-1)}, \sqrt{b(r-1)}, r-1), \; x^- = (-\sqrt{b(r-1)}, -\sqrt{b(r-1)}, r-1)$$

are born.
– At $r = \frac{s(s+b+30)}{s-b-1}$ (we assume $s > 1 + b$) the two nontrivial fixed points $x^\pm$ become unstable and undergo a (subcritical) Hopf bifurcation.

We now look at the case where the bifurcation parameter r is perturbed by white noise (say), i. e. we study the SDE

$$dx = \begin{pmatrix} -s & s & 0 \\ r & -1 & 0 \\ 0 & 0 & -b \end{pmatrix} x\, dt + \begin{pmatrix} 0 \\ -x_1 x_3 \\ x_1 x_2 \end{pmatrix} dt + \sigma \begin{pmatrix} 0 & 0 & 0 \\ 1 & 0 & 0 \\ 0 & 0 & 0 \end{pmatrix} x \circ dW_t.$$

(9.5.23)

The bifurcation theory of this SDE was studied by Sri Namachchivaya [322] and Keller [202]. Keller proved the existence of a stationary measure and of a random attractor, also for more general stochastic perturbations, and made extensive numerical studies.

Let us calculate the stochastic D-bifurcation point corresponding to the deterministic pitchfork bifurcation point $r = 1$. The linearization of (9.5.23) at $x = 0$ is

$$dv_t = \begin{pmatrix} -s & s & 0 \\ r & -1 & 0 \\ 0 & 0 & -b \end{pmatrix} v_t dt + \sigma \begin{pmatrix} 0 & 0 & 0 \\ 1 & 0 & 0 \\ 0 & 0 & 0 \end{pmatrix} v_t \circ dW_t. \quad (9.5.24)$$

The deterministic eigenvalues are $a_3 = -b$ and

$$a_{1,2} = -\frac{s+1}{2} \pm \frac{1}{2}\sqrt{(s+1)^2 - 4s(r-1)}.$$

The third equation in (9.5.24) is always stable and decoupled from the first two equations which, after transformation of the drift matrix to diagonal form, reads

$$du_t = \begin{pmatrix} a_1 & 0 \\ 0 & a_2 \end{pmatrix} u_t dt + \sigma \frac{s}{a_1 - a_2}\begin{pmatrix} 1 & 1 \\ -1 & -1 \end{pmatrix} u_t \circ dW_t. \quad (9.5.25)$$

For small intensity parameter σ Theorem 9.4.11 yields for the top Lyapunov exponent of (9.5.25) and of (9.5.24)

$$\lambda_1(r, \sigma) = a_1 - \frac{\sigma^2}{2}\frac{s^2}{(s+1)^2 - 4s(1-r)} + O(\sigma^4). \quad (9.5.26)$$

For $r = 1$, $\lambda_1(1, \sigma) = -\frac{\sigma^2}{2}\frac{s^2}{(s+1)^2} < 0$, i.e. the origin is still exponentially stable, and the pitchfork bifurcation is delayed to the parameter value r_D for which $\lambda_1(r_D, \sigma) = 0$ which is, using (9.5.26),

$$r_D = 1 + \sigma^2 \frac{s}{2(s+1)} + O(\sigma^4).$$

Since the moment Lyapunov exponent function $g(p)$ of (9.5.25) satisfies $g(-2) = -\text{trace} A_0 = s + 1 > 0$, there will be no parameter value $r > r_D$ for which we have a P-bifurcation of the new stationary measure in a neighborhood of $x = 0$.

The Stochastic Brusselator

After the discovery of oscillatory chemical reactions by Belousov and Zhabotinskii, the Brussels school (see Nicolis and Prigogine [263: Chap. 7]) proposed the following "simplest possible" differential equation for the concentrations x_1 and x_2 of reactants of a chemical reaction scheme which undergoes a Hopf bifurcation:

$$\begin{aligned} \dot{x}_1 &= a - (b+1)x_1 + x_1^2 x_2, \\ \dot{x}_2 &= bx_1 - x_1^2 x_2. \end{aligned} \quad (9.5.27)$$

Here $a, b > 0$ are parameters, of which b is the bifurcation parameter. This model became known as the *Brusselator*, and has been one of the paradigms of nonequilibrium dynamics.

9.5 General Dimension. Further Studies

The system (9.5.27) has a unique fixed point at $x_0 = (a, \frac{b}{a})$ which is stable for $0 < b < a^2 + 1$ and unstable for $b > a^2 + 1$. At $b = a^2 + 1$, the fixed point x_0 undergoes a Hopf bifurcation.

If we perturb the bifurcation parameter b by white noise (say), we obtain the SDE

$$dx = \begin{pmatrix} a - (b+1)x_1 + x_1^2 x_2 \\ bx_1 - x_1^2 x_2 \end{pmatrix} dt + \sigma \begin{pmatrix} -x_1 \\ x_1 \end{pmatrix} \circ dW_t. \qquad (9.5.28)$$

Note that the noise on b is additive noise with respect to the fixed point x_0 (see Subsect. 9.2.2 for the definition of additive and multiplicative noise), hence the mathematically rigorous methods developed so far cannot be applied.

The effect of noise on the Hopf scenario of the Brusselator was studied by Lefever and Turner [232, 233] using asymptotic methods. They showed that the stationary density undergoes a delayed transition from a bell-shaped to a crater-type form.

The latter was made more precise in the systematic numerical study by Arnold, Bleckert and Schenk-Hoppé [9]. They also demonstrated that "parametric noise destroys the deterministic Hopf bifurcation" in the following sense: For all values of the bifurcation parameter b and the noise intensity $\sigma > 0$ the local RDS generated by (9.5.28) has a unique invariant measure. This measure is a random Dirac measure whose top Lyapunov exponent is negative and whose support is the global random attractor. See Example (9.2.5) for a similar phenomenon in dimension 1.

Part IV

Appendices

Appendix A. Measurable Dynamical Systems

A.1 Ergodic Theory

This is a summary of some well-known facts on ergodic theory which can be found in most of the many excellent books on the subject, e.g. in the books by Cornfeld, Fomin and Sinai [106], Krengel [220], Mañé [249], Petersen [276], Rudolph [291], Sinai [318], Walters [336].

Probability Space $(\Omega, \mathcal{F}, \mathbb{P})$**.** Let $\Omega \neq \emptyset$ be an abstract set, $\mathcal{F}$ a σ-algebra of subsets of Ω and $\mathbb{P}$ a probability measure on $\mathcal{F}$. The pair $(\Omega, \mathcal{F})$ is called a *measurable space* and the triple $(\Omega, \mathcal{F}, \mathbb{P})$ a *probability space*. A probability space is said to be *complete* if the σ-algebra $\mathcal{F}$ contains all subsets of sets of probability 0. In this book we do in general not assume completeness of our probability space, unless explicitly stated.

A σ-algebra $\mathcal{F}$ is called *countably generated* if there exists a countable family $\mathcal{E} = (A_n)_{n \in \mathbb{N}} \subset \mathcal{F}$ such that the σ-algebra $\sigma(\mathcal{E})$ generated by it equals $\mathcal{F}$. A probability space $(\Omega, \mathcal{F}, \mathbb{P})$ is said to be countably generated if there exists a countable family $\mathcal{E} \subset \mathcal{F}$ such that for each $A \in \mathcal{F}$ and $\varepsilon > 0$ there exists an $A_\varepsilon \in \mathcal{E}$ for which $\mathbb{P}(A \triangle A_\varepsilon) < \varepsilon$. The latter is equivalent to $L^p(\Omega, \mathcal{F}, \mathbb{P})$ being separable for all $1 \leq p < \infty$. If $\mathcal{F}$ is countably generated, then $(\Omega, \mathcal{F}, \mathbb{P})$ is countably generated for any $\mathbb{P}$.

A set $\tilde{\Omega} \in \mathcal{F}$ with $\mathbb{P}(\tilde{\Omega}) = 1$ is called a *set of full measure* and a set $N \in \mathcal{F}$ with $\mathbb{P}(N) = 0$ is called an *exceptional set* or *null set*. A function $f : \Omega_1 \to \Omega_2$ of two measurable spaces $(\Omega_1, \mathcal{F}_1)$ and $(\Omega_2, \mathcal{F}_2)$ is called *measurable* if $f^{-1}(\mathcal{F}_2) \subset \mathcal{F}_1$. It is called a *bimeasurable bijection* if it is measurable and measurably invertible.

Time $\mathbb{T}$. We will study families of mappings indexed by the elements t of a set $\mathbb{T}$ called *time*. The most general time could be a measurable (semi)group $\mathbf{T}$ (i.e. $\mathbb{T}$ is endowed with a σ-algebra $\mathcal{T}$ rendering the (semi)group operations measurable). However, in this book $\mathbb{T}$ exclusively stands for the following (additive topological) (semi)groups:

- $\mathbb{T} = \mathbb{R}$: *two-sided continuous time*,
- $\mathbb{T} = \mathbb{R}^+$ (sometimes $\mathbb{T} = \mathbb{R}^- := -\mathbb{R}^+$): *one-sided continuous time*,
- $\mathbb{T} = \mathbb{Z} := \{0, \pm 1, \pm 2, \ldots\}$: *two-sided discrete time*,
- $\mathbb{T} = \mathbb{Z}^+ := \{0, 1, 2, \ldots\}$ (sometimes $\mathbb{T} = \mathbb{Z}^- := -\mathbb{Z}^+$ or $T = \mathbb{N} := \{1, 2, \ldots\}$): *one-sided discrete time*.

$\mathbb{T}$ is always endowed with its Borel σ-algebra $\mathcal{B}(\mathbb{T})$.

Measurable Dynamical Systems[1]. A family $(\theta(t))_{t \in \mathbb{T}}$ of mappings of $(\Omega, \mathcal{F})$ into itself (such self-mappings are also called *transformations*) is called a *measurable dynamical system* with time $\mathbb{T}$ (or *measurable (semi)flow* if $\mathbb{T}$ is a (semi)group) if it satisfies the following three conditions:

1. $(\omega, t) \mapsto \theta(t)\omega$ is measurable[2],
2. $\theta(0) = \mathrm{id}_\Omega =$ identity on Ω (if $0 \in \mathbb{T}$),
3. *(Semi)flow property*: $\theta(s+t) = \theta(s) \circ \theta(t)$ for all $s, t \in \mathbb{T}$.[3]

It follows from condition 1 that $\theta(t) : \Omega \to \Omega$ is measurable for all $t \in \mathbb{T}$ as a section of a measurable mapping (for discrete time this is equivalent to condition 1). If $\mathbb{T}$ is a group, conditions 2 and 3 imply that all $\theta(t)$ are measurably invertible with $\theta(t)^{-1} = \theta(-t)$.

If $\mathbb{T}$ is discrete then $\theta(n) = \theta^n$, $n \in \mathbb{T}$, where $\theta := \theta(1)$ is the time one mapping. In this case the measurable DS consists of the iterates of a measurable mapping (in case $\mathbb{T} = \mathbb{Z}^+$ or $\mathbb{N}$) or a bimeasurable bijection θ (in case $\mathbb{T} = \mathbb{Z}$). Conversely, every such mapping generates through its iterates a measurable DS.

For continuous time $\mathbb{T} = \mathbb{R}$ or $\mathbb{R}^+$ we often use the less clumsy notation θ_t instead of $\theta(t)$.

Measure Preserving Transformations. Let θ be a measurable mapping of $(\Omega_1, \mathcal{F}_1, \mathbb{P}_1)$ to $(\Omega_2, \mathcal{F}_2)$. The measure $\theta\mathbb{P}_1$ on $\mathcal{F}_2$ defined by $\theta\mathbb{P}_1(A) := \mathbb{P}_1\{\theta^{-1}(A)\}$, $A \in \mathcal{F}_2$, is the image of $\mathbb{P}_1$ with respect to θ. The mapping $f \mapsto f \circ \theta =: Uf$ is an isometry of $L^p(\Omega_2, \mathcal{F}_2, \theta\mathbb{P}_1)$ to $L^p(\Omega_1, \mathcal{F}_1, \mathbb{P}_1)$, $1 \le p \le \infty$, and $\int f \circ \theta \, d\mathbb{P}_1 = \int f d(\theta\mathbb{P}_1)$. A measurable mapping θ of $(\Omega_1, \mathcal{F}_1, \mathbb{P}_1)$ to $(\Omega_2, \mathcal{F}_2, \mathbb{P}_2)$ with $\theta\mathbb{P}_1 = \mathbb{P}_2$ is called a *homomorphism* of the corresponding probability spaces. It is called an *isomorphism* if, in addition, it is measurably invertible. A homomorphism of $(\Omega, \mathcal{F}, \mathbb{P})$ to itself (i.e. a measurable map θ with $\theta\mathbb{P} = \mathbb{P}$) is called an *endomorphism* and $\mathbb{P}$ is said to be *invariant* with respect to θ. An endomorphism which is measurably invertible is called an *automorphism*. For an automorphism we also have $\theta^{-1}\mathbb{P} = \mathbb{P}$.

---

[1] "Dynamical System(s)" is henceforth abbreviated as "DS".
[2] σ-algebras with respect to which measurability is to be understood are not mentioned in case they are clear from the context, e.g. in product spaces we take the product σ-algebra and in topological spaces we take the Borel σ-algebra generated by the open sets.
[3] $\circ$ means composition.

A measurable DS $(\theta(t))_{t\in\mathbb{T}}$ on a probability space $(\Omega, \mathcal{F}, \mathbb{P})$ for which each $\theta(t)$ is an endomorphism is called a measure preserving or *metric DS* and is denoted by $\Sigma = (\Omega, \mathcal{F}, \mathbb{P}, (\theta(t))_{t\in\mathbb{T}})$ or, for short, by $\theta(\cdot)$ or θ. If $\mathbb{T}$ is discrete, $\mathbb{P}$ is invariant with respect to $(\theta(t))_{t\in\mathbb{T}}$ if and only if it is invariant with respect to $\theta := \theta(1)$.

Let Σ be a metric DS with continuous time. Then the (semi)group $(U_t)_{t\in\mathbb{T}}$ of isometries of $L^p(\Omega, \mathcal{F}, \mathbb{P})$, $1 \leq p < \infty$, defined by $f \mapsto U_t f := f \circ \theta_t$, is strongly continuous with respect to $t \in \mathbb{T}$ (Krengel [220: Theorem 6.13]). As a consequence, for each measurable f, the measurable stationary stochastic process (see Appendix A.3) $t \mapsto f(\theta_t \cdot)$ is continuous in probability.

Invariant Functions and Sets. Let $(\Omega, \mathcal{F})$ be a measurable space. A *function* f is called *invariant* with respect to the measurable self-mapping θ of $(\Omega, \mathcal{F})$ (with respect to the measurable DS $(\theta(t))_{t\in\mathbb{T}}$) if $f(\theta\omega) = f(\omega)$ for all $\omega \in \Omega$ ($f(\theta(t)\omega) = f(\omega)$ for all $t \in \mathbb{T}$ and all $\omega \in \Omega$). A *set* is called *invariant* with respect to θ (or $(\theta(t))_{t\in\mathbb{T}}$) if 1_A is invariant, i.e. if $\theta^{-1}A = A$ (or $\theta(t)^{-1}A = A$ for all $t \in \mathbb{T}$). An invariant set of a measurable DS consists of whole *orbits* or *trajectories*, i.e. $(\theta(t)\omega)_{t\in\mathbb{T}} \subset A$ if $\omega \in A$. The family of measurable invariant sets of θ (or $(\theta(t))_{t\in\mathbb{T}}$) forms a sub-$\sigma$-algebra $\mathcal{I} \subset \mathcal{F}$. A set A is called *forward invariant* with respect to θ (or $(\theta(t))_{t\in\mathbb{T}}$) if $A \subset \theta^{-1}A$ (or $A \subset \theta(t)^{-1}A$ for all $t \in \mathbb{T}^+$, where $\mathbb{T}^+ := \mathbb{T} \cap \mathbb{R}^+$), equivalently[4] $\theta A \subset A$ (or $\theta(t)A \subset A$ for all $t \in \mathbb{T}^+$). An invariant set is clearly forward invariant.

Notions Mod $\mathbb{P}$. If for two measurable functions f_1, f_2 on a probability space $(\Omega, \mathcal{F}, \mathbb{P})$ the set $\{\omega : f_1(\omega) = f_2(\omega)\} \in \mathcal{F}$ has full measure we say that $f_1 = f_2$ mod $\mathbb{P}$. A measurable function f is called *invariant* mod $\mathbb{P}$ with respect to the endomorphism θ (the metric DS $(\Omega, \mathcal{F}, \mathbb{P}, (\theta(t))_{t\in\mathbb{T}})$) if $f \circ \theta = f$ mod $\mathbb{P}$ ($f \circ \theta(t) = f$ mod $\mathbb{P}$ for all $t \in \mathbb{T}$ where the exceptional set N_t can depend on $t \in \mathbb{T}$). A set $A \in \mathcal{F}$ is called invariant mod $\mathbb{P}$ with respect to the endomorphism θ (the metric DS $(\Omega, \mathcal{F}, \mathbb{P}, (\theta(t))_{t\in\mathbb{T}})$) if 1_A is invariant mod $\mathbb{P}$, equivalently if $\mathbb{P}(A \triangle \theta^{-1}A) = 0$ (if $\mathbb{P}(A \triangle \theta(t)^{-1}A) = 0$ for all $t \in \mathbb{T}$). The family $\mathcal{I}_\mathbb{P}$ of sets invariant mod $\mathbb{P}$ is a σ-algebra (depending on $\mathbb{P}$) with $\mathcal{I} \subset \mathcal{I}_\mathbb{P} \subset \mathcal{F}$, where $\mathcal{I}_\mathbb{P}$ and $\mathcal{I}$ are related as follows:

$$\mathcal{I}_\mathbb{P} = \{B \in \mathcal{F} : \text{ there is an } A \in \mathcal{I} \text{ with } \mathbb{P}(A \triangle B) = 0\}.$$

For the proof of this relation one needs the fact that to each real-valued measurable function f invariant mod $\mathbb{P}$ with respect to the metric DS $(\Omega, \mathcal{F}, \mathbb{P}, (\theta(t))_{t\in\mathbb{T}})$ there is a function $\tilde{f} = f$ mod $\mathbb{P}$ which is invariant. In particular, to each set $A \in \mathcal{F}$ invariant mod $\mathbb{P}$ with respect to a metric DS there exists a set $\tilde{A} = A$ mod $\mathbb{P}$ which is invariant. In case of discrete time take $\tilde{f} := \limsup_{n\to\infty} f \circ \theta^n$ and in the case of continuous time set

---

[4] Here and at many other occasions we use the following elementary relations: $f^{-1}(f(A)) \supset A$, $f(f^{-1}(A)) \subset A$, $f^{-1}(f(A)) = A$ if f is injective, and $f(f^{-1}(A)) = A$ if f is surjective.

$$\tilde{f}(\omega) := \lim_{t\to\infty} \text{ess sup}\{f(\theta(s)\omega) : s \geq t\},$$

the "ess sup" taken for fixed ω with respect to Lebesgue measure on $\mathbb{T}$.

A forward invariant set is clearly invariant mod $\mathbb{P}$ for any $\mathbb{P}$.

Ergodic Theorems. Let $(\Omega, \mathcal{F}, (\theta(t))_{t\in\mathbb{T}})$ be a measurable DS and let f be a real-valued measurable function. Define

(i) for $\mathbb{T} = \mathbb{Z}^+$:

$$\Omega_f := \{\omega : \lim_{n\to\infty} \frac{1}{n} \sum_{k=0}^{n-1} f(\theta^k \omega) =: \bar{f}(\omega) \text{ exists}\},$$

(ii) for $\mathbb{T} = \mathbb{Z}$

$$\Omega_f := \{\omega : \lim_{n\to\infty} \frac{1}{n} \sum_{k=0}^{n-1} f(\theta^k \omega) =: \bar{f}_+(\omega),$$

$$\lim_{n\to\infty} \frac{1}{n} \sum_{k=0}^{n-1} f(\theta^{-k}\omega) =: \bar{f}_-(\omega) \text{ both exist, and}$$

$$\bar{f}_+(\omega) = \bar{f}_-(\omega) =: \bar{f}(\omega)\}.$$

Then $\Omega_f \in \mathcal{I}$ and $\bar{f}$ is invariant on Ω_f.

(iii) For $\mathbb{T} = \mathbb{R}^+$: Let

$$F(\omega) := \int_0^1 f(\theta(t)\omega)\, dt, \quad F_0(\omega) := \int_0^1 |f(\theta(t)\omega)|\, dt,$$

$$S_n(\omega) := \int_0^n f(\theta(t)\omega)\, dt = \sum_{k=0}^{n-1} F(\theta^k \omega),$$

and

$$\Omega_f := \{\omega : f(\theta(\cdot)\omega) \text{ locally integrable}, \lim_{n\to\infty} \frac{1}{n} F_0(\theta^n \omega) = 0,$$

$$\lim_{n\to\infty} \frac{1}{n} S_n(\omega) =: \bar{f}(\omega) \text{ exists}\}.$$

Then $\Omega_f \in \mathcal{F}$ is forward invariant, $\bar{f}$ is invariant on Ω_f and for all $\omega \in \Omega_f$

$$\lim_{t\to\infty} \frac{1}{t} \int_0^t f(\theta(s)\omega)\, ds = \bar{f}(\omega).$$

(iv) For $T = \mathbb{R}$: Let, with the notations of the case $\mathbb{T} = \mathbb{R}^+$

$$\Omega_f := \{\omega : f(\theta(\cdot)\omega) \text{ is locally integrable}, \lim_{n\to\pm\infty} \frac{1}{n} F_0(\theta^n \omega) = 0,$$
$$\lim_{n\to\infty} \frac{1}{n} S_n(\omega) =: \bar{f}_+(\omega) \text{ and } \lim_{n\to\infty} \frac{1}{n} S_{-n}(\omega) =: \bar{f}_-(\omega)$$
$$\text{both exist, and } \bar{f}_+(\omega) = \bar{f}_-(\omega) =: \bar{f}(\omega)\}.$$

Then $\Omega_f \in \mathcal{I}$, $\bar{f}$ is invariant on Ω_f and for all $\omega \in \Omega_f$

$$\lim_{t\to\infty} \frac{1}{t} \int_0^t f(\theta(s)\omega)\, ds = \lim_{t\to\infty} \frac{1}{t} \int_{-t}^0 f(\theta(s)\omega)\, ds = \bar{f}(\omega).$$

The *(Birkhoff-Chintchin) ergodic theorem* states that if $(\theta(t))_{t\in\mathbb{T}}$ is a measurable DS on $(\Omega, \mathcal{F})$, then for any θ-invariant probability $\mathbb{P}$ on $(\Omega, \mathcal{F})$ and any $f \in L^1(\Omega, \mathcal{F}, \mathbb{P})$

1. $\mathbb{P}(\Omega_f) = 1$,
2. $\bar{f}$ (defined outside Ω_f by $\bar{f}(\omega) = 0$) is a version of $\mathbb{E}(f|\mathcal{I})$ for $\mathbb{T} = \mathbb{Z}$, $\mathbb{T} = \mathbb{Z}^+$ and $\mathbb{T} = \mathbb{R}$, and a version of $\mathbb{E}(f|\mathcal{I}_\mathbb{P})$ for $\mathbb{T} = \mathbb{R}^+$,
3. if $f \in L^p$ for some $1 \leq p < \infty$, then $\bar{f} \in L^p$ and convergence to $\bar{f}$ also holds in L^p.

We also need the following *extended version* of the ergodic theorem: If only $f^+ \in L^1$, then still $\mathbb{P}(\Omega_f) = 1$, provided we allow $\bar{f}(\omega) \in \mathbb{R} \cup \{-\infty\}$. But we only have $\bar{f}^+ \in L^1$.

Ergodic Dynamical Systems. A metric DS $(\Omega, \mathcal{F}, \mathbb{P}, (\theta(t))_{t\in\mathbb{T}})$ is called *ergodic* if all sets in $\mathcal{I}$ (equivalently, all sets in $\mathcal{I}_\mathbb{P}$) have probability 0 or 1. Equivalently, a metric DS is ergodic if and only if all invariant functions (invariant functions mod $\mathbb{P}$) are constant mod $\mathbb{P}$. As a result, the limit $\bar{f}$ in the ergodic theorem will be constant on an invariant set (invariant set mod $\mathbb{P}$ only for $\mathbb{T} = \mathbb{R}^+$) of full measure if the system is ergodic.

Let for a measurable DS $(\Omega, \mathcal{F}, (\theta(t))_{t\in\mathbb{T}})$, $\mathcal{I}(\theta)$ denote the convex set of θ-invariant probabilities and let $\mathcal{E}(\theta)$ denote the ergodic measures in $\mathcal{I}(\theta)$. Then two elements $\mathbb{P}_1, \mathbb{P}_2 \in \mathcal{I}(\theta)$ are either equal or differ already on the σ-algebra $\mathcal{I}$. In particular, different elements of $\mathcal{E}(\theta)$ are mutually singular and $\mathcal{E}(\theta)$ coincides with the extreme points of the convex set $\mathcal{I}(\theta)$.

We will often restrict ourselves to the ergodic case on the grounds of the existence of a decomposition of an arbitrary invariant measure into ergodic components. For example, we have the following *ergodic decomposition theorem* (see Deuschel and Stroock [126: Theorem 5.2.16]): Let $(\Omega, \mathcal{F}, (\theta(t))_{t\in\mathbb{T}})$ be a measurable DS. Assume in addition that Ω is a Polish space and that for each fixed $t \in \mathbb{T}$, $\theta(t) : \Omega \to \Omega$ is continuous. Then for each $\mathbb{P} \in \mathcal{I}(\theta)$ there is a probability $\rho_\mathbb{P}$ on $\mathcal{E}(\theta)$ for which

$$\mathbb{P}(\cdot) = \int_{\mathcal{E}(\theta)} Q(\cdot)\, \rho_\mathbb{P}(dQ).$$

An ergodic decomposition theorem for compact metric Ω and measurable θ is given by Mañé [249: Chap. II.6]. For Lebesgue spaces see Rokhlin [289].

Isomorphisms of Metric DS. Two probability spaces $(\Omega_i, \mathcal{F}_i, \mathbb{P}_i)$, $i = 1, 2$, are said to be *metrically isomorphic* if there exist sets $\tilde{\Omega}_i$ of full $\mathbb{P}_i$ measure and an isomorphism ψ from $(\tilde{\Omega}_1, \tilde{\Omega}_1 \cap \mathcal{F}_1, \mathbb{P}_1)$ to $(\tilde{\Omega}_2, \tilde{\Omega}_2 \cap \mathcal{F}_2, \mathbb{P}_2)$. Two metric DS $(\Omega_i, \mathcal{F}_i, \mathbb{P}_i, (\theta_i(t))_{t \in \mathbb{T}})$, $i = 1, 2$ (with the same time) are *metrically isomorphic* if there exists a metric isomorphism ψ of the corresponding probability spaces with sets $\tilde{\Omega}_i$ of full measure which are forward invariant for one-sided time, and invariant for two-sided time, such that on $\tilde{\Omega}_1$

$$\theta_2(t) \circ \psi = \psi \circ \theta_1(t), \text{ all } t \in \mathbb{T}.$$

Factors and Extensions. The probability space $(\Omega, \mathcal{F}, \mathbb{P})$ is a *factor* of $(\Omega_1, \mathcal{F}_1, \mathbb{P}_1)$ (and the latter is called an *extension* of the former) if there exist sets $\tilde{\Omega}$, $\tilde{\Omega}_1$ of full measure and a homomorphism ψ of $(\tilde{\Omega}_1, \tilde{\Omega}_1 \cap \mathcal{F}_1, \mathbb{P}_1)$ to $(\tilde{\Omega}, \tilde{\Omega} \cap \mathcal{F}, \mathbb{P})$.

Given a metric DS $(\Omega, \mathcal{F}, \mathbb{P}, (\theta(t))_{t \in \mathbb{T}})$, it is called a factor of the metric DS $(\Omega_1, \mathcal{F}_1, \mathbb{P}_1, (\theta_1(t))_{t \in \mathbb{T}})$ (and the latter is called an extension of the former) if this holds for the corresponding probability spaces with sets $\tilde{\Omega}$, $\tilde{\Omega}_1$ of full measure which are forward invariant for one-sided time, and invariant for two-sided time, and a homomorphism $\psi : \tilde{\Omega}_1 \to \tilde{\Omega}$ satisfying on $\tilde{\Omega}_1$

$$\psi \circ \theta_1(t) = \theta(t) \circ \psi, \text{ all } t \in \mathbb{T}.$$

We say that $\theta_1(\cdot)$ *covers* $\theta(\cdot)$. Two metric DS are isomorphic if and only if they are factors of each other. Factors of metric DS on Lebesgue spaces are determined by invariant sub-σ-algebras, and a metric DS with a factor allows a skew product representation.

Natural Extension. In the theory of random dynamical systems it is very advantageous to have a model of the noise which has two-sided time. Two-sided time can be achieved basically without loss of generality by extending a metric DS with one-sided time. This procedure is known as natural extension and was carried out for discrete time and a Lebesgue space $(\Omega, \mathcal{F}, \mathbb{P})$ by Rokhlin [290: §3]. See also Cornfeld, Fomin and Sinai [106: p. 240] and Sinai [318: p. 27].

Let $\mathbb{T} = \mathbb{Z}$ or $\mathbb{R}$, put $\mathbb{T}^+ = \mathbb{Z}^+$ or $\mathbb{R}^+$ and let

$$\Sigma^+ = (\Omega^+, \mathcal{F}^+, \mathbb{P}^+, (\theta^+(t))_{t \in \mathbb{T}^+})$$

be a metric DS with one-sided time. The metric DS $\Sigma = (\Omega, \mathcal{F}, \mathbb{P}, (\theta(t))_{t \in \mathbb{T}})$ with two-sided time is called a *natural extension* of Σ^+ if

(i) Σ^+ is a factor of the DS Σ restricted to $\mathbb{T}^+$,

(ii) the σ-algebra $\mathcal{G} := \psi^{-1}(\mathcal{F}^+)$ is *exhaustive* in the following sense: $\sigma(\theta(t)\mathcal{G} : t \in \mathbb{T}^+) = \mathcal{F} \mod \mathbb{P}$. Here ψ is the homomorphism occuring in (i).

A natural extension exists and is unique up to isomorphisms for discrete time and $(\Omega^+, \mathcal{F}^+)$ a standard measurable space or $(\Omega^+, \mathcal{F}^+, \mathbb{P}^+)$ a Lebesgue space. Further, Σ is ergodic if and only if Σ^+ is, and Σ and Σ^+ have the same entropy (see Rokhlin [290: pp. 23 and 26]).

If $\Sigma = (\Omega, \mathcal{F}, \mathbb{P}, (\theta(t))_{t \in \mathbb{T}})$ is a two-sided metric DS, then $\Sigma_1^+ = (\Omega, \mathcal{F}, \mathbb{P}, (\theta(t))_{t \in \mathbb{T}^+})$ is a one-sided metric DS and Σ is the natural extension of Σ_1^+ with associated homomorphism $\psi = \mathrm{id}_\Omega$. A new DS with one-sided time on Ω can be defined by a sub-σ-algebra $\mathcal{G}^+ \subset \mathcal{F}$ with $\theta(t)^{-1}\mathcal{G}^+ \subset \mathcal{G}^+$ for $t \in \mathbb{T}^+$. Now $\Sigma_2^+ = (\Omega, \mathcal{G}^+, \mathbb{P}, (\theta(t))_{t \in \mathbb{T}^+})$ is a factor of Σ_1^+, again with associated homomorphism $\psi = \mathrm{id}_\Omega$. Its natural extension is the two-sided DS with $\mathcal{F}$ replaced with $\mathcal{G} := \sigma(\theta(t)\mathcal{G}^+ : t \in \mathbb{T}^+)$, which is a factor of Σ.

For the canonical metric DS on path space with shift corresponding to a stationary stochastic process there is a more direct extension from one-sided to two-sided time, see Appendix A.2.

Lebesgue Spaces. A probability space $(\Omega, \mathcal{F}, \mathbb{P})$ is called *Lebesgue space* (see the pioneering work [288, 289] of Rokhlin) if it is metrically isomorphic to a probability space which is the disjoint union of an at most countable (possibly empty) set $\{x_1, x_2, \ldots\}$ of points each of positive measure and the space $([0, s), \mathcal{L}, \lambda)$ (possibly absent), where $\mathcal{L}$ is the σ-algebra of Lebesgue measurable subsets of the interval $[0, s)$ and λ is Lebesgue measure. Here $s = 1 - \sum p_n$, $p_n = $ measure of the point x_n.

As ergodic theory becomes more satisfactory for Lebesgue spaces and since this class covers many applications some books on ergodic theory deal exclusively with Lebesgue spaces, e.g. Petersen [276] and Rudolph [291].

Standard Measurable Spaces. A measurable space $(\Omega, \mathcal{F})$ is called *standard measurable space* (sometimes also called *Borel space*) if it is isomorphic (by means of a bimeasurable bijection) to a Borel subset of a Polish space (see Zimmer [353: Appendix A] or Parthasarathy [272]). In particular, the σ-algebra $\mathcal{F}$ of a standard measurable space is countably generated and countably separated. For any probability $\mathbb{P}$ on a standard space $(\Omega, \mathcal{F})$, define $\mathcal{F}^\mathbb{P}$ to be the $\mathbb{P}$-completion of $\mathcal{F}$. Then $(\Omega, \mathcal{F}^\mathbb{P}, \mathbb{P})$ is a Lebesgue space.

Almost Periodic Functions and DS. A subset of $\mathbb{R}$ is called relatively dense if there is an $L > 0$ such that each interval of length L contains an element of the set. For a function $f : \mathbb{R} \to \mathbb{R}^d$ and $\varepsilon > 0$ a number $\tau \in \mathbb{R}$ is called an ε-almost period of f, if $|f(t + \tau) - f(t)| < \varepsilon$ for all $t \in \mathbb{R}$.

A function $f : \mathbb{R} \to \mathbb{R}^d$ is called *almost periodic* (in the sense of H. Bohr) if (i) f is continuous and (ii) for each $\varepsilon > 0$ the set of ε-almost periods of f is relatively dense. For details see Dunford and Schwartz [132: IV.8].

An almost periodic function is bounded and uniformly continuous. Let $H(f) := \{f(\cdot + t) : t \in \mathbb{R}\}$ be the *hull*, i.e. the set of all translates of $f \in \mathcal{C}_b(\mathbb{R}, \mathbb{R}^d)$. Then f is almost periodic if and only if the closure $\overline{H(f)}$ of $H(f)$ is compact[5]. If this is the case, $\overline{H(f)}$ consists of almost periodic functions and has the structure of a compact Abelian Polish group G with unit $e = f$. The group operation is defined as follows: For $g = f(\cdot + t)$ and $h = f(\cdot + s)$ put $g * h = f(\cdot + s + t)$, for $g = \lim f(\cdot + t_n)$ and $h = \lim f(\cdot + s_n)$ put $g * h = \lim f(\cdot + t_n + s_n)$.

We associate to an almost periodic function f the following canonical metric DS: Let $\mathbb{T} = \mathbb{R}$, $\Omega := G = \overline{H(f)}$, $\mathcal{F}$ the Borel σ-algebra of G, $\theta(t)\omega := \omega(\cdot + t)$ the translation of ω by t. Then $(t, \omega) \mapsto \theta(t)\omega$ is continuous and hence measurable. The normalized Haar measure of G is the unique θ-invariant probability. Under $\mathbb{P}$, θ is ergodic. Particular cases: (i) $G = T^1$ (1-torus for f periodic), (ii) $G = T^n$ (n-torus for f quasi-periodic).

Noise modeled by this DS is of particular interest as it is in the intersection of topological dynamics and ergodic theory and is a first step beyond periodic excitation.

A.2 Stochastic Processes and Dynamical Systems

Several classes of stochastic processes (e.g. stationary processes, processes with stationary increments, in particular Brownian motion (white noise)) which occupy strategic positions in probability theory are linked to metric DS, which will be briefly surveyed in this section. Our motivation is that metric DS constructed in this section enter as *noise* into parameters of difference or differential equations, this way generating a *random dynamical system*, see Chap. 2.

There is an abundance of good texts from which we quote freely, and of which we only mention Bauer [52], Breiman [81], Doob [128], Dudley [131], Durrett [133], Gänssler and Stute [156], Gihman and Skorohod [160], Karlin and Taylor [198], Meyn and Tweedie [253], Wentzell [345].

Basic Definitions. Let $(\Omega, \mathcal{F}, \mathbb{P})$ be a probability space, $(E, \mathcal{E})$ a measurable space and $\mathbb{T} \neq \emptyset$ a set. A family $\xi = (\xi_t)_{t \in \mathbb{T}}$ of random variables $\xi_t : \Omega \to E$ is called a *stochastic* or *random process* with parameter set or *time* $\mathbb{T}$ and *state space* E. In this book time $\mathbb{T}$ is equal to one of the additive (semi)groups $\mathbb{N}$, $\mathbb{Z}^+$, $\mathbb{R}^+$, $\mathbb{Z}^-$, $\mathbb{R}^-$, $\mathbb{Z}$, $\mathbb{R}$. For fixed ω the function $\xi.(\omega)$ given by $t \mapsto \xi_t(\omega)$ is called a *sample function* (*trajectory*, *path*). The mapping ξ given by $\omega \mapsto \xi.(\omega)$ into $(E^\mathbb{T}, \mathcal{E}^\mathbb{T})$, where

---

[5] $H(f)$ is compact if and only if f is periodic (exercise).

$$E^{\mathbb{T}} := \prod_{t \in \mathbb{T}} E, \quad \mathcal{E}^{\mathbb{T}} := \bigotimes_{t \in \mathbb{T}} \mathcal{E},$$

is measurable. Note that $\mathcal{E}^{\mathbb{T}}$ is generated by the algebra of *cylinder sets* in $E^{\mathbb{T}}$, i.e. sets of the form

$$Z = A_S \times \prod_{t \in \mathbb{T} \setminus S} E = \{\omega \in E^{\mathbb{T}} : (\omega(t_1), \ldots, \omega(t_r)) \in A_S\},$$

where $S = \{t_1, \ldots, t_r\} \in \mathfrak{P}_\circ(\mathbb{T})$, the family of non-void finite subsets of $\mathbb{T}$, and $A_S \in \mathcal{E}^S$. Put $\mathbb{P}_\xi := \xi \mathbb{P}$. The stochastic process $\bar{\xi}_t$ given by the coordinate functions $\bar{\xi}_t(\bar{\omega}) = \bar{\omega}(t)$ on $(\bar{\Omega}, \bar{\mathcal{F}}, \bar{\mathbb{P}}) := (E^{\mathbb{T}}, \mathcal{E}^{\mathbb{T}}, \mathbb{P}_\xi)$ is called the *canonical realization* of ξ. We have $\mathbb{P}_\xi = \mathbb{P}_{\bar{\xi}}$, in which case the two stochastic processes ξ and $\bar{\xi}$ are called *equivalent*. Two processes are equivalent if and only if their *finite-dimensional distributions* (i.e. the distributions $\mathbb{P}_S$ of ξ_S on $(E^S, \mathcal{E}^S)$) are the same for any $S \in \mathfrak{P}_\circ(\mathbb{T})$.

For two processes $\xi, \bar{\xi}$ on the same probability space, $\bar{\xi}$ is called a *version* or *modification* of ξ if $\mathbb{P}(\xi_t = \bar{\xi}_t) = 1$ for all $t \in \mathbb{T}$. They are called *indistinguishable* if $\{\omega : \xi_t(\omega) \neq \bar{\xi}_t(\omega) \text{ for some } t \in \mathbb{T}\} \subset N$ with $N \in \mathcal{F}$ and $\mathbb{P}(N) = 0$. Indistinguishable processes have $\mathbb{P}$-a.s. the same trajectories, while versions can have disjoint sets of trajectories. Versions are indistinguishable for discrete time and also in the case of continuous time if E is a Hausdorff space and the processes have $\mathbb{P}$-a.s. continuous (or right/left continuous) trajectories.

Let $(\mathbb{P}_S)_{S \in \mathfrak{P}_\circ(\mathbb{T})}$ be a projective family of finite-dimensional distributions (*projective* meaning that if $S_1 \subset S_2$ then $\mathbb{P}_{S_1} = \pi_{S_1}^{S_2} \mathbb{P}_{S_2}$, $\pi_{S_1}^{S_2}$ the canonical projection from E^{S_2} to E^{S_1}). Then Kolmogorov's celebrated *fundamental theorem* states that there is a stochastic process $\xi = (\xi_t)_{t \in \mathbb{T}}$ on some probability space $(\Omega, \mathcal{F}, \mathbb{P})$ whose finite-dimensional distributions are the prescribed ones – provided $(E, \mathcal{E})$ is a Polish space with its Borel σ-algebra (or more generally: $(E, \mathcal{E})$ is a standard space).

Canonical Realization on Path Space. Shift. For any stochastic process we can switch to the equivalent canonical realization $(\Omega, \mathcal{F}, \mathbb{P}) = (E^{\mathbb{T}}, \mathcal{E}^{\mathbb{T}}, \mathbb{P})$ and $\xi_t(\omega) = \omega(t)$.

Define the transformations $\theta(t) : E^{\mathbb{T}} \to E^{\mathbb{T}}$, $t \in \mathbb{T}$, by

$$\theta(t)\omega(s) := \omega(t+s), \quad s, t \in \mathbb{T}.$$

The mapping $\theta(t)$ is $\mathcal{E}^{\mathbb{T}}, \mathcal{E}^{\mathbb{T}}$ measurable. More precisely: If

$$\mathcal{E}_s^t := \sigma(\omega(u) : s \leq u \leq t), \quad s \leq t,$$

then $\theta(u)^{-1}\mathcal{E}_s^t = \mathcal{E}_{s+u}^{t+u}$, so that $\theta(u)$ is $\mathcal{E}_{s+u}^{t+u}, \mathcal{E}_s^t$ measurable for all $s \leq t$. θ is thus *filtered* with respect to the filtration $\mathcal{E}_s^t$. Further, $\theta(0) = \text{id}$ (if $0 \in \mathbb{T}$), $\theta(s+t) = \theta(s) \circ \theta(t)$, from which for two-sided time $\theta(t)^{-1} = \theta(-t)$ follows. The transformations are called the *unilateral* (for one-sided time) or *bilateral* (for two-sided time) *shift transformations*.

If $(E, \mathcal{E})$ is a topological space with its Borel σ-algebra, then the unilateral shift $\omega \mapsto \theta(t)\omega$ is continuous for each fixed $t \in \mathbb{T}$ in the product topology of $E^{\mathbb{T}}$ and the bilateral shift is even a homeomorphism. Note, however, that in the uncountable case $\mathcal{E}^{\mathbb{T}} \subsetneq \mathcal{B}(E^{\mathbb{T}})$.

For discrete time $\mathbb{T} = \mathbb{N}, \mathbb{Z}^+, \mathbb{Z}$, the mapping $(t, \omega) \mapsto \theta(t)\omega$ is trivially measurable, hence $(E^{\mathbb{T}}, \mathcal{E}^{\mathbb{T}}, (\theta(t))_{t \in \mathbb{T}})$ is a measurable DS.

For continuous time $\mathbb{T} = \mathbb{R}$ or $\mathbb{R}^+$ the canonical realization has certain defects which make the model useless for our purposes, e.g.

(i) many interesting sets (namely those whose description depends on an uncountable number of t's) are not in $\mathcal{E}^{\mathbb{T}}$,

(ii) the mapping $(t, \omega) \mapsto \theta(t)\omega$ is not $\mathcal{B}(\mathbb{T}) \otimes \mathcal{E}^{\mathbb{T}}, \mathcal{E}^{\mathbb{T}}$ measurable whenever $\mathcal{E}$ contains a non-trivial set.

The following is a procedure for obtaining an equivalent realization for continuous time without the above deficiencies: If $\Omega_0 \notin \mathcal{E}^{\mathbb{T}}$, but $\mathbb{P}^*(\Omega_0) = 1$, where

$$\mathbb{P}^*(\Omega_0) := \inf\{\mathbb{P}(A) : \Omega_0 \subset A, A \in \mathcal{E}^{\mathbb{T}}\}$$

is the *outer measure* of Ω_0, we can go down from $(\Omega, \mathcal{F}, \mathbb{P}) = (E^{\mathbb{T}}, \mathcal{E}^{\mathbb{T}}, \mathbb{P})$ to the space $(\Omega_0, \Omega_0 \cap \mathcal{E}^{\mathbb{T}}, \mathbb{P}_0)$, where $\mathbb{P}_0(\Omega_0 \cap A) := \mathbb{P}(A)$ is the unique probability measure on $\Omega_0 \cap \mathcal{E}^{\mathbb{T}}$ which has the same finite-dimensional distributions as $\mathbb{P}$. The coordinate functions $\xi_t(\omega) := \omega(t)$ on both spaces define equivalent stochastic processes, but all trajectories of the second process are in Ω_0. We will now describe some important cases for which there are also convenient criteria for $\mathbb{P}^*(\Omega_0) = 1$.

Canonical Spaces $\mathcal{C}(\mathbb{R}, \mathbb{R}^m)$ and $\mathcal{C}(\mathbb{R}^+, \mathbb{R}^m)$. First consider the case $\mathbb{T} = \mathbb{R}$. Let $\Omega := \mathcal{C}(\mathbb{R}, \mathbb{R}^m) \subset (\mathbb{R}^m)^{\mathbb{R}}$ ($\mathbb{R}^m$ could be replaced by a Polish space E). Clearly $\Omega \notin (\mathcal{B}^m)^{\mathbb{R}}$. Endow Ω with the compact open topology given by the complete metric

$$d(\omega, \omega') := \sum_{n=1}^{\infty} \frac{1}{2^n} \frac{\|\omega - \omega'\|_n}{1 + \|\omega - \omega'\|_n}, \quad \|\omega - \omega'\|_n := \sup_{-n \le t \le n} |\omega(t) - \omega'(t)|.$$

This makes Ω a Polish space (in fact, a Fréchet space). The Borel σ-algebra $\mathcal{F}$ is the trace in Ω of the product σ-algebra $(\mathcal{B}^m)^{\mathbb{R}}$, so $\mathcal{F} = \sigma(\omega(t) : t \in \mathbb{R})$. The set $\Omega \subset (\mathbb{R}^m)^{\mathbb{R}}$ is invariant with respect to the group of shifts $(\theta(t))_{t \in \mathbb{R}}$. For fixed $t \in \mathbb{T}$, $\theta(t)$ is a homeomorphism and $(t, \omega) \mapsto \theta(t)\omega$ is continuous, thus (see Dudley [131: Proposition 4.1.7]) measurable. Therefore $(\Omega, \mathcal{F}, (\theta(t))_{t \in \mathbb{R}})$ is a *measurable DS*. We have the natural filtration

$$\mathcal{F}_s^t := \sigma(\omega(u) : s \le u \le t), \quad s \le t,$$

with $\theta(u)^{-1} \mathcal{F}_s^t = \mathcal{F}_{s+u}^{t+u}$, hence θ is filtered with respect to $\mathcal{F}_s^t$.

A criterion for $\mathbb{P}^*(\Omega) = 1$ is *Kolmogorov's criterion* (see e.g. Kunita [224: p. 31]).

For $\mathbb{T} = \mathbb{R}^+$, analogous statements hold for $\Omega = \mathcal{C}(\mathbb{R}^+, \mathbb{R}^m)$ with obvious changes.

Canonical Spaces $\mathcal{D}(\mathbb{R}, \mathbb{R}^m)$ and $\mathcal{D}(\mathbb{R}^+, \mathbb{R}^m)$. Many important classes of possibly discontinuous stochastic processes (like (semi)martingales, Markov processes, processes with independent increments) have trajectories which are *cadlag* (or "càdlàg" after the French "continu à droite avec des limites à gauche"). The set of cadlag functions on $\mathbb{T} = \mathbb{R}$ with values in $\mathbb{R}^m$ (which could be replaced by a Polish space E) is defined by

$$\Omega = \mathcal{D}(\mathbb{R}, \mathbb{R}^m) :=$$
$$\{\omega \in (\mathbb{R}^m)^{\mathbb{R}} : \text{For all } t \in \mathbb{R} \lim_{s \uparrow t} \omega(s) =: \omega(t-),\ \lim_{s \downarrow t} \omega(s) = \omega(t) \text{ exist}\}.$$

Clearly $\Omega \notin (\mathcal{B}^m)^{\mathbb{R}}$. Each $\omega \in \Omega$ is measurable, bounded on bounded intervals and has only (at most countably many) discontinuities of the first kind (jumps). It is regularized at jumps to be right-continuous.

$\Omega = \mathcal{D}(\mathbb{R}, \mathbb{R}^m)$ can be made a Polish space (the so-called Skorokhod space) (for details see e.g. Billingsley [69: Chap. 3], Jacod and Shiryaev [186: Chap. IV], or Ethier and Kurtz [147: Chap. 3]). The Borel σ-algebra $\mathcal{F}$ is the trace in Ω of the product σ-algebra $(\mathcal{B}^m)^{\mathbb{R}}$, so $\mathcal{F} = \sigma(\omega(t) : t \in \mathbb{R})$. $\mathcal{C}(\mathbb{R}, \mathbb{R}^m)$ is continuously imbedded in $\mathcal{D}(\mathbb{R}, \mathbb{R}^m)$ and the relative Skorokhod topology in $\mathcal{C}$ is the $\mathcal{C}$ topology.

The set $\Omega \subset (\mathbb{R}^m)^{\mathbb{R}}$ is invariant with respect to the group of shifts $(\theta(t))_{t \in \mathbb{R}}$. For fixed $t \in \mathbb{R}$, $\theta(t) : \Omega \to \Omega$ is a homeomorphism and $(t, \omega) \mapsto \theta(t)\omega$ is continuous, thus measurable. Therefore $(\Omega, \mathcal{F}, (\theta(t))_{t \in \mathbb{R}})$ is a *measurable DS*. We again have the natural filtration $\mathcal{F}_s^t := \sigma(\omega(u) : s \leq u \leq t)$, $s \leq t$, with $\theta(u)^{-1} \mathcal{F}_s^t = \mathcal{F}_{s+u}^{t+u}$, so that θ is filtered with respect to $\mathcal{F}_s^t$.

For criteria for $\mathbb{P}^*(\Omega) = 1$ see e.g. Billingsley [69: Chap. 3] or Gänssler and Stute [156: pp. 283ff.].

For $\mathbb{T} = \mathbb{R}^+$, analogous statements hold for $\Omega = \mathcal{D}(\mathbb{R}^+, \mathbb{R}^m)$ with obvious changes.

A.3 Stationary Processes

Canonical DS Corresponding to a Stationary Process.. A stochastic process ξ with time $\mathbb{T}$ and state space $(E, \mathcal{E})$ is called *stationary* if for all $t, t_1, \ldots, t_r \in \mathbb{T}$ we have $\mathbb{P}_{t_1+t,\ldots,t_r+t} = \mathbb{P}_{t_1,\ldots,t_r}$. An equivalent, but more concise way of saying this is that on $(E^{\mathbb{T}}, \mathcal{E}^{\mathbb{T}}, \mathbb{P}_\xi)$

$$\theta(t) \mathbb{P}_\xi = \mathbb{P}_\xi \quad \text{for all } t \in \mathbb{T},$$

i.e. the shifts are measure preserving. For discrete time, a stationary stochastic process gives rise to the canonical metric DS

$$\Sigma := (E^{\mathbb{T}}, \mathcal{E}^{\mathbb{T}}, \mathbb{P}_\xi, (\theta(t))_{t \in \mathbb{T}}).$$

For continuous time, this model is inadequate (see above). However, in many relevant cases the stationary process ξ can be realized on $\mathcal{C}(\mathbb{R}, \mathbb{R}^m)$ ($\mathcal{C}(\mathbb{R}^+, \mathbb{R}^m)$) or $\mathcal{D}(\mathbb{R}, \mathbb{R}^m)$ ($\mathcal{D}(\mathbb{R}^+, \mathbb{R}^m)$), and in those cases we consider the canonical metric DS on the respective space introduced above.

Conversely, if Σ is a metric DS and $f : (\Omega, \mathcal{F}) \to (E, \mathcal{E})$ is measurable then $\xi_t(\omega) := f(\theta(t)\omega)$ is a measurable stationary stochastic process. A stochastic process is called *measurable* if the mapping $(t, \omega) \mapsto \xi_t(\omega)$ is measurable.

Canonical DS for Processes with Stationary Increments. A process ξ with continuous time $\mathbb{T} = \mathbb{R}$ or $\mathbb{R}^+$, state space $\mathbb{R}^m$ and with $\xi_0 = 0$ is said to have *stationary increments* if for any $t_1 \leq \ldots \leq t_r$ the distribution of $(\xi_{t_2+t} - \xi_{t_1+t}, \ldots, \xi_{t_r+t} - \xi_{t_{r-1}+t})$ is independent of $t \in \mathbb{T}$. In case ξ can be realized in $\mathcal{C}(\mathbb{R}, \mathbb{R}^m)$ (or $\mathcal{C}(\mathbb{R}^+, \mathbb{R}^m)$, $\mathcal{D}(\mathbb{R}, \mathbb{R}^m)$, $\mathcal{D}(\mathbb{R}^+, \mathbb{R}^m)$) we consider the Borel set

$$\Omega = \mathcal{C}_0(\mathbb{R}, \mathbb{R}^m) := \{\omega \in \mathcal{C}(\mathbb{R}, \mathbb{R}^m) : \omega(0) = 0\}$$

with its relative topology and Borel σ-algebra $\mathcal{F}$. We redefine the shift so that it leaves Ω invariant by

$$\theta(t) : \Omega \to \Omega, \quad \theta(t)\omega(s) := \omega(s+t) - \omega(t), \quad s, t \in \mathbb{R}.$$

Then $\theta(t)$ is a homeomorphism for each t and $(t, \omega) \mapsto \theta(t)\omega$ is continuous, hence measurable. With the natural filtration now defined as

$$\mathcal{F}_s^t := \sigma(\omega(u) - \omega(v) : s \leq u, v \leq t),$$

we have $\theta(u)^{-1} \mathcal{F}_s^t = \mathcal{F}_{s+u}^{t+u}$, so θ is filtered with respect to $\mathcal{F}_s^t$.

Similarly for the cases $\mathcal{C}_0(\mathbb{R}^+, \mathbb{R}^m)$, $\mathcal{D}_0(\mathbb{R}, \mathbb{R}^m)$ and $\mathcal{D}_0(\mathbb{R}^+, \mathbb{R}^m)$.

ξ has stationary increments if and only if its distribution $\mathbb{P}_\xi$ on Ω is shift invariant. The corresponding metric DS is the canonical one for ξ. The canonical process $\xi_t(\omega) := \omega(t)$ is a *helix* over θ, i.e. satisfies identically

$$\xi_{t+s}(\omega) = \xi_s(\omega) + \xi_t(\theta(s)\omega).$$

Every stochastically continuous process with stationary independent increments and $\xi_0 = 0$ can be realized on $\mathcal{D}_0(\mathbb{R}, \mathbb{R}^m)$ and gives thus rise to a canonical metric DS (see Gänssler and Stute [156: p. 298]). Since by the zero-one law the tail σ-algebra $\mathcal{T}^\infty$ is trivial mod $\mathbb{P}$, the DS is ergodic (see below).

Canonical DS for Brownian Motion/Wiener Process/White Noise. The only processes with stationary and independent increments which can be realized on $\mathcal{C}_0(\mathbb{R}, \mathbb{R}^m)$ have the form

$$\xi_t = t\lambda + G_t,$$

where $\lambda \in \mathbb{R}^m$ and G_t is a Gaussian process with stationary independent increments, $G_0 = 0$ and $G_t - G_s \sim \mathcal{N}(0, |t-s|\sigma)$, with $\sigma \in \mathbb{R}^{m \times m}$ non-negative definite.

A *standard Brownian motion* or *Wiener process* $(W_t)_{t\in\mathbb{R}}$ (i.e. with two-sided time) in $\mathbb{R}^m$ is a process with $W_0 = 0$ and stationary independent increments satisfying $W_t - W_s \sim \mathcal{N}(0, |t-s|I)$. The corresponding measure $\mathbb{P}$ on $(\Omega, \mathcal{F})$, where $\Omega = \mathcal{C}_0(\mathbb{R}, \mathbb{R}^m)$ and $\mathcal{F}$ is the (uncompleted!) Borel σ-algebra on Ω, is called *Wiener measure*, the probability space $(\Omega, \mathcal{F}, \mathbb{P})$ is called *Wiener space*. The corresponding canonical (ergodic) metric DS describes Brownian motion or *(Gaussian) white noise* as a metric DS.

We stress that $(t,\omega) \mapsto \theta(t)\omega$ is not $\mathcal{B} \otimes \mathcal{F}^\mathbb{P}, \mathcal{F}^\mathbb{P}$ measurable, where $\mathcal{F}^\mathbb{P}$ is the completion of $\mathcal{F}$ with respect to Wiener measure.[6]

This canonical DS describing Brownian motion/Wiener process/white noise is one of the fundamental objects of this book since it drives stochastic differential equations.

Invariant and Tail σ-Algebra. Let $\Sigma = (\Omega, \mathcal{F}, \mathbb{P}, (\theta(t))_{t\in\mathbb{T}})$ be one of the canonical DS introduced above, with the canonical filtration $\mathcal{F}_s^t$, $s \leq t$. Let

$$\mathcal{T}^\infty := \cap_{t\in\mathbb{T}} \mathcal{F}_t^\infty,$$

and for two-sided time

$$\mathcal{T}_{-\infty} := \cap_{t\in\mathbb{T}} \mathcal{F}_{-\infty}^t,$$

be the *tail* σ-algebras ($\mathcal{T}_{-\infty}$: remote past, $\mathcal{T}^\infty$: remote future). These σ-algebras are θ-invariant, $\theta(t)^{-1}\mathcal{T}^\infty = \mathcal{T}^\infty$, $\theta(t)^{-1}\mathcal{T}_{-\infty} = \mathcal{T}_{-\infty}$. Recall that $\mathcal{I} \subset \mathcal{I}_\mathbb{P} \subset \mathcal{F}$ are the σ-algebras of invariant sets and mod $\mathbb{P}$ invariant sets of the DS. Then

(i) if $\mathbb{T}$ is one-sided: $\mathcal{I} \subset \mathcal{T}^\infty$ (hence $\mathcal{I}_\mathbb{P} \subset \mathcal{T}^\infty$ mod $\mathbb{P}$ [7]),
(ii) if $\mathbb{T}$ is two-sided: $\mathcal{I}_\mathbb{P} \subset \mathcal{T}^\infty$ mod $\mathbb{P}$, and $\mathcal{I}_\mathbb{P} \subset \mathcal{T}_{-\infty}$ mod $\mathbb{P}$.

Consequently, if one of the tail σ-algebras is trivial mod $\mathbb{P}$, Σ is ergodic. For example, if $\mathbb{T}$ is discrete and $\mathbb{P} = \rho^\mathbb{T}$ is a product measure or if $\mathbb{P}$ corresponds to a process with stationary independent increments, then $\mathcal{T}^\infty$ is trivial mod $\mathbb{P}$ by Kolmogorov's zero-one law.

Natural Extension of Canonical DS. We can always assume that the canonical DS described above have two-sided time thanks to the existence of the following natural extension from one- to two-sided time (for the general method see Appendix A.1).

(i) Discrete time case: Suppose $\Sigma^+ = (E^{\mathbb{Z}^+}, \mathcal{E}^{\mathbb{Z}^+}, \mathbb{P}^+, (\theta^+(n))_{n\in\mathbb{Z}^+})$ is a canonical one-sided time metric DS, where $(E, \mathcal{E})$ is a standard measurable space, and $(\theta^+(n)\omega^+)_k = \omega_{k+n}^+$ is the one-sided shift. Define a θ-invariant probability $\mathbb{P}$ on the measurable DS $(E^\mathbb{Z}, \mathcal{E}^\mathbb{Z}, (\theta(n))_{n\in\mathbb{Z}})$ with two-sided time as follows: On a typical cylinder put

---

[6] The mapping $(t,\omega) \mapsto \theta(t)\omega$ is, however, $(\mathcal{B} \otimes \mathcal{F})^{\mu\times\mathbb{P}}, \mathcal{F}^\mathbb{P}$ measurable, where μ is any σ-finite measure on $\mathcal{B}$ (e.g. Lebesgue measure). See Kager [193: Lemma 3.35].

[7] $\mathcal{A} \subset \mathcal{B}$ mod $\mathbb{P}$ means that for each $A \in \mathcal{A}$ there is a $B \in \mathcal{B}$ with $\mathbb{P}(A \triangle B) = 0$.

$$\mathbb{P}\{\omega : \omega_{n_1} \in A_1, \ldots, \omega_{n_r} \in A_r\} :=$$
$$\mathbb{P}^+\{\omega^+ : \omega^+_{n_1+N} \in A_1, \ldots, \omega^+_{n_r+N} \in A_r\},$$

where we have chosen $N \geq 0$ so big that $n_i + N \geq 0$ for $i = 1, \ldots, r$. This projective family of finite-dimensional distributions uniquely extends by Kolmogorov's theorem to a probability $\mathbb{P}$ on $\mathcal{E}^{\mathbb{Z}}$ which by construction is θ-invariant. The resulting two-sided canonical metric DS $\Sigma = (E^{\mathbb{Z}}, \mathcal{E}^{\mathbb{Z}}, \mathbb{P}, (\theta(n))_{n \in \mathbb{Z}})$ is the natural extension of Σ^+, with homomorphism $\psi : E^{\mathbb{Z}} \to E^{\mathbb{Z}^+}$ given by $\omega = (\ldots, \omega_{-1}, \omega_0, \omega_1, \ldots) \mapsto \psi(\omega) := (\omega_0, \omega_1, \ldots) = \omega^+$. The σ-algebra $\mathcal{G} = \psi^{-1}(\mathcal{E}^{\mathbb{Z}^+}) = \sigma(\omega_n : n \in \mathbb{Z}^+)$ is exhaustive since $\sigma(\theta(n)\mathcal{G} : n \in \mathbb{Z}^+) = \sigma(\omega_n : n \in \mathbb{Z}) = \mathcal{E}^{\mathbb{Z}}$.

(ii) Continuous time case: Now our one-sided metric DS Σ^+ is one of the canonical DS desribed above on $\Omega^+ = \mathcal{C}(\mathbb{R}^+, \mathbb{R}^m), \mathcal{D}(\mathbb{R}^+, \mathbb{R}^m), \mathcal{C}_0(\mathbb{R}^+, \mathbb{R}^m)$ or $\mathcal{D}_0(\mathbb{R}^+, \mathbb{R}^m)$. By the same procedure as for discrete time, using Kolmogorov's theorem, we can naturally extend it to a canonical metric DS Σ on $\Omega = \mathcal{C}(\mathbb{R}, \mathbb{R}^m)$, etc. Here the homomorphism $\psi : \Omega \to \Omega^+$ is the restriction of a function ω from $\mathbb{R}$ to $\mathbb{R}^+$.

A.4 Markov Processes

Basic Definitions. Let $\mathbb{T} = \mathbb{Z}^+$ or $\mathbb{R}^+$, let $(E, \mathcal{E})$ be a standard space and let $(P_t)_{t \in \mathbb{T}}$ be a *transition function* on $(E, \mathcal{E})$, i.e. $P_t(x, B)$ is a *kernel* for each fixed t ($P_t(\cdot, B)$ is measurable for each fixed $B \in \mathcal{E}$ and $P_t(x, \cdot)$ is a probability on $\mathcal{E}$ for each fixed $x \in E$) with $P_0(x, B) = \delta_x(B)$ and

$$P_{s+t}(x, B) = \int_E P_s(y, B) P_t(x, dy), \quad \text{for all } s, t \in \mathbb{T}$$

(*Chapman-Kolmogorov equation*). Let ρ be a probability on $\mathcal{E}$. Then there exists a unique probability $\mathbb{P}_\rho$ on the canonical space $(\Omega, \mathcal{F}) = (E^{\mathbb{T}}, \mathcal{E}^{\mathbb{T}})$ such that the coordinate process $\xi_t(\omega) = \omega(t)$ is a (homogeneous) *Markov process* with respect to the natural filtration $\mathcal{F}_0^t = \sigma(\xi_s : 0 \leq s \leq t)$ with transition function $(P_t)_{t \in \mathbb{T}}$ and initial distribution ρ, i.e. it satisfies for any $s \leq t$ and any measurable $f : E \to \mathbb{R}^+$

$$\mathbb{E}_\rho(f(\xi_t)|\mathcal{F}_0^s) = P_{t-s}f(\xi_s) \quad \mathbb{P}_\rho\text{-a.s.},$$

where $P_t f(x) := \int_E f(y) P_t(x, dy)$ is the semigroup of positive linear operators corresponding to the kernel (P_t) and $\mathbb{P}_\rho$ is given on cylinders by

$$\mathbb{P}_\rho(\xi_{t_1} \in B_1, \ldots, \xi_{t_r} \in B_r) =$$
$$\int_E \int_{B_1} \cdots \int_{B_r} \rho(dx_0) P_{t_1}(x_0, dx_1) P_{t_2-t_1}(x_1, dx_2) \ldots P_{t_r-t_{r-1}}(x_{r-1}, dx_r),$$

where $t_1 \leq t_2 \leq \ldots \leq t_r$ and $B_1, \ldots, B_r \in \mathcal{E}$, and extended to $\mathcal{E}^{\mathbb{T}}$ by Kolmogorov's theorem (for $\mathbb{T} = \mathbb{Z}^+$, Ionescu Tulcea's theorem could be used which works for any measurable space $(E, \mathcal{E})$).

With $\mathbb{P}_x := \mathbb{P}_{\delta_x}$ the *Markov property* on the canonical space reads: For all $t \geq 0$ and measurable $f : \Omega \to \mathbb{R}^+$

$$\mathbb{E}_\rho(f \circ \theta(t)|\mathcal{F}_0^t) = \mathbb{E}_{\xi_t} f \quad \mathbb{P}_\rho\text{-a.s.}$$

In particular, for any $x \in E$, $t \geq 0$ and measurable $f \geq 0$

$$\mathbb{E}_x(f \circ \theta(t)|\mathcal{F}_0^t) = \mathbb{E}_{\xi_t} f \quad \mathbb{P}_x\text{-a.s.},$$

i.e. $(\Omega, \mathcal{F}, (\mathbb{P}_x)_{x \in E}, (\xi_t)_{t \in \mathbb{T}})$ forms a *Markov family* with transition function $(P_t)_{t \in \mathbb{T}}$.

Stationary Measures. Feller Processes. The measure $\mathbb{P}_\rho$ is $\theta(t)$ invariant (i.e. the Markov process is stationary) if and only if ρ is P_t-invariant or *stationary*[8], i.e.

$$\rho(\cdot) = \int_E P_t(x, \cdot) \rho(dx) \quad \text{for all} \quad t \geq 0$$

($t = 1$ suffices if $\mathbb{T} = \mathbb{Z}^+$). When does P_t admit a stationary ρ?

Suppose $(E, \mathcal{E})$ is a Polish space with its Borel σ-algebra and let $\mathcal{C}_b(E)$ be the Banach space of bounded continuous real-valued functions on E. If
(i) $P_t \mathcal{C}_b(E) \subset \mathcal{C}_b(E)$ for all $t \geq 0$,
(ii) $\lim_{t \downarrow 0} \|P_t f - f\| = 0$ for every $f \in \mathcal{C}_b(E)$ (only for $\mathbb{T} = \mathbb{R}^+$),
the semigroup P_t and the associated Markov process are called *Feller*.

A Feller transition function P_t admits a stationary ρ if and only if the sequence

$$\rho_T^x(\cdot) = \begin{cases} \frac{1}{T} \sum_{k=1}^T P_k(x, \cdot), & \mathbb{T} = \mathbb{Z}^+, \\ \frac{1}{T} \int_0^T P_t(x, \cdot) \, dt, & \mathbb{T} = \mathbb{R}^+, \end{cases}$$

has a limit point in the topology of weak convergence for $T \to \infty$ for some $x \in E$, and every such limit point is a stationary probability.

For discrete time $\mathbb{T} = \mathbb{Z}^+$, a standard state space $(E, \mathcal{E})$, a transition probability P_t and a stationary ρ we naturally extend (see above) to two-sided time $\mathbb{T} = \mathbb{Z}$ and obtain a metric DS $\Sigma_\rho := (E^\mathbb{Z}, \mathcal{E}^\mathbb{Z}, \mathbb{P}_\rho, (\theta(t))_{t \in \mathbb{Z}})$ where the coordinate process $\xi_n(\omega) = \omega_n$ satisfies the Markov property on all of $\mathbb{Z}$. Further, the sub-σ-algebra $\mathcal{I}_{\mathbb{P}_\rho}$ of mod $\mathbb{P}_\rho$ θ-invariant sets satisfies $\mathcal{I}_{\mathbb{P}_\rho} \subset \sigma(\xi_0)$ mod $\mathbb{P}_\rho$, i.e. for Markov processes, invariant sets can be "seen" in state space. This is the basis of the equivalence of the concept of invariance and ergodicity for Σ_ρ introduced in Appendix A.1 and the following state space concept:

Call a bounded measurable function g on E ρ-*invariant* if $P_t(g) = g$ ρ-a.s. for all $t \in \mathbb{Z}^+$. A stationary ρ is called *ergodic* if all ρ-invariant functions are constant ρ-a.s. By the above, there is a one-to-one correspondence between ρ-invariant functions g and mod $\mathbb{P}_\rho$ θ-invariant functions h via $g \circ \xi_0 = h$. Hence ρ is ergodic if and only if $\mathbb{P}_\rho$ is ergodic (see Meyn and Tweedie [253: Chap. 17] or Kifer [207: Sect. 1.2]).

---

[8] It is important to distinguish between P_t-invariant measures and invariant measures for dynamical systems. This is the reason why we introduce the name "stationary" for the former.

Every stationary ρ on a standard space has an ergodic decomposition $\rho(\cdot) = \int \eta(\cdot) \, d\nu_\rho(\eta)$, where ν_ρ is a probability measure on the space of all stationary measures concentrated on the ergodic stationary measures (see Kifer [207: Appendix A.1] for $\mathbb{T} = \mathbb{Z}^+$ and Skorokhod [320: §1.2] for $\mathbb{T} = \mathbb{R}^+$).

When does a Markov process with continuous time $\mathbb{T} = \mathbb{R}^+$ admit a cadlag version? Suppose E is locally compact with a countable base (thus Polish), let $\mathcal{C}_0(E)$ be the Banach space of continuous real-valued functions vanishing at infinity. Then a Markov process which is Feller in $\mathcal{C}_0(E)$ (see the above definition with $\mathcal{C}_0(E)$ instead of $\mathcal{C}_b(E)$) admits a cadlag version, more precisely a version on $\mathcal{D}(\mathbb{R}^+, \hat{E})$, where $\hat{E}$ is the one-point compactification of E by the point ∞.

If d is a metric on E and $\lim_{t \downarrow 0} \sup_{x \in K} \frac{1}{t} P_t(x, \{y : d(x,y) > \varepsilon\}) = 0$ for every $\varepsilon > 0$ and every compact set $K \subset E$, then ξ can be realized on $\mathcal{C}(\mathbb{R}, \hat{E})$.

If thus ρ is stationary for the semigroup P_t which is Feller in $\mathcal{C}_0(E)$ we obtain by natural extension a canonical metric DS Σ_ρ on $\mathcal{D}(\mathbb{R}, \hat{E})$.

Appendix B. Smooth Dynamical Systems

Throughout Appendix B, time $\mathbb{T}$ is always equal to $\mathbb{R}$.

B.1 Two-Parameter Flows on a Manifold

B.1.1 Definition (Smooth manifold). A d-dimensional *manifold* M is

(i) a *topological manifold* in the sense that M is a Hausdorff topological space which is locally Euclidean of dimension d and has a countable base of open sets. It follows that M is locally connected, locally compact, normal, metrizable by a complete metric, paracompact and separable (hence Polish) with countably many connected components (see e.g. Boothby [74: pp. 6–11]).

We assume once and for all that M is connected.

(ii) It is *smooth* in the sense that the changes between local coordinates are C^∞ functions. The choice of a C^∞ structure is without loss of generality for us, since in most situations we need at least a C^1 structure on M. But on a paracompact manifold M there is a C^∞ structure subordinate to a C^k structure of M for $1 \leq k < \infty$ (Whitney). ■

B.1.2 Definition. A (two-parameter) *continuous local flow* on a manifold M is a continuous mapping

$$\varphi : D \to M, \quad (s,t,x) \mapsto \varphi_{s,t}(x), \quad D \subset \mathbb{R} \times \mathbb{R} \times M \quad \text{an open set,}$$

such that:

(i) For each $s \in \mathbb{R}$, $x \in M$

$$D(s,x) := \{t \in \mathbb{R} : (s,t,x) \in D\} = (t^-(s,x), t^+(s,x))$$

is an open interval of $\mathbb{R}$ containing s,

(ii) φ is a *local flow* in the following sense:

1. $\varphi_{s,s} = \mathrm{id}_M$ for all $s \in \mathbb{R}$,
2. for all $s \in \mathbb{R}$, $x \in M$ and $u \in D(s,x)$, we have $t \in D(u, \varphi_{s,u}(x))$ if and only if $t \in D(s,x)$, and in this case the two-parameter local flow property holds:

$$\varphi_{u,t}(\varphi_{s,u}(x)) = \varphi_{s,t}(x).$$

If the mappings

$$\varphi_{s,t} : D_{s,t} \to M, \quad D_{s,t} := \{x \in M : (s,t,x) \in D\}, \quad x \mapsto \varphi_{s,t}(x),$$

(which are continuous) are C^k, $k \geq 1$ (i.e. k times differentiable with respect to x and the derivatives are continuous with respect to (s,t,x)), then the local flow is called C^k.

The local flow is called a *global flow*, or just a *flow*, if $D = \mathbb{R} \times \mathbb{R} \times M$ (or, equivalently, if $D(s,x) = \mathbb{R}$ for all $(s,x) \in \mathbb{R} \times M$, or if $D_{s,t} = M$ for all $s,t \in \mathbb{R}$). ∎

B.1.3 Theorem. *Let $\varphi_{s,t}(x)$ be a local continuous/C^k flow. Then*

(i) the function $(s,x) \mapsto t^+(s,x) \in (s,+\infty]$ is lower semicontinuous, and the function $(s,x) \mapsto t^-(s,x) \in [-\infty,s)$ is upper semicontinuous,

(ii) for all $t \in D(s,x)$

$$D(s,x) = D(t, \varphi_{s,t}(x)),$$

(iii) for all $(s,t) \in \mathbb{R} \times \mathbb{R}$ the mapping

$$\varphi_{s,t} : D_{s,t} \to D_{t,s}, \quad x \mapsto \varphi_{s,t}(x),$$

is a local homeomorphism/C^k diffeomorphism and

$$\varphi_{s,t}^{-1} = \varphi_{t,s} : D_{t,s} \to D_{s,t}.$$

(iv) The mapping

$$(s,t,x) \mapsto \varphi_{s,t}^{-1}(x) = \varphi_{t,s}(x)$$

is continuous/C^k.

A proof of these facts can be basically found in Amann [1: p. 134ff.] or Hartman [168: p. 94ff.].

B.2 Spaces of Functions in $\mathbb{R}^d$

B.2.1 Definition. Let $k \in \mathbb{Z}^+$ and $0 \leq \delta \leq 1$. Let $\mathcal{C}^{k,\delta}$ be the Fréchet space of functions $f : \mathbb{R}^d \to \mathbb{R}^d$ which are k times continuously differentiable and (for $\delta > 0$) whose k-th derivative is locally δ-Hölder continuous (for $\delta = 1$: locally Lipschitz continuous), with seminorms

$$\|f\|_{k,0;K} := \sum_{0 \leq |\alpha| \leq k} \sup_{x \in K} |D^\alpha f(x)|,$$

$$\|f\|_{k,\delta;K} := \|f\|_{k,0;K} + \sum_{|\alpha|=k} \sup_{x,y \in K, x \neq y} \frac{|D^\alpha f(x) - D^\alpha f(y)|}{|x-y|^\delta}, \quad 0 < \delta \leq 1,$$

where K is a compact convex subset of $\mathbb{R}^d$, $\alpha = (\alpha_1, \ldots, \alpha_d) \in (\mathbb{Z}^+)^d$ is a multi-index, $|\alpha| = \alpha_1 + \ldots + \alpha_d$, and

$$D^\alpha f(x) = D_x^\alpha f(x) = \frac{\partial^{|\alpha|}}{(\partial x_1)^{\alpha_1} \ldots (\partial x_d)^{\alpha_d}} f(x).$$

A complete metric is given by

$$\rho(f,g) := \sum_{n=1}^{\infty} \frac{1}{2^n} \frac{\|f-g\|_{k,\delta;K_n}}{1+\|f-g\|_{k,\delta;K_n}},$$

where (K_n) is some increasing sequence of compact convex sets exhausting $\mathbb{R}^d$. The topology of $\mathcal{C}^{k,\delta}$ is called the *compact-open topology*. While $\mathcal{C}^{k,0}$ is separable (hence Polish), the space $\mathcal{C}^{k,\delta}$ for $0 < \delta \leq 1$ is not separable (see Kufner, John and Fučik [221: Chap. 1]). We have for any $k \in \mathbb{Z}^+$ resp. $k-1 \in \mathbb{Z}^+$ and $0 \leq \varepsilon, \delta \leq 1$ the continuous inclusions

$$\mathcal{C}^{k,\delta} \hookrightarrow \mathcal{C}^{k,0} \hookrightarrow \mathcal{C}^{k-1,\varepsilon}.$$

We will sometimes write $\mathcal{C}^k := \mathcal{C}^{k,0}$ and will say that $f \in \mathcal{C}^\infty$ if $f \in \mathcal{C}^k$ for all $k \geq 1$. Composition $(f,g) \mapsto f \circ g$ is continuous in $\mathcal{C}^k$ for $k \in \{0, 1, \ldots, \infty\}$, and $\mathcal{C}^k$ is a Polish semigroup. ∎

B.2.2 Definition. Let

$$\text{Diff}^k(\mathbb{R}^d) := \{f \in \mathcal{C}^k : f \text{ bijective and } f^{-1} \in \mathcal{C}^k\},$$

put for $k = 0$

$$\text{Homeo}(\mathbb{R}^d) := \text{Diff}^0(\mathbb{R}^d),$$

endowed with its relative compact-open topology. A complete metric generating the same topology is given by

$$d(f,g) := \rho(f,g) + \rho(f^{-1}, g^{-1}),$$

where ρ is any complete metric of $\mathcal{C}^k$, making $\text{Diff}^k(\mathbb{R}^d)/\text{Homeo}(\mathbb{R}^d)$ a Polish group with respect to composition, the group of $\mathcal{C}^k$ diffeomorphisms/homeomorphisms of $\mathbb{R}^d$. ∎

B.2.3 Definition. Let $k \in \mathbb{Z}^+$ and $0 \leq \delta \leq 1$. Let $\mathcal{C}_b^{k,\delta}$ be the Banach space of functions $f : \mathbb{R}^d \to \mathbb{R}^d$ which are in $\mathcal{C}^{k,\delta}$ and for which the norm

$$\|f\|_{k,0} := \sup_{x \in \mathbb{R}^d} \frac{|f(x)|}{1+|x|} + \sum_{1 \leq |\alpha| \leq k} \sup_{x \in \mathbb{R}^d} |D^\alpha f(x)|,$$

$$\|f\|_{k,\delta} := \|f\|_{k,0} + \sum_{|\alpha|=k} \sup_{x \neq y} \frac{|D^\alpha f(x) - D^\alpha f(y)|}{|x-y|^\delta}, \quad 0 < \delta \leq 1,$$

is finite. We have for any $k \in \mathbb{Z}^+$, respectively $k-1 \in \mathbb{Z}^+$, and $0 \leq \varepsilon, \delta \leq 1$ the continuous inclusions

$$\mathcal{C}_b^{k,\delta} \hookrightarrow \mathcal{C}_b^{k,0} \hookrightarrow \mathcal{C}_b^{k-1,\varepsilon},$$

and sometimes write $\mathcal{C}_b^k := \mathcal{C}_b^{k,0}$. ∎

B.2.4 Definition. Let $k \in \mathbb{Z}^+$ and $0 \leq \delta \leq 1$. Let $L_{\text{loc}}(\mathbb{R}, \mathcal{C}^{k,\delta})$ be the set of measurable functions $f : \mathbb{R} \times \mathbb{R}^d \to \mathbb{R}^d$ for which

- $f(t, \cdot) \in \mathcal{C}^{k,\delta}$ for every $t \in \mathbb{R}$ ("for Lebesgue-almost all t" would suffice),
- for every compact set $K \subset \mathbb{R}^d$ and every bounded interval $[a,b] \subset \mathbb{R}$

$$\int_a^b \|f(t,\cdot)\|_{k,\delta;K}\, dt < \infty. \tag{B.2.1}$$

With the seminorms given by (B.2.1), $L_{\text{loc}}(\mathbb{R}, \mathcal{C}^{k,\delta})$ is a Fréchet space. We have the continuous inclusions

$$L_{\text{loc}}(\mathbb{R}, \mathcal{C}^{k,\delta}) \hookrightarrow L_{\text{loc}}(\mathbb{R}, \mathcal{C}^{k,0}) \hookrightarrow L_{\text{loc}}(\mathbb{R}, \mathcal{C}^{k-1,\varepsilon}).$$

The Fréchet spaces $L_{\text{loc}}(\mathbb{R}, \mathcal{C}_b^{k,\delta})$ are defined analogously and obey analogous inclusion properties. ∎

B.2.5 Definition. Let $k \in \mathbb{Z}^+$, $0 \leq \delta \leq 1$, $f : \mathbb{R} \times \mathbb{R}^d \to \mathbb{R}^d$ and let $(t,x) \mapsto f(t,x)$ be continuous. We say that $f \in \mathcal{C}^{0;k,\delta}$ if

- for each $t \in \mathbb{R}$, $f(t,\cdot) \in \mathcal{C}^{k,\delta}$,
- in case $k \geq 1$ the derivatives $D_x^\alpha f(t,x)$ are continuous with respect to (t,x) for all $1 \leq |\alpha| \leq k$,
- in case $\delta > 0$, for $|\alpha| = k$ the derivatives are locally δ-Hölder continuous (for $\delta = 1$ locally Lipschitz continuous) with respect to x.

The spaces are Fréchet with seminorms $\sup_{a \leq t \leq b} \|f(t,\cdot)\|_{k,\delta;K}$ and we have the continuous inclusions

$$\mathcal{C}^{0;k,\delta} \hookrightarrow \mathcal{C}^{0;k,0} \hookrightarrow \mathcal{C}^{0;k-1,\varepsilon}.$$

Further,

$$\mathcal{C}^{0;k,\delta} \hookrightarrow L_{\text{loc}}(\mathbb{R}, \mathcal{C}^{k,\delta}).$$

The last inclusion is clear by the fact that local Hölder continuity of $f(t,x)$ with respect to x is equivalent to uniform Hölder continuity on compact sets $[a,b] \times K$ (see e.g. Amann [1: p. 102]).

We say that $f \in \mathcal{C}_b^{0;k,\delta}$ if $f \in \mathcal{C}^{0;k,\delta}$ and for each $t \in \mathbb{R}$, $f(t,\cdot) \in \mathcal{C}_b^{k,\delta}$. ∎

B.2.6 Remark. Observe that we have (following Kunita [224]) built into the definition of our spaces $\mathcal{C}_b^{k,\delta}$ that its elements are permitted to be unbounded, but with at most linear growth, i.e. for each $f \in \mathcal{C}_b^{k,\delta}$ there are constants α and β such that for all $x \in \mathbb{R}^d$

$$|f(x)| \leq \alpha |x| + \beta.$$

It implies that for $f \in L_{\text{loc}}(\mathbb{R}, \mathcal{C}_b^{k,\delta})$

$$|f(t,x)| \le \alpha(t)|x| + \beta(t)$$

with locally integrable $\alpha(\cdot)$, $\beta(\cdot)$. This "trick" allows a more elegant formulation of the conditions for the existence of global flows. ∎

B.3 Differential Equations in $\mathbb{R}^d$

In studying differential equations $\dot{x}_t = f(t, x_t)$ and their solutions we will adopt the dynamical systems or flow point of view: Rather than studying a particular solution for given initial values, we consider the totality of solutions starting at arbitrary times $s \in \mathbb{R}$ and at arbitrary points $x \in \mathbb{R}^d$. We then look at the mappings defined by "initial value x at time s goes to solution of $\dot{x}_t = f(t, x_t)$ at time t". There is a basically one-to-one correspondence between (local) continuous/C^k two-parameter flows $\varphi_{s,t}$, which are absolutely continuous with respect to t, and differential equations.

We say that $t \mapsto \varphi_{s,t}(x)$ solves or is a *solution* of $\dot{x}_t = f(t, x_t)$ (sometimes called *solution in the sense of Carathéodory*, see Aulbach and Wanner [47]), or that the differential equation *generates* the flow $\varphi_{s,t}$, if it satisfies

$$\varphi_{s,t}(x) = x + \int_s^t f(u, \varphi_{s,u}(x))\, du \tag{B.3.1}$$

for all $t \in D(s, x)$, an open interval of $\mathbb{R}$ containing s (this interval is all of $\mathbb{R}$ in the global case). Note that (B.3.1) implies $\varphi_{s,s}(x) = x$ for all s and x.

If the solution is in fact differentiable with respect to t and satisfies for $t \in D(s, x)$

$$\frac{d}{dt}\varphi_{s,t}(x) = f(t, \varphi_{s,t}(x)), \quad \varphi_{s,s}(x) = x, \tag{B.3.2}$$

it is called a *classical solution* of $\dot{x}_t = f(t, x_t)$.

B.3.1 Theorem (Local Flow from Differential Equation). *Consider the differential equation*

$$\dot{x}_t = f(t, x_t) \tag{B.3.3}$$

in $\mathbb{R}^d$.

(i) If $f \in L_{\mathrm{loc}}(\mathbb{R}, C^{0,1})$, then (B.3.3) uniquely generates through its maximal solution a local continuous two-parameter flow $\varphi_{s,t}$.

(ii) If $f \in L_{\mathrm{loc}}(\mathbb{R}, C^{k,0})$ for some $k \ge 1$, then (B.3.3) uniquely generates a local C^k flow. The Jacobian of $\varphi_{s,t}$ at x,

$$D\varphi_{s,t}(x) := \left(\frac{\partial(\varphi_{s,t}(x))_i}{\partial x_j}\right) \in Gl(d, \mathbb{R}),$$

uniquely solves the variational equation $\dot{v}_t = Df(t, x_t)v_t$, i.e. we have for all $t \in D(s, x)$

$$D\varphi_{s,t}(x) = I + \int_s^t Df(u, \varphi_{s,u}(x)) \, D\varphi_{s,u}(x) \, du,$$

where $Df(t,x) := (\frac{\partial f_i(t,x)}{\partial x_j}) \in \mathbb{R}^{d \times d}$ is the Jacobian of $f(t,x)$. Finally, the determinant $\det D\varphi_{s,t}(x)$ satisfies Liouville's equation

$$\det D\varphi_{s,t}(x) = \exp \int_s^t \operatorname{trace} Df(u, \varphi_{s,u}(x)) \, du, \quad t \in D(s,x).$$

(iii) If $f \in \mathcal{C}^{0;0,1}$, then the solution of (i) is classical. More precisely, $\varphi_{s,t}(x)$ is Lipschitz continuous in the variable (s,t,x) and is C^1 with respect to t.

(iv) If $f \in \mathcal{C}^{0;k,0}$ for some $k \geq 1$, then the solution of (ii) is classical. The Jacobian and its determinant satisfy the classical versions of the variational and Liouville's equation, respectively.

The proof of this theorem can be basically found in many textbooks, see in particular Amann [1], Coddington and Levinson [102] or Walter [335].

B.3.2 Remark. (i) Even if $f(t,\cdot)$ is C^∞ with respect to x the solution $\varphi_{s,t}(x)$ is in general only absolutely continuous with respect to t.

(ii) To assume that $f(t,\cdot) \in \mathcal{C}^k$ for each fixed $t \in \mathbb{R}$ is in general not enough to guarantee the result.

(iii) The integral equation (B.3.1) and the differential equation (B.3.2) are always with respect to the argument t for fixed s. Initial time s plays the rôle of a parameter and is fixed. In $\varphi_{s,t}$, s always denotes the initial time no matter whether $s \leq t$ or $s \geq t$.

(iv) The maximal solution $\varphi_{s,t}(x)$ has the property that if $t^+(s,x) < \infty$ then

$$\lim_{t \uparrow t^+(s,x)} |\varphi_{s,t}(x)| = \infty,$$

similarly for $t^-(s,x) > -\infty$. ∎

B.3.3 Theorem (Global Flow from Differential Equation).
Consider the differential equation (B.3.3) in $\mathbb{R}^d$.

(i) Suppose $f \in L_{\mathrm{loc}}(\mathbb{R}, \mathcal{C}_b^{0,1})/L_{\mathrm{loc}}(\mathbb{R}, \mathcal{C}_b^{k,0})$ or, more generally, suppose $f \in L_{\mathrm{loc}}(\mathbb{R}, \mathcal{C}^{0,1})/L_{\mathrm{loc}}(\mathbb{R}, \mathcal{C}^{k,0})$ and

$$|f(t,x)| \leq \alpha(t)|x| + \beta(t) \tag{B.3.4}$$

with $\alpha \geq 0$, $\beta \geq 0$ locally integrable. Then (B.3.3) uniquely generates a continuous/C^k flow.

(ii) Suppose $f \in \mathcal{C}_b^{0;0,1}/\mathcal{C}_b^{0;k,0}$ or, more generally, suppose $f \in \mathcal{C}^{0;0,1}/\mathcal{C}^{0;k,0}$ and (B.3.4) holds. Then (B.3.3) uniquely generates a classical continuous/C^k flow.

The fact that the linear growth condition (B.3.4) implies a global solution follows from the Gronwall-Bellman lemma which we state for convenience.

B.3.4 Lemma (Gronwall-Bellman Lemma). *Let* $a, x : [t_1, t_2] \to \mathbb{R}^+$ *be continuous and let* $b : [t_1, t_2] \to \mathbb{R}^+$ *be integrable. If*

$$x(t) \leq a(t) + \int_{t_1}^{t} b(s)x(s)ds \quad \text{for all} \quad t \in [t_1, t_2],$$

then

$$x(t) \leq a(t) + \int_{t_1}^{t} a(s)b(s) \exp\left(\int_s^t b(u)du\right) ds \quad \text{for all} \quad t \in [t_1, t_2],$$

and the right-hand side is finite. If $a(\cdot)$ *is increasing, then*

$$x(t) \leq a(t) \exp\left(\int_{t_1}^{t} b(s)ds\right) \quad \text{for all} \quad t \in [t_1, t_2].$$

We now deal with the inverse problem of when for a given flow $\varphi_{s,t}$ there exists a differential equation $\dot{x}_t = f(t, x_t)$ which generates it.

B.3.5 Theorem (Differential Equation from Flow). *(i) Let* $\varphi_{s,t}$ *be a local continuous flow for which* $t \mapsto \varphi_{s,t}(x)$ *is differentiable at* $t = s$ *for all* x. *Put*

$$f(s, x) := \frac{d}{dt}\varphi_{s,t}(x)|_{t=s}.$$

Then f *is measurable,* $t \mapsto \varphi_{s,t}(x)$ *is differentiable for all* $t \in D(s, x)$ *and*

$$\frac{d}{dt}\varphi_{s,t}(x) = f(t, \varphi_{s,t}(x)), \quad \varphi_{s,s}(x) = x.$$

(ii) Let $\varphi_{s,t}$ *be a local continuous flow for which* $t \mapsto \varphi_{s,t}(x)$ *is absolutely continuous with respect to* $t \in D(s, x)$ *for all* x. *Then there exists a measurable function* $f : \mathbb{R} \times \mathbb{R}^d \to \mathbb{R}^d$ *for which for all* x

$$\varphi_{s,t}(x) = x + \int_s^t f(u, \varphi_{s,u}(x))\, du, \quad t \in D(s, x).$$

The result is that we have a basically one-to-one correspondence between two-parameter flows and non-autonomous differential equations.

B.4 Autonomous Case: Dynamical Systems

If the differential equation is autonomous, $\dot{x}_t = f(x_t)$, then clearly $f \in L_{\text{loc}}(\mathbb{R}, C_b^{k,\delta})$ if and only if $f \in C_b^{0;k,\delta}$ if and only if $f \in C_b^{k,\delta}$, similarly for $C^{k,\delta}$. This means that we are always in the case of classical solutions.

Hence, if e.g. $f \in C_b^{0,1}$ the differential equation uniquely generates a continuous flow $\varphi_{s,t}$. By substituting $v = u - s$ we obtain

$$\varphi_{s,t}(x) = x + \int_s^t f(\varphi_{s,u}(x))du = x + \int_0^{t-s} f(\varphi_{s,s+v}(x))dv,$$

thus by uniqueness of solution

$$\varphi_{s,t}(x) = \varphi_{0,t-s}(x), \tag{B.4.1}$$

hence the one-parameter family $\varphi(t) := \varphi_{0,t}$ satisfies the one-parameter flow property $\varphi(t)\varphi(s)x = \varphi(t+s)x$.

A two-parameter flow with the extra property (B.4.1) obtains the name dynamical system.

B.4.1 Definition (Dynamical system). A *local continuous dynamical system*[1] on a manifold M is a continuous mapping

$$\varphi : D \to M, \quad (t,x) \mapsto \varphi(t)x,$$

where $D \subset \mathbb{R} \times M$ is open, such that for each $x \in M$

$$D(x) := \{t \in \mathbb{R} : (t,x) \in D\} =: (\tau^-(x), \tau^+(x))$$

is an open interval of $\mathbb{R}$ containing 0 and φ satisfies the following conditions:

1. $\varphi(0) = \mathrm{id}_M$,
2. For all $x \in M$ and all $s \in D(x)$ the local flow property holds: We have $t \in D(\varphi(s)x)$ if and only if $t+s \in D(x)$, and in that case

$$\varphi(t)(\varphi(s)x) = \varphi(t+s)x.$$

If the mapping $(t,x) \mapsto \varphi(t)x$ is k times continuously differentiable with respect to x, where $k \geq 1$, the local continuous dynamical system is called C^k.

If $D = \mathbb{R} \times M$ (equivalently, if $D(x) = \mathbb{R}$ for all $x \in M$ or if $D(t) = M$ for all $t \in \mathbb{R}$, where $D(t) := \{x \in M : (t,x) \in D\}$) then the local dynamical system is called *(global) dynamical system*. ∎

The sets $D(x) \subset \mathbb{R}$ and $D(t) \subset M$ from Definition B.4.1 are open as sections of the open set D. We have $t \in D(x) \iff x \in D(t)$, implying $D(x) = \{t : x \in D(t)\}$ and $D(t) = \{x : t \in D(x)\}$.

B.4.2 Theorem. *Let φ be a local continuous/C^k DS on a manifold M. Then*

(i) $x \mapsto \tau^+(x) \in (0, +\infty]$ *is lower semicontinuous and* $x \mapsto \tau^-(x) \in [-\infty, 0)$ *is upper semicontinuous,*

(ii) for all $(t,x) \in D$

$$D(x) = t + D(\varphi(t)x),$$

(iii) for all $t \in \mathbb{R}$

---
[1] "Dynamical system" is abbreviated as "DS".

$$\varphi(t) : D(t) \to D(-t)$$

is a local homeomorphism/C^k diffeomorphism and

$$\varphi(t)^{-1} = \varphi(-t) : D(-t) \to D(t),$$

(iv) the mapping $(t, x) \mapsto \varphi(t)^{-1}x = \varphi(-t)x$ is continuous/C^k.

For a proof see e. g. Amann [1: pp. 100ff.].

B.4.3 Theorem (Local DS from Vector Field). *(i) If $f \in C^{0,1}$, then the differential equation $\dot{x}_t = f(x_t)$ in $\mathbb{R}^d$ generates through its unique maximal classical solution $(t, x) \mapsto \varphi(t)x$ a local continuous DS which is Lipschitz with respect to (t, x) and C^1 with respect to t.*
(ii) If $f \in C^k$, $k \geq 1$, then the local DS is C^k. In this case $D\varphi(t)x$ satisfies locally the non-autonomous variational equation

$$\dot{v}_t = Df(\varphi(t)x)v_t, \quad v_0 = \mathrm{id}_{\mathbb{R}^d},$$

and Liouville's equation

$$\det D\varphi(t)x = \exp \int_0^t \mathrm{trace} Df(\varphi(s)x) \, ds.$$

If for the maximal local solution $\tau^+(x) < \infty$ or $\tau^-(x) > -\infty$ for some x, then

$$\lim_{t \uparrow \tau^+(x)} |\varphi(t)x| = \infty \quad \text{or} \quad \lim_{t \downarrow \tau^-(x)} |\varphi(t)x| = \infty.$$

B.4.4 Theorem (Global DS from Vector Field). *Suppose $f \in C_b^{0,1}/C_b^{k,0}$ for some $k \geq 1$. More generally, suppose $f \in C^{0,1}/C^{k,0}$ and for some constants $\alpha, \beta \geq 0$*

$$|f(x)| \leq \alpha|x| + \beta \quad \text{for all} \quad x \in \mathbb{R}^d.$$

Then the differential equation $\dot{x}_t = f(x_t)$ in $\mathbb{R}^d$ uniquely generates a continuous/C^k DS. If $f \in C_b^{k,0}$, then $(\varphi, D\varphi)$ is a C^{k-1} DS on $\mathbb{R}^d \times \mathbb{R}^d$.

B.4.5 Theorem (Vector Field from DS). *Let φ be a local continuous DS.*
(i) If $t \mapsto \varphi(t)x$ is differentiable at $t = 0$ for all x, then, putting

$$f(x) := \frac{d}{dt}\varphi(t)x|_{t=0},$$

φ is generated by $\dot{x}_t = f(x_t)$ and $t \mapsto \varphi(t)x$ is differentiable for all $t \in D(x)$ and all x.
(ii) If $t \mapsto \varphi(t)x$ is absolutely continuous for $t \in D(x)$ and all x, then there exists a measurable function $f : \mathbb{R}^d \to \mathbb{R}^d$ for which for all x

$$\varphi(t)x = x + \int_0^t f(\varphi(s)x)\,ds, \quad t \in D(x).$$

A consequence of the above is that the classes of DS and autonomous differential equations (vector fields) are basically the same, symbolically written as

$$\text{dynamical systems} = \exp(\text{vector fields}).$$

B.5 Vector Fields and Flows on Manifolds

The definitions B.1.2 and B.4.1 above are already formulated for the manifold case. We claim that all local results stated above for $\mathbb{R}^d$ will remain valid for the case where $\mathbb{R}^d$ is replaced by a manifold M. As most things are obvious, we will be quite brief.

Let now M be a d-dimensional manifold. For $1 \leq k \leq \infty$ let $\mathfrak{X}^k(M)$ be the separable Fréchet space of C^k vector fields. We will consider non-autonomous differential equations $\dot{x}_t = f(t, x_t)$ on M with right-hand side

$$f : \mathbb{R} \to \mathfrak{X}^k(M), \quad t \mapsto f(t, \cdot) = f(t). \tag{B.5.1}$$

The spaces $\mathcal{C}^{0;k,\delta} = \mathcal{C}^{0;k,\delta}(\mathbb{R} \times M, TM)$ can now be defined (see Kunita [224: pp. 185ff.]) as those functions (B.5.1) for which $\mathbb{R} \times M \ni (t, x) \mapsto f(t)(x) \in T_x M$ is, for any chart over which TM is locally trivialized, in $\mathcal{C}^{0;k,\delta}$ (the space introduced in Definition B.2.5).

The spaces $L_{\text{loc}}(\mathbb{R}, \mathcal{C}^{k,\delta}) = L_{\text{loc}}(\mathbb{R}, \mathcal{C}^{k,\delta}(M, TM))$ can also be defined as those functions (B.5.1) for which $t \mapsto \|f(t)\|_{k,\delta;K}$ is locally integrable for any compact $K \subset M$. Here the seminorms $\|f\|_{k,\delta;K}$ of a vector field $f \in \mathfrak{X}^k(M)$ are defined in an obvious manner by covering K by finitely many charts over which TM is trivialized, and then using Definition B.2.1.

The space $\mathcal{C}^k(M, N)$ of C^k mappings $f : M \to N$, M, N manifolds, $0 \leq k \leq \infty$, with its C^k compact-open topology can be similarly defined (see e.g. Baxendale [54: Sect. 2]). $\mathcal{C}^k(M, N)$ is a Polish space, and $\mathcal{C}^k(M, M)$ is a Polish semigroup.

Further,

$$\text{Diff}^k(M) := \{f \in \mathcal{C}^k(M, M) : f \text{ bijective and } f^{-1} \in \mathcal{C}^k(M, M)\},$$

with

$$\text{Homeo}(M) := \text{Diff}^0(M),$$

given the relative compact-open topology, are Polish groups. In fact, if ρ is a complete metric of $\mathcal{C}^k(M, M)$, then

$$d(f, g) := \rho(f, g) + \rho(f^{-1}, g^{-1})$$

is a complete metric of $\text{Diff}^k(M)/\text{Homeo}(M)$, still generating the compact-open topology.

B.5 Vector Fields and Flows on Manifolds

Let f be a function as in (B.5.1). We call a function $\varphi : \mathbb{R} \times \mathbb{R} \times M \to M$, $(s,t,x) \mapsto \varphi_{s,t}(x)$ a *solution* of $\dot{x}_t = f(t, x_t)$ if for each test function $h \in \mathcal{C}_K^\infty(M, \mathbb{R})$, the space of real-valued C^∞ functions with compact support,

$$h(\varphi_{s,t}(x)) = h(x) + \int_s^t (f(u, \cdot)h)(\varphi_{s,u}(x))du. \tag{B.5.2}$$

It will be called a *classical solution* if for each $h \in \mathcal{C}_K^\infty(M)$

$$\frac{d}{dt} h(\varphi_{s,t}(x)) = (f(t, \cdot)h)(\varphi_{s,t}(x)), \quad h(\varphi_{s,s}(x)) = h(x). \tag{B.5.3}$$

With those notions the existence and uniqueness part of Theorem B.3.1 holds verbatim.

Let φ be a local C^1 flow on M. Denote by $T\varphi$, the derivative of φ, the local continuous flow on the tangent bundle TM which covers φ and is on fibers defined by the derivative of $\varphi_{s,t}$ at $x \in D_{s,t}$,

$$T\varphi_{s,t}(x) : T_x M \to T_{\varphi_{s,t}(x)} M,$$

where for a local diffeomorphism φ for each x in the domain of φ the linear isomorphism

$$T\varphi(x) : T_x M \to T_{\varphi(x)} M, \quad v \mapsto w = T\varphi(x)v,$$

is defined by $w(h) := v(h(\varphi(x)))$, $h \in \mathcal{C}_K^\infty(M)$.

If φ solves $\dot{x}_t = f(t, x_t)$ with $f \in L_{\text{loc}}(\mathbb{R}, \mathcal{C}^{k,0})$, then $T\varphi$ is a local C^{k-1} flow which uniquely solves

$$\dot{v}_t = Tf(t, v_t), \quad v_s = v \in TM, \tag{B.5.4}$$

where $Tf(t, \cdot)$ is the natural lift of the vector field $f(t, \cdot)$ on M to a vector field on TM, in local coordinates $Tf(t, x, v) = (x, v, f(t, x), Df(t, x)v)$.

If M is a Riemannian manifold, we can use the Riemannian connection to decompose $T_v TM$ into horizontal and vertical components. Then $Tf(t, v)$ has horizontal component (identified with) $f(t, x)$ and vertical component $\nabla f(t, x)v$, where ∇ denotes the covariant derivative of the vector field $f(t, \cdot)$ at x in direction $v \in T_x M$. Equation (B.5.4) is then equivalent to the coupled system of the original equation $\dot{x}_t = f(t, x_t)$ and the *variational equation*

$$v'_t = \nabla f(t, \varphi_{s,t}(x))v_t, \quad v_s = v \in T_x M. \tag{B.5.5}$$

Here the *absolute* (or covariant) *derivative* v'_t is taken along the integral curve $t \mapsto \varphi_{s,t}(x)$ in the direction of the vector field $f(t, \cdot)$ (see Elworthy [138: Appendix B] or [139: p. 300], Klingenberg [213: Sect. 1.5]).

The expression for the variational equation (B.5.5) in a chart around $x \in M$ is

$$v'^k_t = \dot{v}^k_t + \Gamma^k_{ij} f^j v^i_t = (\nabla_i f^k) v^i_t, \quad \nabla_i f^k = \frac{\partial f^k}{\partial x^i} + \Gamma^k_{ij} f^j,$$

with summation of repeated indices.

Moreover,

$$\operatorname{div} f(t, x) = \operatorname{trace} \nabla f(t, x) := \sum_{i=1}^{d} \langle \nabla f(t, x) v_i, v_i \rangle_x,$$

where (v_i) is an orthonormal basis of $(T_x M, \langle \cdot, \cdot \rangle_x)$, and we have *Liouville's equation*: For $t \in D(s, x)$

$$\det T\varphi_{s,t}(x) = \exp \int_s^t \operatorname{div} f(u, \varphi_{s,u}(x)) \, du = \exp \int_s^t \operatorname{trace} \nabla f(u, \varphi_{s,u}(x)) \, du.$$

A local C^k flow on M generated by $\dot{x}_t = f(t, x_t)$ with $f \in L_{\text{loc}}(\mathbb{R}, \mathcal{C}^{k,0})$ is global if M is compact, or if, by embedding M into an $\mathbb{R}^N$, $f(t, \cdot)$ is the restriction of a vector field $\tilde{f}(t, \cdot)$ on $\mathbb{R}^N$ for which $\tilde{f} \in L_{\text{loc}}(\mathbb{R}, \mathcal{C}_b^{k,0})$.

Dynamical Systems on Manifolds

If $f \in \mathfrak{X}^k(M)$, then $\dot{x}_t = f(x_t)$ generates a local C^k DS. The derivative $T\varphi$ satisfies

$$\dot{v}_t = Tf(v_t), \quad v_0 = v \in TM, \tag{B.5.6}$$

and $T\varphi$ is a C^{k-1} DS as well as a continuous linear bundle DS on TM over the DS φ. If M is a Riemannian manifold, then (B.5.6) is equivalent to $\dot{x}_t = f(x_t)$ and the variational equation $v'_t = \nabla f(\varphi(t)x) v_t$, and

$$\det T\varphi(t)x = \exp \int_0^t \operatorname{div} f(\varphi(s)x) \, ds = \exp \int_0^t \operatorname{trace} \nabla f(\varphi(s)x) \, ds.$$

References

1. H. Amann. *Gewöhnliche Differentialgleichungen*. Walter de Gruyter, Berlin, 1983.
2. D. V. Anosov and V. I. Arnold. *Dynamical systems I*. Springer-Verlag, Berlin Heidelberg New York, 1988.
3. R. Arens. Topologies for homeomorphism groups. *Amer. J. Math.*, 68:593–610, 1946.
4. S. T. Ariaratnam. Some illustrative examples of stochastic bifurcation. In J. M. T. Thompson and S. R. Bishop, editors, *Nonlinearity and chaos in engineering dynamics (IUTAM Symposium 1994)*, pages 267–274. Wiley, New York, 1994.
5. S. T. Ariaratnam and N. Sri Namachchivaya. Stochastically perturbed Hopf bifurcation. In F. M. A. Salam and G. I. L. Levi, editors, *Dynamical Systems Approaches to Nonlinear Problems in Systems and Circuits*, pages 39–52. SIAM, 1988.
6. S. T. Ariaratnam and W. C. Xie. Stochastic bifurcations in engineering mechanics. In *Proceedings of the 47th Session of the International Statistical Institute, Paris, France, Volume L III.3*, pages 479–496, 1989.
7. S. T. Ariaratnam and W. C. Xie. Lyapunov exponents in stochastic structural dynamics. In Arnold et al. [14], pages 271–289.
8. L. Arnold. The unfolding of dynamics in stochastic analysis. *Matemática Aplicada e Computacional*, 16:3–25, 1997.
9. L. Arnold, G. Bleckert, and K.-R. Schenk-Hoppé. The stochastic Brusselator: parametric noise destroys Hopf bifurcation. Report 425, Institut für Dynamische Systeme, Universität Bremen, 1998.
10. L. Arnold and P. Boxler. Stochastische Bifurkationstheorie. Report 163, Institut für Dynamische Systeme, Universität Bremen, 1986.
11. L. Arnold and P. Boxler. Additive noise turns a hyperbolic fixed point into a stationary solution. In Arnold et al. [14], pages 159–164.
12. L. Arnold and P. Boxler. Stochastic bifurcation: instructive examples in dimension one. In M. Pinsky and V. Wihstutz, editors, *Diffusion processes and related problems in analysis, volume II: Stochastic flows*, volume 27 of *Progress in Probability*, pages 241–255. Birkhäuser, Boston Basel Stuttgart, 1992.
13. L. Arnold and H. Crauel. Iterated function systems and multiplicative ergodic theory. In M. Pinsky and V. Wihstutz, editors, *Diffusion processes and related problems in analysis, volume II: Stochastic flows*, volume 27 of *Progress in Probability*, pages 283–305. Birkhäuser, Boston Basel Stuttgart, 1992.
14. L. Arnold, H. Crauel, and J.-P. Eckmann, editors. *Lyapunov Exponents. Proceedings, Oberwolfach 1990*, volume 1486 of *Springer Lecture Notes in Mathematics*. Springer-Verlag, Berlin Heidelberg New York, 1991.

15. L. Arnold, L. Demetrius, and M. V. Gundlach. Evolutionary formalism for products of positive random matrices. *Annals of Applied Probability*, 4:859–901, 1994.
16. L. Arnold, M. Doyle, and N. Sri Namachchivaya. Small noise expansion of moment Lyapunov exponents for two–dimensional systems. *Dynamics and Stability of Systems*, 12:187–211, 1997.
17. L. Arnold, A. Eizenberg, and V. Wihstutz. Large noise asymptotics of invariant measures, with applications to Lyapunov exponents. *Stochastics and Stochastics Reports*, 59:71–142, 1996.
18. L. Arnold and P. Imkeller. Furstenberg-Khasminskii formulas for Lyapunov exponents via anticipative calculus. *Stochastics and Stochastics Reports*, 54:127–168, 1995.
19. L. Arnold and P. Imkeller. Stratonovich calculus with spatial parameters and anticipative problems in multiplicative ergodic theory. *Stochastic Processes and their Applications*, 62:19–54, 1996.
20. L. Arnold and P. Imkeller. Rotation numbers for linear stochastic differential equations. Report 415, Institut für Dynamische Systeme, Universität Bremen, 1997.
21. L. Arnold and P. Imkeller. Normal forms for stochastic differential equations. *Probab. Theory Relat. Fields*, 1998. To appear.
22. L. Arnold and P. Imkeller. On the integrability condition in the multiplicative ergodic theorem for stochastic differential equations. *Stochastics and Stochastics Reports*, 1998. To appear.
23. L. Arnold and R. Khasminskii. Stability index for nonlinear stochastic differential equations. In M. C. Cranston and M. A. Pinsky, editors, *Stochastic Analysis (Proceedings of Symposia in Pure Mathematics, Volume 57)*, pages 543–551. American Mathematical Society, Providence, Rhode Island, 1995.
24. L. Arnold and W. Kliemann. Qualitative theory of stochastic systems. In A. T. Bharucha-Reid, editor, , volume 3 of *Probabilistic Analysis and Related Topics*, pages 1–79. Academic Press, New York, 1983.
25. L. Arnold and W. Kliemann. Large deviations of linear stochastic differential equations. In H. J. Engelbert and W. Schmidt, editors, *Stochastic Differential Systems*, pages 117–151, 1987. Springer Lecture Notes in Control and Information Sciences, Volume 96.
26. L. Arnold and W. Kliemann. On unique ergodicity for degenerate diffusions. *Stochastics*, 21:41–61, 1987.
27. L. Arnold, W. Kliemann, and E. Oeljeklaus. Lyapunov exponents of linear stochastic systems. In L. Arnold and V. Wihstutz, editors, *Lyapunov Exponents. Proceedings, Bremen 1984*, pages 85–125. Springer-Verlag, Berlin Heidelberg New York, 1986.
28. L. Arnold and P. Kloeden. Discretization of a random dynamical system near a hyperbolic point. *Mathematische Nachrichten*, 181:43–72, 1996.
29. L. Arnold and Nguyen Dinh Cong. Generic properties of Lyapunov exponents. *Random & Computational Dynamics*, 2(3&4):335–345, 1994.
30. L. Arnold and Nguyen Dinh Cong. On the simplicity of the Lyaponov spectrum of products of random matrices. *Ergodic Theory and Dynamical Systems*, 17:1005–1025, 1997.
31. L. Arnold, Nguyen Dinh Cong, and V. I. Oseledets. Jordan normal form for linear cocycles. Report 408, Institut für Dynamische Systeme, Universität Bremen, 1997.

32. L. Arnold, E. Oeljeklaus, and E. Pardoux. Almost sure and moment stability for linear Itô equations. In L. Arnold and V. Wihstutz, editors, *Lyapunov Exponents. Proceedings, Bremen 1984*, pages 129–159. Springer-Verlag, Berlin Heidelberg New York, 1986.
33. L. Arnold and L. San Martin. A control problem related to the lyapunov spectrum of stochastic flows. *Matemática Aplicada e Computacional*, 5:31–64, 1986.
34. L. Arnold and L. San Martin. A multiplicative ergodic theorem for rotation numbers. *Journal of Dynamics and Differential Equations*, 1:95–119, 1989.
35. L. Arnold and M. Scheutzow. Perfect cocycles through stochastic differential equations. *Probab. Theory Relat. Fields*, 101:65–88, 1995.
36. L. Arnold and B. Schmalfuß. Fixed points and attractors for random dynamical systems. In A. Naess and S. Krenk, editors, *IUTAM Symposium on Advances in Nonlinear Stochastic Mechanics*, pages 19–28. Kluwer, Dordrecht, 1996.
37. L. Arnold, N. Sri Namachchivaya, and K. R. Schenk–Hoppé. Toward an understanding of stochastic Hopf bifurcation: a case study. *International Journal of Bifurcation and Chaos*, 6:1947–1975, 1996.
38. L. Arnold and V. Wihstutz. Stationary solutions of linear systems with additive and multiplicative noise. *Stochastics*, 7:133–155, 1982.
39. L. Arnold and V. Wihstutz, editors. *Lyapunov Exponents. Proceedings, Bremen 1984*, volume 1186 of *Springer Lecture Notes in Mathematics*. Springer-Verlag, Berlin Heidelberg New York, 1986.
40. L. Arnold and Xu Kedai. Normal forms for random diffeomorphisms. *J. Dynamics and Differential Equations*, 4:445–483, 1992.
41. L. Arnold and Xu Kedai. Simultaneous normal form and center manifold reduction for random differential equations. In C. Perelló, C. Simó, and J. Solà-Morales, editors, *EQUADIFF–91*, volume 1, pages 68–80. World Scientific, Singapore, 1993.
42. L. Arnold and Xu Kedai. Invariant measures for random dynamical systems, and a necessary condition for stochastic bifurcation from a fixed point. *Random & Computational Dynamics*, 2:165–182, 1994.
43. L. Arnold and Xu Kedai. Normal forms for random differential equations. *Journal of Differential Equations*, 116:484–503, 1995.
44. V. I. Arnold. *Geometrical methods in the theory of ordinary differential equations*. Springer-Verlag, Berlin Heidelberg New York, 1983.
45. V. I. Arnold. *Dynamical systems V*. Springer-Verlag, Berlin Heidelberg New York, 1994.
46. B. Aulbach and T. Wanner. *Invariant fiber bundles in dynamical processes*. Preliminary version of a book, 1995.
47. B. Aulbach and T. Wanner. Integral manifolds for Carathéodory type differential equations in Banach spaces. In B. Aulbach and F. Colonius, editors, *Six lectures on dynamical systems*, pages 45–119. World Scientific, Singapore, 1996.
48. M. Barlow and M. Yor. Semi–martingale inequalities via the Garsia–Rodemich–Rumsey lemma and applications to local times. *Journal of Functional Analysis*, 49:198–229, 1982.
49. M. F. Barnsley. *Fractals everywhere*. Academic Press, New York, 1988.
50. M. F. Barnsley and J. Elton. A new class of Markov processes for image encoding. *Adv. Appl. Probab.*, 20:14–32, 1988.

51. H. Bauer. *Maß- und Integrationstheorie*. Walter de Gruyter, Berlin, 1990.
52. H. Bauer. *Wahrscheinlichkeitstheorie*. Walter de Gruyter, Berlin, 1991.
53. P. Baxendale. Wiener processes on manifolds of maps. *Proc. Royal Soc. Edinburgh*, 87A:127–152, 1980.
54. P. Baxendale. Brownian motion in the diffeomorphism group I. *Compositio Mathematica*, 53:19–50, 1984.
55. P. Baxendale. Asymptotic behavior of stochastic flows of diffeomorphisms: two case studies. *Probab. Theory Relat. Fields*, 73:51–85, 1986.
56. P. Baxendale. Asymptotic behaviour of stochastic flows of diffeomorphisms. In K. Itô and T. Hida, editors, *Stochastic processes and their applications*, volume 1203 of *Springer Lecture Notes in Mathematics*, pages 1–19. Springer-Verlag, Berlin Heidelberg New York, 1986.
57. P. Baxendale. The Lyapunov spectrum of a stochastic flow of diffeomorphisms. In L. Arnold and V. Wihstutz, editors, *Lyapunov exponents. Proceedings, Bremen 1984*, pages 322–337. Springer-Verlag, Berlin Heidelberg New York, 1986.
58. P. Baxendale. Moment stability and large deviations for linear stochastic differential equations. In N. Ikeda, editor, *Proc. Taniguchi Symposium on probabilistic methods in mathematical physics*, pages 31–54. Kinokuniya, Tokyo, 1987.
59. P. Baxendale. Lyapunov exponents and relative entropy for a stochastic flow of diffeomorphisms. *Probab. Theory Relat. Fields*, 81:521–554, 1989.
60. P. Baxendale. Invariant measures for nonlinear stochastic differential equations. In Arnold et al. [14], pages 123–140.
61. P. Baxendale. Statistical equilibrium and two–point motion for a stochastic flow of diffeomorphisms. In K. S. Alexander and J. C. Watkins, editors, *Spatial stochastic processes*, pages 189–218. Birkhäuser, Boston, 1991.
62. P. Baxendale. Stability and equilibrium properties of stochastic flows of diffeomorphisms. In M. Pinsky and V. Wihstutz, editors, *Diffusion processes and related problems in analysis, volume II: Stochastic flows*, volume 27 of *Progress in Probability*, pages 3–35. Birkhäuser, Boston Basel Stuttgart, 1992.
63. P. Baxendale. A stochastic Hopf bifurcation. *Probab. Theory Relat. Fields*, 99:581–616, 1994.
64. P. Baxendale and D. Stroock. Large deviations and stochastic flows of diffeomorphisms. *Probab. Theory Relat. Fields*, 80:169–215, 1988.
65. Y. I. Belopolskaya and Y. L. Dalecky. *Stochastic equations and differential geometry*. Kluwer, Dordrecht, 1990.
66. E. Benoît, editor. *Dynamic bifurcations*, volume 1493 of *Springer Lecture Notes in Mathematics*. Springer-Verlag, Berlin Heidelberg New York, 1991.
67. M. Berger. *Introduction to probability and stochastic processes*. Springer-Verlag, Berlin Heidelberg New York, 1993.
68. N. Bhatia and G. Szegö. *Stability theory of dynamical systems*. Springer-Verlag, Berlin Heidelberg New York, 1970.
69. P. Billingsley. *Convergence of probability measures*. Wiley, New York, 1968.
70. J.-M. Bismut. Flots stochastiques et formule de Ito-Stratonovitch généralisée. *C. R. Acad. Sci., Paris, Ser. I*, 290:483–486, 1980.
71. J.-M. Bismut. *Mécanique aléatoire*, volume 866 of *Springer Lecture Notes in Mathematics*. Springer-Verlag, Berlin Heidelberg New York, 1981.

72. J.-P. Bismut. A generalized formula of Itô and some other properties of stochastic flows. *Z. Wahrscheinlichkeitstheorie verw. Gebiete*, 55:331–350, 1981.
73. T. Bogenschütz. *Equilibrium states for random dynamical systems*. PhD thesis, Institut für Dynamische Systeme, Universität Bremen, 1993.
74. W. M. Boothby. *An introduction to differentiable manifolds and Riemannian geometry*. Academic Press, New York, 1975.
75. P. Bougerol. Comparaison des exposants de Lyapounov des processus markoviens multiplicatifs. *Ann. Inst. H. Poincaré Probab. Statist.*, 24:439–489, 1988.
76. P. Bougerol. Théorèmes limite pour les systèmes linéaires à coefficients markovien. *Probab. Theory Relat. Fields*, 78:193–221, 1988.
77. P. Bougerol and J. Lacroix. *Products of random matrices with applications to Schrödinger operators*. Birkhäuser, Boston Basel Stuttgart, 1985.
78. N. Bourbaki. *Eléments de mathématique. Fascicule XXV, Livre Intégration*. Hermann, Paris, 1959.
79. P. Boxler. A stochastic version of center manifold theory. *Probab. Theory Relat. Fields*, 83:509–545, 1989.
80. P. Boxler. How to construct stochastic center manifolds on the level of vector fields. In Arnold et al. [14], pages 141–158.
81. L. Breiman. *Probability*. SIAM, Philadelphia, 1992.
82. M. Brin and Y. Kifer. Dynamics of Markov chains and stable manifolds for random diffeomorphisms. *Ergodic Theory and Dynamical Systems*, 7:351–374, 1987.
83. I. Y. Bronshtein. *Non-autonomous dynamical systems*. Akademia Nauk Moldavskoi SSR, Kishinev, 1984.
84. H. Bunke. *Gewöhnliche Differentialgleichungen mit zufälligen Parametern*. Akademie-Verlag, Berlin, 1972.
85. B. F. Bylov, R. E. Vinograd, D. M. Grobman, and V. V. Nemytskii. *Theory of Lyapunov exponents*. Nauka, Moscow, 1966 (in Russian).
86. R. Carmona and J. Lacroix. *Spectral theory of random Schrödinger operators* Birkhäuser, Boston Basel Stuttgart, 1990.
87. R. Carmona and D. Nualart. *Nonlinear stochastic integrators, equations and flows*. Gordon and Breach, New York, 1990.
88. A. Carverhill. A pair of stochastic dynamical systems which have the same infinitesimal generator, but of which one is strongly complete, and the other is not. Technical report, Mathematics Institute, University of Warwick, 1981.
89. A. Carverhill. Flows of stochastic dynamical systems: ergodic theory. *Stochastics*, 14:273–317, 1985.
90. A. Carverhill. A formula for the Lyapunov numbers of a stochastic flow. Application to a perturbation theorem. *Stochastics*, 14:209–226, 1985.
91. A. Carverhill. A nonrandom Lyapunov spectrum for nonlinear stochastic dynamical systems. *Stochastics*, 17:253–287, 1986.
92. A. Carverhill. Survey: Lyapunov exponents for stochastic flows on manifolds. In L. Arnold and V. Wihstutz, editors, *Lyapunov Exponents. Proceedings, Bremen 1984*, volume 1186 of *Springer Lecture Notes in Mathematics*, pages 292–307. Springer-Verlag, Berlin Heidelberg New York, 1986.

93. A. Carverhill, M. Chappell, and D. K. Elworthy. Characteristic exponents for stochastic flows. In S. Albeverio, P. Blanchard, and L. Streit, editors, *Stochastic processes — mathematics and physics*, volume 1158 of *Springer Lecture Notes in Mathematics*, pages 52–80. Springer-Verlag, Berlin Heidelberg New York, 1986.
94. A. Carverhill and K. D. Elworthy. Flows of stochastic dynamical systems: the functional analytic approach. *Probab. Theory Relat. Fields*, 65:245–267, 1983.
95. A. Carverhill and K. D. Elworthy. Lyapunov exponents for a stochastic analogue of the geodesic flow. *Trans. Amer. Math. Soc.*, 295:85–105, 1986.
96. C. Castaing and M. Valadier. *Convex analysis and measurable multifunctions*, volume 580 of *Springer Lecture Notes in Mathematics*. Springer-Verlag, Berlin Heidelberg New York, 1977.
97. L. Cesari. *Asymptotic behavior and stability problems in ordinary differential equations*. Springer-Verlag, Berlin Heidelberg New York, 1963.
98. M. Chappell. Bounds for average Lyapunov exponents of gradient stochastic systems. In L. Arnold and V. Wihstutz, editors, *Lyapunov Exponents. Proceedings, Bremen 1984*, volume 1186 of *Springer Lecture Notes in Mathematics*, pages 308–321. Springer-Verlag, Berlin Heidelberg New York, 1986.
99. S. Chow and Y. Yi. Center manifold and stability for skew product flows. *J. Dynamics and Differential Equations*, 6:543–582, 1994.
100. S.-N. Chow and J. K. Hale. *Methods of bifurcation theory*. Springer-Verlag, Berlin Heidelberg New York, 1996.
101. S.-N. Chow, K. Lu, and Y.-Q. Shen. Normal forms for quasiperiodic evolutionary equations. *Discrete and continuous dynamical systems*, 2:65–94, 1996.
102. E. Coddington and N. Levinson. *Theory of ordinary differential equations*. McGraw-Hill, New York, 1955.
103. J. E. Cohen, H. Kesten, and C. M. Newman. Oseledec's multiplicative ergodic theorem: a proof. In *Random matrices and their applications* [104], pages 23–30.
104. J. E. Cohen, H. Kesten, and C. M. Newman, editors. *Random Matrices and Their Applications*, volume 50 of *Contemporary Mathematics*. American Mathematical Society, Providence, Rhode Island, 1986.
105. F. Colonius and W. Kliemann. *Dynamics of control systems*. Birkhäuser, Boston Basel Stuttgart, 1998.
106. I. P. Cornfeld, S. V. Fomin, and Y. G. Sinai. *Ergodic theory*. Springer-Verlag, Berlin Heidelberg New York, 1982.
107. P. H. Coullet, C. Elphick, and E. Tirapegui. Normal form of a Hopf bifurcation with noise. *Physics Letters*, 111A:277–282, 1985.
108. H. Crauel. Ergodentheorie linearer stochastischer Systeme. Report 59, Institut für Dynamische Systeme, Universität Bremen, 1981.
109. H. Crauel. *Random dynamical systems — positivity of Lyapunov exponents, and Markov systems*. PhD thesis, Institut für Dynamische Systeme, Universität Bremen, 1988.
110. H. Crauel. Extremal exponents of random dynamical systems do not vanish. *Journal of Dynamics and Differential Equations*, 2:245–291, 1990.
111. H. Crauel. Lyapunov exponents of random dynamical systems on Grassmannians. In Arnold et al. [14], pages 38–50.
112. H. Crauel. Markov measures for random dynamical systems. *Stochastics and Stochastics Reports*, 37:153–173, 1991.

113. H. Crauel. Non-Markovian invariant measures are hyperbolic. *Stochastic Processes and their Applications*, 45:13–28, 1993.
114. H. Crauel. Random probability measures on Polish spaces. Habilitationsschrift, Universität Bremen, 1995.
115. H. Crauel. Global random attractors are uniquely determined by attracting deterministic compact sets. *Annali di Matematica*, 1998. To appear.
116. H. Crauel, A. Debussche, and F. Flandoli. Random attractors. *Journal of Dynamics and Differential Equations*, 9:307–341, 1997.
117. H. Crauel and F. Flandoli. Attractors for random dynamical systems. *Probab. Theory Relat. Fields*, 100:365–393, 1994.
118. H. Crauel and F. Flandoli. Additive noise destroys a pitchfork bifurcation. *Journal of Dynamics and Differential Equations*, 10:259–274, 1998.
119. H. Crauel and F. Flandoli. Hausdorff dimension of invariant sets for random dynamical systems. *Journal of Dynamics and Differential Equations*, 1998. (to appear).
120. S. Dahlke. *Invariante Mannigfaltigkeiten für Produkte zufälliger Diffeomorphismen*. PhD thesis, Institut für Dynamische Systeme, Universität Bremen, 1989.
121. S. Dahlke. Invariant manifolds for products of random diffeomorphisms. *J. Dynamics and Differential Equations*, 9:157–210, 1997.
122. J. de Sam Lazaro and P. A. Meyer. Méthodes de martingales et théorie des flots. *Z. Wahrscheinlichkeitstheorie verw. Gebiete*, 18:116–140, 1971.
123. J. de Sam Lazaro and P. A. Meyer. Questions de théorie des flots. *Séminaire de Probabilités*, IX:1–153, 1975. Springer Lecture Notes in Mathematics, Volume 465.
124. A. Debussche. On the finite dimensionality of random attractors. *Stochastic Analysis and Applications*, 15:473–492, 1997.
125. F. Delyon and P. Foulon. Complex entropy for dynamical systems. Preprint, 1987.
126. J.-D. Deuschel and D. Stroock. *Large Deviations*. Academic Press, New York, 1989.
127. J. Dieudonné. *Grundzüge der modernen Analysis, Volume 9*. VEB Deutscher Verlag der Wissenschaften, Berlin, 1987.
128. J. Doob. *Stochastic processes*. Wiley, New York, 1953.
129. J. Doob. *Classical potential theory and its probabilistic counterpart*. Springer-Verlag, Berlin Heidelberg New York, 1984.
130. D. Dubischar. Representation of Markov processes by random dynamical systems. Report 393, Institut für Dynamische Systeme, Universität Bremen, 1997.
131. R. M. Dudley. *Real Analysis and Probability*. Wadsworth & Brooks/Cole, Pacific Grove, California, 1989.
132. N. Dunford and J. T. Schwartz. *Linear Operators. Part I: General theory*. Interscience, New York, 1958.
133. R. Durrett. *Probability: theory and examples*. Wadsworth & Brooks/Cole, Pacific Grove, California, 1991.
134. W. Ebeling, H. Herzel, W. Richert, and L. Schimansky-Geier. Influence of noise on Duffing-Van der Pol oscillators. *Zeitschrift f. Angew. Math. u. Mechanik*, 66:141–146, 1986.

135. C. Elphick, E. Tirapegui, M. E. Brachet, P. H. Coullet, and G. Iooss. A simple global characterization for normal forms of singular vector fields. *Physica D*, 29:95–127, 1987.
136. J. Elton. A multiplicative ergodic theorem for Lipschitz maps. *Stochastic Processes and their Applications*, 34:39–47, 1990.
137. K. D. Elworthy. Stochastic dynamical systems and their flows. In A. Friedman and M. Pinsky, editors, *Stochastic Analysis*, pages 79–95. Academic Press, New York, 1978.
138. K. D. Elworthy. *Stochastic differential equations on manifolds*. Cambridge University Press, Cambridge, 1982.
139. K. D. Elworthy. Geometric aspects of diffusions on manifolds. *École d'Été de Probabilités de Saint-Flour*, pages 277–425, 1988. Springer Lecture Notes in Mathematics, Volume 1362.
140. K. D. Elworthy. Stochastic flows on Riemannian manifolds. In M. Pinsky and V. Wihstutz, editors, *Diffusion processes and related problems in analysis, volume II: Stochastic flows*, volume 27 of *Progress in Probability*, pages 37–72. Birkhäuser, Boston Basel Stuttgart, 1992.
141. K. D. Elworthy. Stochastic differential geometry. *Bull. Sc. math.*, 2e série, 117:7–28, 1993.
142. K. D. Elworthy, Y. Le Jan, and X.-M. Li. Concerning the geometry of stochastic differential equations and stochastic flow. Preprint 56, Warwick University, 1994.
143. K. D. Elworthy and X.-M. Li. Derivative flows of stochastic differential equations: moment exponents and geometric properties. In M. C. Cranston and M. A. Pinsky, editors, *Stochastic Analysis (Proceedings of Symposia in Pure Mathematics, Volume 57)*, pages 565–574. American Mathematical Society, Providence, Rhode Island, 1995.
144. K. D. Elworthy and S. Rosenberg. Homotopy and homology vanishing theorems and the stability of stochastic flows. *Geometric and Functional Analysis*, 6:51–78, 1996.
145. K. D. Elworthy and M. Yor. Conditional expectations for derivatives of certain stochastic flows. *Séminaire de Probabilités*, XXVII:159–172, 1993. Springer Lecture Notes in Mathematics, Volume 1557.
146. M. Emery. *Stochastic calculus in manifolds*. Springer-Verlag, Berlin Heidelberg New York, 1989.
147. S. Ethier and T. Kurtz. *Markov processes. Characterization and convergence*. Wiley, New York, 1986.
148. M. Farkas. *Periodic motion*. Springer-Verlag, Berlin Heidelberg New York, 1994.
149. A. Fathi, M. R. Herman, and J. C. Yoccoz. A proof of Pesin's stable manifold theorem. In J. Palis, editor, *Geometric Dynamics*, pages 177–215, 1983. Springer Lecture Notes in Mathematics, Volume 1007.
150. F. Flandoli and K.-U. Schaumlöffel. A multiplicative ergodic theorem with applications to a first order stochastic hyperbolic equation in a bounded domain. *Stochastics and Stochastics Reports*, 34:241–255, 1991.
151. F. Flandoli and B. Schmalfuß. Random attractors for the 3D stochastic Navier–Stokes equation with multiplicative white noise. *Stochastics and Stochastics Reports*, 59:21–45, 1996.
152. M. I. Freidlin and A. Wentzell. *Random perturbations of dynamical systems*. Springer-Verlag, Berlin Heidelberg New York, 1984.

153. H. Furstenberg. Noncommuting random products. *Trans. Amer. Math. Soc.*, 108:377–428, 1963.
154. H. Furstenberg and H. Kesten. Products of random matrices. *Ann. Math.*, 31:457–469, 1960.
155. H. Furstenberg and Y. Kifer. Random matrix products and measures on projective spaces. *Israel Journal of Mathematics*, 46:12–32, 1983.
156. P. Gänssler and W. Stute. *Wahrscheinlichkeitstheorie*. Springer-Verlag, Berlin Heidelberg New York, 1977.
157. F. R. Gantmacher. *The theory of matrices*. Chelsea, New York, 1960.
158. T. C. Gard. *Introduction to stochastic differential equations*. Marcel Dekker, New York, 1988.
159. I. Gihman and A. Skorohod. *Stochastic differential equations*. Springer-Verlag, Berlin Heidelberg New York, 1972.
160. I. Gihman and A. Skorohod. *The theory of stochastic processes I,II,III*. Springer-Verlag, Berlin Heidelberg New York, 1974–1979.
161. I. Y. Goldsheid and G. A. Margulis. Lyapunov indices of a product of random matrices. *Russian Mathematical Surveys*, 44:11–71, 1989.
162. J. Guckenheimer and P. Holmes. *Nonlinear oscillations, dynamical systems, and bifurcation of vector fields*. Springer-Verlag, Berlin Heidelberg New York, 1983.
163. Y. Guivarc'h and A. Raugi. Frontière de Furstenberg, propriétés de contraction et théorème de convergence. *Z. Wahrscheinlichkeitstheorie Verw. Gebiete*, 69:187–242, 1985.
164. V. M. Gundlach. Random homoclinic orbits. *Random & Computational Dynamics*, 3:1–33, 1995.
165. W. Hackenbroch and A. Thalmaier. *Stochastische Analysis*. Teubner, Stuttgart, 1994.
166. W. Hahn. *Stability of motion*. Springer-Verlag, Berlin Heidelberg New York, 1967.
167. J. Hale and H. Koçak. *Dynamics and bifurcations*. Springer-Verlag, Berlin Heidelberg New York, 1991.
168. P. Hartman. *Ordinary differential equations*. Birkhäuser, Boston Basel Stuttgart, 1982.
169. H. Hennion. Loi de grands nombres et perturbations pour des produits réductibles de matrices aléatoires indépendantes. *Z. Wahrscheinlichkeitstheorie verw. Gebiete*, 67:265–278, 1984.
170. B. Hess, M. Kiwi, M. Markus, and J. Rössler. Periodically and randomly modulated non linear processes. In E. Tirapegui and D. Villarroel, editors, *Instabilities and Nonequilibrium Structures*, pages 67–74. Kluwer, Dordrecht, 1989.
171. S. Hilger. Smoothness of invariant manifolds. *J. Functional Analysis*, 106:95–129, 1992.
172. G. Högnäs and A. Mukherjea. *Probability measures on semigroups: convolution products, random walks, and random matrices*. Plenum Press, 1995.
173. P. Holmes and D. Rand. Phase portraits and bifurcations of the nonlinear oscillator $\ddot{x} + (\alpha + \gamma x^2)\dot{x} + \beta x + \delta x^3 = 0$. *International Journal of Non–Linear Mechanics*, 15:449–458, 1980.

174. W. Horsthemke and D. K. Kondepudi, editors. *Fluctuations and sensitivity in nonequilibrium systems*, volume 1 of *Springer Proceedings in Physics*. Springer-Verlag, Berlin Heidelberg New York, 1984.
175. W. Horsthemke and R. Lefever. *Noise-induced transitions*. Springer-Verlag, Berlin Heidelberg New York, 1984.
176. D. Husemoller. *Fibre bundles*. Springer-Verlag, Berlin Heidelberg New York, third edition edition, 1994.
177. J. E. Hutchinson. Fractals and self similarity. *Indiana Univ. Math. J.*, 30:713–747, 1981.
178. N. Ikeda and S. Watanabe. *Stochastic differential equations and diffusion processes*. North Holland-Kodansha, Tokyo, 1981.
179. P. Imkeller. The smoothness of laws of random flags and Oseledets spaces of linear stochastic differential equations. *Potential Analysis*, 1998. To appear.
180. G. Iooss. Global characterization of the normal form for a vector field near a closed orbit. *Journal of Differential Equations*, 76:47–76, 1988.
181. M. C. Irwin. *Smooth dynamical systems*. Academic Press, New York, 1981.
182. K. Itô. Differential equations determining a Markov process. *Journal of the Pan-Japanese Mathematical Colloquium*, 1077, 1942.
183. K. Itô. *On stochastic differential equations*. Memoirs of the American Mathematical Society, Volume 4, 1951.
184. K. Itô. Stochastic integral. *Proc. Imp. Acad. Tokyo*, 1944:519–524, 20.
185. J. Jacod. *Calcul stochastique et problèmes de martingales*. Springer-Verlag, Berlin Heidelberg New York, 1979. Lecture Notes in Mathematics Volume 714.
186. J. Jacod and A. N. Shiryaev. *Limit theorems for stochastic processes*. Springer-Verlag, Berlin Heidelberg New York, 1987.
187. R. Johnson. On a Floquet theory for almost periodic, two-dimensional linear systems. *Journal of Differential Equations*, 37:184–205, 1980.
188. R. Johnson. m–functions and Floquet exponents for linear differential systems. *Ann. Mat. Pura Appl.*, 147:211–248, 1987.
189. R. Johnson. The Oseledec and Sacker–Sell spectra for almost periodic linear systems: an example. *Proceedings of the Amer. Math. Soc.*, 99:261–267, 1987.
190. R. Johnson. Exponential dichotomy and rotation number for linear Hamiltonian systems. *J. of Differential Equations*, 108:201–216, 1994.
191. R. Johnson and J. Moser. The rotation number for almost periodic potentials. *Commun. Math. Phys*, 84:403–438, 1982.
192. R. Johnson, K. Palmer, and G. R. Sell. Ergodic properties of linear dynamical systems. *SIAM J. Math. Anal.*, 18:1–33, 1987.
193. G. Kager. *Zur Perfektionierung nicht invertierbarer grober Kozykel*. PhD thesis, Fachbereich Mathematik, Technische Universität Berlin, 1996.
194. G. Kager and M. Scheutzow. Generation of one–sided random dynamical systems by stochastic differential equations. *Electronic J. Prob.*, 2:1–17, 1997.
195. V. A. Kaimanovich. Lyapunov exponents, symmetric spaces, and a multiplicative ergodic theorem for semisimple Lie groups. *Zapiski Nauchn. Sem. Leningr. Otdel. Matemat. Inst. Steklova*, 164:30–46, 1987.
196. J. Kao and V. Wihstutz. Stabilization of companion form systems by mean zero noise. *Stochastics and Stochastics Reports*, 49:1–25, 1994.
197. I. Karatzas and S. E. Shreve. *Brownian motion and stochastic calculus*. Springer-Verlag, Berlin Heidelberg New York, 1988.

198. S. Karlin and H. Taylor. *A second course in stochastic processes*. Academic Press, New York, 1981.
199. T. Kato. *Perturbation theory for linear operators*. Springer-Verlag, Berlin Heidelberg New York, second edition edition, 1980.
200. A. Katok and B. Hasselblatt. *Introduction to the modern theory of dynamical systems*. Cambridge University Press, 1995.
201. A. Katok and J.-M. Strelcyn. *Invariant manifolds, entropy and billiards; smooth maps with singularities*, volume 1222 of *Springer Lecture Notes in Mathematics*. Springer-Verlag, Berlin Heidelberg New York, 1986.
202. H. Keller. Attractors and bifurcations of the stochastic Lorenz system. Report 389, Institut für Dynamische Systeme, Universität Bremen, 1996.
203. E. S. Key. Lyapunov exponents for matrices with invariant subspaces. *Annals of Probability*, 16:1721–1728, 1988.
204. R. Z. Khasminskii. A limit theorem for solutions of differential equations with random right–hand side. *Theory of Probability and its Applications*, 11:390–406, 1966.
205. R. Z. Khasminskii. Necessary and sufficient conditions for the asymptotic stability of linear stochastic systems. *Theory of Probability and its Applications*, 12:144–147, 1967.
206. R. Z. Khasminskii. *Stochastic stability of differential equations*. Sijthoff and Noordhoff, Alphen, 1980. (Translation of the Russian edition, Nauka, Moscow 1969).
207. Y. Kifer. *Ergodic theory of random transformations*. Birkhäuser, Boston Basel Stuttgart, 1986.
208. Y. Kifer. A note on integrability of C^r-norms of stochastic flows and applications. In A. Truman and I. M. Davies, editors, *Stochastic Mechanics and Stochastic Processes*, pages 125–131, 1988. Springer Lecture Notes in Mathematics, Volume 1325.
209. J. F. C. Kingman. The ergodic theory of subadditive stochastic processes. *J. Royal Statist. Soc. Ser. B*, 30:499–510, 1968.
210. J. F. C. Kingman. Subadditive ergodic theory. *Annals of Probability*, 1:883–904, 1973.
211. J. F. C. Kingman. Subadditive processes. *École d'Été de Probabilités de Saint-Flour*, pages 167–223, 1976. Springer Lecture Notes in Mathematics, Volume 539.
212. W. Kliemann. Recurrence and invariant measures for degenerate diffusions. *Annals of Probability*, 15:690–707, 1987.
213. W. Klingenberg. *Riemannian geometry*. Walter de Gruyter, Berlin, 1982.
214. P. E. Kloeden and E. Platen. *Numerical solution of stochastic differential equations*. Springer-Verlag, Berlin Heidelberg New York, 1992.
215. O. Knill. The upper Lyapunov exponent of Sl(2,R) cocycles: Discontinuity and the problem of positivity. In Arnold et al. [14], pages 86–97.
216. O. Knill. Positive Lyapunov exponents for a dense set of bounded measurable Sl(2,R) cocycles. *Ergodic Theory and Dynamical Systems*, 12:319–331, 1992.
217. S. Kobayashi and K. Nomizu. *Foundations of differential geometry, volume I*. Wiley, New York, 1963.
218. A. N. Kolmogorov. The Wiener helix, and other interesting curves in Hilbert space. *Dokl. Akad. Nauk*, 26:115–118, 1940.

219. H.-J. Kowalsky. *Lineare Algebra*. Walter de Gruyter, Berlin, 9th edition edition, 1979.
220. U. Krengel. *Ergodic theorems*. Walter de Gruyter, Berlin, 1985.
221. A. Kufner, O. John, and S. Fučik. *Function spaces*. Academia, Publishing House of the Czechoslovak Academy of Sciences, Prague, 1977.
222. H. Kunita. On the decomposition of solutions of stochastic differential equations. In D. Williams, editor, *Stochastic Integrals*, pages 213–255, 1981. Springer Lecture Notes in Mathematics, Volume 851.
223. H. Kunita. Stochastic differential equations and stochastic flows of diffeomorphisms. *École d'Été de Probabilités de Saint-Flour*, pages 143–303, 1981. Springer Lecture Notes in Mathematics, Volume 1097.
224. H. Kunita. *Stochastic flows and stochastic differential equations*. Cambridge University Press, Cambridge, 1990.
225. H. Kushner. *Stochastic stability and control*. Academic Press, New York, 1967.
226. A. Lasota and M. Mackey. *Chaos, fractals, and noise*. Springer-Verlag, Berlin Heidelberg New York, 1994.
227. Y. Le Jan. Équilibre statistique pour les produits de difféomorphismes aléatoires indépendants. *Ann. Inst. H. Poincaré Probab. Statist.*, 23:111–120, 1987.
228. Y. Le Jan and S. Watanabe. Stochastic flows of diffeomorphisms. In K. Itô, editor, *Stochastic Analysis (Proceedings of the Taniguchi Symposium 1982)*, pages 307–332. North Holland, Amsterdam, 1984.
229. R. Léandre. Un exemple en theorie des flots stochastiques. *Séminaire de Probabilités*, XVII:158–161, 1981. Lecture Notes in Mathematics Volume 986.
230. F. Ledrappier. Quelques propriétés des exposants caractéristiques. *École d'Été de Probabilités de Saint-Flour*, pages 305–396, 1984. Springer Lecture Notes in Mathematics, Volume 1097.
231. F. Ledrappier and L.-S. Young. Stability of Lyapunov exponents. *Ergod. Th. & Dynam. Sys.*, 11:469–484, 1991.
232. R. Lefever and J. Turner. Sensitivity of a Hopf bifurcation to external multiplicative noise. In W. Horsthemke and D. K. Kondepudi, editors, *Fluctuations and sensitivity in nonequilibrium systems*, pages 143–149. Springer-Verlag, Berlin Heidelberg New York, 1984.
233. R. Lefever and J. Turner. Sensitivity of a Hopf bifurcation to multiplicative colored noise. *Physical Review Letters*, 56:1631–1634, 1986.
234. Li Xue–Mei. Strong p–completeness of stochastic differential equations and the existence of smooth flows on noncompact manifolds. *Probab. Theory Relat. Fields*, 100:485–511, 1994.
235. Li Xue-Mei. *Stochastic flows on noncompact manifolds*. PhD thesis, University of Warwick, 1993.
236. M. Liao. Stochastic flows on the boundaries of Lie groups. *Stochastics and Stochastics Reports*, 39:213–237, 1992.
237. M. Liao. Stochastic flows on the boundaries of SL(n,R). *Probab. Theory Relat. Fields*, 96:261–281, 1993.
238. M. Liao. The Brownian motion and the canonical stochastic flow on a symmetric space. *Trans. Amer. Math. Soc.*, 341:253–274, 1994.
239. M. Liao. Lyapunov exponents of stochastic flows. *Annals of Probability*, 25:1241–1256, 1997.

240. M. A. Liapounoff. *Problème général de la stabilité du mouvement.* Annales Fac. Sciences Toulouse 9 (1907) (Translation of the Russian edition, Kharkcv 1892). Reprinted by Princeton University Press, Princeton, N. J., 1949 and 1952.
241. I. Lindemann. Stability of hyperbolic PDE's with random loads. *SIAM J. Appl. Math.*, 52:347–367, 1992.
242. I. Lindemann. Lyapunov exponents of random block diagonal systems in infinite dimensions. *Stochastics and Stochastics Reports*, 50:245–271, 1994.
243. R. Liptser. A strong law of large numbers for local martingales. *Stochastics*, 3:217–228, 1980.
244. P.-D. Liu and M. Qian. *Smooth ergodic theory of random dynamical systems*, volume 1606 of *Springer Lecture Notes in Mathematics*. Springer-Verlag, Berlin Heidelberg New York, 1995.
245. M. Lücke. Bifurcation behavior under modulated control parameters. In F. Moss and P. McClintock, editors, *Noise in nonlinear dynamical systems*, volume 2, pages 100–144. Cambridge University Press, 1989.
246. M. Lücke and F. Schank. Response to parametric modulation near an instability. *Phys. Rev. Letters*, 54:1465–1468, 1985.
247. P. Malliavin. *Stochastic analysis.* Springer-Verlag, Berlin Heidelberg New York, 1997.
248. R. Mañé. Lyapunov exponents and stable manifolds for compact transformations. In J. Palis, editor, *Geometric Dynamics*, pages 522–577, 1983. Springer Lecture Notes in Mathematics, Volume 1007.
249. R. Mañé. *Ergodic theory and differentiable dynamics.* Springer-Verlag, Berlin Heidelberg New York, 1987.
250. M. S. Marasimhan and S. Ramanan. Existence of universal connections. *American J. Math.*, 83:563–572, 1961.
251. G. A. Margulis. *Discrete subgroups of semisimple Lie groups.* Springer-Verlag, Berlin Heidelberg New York, 1991.
252. P. A. Meyer. Flot d'une équation differentielle stochastique. *Séminaire de Probabilités*, pages 103–117, 1981. Springer Lecture Notes in Mathematics, Volume 850.
253. S. Meyn and R. Tweedie. *Markov chains and stochastic stability.* Springer-Verlag, Berlin Heidelberg New York, 1993.
254. V. M. Millionshchikov. Metric theory of linear systems of differential equations. *Math. USSR Sbornik*, 6:149–158, 1968.
255. S.-E. Mohammed. The Lyapunov spectrum and stable manifolds for stochastic linear delay equations. *Stochastics and Stochastics Reports*, 29:89–131, 1990.
256. S.-E. Mohammed. Lyapunov exponents and stochastic flows of linear and affine hereditary systems. In M. Pinsky and V. Wihstutz, editors, *Diffusion processes and related problems in analysis, volume II: Stochastic flows*, volume 27 of *Progress in Probability*, pages 141–169. Birkhäuser, Boston Basel Stuttgart, 1992.
257. S.-E. Mohammed and M. Scheutzow. Lyapunov exponents of linear stochastic functional differential equations driven by semimartingales. Part I: the multiplicative ergodic theory. *Ann. Inst. H. Poincaré Probab. Statist.*, 32:69–105, 1996.
258. S.-E. Mohammed and M. Scheutzow. The stable manifold theorem for stochastic differential equations. Report, University of Southern Illinois, 1997.

259. F. Moss and P. V. E. McClintock. *Noise in nonlinear dynamical systems, Volumes 1–3*. Cambridge University Press, 1989.
260. Nguyen Dinh Cong. Structural stability and topological classification of continuous–time linear hyperbolic cocycles. *Random & Computational Dynamics*, 5:19–63, 1997.
261. Nguyen Dinh Cong. *Topological dynamics of random dynamical systems*. Oxford University Press, 1997.
262. C. Nicolis and G. Nicolis. Normal form analysis of stochastically forced dynamical systems. *Dynamics and Stability of Systems*, 1:249–253, 1986.
263. G. Nicolis and I. Prigogine. *Self-organization in non–equilibrium systems*. Wiley, New York, 1977.
264. G. L. O'Brien. The occurence of large values in stationary sequences. *Z. Wahrscheinlichkeitstheorie verw. Gebiete*, 61:347–353, 1982.
265. G. Ochs. On the stability of Oseledets spaces under random perturbations. Report 368, Institut für Dynamische Systeme, Universität Bremen, 1996.
266. T. Ohno. Asymptotic behaviors of dynamical systems with random parameters. *Publ. RIMS Kyoto Univ.*, 19:83–98, 1983.
267. S. Orey. Stationary solutions for linear systems with additive noise. *Stochastics*, 5:241–252, 1981.
268. V. I. Oseledets. A multiplicative ergodic theorem. Lyapunov characteristic numbers for dynamical systems. *Trans. Moscow Math. Soc.*, 19:197–231, 1968.
269. K. J. Palmer. A proof of Oseledec's multiplicative ergodic theorem. Preprint, 1984.
270. E. Pardoux and V. Wihstutz. Lyapunov exponent and rotation number of two–dimensional linear stochastic systems with small diffusion. *SIAM J. Appl. Math.*, 48:442–457, 1988.
271. E. Pardoux and V. Wihstutz. Lyapunov exponent of linear stochastic systems with large diffusion term. *Stochastic Processes and their Applications*, 40:289–308, 1992.
272. K. P. Parthasarathy. *Probability measures on metric spaces*. Academic Press, New York, 1967.
273. L. Pastur and A. Figotin. *Spectra of random and almost–periodic operators*. Springer-Verlag, Berlin Heidelberg New York, 1992.
274. Y. B. Pesin. Families of invariant manifolds corresponding to nonzero characteristic exponents. *Math. USSR Izvestija*, 10:1261–1305, 1976.
275. Y. B. Pesin. Characteristic Lyapunov exponents and smooth ergodic theory. *Russian Mathematical Surveys*, 32:55–114, 1977.
276. K. Petersen. *Ergodic theory*. Cambridge University Press, Cambridge, 1983.
277. R. Phelps. *Lectures on Choquet's theorem*. Van Nostrand, Princeton, N.J., 1966.
278. M. Pinsky. Extremal character of the Lyapunov exponent of the stochastic harmonic oscillator. *The Annals of Applied Probability*, 4:942–950, 1992.
279. M. Pinsky and W. Wihstutz. Lyapunov exponents of nilpotent Itô systems. *Stochastics*, 25:43–57, 1989.
280. P. Protter. Semimartingales and measure preserving flows. *Ann. Inst. H. Poincaré Probab. Statist.*, 22:127–147, 1986.
281. P. Protter. *Stochastic integration and differential equations*. Springer-Verlag, Berlin Heidelberg New York, 1990.

282. C. Pugh and M. Shub. Ergodic attractors. *Trans. Amer. Math. Soc.*, 312:1–54, 1989.
283. A. Quas. On representations of Markov chains by random smooth maps. *Bull. London Math. Soc.*, 23:487–492, 1991.
284. A. Quas. Representation of Markov chains on tori. *Random & Computational Dynamics*, 1:261–276, 1992–93.
285. M. S. Ragunathan. A proof of Oseledec's multiplicative ergodic theorem. *Israel J. Math.*, 32:356–362, 1979.
286. D. Revuz and M. Yor. *Continuous martingales and Brownian motion*. Springer-Verlag, Berlin Heidelberg New York, 1991.
287. L. C. G. Rogers and D. Williams. *Diffusions, Markov processes, and martingales. Volume 2: Itô calculus*. Wiley, New York, 1987.
288. V. A. Rokhlin. On the fundamental ideas of measure theory. *Mat. Sb. (N.S.)*, 67:107–150, 1949. English translation: AMS Transl. (1) 10 (1962), 1–54.
289. V. A. Rokhlin. Selected topics from the metric theory of dynamical systems. *Usp. Mat. Nauk*, 4:57–128, 1949. English translation: AMS Transl. (2) 49 (1966), 171–240.
290. V. A. Rokhlin. Exact endomorphisms of a Lebesgue space. *Izv. Acad. Sci. USSR, Ser. Mat.*, 25:499–530, 1961. English translation: AMS Transl. (2) 39 (1964), 1–36.
291. D. J. Rudolph. *Fundamentals of measurable dynamics*. Oxford University Press, Oxford, 1990.
292. D. Ruelle. Ergodic theory of differentiable dynamical systems. *Publ. Math., Inst. Hautes Etud. Sci.*, 50:275–306, 1979.
293. D. Ruelle. Characteristic exponents and invariant manifolds in Hilbert space. *Annals of Mathematics*, 115:243–290, 1982.
294. D. Ruelle. Rotation numbers for diffeomorphisms and flows. *Ann. Inst. H. Poincaré*, 42:109–115, 1985.
295. D. Ruelle. *Elements of differentiable dynamics and bifurcation theory*. Academic Press, New York, 1989.
296. P. Ruffino. *Rotation numbers for stochastic dynamical systems*. PhD thesis, University of Warwick, 1995.
297. P. Ruffino. Rotation numbers for stochastic dynamical systems. *Stochastics and Stochastics Reports*, 60:289–318, 1997.
298. R. Sacker and G. Sell. Lifting properties in skew product flows with applications to differential equations. Memoirs of the American Mathematical Society, Volume 190, 1977.
299. R. Sacker and G. Sell. A spectral theory for linear differential systems. *Journal of Differential Equations*, 27:320–358, 1978.
300. L. San Martin. Rotation numbers in higher dimensions. Report 199, Institut für Dynamische Systeme, Universität Bremen, 1988.
301. K.-U. Schaumlöffel. Multiplicative ergodic theorems in infinite dimensions. In Arnold et al. [14], pages 187–195.
302. K. R. Schenk–Hoppé. Bifurcation scenarios of the noisy Duffing–van der Pol oscillator. *Nonlinear Dynamics*, 11:255–274, 1996.
303. K. R. Schenk–Hoppé. Deterministic and stochastic Duffing–van der Pol oscillators are non–explosive. *ZAMP*, 47:740–759, 1996.
304. K. R. Schenk–Hoppé. *The stochastic Duffing–van der Pol equation*. PhD thesis, Institut für Dynamische Systeme, Universität Bremen, 1996.

305. K. R. Schenk–Hoppé. Stochastic Hopf bifurcation: an example. *Int. J. Non-Linear Mechanics*, 31:685–692, 1996.
306. K. R. Schenk-Hoppé. Random attractors – general properties, existence, and applications to stochastic bifurcation theory. *Discrete and Continuous Dynamical Systems*, 4:99–130, 1998.
307. B. Schmalfuß. Backward cocycles and attractors of stochastic differential equations. In V. Reitmann, T. Riedrich, and N. Koksch, editors, *International Seminar on Applied Mathematics – Nonlinear Dynamics: Attractor Approximation and Global Behaviour*, pages 185–192. Teubner, Leipzig, 1992.
308. B. Schmalfuß. Attractors for the non–autonomous Navier Stokes equation. Submitted, 1996.
309. B. Schmalfuß. Measure attractors and stochastic attractors for stochastic partial differential equations. *Stochastic Analysis and Applications*, 1997. To appear.
310. B. Schmalfuß. The random attractor of the stochastic Lorenz system. *ZAMP*, 48:951–975, 1997.
311. B. Schmalfuß. A random fixed point theorem and the random graph transform. Report 402, Institut für Dynamische Systeme, Universität Bremen, 1997.
312. K. Schmidt. *Cocycles on ergodic transformation groups*. MacMillan, Delhi, 1977.
313. K. Schmidt. *Algebraic ideas in ergodic theory*. American Mathematical Society, Providence, R. I., 1990. Regional Conference Series in Mathematics, Number 76.
314. G. Schöner and H. Haken. The slaving principle for Stratonovich stochastic differential equations. *Z. Phys. B*, 63:493–504, 1986.
315. G. Schöner and H. Haken. A systematic elimination procedure for Ito stochastic differential equations and the adiabatic approximation. *Z. Phys. B*, 68:89–103, 1987.
316. G. Sell. The structure of a flow in the vicinity of an almost periodic motion. *Journal of Differential Equations*, 27:359–393, 1978.
317. S. Siegmund. Zur Differenzierbarkeit von Integralmannigfaltigkeiten. Diplomarbeit, Universität Augsburg, 1996.
318. Y. G. Sinai, editor. *Dynamical systems II*. Encyclopaedia of Mathematical Sciences, Volume 2. Springer-Verlag, Berlin Heidelberg New York, 1989.
319. A. V. Skorokhod. *Random linear operators*. Reidel Dordrecht, 1984. (Translation of the Russian edition, Naukova Dumka, Kiev 1978).
320. A. V. Skorokhod. *Asymptotic methods in the theory of stochastic differential equations*. Volume 78, Transl. Amer. Math. Soc., Providence, R. I., 1989. (Translation of the Russian edition, Nauka, Moscow 1987).
321. C. Sparrow. *The Lorenz equations: bifurcations, chaos, and strange attractors*. Springer-Verlag, Berlin Heidelberg New York, 1982.
322. N. Sri Namachchivaya. Stochastic bifurcation. *Applied Mathematics and Computation*, 38:101–159, 1990.
323. N. Sri Namachchivaya and Y. K. Lin. Methods of stochastic normal forms. *Int. J. Nonlin. Mech.*, 26:931–943, 1991.
324. J. M. Steele. Kingman's subadditive ergodic theorem. *Ann. Inst. H. Poincaré Probab. Statist.*, 25:93–98, 1989.

325. D. Stroock. On the rate at which a homogeneous diffusion approaches a limit, an application of the large deviation theory of certain stochastic integrals. *Annals of Probability*, 14:840–859, 1986.
326. D. W. Stroock and S. R. S. Varadhan. *Multidimensional diffusion processes*. Springer-Verlag, Berlin Heidelberg New York, 1979.
327. D. Tanny. A zero-one law for stationary sequences. *Z. Wahrscheinlichkeitstheorie verw. Gebiete*, 30:139–148, 1974.
328. R. Temam. *Infinite-dimensional dynamical systems in mechanics and physics*. Springer-Verlag, Berlin Heidelberg New York, 1988.
329. P. Thieullen. Fibres dynamiques asymptotiquement compacts — exposants de Lyapunov. Entropie. Dimension. *Ann. Inst. H. Poincaré, Anal. Non Linéaire*, 4(1):49–97, 1987.
330. C. W. S. To and D. M. Li. Largest Lyapunov exponents and bifurcations of stochastic nonlinear systems. *Shock and Vibration*, 3:313–320, 1996.
331. C. W. S. To, D. M. Li, and K. L. Huang. Supercritical and subcritical Hopf bifurcations in a stochastically excited system. *Journal of Sound and Vibration*, 201:648–656, 1997.
332. A. Vanderbauwhede. Center manifolds, normal forms and elementary bifurcations. In U. Kirchgraber and H. O. Walter, editors, *Dynamics Reported*, volume 2, pages 89–169. Teubner and Wiley, New York, 1989.
333. A. Vanderbauwhede and S. Van Gils. Center manifolds and contractions on a scale of banach spaces. *Journal of Functional Analysis*, 72:209–224, 1987.
334. J. B. Walsh. The perfection of multiplicative functionals. *Séminaire de Probabilités*, pages 233–242, 1972. Springer Lecture Notes in Mathematics, Volume 258.
335. W. Walter. *Gewöhnliche Differentialgleichungen*. Springer-Verlag, Berlin Heidelberg New York, 5th edition, 1993.
336. P. Walters. *An introduction to ergodic theory*. Springer-Verlag, Berlin Heidelberg New York, 1982.
337. P. Walters. A dynamical proof of the multiplicative ergodic theorem. *Trans. Am. Math. Soc.*, 335:245–257, 1993.
338. T. Wanner. Invariante Faserbündel und topologische Äquivalenz bei dynamischen Prozessen. Diplom–Arbeit, Universität Augsburg, 1991.
339. T. Wanner. *Zur Linearisierung zufälliger dynamischer Systeme*. PhD thesis, Universität Augsburg, 1993.
340. T. Wanner. Linearization of random dynamical systems. In C. Jones, U. Kirchgraber, and H. O. Walther, editors, *Dynamics Reported*, volume 4, pages 203–269. Springer-Verlag, Berlin Heidelberg New York, 1995.
341. T. Wanner. Qualitative behavior of random differential equations. In D. Bainov and A. Dishliev, editors, *Proceedings of the Fifth International Colloquium on Differential Equations*, pages 242–257. SCT Publishing, 1995.
342. W. Wedig. Vom Chaos zur Ordnung. *GAMM-Mitteilungen*, 2:3–31, 1989.
343. W. Wedig. Lyapunov exponents and invariant measures of equilibria and limit cycles. In Arnold et al. [14], pages 309–321.
344. W. Wedig. Simulation and analysis of mechanical systems with parameter fluctuations. *International Series of Numerical Mathematics*, 102:159–165, 1991.
345. A. D. Wentzell. *Theorie zufälliger Prozesse*. Akademie-Verlag, Berlin, 1979.

346. S. Wiggins. *Introduction to applied nonlinear dynamical systems and chaos.* Springer-Verlag, Berlin Heidelberg New York, 1990.
347. Xu Kedai. Bifurcations of random differential equations in dimension one. *Random & Computational Dynamics*, 1:277–305, 1993.
348. Y. Yi. A generalized integral manifold theorem. *Journal of Differential Equations*, 102:153–187, 1993.
349. Y. Yi. Stability of integral manifold and orbital attraction of quasi–periodic motion. *Journal of Differential Equations*, 103:278–322, 1993.
350. M. I. Zakharevich. Characteristic exponents and a vector ergodic theorem. *Vestn. Leningr. Gos. Univ.*, 7:28–34, 1978.
351. E. C. Zeeman. On the classification of dynamical systems. *Bull. London Math. Soc.*, 20:545–557, 1988.
352. E. C. Zeeman. Stability of dynamical systems. *Nonlinearity*, 1:115–155, 1988.
353. R. J. Zimmer. *Ergodic theory and semisimple groups.* Birkhäuser, Boston Basel Stuttgart, 1984.

Index

absolute derivative 561
absorbing set 483
abstract bifurcation point 469
- local 469
action of group on space 238
additive noise 473
affine cocycle 222
affine RDE 63, 222
affine RDS 51, 222
- invariant measure 223, 228, 232
affine SDE 94, 222
almost periodic function 541
angle
- between subspaces 216
- between vectors 215
attractor 483
average Lyapunov exponent 132, 276

backward cocycle 10, 239
backward Markov measure 39
backward semimartingale 71
bifurcation
- Hopf 497
- pitchfork 481, 484, 489, 511
- saddle-node 482, 490
- transcritical 478, 489
bifurcation point
- abstract 469
- dynamical 469
- phenomenological 472
Borel space 541
Brownian motion 547
Brusselator 530
bundle RDS 43
- continuous 45
- linear 45, 174
- smooth 45

C^k RDS 6
cadlag 545
canonical realization 543

center manifold 340, 344
characteristic exponent 115
coboundary 231
cocycle 5
- affine 222
- backward 10, 239
- crude 5
- group-valued 17, 73, 237
- induced by action 239
- local 12
- on homogeneous space 240
- on principal bundle 241
- perfect 5
- semimartingale 82
- very crude 5
cocycle property 5
cohomological equation 231, 411, 413
- RDE case 424
- SDE case 449
cohomological operator 231, 408, 413
- RDE case 424
- SDE case 449
cohomology 46
- Lyapunov 167
compact-open topology 553
complete RDS 95
continuous DS
- global 558
- local 558
continuous RDS 5
contraction theorem 357
countably generated 535
covariant derivative 561

D-bifurcation 469
- local 469
- sufficient condition 522
determinant 213
diagonalization
- of linear RDE 253
- of linear SDE 260

dichotomy spectrum   178
differential equation   555
– classical solution   555
– solution in the sense of Carathéodory   555
disintegration   *see* factorization
domain of attraction   483
DS   *see* dynamical system
Duffing-van der Pol oscillator   442, 462, 491
– averaged   526
– normal form   444, 462, 509, 515
dynamical bifurcation   469
dynamical characterization   320, 387
dynamical system   536
– C^k 558
– continuous   558
– ergodic   539
– filtered   72, 543
– metric   537
– on manifold   562

ε-slowly varying   187
equivalence of RDS
– local topological   385
– smooth   47, 412, 422
– topological   47, 369
equivariant measure   25
ergodic decomposition   539
ergodic DS   539
ergodic theorem   539
– extended version   539
exhaustive σ-algebra   37
explosion time   12
extension of DS   540
exterior power   118

φ-invariant measure   22
factor of DS   540
factorization
– of image measure   23
– of measure   22
Feller Markov process   549
filtered DS   72, 543
filtration   71, 115, 134
– measurability of   159, 212
flag
– of stable manifolds   338
– of unstable manifolds   337
flag manifold   145
Floquet exponent   117
flow   536, 552
– global   552

– local   551
formal linearization
– of RDE   425
– of RDS   419
– of SDE   450
forward invariant set   34, 537
forward Markov measure   39
forward semimartingale   71
Furstenberg-Kesten theorem
– for backward cocycles   128
– for one-sided time   123
– for two-sided time   131
Furstenberg-Khasminskii formulas
– for discrete time   251
– for linear RDE   252
– for linear SDE   258
– – for top exponent   259
– for SDE on manifold   272, 275
– for sum of exponents   268
future of RDS   37

gap conditions   348
generation of RDS
– by Itô SDE   97
– by RDE   58
– by Stratonovich SDE   83, 92
– one-sided discrete time   50
– two-sided discrete time   52
global flow   552
global RDS
– from Itô SDE   97
– from RDE   60
– from scalar SDE   96
– from Stratonovich SDE   92, 93
globalization of local manifold   390
gradient Brownian flow   400
Gronwall-Bellman lemma
– continuous time   556
– discrete time   327
group-valued cocycle   17, 73, 237

Hartman-Grobman theorem   377
– hyperbolic case   377
– local   385, 394
– non-hyperbolic case   378
helix   73, 546
– semimartingale   74
Hopf bifurcation   497

independent increments of RDS   94, 105
indistinguishable   543
invariant manifolds
– center   307

– unstable   307
invariant foliations   362
invariant function   537
invariant manifolds   306
– C^k smoothness   347
– classical vs. generalized   306
– dynamical characterization of   320, 387
– foliation by   362
– for RDE   395
– global   360
– global center   340, 344
– global stable   337
– global unstable   318, 331
– globalization of local   390
– local   382, 385, 394
– Oseledets   307
– stable   307, 391
invariant measure   22, 536
– factorization of   22
– for affine RDS   223, 228, 232
– for continuous RDS   26
– for local RDS   40
– for RDS induced on $G_k(d)$ 265
– for RDS induced on S^{d-1} 247
– for state space $\mathbb{R}$ 41
– lift of   247, 270, 287
– on random set   36
– on Stiefel bundle   287
– on unit sphere bundle   270
invariant mod $\mathbb{P}$ 537
invariant set   35, 537
– forward   34, 537
isomorphism
– metric   540
isomorphism of RDS   45
– linear   47
– metric   47
iterated function system   50, 231

Kingman's theorem   *see* subadditive ergodic theorem
Kolmogorov's fundamental theorem   543
Kronecker product   219
Krylov-Bogolyubov procedure   29

Langevin equation   103
Lebesgue space   541
lift of measure   247, 270, 287
linear bundle RDS   174
linear isomorphism of RDS   47
linear RDE   62, 159, 202

– diagonalization   253
linear RDS   6, 51
– on invariant bundle   206
– on quotient space   206
linear SDE   94, 162, 202
– diagonalization   260
linearization of RDS   174
Liouville's equation   556
– for RDE   60
– for RDE on manifold   68
– for SDE   93
– for SDE on manifold   102
– on manifold   562
local characteristics   77, 79
local cocycle property   12
local flow   551
local RDS   11
– from RDE   58
– from RDE on manifold   67
– from SDE   86, 95
– from SDE on manifold   102
local topological equivalence of RDS   385
Lorenz equation   529
Lyapunov cohomology   167
– for different bases   196
– for random norms   195
– for random Riemannian metric   199
Lyapunov exponent   114, 129
– average   132, 276
– backward   114
– forward   114
– multiplicity of   115, 129
– of a system   115
– small noise expansion   499, 512
– top   115
Lyapunov index   114
– and scale of Banach spaces   315
– of angle between subspaces   217
– of angle between vectors   215
– of determinant   214
– of stationary process   166
– of volume   213
Lyapunov metric, norm   194
Lyapunov spectrum   115, 129, 135
– of exterior power   211
– of inverse and adjoint   202
– of RDS induced on S^{d-1} 262
– of tensor product   220
– simplicity of   259

manifold   551
Markov measure   39, 107

- and stationary measure 56, 107
- backward 39
- forward 39
- on S^{d-1} 249

Markov process 548
Markov property 549
Markovian RDS 104
measurability of filtration 159, 212
measurable bundle 44
measurable isomorphism of RDS 45
measurable RDS 5
measurable selection theorem 33
measurable space 535
memoryless RDE 64
MET see multiplicative ergodic theorem
metric DS 537
metric isomorphism 540
metric isomorphism of RDS 47
multiplicative ergodic theorem 134, 153
- for exterior power 211
- for inverse and adjoint 202
- for linear RDE 159
- for linear SDE 162
- for local RDS 175
- for one-sided time 134
- for RDE 180
- for RDS on manifolds 174
- for rotation numbers 296, 299
- for SDE 181, 183, 184
- for subbundle and quotient 207
- for tensor product 220
- for two-sided time 153
- independent of norm 153
- Lyapunov spectrum 135
- on Euclidean bundle 170
- strong version of 196
- total measure 177
multiplicative noise 473

n-point motion 104
- backward 104
- forward 104
natural extension 540
- of canonical DS 547
Nemytskii operator 350
- regularity of 352
never exploding initial values 40
noisy parameters 65
nonresonant of order n 417, 425
normal form
- for discrete time 412
- for RDE 428
- for SDE 446
normal form and center manifold
- for RDE 435
- with parameters 439
normed linear bundle 167

orbit 537
Oseledets manifolds 307
Oseledets spaces 153
Oseledets splitting 154, 175
- for exterior power 212
- for inverse and adjoint 202
- for RDS induced on S^{d-1} 262
- for tensor product 220
Oseledets's theorem see multiplicative ergodic theorem

P-bifurcation 472
- sufficient condition 523
past of RDS 37
perfection of a cocycle 15
- for continuous time 17
- for discrete time 16
periodic linear system 117, 171
phenomenological bifurcation 472
pitchfork bifurcation 481, 484, 489, 511
prepared RDS 316
principal fiber bundle 241
probability space 535
- complete 535
product of random mappings 50
product of random matrices 51

random attractor 483
random averaging 429
random coordinate transformation 412
random differential equation 58
- affine 63
- classical solution 58
- linear 62
- solution in the sense of Carathéodory 58
random Dirac measure 25, 310
random dynamical system 5
- affine 51
- C^k 6
- complete 95
- continuous 5
- future of 37
- generation of
-- by RDE 58

– – by SDE 83, 92, 97
– – for discrete time 50
– independent increments 53, 93
– linear 6
– local 11
– Markov chain for 53
– Markovian 104
– measurable 5
– on bundle 44
– on subset 43
– past of 37
– prepared 316
– smooth 6
– stationary increments 9
– strictly complete 95
– topological 5
random eigenvalue 249
random eigenvectors 249
– basis of 250
random homeomorphism 369
random norm 191
random Riemannian metric 198
random scalar product 191
random set 32
RDE *see* random differential equation
RDS *see* random dynamical system
reduction principle 372
regular linear system 115, 116
resonant of order n 417, 425
right-invariant
– RDE 238
– SDE 238
rotation number
– for canonical plane 290
– for invariant measure 289
– for linear differential equation 282
– for linear SDE 298
– for nonlinear differential equation 284
– for nonlinear RDE 286
– MET 296, 299

Sacker-Sell spectrum 178
saddle-node bifurcation 482, 490
sample measure *see* factorization
scale of Banach spaces 314
– contraction theorem 357
SDE *see* stochastic differential equation
semimartingale 71
– backward 71
– forward 71
semimartingale cocycle 82

semimartingale helix 74
– $\mathcal{C}_b^{k,\delta}, \mathcal{C}^{k,\delta}$ 78
– local characteristics of 77, 79
– with spatial parameters 78
shift 543
simple Lyapunov spectrum 259
singular value 117
skew product 9
slowly varying 187
smooth equivalence of RDS 47, 412, 422
smooth RDS 6
spectral theory of linear RDS 168
stable manifolds 337
stable reference solution 372
– asymptotically 372, 393
– exponentially 372
standard measurable space 541
stationary increments of RDS 9
stationary measure 55, 107, 549
– and Markov measure 56, 107
– for extreme exponents 257
stationary process 545
Stiefel bundle 283
Stiefel manifold 279
stochastic differential equation 83
– Itô 97
– Stratonovich 83, 92
stochastic integral
– backward Itô 97
– backward Stratonovich 81
– forward Itô 96
– forward Stratonovich 81
stochastic process 542
– equivalent 543
– indistinguishable 543
– measurable 546
– stationary 545
– with stationary increments 546
strictly (forward) invariant set 35
strictly complete RDS 95
– backward 95
– forward 95
subadditive ergodic theorem 122
subadditive process 122
support of a measure 33

tail σ-algebra 547
tangent bundle 173
tempered 164
– ball 386
– from above 164
– from below 164

– set   508
tensor product   218
time $\mathbb{T}$ 4, 535
time reversible   205, 232
topological decoupling   369, 371
topological equivalence of RDS   47, 369
– local   385
topological linearization   375, 376
topology of weak convergence   27
transcritical bifurcation   478, 489
two-parameter flow   551

universally measurable set   32
universe of sets   483

unstable manifolds   318
– graph of   331

variational equation   555
– for RDE   60
– for RDE on manifold   68
– for SDE   93
– for SDE on manifold   102
– on manifold   561

weak convergence   27
white noise   547
Wiener process   547

Springer and the environment

At Springer we firmly believe that an international science publisher has a special obligation to the environment, and our corporate policies consistently reflect this conviction.

We also expect our business partners – paper mills, printers, packaging manufacturers, etc. – to commit themselves to using materials and production processes that do not harm the environment. The paper in this book is made from low- or no-chlorine pulp and is acid free, in conformance with international standards for paper permanency.

Printing: Mercedesdruck, Berlin
Binding: Buchbinderei Lüderitz & Bauer, Berlin